Antimicrobial Resistance of the Human Eye

The book covers antimicrobial resistance in ocular diseases, including the microbiology of the ocular surface, the history and origin of antimicrobials, methods to detect antimicrobial resistance and antimicrobial resistance genes, and the impact of antimicrobial resistance on a variety of ocular diseases. The inclusion of chapters covers bacterial keratitis, fungal keratitis, viral keratitis, acanthamoeba keratitis, endophthalmitis, dry eye disease, post fever retinitis, and uveitis including management and prevention of antimicrobial resistance.

Key Features:

- Focusses on various ocular diseases and their association with antimicrobial resistance;
- Includes data relevant to drug industry to develop drugs for specific ophthalmic use;
- Illustrates ocular surface microbiome under various diseased conditions;
- Highlights the microbes associated with the conjunctiva and cornea of the human eye;
- Enumerates the changes in the abundance and diversity of the antimicrobial resistant (AMR) microorganisms in the diseased eye.

This book is aimed at professionals and researchers in ophthalmology, microbiologists, infectious disease specialists, and public health.

Antimicrobial Resistance of the Human Eye

Edited by

S. Shivaji

CRC Press
Taylor & Francis Group
Boca Raton London New York

CRC Press is an imprint of the
Taylor & Francis Group, an **informa** business

Designed cover image: Cover illustration created by Harsh Kapoor, a researcher depicting the three components of antimicrobial resistance: the eye, the microbes and the drugs.

First edition published 2025
by CRC Press
2385 NW Executive Center Drive, Suite 320, Boca Raton FL 33431

and by CRC Press
4 Park Square, Milton Park, Abingdon, Oxon, OX14 4RN

CRC Press is an imprint of Taylor & Francis Group, LLC

ISBN: 9781032583495 (hbk)
ISBN: 9781032586991 (pbk)
ISBN: 9781003451105 (ebk)

DOI: 10.1201/9781003451105

Typeset in Times
by Newgen Publishing UK

To all scientists and clinicians who recognised antimicrobial resistance as a global menace with devastating effects on life forms.

Contents

Foreword

Antimicrobial resistance (AMR) has emerged as a global menace which is likely to escalate to ten million annual deaths by the year 2050 due to AMR complications, accompanied by staggering economic burdens stemming from healthcare impact and reduced productivity, the urgency to address this issue cannot be overstated.

While AMR in various regions of the human body has been extensively studied, our understanding of AMR in the human eye remains limited. The ocular surface, which hosts a diverse range of bacteria and fungi in a commensal relationship with the eye, can also become a site of infection following trauma or injury. These ocular infections can lead to visual impairment and even blindness. Compounding the issue is the development of antimicrobial resistance among these eye-associated microorganisms, making their eradication challenging.

This book, *Antimicrobial Resistance of the Human Eye*, addresses the specific challenges posed by antimicrobial resistance in the human eye. It aims to consolidate the studies conducted on ocular specific antimicrobial resistance and its impact on ocular diseases. By compiling the available information, this book endeavours to shed light on this important but relatively understudied aspect of AMR.

The book begins by exploring the ocular surface microbiota and its role in ocular diseases. It highlights the delicate balance of commensal bacteria that defend against pathogens through adaptive and innate immunity. However, under conditions of trauma or injury, this balance is disrupted, leading to an increase in virulence and the onset of various ocular infectious diseases. Traditionally, these diseases have been treated with antimicrobial agents and anti-inflammatory or immunosuppressive drugs.

This book delves into various aspects of AMR within the human eye to address the knowledge gaps surrounding this area, which has only recently begun to be studied. It explores the unique profile of antimicrobial-resistant bacteria on the ocular surface, examining differences between regions such as the conjunctiva, limbus, and cornea. The book also investigates how the profile of antimicrobial resistance varies between healthy and diseased eyes, different ocular diseases, and factors such as age, gender, ethnicity, and diet. Additionally, the possibility is explored of effectively treating the ocular surface to eliminate antimicrobial-resistant bacteria.

The comprehensive structure of the book ensures a thorough examination of the topic. It covers chapters dedicated to different ocular diseases affected by antimicrobial resistance, including bacterial keratitis, fungal keratitis, viral keratitis, dry eye disease, conjunctivitis, scleritis, uveitis, post-fever retinitis, endophthalmitis, and dacryocystitis. Each chapter provides insights into the abundance, diversity, and resistance mechanisms of antimicrobial-resistant microorganisms in the affected eye. Moreover, the book emphasizes strategies to overcome this growing menace.

Antimicrobial Resistance of the Human Eye serves as a valuable resource for ophthalmologists, ocular microbiologists, general microbiologists, and students in medicine and biology. By bridging the gap between conventional microbiology and molecular microbiology, it brings together a wealth of knowledge to address the health and economic burden of AMR.

I extend my appreciation to the authors who have worked diligently to compile this valuable resource. Their collective expertise and dedication under the editorship of Dr. Shivaji will undoubtedly contribute to our understanding of antimicrobial resistance in the human eye, paving the way for improved strategies to combat this global menace. I am confident that this book will contribute significantly to ongoing research, inform clinical practice, and stimulate further investigations to combat the global menace of antimicrobial resistance.

Gullapalli N. Rao
LV Prasad Eye Institute

Preface

Antimicrobial Resistance: A Global Menace

Antimicrobial resistance (AMR), the ability of a microorganism to resist the killing effect of a drug, is a phenomenon that knows no geographical limits, that affects the young and the aged and does not discriminate between genders. A conservative estimate is that by the year 2050 ten million people would die annually from AMR complications and the resultant economic burden would amount to billions of dollars due to the impact on healthcare and loss of productivity. AMR also impacts the agricultural and veterinary sectors and thus it is recognised as a menace to human beings, animals, and plants. In human beings, microbes resistant to antimicrobial drugs have been reported from those regions of the body which support the colonisation of microorganisms like the skin, nasal cavity, oral cavity, dental plaque, blood, urine/UTI, digestive tract and so forth, and these niches have been explored with respect to the microbes and their AMR profiles. Comparatively little is known about AMR of the human eye, though the ocular surface harbours a diversity of bacteria and fungi, which have a commensal relationship with the eye. But under conditions of trauma and injury these eye-associated microorganisms cause ocular infections that could result in visual impairment and blindness. An additional complication is that these ocular microorganisms over time become resistant to several classes of antimicrobial agents, thus posing a challenge to overcoming their infections. Pioneering studies on ocular surface niche-specific antimicrobial resistant microorganisms and their ability to cause ocular diseases that could result in visual impairment and blindness was an impetus to put together the present book, *Antimicrobial Resistance of the Human Eye*.

Ocular Surface Microbiota and Ocular Disease

The healthy normal human eye fosters a paucibacterial microbiome which predominantly includes *Propionibacterium* sp., *Corynebacterium* sp., *Staphylococcus aureus*, *Micrococcus* sp. and so forth. These commensal bacteria on the ocular surface defend the host against pathogens through both adaptive and innate immunity. But under conditions of trauma or injury to the eye, the composition and relative abundance of the microbiota changes and, concomitantly, an increase in virulence is observed, which may trigger ocular infectious diseases like keratitis, uveitis, endophthalmitis, blepharitis, conjunctivitis, orbital cellulitis and so forth. Normally, these infectious diseases of the eye are treated using antimicrobials (antibacterial, antifungals, antivirals) and (or) with an anti-inflammatory or an immunosuppressive drug as deemed appropriate.

Emergence of Antimicrobial Resistance on the Ocular Surface

AMR in bacteria predates our present-day use of antibiotics and has been studied extensively in various systemic infections. However, the emergence of AMR on the ocular surface is more recent and was reported by the TRUST (Tracking Resistance in the U.S. Today) program about two decades ago, in 2000, when ocular methicillin-resistant *Staphylococcus aureus* was detected. The factors that facilitate AMR could be extrinsically attributed to improper dosing, excess consumption, use of sub-standard antibiotics, use against non-target organisms, improper disposal, and lack of effective methods to prevent transmission of resistant bacteria. In contrast, the intrinsic factors that influence AMR, are microbe centric and include the ability of the microbe to inactivate the antimicrobial agent, to enzymatically modify the targets of the antibiotic, induce active efflux of the antibiotic, acquire genes that confer resistance to the antimicrobial agent and guard against the lethal effects of the drug by biofilm formation. Thus, AMR could be overcome by overcoming the above defensive strategies of the microorganisms.

Antimicrobial Resistance of the Human Eye

Several questions need to be addressed with respect to the human ocular surface and associated antimicrobial resistant microorganisms, questions such as: Does the ocular surface have a unique profile of antimicrobial resistant bacteria? Does the profile of antimicrobial resistant bacteria differ between regions of the eye such as conjunctiva, limbus, and cornea? Is the profile of antimicrobial resistant bacteria specific to the ocular surface? Does the profile of antimicrobial resistant bacteria differ between the healthy and diseased eye and between different ocular diseases? Does it differ with age, gender, ethnicity, diet, and other factors? Is it possible to treat the ocular surface free of antimicrobial resistant bacteria? And so forth. A reasonable number of studies have addressed some of the above concerns related to AMR, and there is a need to capture all this information in a comprehensive book.

Contents of the Book *Antimicrobial Resistance of the Human Eye*

Over the years there has been an exponential growth in the number of research papers and reviews on AMR and Resistome analysis. Further, several books have been published on AMR as a global challenge and threat, on development and spread of AMR, on methods to combat and overcome AMR, on clinical and epidemiological aspects of AMR, on antibiotic resistance protocols and so forth. A good proportion of these books are also related to dissemination of AMR in the environment, in food bacteria and in livestock and companion animals. But books on AMR of a given body niche are limited to the gut and skin. Thus, a book dedicated to the human ocular surface AMR would add to the characterisation of AMR in yet another niche of the human body and help unravel its impact on ocular health.

In Chapter 1, the book introduces the human ocular surface as a habitat for multitude of commensal microorganisms covering the range of prokaryotes to eukaryotes.

These commensals, especially the prokaryotes, under conditions of trauma get transformed into ocular disease-causing organisms which are susceptible to drugs but could develop antimicrobial resistance (AMR). As a logical follow up in Chapter 2 the global menace of AMR and the national and international efforts to overcome it are discussed. Further, Chapter 3 highlights the various phenotypic and DNA-based methods that are in place to detect AMR in microorganisms of the eye. The chapter also highlights the fact that phenotypic methods are the gold standards for AMR in ocular samples.

Chapters 1–3 are considered as the foundation for the subsequent Chapters 4 to Chapter 14, with each individual chapter dedicated to a specific ocular disease like bacterial keratitis, uveitis, endophthalmitis, dry eye, post-fever retinitis and so forth. These chapters delve on the symptoms, aetiology, associated risk factors, drugs used, AMR with respect to the specific antibacterial, antifungal, antiviral or anti-amoebic drug, the molecular mechanism of resistance, host immune factors and surgical procedures when the disease is refractory to drugs and so forth. Finally, Chapter 15 is the concluding chapter, highlighting the contents of the book and concluding that it is necessary to take a step back to comprehend the importance of AMR and tackle the global health threat of AMR.

Overall, the book highlights the microbes associated with the human eye with specific reference to the emergence of antimicrobial resistance as detected by conventional and metagenomic approaches and unravels the factors that trigger AMR. Emphasis is on enumerating the changes in the abundance and diversity of the antimicrobial resistant microorganisms in the diseased eye impacted with ocular diseases like keratitis, conjunctivitis, blepharitis, endophthalmitis, uveitis, post fever retinitis and dry eye disease and so forth. All these chapters, apart from unravelling AMR on the ocular surface, would also provide insight into the possible strategies to overcome this worldwide menace.

Target Audience and Importance

This book will cater to the needs of ophthalmologists, ocular microbiologists, general microbiologists, students of medicine and biology and help to bridge the gap between conventional microbiologists and molecular microbiologists. Additionally, this book will be useful to International and National organisations concerned with the health and economic burden of AMR.

Acknowledgements

Thanks to Drs. D. Balasubramanian and Gullapalli Nageswara Rao who, with their relentless commitment to science for the welfare of the common man, have inspired and motivated me to do what I could do best—that is involve myself in academics, especially research. The clinicians and scientists of LV Prasad Eye Institute, Hyderabad, India, were a major influence in carving the contents of the book. Specific thanks to Drs. Arunasri, Joveeta Joseph, Bhupesh Bagga, Ponnari Gottipati and Shalem Raj for their specific contributions to the book. Several students at the Institute did a meticulous job in compiling the references and in formatting the chapters. The affection and moral support of my granddaughter, my sons, daughters-in-law, and my wife made my task all the easier.

Editor Biography

Dr. S. Shivaji's research interests include conservation of endangered animals, biodiversity of Antarctica, Arctic, Himalayas and so forth, and microbiome of the gut and eye with special reference to ocular diseases. The Government of India has conferred on Dr. Shivaji the prestigious National Award for research in Cryosphere and Polar sciences and the Antarctic Award for excellence in the field of Biological Sciences. He was a visiting scientist at Max-Planck-Institut für biophysikalische Chemie and Institute for Medicine, Germany, and National Institute of Basic Biology, Japan. Dr. Shivaji is a fellow of the National Academy of Sciences, India; Indian Academy of Sciences, and the Zoological Society of London, 2002.

Contributors

Bhupesh Bagga
Consultant, Shantilal Shanghvi Cornea Institute Head, The Ramoji Foundation Centre for Ocular Infections, L V Prasad Eye Institute, Hyderabad, India

Ponnari Gottipati
Associate Director, Research Management, L V Prasad Eye Institute Co-Founder and Lead—Superheroes against Superbugs (https://sasuperbugs.org/) Kallam Anji Reddy Campus, Hyderabad, India

Joveeta Joseph
Head of Microbiology—LVPEI Network, Jhaveri Microbiology Centre, Dr. Savitri Sharma Chair of Ocular Microbiology, L V Prasad Eye Institute, Hyderabad, India

Arunasri Anakala Kotakonda
Scientist, Prof. Brien Holden Eye Research Centre, L V Prasad Eye Institute, Kallam Anji Reddy Campus, Hyderabad, India

Shalem Raj Padakandla
Postdoctoral Fellow, Prof. Brien Holden Eye Research Centre, L V Prasad Eye Institute, Kallam Anji Reddy Campus, Hyderabad, India

S. Shivaji
Distinguished Scientist and Director Emeritus, Prof. Brien Holden Eye Research Centre, L V Prasad Eye Institute, Kallam Anji Reddy Campus, Hyderabad, India

Abbreviations

AAO	American Academy of Ophthalmology
ACV	Acyclovir
AFB	Acid-fast bacilli
AFM-IR	Atomic force microscopy-infrared spectroscopy
AFR	Antifungal resistance
AgNP	Silver nanoparticles
AHC	Acute haemorrhagic conjunctivitis
AIDS	Acquired immunodeficiency syndrome
AK	*Acanthamoeba* (amoebic) keratitis
AKC	Atopic Keratoconjunctivitis
AmB-CH-NLC	Amphotericin B-loaded chitosan-modified nanostructured lipid carriers
AmB/PLA-g-CS	Amphotericin B-loaded self-aggregating nanoparticles
AMD	Age-related Macular Degeneration
AMP	Antimicrobial peptide
AMR	Antimicrobial resistance
AMRFinderPlus	Identifies AMR genes
AMR MPTF	Antimicrobial Resistance Multi Partner Trust Fund
AMT	Amniotic membrane transplantation
ANN	Artificial neural networks
ARG	Antibiotic resistance genes
ARG-ANNOT	Bioinformatic tool to discover antibiotic resistance genes in bacteria
ARGs-OAP v2.0	Database for quantification of ARGs in metagenomes
ARIBA	A tool to identify ARGs from reference sequences
ARMOR	Antibiotic resistance monitoring in ocular microorganisms
AST	Antimicrobial susceptibility testing
ATP	Adenosine triphosphate
ATT	Antitubercular therapy
AZA	Azathioprine
BCR	B-cell receptor
BK	Bacterial keratitis
BRiSK	Biome representational in silico karyotyping
CARD	Comprehensive antibiotic resistance database
Cas9	CRISPR associated protein 9
CDC	Centre for Disease Control
CDC	Contact dermato-conjunctivitis
cDNA	Complementary DNA
CFU	Colony forming unit
CHIKV	Chikungunya virus

CL	Contact lens
CLAIR	Cross-Linking Assisted Infection Reduction
CLSI	Clinical and Laboratory Standards Institute
CMV	Cytomegalovirus
CNS	Central nervous system
CoNS	Coagulase-negative staphylococci
cRNA	Complementary RNA
CRISPR	Clustered regularly interspaced short palindromic repeats in bacteria
CSF	Cerebrospinal fluid
CTF	Colorado tick fever
CXL	Cross-linking of cornea
CZS	Congenital Zika syndrome
DEC	Diethylcarbamazine
DED	Dry eye disease
DeepArgs	Applies deep learning to identify ARGs
DENV	Dengue virus
DEP	Dielectrophoretic properties
DMARDS	Disease-modifying anti-rheumatic drugs
DNA	Deoxyribonucleic acid
EBV	Epstein-Barr virus
ECDC	European Centre for Disease Prevention and Control
EKC	Epidemic keratoconjunctivitis
ELISA	Enzyme-linked immunosorbent assay
ERK	Extracellular signal-regulated kinases
ESBLs	Extended-spectrum β-lactamases
ESKAPE	Includes six nosocomial pathogens, *Enterococcus faecium, S. aureus, K. pneumoniae, A. baumannii, P. aeruginosa,* and *Enterobacter* spp.
EV	Enterovirus
E-test	Epsilometer test
FAO	Food and Agriculture Organization
FDA	US Food and Drug Administration
FFA	Fundus fluorescein angiography
FK	Fungal Keratitis
Fk01-37	Viable fungi from corneal surface of FK patients
FK-CR	Corneal scraping of fungal keratitis patients
FK-SW	Conjunctival swabs of fungal keratitis patients
FT-IR	Fourier transform infrared
GAE	Granulomatous amoebic encephalitis
GAIN	Generating Antibiotics Incentives Now
GAP	Global Action Plan
GARDP	Global Antibiotic Research & Development Partnership
GC	Gonococcal conjunctivitis
gD	Glycoprotein D
GLASS	Global Antimicrobial Resistance and Use Surveillance System
GN	Gram-negative
GO-AuNP	Graphene oxide-gold nanoparticle

GP	Gram-positive
GPA	Granulomatosis with polyangiitis
GPC	Giant papillary conjunctivitis
gRNA	Guide RNA
GROOT	Tool for identifying acquired resistance genes from gene data base
GVHD	Graft-versus-host disease
hAD	Human adenovirus
hBD	Human β-defensins
HC001-049	Viable fungi from conjunctiva of healthy individuals
HCC1-10	Viable fungi from the corneal surface of healthy individuals;
HEDS	Herpetic Eye Disease Study
HI	Harlequin ichthyosis
HLA	Human leukocyte antigen
hNP	Human neutrophil peptides
HC-SW	Conjunctival swabs of healthy controls
HERV-K	Human Endogenous Retrovirus K
HGA	Human granulocytotropic anaplasmosis
HIV	Human immunodeficiency virus
HME	Human monocytotropic ehrlichiosis
HPV	Human papillomavirus
HSK	Keratitis due to HSV
HSP	Heat shock protein
HSV	Herpes simplex virus
HTLV	Human T-lymphotropic virus
HT-qPCR	High throughput quantitative PCR
HZO	Herpes zoster ophthalmicus
IACG	Interagency Coordination Group on AMR
ICUs	Intensive care units
IDA	Isothermal DNA amplification
IFA	Immunofluorescence assay
IGRA	Interferon-gamma release assay
IL	Interleukins
INF	Interferons
IR	Infrared
ITS2	Internal transcribed spacer region of the nuclear ribosomal gene
IUSG	International Uveitis Study Group
IVCM	*In vivo* confocal microscopy
KD	Kawasaki disease
Kmer Resistance	Identifies SRGs in bacterial whole genome data
KP	Keratic precipitate
LAMP	Loop-mediated isothermal amplification
LC	Ligneous conjunctivitis
LEAP	Liver like antimicrobial peptide
LPS	Lipopolysaccharide
LVPEI	L V Prasad Eye Institute
MALDI-TOF	Matrix-assisted laser desorption/ionization time-of-flight mass spectrometry

MAPK	MAPK signal transduction pathways
MARDy	Mycology Antifungal Resistance Database (http://www.mardy.net)
MC	*Molluscum contagiosum*
MCV	Merkel Cell Polyomavirus
MDR	Multidrug resistant
MDR-PA	Multidrug resistant *Pseudomonas aeruginosa*
MEGARes 2.0	Database of published resistance sequences for antimicrobial drugs
MGD	Meibomian gland dysfunction patients
MIC	Minimum inhibitory concentration
MK	Microbial keratitis
MMP	Matrix metallo-proteinase
mNPS	Metagenomic nanopore sequencing
MRCNS	Methicillin-resistant coagulase-negative staphylococci
MRSA	Methicillin resistant *S. aureus*
MSRV	Multiple Sclerosis – associated retrovirus
MSSA	Methicillin-susceptible *S. aureus*
MTB	*Mycobacterium tuberculosis*
MTBBDRplus	A strip-based platform that simultaneously detects *M. tuberculosis* and mutations in rifampicin (rpoB gene)
MTB/RIF	Real-time PCR to detect *M. tuberculosis* and rifampicin resistance
MUTT	Mycotic Ulcer Treatment Trial
NAAT	Nucleic acid amplification test
NDM1	New Delhi metallo-beta-lactamase 1 gene
NGS	Next generation sequencing
NK	Neurotrophic keratitis
NLDO	Nasolacrimal duct obstruction
NMR	Nuclear magnetic resonance
NSAIDs	Non-steroidal anti-inflammatory drugs
NTM	Nontuberculous mycobacteria
NTS	Non-typhoidal *Salmonella*
OASIS	Ocular Autoimmune Systemic Inflammatory Infectious Study
OCP	Ocular cicatricial pemphigoid
OCT	Optical coherence tomography
OD	Optical density
OIE	Office International des Epizooties, global authority on animal health
OLM	Ocular larva migrans
OMCD	Optical Microscopy-based Microcolony Detection method
OTUs	Operational taxonomic units
PAC	Perennial allergic conjunctivitis
PAD	Paediatric acute dacryocystitis
PATRIC	Bacterial bioinformatics database and analysis resource
PCF	Pharyngoconjunctival fever
PCR	Polymerase chain reaction
PDAT	Photodynamic antimicrobial therapy
PDMS	Polydimethylsiloxane

PFR	Post-fever retinitis
PHMB	Polyhexamethylene biguanide
PLA-g-CS	Poly- (lactic acid) grafted-chitosan
PLSDA	Partial least square discriminant analysis
PLSR	Partial least square regression
PNA	Prednisolone
PNSP	Penicillin-non-susceptible *S. pneumoniae* (PNSP)
POC	Point-of-care
Point-finder	Detects chromosomal point mutations associated with antimicrobial resistance in bacterial WGS data
POS	Parinaud oculo-glandular syndrome
POW	Powassan encephalitis
PRISM	Phase-shift reflectometric interference spectroscopic measurement
PRK	Photorefractive keratectomy
PSSP	Penicillin-sensitive *S. pneumoniae*
PTFE	Polytetrafluoroethylene
PTK	Phototherapeutic keratectomy
QCM	Quartz-crystal microbalance
QOL	Quality of life
qPCR	Real-time polymerase chain reaction
RA	Rheumatoid arthritis
RAPD	Randomly amplified polymorphic DNA
ResFinder	Identifies antimicrobial resistance genes in NGS data
ReLAVRA	Latin American Antimicrobial Resistance Surveillance Network
RFLP	Restriction fragment length polymorphism
RGI	Resistance Gene Identifier from genome sequence data
RMSF	Rocky Mountain Spotted Fever
RNA	Ribonucleic acid
RP	Relapsing polychondritis
RPA	Recombinase polymerase amplification
rRNA	Ribosomal RNA
SAC	Seasonal allergic conjunctivitis
SARG	Structured Antibiotic Resistance Gene database
SCUT	Steroids for Corneal Ulcers Trial
SERS	Surface enhanced Raman spectroscopy or surface-enhanced Raman Scattering
ShortBRED	Profiles protein families in shotgun metagenomic sequencing data
SJS	Stevens-Johnson syndrome
SLE	Systemic lupus erythematosus
SLK	Superior limbic keratoconjunctivitis
SRST2	Short read sequence typing for bacterial pathogens
SSTAR	A stand-alone easy-to-use AMR gene predictor
SUN	Standardization of Uveitis Nomenclature
TB	Tuberculosis
Tβ4	Thymosin beta-4
TBRF	Tick-borne relapsing fever
TEM	TEM-1 β-lactamase is the most common plasmid-encoded β-lactamase

TFOS	Tear film and ocular surface society
TG	Trigeminal ganglion
TGF	Tumor growth factor
TLR	Toll-like receptor
TNF	Tumor necrosis factor
tRNA	Transfer RNA
TRUST	Tracking resistance in the United States today
TTV	Torque teno virus
UN	United Nations
UTI	Urinary tract infections
UV	Ultraviolet
VEGF	Vascular endothelial growth factor
VITEK	Vitek 2 Compact system for antimicrobial susceptibility testing
VK	Vernal keratoconjunctivitis
VKC	Viral keratitis
VLM	Visceral larva migrans
VMA	A vacuolar ATPase subunit
VZV	Varicella-zoster virus
VZV-E	VZV endotheliitis
VZV-SK	VZV stromal keratitis
WAA	World Antimicrobial Awareness Week
WGS	Whole genome sequencing
WHO	World Health Organisation
WMS	Whole metagenome sequencing
XDR	Extensively-drug resistant
XP	Xeroderma pigmentosum

1

Microbes of the Ocular Surface of the Human Eye

S. Shivaji

1.1 Introduction

Robert Hooke, in 1665, a Fellow of The Royal Society, was undoubtedly the first to describe the existence of a microorganism, the microfungus *Mucor* (Gest, 2004). Subsequently, bacteria and protozoa were discovered by Antoni van Leeuwenhoek in 1676. These path-breaking discoveries laid the foundations to microbiology and led to the discovery of microbes (bacteria, fungi and protozoa) from the environment, both from the biotic and abiotic habitats, including the freezing lakes of Antarctica (Fox, 2014), the hydrothermal vents of the sea-floor (Dick, 2019), the stratosphere (Shivaji et al., 2009) and so forth, and were also found associated with plants (Turner et al., 2013), animals (O'Hara et al., 2020) and human beings, including the surface and inside of the human body (Heintz-Buschart et al., 2018; Shivaji et al., 2021a; 2021b).

1.2 Discovery of Microbes of the Eye: The Cultivable Approach

Keilty in 1930, was the first to cultivate several ocular microorganisms, including haemolytic *Staphylococcus* using conjunctival swabs of healthy individuals. This also led to the discovery of other bacteria on the surface of the eye such as *Staphylococcus aureus*, an aerobic pathogen and *Propionibacterium acnes* and *Peptostreptococcus* spp., which are anaerobic bacteria (Perkins et al., 1975). It was also demonstrated that the bacterial abundance of the ocular surface was less compared to other habitats of the human body. The reduced abundance was probably because of antimicrobial factors in the tear film (McDermott, 2013) and the movement of the eyelids (Shovlin et al., 2013). Nevertheless, studies were undertaken to realise the abundance and diversity of bacteria associated with the eye since these microbes may have health implications. Several Gram-positive (*Staphylococcus epidermidis*, coagulase-negative *Staphylococcus* spp., *Corynebacterium* spp., *Micrococcus* spp., *Streptococcus* spp., and *Propionibacterium acnes*) and Gram-negative bacteria (*Escherichia coli*, *Pseudomonas aeruginosa, Neisseria* spp., and *H. influenzae*) (Sharma, 2012) were identified on the surface of the eye. These studies also indicated that the organisms recovered from the tear film were less in abundance compared to that recovered

DOI: 10.1201/9781003451105-1

from the conjunctival and lid swabs (Willcox, 2013). The fungi *Fusarium solani, Acremonium implicatum, Cladosporium sphaerospermum, Candida albicans* and *Aspergillus fumigatus* were also identified in some of these studies (Ramage et al., 2009, 2011; Shivaji et al., 2019).

1.3 Discovery of Microbes of the Eye: The Microbiome Approach

The cultivation method results in culturing only a small, miniscule fraction of the total community (<1%) of microbes derived from a biotic or abiotic sample and consequently the number of cultivable bacteria enumerated on a media plate is <90% compared to the number observed under the microscope (Harwani, 2012). This anomaly is referred to as 'The great plate count anomaly' (Staley and Konopka, 1985). To overcome this anomaly the microbiome approach is the preferred approach to establish the diversity and abundance of microorganisms (bacteria, fungi, viruses, protozoans, etc.). The microbiome encompasses a community of commensal, symbiotic, and pathogenic microorganisms associated with biotic sources like the human or animal body space or the environment (Lederberg and McCray, 2001) and their genomes, gene expression patterns and proteomes (Marchesi and Ravel, 2015). Compared to the conventional cultivation-dependent approach, the microbiome approach is a culture-independent method of detection of microbes which does not rely on culturing of microbes to identify them but it depends on the ability to extract DNA from a biological source, PCR amplification of a specific conserved phylogenetic marker like the V3-V4 region of 16S rRNA gene for prokaryotes and the ITS2 of the internal transcribed spacer region of the nuclear ribosomal gene for the eukaryotes and finally assigning the sequence of the amplicon to a taxa (Graham et al., 2007; Dong et al., 2011;Doan et al., 2016). This approach was extensively used to unravel the microbiomes of the gut (Turnbaugh et al., 2007; Jayasudha et al., 2019; Chakravarthy et al., 2018a; 2018b; Shivaji, 2017; 2019), skin, oral cavity, vagina, dental plaque (Liu et al., 2021; Peterson et al., 2013), ocular microbiomes (Dong et al., 2011; Doan et al., 2016; Lee et al., 2012; Graham et al., 2007; Shivaji et al., 2019; 2021a; 2021b; 2022; Konduri et al., 2021) and environmental samples of soil, water and so forth (Rappe and Giovannoni 2003; Achtman and Wagner 2008). A major outcome of this non-conventional microbiome-approach was the identification of a vast gene pool of sequences which were categorised as new uncultured and unexplored bacteria (Harwani, 2012).

1.3.1 Amplicon Sequencing: A Culture-independent Method of Detection of Microbes

In amplicon sequencing, total genomic DNA (Gallon et al., 2019) (Figure 1.1) from a given sample is extracted under aseptic conditions, quality checked, quantified and a specific phylogenetic marker (V3-V4 or ITS2 region) is amplified and an amplicon library prepared. Subsequently, following purification and quantification the amplicons are sequenced and the resulting sequences analyzed using tools of bioinformatics to annotate the sequences to a microorganism or a gene (Figure 1.2). In this approach the diversity of the microorganisms is revealed and is thus appropriate to unravel microbial diversity.

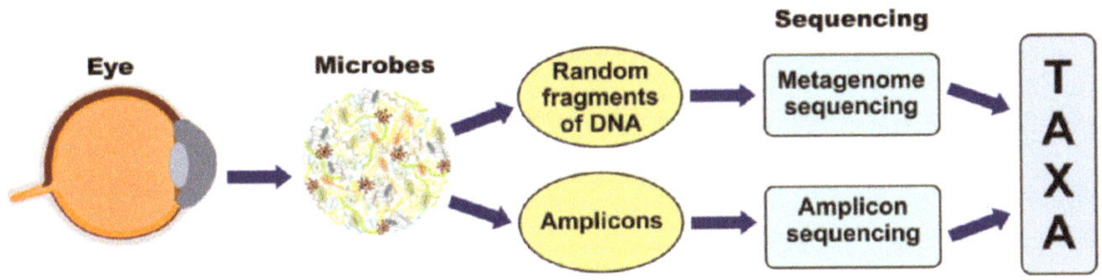

FIGURE 1.1 Schematic representation of the two different sequencing strategies of the microbiome approach namely amplicon sequencing or targeted sequencing and whole metagenome sequencing. (Modified from Shivaji, 2022.)

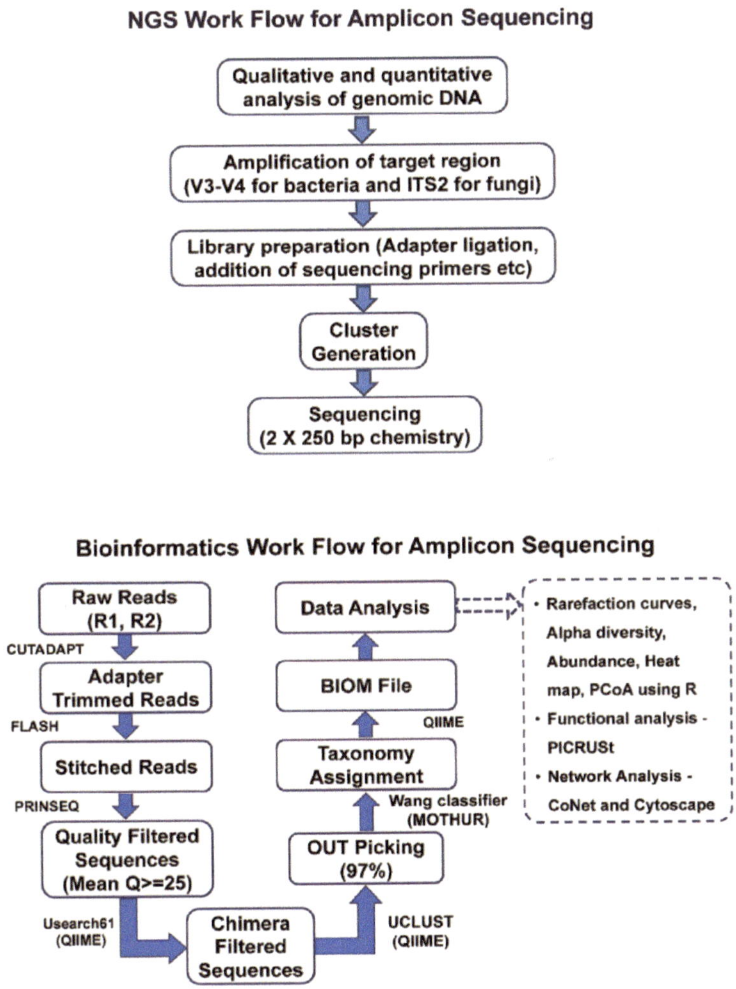

FIGURE 1.2 Steps involved in amplicon sequencing and bioinformatics workflow. The various programmes used for the bioinformatics analyses are adjacent to the arrow. (Modified from Shivaji, 2022.)

1.3.2 Metagenome Sequencing: A Culture-independent Method of Detection of Microbes and Functional Analysis

Total genomic DNA was fragmented, adapter-tagged and used for the preparation of the metagenomic library (Nextera reagent kit; Illumina Inc., USA). Libraries were quality checked prior to sequencing and processed to remove adapter sequences, low quality reads and sequences corresponding to host DNA to obtain high-quality reads (Li et al., 2015) and sequences affiliated to fungi, bacteria, viruses, protozoa, and archaea were separated using the Kraken database v 2.0.9 (Wood et al., 2019) (Figure 1.3). The sequences could be used for α-diversity analysis and for functional analysis such as antimicrobial resistance using gene tools of bioinformatics such as MEGARes 2.0 database (Doster et al., 2020), Resistance Gene Identifier tool v 5.1.0, a collation of multiple databases (CARD, ARG-ANNOT and ResFinder) and the specific Mycology Antifungal Resistance Database (MARDy) (http://www.mardy.net). Thus, the metagenome approach unveils the taxonomy of the microbes and could also be analysed to identify genomes and genes of particular interest such as virulence genes, antibiotic resistance genes and so forth (Figure 1.3) (Thomas et al., 2012).

1.3.3 Identification of Ocular Bacteria by the Microbiome-approach in the Normal Eye

Several studies have characterised commensal bacteria on the human ocular surface using NGS (Doan et al., 2016; Dong et al., 2011; Graham et al., 2007; Lee et al., 2012; Shin et al., 2016). At the phyla level the bacteria were affiliated to Firmicutes, Actinobacteria and Proteobacteria (Dong et al., 2011; Huang et al., 2016; Ozkan et al., 2017; Zhou et al., 2014) and could be assigned to several genera. In the human conjunctivae the number of genera varied from study to study and ranged from a minimum of fifty-nine bacterial genera (Dong et al., 2011) to a maximum of six hundred genera (Zhou et al., 2014). Hamady and Knight, (2009) referred to the genera that occurred in all the microbiomes as the "core genera" and the number of genera in the 'core' was not constant and varied. Huang et al., (2016) analysed the conjunctival microbiome in healthy human conjunctivae and identified 10 'core' genera with >1% mean abundance; these included *Corynebacterium* spp., *Staphylococcus* spp., *Pseudomonas* spp., *Millisia* spp., *Acinetobacter* spp., *Streptococcus* spp., *Anaerococcus* spp., *Simonsiella* spp., *Finegoldia* spp., and *Veillonella* spp. Five of these 'core' genera namely *Pseudomonas* spp., *Staphylococcus* spp., *Corynebacterium* spp., *Acinetobacter* spp., *Staphylococcus* spp., and *Streptococcus* spp., along with 7 other 'core' genera that included *Propionibacterium* spp., *Bradyrhizobium* spp., *Brevundimonas* spp., *Aquabacterium* spp., *Sphingomonas* spp., *Simonsiella* spp., and *Methylobacterium* spp. were reported by Dong et al., (2011) in the conjunctivae. In contrast, Zhou et al., (2014) identified *Corynebacterium* as the only 'core' genus in the conjunctiva. Ozkan et al., (2017) did not identify any 'core' genera but referred to a "putative core" consisting of genera with >1% abundance in approximately 80% of the conjunctival microbiomes. Compared to the ocular surface microbiome of the healthy eye studies have indicated that the eye surface microbiome varied in composition in various eye diseases, like dry eye disease, blepharitis, and trachomatous disease (Graham et al., 2007; Lee et al., 2012). But whether the ocular surface microbiome caused the disease or whether the disease altered the microbiome was not evident (Zilliox et al., 2020).

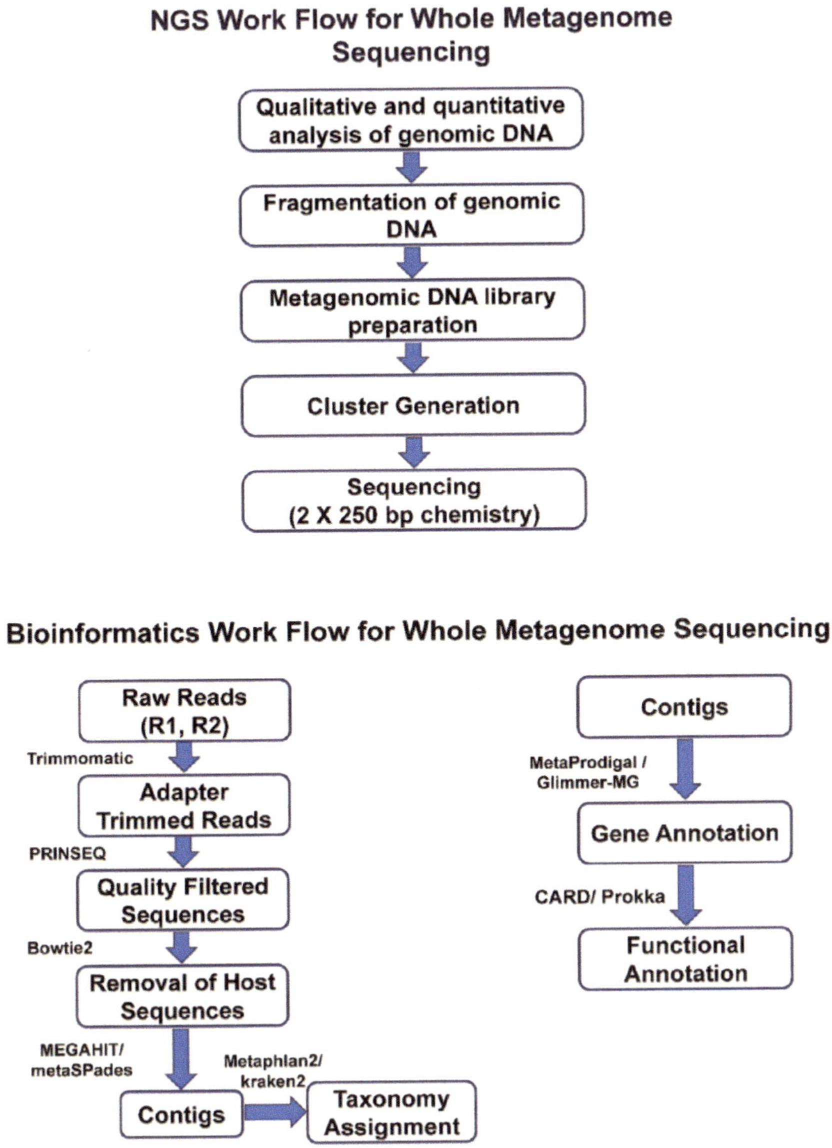

FIGURE 1.3 Steps in metagenome sequencing and bioinformatics workflow. The various programmes used for the bioinformatics analyses are adjacent to the arrow. (Modified from Shivaji, 2022.)

It should also be cautioned that the differences in the 'core' genera of the conjunctivae may also be due to various factors such as ethnic variations between individuals, the mode of sample collection, the region of the conjunctiva sampled (Huang et al., 2016), differences in environmental exposure, and due to the use of tetracaine during sample collection (Dong et al., 2011; Ozkan et al., 2018; Shin et al., 2016). Further, NGS cannot confirm whether the reported species are viable and resident of the ocular surface and were not acquired from the environment (Zegans and Van Gelder, 2014) and (or) the regions surrounding the eye. The detected microorganism in the conjunctival tissue microbiome could also be a bonafide commensal like *Corynebacterium mastitidis* on the conjunctival tissue of C57BL/6 mice (St. Leger and Caspi, 2018).

1.3.4 Identification of Ocular Fungi by the Microbiome-approach in the Normal Eye

In the conjunctival swabs of healthy control individuals, the predominant fungi were affiliated to Ascomycota and Basidiomycota and thy constituted 80% of the abundance and the Zygomycota fungi were a minority population (mean abundance, 0.05%). At the genera level, 80 different genera were identified in the conjunctival swabs of healthy individuals and 11 genera were present in majority of the eyes; those that were common to the eyes of all the individuals included *Aspergillus* spp., *Malassezia* spp., *Setosphaeria* spp., *Haematonectria* spp., *Penicillium* spp., *Candida* spp., *Emericella* spp., *Fusarium* spp., *Choiromyces* spp., *Cladosporium* spp., and *Cochliobolus* spp. (Sai Prashanthi et al., 2019; Shivaji et al., 2019). Sai Prashanthi et al., (2019) also confirmed that in the conjunctivae of healthy individuals *Aspergillus flavus* and *A. niger* were the only fungi identified whereas the corneal scrapings of fungal keratitis individuals were more diverse and in addition to *A. flavus*, *A. fumigatus*, *Fusarium spp.*, *Fusarium solani*, *Colletotrichum coccodes*, *Exserohilum rostratum*, *Cladosporium* sp., and *Curvularia lunata*, were identified. It was also observed that several sequences did not identify with a known fungal sequence and were categorised as unidentified septate and aseptate fungal filaments and dematiaceous fungi. This would imply that the fungal mycobiome of the human ocular surface is yet to be completely characterised.

1.3.5 Identification of Viruses by the Microbiome-approach in the Normal Eye

Culture independent approach has been used to characterise the viruses associated with the eye. For this purpose, conjunctival DNA was used and analysed using BRiSK (biome representational in silico karyotyping), a culture independent approach. In this method all the DNA present in a sample is restricted using a Type IIB restriction enzyme (BsaXI) and subsequently the fragments were sequenced. Doan et al., (2016) used the BRiSK method to identify Torque teno virus (TTV), Human Endogenous Retrovirus K (HERV-K) and Multiple Sclerosis–associated retrovirus (MSRV) in the conjunctiva. In a few ocular samples the Merkel Cell Polyomavirus (MCV), human papillomavirus (HPV) and Abelson murine leukaemia virus were also detected (Spurgeon et al., 2015). Of these viruses MSRV and HERV-K are endogenous to human beings and may have originated from the host DNA. But, in contrast TTV is known to be present on the ocular surface and associated with seasonal hyperacute pan uveitis and culture-negative endophthalmitis. Thus, it is not surprising that TTV was detected by using quantitative PCR in healthy conjunctiva and its presence was not influenced by gender and age (Lee et al., 2015; Smits et al., 2012).

1.4 Identification of Bacteria in the Diseased Eye

Several Gram-positive and Gram-negative bacteria (Ramage et al., 2009; Shivaji et al., 2019) are the etiological agents of ocular diseases like keratitis, conjunctivitis, blepharitis, keratitis, endophthalmitis, orbital cellulitis, retinitis, post fever retinitis, dacryocystitis and so forth (Table 1.1). Lee et al., (2012) observed increase in the relative abundance of *Staphylococcus* spp., *Corynebacterium* spp., *Streptophyta* spp., and *Enhydrobacter* spp. and a decrease in *Propionibacterium* spp. in blepharitis subjects. In Meibomian gland dysfunction (MGD) patients the abundance of *Pseudomonas acnes* increased whereas the abundance of *Staphylococcus* spp. decreased (Watters et al., 2016). Changes in the relative abundances of *Staphylococcus aureus,* coagulase negative *Staphylococcus* spp., *Corynebacterium* spp., *Propionibacterium* spp., *Rhodococcus* spp. *and Klebsiella oxytoca* were also seen in the ocular surface microbiota of individuals with dry eye disease (DED) (Graham et al., 2007). In individuals with Stevens-Johnson syndrome (SJS) (Kohanim et al., 2016) the inferior conjunctiva revealed growth of more pathogenic bacteria such as *Serratia nonliquefaciens, Enterobacter* spp., *Escherichia coli, Proteus mirabilis Morganella morganii,* and *Haemophilus* spp. and commensals like *Micrococcus* spp., and *Streptococcus viridians* compared to the healthy controls (Venugopal et al., 2016). In patients with bacterial conjunctivitis several bacteria like *Haemophilus influenza*, *Streptococcus pneumoniae* and *Staphylococcus aureus* were commonly reported (Sheikh et al., 2006); in contact lens wearers with bacterial keratitis coagulase-negative staphylococci, *S. aureus*, *S. pneumoniae,* and *Pseudomonas aeruginosa* were commonly detected (Green et al., 2008) whereas streptococci, coagulase-negative staphylococci, *S. aureus*, and Gram-negative bacilli were common in endophthalmitis patients (Watson et al., 2018) (Table 1.1).

1.5 Identification of Fungi in the Diseased Eye

Several fungal infections of the eye were identified by Klotz et al., (2000) (Table 1.2). Several species of fungi (>70) cause fungal keratitis (Thomas, 1994) and the incidence varied from 6 to 53% of all ulcerative keratitis in different countries (Thomas, 1994). Most of the cases were detected after trauma to the cornea in agricultural workers and gardeners from contaminated plant material (leaves, grain, branches, or wood), implements or contact lens wear. In fungal keratitis patients septate, filamentous, and saprophytic fungi are common and occasionally zygomycetes such as *Absidia* (Marshall, et al.1997) or *Rhizopus* spp. (Schwartz et al., 1978) may be implicated in keratitis. Filamentous fungal keratitis occurs less often with lens wear (Strelow et al., 1992; Yamamoto, et al., 1978). *Candida* keratitis is also common in patients with abnormal cornea as in dry eye syndrome, chronic ulceration, erythema multiforme, and in HIV patients (Hemady,1995) and in patients who chronically abused corneal anaesthetics (Chen, et al., 1998). Fungi like *Paracoccidioides brasiliensis* cause eyelid lesions (Slack et al., 1992). Endophthalmitis is a rare consequence of fungal keratitis (Borderie, et al., 1997). Fungal infections have also been reported from the adjacent structures of the eye such as the eyelids (Bloom et al., 1996), lacrimal system (Bloom et al., 1996), eye lashes (Creach et al., 1995), eyebrows, orbit (Rippon, 1988) and fungi

TABLE 1.1

Bacteria Associated with Ocular Diseases

Ocular disease	Disease-causing bacteria	Reference
Keratitis	*Actinomyces spp., Corynebacterium spp., Corynebacterium diphtheriae, Escherichia coli, Citrobacter spp., Enterobacter spp., Enterococcus spp., Moraxella spp., Moraxella lacunata, M. liquefaciens, Mycobacterium spp., Nocardia spp., Propionibacterium acne, Proteus spp., Pseudomonas aeruginosa, Serratia spp., S. marcescens; S. pneumoniae, S. viridians, Staphylococcus epidermidis, Staphylococcal spp., Staphylococcus aureus, Stenotrophomonas spp., Stenotrophomonas maltophilia, Klebsiella spp., Streptococcus spp., Streptococcus pneumoniae, Neisseria gonorrhoeae, Haemophilus ducreyi, Listeria monocytogenes, H. influenzae, H. parainfluenza, Alcaligenes spp., Achromobacter spp., Flavobacterium spp., Abiotrophia spp. and Bacillus spp.*	Egrilmez and Yildirim-Theveny, 2020; Turner and Stinson, 1965; Zhai et al., 2018; Kowalski and Dhaliwal, 2005.
Conjunctivitis	*Chlamydia trachomatis, H. influenzae, Klebsiella spp., Staphylococcus aureus, Neisseria gonorrhoeae, Staphylococcus epidermidis, Moraxella catarrhalis, Moraxella lacunata, M. liquefaciens, Streptococcus pneumoniae, Streptococcus viridians, -haemolytic and nonhemolytic streptococci, Enterococcus spp., Pseudomonas aeruginosa, Escherichia coli, Serratia marcescens and Proteus mirabilis.*	Alfonso et al., 2015; Ghosh et al., 1995; Quinet et al., 2010; Kowalski and Dhaliwal, 2005.
Blepharitis	*Streptophyta spp., Corynebacterium spp., Enhydrobacter spp., Bacillus oleronius, Streptococcus pyogenes, coagulase-negative staphylococci, S. aureus and Propionibacterium spp.*	Lee et al., 2012; Kowalski and Dhaliwal, 2005.
Scleritis	*Pseudomonas aeruginosa, Corynebacterium diphtheriae, Mycobacterium spp., M. tuberculosis, Staphylococcus pneumoniae, Nocardia spp., Stenotrophomonas maltophilia and Streptococcus spp.*	Jain et al., 2009; Reddy et al., 2015.

Endophthalmitis	*Enterococcus faecalis, E. coli, Neisseria meningitidis, Klebsiella spp., Micrococcus spp., S. aureus, S. pnuemoniae, S. viridians, S. epidermidis, Moraxella spp., Staphylococcus pyogenes, Serratia spp., Streptococcus viridians, Streptococcus spp., Propionibacterium acnes, Bacillus cereus, Pseudomonas spp., Proteus spp., Haemophilus influenza, Klebsiella spp., and coliform bacteria.*	Callegan et al., 2002; Driebe et al., 1986; Maitray et al., 2019; Tseng et al., 1996; Kowalski and Dhaliwal, 2005.
Orbital cellulitis	*Staphylococcus spp., S. aureus, Veillonella spp., Streptococcus spp., Haemophilus influenzae, Propionibacterium acne, Pseudomonas spp., E. coli, Fusobacterium lentum, F. necrophorum, Peptostreptococcus spp., Eikenellacorrodens, Arcanobacterium haemolyticum, Moraxella catarrhalis, Bacteroides spp., and Prevotella spp.*	Malik et al., 2004; Partamian et al., 1983; Sharma et al., 2012;
Dacryocystitis	*Citrobacter spp., Enterobacter spp., E. coli, Haemophilus influenzae, Pseudomonas vulgaris, Pseudomonas spp., Staphylococcus spp., S. aureus, S. epidermidis, Proteus mirabilis, Klebsiella pneumoniae, Streptococcus pneumonia, Streptococcus pyogenes, Propionibacterium acne, Peptostreptococcus spp., and Pseudomonas aeruginosa.*	Assefa et al., 2015; Chaudhry et al., 2005; Eslami et al., 2018; Mills et al., 2007.
Retinitis	*Treponema pallidum, Bartonella henselea, Mycobacterium tuberculosis, Borrelia burgdorferi* and *Toxoplasma gondii.*	Gupta et al., 2007; Sathiamoorthi and Smith, 2016.
Post-fever retinitis	*Rickettsia species* and *Salmonella typhi.*	Mahendradas et al., 2020; Prabhushanker et al., 2017.
Uveitis	*Mycobacterium tuberculosis, Chlamydia trachomatis, Treponema pallidum spp. and Borrelia burgdorferi.*	Barisani-Asenbauer et al., 2012.
Dry eye disease	*Coagulase negative Staphylococci, Corynebacterium spp., Propionibacterium spp., Rhodococcus erythropolis, Klebsiella oxytoca and Erwinia spp.*	Graham et al., 2007; Kohanim et al., 2016.

(continued)

TABLE 1.1 (Continued)

Bacteria Associated with Ocular Diseases

Ocular disease	Disease-causing bacteria	Reference
Meibomian gland dysfunction	*Staphylococcus* spp., *S. lugdunesis, S. aureus*, *Lysinibacillus* spp., *Microbacteriaceae*, *Bacillus* spp., *Moraxella osloensis*, *Xanthomonadaceae*, *Corynebacterium* spp., *Paenibacillus* spp., *Corynebacterium macginleyi, Acinetobacter calcoaceticus*, *Haemophilus influenzae*, *Actinomyces* spp., *Stenotrophomonas maltophilia, Pantoea agglomerans* and *Cladosporium* spp.	Jiang et al., 2018; Watters et al., 2016
Chronic Stevens-Johnson syndrome	*Pseudomonas* spp., *Staphylococcus* spp., *S. aureus*, S. *epidermidis*, *Streptococcus* spp., and *Acinetobacter* spp.	Kittipibul et al., 2020; Moeller et al., 2005; .
Sjorgens syndrome	Firmicutes, Actinobacteria, Proteobacteria, Bacteroidetes	de Paiva et al., 2016.
Lax eyelid syndrome	*Corynebacterium* spp., *Hydrotalea* spp., *Ralstonia* spp., *Clostridium_*XI, *Staphylococcus* spp. and *Rothia* spp.	Zilliox et al., 2020.
Ocular graft-vs host disease	*Staphylococcus* species, *alpha-hemo Streptococcus* spp., *Corynebacterium* spp., *Propionibacterium acnes*, *Haemophilus influenza*, *Enterobacteriaceae*, *Selenomonas* spp., *Pasteurellaceae* and *Lactobacillus/Streptococcus* spp.	Zilliox et al., 2020.

Source: Adapted and modified from Konduri et al., 2021 and Shivaji, 2022.

TABLE 1.2

Fungi Associated with an Ocular Disease

Ocular disease	Disease causing fungi
Keratitis	*Fusarium solani, Fusarium spp., Alternaria spp., Aspergillus* spp., *Aspergillus flavus, Candida* spp., *Paecilomyces lilacinus (*Okhravi, 1997*), Absidia* spp., *(*Marshall, et al., 1997*), Rhizopus spp.* (Schwartz et al., 1978), *Candida albicans* (Butrus and Klotz. 1986), *Candida lipolytica, Candida humicola* (Nitzulescu and Niculescu. 1975), *Candida guilliermondii* (Ainbinder, 1998), *Curvularia* spp. (Thomas,1994*), Cryptococcus laurentii* (Ritterband et al., 1998*), Acremonium, Phialophora* spp. (Liesegang, and Forster. 1980) *and Pseudomonas aeruginosa* (Butrus et al., 1987).
Conjunctivitis	*Pneumocystis* spp. (Ruggli, et al., 1997*), Candida* spp. (Gumbel et al., 1990)*, Cryptococcus* spp. (Saran and Pomilla. 1996)*, Pneumocystis* spp.*, P. carinii* (Sha et al., 1992), *Histoplasma* spp., *Aspergillus* spp. (Morinelli, et al., 1993), *Fusarium* spp. (Glasgow et al., 1996) *and Bipolaris* spp. (Pavan and Margo,1993).
Endogenous Endophthalmitis	*Candida spp., Candida albicans, Aspergillus spp., Aspergillus flavus* (Roney, et al., 1986), *Aspergillus fumigatus, Aspergillus niger, Aspergillus terreus, Aspergillus glaucus* (Weishaar et al., 1998), *Aspergillus nidulans* (Tong et al., 1990), *Fusarium spp., (*Glasgow et al., 1996*) Penicillium spp., H. capsulatum* (Magie,1995*), B. dermatitidis* (Logan and O'Day 1995*), Sporothrixschenckii* (Agger, 1995), *C. immitis* (Ganley and Klotz, 1995), and *Cryptococcus neoformans* (Crump et al., 1992).
Exogenous Endophthalmitis	*Candida* spp., *Candida parapsilosis, Candida glabrata* (Cameron et al., 1998), *Candida famata* (Rao et al., 1991), *Fusarium* species (Pflugfelder et al., 1988), *Paecilomyces lilacinus* (Okhravi et al., 1997), *Fusarium moniliforme* (Srdic, et al., 1993), *Exophiala jeanselmei* (Hammer et al., 1983), *P. boydii* (Carney et al., 1996), *A. Niger* (Jager et al., 1994), *Scytalidium imidiatum* (al-Rajhi et al., 1993), *Helminthosporium* spp. (Das et al., 1994), *S. schenckii* (Witherspoon et al., 1990), *Penicillium chrysogenum* (Eschete et al., 1981), and *Lasiodiplodia theobramae* (Borderie et al., 1997).
Optic neuropathy	*Zygomycetes* spp., *Histoplasma* spp. (Yau et al., 1996) and *Cryptococcus* spp. (Garrity et al., 1993).
Sino-orbital disease	*Aspergillus* spp. (Johnson et al., 1999).
Orbital cellulitis	*Bipolaris* spp. (Jacobson et al., 1992)
Chorioretinitis	*Pneumocystis carinii* (Klotz et al., 2000)
Neuro-retinitis	coagulase-negative *Staphylococcus*, diphtheroids, and *Propionibacteriumacnes* (Wilhelmus 1987).
Blepharitis	*Microsporum canis* (Creach et al., 1995), *Malassezia furfur*, and *Candida* species (Parunovic and Halde 1967).

Source: Compiled by the author.

could cause blepharitis (Parunovic, and Halde, 1967), proptosis of the eye (Ammon et al., 1992; Ruggli et al., 1997) and dacryocystitis (Brook and Frazier. 1998). Fungi do not appear to pose a problem in patients with anophthalmus. A combination of surgery and antifungal drugs is the usual therapeutic approach to most fungal eye infections (Klotz et al., 2000).

In a recent study, NGS detected 65 distinct fungal genera with *Aspergillus* spp., *Setosphaeria* spp., *Malassezia* spp., and *Haematonectria* spp. present in all the 25 eyes that were sampled (Shivaji et al., 2019). In a subsequent study it was demonstrated that compared to the conjunctivae of healthy controls, the conjunctiva and corneal scrapings of fungal keratitis individuals had a greater abundance of opportunistic pathogens or pathogens which could be related to ocular disease (Sai Prashanthi et al., 2019). In this study using the mycobiome approach a total of 111 genera were identified in the conjunctivae of healthy controls and in the conjunctivae and corneal scrapings of fungal keratitis patients (Table 1.3). Further, the Venn diagram (Figure 1.4) based on the distribution of the 111 different genera showed that the conjunctivae of the healthy controls were more similar with the conjunctivae (56 genera were common) rather than to corneal scrapings from fungal keratitis individuals (37 genera were common). In addition, genera unique to a particular group are shown in Table 1.4. Three fungal genera could discriminate between conjunctivae of healthy controls and keratitis individuals, 33 genera between the conjunctivae of healthy controls and corneal scrapings of keratitis individuals and 27 genera between the conjunctivae and corneal scrapings of keratitis individuals (Sai Prashanthi et al., 2019). These observations agree with a review published by Karsten et al., (2012) which had indicated based on published literature between 1950–2012 that 144 species of fungi from 92 genera were implicated in keratitis (Karsten et al., 2012).

1.6 Identification of Protozoans, Nematodes, Cestodes, Trematodes, and Ectoparasites in the Diseased Eye

Several other organisms such as protozoa, nematodes, cestodes, trematodes, and ectoparasites are associated with the diseased eye (Tables 1.5 and 1.6). These infections may be confined to a specific region of the world or may be common to all countries. Further some of them could be congenital and occur more frequently in AIDS patients. Tables1.5 and 1.6 summarise the above pathogens involved in ocular infection and their characteristics such as occurrence, infection source, symptoms of the disease, diagnosis, and therapy.

1.7 Ocular Infectious Diseases and Treatment

Bacteria, fungi, viruses, protozoans, ectoparasites and so forth could cause a multitude of eye diseases (Tables 1.1, 1.2, 1.5 and 1.6). Infections could affect the cornea, the conjunctiva, and the inside of the eye, as in bacterial and fungal endophthalmitis or infect the tear glands causing inflammatory dacryostenosis and or block the tear drainage system as in dacryocystitis (Chou et al., 2022). Infection can also be due to a corneal ulcer, and this could lead to severe vision loss. In orbital cellulitis, infection

TABLE 1.3

Fungal Genera Associated with the Human Ocular Surface

S.No.	Genus	HC-SW Median abundance	HC-SW Present out of 37 eyes sampled	FK-SW Median abundance	FK-SW Present out of 23 eyes sampled	FK-CR Median abundance	FK-CR Present out of 29 eyes sampled
1	*Aspergillus*	3.5886	37	7.4317	23	0.4797	29
2	*Malassezia*	2.7353	37	1.1787	23	0.0137	20
3	*Emericella*	0.1851	32	0.0056	15	0	5
4	*Haematonectria*	0.1693	35	0.2046	23	1.7626	28
5	*Candida*	0.0572	28	0.0589	17	0.0315	18
6	*Cladosporium*	0.0208	32	0.0049	19	0	5
7	*Fusarium*	0.014	28	0.0036	21	0.3567	24
8	*Setosphaeria*	0.01	25	0	6	0	9
9	*Penicillium*	0.0022	25	0	7	0	3
10	*Choiromyces*	0.0007	27	0.0003	16	0	2
11	*Cochliobolus*	0.0004	30	0.0007	16	0.0101	25
12	*Neosartorya*	0.0004	19	0	2	0	5
13	*Xeromyces*	0.0002	22	0	9	0	13
14	*Pichia*	0.0001	19	0.0002	13	0	10
15	*Clavispora*	0	7	0.0002	13	0	0
16	*Nectria*	0	18	0.0001	12	0.001	18
17	*Pleurostomophora*	0	3	0	6	0.0011	17
18	*Neocosmospora*	0	9	0	7	0.0009	17
19	*Botryosphaeria*	0	0	0	1	0.0005	15
20	*Acremonium*	0	3	0	1	0	0
21	*Alternaria*	0	6	0	1	0	1

(continued)

TABLE 1.3 (Continued)

Fungal Genera Associated with the Human Ocular Surface

S.No.	Genus	HC-SW		FK-SW		FK-CR	
		Median abundance	Present out of 37 eyes sampled	Median abundance	Present out of 23 eyes sampled	Median abundance	Present out of 29 eyes sampled
22	*Anthracophyllum*	0	0	0	1	0	0
23	*Arthrinium*	0	9	0	0	0	0
24	*Arthrobotrys*	0	0	0	1	0	0
25	*Arxula*	0	1	0	0	0	0
26	*Aureobasidium*	0	16	0	4	0	3
27	*Auricularia*	0	3	0	1	0	0
28	*Bartalinia*	0	0	0	3	0	0
29	*Bipolaris*	0	6	0	2	0	12
30	*Bullera*	0	1	0	3	0	1
31	*Calcarisporiella*	0	1	0	0	0	0
32	*Cephaliophora*	0	0	0	0	0	11
33	*Ceratobasidium*	0	0	0	2	0	0
34	*Cercospora*	0	14	0	9	0	0
35	*Chaetomium*	0	1	0	10	0	1
36	*Chrysosporium*	0	2	0	1	0	0
37	*Cladorrhinum*	0	0	0	1	0	1
38	*Colletotrichum*	0	6	0	7	0	5
39	*Coniothyrium*	0	0	0	5	0	0
40	*Coriolopsis*	0	3	0	5	0	0
41	*Cryptococcus*	0	6	0	1	0	2
42	*Curvularia*	0	0	0	2	0	6
43	*Cyberlindnera*	0	0	0	2	0	3

44	*Cytospora*	0	0	0	1	0	0
45	*Dactylella*	0	1	0	0	0	0
46	*Diaporthe*	0	0	0	0	0	11
47	*Ellurema*	0	0	0	1	0	0
48	*Epicoccum*	0	0	0	1	0	0
49	*Eupenicillium*	0	2	0	0	0	1
50	*Eurotium*	0	13	0	6	0	0
51	*Filobasidiella*	0	0	0	0	0	3
52	*Flavodon*	0	1	0	6	0	0
53	*Galactomyces*	0	1	0	1	0	0
54	*Gibberella*	0	0	0	0	0	13
55	*Gloeomyces*	0	0	0	1	0	0
56	*Hanseniaspora*	0	6	0	1	0	2
57	*Hortaea*	0	0	0	3	0	0
58	*Humicola*	0	0	0	1	0	0
59	*Hypoxylon*	0	1	0	2	0	0
60	*Kluyveromyces*	0	1	0	2	0	0
61	*Lasiodiplodia*	0	4	0	0	0	13
62	*Leptosphaerulina*	0	14	0	8	0	0
63	*Macrophomina*	0	2	0	2	0	11
64	*Musicillium*	0	1	0	1	0	0
65	*Mycosphaerella*	0	2	0	1	0	0
66	*Myrothecium*	0	2	0	0	0	11
67	*Nakaseomyces*	0	0	0	1	0	5
68	*Nigrospora*	0	7	0	0	0	0
69	*Paecilomyces*	0	2	0	0	0	2
70	*Passalora*	0	4	0	0	0	0

(continued)

TABLE 1.3 (Continued)

Fungal Genera Associated with the Human Ocular Surface

		HC-SW		FK-SW		FK-CR	
S.No.	**Genus**	**Median abundance**	**Present out of 37 eyes sampled**	**Median abundance**	**Present out of 23 eyes sampled**	**Median abundance**	**Present out of 29 eyes sampled**
71	*Penidiella*	0	0	0	0	0	1
72	*Peniophora*	0	3	0	6	0	0
73	*Pestalotiopsis*	0	4	0	0	0	0
74	*Phaeoacremonium*	0	0	0	2	0	0
75	*Phaeosphaeriopsis*	0	1	0	0	0	0
76	*Phanerochaete*	0	1	0	0	0	0
77	*Phialophora*	0	1	0	0	0	0
78	*Phlebiopsis*	0	1	0	3	0	0
79	*Phoma*	0	7	0	5	0	0
80	*Podospora*	0	12	0	5	0	0
81	*Psathyrella*	0	4	0	2	0	0
82	*Pseudozyma*	0	2	0	0	0	0
83	*Pyricularia*	0	2	0	0	0	0
84	*Radulidium*	0	0	0	1	0	0
85	*Ramichloridium*	0	2	0	0	0	0
86	*Retroconis*	0	1	0	0	0	11
87	*Rhinocladiella*	0	0	0	0	0	4
88	*Rhizoctonia*	0	0	0	1	0	11
89	*Rhizopus*	0	2	0	1	0	0
90	*Rhizopycnis*	0	0	0	2	0	0
91	*Rhodotorula*	0	2	0	3	0	0
92	*Saccharomyces*	0	5	0	10	0	3

93	*Sagenomella*	0	2	0	0	0	0
94	*Sarcinomyces*	0	2	0	0	0	0
95	*Scedosporium*	0	0	0	1	0	0
96	*Scolecobasidium*	0	0	0	1	0	0
97	*Stachybotrys*	0	2	0	0	0	0
98	*Stagonosporopsis*	0	5	0	7	0	1
99	*Strelitziana*	0	1	0	1	0	0
100	*Talaromyces*	0	4	0	1	0	3
101	*Thielavia*	0	0	0	4	0	7
102	*Torulaspora*	0	2	0	1	0	0
103	*Toxicocladosporium*	0	0	0	1	0	0
104	*Trichoderma*	0	4	0	0	0	1
105	*Trichomonascus*	0	2	0	1	0	0
106	*Ustilaginoidea*	0	2	0	3	0	2
107	*Villosiclava*	0	2	0	1	0	0
108	*Wallemia*	0	5	0	5	0	6
109	*Yamadazyma*	0	0	0	1	0	0
110	*Zasmidium*	0	4	0	0	0	0
111	*Zygoascus*	0	1	0	0	0	0
112	Unclassified	59.2213	37	20.7464	23	3.5052	29
113	*Others	0.7041	37	1.583	23	1.2312	29
	Total Genera	**80**		**81**		**50**	

Source: Modified from Sai Prashanthi et al. 2019.

Notes: Fungal microbiomes from conjunctivae of healthy controls (HC-SW), conjunctival swabs of fungal keratitis patients (FK-SW) and corneal scraping of fungal keratitis patients (FK-CR). Others include singletons and sparse
OTUs (with <0.001% of total high-quality reads). Sparse OTUs with < 0.001% of total high-quality reads were not included.

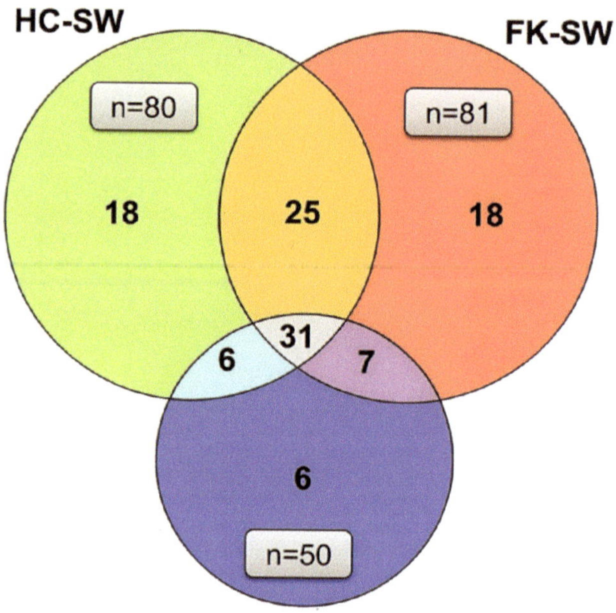

FIGURE 1.4 Venn diagram representing the number of fungal genera in the conjunctivae of healthy controls (HC-SW), conjunctivae of keratitis individuals (FK-SW), and corneal scrapings of (FK-CR) of keratitis individuals. (Modified from Sai Prashanthi et al., 2019.)

involves the soft tissue of the eyelids and could spread in the absence of treatment. Most ocular infections are self-resolving and need no medication but in the remaining cases the infection could be profoundly serious and requires immediate medical attention (Chou et al., 2022). Diagnosis is by physical examination of the affected area and by culturing a sample from the affected area to detect the infective agent and determine the most effective treatment, such as an antibiotic, an antifungal or antiviral agent that selectively targets the infective agent. Viral infections require administration of steroid eye drops to reduce ocular inflammation. Ocular bacterial infections are treated with several classes of antimicrobial agents including cephalosporins, β-lactams, fluoroquinolones and so forth, which were developed for systemic use. Treatment of ocular infections with systemic antibiotics is slightly different. For instance, ocular infections are treated using topical therapy of antimicrobial drugs as antibiotic eye drops. Topical therapy using antibacterial agents for the treatment of superficial ocular infections including conjunctivitis, blepharitis and corneal ulcers have several advantages when compared with systemic therapy. The most obvious advantage is that high concentrations of the antibiotic can be delivered to the site of infection. In addition, antibacterial agents are frequently used to augment treatment for intraocular infections.

TABLE 1.4

Unique Fungal Genera Associated with the Human Ocular Surface

	Present exclusively in HC-SW		Present exclusively in FK-SW		Present exclusively in FK-CR	
S.No.	**Genera**	**Habit***	**Genera**	**Habit**	**Genera**	**Habit***
1	*Arthrinium*		*Coniothyrium*	P	*Gibberella*	P
2	*Nigrospora*	P/H	*Hortaea*	H	*Diaporthe*	P
3	*Zasmidium*		*Bartalinia*	P	*Cephaliophora*	P
4	*Passalora*	P	*Rhizopycnis*	P	*Rhinocladiella*	H
5	*Pestalotiopsis*	P	*Phaeoacremonium*	P/H	*Filobasidiella*	H
6	*Pseudozyma*	H	*Ceratobasidium*	P	*Penidiella*	P
7	*Sagenomella*	A/H	*Epicoccum*	P/A		
8	*Sarcinomyces*		*Toxicocladosporium*	P/A		
9	*Pyricularia*	P	*Gloeomyces*			
10	*Ramichloridium*	H	*Yamadazyma*	P/H		
11	*Stachybotrys*	A/H	*Anthracophyllum*			
12	*Dactylella*	P	*Cytospora*	P		
13	*Phaeosphaeriopsis*	P	*Ellurema*			
14	*Zygoascus*	H	*Scedosporium*	H		
15	*Phanerochaete*	P	*Scolecobasidium*	A/H		
16	*Arxula*	H	*Arthrobotrys*	P		
17	*Phialophora*	H	*Humicola*	P/H		
18	*Calcarisporiella*		*Radulidium*			

Source: Modified from Sai Prashanthi et al. 2019.

Note: Fungal genera unique to the conjunctivae of healthy controls (HC-SW), and conjunctivae (FK-SW) and corneal scrapings (FK-CR) of keratitis individuals.

TABLE 1.5
Protozoans Associated with an Ocular Disease

Organism/ disease	Occurrence, infection sources, symptoms, diagnosis and therapy
	Protozoans
Toxoplasmosis: *Toxoplasma gondii*	**Occurrence:** *Toxoplasma gondii* affects approximately one-third of the adult human population (Frenkel, 1971) and occurs in warm and humid climates as in Caribbean and Central America (Feldman, 1982) and also in Europe (McCabe and Remmington 1988). **Infection sources:** raw or undercooked meat, contaminated water, feces of cat, the definitive host. **Symptoms:** necrotizing chorioretinitis (Feldman, 1982) which may include strabismus, nystagmus, and blindness (Perry and Merritt, 1983); acute disease is associated with scotoma, photophobia, and loss of central vision. **Diagnosis:** serologic testing, ELISA, agglutination assay, indirect immunofluorescence, immunosorbent agglutination assay, immunocapture and immunoblot assays (Garcia and Bruckner, 1997). **Therapy:** Pyrimethamine with sulfadiazine are the first-line therapy of *T. gondii*-related chorioretinitis (Engstrom et al., 1991).
Trypanosomiasis: *Trypanosoma cruzi*	**Occurrence:** endemic to Central and South America and may cause 10% of all deaths. **Infection sources:** 'Chagas' disease, or American trypanosomiasis, in which *Trypanosoma cruzi* bites and transfers trypomastigotes into the blood stream which then move to muscle cells, neural tissue, and the reticuloendothelial system. Trypanosomiasis may also occur through blood transfusion, organ transplantation, placental transfer, and accidental ingestion (Garcia and Bruckner, 1997). **Symptoms:** *T. cruzi* causes chagoma and if the protozoan bites near the orbit the patient experiences palpebral and periorbital painless edema (Romana's sign) accompanied by fever, malaise, and anorexia and may also lead to cardiomyopathy, megaesophagus, and megacolon. **Diagnosis:** detection of trypomastigotes in the bloodstream by culture (Albesa and Eraso. 1983) or PCR. **Therapy:** in the acute stage nifurtimox and benznidazole are administered.
Plasmodium vivax, *P. ovale*, *P. malariae*, and *P. falciparum*	**Occurrence:** affects over 3 million humans. Infection sources: *Anopheles* mosquito transfers the *Plasmodium* sporozoites to humans the intermediate host. **Symptoms:** microvascular obstruction, hemolysis, tissue hypoxia, hypoglycaemia, retinal haemorrhage (Kochar et al., 1998). Other ocular effects include amaurosis fugax, optic neuritis, glaucoma, panuveitis, oculomotor paralysis, and cortical blindness. **Diagnosis:** Microscopic examination.

	Therapy: Chloroquine is the standard of treatment, but chloroquine and its derivatives cause several ocular side effects such as blurred vision, scotoma, visual-field defects, disturbed retinal pigmentation and sensitivity to red light (Klotz et al., 2000).
Leishmaniasis: *Leishmania* spp.	**Occurrence**: occurs worldwide and is more prevalent in Asia, the Middle East, and Africa. **Infection sources**: Leishmaniasis occurs when promastigotes are injected into humans by the sandfly. Leishmaniasis could be visceral, cutaneous or mucocutaneous leishmaniasis. Visceral leishmaniasis, is a systemic disease and is also known as kala-azar. **Symptoms:** Visceral leishmaniasis (Kala-azar) causes chorioretinitis, central retinal vein thrombosis, iritis, papillitis, keratitis (Kean et al., 1991), retinal hemorrhages and glaucoma. Cutaneous leishmaniasis affects the eye and the lacrimal duct (Chaudhry et al., 1998; Kean et al., 1991) and causes inflammation and necrosis of periorbital structures (Sodafy et al., 1981). Mucocutaneous leishmaniasis affects the conjunctival mucosa leading to ulceration and possible loss of the eye. **Diagnosis:** ocular manifestations of kala-azar are rare and when it occurs it causes chorioretinitis, central retinal vein thrombosis, iritis, papillitis, and keratitis. Additionally, retinal haemorrhage and glaucoma may occur after the treatment of kala-azar. In cutaneous leishmaniasis infection occurs near the eye and occasionally spreads to the lacrimal duct (Chaudhry et al., 1998), and inflammation and necrosis of periorbital structures, including the eyelid is seen (Sodafy et al., 1981). If the bite occurs on the conjunctival mucosa, the disease is termed mucocutaneous leishmaniasis and causes ulceration and possible loss of the eye. Leishmaniasis is diagnosed by direct demonstration of organisms on tissue smears or biopsy or by culturing. **Therapy**: No universally applicable treatment has been identified for Leishmaniasis. Parasite and host factors, as well as clinical characteristics must be considered (Aronson et al., 2017). Some the drugs used include Ampotericin B deoxycholate, pentavalent antimonials, azoles, paromycin preparations, hetatherapoy, cryo-therapy, pentamidine isethionate, miltefosine, (Aronson et al., 2017).

(*continued*)

TABLE 1.5 (Continued)

Protozoans Associated with an Ocular Disease

Organism/ disease	Occurrence, infection sources, symptoms, diagnosis and therapy
	Protozoans
Acanthamoeba keratitis: *Acanthamoeba* spp., *Acanthamoeba castellani* and *Acanthamoeba polyphaga*	**Occurence**:Acanthamoeba keratitis is a rare corneal infection and the common causative species are *Acanthamoeba castellani* and *A. polyphaga*. **Infection sources:** Transmission is through the pathogen contact with the cornea through contaminated substances, water, soil, or trauma to the eye. The amoebae feed on the keratocytes and destroy the corneal stroma. **Symptoms**: corneal injury (Kean et al., 1991), severe pain, conjunctival edema, loss of vision (Garcia and Bruckner, 1997), a ring infiltrate around the cornea, corneal penetration (Harwood, et al., 1988), intense inflammation (Kean et al., 1991), hypopyon, hyphema, and uveitis (Heffler, et al., 1996). Advanced stages of the disease result in vision loss and the need for corneal transplants. **Diagnosis**: detection of trophozoites and cysts in corneal scrape/biopsy, polymerase chain reactions, impression cytology, or confocal microscopy and culturing of the *Acanthamoeba* (Visvesvara, 1995). **Therapy**: combination therapy with biguanide and aromatic diadine eye drops for several months (Gilbert et al.1998).
Microsporidiosis	**Occurrence:**Microsporidia are observed in particularly in AIDS and immunocompetent individuals (Han and Weiss, 2017). **Infection sources**: different species of microsporidia like *Enterocytozoon bieneusi* and *Encephalitozoon intestinalis*. **Symptoms:** keratoconjunctivitis (Lowder et al., 1996), conjunctival hyperemia, punctate epithelial keratitis, necrotizing keratitis, and corneal ulcers accompanied by photophobia, foreign body sensation, and decreased visual acuity (Didier, et al., 1991). **Diagnosis:** detection of spores in stool and biopsy material, microscopic examination of tissue sections, immunofluorescence or electron microscopy (Weber et al., 1994) and PCR (Kock et al., 1997). **Therapy**: Fumagillin and albendazole are used (Didier et al., 2005; Han and Weiss, 2018).

Giardiasis: *Giardia lamblia*	**Occurrence**:*Giardia* parasites have been detected in most countries and continents and is a serious problem in countries with poor sanitation. **Infection sources**: In the United States, giardiasis is the most common parasitic infection of the intestines. **Symptoms:** manifests as intestinal illness, but ocular findings manifest as chorioretinitis, iridocyclitis, retinal haemorrhage, retinal arteritis, vitreous haemorrhage, and uveitis (Klotz et al., 2000). **Diagnosis**; stool examination for the *Giardia* parasite **Therapy:** may resolve by itself, using antibiotics like Metronidazole, Tinidazole and Nitazoxanide (Klotz et al., 2000).
Rhinosporidiosis: *Rhinoporidium seeberi*	**Occurrence:** Rhinosporidiosis is a chronic granulomatous tropical disease of the mucous membranes by *Rhinoporidium seeberi* (Izimukwiye, et al., 2019) and ocular manifestations account for 15% of cases. Frequently seen in India and tropical South America. In India, conjunctival rhinosporidiosis occurs in desert regions. **Infection sources**: caused by *Rhinoporidium seeberi*, an endosporulating microorganism. **Symptoms:** *Rhinoporidium seeberi* causes mucocutaneous disease involving the palpebral conjunctiva (Rippon, 1988). **Diagnosis**: history, clinical findings, examination of smears of macerated tissue, histopathology of the biopsied tissues using Gomori methenamine silver, periodic acid-Schiff, or hematoxylin and eosin staining, potassium hydroxide preparation and microscopic identification of *Rhinoporidium seeberi*. Serologic testing by immunoblot or enzyme-linked immunosorbent assay identification of antirhinosporidial antibody has been developed (Sudasinghe et al., 2011). **Therapy:** surgical excision of the rhinosporidiosis lesions, followed by electro-cautery of the lesion's base (Godwins et al., 2008). Drugs like griseofluvin, amphoterecin B and dapsone (4, 4-diaminodiphenyl sulphone) (Madke et al., 2011) are useful in individuals with multisite rhinosporidiosis but no tangible success was observed with medical therapy (Nayak et al., 2007).

Source: Compiled by the author.

TABLE 1.6

Nematodes, Cestodes and Ectoparasites Associated with an Ocular Disease

	Nematodes
Onchocerciasis: *Onchocercus volvulus*	**Occurrence**: *Onchocercus volvulus* is estimated to have infected forty million humans out of which 5% are blind due to the infection (river blindness) (Connor and Neafie,1976). Onchocerciasis is endemic in tropical Africa, South America, and the Arabian Peninsula. Infection source: infection occurs when an infected female blackfly, *Simulium* spp. injects mature larvae into a human and produce adult worms capable of producing microfilariae which manifest as dermatitis (Connor and Neafie, 1976). **Symptoms:** *Onchocercus volvulus* cause punctate keratitis, sclerosing keratitis, iridocyclitis, chorioretinitis (Anderson and Front. 1976; Paul and Zimmerman. 1970), and optic atrophy (Kean et al., 1991). **Diagnosis:** based on clinical symptoms, histo-pathology, slit lamp examination, a Sclerocorneal punch biopsy and Xenodiagnosis, using laboratory-bred black flies. PCR may aid in detecting low burden of microfilariae. **Therapy:** diethylcarbamazine orally, manual removal of adult worms and killing microfilariae using pharmacological agents like Ivermectin. Suramin could also be used to eliminate both microfilariae and adult worms (Klotz et al., 2000) but it is toxic.
Loiasis: *Loa loa*	**Occurrence**: Loiasis is caused by the African eyeworm prevalent in Central and West Africa and causes parasitic ocular disease. **Infection source**: tabinid flies, including the mango fly and horsefly transfer the larvae to the humans where it matures into microfilariae causing tingling sensation at the nose, the bulbar conjunctiva, and the eyelid (Gibbs,1979; Garcia and Bruckner, 1997). Additionally, it could form a localized calabar swelling. **Symptoms:** the microfilariae and the adult worm is associated with conjunctival injection, pain, perivascular inflammation and may affect vision transiently. Worms in the vitreous, eyelid and anterior chamber cause retinal hemorrhages resulting from aneurysmal dilatation of retinal vessels (Toussaint and Danis, 1965). **Diagnosis:** identification of the adult worm by a microbiologist or pathologist or health care provider from under the skin or eye, identification of the microfilariae on a blood smear or using Giemsa-stained slides, by biopsy analysis (Kean et al., 1991) and identification of antibodies against *L. loa*. **Therapy:** surgical removal of adult worms from the eye, treatment with diethylcarbamazine and albendazole to kill the worms.

Dirofilariasis: *Dirofilariaimmitis*,	**Occurrence:** Dirofilariasis occurs throughout the world (Grieve et al., 1983). **Infection source:** infected mosquitoes transfer the larvae to human beings where they mature into the adult worm. **Symptoms:** Dirofilariasis affects the eyelid causing pruritis, pain, edema, and congestion of the conjunctiva. Manifests as ocular photophobia, diplopia, foreign body sensation, and floaters (Kean et al., 1991). **Diagnosis:** by histology and extraction from the conjunctiva (Arvantis et al., 1997) and by ophthalmoscopy. **Therapy:** Surgical excision of the adult worm (Kean et al., 1991) and a course of ivermectin and diethylcarbamazine to prevent further invasive surgery.
Gnathostomiasis: *Gnathostoma* spp., *Gnathostomaspinigerum*	**Occurrence:** *Gnathostoma* spp., and *Gnathostoma spinigerum* are the most-common ocular parasites causing Gnathostomiasis in Thailand (Teckhasaenee et al., 1986). **Infection:** due to the consumption of contaminated fish, pork, chicken, frog, or snake through third-stage larvae (Daengrang,1981). **Symptoms**: orbital symptoms include erythema, pruritis, and localized warmth (Daengrang, 1981). Both intraocular and corneal or conjunctival invasion may occur, leading to corneal ulceration, orbital cellulitis, uveitis, cataract, glaucoma, central retinal vein occlusion, retinitis, vitreous haemorrhage (Kean et al., 1991) and retinal detachment (Tudor and Blair, 1971). **Diagnosis:** is difficult (Kean et al., 1991). **Therapy:** surgical treatment remains the only treatment (Klotz et al., 2000).
Cestodes	
Cysticercosis: *Taenia solium*	**Occurrence:** caused by the tapeworm *Taenia solium*. **Infection:** infection caused by the adult worm is known as taeniasis whereas infection by the larvae is referred to as cysticercosis. Only Cysticercosis is associated with ocular diseases and acquired by the consumption of *Taenia* eggs in food or water which mature to larvae and migrate to the brain and eyes. **Symptoms:** the eyelid, conjunctiva, vitreous, anterior chamber, and subretina are infected and the symptoms include periorbital pain, diplopia, ptosis, blurring or loss of vision, distortion of images, and the sensation of light flashes (Messner and Kaminerer. 1979; Sen 1980). **Diagnosis:** by direct ophthalmoscopy, by ocular ultrasonography (Wei, 1990), by CT and histology. **Therapy:** vitrectomy along with photocoagulation to remove cystercerci from the vitreous (Seo et al., 1996). Metrifonate has been used for treatment (Kean et al., 1991). Interestingly, spontaneous extrusion of cystercerci from the eye may occur (Raina et al., 1996).

(continued)

TABLE 1.6 (Continued)

Nematodes, Cestodes and Ectoparasites Associated with an Ocular Disease

	Nematodes
Toxocariasis: *Toxocara* spp., *Toxocaracanis*	**Occurrence:** *Toxocara* is common *in* children and owners of dogs or cats. Approximately 5% of the U.S. population has antibodies to *Toxocara*. **Infection source:** toxocariasis is the result of infection by *Toxocara canis* found in soil contaminated with canine feces. After ingestion, the eggs hatch in the small intestine, the resulting larvae penetrate the mucosa and migrate to the liver, lung, and trachea causing visceral larva migrans (VLM) and when it migrates to the eye it is referred to as ocular larva migrans (OLM) (Zinkham, 1978). **Symptoms:** OLM manifests as retinal detachment, posterior pole granuloma, uveitis, vitreous abscess, optic neuritis, keratitis, iritis, and hypopyon (Krukar-Baster, et al., 1996; Molk, 1983; Zinkham, 1978), strabismus, decreased or loss of vision, esotropia, and leukocoria (Molk, 1983; Schantz, et al., 1979; Zinkham, 1978). **Diagnosis**: diagnosis of OLM is based on eosinophilia, serological testing, enzyme immunoassay using larval antigen and presence of antibody in vitreous humor (Molk, 1983). **Therapy:** Diethylcarbamazine (DEC), albendazole, and mebendazole were not effective. Oral steroids have been suggested along with photocoagulation.
Echinococcosis: *Echinococcus* spp. and *Echinococcus granulosus*	**Occurrence:** ocular echinococcosis is caused by the larvae of *Echinococcus granulosus.* Disease is common in Australia, East Africa, India, the Mediterranean, the Middle East, South America, and Russia, with the highest incidence reported in Africa. Echinococcosis is also known as Hydatid disease and humans are the intermediate hosts and approximately 1% of infections manifest ocular findings. **Infection sources:** Consumption of contaminated food or water contaminated with canine feces containing the eggs of the parasite. These eggs, hatch in the human duodenum and then travel to other organs. **Symptoms:** Proptosis, keratitis, conjunctival congestion, ulceration of the cornea (Kean et al., 1991), erosion of the orbital wall, optic atrophy, optic neuritis and subretinal hydatid disease has been reported. Severe cases may lead to blindness (Kean et al.1991). **Diagnosis:** ultrasonography, magnetic resonance imaging, serological testing, indirect hemagglutination assay and antigen detection. **Therapy**: surgical removal of the cyst (Akham et al., 1998) and medical therapy with mebendazole or albendazole (Kean et al., 1991).

Ectoprasites Myiasis: larvae of *Dermatobia*, *Gasterophilus*, *Oestra*, *Cordylobia*, *Chrysomia*, *Wohlfahrtia*, *Cochliomyia*, and *Hypoderma*	**Occurrence**:Ocular myiasis is caused by the larvae of several different genera of flies which include *Dermatobia*, *Gasterophilus*, *Oestra*, *Cordylobia*, *Chrysomia*, *Wohlfahrtia*, *Cochliomyia*, and *Hypoderma*. **Infection sources**: Geographic location, shepherding and sanitation are important determining factors of infection. Myiasis is considered ophthalmomyiasis externa when the larvae remain outside of the eye (Heyde, et al., 1986) and infect the eyelid, conjunctiva, lacrimal sac, and nasolacrimal ducts (Kean et al., 1991; Cogen et al., 1987). **Symptoms:** Conjunctivitis, foreign body sensation with watery or mucopurulent drainage, crawling or wriggling sensation, swelling and cellulitis in palpebral myiasis. Ophthalmomyiasis interna leads to uveitis, lens dislocation and retinal detachment (Mason, 1981; Syrdalen, et al., 1982). Orbital myiasis infection is due to several fly species and causes local destruction of tissue by the maggots. (Kersten et al., 1986). Foul odour is the most likely attraction of the flies, **Diagnosis**: by demonstration of maggots by the detection of larval tracks across the subretina and histological examination which may reveal a granuloma **Therapy:** removal of larvae, anticholinesterase ointment to kill or paralyze the larvae, steroids and antibiotics to control inflammation (Klotz et al., 2000).
Lice: *Pediculus humanus* var. *corporis*, *Pediculus humanus* var. *capitis* and *Phthirus pubis*	**Occurrence:** *P. pubis* affects the eyebrows and eye lashes. **Infection sources**: direct transfer of lice from the hair of one person to the hair of another. **Symptoms**: pruritis, small erythematous papules with excoriation and crusting of the lid margins. **Diagnosis:** itching at the root of the eyelashes. **Therapy:** a thick layer of petrolatum along the eyelid or the application of 1% yellow oxide of mercury (Klotz et al., 2000).
Ticks	Tick-borne diseases with ocular manifestations include Lyme disease, babesiosis, Rocky Mountain spotted fever, Colorado tick fever, tick-borne relapsing fever, Powassan encephalitis, Tularemia, Ehrlichiosis and Anaplasmosis (Sathiamoorthi and Smith, 2016). **Lyme disease** **Occurrence**: is a tick-borne disease **Infection source**: caused by *Borrelia burgdorferi* transmitted by the Ixodes tick. *B. burgdorferi, B. garinii*and *B. afzelii* are implicated in European cases (Stanek et al., 2012).

(*continued*)

TABLE 1.6 (Continued)

Nematodes, Cestodes and Ectoparasites Associated with an Ocular Disease

	Nematodes
	Symptoms: Ocular involvement may cause redness or tearing, follicular conjunctivitis, periorbital edema and mild photophobia (Steere et al., 1983), optic neuritis (Träisk and Lindquist, 2012), neuroretinitis (Sauer et al., 2009), eyelid edema (Raja et al., 2016), keratitis (Raja et al., 2016), anterior and posterior scleritis (Krist and Wenkel, 2002) and uveitis (Rosenbaum and Rahn, 1997).
	Therapy: Macular edema, chorioretinitis and optic neuropathy during Lyme disease require treatment with corticosteroids along with the recommended antimicrobial regimen. Delays in therapy may result in vision impairment and even blindness (Sathiamoorthi and Smith, 2016).
	Babesiosis
	Occurrence: Ixodes tick transmits *Babesia microti*, a protozoan parasite.
	Symptoms:
	Most infections are asymptomatic, some patients experience gradual onset of malaise and fatigue with fevers, anorexia, headache, myalgia, arthralgia, and nausea. Hemolyticanemia and thrombocytopenia are common (Vannier and Krause, 2012). Ocular findings include conjunctival injection, retinal hemorrhages, papillitis and nerve fiber layer infarcts (Ortiz and Eagle,1982).
	Tick-borne relapsing fever (TBRF)
	Infection: caused by spirochetes of the *Borrelia* genus and transmitted to humans by the ticks.
	Symptoms: febrile episodes, headache, myalgia and arthralgia (Dworkin et al., 2008). Photophobia, eye pain, anterior and intermediate uveitis and optic neuritis (Cutler, 2015).
	Powassan encephalitis (POW)
	Occurrence: POW encephalitis is a very rare neurologic infection.
	Infection source: caused by POW virus, an RNA virus transmitted by several different ticks including Ixodes scapularis.
	Symptoms: fever, headache, weakness, confusion, seizures, and memory loss (Piantadosi et al., 2016).

Ophthalmic manifestations include retinal vein tortuosity, optic disk edema and ophthalmoplegia (Lessell and Collins, 2003; McLean and Donohue 1959).

Erhlichiosos and Anaplasmosis

Occurrence: Human monocytotropic ehrlichiosis (HME) and human granulocytotropic anaplasmosis (HGA) have similar clinical presentations.

Infection source: HME is caused by *Ehrlichia chaffeensis*, and HGA is associated with *Anaplasma phagocytophilum* transmitted by ticks.

Symptoms: both diseases are mild, self-limiting and manifestations include fever, headache, fatigue, nausea and myalgias. Ocular findings with HGA are very rare and have included pupil abnormalities, facial nerve palsy, ophthalmoplegia and optic neuritis (Lee et al., 2003). HME is associated with optic neuritis and orbital myositis (Pendse et al., 2006).

Rocky Mountain Spotted Fever (RMSF)

Occurrence: Common (2000 cases) during summer in the south Atlantic and the south-central United States among children (5–9 years) and in middle-aged adults (CDC, 2015a-d).

Infection: caused by *Rickettsia rickettsia*, an intracellular bacterium transmitted primarily by the dog tick.

Symptoms: classic clinical triad includes headache, fever, and petechial rash. Other symptoms include myalgias, arthralgias, abdominal pain, nausea and vomiting. Severe cases may involve renal failure, hepatomegaly, meningismus, neurologic abnormalities and can result in death.

Ocular manifestations of RMSF are common and involve vasculitis of the conjunctiva, retina and choroid, retinal cotton wool spots, macular edema, vascular sheathing, venous tortuosity, retinal artery or vein occlusion and intraretinal hemorrhages and exudates (Khairallah et al., 2004). Optic disk edema, optic neuritis, ischemic optic neuropathy and neuroretinitis have also been reported (Sulewski and Green, 1986). Less often, anterior uveitis has been described. Typically, the disease resolves in a few weeks, although significant retinovascular involvement can lead to vision loss (Raja et al., 2016).

Tulremia

Occurrence: approximately 200 cases are reported annually (CDC, 2015a-d) and Tularemia affects all ages but is most common in children.

(*continued*)

TABLE 1.6 (Continued)

Nematodes, Cestodes and Ectoparasites Associated with an Ocular Disease

Nematodes
Infection source: Tularemia, also known as rabbit fever, is caused by the bacterium *Francisella tularensis* and ticks, mosquitoes and fleas are the carriers (CDC, 2015a). Other routes of transmission include direct contact with infected animals, ingestion, inhalation and infected meat.
Symptoms: Oculo-glandular tularemia is rare (Kosker et al., 2013). Signs of ocular infection may include conjunctivitis, painful preauricular or submandibular lymphadenopathy, lid edema and occasionally ulcerations of the conjunctiva. Dacryocystitis, corneal ulcers, corneal perforation, corneal edema and optic nerve atrophy have also been described (Celik et al., 2013).
Colorado tick fever (CTF)
Occurrence: CTF or mountain fever, is the second most common arbovirus infection in the United States. The virus occurs t in the Rocky Mountain region and Southwest Canada. In total, 200–400 cases are reported each year, mainly in men between the ages of 20 and 49 years (Romero and Simonsen, 2008).
Infection source: CTF, is an RNA virus transmitted by the primary vector *Dermacentor andersoni*.
Symptoms: CTF may present with the abrupt onset of fever, chills, headache, retro-orbital pain, photophobia, myalgias and generalized weakness. The febrile pattern, termed 'saddleback' fever, can be biphasic, improving for 24–48 h and then recurring. Ophthalmic involvement is limited to retro-orbital pain and conjunctivitis (Spruance and Bailey, 1973).

Source: Compiled by the author.

1.7.1 Fate of Ocular Drug Therapy

Several studies have also addressed specifically the fate of drugs used for ocular therapy especially those used as eye drops. A few barriers like the tear film, the reflex tears and blinking of the eyelids contribute to the dissipation of the drug. For instance, the tear film initially facilitates rapid delivery of high concentrations of the drug to the cornea and conjunctiva, but the drug is rapidly dissipated (Patton and Francoeur, 1978). Reflex tearing and blinking also contribute to spillage of the drug onto the skin and drainage through the puncta (Maurice, 1973). Several other factors increase ocular availability of the drug used as an eye drop such as nasolacrimal occlusion (Zimmerman et al., 1984), instilling drops more frequently (Chrai et al., 1974; Davis et al., 1978 a and b), using high viscosity non-aqueous ointments and aqueous gels (Robin and Ellis, 1978) and using collagen shields impregnated with antibacterial agents (Callegan et al., 1994a; 1994b). It is also known that when 2 agents are applied to the eye the agent instilled second displaces the first drug from the eye (Chrai et al., 1974; Davis et al., 1978 a and b); immiscible agents should be instilled rapidly within 5 minutes to maximise drug availability in the eye and tear film concentrations of agents delivered in similar vehicles decay at similar rates (Richman et al., 1990). In some drugs the increase in intraocular penetration is not directly proportional to the concentrations applied (Benson, 1974).

1.7.2 Absorption into the Eye

Apart from the above factors that decide the fate of the drug on the ocular surface several other factors such as the route of administration (topical or intraocular) also decides on the ultimate efficacy of the drug such as the absorption of the drug into the eye (Weng et al., 2017). The epithelium is the first ocular barrier which topical agents encounter and penetration is increased by physical damage caused mechanically or because of infection. Factors such as hydrophobicity, lipophilicity, size and ionisation potential influence the diffusion of the drug across the corneal epithelium and stroma (Grass and Robinson, 1984) and thus different concentrations of the drug have been observed in the different layers of the cornea (Mindel et al., 1984). Iontophoresis is a technique used to increase the penetration of charged molecules into the eye by applying an electric current (Callegan et al., 1994a; 1994b) and superior penetration of gentamicin was observed in aphakic rabbit eyes, subjected to iontophoresis (Fishman et al., 1984). Chloramphenicol and some fluoroquinolones, efficiently penetrate the cornea and therapeutic concentrations are achieved in the aqueous humour. But the flow of aqueous humour from the ciliary processes through the posterior and anterior chambers and out of the eye via Schlemm's canal prevents the accumulation of high concentrations of antibacterial at this site. Topical administration results in antibacterials in the aqueous humour which are then absorbed by the anterior uvea and lens but the concentrations achieved in the vitreous body are generally subtherapeutic, thus necessitating systemic or local injection (preferably intravitreal) for patients with infective endophthalmitis. The pharmacokinetic data available on antibacterials applied topically to the eye (Benson,1974) are collected from animal studies and therefore caution is necessary while extrapolating these data for predicting the clinical efficacy or potential toxicity of antibacterials to the human eye because of anatomical differences (Lee and Robinson, 1986). Furthermore, individual variation between individuals is large as observed for the penetration of topical fluoroquinolones into the human eye (Diamond et al., 1995; Leeming et al., 1994). It is also relevant that the concentrations required to kill ocular pathogens would be significantly increased compared to when the antibiotic is used to clear systemic infections (Kowalski and Dhaliwal 2005; Kowalski et al., 2004a; 2004b; Mah et al., 2004).

1.8 Conclusions

The ocular surface harbours a multitude of microorganisms which include viruses, prokaryotes like bacteria, eukaryotes like fungi and protozoans and multicellular eukaryotes like nematodes, cestodes and so forth, which may be associated with the eye as commensals and are thus harmless (Tables 1.1, 1.2, 1.5 and 1.6). But, under conditions of trauma many of these become pathogenic and cause several diseases of the eye like keratitis, conjunctivitis, blepharitis, endophthalmitis, orbital cellulitis, retinitis, post fever retinitis, dacryocystitis and so forth. A few of the ocular diseases could be self-limiting like Giardiasis caused by *Giardia lamblia*, ocular toxoplasmosis caused by *Toxoplasma gondii*, Rocky Mountain Spotted Fever caused by *Rickettsia rickettsia,* viral epidemic keratoconjunctivitis, viral pharyngoconjunctival fever and so forth. Ocular infectious diseases are susceptible to various classes of dugs (antibiotics, antifungals, antivirals and so forth) and thus are manageable. But over the years these above drugs have become ineffective due to the emergence of antimicrobial resistance influenced by several biotic and abiotic factors thus posing a challenge to the treatment of the eye and restoration of normal vision.

The objective of this introductory chapter is to introduce the vast heterogeneity of ocular cultivable and the gross non-cultivable bacteria, fungi, viruses, and so forth. associated with the normal eye and the diseased eye, the modes of treatment, the fate of the drugs and to allude to the phenomenon of antibacterial resistance and resistome analysis of the ocular bacteria. Antibacterial resistance of the eye is the focus of the subsequent chapters.

REFERENCES

Achtman, M., Wagner, M. 2008. Microbial diversity and the genetic nature of microbial species. *Nat Rev Microbiol.* 6:431–440. doi:10.1038/nrmicro1872

Agger, W.A. 1995. Sporotrichosis. In F. T. Fraunfelder and F. H.Roy (ed.), *Current ocular therapy.* W. B. Saunders Company, Philadelphia,Pa, p. 79–80.

Ainbinder, D., Parmley, V.C., Mader, T.H., Nelson, M.L. 1998. Infectious crystalline keratopathy caused by *Candida guilliermondii. Am. J. Ophthalmol.* 125:723–725.

Akham, O., Bilgi, S., Akata, D., Kiratli, H., Ozmen, M. 1998. Percutaneous treatment of an orbital hydatic cyst: A new therapeutic approach. *Am. J. Ophthalmol.* 125:827–829.

Albesa, I., Eraso, A. 1983. Primary isolation of *Trypanosoma cruzi*by hemoculture: Method of glucose concentration. *Am. J. Trop. Med. Hyg.* 32:963–967.

Alfonso, S.A., Fawley, J.D., Alexa Lu, X. 2015. Conjunctivitis. *Prim Care.* 42, 325–345. https://doi.org/10.1016/j.pop.2015.05.001

al-Rajhi, A., Awad, A.H., al-Hedaithy, S.S., Forster, R.K., Caldwell, K.C. 1993. *Scytalidium dimidiatum* fungal endophthalmitis. *Br. J. Ophthalmol.* 77:388–390.

Ammon, A., Rumpf, K.W., Hommerich, C.P., Behrens-Baumann, W., Ruchel, R. 1992. Rhinocerebral mucormycosis during deferoxamine therapy. *Dtsch. Med. Wochenschr.* 117:1434–1438.

Anderson, J., Front, R. 1976. Ocular onchocerciasis. In C. Binford and D. Connor (ed.), *Pathology of tropical and extraordinary diseases*. Armed Forces Institute of Pathology, Washington, D.C., pP. 373–381.

Aronson, N., Herwaldt, B.L., Libman, M., Pearson, R., Lopez-Velez, R., Weina, P., Carvalho, E., Ephros, M., Jeronimo, S., Magill, A. 2017 Jan 11. Diagnosis and treatment of Leishmaniasis: Clinical practice guidelines by the Infectious Diseases Society of America (IDSA) and the American Society of Tropical Medicine and Hygiene (ASTMH). *Am J Trop Med Hyg*. 96(1):24–45. doi: 10.4269/ajtmh.16-84256. Epub 2016 Dec 7. PMID: 27927991; PMCID: PMC5239701

Arvantis, P., Vakalis, N., Damanakis, A., Theodossiadis, G. 1997. Ophthalmic dirofilariasis. *Am. J. Ophthalmol*. 123:689–691.

Assefa, Y., Moges, F., Endris, M., Zereay, B., Amare, B., Bekele, D., Tesfaye, S., Mulu, A., Belyhun, Y. 2015. Bacteriological profile and drug susceptibility patterns in dacryocystitis patients attending Gondar University Teaching Hospital, Northwest Ethiopia. *BMC Ophthalmol*. 2015 Apr 2; 15:34. doi: 10.1186/s12886-015-0016-0. PMID: 25880996; PMCID: PMC4396718.

Barisani-Asenbauer, T., Maca, S.M., Mejdoubi, L., Emminger, W., Machold, K., Auer, H. 2012. Uveitis- a rare disease often associated with systemic diseases and infections- a systematic review of 2619 patients. *Orphanet J. Rare Dis*. 7, 57. https://doi.org/10.1186/1750-1172-7-57

Benson, H. Permeability of the cornea to topically applied drugs. *Arch Ophthalmol*. 1974;91:313–27.

Bloom, J., Hamor, R.E., Gerding, P.A. Jr. 1996. Ocular blastomycosisin dogs: 73 cases, 108 eyes (1985–1993). *J. Am. Vet. Med*. 209:1271–1274.

Borderie, V., Bourcier, T.M., Poirot, J.L., Baudrimont, M., Prudhommede Saint-Maur, P., Laroche, L. 1997. Endophthalmitis after *Lasiodiplodiatheobromae*corneal abscess. *Graefes Arch. Clin. Exp. Ophthalmol*. 235:259–261.

Brook, I., Frazier, E.H. 1998. Aerobic and anaerobic microbiology ofdacryocystitis. *Am. J. Ophthalmol*. 125:552–554.

Butrus, S.I., Klotz, S.A. 1986. Blocking *Candida* adherence to contact lenses. *Curr. Eye Res*. 5:745–750.

Butrus, S.I., Klotz, S.A., Misra, R.P. 1987. The adherence of *Pseudomonas aeruginosa* to soft contact lenses. *Ophthalmol*. 94:1310–1314.

Callegan, M.C., Engel, L.S., Clinch, T.E., et al., 1994a. Efficacy of tobramycin drops applied to collagen shields for experimental staphylococcal keratitis. *Curr Eye Res*. 13:875–8.

Callegan, M.C., Engelbert, M., Parke 2nd, D.W., Jett, B.D., Gilmore, M.S. 2002. Bacterial endophthalmitis: Epidemiology, therapeutics, and bacterium-host interactions. *Clin. Microbiol. Rev*. 15, 111–124. https://doi.org/10.1128/cmr.15.1.111-124.2002

Callegan, M.C., O'Callaghan, R.J., Hill, J.M. 1994b. Pharmacokinetic considerations in the treatment of bacterial keratitis. *Clin Pharmacokinet*. 27:129–49.

Cameron, J., Badr I.A., Miguel Risco, J., Abboud, E., Gaonnah, el S. 1998. Endophthalmitis cluster from contaminated donor corneas following penetrating keratoplasty. *Can. J. Ophthalmol*. 33:8–13.

Carney, M.D., Tabassian, A., Guerry, R.K. 1996. Pseudo-*Allescheria boydii* endophthalmitis. *Retina*. 16:263–264.

CDC. Statistics and Epidemiology j Rocky Mountain Spotted Fever (RMSF)j CDC. Retrieved from www.cdc.gov/rmsf/stats/index.html. (Accessed 3 September 2015a).

CDC. Statistics j Tularemiaj CDC. Retrieved www.cdc.gov/tularemia/statistics/index.html. (Accessed 26 October 2015b).

CDC. Transmission j Tularemiaj CDC. Retrieved www.cdc.gov/tularemia/transmission/index.html. (Accessed 26 October 2015c).

CDC. Statistics & Maps j Colorado Tick Feverj CDC. Retrieved www.cdc.gov/coloradoti ckfever/statistics.html. (Accessed 13 January 2015d).

Celik, T., Yuksel, D., Kosker, M., Turkoglu, E.B. 2013. Unilateral acute dacryocystitis associated with oculoglandulartularemia: A case report. *Semin Ophthalmol.* 28:91–93.

Chakravarthy, K.S., Jayasudha, R., Ranjith, K., Dutta, A., Pinna, N.K., Mande, S.S., Sharma, S., Shivaji, S. 2018a. Alterations in the gut bacterial microbiome in fungal Keratitis patients. *PLoS ONE* 13(6):e0199640.

Chakravarthy, K.S., Jayasudha, R., Sai Prashanthi, G. et al., 2018b. Dysbiosis in the gut bacterial microbiome of patients with Uveitis, an inflammatory disease of the eye. *Indian J Microbiol.* 58:457–469. https://doi.org/10.1007/s12088-018-0746-9

Chaudhry, I., Hylton, C., DesMarchais, B. 1998. Bilateral ptosis and lower eyelid ectropion secondary to cutaneous leishmaniasis. *Arch. Ophthalmol.* 116:1244–1245.

Chaudhry, I.A., Shamsi, F.A., Al-Rashed, W. 2005. Bacteriology of chronic dacryocystitis in a tertiary eye care center. *Ophthalmic Plast. Reconstr. Surg.* 21, 207–210. https://doi.org/10.1097/01.iop.0000161718.54275.7d

Chen, S., Chung, Y.M., Liu, J.H. 1998. Endogenous *Candida* endophthalmitis after induced abortion. *Am. J. Ophthalmol.* 125:873–875.

Chou, B. Eye Infections: Causes, Symptoms, Treatment - All About Vision. www.allabou tvision.com In: All about vision March 2022.

Chrai, S.S., Makoid, M.C., Eriksen, S.P., et al., 1974. Drop size and initial dosing frequency problems of topically applied ophthalmic drugs. *J Pharm Sci.* 63:333–338.

Cogen, M., Hays, S., Dixon, J. 1987. Cutaneous myiasis of the eyelid due to *Cuterebra*larva. *JAMA.* 258:1795–1796.

Connor, D., Neafie, R. 1976. Onchocerciasis. In C. Binford and D. Connor (ed.), *Pathology of tropical and extraordinary diseases.* Armed Forces Institute of Pathology, Washington, D.C., p. 360–372.

Creach, P., Auffret, N., Buot, G., Binet, O. 1995. *Microsporum canis* tinea ciliaris and blepharitis. *Ann. Dermatol. Venereol.* 122:773–774.

Crump, J.R., Elner, S.G., Elner, V.M., Kaufmann, C.A. 1992. Cryptococcal endophthalmitis: Case report and review. *Clin. Infect. Dis.* 14:1069–1073.

Cutler, S.J. 2015. Relapsing fever Borreliae: A global review. *Clin Lab Med.* 35:847–865.

Daengrang, S. 1981. Gnathostomiasis in Southeast Asia. *Southeast Asian J.Trop. Med. Public Health.* 12:319–332.

Das, T., Gopinathan, U., Sharma, S. 1994. Exogenous *Helminthosporium* endophthalmitis. *Br. J. Ophthalmol.* 78:492–493.

Davis, S.D., Sarff, L.D., Hyndiuk, R.A. 1978a. Corticosteroid in experimentally induced *Pseudomonas* keratitis: Failure of prednisolone to impair the efficacy of tobramycin and carbenicillin therapy. *Arch Ophthalmol.* 96:126–8.

Davis, S.D., Sarff, L.D., Hyndiuk, R.A. 1978b Topical tobramycin therapy of experimental *Pseudomonas* keratitis. *Arch Ophthalmol.* 96:123–5.

de Paiva, C.S., Jones, D.B., Stern, M.E., Bian, F., Moore, Q.L., Corbiere, S., Streckfus, C.F., Hutchinson, D.S., Ajami, N.J., Petrosino, J.F., Pflugfelder, S.C. 2016. Altered mucosal microbiome diversity and disease severity in sjogren syndrome. *Sci. Rep.* 6, 23561. https://doi.org/10.1038/srep23561

Diamond, J.P., White, L., Leeming, J.P. et al., 1995. Topical 0.3% ciprofloxacin norfloxacin and ofloxacin in the treatment of bacterial keratitis: a new method for comparative evaluation of ocular drug penetration. *Br J Ophthalmol.* 79:606–9.

Dick, G.J. 2019. The microbiomes of deep-sea hydrothermal vents: Distributed globally, shaped locally. *Nat Rev Microbiol.* 17:271–283. https://doi.org/10.1038/s41579-019-0160-2

Didier, E., Shadduck, J., Didier, P., Millichanp, N., Vossbrinck, C. 1991. Studies on ocular microsporidia. *J. Protozool.* 38:635–638.

Didier, E.S., Maddry, J.A., Brindley, P.J., Stovall, M.E., Didier, P.J. 2005 June. Therapeutic strategies for human microsporidia infections. *Expert Rev Anti Infect Ther.* 3(3):419–34. doi:10.1586/14787210.3.3.419. PMID 15954858. S2CID 42354627

Doan, T., Akileswaran, L., Andersen, D. et al., 2016. Paucibacterial microbiome and resident DNA virome of the healthy conjunctiva. *Invest Ophthalmol Vis Sci.* 57:5116–5126. doi:10.1167/iovs.16-19803

Dong, Q., Brulc, J.M., Iovieno, A. et al., 2011. Diversity of bacteria at healthy human conjunctiva. *Invest Ophthalmol Vis Sci.* 52:5408–5413. doi:10.1167/iovs.10-6939

Doster, E., Lakin, S.M., Dean, C.J. et al., 2020. MEGARes 2.0: A database for classification of antimicrobial drug, biocide and metal resistance determinants in metagenomic sequence data. *Nucleic Acids Res.* 48:D561–D569. doi:10.1093/nar/gkz1010

Driebe Jr., W.T., Mandelbaum, S., Forster, R.K., Schwartz, L.K., Culbertson, W.W. 1986. Pseudophakic endophthalmitis. Diagnosis and management. *Ophthalmol.* 93, 442–448. https://doi.org/10.1016/s0161-6420(86)33722-9

Dworkin, M.S., Schwan, T.G., Anderson, D.E. Jr, Borchardt, S.M. 2008. Tick-borne relapsingfever. *Infect Dis Clin North Am.* 22:449–468.

Echejoh, G.O., Manasseh, A.N., Tanko M.N., Ogala-Echejoh, S.E., Silas, O.A., Nimkur, T.L., Mandong, B.M. 2008. Nasal rhinosporidiosis. *J Natl Med Assoc.* 100:713–715. https:// doi: 10.1016/s0027-9684(15)31347-x

Egrilmez, S., Yildirim-Theveny, S. 2020. Treatment-resistant bacterial keratitis: Challenges and solutions. *Clin. Ophthalmol.* 14, 287–297. https://doi.org/10.2147/ OPTH.S18199

Engstrom, R.J., Holland, G., Nussenblatt, R., Jabs, D. 1991. Currentpractices in the management of ocular toxoplasmosis. *Am. J. Ophthalmol.*111:601–610.

Eschete, M., King, J.W., West, B.C., Oberle, A. 1981. *Penicillium chrysogenum* endophthalmitis. *Mycopathologia.* 74:125–127.

Eslami, F., Ghasemi Basir, H.R., Moradi, A., Heidari Farah, S. 2018. Microbiological study of dacryocystitis in northwest of Iran. *Clin. Ophthalmol.* 12, 1859–1864. https://doi.org/10.2147/OPTH.S175463.

Feldman, H. 1982. Epidemiology of toxoplasma infections. *Epidemiol. Rev.* 4:204–213.

Fishman, P.H., Jay, W.M., Rissing, P. et al., 1984. Iontophoresis of gentamicin into aphakic rabbit eyes. *Invest Ophthalmol Vis Sci.* 25:343–5.

Fox, D. 2014. Lakes under the ice: Antarctica's secret garden. *Nature.* 512:244–246. https://doi.org/10.1038/512244a

Frenkel, J. 1971. Toxoplasmosis. In R. Marcial-Rojas (ed.), *Pathology of protozoal and helminthic diseases.* Williams and Wilkins,Baltimore, Md, p. 254–290.

Gallon, P., Parekh, M., Ferrari, S. et al., 2019. Metagenomics in ophthalmology: Hypothesis or real prospective? *Biotechnol Rep (Amst).* 23:e00355. doi:10.1016/j.btre.2019.e00355

Ganley, J.P., Klotz, S.A. 1995. Coccidioidomycosis. In F. T. Fraunfelder and F. H. Roy (ed.), *Current ocular therapy.* W. B. Saunders Company, Philadelphia, Pa, p. 70–73.

Garcia, L., Bruckner, D. 1997. *Diagnostic medical parasitology*, 3rd ed. ASM Press, Washington, D.C.

Garrity, J., Herman, D.C., Imes, R., Fries, P., Hughes, C.F., Campbell, R.J. 1993. Optic nerve sheath decompression for visual loss in patients with acquired immunodeficiency syndrome and cryptococcal meningitis with papilledema. *Am. J. Ophthalmol.* 116:472–478.

Gest, H. 2004. The discovery of microorganisms by Robert Hooke and Antoni van Leeuwenhoek, Fellows of The Royal Society. *Notes Rec. R. Soc. Lond.* 58187–201. http://doi.org/10.1098/rsnr.2004.0055

Ghosh, S., Chatterjee, B.D., Chakraborty, C.K., Chakravarty, A., Khatua, S.P. 1995. Bacteria in surface infections of neonates. *J. Indian Med. Assoc.* 93:132–135.

Gibbs, R. 1979. Loiasis: Report of three cases and literature review. *J. Nat. Med. Assoc.* 71:853–84.

Gilbert, D., Moellering, R., Sande, M. 1998. *The Sanford guide to antimicrobial therapy*, 28th ed. Antimicrobial Therapy Inc., Vienna, Va.

Glasgow, B., Engstrom, R.E. Jr., Holland, G.N., Kreiger, A.E., Wool, M.G. 1996. Bilateral endogenous *Fusarium* endophthalmitis associated with acquired immunodeficiency syndrome. *Arch. Ophthalmol.* 114:873–877.

Graham, J.E., Moore, J.E., Jiru, X. et al., 2007. Ocular pathogen or commensal: A PCR-based study of surface bacterial flora in normal and dry eyes. *Invest Ophthalmol Vis Sci.* 48:5616–5623. doi:10.1167/iovs.07-0588

Grass, G.M., Robinson, J.R. 1984. Relationship of chemical structure to corneal penetration and influence of low-viscosity solution on ocular bioavailability. *J Pharm Sci.* 73:1021–7.

Green, M., Apel, A., Stapleton, F. 2008. Risk factors and causative organisms in microbial keratitis. *Cornea.* 27:22–7. 10.1097/ICO.0b013e318156caf2

Grieve, R., Lok, J., Glickman, L. 1983. Epidemiology of canine heartworm infection. *Epidemiol. Rev.* 5:220–246.

Gumbel, H., Ohrloff, C., Shah, P.M. 1990. The conjunctival flora of HIV-positive patients in an advanced stage. *Fortschr. Ophthalmol.* 87:382–383.

Gupta, V., Gupta, A., Rao, N.A. 2007. Intraocular tuberculosis–an update. *Surv. Ophthalmol.* 52, 561–587. https://doi.org/10.1016/j.survophthal.2007.08.015

Hamady, M., Knight, R. 2009. Microbial community profiling for human microbiome projects: Tools, techniques, and challenges. *Genome Res.* 19:1141–1152. doi:10.1101/gr.085464.108

Hammer, M., Harding, S., Wynn, P. 1983. Post-traumatic fungal endophthalmitis caused by *Exophialajeanselmei*. *Ann. Ophthalmol.* 15:853–855.

Han, B., Weiss, L.M. 2017. Microsporidia: Obligate intracellular pathogens within the fungal kingdom. *Microbiology Spectrum*, 5(2):10.1128/microbiolspec.FUNK-0018-2016. doi: 10.1128/microbiolspec.FUNK-0018-2016. PMID: 28944750; PMCID: PMC5613672.

Han, B., Weiss, L.M. 2018 Nov. Therapeutic targets for the treatment of microsporidiosis in humans. *Expert OpinTher Targets.* 22(11):903–915. doi: 10.1080/14728222.2018.1538360. Epub 2018 Nov 1. PMID: 30336698; PMCID: PMC6300147

Harwani, D. 2012. The great plate count anomaly and the unculturable bacteria. *Int. J. Sci. Res.* 2:350–351. doi:10.15373/22778179/sep2013/122

Harwood, C., Rich, G., McAleer, R., Cherian, G. 1988. Isolation of *Acanthamoeba* from a cerebral abscess. *Med. J. Austr.* 148:47–49.

Heffler, K., Echhardt, T., Reboli, A., Stieritz, D. 1996. *Acanthamoeba*endophthalmitis in acquired immunodeficiency syndrome. *Am. J.Ophthalmol.* 122:584–586.

Heintz-Buschart, A., Wilmes, P. 2018 Jul. Human Gut Microbiome: Function matters. *Trends Microbiol.* 26(7):563–574. doi: 10.1016/j.tim.2017.11.002. Epub 2017 Nov 22. PMID: 29173869

Hemady, R. 1995. Microbial keratitis in patients infected with the human immunodeficiency virus. *Ophthalmol.* 102:1026–1030.

Heyde, R., Seiff, S., Mucia J. 1986. Ophthalmomyiasis externa in California. *W. J Med.* 144:80–81.

Huang, Y., Yang, B., Li, W. 2016. Defining the normal core microbiome of conjunctival microbial communities. *Clin Microbiol Infect.* 22:643 e647–643 e612. doi:10.1016/j.cmi.2016.04.008

Izimukwiye, A.I., Mbarushimana, D., Ndayisaba, M.C., Bigirimana, V., Rugwizangoga, B., Laga, A.C. 2019. Cluster of Nasal Rhinosporidiosis, eastern province, Rwanda. *Emerg. Infect. Dis.* 25(9):1727–1729. https://dx.doi.org/10.3201/eid2509.190021

Jacobson, M., Galetta, S.L., Atlas, S.W., Curtis, M.T., Wulc, A.W. 1992 Dec. Bipolaris-induced orbital cellulitis. *J Clin Neuroophthalmol.* 12(4):250–6. PMID: 1287050

Jager, M., Chodosh, J., Huang, A.J., Alfonso, E.C., Culbertson, W.W., Forster, R.K. 1994. *Aspergillus niger*as an unusual cause of scleritis and endophthalmitis. *Br. J. Ophthalmol.* 78:584–586.

Jain, V., Garg, P., Sharma, S. 2009. Microbial scleritis-experience from a developing country. *Eye.* 23, 255–261. https://doi.org/10.1038/sj.eye.6703099

Jayasudha, R., Kalyana Chakravarthy, S., Sai Prashanthi, G., Sharma, S., Tyagi, M., Shivaji, S. 2019. Implicating dysbiosis of the gut fungal microbiome in Uveitis, an inflammatory disease of the eye. *Invest Ophthalmol Vis Sci.* 60:1384–1393. doi:10.1167/iovs.18-26426

Jiang, X., Deng, A., Yang, J., Bai, H., Yang, Z., Wu, J., Lv, H., Li, X., Wen, T. 2018. Pathogens in the Meibomian gland and conjunctival sac: Microbiome of normal subjects and patients with Meibomian gland dysfunction. *Infect. Drug Resist.* 11, 1729–1740. https://doi.org/10.2147/IDR.S162135

Johnson, T.E., Casiano, R.R., Kronish, J.W., Tse, D.T., Meldrum, M., Chang, W. 1999. Sino-orbital aspergillosis in acquired immunodeficiencysyndrome. *Arch. Ophthalmol.* 117:57–64.

Karsten, E., Watson, S.L., Foster, L.J.R. 2012. Diversity of microbial species implicated in keratitis: A review. *Open Ophthalmol. J.* 6:110–124.

Kean, B., Sun, T., Ellsworth, R. 1991. *Color atlas/text of ophthalmologicparasitology.* 1st ed. Igaku-Shoin, New York, N.Y.

Keilty, R.A. 1930. The bacterial flora of the normal conjunctiva with comparative nasal culture study. *Am. J. Ophthalmol.* 13, 876–879. https://doi.org/10.1016/s0002- 9394(30)92437-3.

Kersten, R., Showkrey, N., Tabbara, K. 1986. Orbital myiasis. *Ophthalmol.* 93:1228–1232.

Khairallah, M., Ladjimi, A., Chakroun, M., et al., 2004. Posterior segment manifestations of Rickettsia conorii infection. *Ophthalmol.* 111:529–534.

Kittipibul, T., Puangsricharern, V., Chatsuwan, T. 2020. Comparison of the ocular microbiome between chronic Stevens-Johnson syndrome patients and healthy subjects. *Sci. Rep.* 10:4353. https://doi.org/10.1038/s41598-020-60794-w

Klotz, S.A., Penn, C.C., Negvesky, G.J., Butrus, S.I. 2000 Oct. Fungal and parasitic infections of the eye. *Clin Microbiol Rev.* 13(4):662–85. doi: 10.1128/CMR.13.4.662. PMID: 11023963; PMCID: PMC88956

Kochar, D., Shubhakaran, B. Kumawat, I. Thanvi, Joshi, A., Vyas, S. 1998. Ophthalmoscopic abnormalities in adults with falciparum malaria. *Q. J. Med.* 91:845–852.

Kock, N.P., Petersen, H., Fenner, T., Sobottka, I., Schmetz, C., Deplazes, P., Pieniazek, N.J., Albrecht, H., Schottelius, J. 1997 May. Species-specific identification of microsporidia in stool and intestinal biopsy specimens by the polymerase chain reaction. *Eur J Clin Microbiol Infect Dis*. 16(5):369–76. doi: 10.1007/BF01726365. PMID: 9228477

Kohanim, S., Palioura, S., Saeed, H.N., Akpek, E.K., Amescua, G., Basu, S., Blomquist, P.H., Bouchard, C.S., Dart, J.K., Gai, X., Gomes, J.A., Gregory, D.G., Iyer, G., Jacobs, D.S., Johnson, A.J., Kinoshita, S., Mantagos, I.S., Mehta, J.S., Perez, V.L., Pflugfelder, S.C., Sangwan, V.S., Sippel, K.C., Sotozono, C., Srinivasan, B., Tan, D.T., Tandon, R., Tseng, S.C., Ueta, M., Chodosh, J. 2016. Acute and chronic ophthalmic involvement in stevens-johnson syndrome/toxic epidermal necrolysis - a comprehensive review and guide to therapy. *II. Ophthalmic disease. Ocul. Surf.* 14:168–188. https://doi. org/10.1016/j.jtos.2016.02.001

Konduri, R., Sharma, S., Shivaji, S. 2021. Microbes of the human eye: Microbiome, antimicrobial resistance, and biofilm formation. *Exp. Eye Res.* 108476. ISSN 0014-4835, https://doi.org/10.1016/j.exer.2021.108476

Kosker, M., Sener, D., Kilic, O., et al., 2013. A case of oculoglandulartularemia resistantto medical treatment. *Scand J Infect Dis*. 45:725–727.

Kowalski, R.P., Deepinder, K. 2005. Dhaliwal Ocular bacterial infections: Current and future treatment options Expert Rev. *Anti. Infect. Ther*. 3(1):131–139.

Kowalski, R.P., Romanowski, E.G., Mah, F.S., Yates, K.A., Gordon, Y.J. 2004a. Gatifloxacin(Zymar™) can be effective in comparison to vancomycin and cefazolin for treatinggatifloxacin-resistant *Staphylococcus aureus* keratitis in a NZW rabbit model. *Invest.Ophthalmol. Vis. Sci.* 45 (E-Abstract 4903).

Kowalski, R.P., Romanowski, E.G., Mah, F.S., Yates, K.A., Gordon, Y.J. 2004b. Topical prophylaxis with moxifloxacin prevents endophthalmitis in a rabbit model. *Am. J. Ophthalmol.* 138:33–37.

Krist, D., Wenkel, H. Posterior scleritis associated with Borrelia burgdorferi (Lyme disease) infection. *Ophthalmol.* 2002;109:143–145.

Krukar-Baster, K., Zygulska-Mach, H., Sajak-Hydzik, K., Kubicka-Trzaska, A., Dymon, M. 1996. Long term observations of ocular toxocariasis in children and youth. *Klin. Oczna*. 98:445–448.

Lederberg, J., McCray, A. 2001. Ome Sweet 'Omics - A genealogical treasury of words. *Scientist* 15:1–2.

Lee, A.Y., Akileswaran, L., Tibbetts, M.D., Garg, S.J., Van Gelder, R.N. 2015. Identification of torque teno virus in culture-negative endophthalmitis by representational deep DNA sequencing. *Ophthalmol.* 122:524–530. https://doi. org/10.1016/ j.ophtha.2014.09.001

Lee, M.S., Goslee, T.E., Lessell, S. 2003. Ehrlichiosis optic neuritis. *Am J Ophthalmol.* 135:412–413.

Lee, S.H., Oh, D.H., Jung, J.Y. et al., 2012. Comparative ocular microbial communities in humans with and without blepharitis. *Invest Ophthalmol Vis Sci.* 53:5585–5593. doi:10.1167/iovs.12-9922

Lee, V.H.L., Robinson, J.R. 1986. Topical ocular drug delivery: Recent developments and future challenges. *J OculPharmacol.* 2:67–108.

Leeming, J.P., Diamond, J.P., Trigg, R., et al., 1994. Ocular penetration of topical ciprofloxacin and norfloxacin drops and their effect upon eyelid flora. *Br J Ophthalmol.* 78:545–8.

Lessell, S., Collins, TE. 2003. Ophthalmoplegia in Powassan encephalitis. *Neurology*. 60:1726–1727.

Li, D., Liu, CM, Luo, R. et al., 2015. MEGAHIT: An ultra-fast single-node solution for large and complex metagenomics assembly via succinct de Bruijn graph. *Bioinformatics*. 31:1674–1676. doi:10.1093/bioinformatics/btv033

Liesegang, T.J., Forster, R.F. 1980. Spectrum of microbial keratitis in South Florida. *Am. J. Ophthalmol*. 90:80.

Liu, F., Zhou, Y., Zhu, L. et al., 2021. Comparative metagenomic analysis of the vaginal microbiome in healthy women. *Synth Syst Biotechnol*. 6:77–84 doi:10.1016/j.synbio.2021.04.002

Logan, R.A., O'Day, D.M. 1995. Blastomycosis. In F. T. Fraunfelder and F. H. Roy (ed.), *Current ocular therapy*. W. B. Saunders Company, Philadelphia, Pa, p. 66–67.

Lowder, C., McMahon, J., Meilser, D., Dodds, E., Calabrese, L., Didier, E., Cali, A. 1996. Microsporidial keratoconjunctivitis caused by *Septataintestinalis* in a patient with acquired immunodeficiency syndrome. *Am. J.Ophthalmol*. 121:715–717.

Madke, B., Mahajan, S., Kharkar, V., Chikhalkar, S., Khopkar, U. 2011 May. Disseminated cutaneous with nasopharyngeal rhinosporidiosis: Light microscopy changes following dapsone therapy. *Australas J Dermatol*. 52(2):e4–6.

Magie, S.K. 1995. Ocular histoplasmosis. In F. T. Fraunfelder and F. H. Roy (ed.), *Current ocular therapy*. W. B. Saunders Company, Philadelphia, Pa, p. 75–77.

Mah, F.S., Romanowski, E.G., Mah, F.S., Yates, K.A., Kowalski, R.P., Gordon, Y.J. 2004. Comparison of a new topical fluoroquinolone, gatifloxacin (Zymar™) with levofloxacin (Quixin™) and ciprofloxacin (Ciloxan™) in the treatment of fluoroquinolone resistant *Staphylococcus aureus* keratitis in a NZW rabbit model. *Invest. Ophthalmol. Vis. Sci*. 45 (E-Abstract 4982).

Mahendradas, P., Kawali, A., Luthra, S., Srinivasan, S., Curi, A.L., Maheswari, S., Ksiaa, I., Khairallah, M. 2020. Post-fever retinitis - newer concepts. *Indian J. Ophthalmol*. 68:1775–1786. https://doi.org/10.4103/ijo.IJO_1352_20

Maitray, A., Rishi, E., Rishi, P., Gopal, L., Bhende, P., Ray, R., Therese, K.L. 2019. Endogenous endophthalmitis in children and adolescents: case series and literature review. *Indian J. Ophthalmol*. 67:795–800. https://doi.org/10.4103/ijo.IJO_710_ 18

Malik, N.N., Goh, D., McLean, C., Huchzermeyer, P. 2004. Orbital cellulitis caused by Peptostreptococcus. *Eye*. 18:643–644. https://doi.org/10.1038/sj.eye.6700657

Marchesi, J.R., Ravel, J. 2015. The vocabulary of microbiome research: A proposal. *Microbiome*. 3:31. doi:10.1186/s40168-015-0094-5

Marshall, D., Browstein, S., Jackson, W.B., Mintsioulis, G., Gilberg, S.M., alZerah, B.F. 1997. Post-traumatic corneal mucormycosis caused by *Absidiacorymbifera*. *Ophthalmol*. 104:1107–1111.

Mason, G. 1981. Bilateral ophthalmomyiasis interna. *Am. J. Ophthalmol*. 91:65–70.

Maurice, D.M. 1973. The dynamics and drainage of tears. *Int Ophthalmol Clin*. 3:103–16.

McCabe, R., Remmington, J. 1988. Toxoplasmosis: The time has come. *N. Engl. J. Med*. 318:313–315.

McDermott, A.M. 2013. Antimicrobial compounds in tears. *Exp. Eye Res*. 117, 53–61. https://doi.org/10.1016/j.exer.2013.07.014

McLean, D.M., Donohue, W.L. 1959. Powassan virus: Isolation of virus from a fatal case of encephalitis. *Can Med Assoc J*. 80:708–711.

Messner, K., Kaminerer, M. 1979. Intraocular cysticercosis. *Arch. Ophthalmol*. 97:1103–1105.

Mills, D.M., Bodman, M.G., Meyer, D.R., Morton 3rd, A.D., Group, A.D.S. 2007. The microbiologic spectrum of dacryocystitis: A national study of acute versus chronic infection. *Ophthalmic Plast. Reconstr. Surg.* 23:302–306. https://doi.org/10.1097/IOP.0b013e318070d237

Mindel, J.S., Smith, H., Jacobs, M., et al., 1984. Drug reservoirs in topical therapy. *Invest Ophthalmol Vis Sci.* 25:346–50.

Moeller, C.T., Branco, B.C., Yu, M.C., Farah, M.E., Santos, M.A., Hofling-Lima, A.L. 2005. Evaluation of normal ocular bacterial flora with two different culture media. *Can. J. Ophthalmol.* 40, 448–453. https://doi.org/10.1139/i05-014

Molk, R. 1983. Ocular toxocariasis: A review of the literature. *Ann. Ophthalmol.* 15:216–231.

Morinelli, E.N., Dugel, P.U., Riffenburgh, R., Rao, N.A. 1993. Infectiousmultifocal choroiditis in patients with acquired immune deficiencysyndrome. *Ophthalmol.* 100:1014–1021.

Nayak, S., Acharjya, B., Devi, B., Sahoo, A., Singh, N. 2007. Disseminated cutenousrhinosporidiosis. *Indian J Dermatol venereolleprol.* 73:185–7.

Nitzulescu, V., Niculescu, M. 1975. Ophthalmopathy determined by*Candida humicola. Arch. Roum. Pathol. Exp. Microbiol.* 34:357–361.

O'Hara, E., Neves, A.L.A., Song, Y., Guan, L.L. 2020 Feb. The role of the gut microbiome in cattle production and health: Driver or Passenger? *Annu Rev AnimBiosci.* 15;8:199–220. doi: 10.1146/annurev-animal-021419-083952. PMID: 32069435

Okhravi, N., Dart, J.K., Towler, H.M., Lightman, S. 1997. *Paecilomyces lilacinus* endophtha lmitis with secondary keratitis: A case report and literaturereview. *Arch. Ophthalmol.* 115:1320–1324.

Ortiz, J.M., Eagle, R.C. Jr. 1982. Ocular findings in human babesiosis (Nantucket fever). *Am J Ophthalmol*;93:307–311.

Ozkan, J., Coroneo, M., Willcox, M. et al., 2018. Identification and visualization of a distinct microbiome in ocular surface conjunctival tissue. *Invest Ophthalmol Vis Sci.* 59:4268–4276. doi:10.1167/iovs.18-24651

Ozkan, J., Nielsen, S., Diez-Vives, C. et al., 2017. Temporal stability and composition of the ocular surface microbiome. *Sci Rep.* 7:9880. doi:10.1038/s41598-017-10494-9

Partamian, L.G., Jay, W.M., Fritz, K.J. 1983. Anaerobic orbital cellulitis. *Ann. Ophthalmol.* 15, 123–126.

Parunovic, A., Halde, C. 1967. *Pityrosporumorbiculare*: Its possible rolein seborrheic blepharitis. *Am. J. Ophthalmol.* 63:815.

Patton, T.F., Francoeur, M. 1978. Ocular bioavailability and systemic loss of topically applied ophthalmic drugs. *Am J Ophthalmol.* 85:225–9.

Paul, E., Zimmerman, L. 1970. Some observations on the ocular pathologyof onchocerciasis. *Hum. Pathol.* 1:581–594.

Pavan, P.R., Margo, C.E. 1993. Endogenousendophthalmitiscausedby*Bipolarishawaiiensis*in a patient with acquired immunodeficiency syndrome. *Am. J. Ophthalmol.* 116:644–645.

Pendse, S., Bilyk, J.R., Lee, M.S. 2006. The ticking time bomb. *SurvOphthalmol.* 51:274–279.

Perkins, R.E., Kundsin, R.B., Pratt, M.V., Abrahamsen, I., Leibowitz, H.M. 1975. Bacteriology of normal and infected conjunctiva. *J. Clin. Microbiol.* 1:147–149. https://doi.org/10.1128/JCM.1.2.147-149.1975

Perry, D., Merritt, J. 1983. Congenital ocular toxoplasmosis. *J. Natl.Med. Assoc.* 75:169–174.

Peterson, S.N., Snesrud, E., Liu, J. et al., 2013. The dental plaque microbiome in health and disease. *PLoS One*. 8:e58487. doi:10.1371/journal.pone.0058487

Pflugfelder, S.C., Flynn, H.W., Zwickey, T.A., Forster, R.K., Tsiligianni, A., Culbertson, W.W., Mandelbaum, S. 1988. Endogenous fungal endophthalmitis. *Ophthalmol.* 95:19–30.

Piantadosi, A., Rubin, D.B., McQuillen, D.P., et al., 2016. Emerging cases of PowassanVirus Encephalitis in New England: Clinical presentation, imaging, and reviewof the literature. *Clin Infect Dis.* 62:707–713.

Prabhushanker, M., Topiwalla, T.T., Ganesan, G., Appandaraj, S. 2017. Bilateral retinitis following typhoid fever. *Int. J. Retina Vitreous* 3, 11. https://doi.org/10.1186/s40942-017-0065-z

Quinet, B., Mitanchez, D., Salauze, B., Carbonne, A., Bingen, E., Fournier, S., Moissenet, D., Vu-Thien, H. 2010. Description and investigation of an outbreak of extended-spectrum beta-lactamase producing Escherichia coli strain in a neonatal unit. *Arch. Pediatr.* 17 (Suppl. 4), S145–S149. https://doi.org/10.1016/S0929- 693X(10)70916-7

Raina, U., Taneja, S., Lamba, P., Bansal, R. 1996. Spontaneous extrusionof extraocular cysticercus cysts. *Am. J. Ophthalmol.* 121:438–441.

Raja, H., Starr, M.R., Bakri, S.J. 2016. Ocular manifestations of tick-borne diseases. *SurvOphthalmol.* pii:S0039–6257(16)30049-2.

Ramage, G., Mowat, E., Jones, B., Williams, C., Lopez-Ribot, J. 2009. Our current understanding of fungal biofilms. *Crit. Rev. Microbiol.* 35, 340–355. https://doi.org/10.3109/10408410903241436

Ramage, G., Rajendran, R., Gutierrez-Correa, M., Jones, B., Williams, C. 2011. Aspergillus biofilms: clinical and industrial significance. *FEMS Microbiol. Lett.* 324, 89–97. https://doi.org/10.1111/j.1574-6968.2011.02381

Rao, N., Nerenberg, A.V., Forster, D.J. 1991. *Torulopsis candida*(*Candida famata*) endophthalmitis simulating *Propionibacterium* syndrome. *Arch. Ophthalmol.* 109:1718–1721.

Rappe, M.S., Giovannoni, S.J. 2003. The uncultured microbial majority. *Annu Rev Microbiol.* 57:369–394. doi:10.1146/annurev.micro.57.030502.090759

Reddy, J.C., Murthy, S.I., Reddy, A.K., Garg, P., 2015. Risk factors and clinical outcomes of bacterial and fungal scleritis at a tertiary eye care hospital. *Middle East Afr. J. Ophthalmol.* 22:203–211. https://doi.org/10.4103/0974-9233.150634

Richman, J., Zolezio, H., Tang-Liu, D. 1990. Comparison of ofloxacin, gentamicin and tobramycin concentrations in tears and in vitro MICs for 90% of test organisms. *Antimicrob Agents Chemother.* 34:1602–4.

Rippon, J.W. 1988. Medical mycology. *The pathogenic fungi and the pathogenic actinomycetes.* W. B. Saunders Company, Philadelphia, Pa, p. 362–372.

Ritterband, D., Seedor, J.A., Shah, M.K., Waheed, S., Schoor, I. 1998. A unique case of *Cryptococcus laurentii*keratitis spread by a rigid gaspermeable contact lens in a patient with onychomycosis. *Cornea* 17:115–118.

Robin, J.S., Ellis, P.P. 1978. Ophthalmic ointments. *SurvOphthalmol.* 22: 335–40.

Romero, J.R., Simonsen, K.A. 2008. Powassan encephalitis and Colorado tick fever. *Infect Dis Clin North Am.* 22:545–559.

Roney, P., Barr, C.C., Chun, C.H. et al., 1986. Endogenous *Aspergillus* endophthalmitis. *Rev. Infect. Dis.* 8:955–958.

Rosenbaum, J.T., Rahn, D.W. Prevalence of Lyme disease among patients withuveitis. *Am J Ophthalmol.* 1991;112:462–463.

Ruggli, G., Weber, R., Messmer, E.P., Font, R.L., Moll, C., Bernauer, W. 1997. *Pneumocystis carinii* infection of the conjunctiva in a patient withacquired immune deficiency syndrome. *Ophthalmol.* 104:1853–1856.

Sai Prashanthi, G.S., Jayasudha, R., Chakravarthy, S.K., Padakandla, S.R., SaiAbhilash, C.R., Sharma, S., Bagga, B., Murthy, S.I., Garg, P., Shivaji, S. 2019. Alterations in the ocular surface fungal microbiome in fungal keratitis patients. *Microorganisms.* 7. https://doi.org/10.3390/microorganisms7090309

Saran, B.R., Pomilla, P.V. 1996. Retinal vascular nonperfusion andretinal neovascularization as a consequence of cytomegalovirus retinitis andcryptococcal choroiditis. *Retina.* 16:510–512.

Sathiamoorthi, S., Smith, W.M. 2016. The eye and tick-borne disease in the United States. *Curr. Opin. Ophthalmol.* 27:530–537. https://doi.org/10.1097/ICU.000000000000030

Sauer, A., Hansmann, Y., Jaulhac, B., et al., 2009. Five cases of paralytic strabismus as a rare feature of Lyme disease. *Clin Infect Dis.* 48:756–759.

Schantz, D., Meyer, D., Glickman, L. 1979. Clinical serologic, and epidemiologic characteristics of ocular toxocariasis. *Am. J. Trop. Med. Hyg.* 28:24–28.

Schwartz, L., Loignon, L.M., Webster, R.G. Jr. 1978. Posttraumaticphycomycosis of the anterior segment. *Arch. Ophthalmol.* 96:860–863.

Sen, D. 1980. *Cysticercuscellulosae*in the lacrimal gland, orbit and eyelid. *Acta Ophthalmol.* 58:144–147.

Seo, M., Woo, J., Park, Y. 1996. Intravitreal cysticercosis. *Korean J. Ophthalmol.* 10:55–59.

Sha, B., Benson, C.A., Deutsch, T., Noskin, G.A., Murphy, R.L., Pottage, J.C., Jr., Finn, W.G., Roth, S.I,. Kessler, H.A. 1992. *Pneumocystiscarinii* choroiditis in patients with AIDS: clinical features, response totherapy, and outcome. *J. Acquir. Immune Defic. Syndr.* 5:1051–1058.

Sharma, S. 2012. Diagnosis of infectious diseases of the eye. *Eye* 26, 177–184. https:// doi.org/10.1038/eye.2011.275

Sheikh, A., Hurwitz, B. 2006. Antibiotics versus placebo for acute bacterial conjunctivitis. *Cochrane Database Syst Rev.* (2):CD001211. 10.1002/14651858.CD001211

Shin, H., Price, K., Albert, L. et al., 2016. Changes in the Eye Microbiota associated with contact lens wearing. *mBio.* 7:e00198. doi:10.1128/mBio.00198-16

Shivaji, S. 2017. We are not alone: A case for the human microbiome in extra intestinal diseases. *Gut Pathogens.* 9:13.

Shivaji, S. 2019. Connect between gut microbiome and diseases of the human eye. *J Biosci.* 44:110. DOI: 10.1007/s12038-019-9931-1

Shivaji, S. 2022. Chapter 3. Microbiome of the human eye. In: *Human ocular microbiome: bacteria, fungi and viruses in the human eye.* Published by Springer, 67–93.

Shivaji, S. et al., 2009. Janibacter hoylei sp. nov., Bacillus isronensis sp. nov. and Bacillus aryabhattai sp. nov., isolated from cryotubes used for collecting air from the upper atmosphere. *Int. J. Syst. Evol.* 59(Pt 12):2977–86. DOI: 10.1099/ijs.0.002527-0

Shivaji, S., Jayasudha, R., Chakravarthy, S.K., SaiAbhilash, C.R., Sai Prashanthi, G., Sharma, S., Garg, P., Murthy, S.I. 2021a Feb. Alterations in the conjunctival surface bacterial microbiome in bacterial keratitis patients. *Exp Eye Res.* 203:108418. Doi 10.1016/j.exer.2020.108418. Epub 2020 Dec 23.PMID: 33359511

Shivaji, S., Jayasudha, R., Chakravarthy, S.K., SaiAbhilash, C.R., Sai Prashanthi, G., Sharma, S., Garg, P., Murthy, S.I. 2021b Feb. Alterations in the conjunctival surface bacterial microbiome in bacterial keratitis patients. *Exp Eye Res.* 203:108418. doi: 10.1016/j.exer.2020.108418. Epub 2020 Dec 23. PMID: 33359511

Shivaji, S., Jayasudha, R., Prashanthi, G. S., Arunasri, K., Das, T. 2022 Apr. Fungi of the human eye: Culture to mycobiome. *Exp. Eye Res.* 217:10892.

Shivaji, S., Jayasudha, R., Sai Prashanthi, G., Kalyana Chakravarthy, S., Sharma, S. 2019. The human ocular surface fungal microbiome. *Invest. Ophthalmol. Vis. Sci.* 60:451–459. https://doi.org/10.1167/iovs.18-26076

Shovlin, J.P., Argueso, P., Carnt, N., Chalmers, R.L., Efron, N., Fleiszig, S.M., Nichols, J.J., Polse, K.A., Stapleton, F., Wiley, L., Willcox, M., Bright, F.V., Efron, N., Jones, L.W., Keir, N., Peterson, R.C., Stapleton, F. 2013. Ocular surface health with contact lens wear. *Contact Lens Anterior Eye* 36 (Suppl. 1):S14–S21. https://doi.org/10.1016/S1367-0484(13)60005-3

Slack, J., Hyndiuk, R.A., Harris, G.J., Simons, K.B. 1992. Blastomycosis of the eyelid and conjunctiva. *Ophthalmol. Plastic Reconstructive Surg.* 8:143–149.

Smits, S.L., Manandhar, A., van Loenen, F.B., van Leeuwen, M., Baarsma, G.S., Dorrestijn, N., Osterhaus, A.D., Margolis, T.P., Verjans, G.M. 2012. High prevalence of anelloviruses in vitreous fluid of children with seasonal hyperacute panuveitis. *J. Infect. Dis.* 205:1877–1884. https://doi.org/10.1093/infdis/jis284

Sodafy, M., Aminlari, A., Resaei, H. 1981. Ophthalmic leishmaniasis.*Clin. Exp. Dermatol.* 6:485–488.

Spruance SL, Bailey A. 1973. Colorado tick fever. A review of 115 laboratory confirmed cases. *Arch Intern Med.* 131:288–293.

Spurgeon, M.E., Cheng, J., Bronson, R.T., Lambert, P.F., DeCaprio, J.A. 2015. Tumorigenic activity of merkel cell polyomavirus T antigens expressed in the stratified epithelium of mice. *Canc. Res.* 75:1068–1079. https://doi.org/10.1158/ 0008-5472.CAN-14-2425

Srdic, N., Radulovic, S., Nonkovic, Z., Velimirovic, S., Cvetkovic, L., Vico, I. 1993. Two cases of exogenous endophthalmitis due to *Fusarium moniliforme*and *Pseudomonas* species as associated aetiological agents. *Mycoses.* 36:441–444.

Staley, J.T., Konopka, A. 1985. Measurement of in situ activities of nonphotosynthetic microorganisms in aquatic and terrestrial habitats. *Annu Rev Microbiol.* 39:321–346 doi:10.1146/annurev.mi.39.100185.001541

Stanek, G., Wormser, G.P., Gray, J., Strle, F. 2012. Lyme borreliosis. *Lancet.* 379:461–473.

Steere, A.C., Bartenhagen, N.H., Craft, J.E., et al., 1983. The early clinical manifestations of Lyme disease. *Ann Intern Med.* 99:76–82.

St. Leger, A.J., Caspi, R.R. 2018. Visions of eye commensals: The known and the unknown about how the microbiome affects eye disease. *Bioessays.* 40:e1800046 doi:10.1002/bies.201800046

Strelow, S.A., Kent, H.D., Eagle, R.C., Cohen, E.J. 1992. A case of contact lens related *Fusarium solani*keratitis. *CLAO J.* 18:125–127.

Sudasinghe, T., Rajapakse, R.P., Perera, N.A., Kumarasiri, P.V., Eriyagama, N.B., Arseculeratne, S.N. 2011 Oct-Nov. The regional sero-epidemiology of rhinosporidiosis in Sri Lankan humans and animals. *Acta Trop.* 120(1–2):72–81.

Sulewski, M.E., Green, W.R. 1986. Ocular histopathologic features of a presumed case of Rocky Mountain spotted fever. *Retina.* 6:125–130.

Syrdalen, P., Nitter, T., Mehl, R. 1982. Ophthalmomyiasis interna posterior. Report of case caused by the reindeer warble fly larva and review of previous reported cases. *Br. J. Ophthalmol.* 66:589–593.

Teckhasaenee, C., Ritch, R., Kanchanaranya, C. 1986. Ocular parasiticinfections in Thailand. *Rev. Infect. Dis.* 8:350–356.

Thomas, P. 1994. Mycotic keratitis—an underestimated mycosis. *J. Med.Vet. Mycol.* 32:235–256.

Thomas, T., Gilbert, J., Meyer, F. 2012. Metagenomics - a guide from sampling to data analysis. *Microb Inform Exp.* 2:3. doi:10.1186/2042-5783-2-3

Tong, Q., Chai, W.X., Wang, Z.F., Kou, J.F., Qi, Z.T., Wang, D.L. 1990. A case of cerebral aspergillosis caused by *Aspergillus nidulans.* Clinical, pathologic and mycologic identifications. *Chinese Med. J.* 103:518–522.

Toussaint, D., Danis, P. 1965. Retinopathy in generalized *Loa loa*filariasis. A clinicopathological study. *Arch. Ophthalmol.* 74:470–476.

Träisk, F., Lindquist, L. 2012. Optic nerve involvement in Lyme disease. *CurrOpinOphthalmol.* 23:485–490.

Tseng, C.Y., Liu, P.Y., Shi, Z.Y., Lau, Y.J., Hu, B.S., Shyr, J.M., Tsai, W.S., Lin, Y.H. 1996. Endogenous endophthalmitis due to Escherichia coli: Case report and review. *Clin. Infect. Dis.* 22:1107–1108. https://doi.org/10.1093/clinids/22.6.1107

Tudor, R., Blair, E. 1971. *G. spinigerum*: An unusual cause of ocular nematodiasis in the Western Hemisphere. *Am. J. Ophthalmol.* 72:185–190.

Turnbaugh, P.J., Ley, R.E., Hamady, M. et al., 2007. The human microbiome project. *Nature.* 449:804–810 doi:10.1038/nature06244

Turner, L., Stinson, I. 1965. Mycobacterium fortuitum; as a cause of corneal ulcer. *Am. J. Ophthalmol.* 60:329–331.

Turner, T.R., James, E.K., Poole, P.S. 2013. The plant microbiome. *Genome Biol.* 14(6):209. Published 2013 Jun 25. doi:10.1186/gb-2013-14-6-209

Vannier, E., Krause, P.J. 2012. Human babesiosis. *N Engl J Med.* 366:2397–2407.

Venugopal, R., Satpathy, G., Sangwan, S., Kapil, A., Aron, N., Agarwal, T., Pushker, N., Sharma, N. 2016. Conjunctival microbial flora in ocular stevens-johnson syndrome sequelae patients at a tertiary eye care center. *Cornea.* 35:1117–1121. https://doi.org/10.1097/ICO.0000000000000857

Visvesvara, G. 1995. Pathogenic and opportunistic free-living amoebae. In P. R. Murray, E. J. Baron, M. A. Pfaller, F. C. Tenover, and R. H. Yolken (ed.), *Manual of clinical microbiology.* ASM Press, Washington, D.C.

Watson, S., Cabrera-Aguas, M., Khoo, P. 2018. Common eye infections. *AustPrescr.* 41(3):67–72. doi:10.18773/austprescr.2018.016280

Watters, C.M., Burton, T., Kirui, D.K., Millenbaugh, N.J. 2016. Enzymatic degradation of in vitro Staphylococcus aureus biofilms supplemented with human plasma. *Infect. Drug Resist.* 9:71–78. https://doi.org/10.2147/IDR.S103101

Weber, R., Bryan, R.T., Schwartz, D.A., Owen, R.L. 1994 Oct. Human microsporidial infections. *Clin Microbiol Rev.* 7(4):426–61. doi: 10.1128/cmr.7.4.426. PMID: 7834600; PMCID: PMC358336

Wei, J. 1990. Ultrasonic diagnosis of intraocular cysticercosis. *Chung-Hua Yen Ko TsaChih* 26:230–231.

Weishaar, P., Flynn, H.W. Jr., Murray, T.G., Davis, J.L., Barr, C.C., Gross, J.G., Mein, C.E., McLean, W.C. Jr., Killian, J.H. 1998. Endogenous *Aspergillus* endophthalmitis. Clinical features and treatment outcomes. *Ophthalmol.* 105:57–65.

Weng, Y., Liu, J., Jin, S., Guo, W., Liang, X., Hu, Z. 2017. Nanotechnology based strategies for treatment of ocular disease. *Acta Pharm Sin B* 7(3), 281–291. doi:10.1016/j.apsb.2016.09.001

Wilhelmus, K. 1987. The pathogenesis of endophthalmitis. *Int. Ophthalmol.Clin.* 27:74–81.

Willcox, M.D., 2013. Characterization of the normal microbiota of the ocular surface. *Exp. Eye Res.* 117, 99–105. https://doi.org/10.1016/j.exer.2013.06.003

Witherspoon, C., Kuhn, F., Owens, S.D., White, M.F., Kimble, J.A., 1990. Endophthalmitis due to *Sporothrixschenckii*after penetrating ocularinjury. *Ann. Ophthalmol.* 22:385–388.

Wood, DE., Lu, J., Langmead, B. 2019. Improved metagenomic analysis with Kraken 2. *Genome Biol.* 20:257. doi:10.1186/s13059-019-1891-0

Yamamoto, G., Pavan-Langston, D., Stowe III, G.C., Albert, D.M. 1978. Fungal invasion of a therapeutic soft contact lens and cornea. *Ann. Ophthalmol.* 11:1731–1735.

Yau, T., Rivera-Velazquez, P.M., Mark, A.S., Cytryn, A.S., Levy, C.S., Shmookler, B.M. Kolsky, M.P. 1996. Unilateral optic neuritis caused by*Histoplasma capsulatum* in a patient with the acquired immunodeficiencysyndrome. *Am. J. Ophthalmol.* 121:324–326.

Zegans, M.E., Van Gelder, R.N. 2014. Considerations in understanding the ocular surface microbiome. *Am J Ophthalmol.* 158:420–422. doi:10.1016/j.ajo.2014.06.014

Zhai, H., Bispo, P.J.M., Kobashi, H., Jacobs, D.S., Gilmore, M.S., Ciolino, J.B. 2018. Resolution of fluoroquinolone-resistant Escherichia coli keratitis with a PROSE device for enhanced targeted antibiotic delivery. *Am. J. Ophthalmol. Case Rep.* 12, 73–75. https://doi.org/10.1016/j.ajoc.2018.09.006

Zhou, Y, Holland, M.J., Makalo, P. et al., 2014. The conjunctival microbiome in health and trachomatous disease: a case control study. *Genome Med.* 6:99. doi:10.1186/s13073-014-0099-x

Zilliox, M.J., Gange, W.S., Kuffel, G. et al., 2020. Assessing the ocular surface microbiome in severe ocular surface diseases. *Ocul Surf.* 18:706–712. doi:10.1016/j.jtos.2020.07.007

Zimmerman, T.J., Kooner, K.S., Kandarakis, A.S., et al., 1984. Improving the therapeutic index of topically applied ocular drugs. *Arch Ophthalmol.*; 102: 551–3.

Zinkham, W. 1978. Visceral larva migrans. *Am. J. Dis. Child.* 32:627–633.

2

Antimicrobial Resistance: History and Origin of Antimicrobials, Targets of Action and Factors Contributing to Antimicrobial Resistance

S. Shivaji

2.1 Introduction

Antimicrobial resistance (AMR) is defined as the ability of a microbe (bacteria, fungi, viruses, and parasites) to resist the killing effect of an antimicrobial drug like antibiotics, antifungals, antivirals and antiparasitic drugs. Therefore, the microbes become resistant to antimicrobial drugs and the infections become increasingly difficult or impossible to treat, thereby contributing to the increase in dissemination of the disease, leading to severe illness and death. AMR, apart from nullifying our ability to treat common infections, is even further alarming since infections are no longer susceptible to the killing effect of the antimicrobial drugs. This menace of AMR is further amplified by the non-availability of new antimicrobials, lack of access to quality antimicrobials and antibiotic shortages (WHO, 2021a-c).

2.2 Antimicrobial Resistance a Global Concern

AMR has become a global phenomenon spreading to every nook and corner of the world and is an economic burden to national economies and their health systems (Founou et al. 2017; Prestinaci et al. 2015). In addition to death and disability it affects productivity of the affected individuals and their dependents due to prolonged hospital stays, expensive treatments with alternate drugs, and due to the complications posed during various surgical procedures, including caesarean sections or hip replacements, cancer chemotherapy, organ transplantation and so forth (WHO, 2021a.c; Prestinaci et al. 2015; ECDC, 2019; Shrestha et al. 2018). It is estimated that annually more than 0.7 million people die world-wide due to diseases that have become drug-resistant (Davies et al., 2013). According to another estimate by the British government, AMR complications would be responsible for the death of ten million people annually by the year 2050 (O'Neill, 2014). But this estimate is considered uncertain, and it is suggested that it be based on more reliable AMR burden estimates, comprehensive

 DOI: 10.1201/9781003451105-2

AMR surveillance data inclusive of low- and middle-income countries, community-acquired infections and to provide AMR-related morbidity and mortality data and demographic data for effective AMR control (de Kraker et al. 2016). In the United States over 2 million individuals are infected by antibiotic-resistant organisms annually, and these infections cause about 23,000 deaths (CDC, 2014). Thus, AMR organisms are a serious threat to public health. In Europe, the economic burden of AMR is more than nine billion Euros annually (Llor et al. 2014; Prestinaci et al. 2015). Furthermore, in the United States the economic cost due to antimicrobial resistance is 20 billion dollars, which does not include 35 billion dollars in loss of productivity annually (CDC, 2014). In the United States the Centre for Disease Control was the first to flag in a report (CDC, 2014) three specific microbes that posed an urgent threat to public health, *Clostridium difficile* (Lawson et al. 2016), carbapenem-resistant Enterobacteriaceae (CRE) and drug-resistant *Neisseria gonorrhoeae* (Liu et al. 2016; Wi et al. 2017). Subsequently, pan-resistant *Klebsiella pneumonia* was also identified in the United States (Chen et al. 2017). Thus, there is an absolute need to develop strategies that could combat AMR including the development of novel antimicrobials and diagnostic tools, and methods for prevention of infections.

2.3 Overcoming Antimicrobial Resistance by Government Interventions

To overcome AMR, governments world over have put in place various agencies such as the Antimicrobial Resistance Multi Partner Trust Fund (AMR MPTF), the Global Antibiotic Research & Development Partnership (GARDP), AMR Action Fund and other funds and initiatives with focus on national surveillance of AMR and to educate the public. In 2001, a global strategy for containment of Antimicrobial Resistance was suggested by the World Health Organization (WHO) to hinder the emergence and spread of AMR. Subsequently, in 2015, the Global Action Plan (GAP) 2015 on AMR was suggested in the World Health Assembly (WHO, 2021a) for implementation of multisectoral national action plans. GAP 2015 was subsequently endorsed by the Food and Agriculture Organization (FAO) and the World Organisation for Animal Health (OIE). In 2015, GLASS (Global Antimicrobial Resistance and Use Surveillance System) was launched as a more comprehensive programme. This programme consolidated progressive data from surveillance of AMR in humans, surveillance of the use of antimicrobial medicines and AMR in the food chain and in the environment (WHO, 2021b). Yet another major driving force to battle AMR was to undertake surveillance of AMR in human, animal, plant and environmental health sector, a worthwhile initiative by heads of state at the United Nations General Assembly in New York in September 2016. Further a "One Health" approach was proposed by WHO along with FAO and OIE to promote best practices to reduce the levels of AMR and slow its development. In the same year in 2016, the Inter agency Coordination Group (IACG) on AMR was convened (United Nations, 2017) and the committee submitted a recommendation titled *No Time to Wait: Securing the Future from Drug-resistant Infections*. The emphasis was to bring together all experts on AMR involving national and international organizations and individuals to formulate a plan for the eradication of AMR (WHO, 2019). As a step in this direction, in 2020 WHO established a tripartite joint secretariat (FAO, OIE and WHO) to drive multi-stakeholder engagement in AMR through the Global Leaders Group on AMR, the Independent Panel on Evidence for Action against AMR, and the Multi-Stakeholder Partnership Platform (WHO, 2021 c). Public participation in AMR was strengthened through World Antimicrobial Awareness Week (WAAW) with a tagline "Antimicrobials: Handle with Care" in 2020 (WHO, 2020), which was conceived as a global campaign.

2.4 History and Evolution of Antimicrobial resistance

The history of AMR (Martin, 2022) is closely linked to the first report describing the antibacterial properties of molds in 1640 by John Parkington (1640). Several centuries later, in 1929, Alexander Fleming demonstrated "penicillin" in the "mould broth filtrate" from a *Penicillium* species (Fleming, 1929), which was antibacterial and effectively controlled infections caused by Gram-positive pathogens such as *Staphylococcus* and *Streptococcus.* It was also demonstrated that penicillin could be used as an antibacterial agent in animals (Chain et al. 1940) and against human infections (Dawson et al. 1943; Florey et al. 1943). Albert Schatz first isolated streptomycin in 1943 from *Streptomyces griseus* (Torok et al. 2009; Renneberg and Demain, 2008). These discoveries together constituted the golden era of antibiotics and was marked by the discovery and designing of newer drugs and modifications of the already existing drugs targeting bacteria and fungi (Table 2.1) (Durand et al. 2018; IDSA, 2016). The above efficacy of the antibiotics revolutionised and transformed medicine such that surgeries, chemotherapy to fight cancer, organ transplantation, joint replacement therapy, corneal transplantation and burn injuries were now controlled and it improved the quality of life for millions of people (Brown and Wright, 2016).

2.4.1 The Post-antibiotic Era

Fleming noted that *Staphylococcus aureus,* which was hitherto susceptible to penicillin (Fleming, 1929), developed penicillin resistance (CDC, 2014). Several other reports also noted that microorganisms often quickly became resistant to antibiotics. Compounding this issue was the realisation that bacteria were evolving resistance more quickly than the introduction of new antibiotics to treat infections (Figure 2.1), and this heralded the emergence of the antibiotic resistant era (Davies and Davies 2010). In this antibiotic resistant era, the quality of life and life expectancy were at risk. We have now entered the post-antibiotic age.

The post-antibiotic era is also marked by rigorous efforts to identify new scaffolds between 1960–1990; but unfortunately, these efforts were not rewarding and resulted in a fresh wave of resistance. The efforts to overcome resistance also involved recombinant DNA approaches, rational drug design, high-throughput screens of biochemical assays (Macarron, et al. 2011), sequencing of thousands of bacterial genomes including the pathogens *Haemophilus influenzae* (Fleischmann et al. 1995), precise deletions at the genome scale of *Staphylococcus aureus, Pseudomonas aeruginosa* (Forsyth et al. 2002; Jacobs et al. 2003) and *Mycobaxcterium tuberculosis* (Jain et al.

TABLE 2.1

Antibiotics, Their Origin, Target of Action and Resistance Mechanism

Antibiotic	Organism	Target	Resistance due to
Aminoglycosides			
Capreomycin	*Streptomyces capreolus*	Translation	Mutations in tly A gene (A33G, T539G and Ins49GC)
Framycetin	*Streptomyces lavendulae*	Translation	Not known
Gentamicin	*Micromonospora purpurea*	Translation	Phosphorylation, acetylation nucleotidylation, efflux and altered target
Kanamycin	*Streptomyces kanamyceticus*	Translation	Bacteriocidal
Natamycin	*Streptomyces natalensis*	Cell membrane	G143A mutation in the cytochrome b gene
Neomycin	*Streptomyces fradiae*	Translation	Bactericidal
Plazomicin	Synthetic	Translation	Modification of the ribosomal binding site
Sisomicin	*Micromonospora inyoensis*	Translation	Enzymatically or non-enzymatically mediated
Spectinomycin	Cyanobacteria and various plants	Translation	Phosphorylation, acetylation nucleotidylation, efflux, altered target and bactericidal
Streptomycin	*Streptomyces griseus*	Translation	Mutations in electron transport chain components and LPS biosynthesis.
Tobramycin	*Streptomyces tenebrarius*	Translation	
Anti-tuberculous drugs			
Ethambutol	Synthetic derivative of ethylenediamine	Peptidoglycan biosynthesis	Unknown
Ethionamide	Nicotinamide derivative	Cell wall	Unknown
Isoniazid	Synthetic	Cell wall	Mutations of *katG* confer isoniazid resistance

(continued)

TABLE 2.1 (Continued)

Antibiotics, Their Origin, Target of Action and Resistance Mechanism

Antibiotic	Organism	Target	Resistance due to
Pyrazinamide	Synthetic	Membrane potential, energy production and translation	Mutations in the *pnc*A gene, which encodes, pyrazinamidase.
β-lactams			
Carbapenem (meropenem)	*Streptomyces cattleya*	Peptidoglycan biosynthesis	Decreased membrane permeability, altered porin channels, transmission of antimicrobial resistance genes, enzymatic inactivation, target site mutations, activation of efflux pumps, presence of various broad spectrum β-lactamases etc.
Cephalosporin (cephamycin)	*Cephalosporium acremonium*	Peptidoglycan biosynthesis	Production of extended-spectrum β-lactamases (ESBLs)
Monobactam (aztreonam)	*Chromobacterium violaceum*	Peptidoglycan biosynthesis	Not known
Penicillin (ampicillin)	*Penicillium notatum, Penicillium chrysogenum*	Peptidoglycan biosynthesis	TEM β-lactamase enzyme
Mupirocin	*Pseudomonas flurorescens*	Isoleucine tRNA synthetase	Mutations in isoleucyl-tRNA synthetase
Cationic peptides			
Colistin	*Bacillus polymyxa*	Cell membrane	Altered target, efflux
Chloramphenicols			
Chloramphenicol	*Penicillium chrysogenum*	Translation	Acetylation, efflux, altered target and bacteriostatic
Fosfomycin			
Fosfomycin	*Streptomyces fradiae*	Cys115 in UDP-N-acetyl-glucosamine- enolpyruvyl-transferase	Mutations of the nonessential glycerophosphate transporter

Glycopeptide			
Dalbavancin	Semisynthetic of naturally occurring compound A-40926,	Peptidoglycan	Mediated through a plasmid gene VanA or VanB from enterococcal species
Oritavancin	Lipoglycopeptide derived from the naturally occurring chloroeremomycin,	Disrupts cell membrane of Gram-positive bacteria resulting in depolarization, permeabilization, and cell death	Not known
Teicoplanin	*Actinoplanes teichomyceticus*	Peptidoglycan biosynthesis	Reprogramming peptidoglycan biosynthesis and bacteriocidal
Vancomycin	*Amycolatopsis orientalis*	Peptidoglycan biosynthesis	
Ketolides			
Telithromycin	Semisynthetic erythromycin derivative	Translation	Mutations in domains II and V of the 23S rRNA, ribosomal proteins L4 and L22 and *erm*(B)
Lincosamide			
Clindamycin	*Streptomyces lincolnensis.*	Translation	Nucleotidylation, efflux, and altered target
Lincomycin	*Streptomyces lincolnensis*	Translation	Methylation of 23S ribosomal RNA, modification of the antibiotic and active efflux
Lipopeptide			
Daptomycin	*Streptomyces roseosporus*	Cell membrane	Altered target
Macrolides			
Azithromycin	Invented as an active pharmaceutical ingredient	Translation	Hydrolysis, glycosylation, phosphorylation, efflux, altered target, and bacteriostatic
Clarithromycin	Synthetic, made from erythromycin	Translation	
Erythromycin	*Streptomyces erythreus*	Translation	

(*continued*)

TABLE 2.1 (Continued)

Antibiotics, Their Origin, Target of Action and Resistance Mechanism

Antibiotic	Organism	Target	Resistance due to
Josamycin	*Streptomyces narbonensis var. josamyceticus*	Translation	Not known
Midecamycin	*Streptomyces mycarofaciens*	Not known	Not known
Spiramycin	*Streptomyces ambofaciens*	Translation	Resistance by several mechanisms
Fidaxomicine	*Dactylosporangium aurantiacum subspecies hamdensis*	Transcription	Mutations in either *rpoB* or *rpoC* genes, Val1143Asp, β' Arg89Gly, β-Gln1074His, β- Val1143Gln and β- Val1143Asp.
Nitrofurant			
Nitrofurantoin	Synthetic, synthesized from hydrazine-acetic acid	Ribosomal proteins, DNA, respiration, pyruvate metabolism and other macromolecules within the cell	Resistance may be chromosomal or plasmid-mediated, and involves inhibition of nitrofuran reductase
Nitroimidazole			
Metronidazole	*Streptomyce ssp.*	DNA synthesis, DNA degradation, cell death.	Decreased uptake of the drug, altered reduction efficiency, over expression of the efflux pumps, inactivation of the drug, and/ or increased DNA damage repair
Ornidazole	Synthetic nitroimidazole	DNA structure and cell death	Not known
Oxazolidinones			
Tedizolid	Synthetic drug	Translation	Mutations in *optrA* and *poxtA*

Polypeptide			
Polymyxin	*Paenibacillus polymyxa*	Cell membrane	Mutations in the lipid A biosynthesis genes lpxA, lpxC or lpxD, increased anionic capsular polysaccharide production, expression of efflux systems and altered expression of outer membrane proteins
Pyrimidines			
Trimethoprim	Synthetic derivative of tri-methoxybenzyl-pyrimidine	C_1 metabolism	Efflux and altered target
Quinolones			
Ciprofloxacin	Synthetic drug	DNA replication	Acetylation, efflux, altered target and bactericidal
Delafloxacin	Synthetic drug	DNA replication	Mutations in *gry*A and *gyr*B
Norfloxacin	Synthetic drug	DNA gyrase, a type II and IV topoisomerase, and cell division	Altered drug target and a reduction in drug uptake
Nalidixicacid	Synthetic drug	DNA replication, DNA gyrase, topoisomerase IV	Mutation in *gyrB*
Rifamycins			
Rifampicin	*Streptomyces mediterranei*	Transcription	ADP-ribosylation, efflux, altered target and bacteriocidal
Steroid			
Fusidicacid	*Fusidium coccinem*	Translation	Alterations in elongation factor G, altered drug permeability, binding by chloramphenicol acetyl-transferase type I and efflux
Streptogramins			
Synercid	A combination drug	Translation	C-O lyase, acetylation, efflux and altered target

(*continued*)

TABLE 2.1 (Continued)

Antibiotics, Their Origin, Target of Action and Resistance Mechanism

Antibiotic	Organism	Target	Resistance due to
Streptogramin B	*Streptomyces graminofaciens*	Translation	C-O lyase, efflux, altered target and bacteriocidal
Pristinamycin	*Streptomyces pristinaespiralis*	Translation	
Sulfonamide			
Sulfamethoxazole	Synthetic drug, structural analog of *para*-aminobenzoic acid	C_1 metabolism	Efflux and altered target
Sulfanilamide	Synthetic, an aniline derivatized with a sulfonamide group	Folate synthesis	Bacteriostatic
Tetracyclines			
Chlortetracycline	*Streptomyces aureofaciens, Streptomyces rimosus*	Translation	Efflux, altered target and enzymatic inactivation
Tetracycline	Semisynthetic drug produced from chlortetracycline,	Translation	Bacteriostatic
Doxycycline	Synthetically derived from oxytetracycline	Translation	Bacteriostatic
Eravacycline	Synthetic drug derived from tetracycline	Translation, efflux	Up-regulation of MDR efflux, modifications of 16S RNA or 30S ribosomal proteins and mutations of *rpsJ*
Minocycline	Synthetic drug derived from tetracycline	Translation	Monoxygenation, efflux, and altered target
Tigecycline	Synthetic glycylcycline drug derived from tetracycline	Translation	

Source: Modified from Durand et al. (2018), Davies and Davies (2010), Morar and Wright (2010), Brown and Wright (2016).

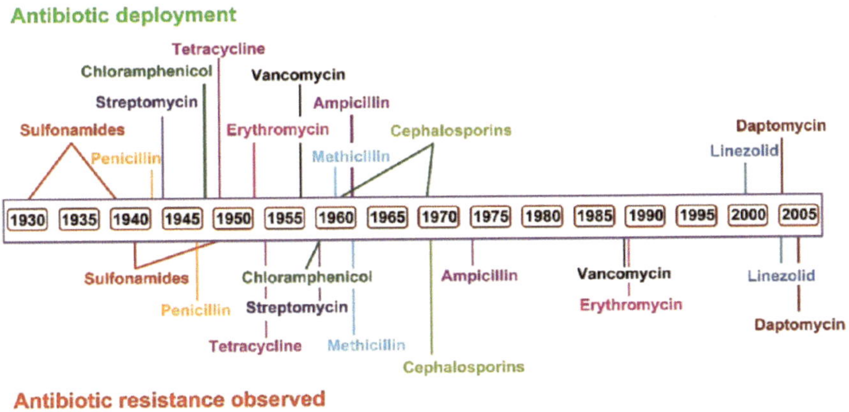

FIGURE 2.1 Timeline representing antibiotic deployment (top of timeline) and antibiotic resistance (bottom of timeline). (Modified from Clatworthy et al. 2007.)

2014) to identify new targets for designing antibiotics that are not susceptible to existing mechanisms of resistance (Brown and Wright 2005). The future challenge is to develop new antibiotics using multiple approaches such as:

(a) an out of the box approach involving a combination of more than one antibiotic as in the case of ceftazidime and avibactam,
(b) incentives to discover new drugs like the GAIN (Generating Antibiotics Incentives Now) act in the United States (Brown, 2013) and the Innovative Medicines Initiative in Europe,
(c) reward companies that successfully launch new antimicrobial drugs (Spellberg et al. 2015),
(d) shift from broad-spectrum antibiotics use to narrow, disease-specific antibiotics (Farha et al. 2013; Ejim et al. 2011) and
(e) improve the accuracy of point-of-care diagnostics to move forward in the field of antibiotic discovery.

2.4.2 Microbial and Human Factors Contributing to Antimicrobial Resistance

The realisation that resistance to antibiotics is of great human concern due to its implications for health and the economy has prompted global efforts to overcome this menace, as discussed in several reviews (Allcock et al., 2017; Castro-Sanchez et al., 2016; Kumarasamy et al., 2010; Muloi et al., 2019; Pai, 2015). The outcome has led to the recognition that resistance to drugs by microbes could be attributed to factors associated with the microbes (intrinsic) and due to human factors (extrinsic) (Konduri et al. 2021). The human factors that facilitate AMR include:

(a) Improper dosing and excess consumption of broad-spectrum antibiotics.
(b) Improper and indiscriminate use of antibiotics for animals and plants (Buckner et al., 2018), which contribute to mobilization of resistance

elements in previously antibiotic-sensitive pathogens thus enabling the enrichment of drug-resistant bacteria through mobilization and horizontal transfer mediated by mobile genetic elements (D'Costa et al. 2006), which harbour genes for resistance (Dias et al. 2020); it is suggested that up to 50% of antibiotics prescribed for humans are unnecessary.

(c) Use of antibiotics in food animals has been linked to antibiotic-resistant bacteria in humans, and thus represents a threat to human health (Landers et al. 2012).

(d) Improper disposal of antibiotics.

(e) Absence of effective methods to prevent transmission of resistant bacteria in healthcare from patient to patient, from healthcare-workers to patients (Mulvey and Simor, 2009; D' Agata et al. 2012), from contaminated equipment (Humphries, et al., 2017) and the biotic and abiotic components of the environment. Hospital drains act as a reservoir for antibiotic-resistant bacteria and need to be disinfected (Weingarten et al. 2018).

(f) Use of sub-standard antibiotics. Resistance in bacteria to antibiotics could be due to presence of antibiotic-resistant genes that encode intrinsic resistance such as to cephalosporins in Enterococci species (Murray, 1990) or intrinsic resistance to ampicillin within *Klebsiella* species (Babini and Livermore, 2000). In addition, several microbial factors that play a role in supporting AMR are attributed to factors that are closely associated with the ability of the microbes to modify the targets of the bacteria that make it susceptible to the killing effects of the drug. Normally, antibiotics kill bacteria by inhibiting processes that are essential for survival like inhibiting cell wall biosynthesis, by inhibiting protein synthesis, by inhibiting nucleic acid metabolism and repair, and by disruption of membrane structure (Figure 2.2) (Li and Webster, 2018).

So, the obvious strategy that bacteria adopt to become resistant to antibiotics is to prevent the antibiotic from interacting with its cellular target (Figure 2.3) by the following strategies:

(a) bacterial cell wall is modified to prevent antibiotics from entering the cell,

(b) bacterial protein targets are altered (Andersson, and Hughes, 2017) so that antibiotics bind with less affinity,

(c) develop methods to inactivate the anti-microbial agent,

(d) induce active efflux of the antibiotic (Cox and Wright, 2013; Fujardo et al. 2008),

(e) upregulate antibiotic-inactivating enzymes (Abraham and Chain, 1940),

(f) acquire AMR genes and plasmids that confer resistance, as in soil bacteria (Manaia 2017; Forsberg et al. 2012). These soil bacteria contain resistance cassettes against five classes of antibiotics (Forsberg et al. 2012). Antibiotic resistance genes (ARGs) were also observed in the human gut microbiome (Forsberg et al. 2012) and the resistomes of the human gastrointestinal tract, soil, and clinical pathogens were all similar (Forsberg et al. 2012). It was also observed that the transmission of resistome genes in a population, was better in pathogens that had higher proportions of mobile genetic elements (Forsberg et al. 2014) and

(g) by their ability to form a biofilm.

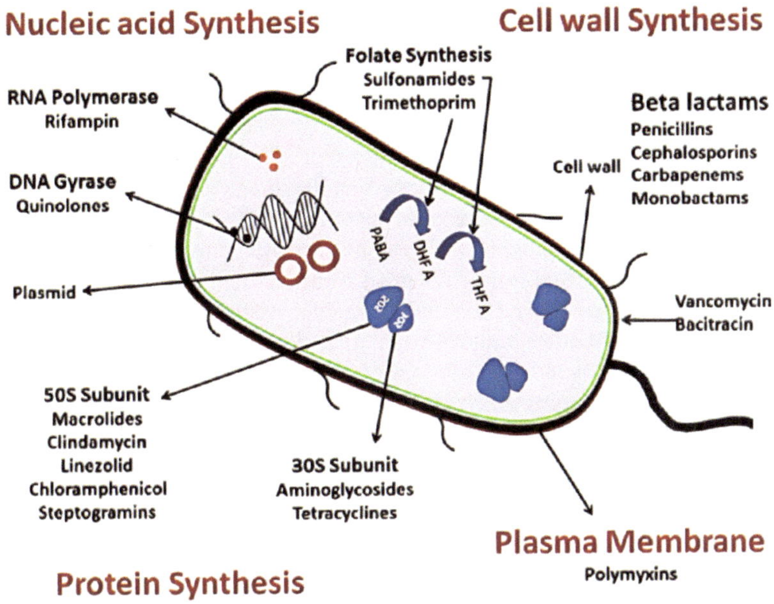

FIGURE 2.2 Targets of antimicrobial agents namely cell wall synthesis, cell membrane integrity, protein synthesis, nucleic acid synthesis and essential enzymes, which render the cells inviable. A few of the antibiotics and their specific targets are indicated. (Modified from Li and Webster (2018) and as in Konduri et al. (2021) and Shivaji (2022).)

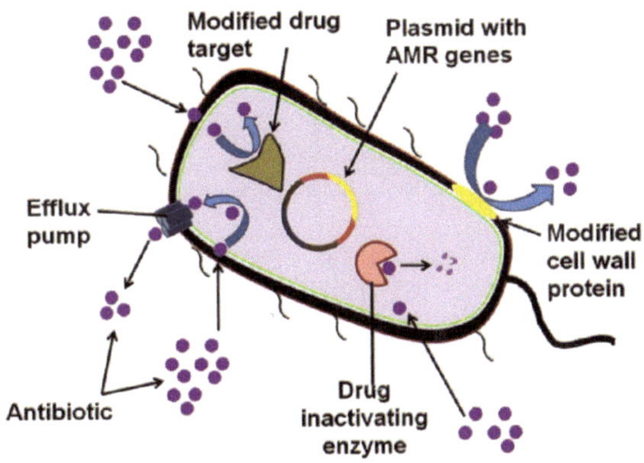

FIGURE 2.3 Multiple microbial strategies that confer resistance to antimicrobial agents. (Modified from Dantas and Sommer (2014) and as in Konduri et al. (2021) and Shivaji (2022).

In 2014 the widespread antibiotic resistance communities prompted the WHO to project the imminence of a post-antibiotic era, where once again minor infections would become potentially deadly (WHO, 2014; Reardon, 2014).

2.5 Conclusions

In this era of antimicrobial resistance, the quality of life and life expectancy are at risk and is eroding our ability to control infections with antibiotics. AMR thus has severe impacts on human health and economy and needs to be addressed at a war-footing to overcome it. And rightly so, apart from international interventions several governments have made active interventions to improve this grave situation. The strategies to overcome AMR are multifold and, apart from developing new antibacterial agents, the strategies include judicious use of antibiotics for treatment of diseases, accurate testing and reporting and administering appropriate antibiotics. The CDC has proposed four core actions to help combat antibiotic resistance, which include: (1) prevention of infections and the spread of antimicrobial resistance; (2) tracking resistance patterns, both domestically and globally; (3) improving antibiotic stewardship; and (4) development of new antibiotics and novel diagnostic tests.

REFERENCES

Abraham, E.P., Chain, E. An enzyme from bacteria able to destroy penicillin. *Nature.* 1940;146, 837.

Allcock, S., Young, E.H., Holmes, M., Gurdasani, D., Dougan, G., Sandhu, M.S., Solomon, L., Torok, M.E. Antimicrobial resistance in human populations: Challenges and opportunities. *Glob. Health Epidemiol. Genom.* 2017;2, e4. https://doi. org/10.1017/gheg.2017.4

Andersson, D.I., Hughes, D. Selection and transmission of antibiotic-resistant bacteria. *Microbiol. Spect.* 2017;5(4).

Babini, G.S., Livermore, D.M. Are SHV β-Lactamases Universal in Klebsiella pneumoniae? *Antimicrob. Agents Chemother.* 2000;44(8), 2230.

Brown, E., Wright, G. Antibacterial drug discovery in the resistance era. *Nature.* 2016;529, 336–343. https://doi.org/10.1038/nature17042

Brown, E.D. Is the GAIN Act a turning point in new antibiotic discovery? *Can. J. Microbiol.* 2013;59, 153–156.

Brown, E.D., Wright, G.D. New targets and screening approaches in antimicrobial drug discovery. *Chem. Rev.* 2005;105, 759–774.

Buckner, M.M.C., Ciusa, M.L., Piddock, L.J.V. Strategies to combat antimicrobial resistance: anti-plasmid and plasmid curing. *FEMS Microbiol. Rev.* 2018;42, 781–804. https://doi.org/10.1093/femsre/fuy031

Castro-Sanchez, E., Moore, L.S., Husson, F., Holmes, A.H. What are the factors driving antimicrobial resistance? Perspectives from a public event in London, England. *BMC Infect. Dis.* 2016;16, 465. https://doi.org/10.1186/s12879-016-1810-x

CDC. *Antibiotic Resistant Threats in the United States*, 2013. 2014: Atlanta, GA.

Chain, E., et al., Penicillin as a chemotherapeutic agent. *The Lancet.* 1940;236(6104), 226–228.

Chen, L., et al., Notes from the Field: Pan-Resistant New Delhi Metallo-Beta-Lactamase-Producing Klebsiella pneumoniae - Washoe County, Nevada, 2016. *MMWR Morb Mortal Wkly Rep*, 2017;66(1), 33.

Clatworthy, A., Pierson, E., Hung, D. Targeting virulence: A new paradigm for antimicrobial therapy. *Nat Chem Biol.* 2007;3, 541–548. https://doi.org/10.1038/nchembio.2007.24

Cox, G., Wright, G.D. Intrinsic antibiotic resistance: Mechanisms, origins, challenges and solutions. *Int. J. Med. Microbiol.* 2013;303, 287–292.

D'Agata, E.M.C., et al., Efficacy of infection control interventions in reducing the spread of multidrug-resistant organisms in the hospital setting. *PLoS ONE*. 2012;7(2), e30170.

Dantas, G., Sommer, M.O.A. How to fight back against antibiotic resistance. *Am. Sci.* 2014;102, 42–51.

Davies J., Davies, D. Origins and evolution of antibiotic resistance. *Microbiol Mol Biol Rev*. 2010 September;74(3), 417–433. 1092-2172/10/$12.00 doi:10.1128/MMBR.00016-10

Davies, S.C., Fowler, T., Watson, J., Livermore, D.M., Walker, D.. Annual report of the chief medical officer: Infection and the rise of antimicrobial resistance. *Lancet.* 2013;381, 1606–1609. https://doi.org/10.1016/S0140-6736(13)60604-2

Dawson, M.H., et al., Penicillin as a chemotherapeutic agent*. *Ann. Intern Med.* 1943;19(5), 707–717.

D'Costa, V.M., McGrann, K.M., Hughes, D.W., Wright, G.D. "Sampling the antibiotic resistome". *Science*. 2006 January;311(5759), 374–377. doi:10.1126/science.1120800. PMID 16424339

de Kraker, M.E.A, Stewardson, A.J., Harbarth, S. Will 10 million people die a year due to antimicrobial resistance by 2050? *PLoS Med.* 2016;13(11), e1002184. https://doi.org/10.1371/journal.pmed.1002184

Dias, M.F., da Rocha Fernandes, G., Cristina de Paiva, M., Christina de Matos Salim, A., Santos A.B., Amaral Nascimento, A.M. "Exploring the resistome, virulome and microbiome of drinking water in environmental and clinical settings". *Water Res.* 2020 May;174, 115630. doi:10.1016/j.watres.2020.115630. PMID 32105997

Durand, G.A., Raoult, D., Dubourg, G. Antibiotic discovery: History, methods and perspectives. *Int. J. Antimicrob. Agents.* 2018. doi: https://doi.org/10.1016/j.ijantimicag.2018.11.010

ECDC. 2019. Surveillance of Antimicrobial Resistance in Europe 2017. Available from: www.ecdc.europa.eu/sites/default/files/documents/EARS-Net-report-2017-update-jan-2019.pdf. Accessed September 16, 2019.

Ejim, L. et al. Combinations of antibiotics and nonantibiotic drugs enhance antimicrobial efficacy. *Nature Chem. Biol.* 2011;7, 348–350.

Fajardo, A., Nadia Martínez-Martín, MaríaMercadillo, Juan C. Galán, Bart Ghysels, Sandra Matthijs,Pierre Cornelis, Lutz Wiehlmann, Burkhard Tümmler, Fernando Baquero, José L. Martínez .The neglected intrinsic resistome of bacterial pathogens. *PLoS ONE*. 2008;3, e1619.

Farha, M. A. et al. Inhibition of WTA synthesis blocks the cooperative action of PBPs and sensitizes MRSA to β-lactams. *ACS Chem. Biol.* 2013;8, 226–233.

Fleischmann, R.D., Adams, M.D., White, O., Clayton, R.A., Kirkness, E.F., Kerlavage, A.R., Bult, C.J., Tomb, J.F., Dougherty, B.A., Merrick, J.M., et al. Whole-genome random sequencing and assembly of Haemophilus influenzae Rd. *Science*. 1995 July 28;269(5223), 496–512. doi: 10.1126/science.7542800. PMID: 7542800

Fleming, A. On the antibacterial action of cultures of a Penicillium, with special reference to their use in the Isolation of B. influenzæ. *Br. J Exp. pathol.* 1929;10(3), 226–236.

Florey, M.E., et al. General and local administration of penicillin. *The Lancet.* 1943;241(6239), 387–397.

Forsberg, K.J., Patel, S., Gibson, M.K., Lauber, C.L., Knight, R., Fierer, N., Dantas, G. "Bacterial phylogeny structures soil resistomes across habitats". *Nature*. 2014 May;509 (7502), 612616. doi:10.1038/nature13377. PMC 4079543. PMID 24847883

Forsberg, K.J., Reyes, A., Wang, B., Selleck, E.M., Sommer, M.O., Dantas, G. "The shared antibiotic resistome of soil bacteria and human pathogens". *Science*. 2012 August;337(6098), 1107.

Forsyth, R.A. et al. A genome-wide strategy for the identification of essential genes in *Staphylococcus aureus*. *Mol. Microbiol.* 2002;43, 1387–1400.

Founou, R.C., Founou, L.L., Essack, S.Y. Clinical and economic impact of antibiotic resistance in developing countries: A systematic review and meta-analysis. *PLoS One*. 2017;12:e0189621. doi: 10.1371/journal.pone.0189621

Humphries, R.M., et al., Duodenoscope-related outbreak of a Carbapenem-Resistant Klebsiella pneumoniae identified using advanced molecular diagnostics. *Clin. Infect. Dis.* 2017;65(7), 1159–1166.

IDSA.2016 *IDSA Antibiotic Resistance Infographic 2016 Final, I.D.S.o.* America, Editor.

Jacobs, M.A. *et al.* Comprehensive transposon mutant library of *Pseudomonas aeruginosa*. *Proc. Natl Acad. Sci. USA*. 2003;100, 14339–14344.

Jain, P., Hsu, T., Arai, M., Biermann, K., Thaler, D.S., Nguyen, A., González, P.A., Tufariello, J.M., Kriakov, J., Chen, B., Larsen, M.H., Jacobs, W.R. Jr.Specialized transduction designed for precise high-throughput unmarked deletions in *Mycobacterium tuberculosis*. *mBio*. 2014;5, e01245–14.

Konduri, R., Sharma, S., Shivaji, S. Microbes of the human eye: Microbiome, antimicrobial resistance and biofilm formation. *Exp. Eye Res*. 2021. 108476. ISSN 0014-4835 https://doi.org/10.1016/j.exer.2021.108476

Kumarasamy, K.K., Toleman, M.A., Walsh, T.R., Bagaria, J., Butt, F., Balakrishnan, R., Chaudhary, U., Doumith, M., Giske, C.G., Irfan, S., Krishnan, P., Kumar, A.V., Maharjan, S., Mushtaq, S., Noorie, T., Paterson, D.L., Pearson, A., Perry, C., Pike, R., Rao, B., Ray, U., Sarma, J.B., Sharma, M., Sheridan, E., Thirunarayan, M.A., Turton, J., Upadhyay, S., Warner, M., Welfare, W., Livermore, D.M., Woodford, N. Emergence of a new antibiotic resistance mechanism in India, Pakistan, and the UK: a molecular, biological, and epidemiological study. *Lancet Infect. Dis*. 2010;10, 597–602. https://doi.org/10.1016/S1473-3099(10)70143-2

Landers, T.F., et al. A review of antibiotic use in food animals: Perspective, policy, and potential. *Public Health Rep*. 2012;127(1), 4–22.

Lawson, P.A., et al. Reclassification of Clostridium difficile as Clostridioides difficile (Hall and O'Toole 1935) Prévot 1938. *Anaerobe*. 2016;40, 95–99.

Li, B., Webster, T.J., Bacteria antibiotic resistance: New challenges and opportunities for implant-associated orthopedic infections. *J. Orthop. Res.* 2018;36, 22–32. https://doi.org/10.1002/jor.23656.

Liu, Y.-Y., et al. Emergence of plasmid-mediated colistin resistance mechanism MCR-1 in animals and human beings in China: A microbiological and molecular biological study. *The Lancet Infectious Diseases*. 2016;16(2), 161–168.

Llor, C., Bjerrum, L. Antimicrobial resistance: Risk associated with antibiotic overuse and initiatives to reduce the problem. *Ther Adv Drug Saf.* 2014;5(6):229–241. doi: 10.1177/2042098614554919

Macarron, R., Banks, M.N., Bojanic, D., Burns, D.J., Cirovic, D.A., Garyantes, T., Green, D.V., Hertzberg, R.P., Janzen, W.P., Paslay, J.W., Schopfer, U., Sittampalam, G.S. Impact of high-throughput screening in biomedical research. *Nat Rev Drug Discov*. March 2011;10(3):188–95. doi: 10.1038/nrd3368. PMID: 21358738

Manaia, C.M. "Assessing the Risk of Antibiotic Resistance Transmission from the Environment to Humans: Non-Direct Proportionality between Abundance and Risk". *Trends in Microbiology*. 2017 March;25(3), 173.

Martin, R.M. Antimicrobial Resistance History and Perspectives Copyright 2018 by. on April 5 2022. http://hwmaint.clsjournal.ascls.org/ Downloaded from 2

Morar, M., Wright, G.D. The genomic enzymology of antibiotic resistance. *Annu. Rev. Genet.* 2010;44, 25–51.

Muloi, D., Kiiru, J., Ward, M.J., Hassell, J.M., Bettridge, J.M., Robinson, T.P., van Bunnik, B.A.D., Chase-Topping, M., Robertson, G., Pedersen, A.B., Fevre, E.M., Woolhouse, M.E.J., Kang'ethe, E.K., Kariuki, S. Epidemiology of antimicrobial-resistant Escherichia coli carriage in sympatric humans and livestock in a rapidly urbanizing city. *Int. J. Antimicrob. Agents.* 2019;54, 531–537. https://doi.org/10.1016/j.ijantimicag.2019.08.014.19

Mulvey, M.R., Simor, A.E. Antimicrobial resistance in hospitals: How concerned should we be? *CMAJ: Canadian Medical Association Journal.* 2009;180(4), 408–415.

Murray, B.E., The life and times of the Enterococcus. *Clin. Microbiol. Rev.* 1990;3(1), 46–65.

O'Neill, J. *Review on antimicrobial resistance antimicrobial resistance: Tackling a crisis for the health and wealth of nations*. London: Review on Antimicrobial Resistance. 2014. Available from: https://amr-review.org/sites/default/files/AMR%20Review%20Paper%20-%20Tackling%20a%20crisis%20for%20the%20health%20and%20wealth%20of%20nations_1.pd

Pai, M.P. Treatment of bacterial infections in obese adult patients: How to appropriately manage antimicrobial dosage. *Curr. Opin. Pharmacol.* 2015;24, 12–17. https://doi.org/10.1016/j.coph.2015.06.004.

Parkington, J. 1640. *Theatrum Botanicum*. London.

Prestinaci, F., Pezzotti, P., Pantosti, A. Antimicrobial resistance: A global multifaceted phenomenon. *Pathog Glob Health.* 2015;109(7):309. doi: 10.1179/2047773215Y.0000000030

Reardon, S. 2014. *WHO warns against 'post-antibiotic' era*. Nature.

Renneberg, R., Demain, A.L. 2008. *Biotechnology for Beginners*. Elsevier. p. 103. ISBN 9780123735812. Archived from the original on 10 September 2017.

Shivaji, S. 2022 *In Human ocular Microbiome; bacteria, fungi and viruses in the human eye*. Springer. 2022.

Shrestha, P., Cooper, B.S., Coast, J., et al. Enumerating the economic cost of antimicrobial resistance per antibiotic consumed to inform the evaluation of interventions affecting their use. *Antimicrob Resist Infect Control.* 2018;7(1):98. doi: 10.1186/s13756-018-0384-3

Spellberg, B., Bartlett, J., Wunderink, R., Gilbert, D. N. Novel approaches are needed to develop tomorrow's antibacterial therapies. *Am. J. Respir. Crit. Care Med.* 2015;191:135–140.

Torok, E., Moran, E., Cooke, F. 2009. *Oxford Handbook of Infectious Diseases and Microbiology*. OUP Oxford. Chapter 2. ISBN 9780191039621. Archived from the original on 8 September 2017.

United nations. 2017. *Interagency Coordination Group on Antimicrobial Resistance*. 17 March 2017.

Weingarten, R.A., Johnson, R.C., Conlan, S., Ramsburg, A.M., Dekker, J.P., Lau, A.F., Khil, P., Odom, R.T., Deming, C., Park, M., Thomas, P.J.; NISC Comparative Sequencing Program, Henderson, D.K., Palmore, T.N., Segre, J.A., Frank, K.M. Genomic analysis of hospital plumbing reveals diverse reservoir of bacterial plasmids conferring

Carbapenem resistance. *mBio*. 2018;9(1):e02011-17. https://doi.org/10.1128/mbio.02011-17.

WHO, 2014 *WHO's first global report on antibiotic resistance reveals serious, worldwide threat to public health*. 2014: Geneva, Switzerland.

WHO, 2019. *No time to Wait: Securing the future from drug-resistant infections. Report to the Secretary-General of the United Nations* 29 April 2019.

WHO, 2020. *World Antimicrobial Awareness Week 2020 - Handle with care: United to preserve*. November 2020.

WHO, 2021a. *Comprehensive Review of the WHO Global Action Plan on Antimicrobial Resistance* – Volume 1: Report 31, August 2021.

WHO, 2021b. *Global Antimicrobial Resistance and Use Surveillance System (GLASS)?*

WHO, 2021c. *Antimicrobial resistance*. 17 November 2021.

Wi, T., Lahra, M.M., Ndowa, F., Bala, M., Dillon, J-AR., Ramon-Pardo, P., et al. (2017) Antimicrobial resistance in *Neisseria gonorrhoeae*: Global surveillance and a call for international collaborative action. PLoS Med 14(7): e1002344. https://doi.org/10.1371/journal.pmed.1002344

3

Methods to Detect Antimicrobial Resistance in Human Ocular Samples

S. Shivaji

3.1 Introduction

Antimicrobial resistance (AMR) is of global concern due to its devastating impact on human health, poultry, animals, plants, and the environment (Friedman et al., 2016). Several organisations across the world are closely monitoring AMR (Álvarez-Molina et al., 2020; Kaprou et al., 2021). AMR is influenced by the delay in detection and characterization of microbial pathogens (Cansizoglu et al., 2019); and most often the delay encourages empirical antibiotic therapy prior to conclusive diagnosis, which may be detrimental to the health of the patient and contribute to AMR. This situation could be overcome by the development of rapid, accurate and low-cost diagnostics that would effectively decrease antibiotic use (Brinkac et al., 2017). In recent years, several reviews have put together methods for AMR diagnostics, which include both phenotypic and genotypic methods and metabolic profiling of bacteria (Leonard et al., 2018; Kaprou et al., 2021; Galhano et al., 2021; Chen and Hong 2021). Detection of AMR in microorganisms has been reviewed by several authors (Sharma 2011; Leonard et al., 2018; Konduri et al., 2021; Thomas et al., 2019; Galhano et al., 2021; Manente et al., 2022). This chapter highlights the various detection methods available including phenotypic, genome based (whole genome and metagenome sequencing) and protein-based methods (MALDI-TOF, MS, and IR spectroscopy) for rapid AMR diagnosis.

3.2 Collection of Ocular Samples

Suitable protocols are used for the collection of ocular samples like conjunctival swabs, corneal scrapings, and ocular fluids like tears. For the collection of conjunctival swab the tarsal and fornix conjunctiva from the lower and upper lids of the eye are swabbed separately and then used to detect the growth of the microorganisms by streak-plating on several different media plates immediately. Corneal scrapings from bacterial keratitis patients are collected using a sterile surgical blade No.15. The cornea are exposed to proparacaine hydrochloride (0.5%) by topical administration of eye drops prior to scraping of the cornea. Corneal epithelia following topical administration of povidone-iodine prior to photorefractive keratectomy (PRK) are collected as controls for corneal scrapings from bacterial keratitis (BK) patients (Shivaji et al., 2021).

DOI: 10.1201/9781003451105-3

3.3 Identification of Microorganisms from Ocular Samples

The conjunctival swabs, corneal scrapings or tears are inoculated onto different types of media to detect the growth of the bacteria in the ocular samples. The popular media that are used include 5% sheep blood agar, chocolate agar, Sabouraud dextrose agar, potato dextrose agar, brain heart infusion broth and thioglycollate broth. Subsequently the media plates were incubated under both aerobic and anaerobic conditions and examined on a day to day basis till the 7th day. Normally bacterial colonies would appear by the seventh day of incubation. Such colonies would be purified and identified based on phenotypic traits using conventional manual microbiological methods or by using an automated identification system (Shivaji et al., 2021) like Vitek 2 Compact System (bioMérieux, France) which help to identify a wide range of clinically relevant organisms including Gram-negative bacilli, Gram-positive bacteria, yeast and yeast-like organisms, *Neisseria*, *Haemophilus* and other fastidious Gram-negative bacteria, anaerobic bacteria and coryneform bacteria.

3.4 AMR screening of Ocular Samples by Phenotypic Methods

Several methods are routinely used in a microbiology laboratory to monitor the growth of the test microorganism in the presence or absence of the drug either by direct colony counting or by quantifying bacteria or fungi using a spectrophotometer or by using a Vitek 2 Compact system (bioMérieux, France) using the VITEK® 2 AST (Antimicrobial susceptibility testing) cards. These cards when loaded with a saline suspension of the organism provide AST results for clinically important Gram-positive cocci, Gram-negative bacilli, and yeasts. The VITEK® 2 Gram-positive (GP) and Gram-negative (GN) cards have 43 and 47 biochemical tests respectively distributed across the 64 wells to monitor carbon source utilization, enzymatic activity and resistance of the organism as indicated by colour change (blue, orange, red, yellow, and green) whereas the controls or negative cultures do not exhibit any change in colour and the wells are colourless (Figure 3.1). Results are available in approximately 10 hours or less. The AST card has the wells prefilled with several antibiotics (Figure 3.1c) to which the bacterial suspension is added and growth is measured by optical density. The results are available in 4 h. Vitek 2 cards are applicable to a wide variety of organisms such as *Enterobacteriaceae,* non-fermenters, *Staphylococci*, *Enterococci*,

Streptococci (including *S. pneumoniae*, *S. viridans* and beta-hemolytic *Streptococci*), yeasts and so forth.

3.4.1 Agar Dilution Method

This time-tested method that has been used to determine the lowest concentration of the antimicrobial agent (MIC) required to inhibit the growth of an organism under specific conditions of temperature, pH, and medium composition (Figure 3.2). It involves preparing a series of agar plates containing increasing concentrations of the drug (i.e., 1, 2, 4, 8, μg ml^{-1}, etc.), on to which 1–5 μl suspension of the organism with a turbidity value of 0.5 McFarland standard [equivalent to about ~ 1×10^8 colony forming units (CFU) ml^{-1}] is spread. Subsequently, the inoculated plates were incubated overnight

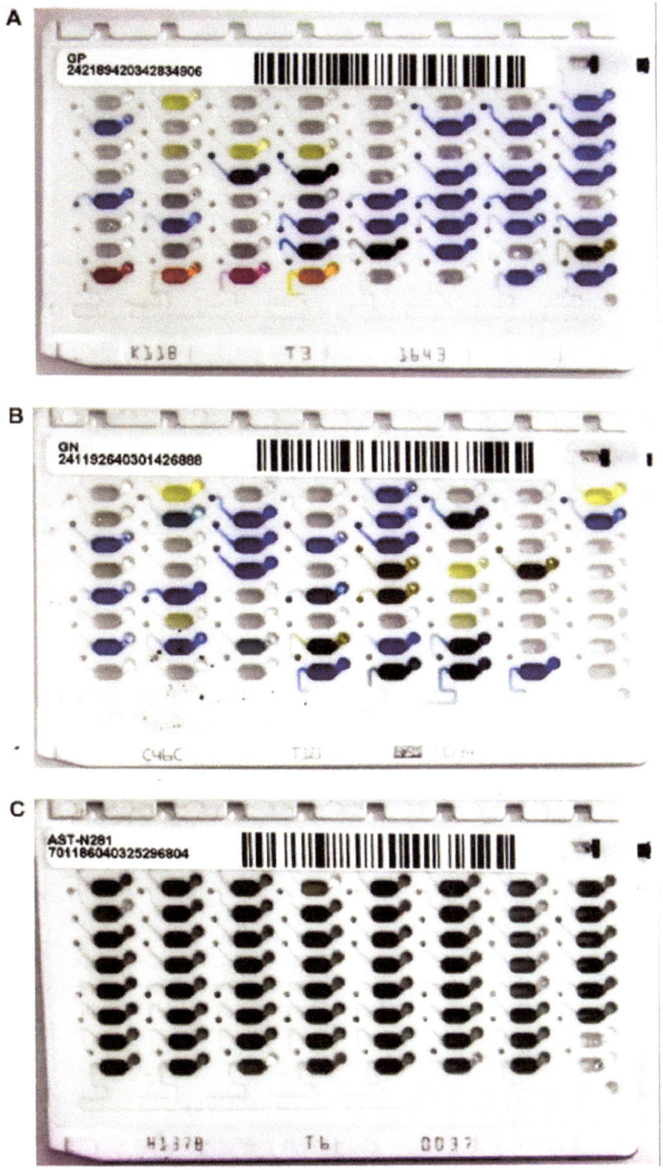

FIGURE 3.1 Identification of Gram-positive and Gram-negative bacteria and antimicrobial susceptibility testing with the Vitek 2 Compact (bioMérieux, France) automated identification system using the Vitek 2 GP card (A), Vitek 2 GN card (B) and the Vitek 2 AST cards respectively (C). The growth of the microorganism is indicated by colour change (blue, orange, red, yellow and green) as in A and B, and the controls or negative cultures do not exhibit any change in colour and the wells are colourless. Results are available in approximately 10 hours or less. The AST card has the wells prefilled with a number of antibiotics (C) to which the bacterial suspension is added and growth is measured by optical density. The results are available in 4 h. (Courtesy Roshni, K., LVPEI, Hyderabad, India.)

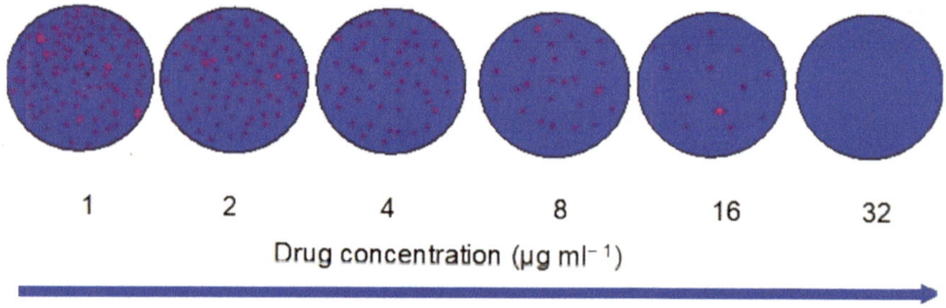

FIGURE 3.2 The agar dilution method. Agar plates containing increasing concentrations of a given drug, usually in doubling dilutions (i.e., 1, 2, 4, 8,16, 32 µg ml^{-1}, etc.) are seeded with 1–5 µl suspension of the organism with a turbidity value of a 0.5 McFarland standard [equivalent to about ~ 1 × 10^8 colony forming units (CFU) ml^{-1}] and incubated overnight. Number of visible colonies are counted at the end of the incubation period and the concentration of the drug on the media plate that exhibited no growth of bacterial colonies was taken to be the MIC of the drug. (Generated by the author.)

at 37^{0}C or at the optimum temperature of growth of the organism. For instance, cold-loving psychrophilic organisms are incubated at 15–20^{0}C. Slow growing organisms are incubated even up to a week. At the end of the incubation period the number of visible colonies are counted and the concentration of the drug on the media plate that exhibited no growth was taken to be the MIC of the drug (Figure 3. 2).

3.4.2 Broth Dilution Method

This method is less laborious and could be carried out in a series of test tubes or a 96 well plate containing graded concentrations of the antimicrobial agent (Figure 3.3). The bacteria to be tested are inoculated into the liquid medium and growth is assessed after incubation for 16–20 h. The concentration of the drug in the tube or well with no visible growth was considered as the MIC (Andrews 2001). In this method, in addition to the MIC, the minimum lethal concentration or minimum bactericidal concentration could be determined (Sharma 2012; Wiegand et al., 2008). This protocol applies only to aerobic bacteria and can be completed in 3 days. Currently several ready to use plates like the Sensititre plates are available commercially. These are microdilution plates that contain serial dilutions of antimicrobial agents for screening bacteria, fungi, and specific organisms (Pujol et al., 2002; Richter et al., 2018).

3.4.3 Kirby-Bauer Disk Diffusion Method

This is an agar diffusion method used for testing rapidly growing ocular and other pathogens (Figure 3.4). The method has been used for bacteria, fungi, oomycetes and so forth.

Briefly, a standardized inoculum of a fresh subculture is swabbed onto the surface of Muller Hinton Agar (MHA) plate and filter paper disks impregnated with a known concentration of an antimicrobial agent, that are commercially available, are placed on the plate. The plates are then incubated according to CLSI guidelines and the size of the zone of inhibition a clear area around the disk is measured after overnight

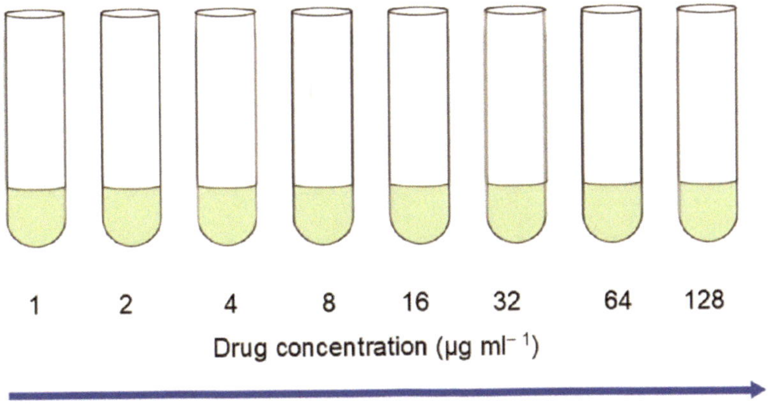

FIGURE 3.3 Broth dilution method. Tubes containing graded concentrations of the antimicrobial agent (i.e., 1, 2, 4, 8,16, 32 μg ml^{-1}, etc.) are inoculated with 1–5 μl suspension of the organism, equivalent to about ~ 1×10^8 colony forming units and assessed for growth after incubation for 16–20 h. The concentration of the drug in the tube with no visible growth was considered as MIC (Andrews 2001). (Generated by the author.)

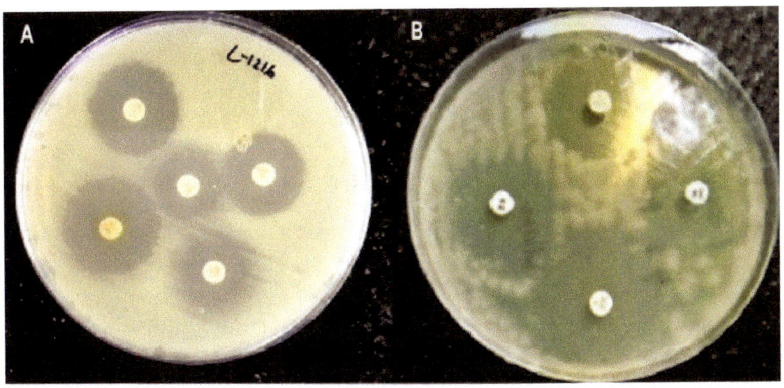

FIGURE 3.4 Kirby-Bauer disk diffusion method for evaluating the susceptibility of ocular *E. coli* to Moxifloxacin, Piperacillin and tazobactam, Gatifloxacin, Cefuroxime and Imipenem (A) and *Pythium insidiosum*, a parasitic oomycete to Tetracycline, Doxycycline, Clarithromycin and Azithromycin (B) showing clear, translucent zones of inhibition which varied in size depending on the antibiotic used. (Courtesy Konduri and Venkatesh, LVPEI, Hyderabad, India.)

incubation (Gy. Horváth et al., 2016). Based on the results the organism is categorised as susceptible, resistant, or intermediately susceptible.

3.4.4 E-test

Epsilometer test or the E- test is a convenient method for MIC determination (Álvarez-Molina et al., 2020) (Figure 3.5). In this method a predefined graded concentrations

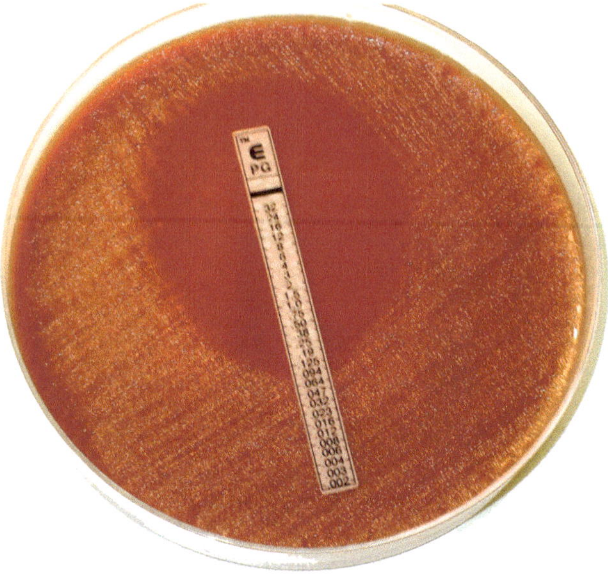

FIGURE 3.5 E- test showing the MIC of Imipenem (IP) and Minocycline (MIN) (A) and the MIC of Tetracycline (TET), Tigecycline (TGC) and Linezolid (LZ) (B). Ocular *E. coli* and *Pythium insidiosum* were used in A and B respectively. The point where the elliptical zone of inhibition, a clear transparent area, intersects the antibiotic strip corresponds to the MIC values of the tested antibiotic. (Courtesy Konduri and Venkatesh, LVPEI, Hyderabad, India.)

of antibiotics or antifungal agents are immobilized on a plastic strip and used to determine the MIC of the antimicrobial agent. The plastic strip is inert and non-porous and about 5mm wide and 60 mm long. This strip is then applied to a plate, which was inoculated with a standard concentration of bacteria or fungi such that a lawn of bacteria or fungi would be visible overnight when incubated at $37^{0}C$. The 'E end' of the strip is placed at the edge of the plate and the concentration gradient was visible (i.e., facing upwards). Following incubation, the drug from the strip diffuses into the agar and with time an antimicrobial concentration gradient is formed in the agar medium and an inhibition zone in the shape of a symmetrical ellipse is visible after overnight incubation. The point where the ellipse intersects the strip corresponds to the MIC value (Figure 3.5). E-test has been used for MIC determination in ocular bacteria and fungi (Sharma 2012; Konduri et al., 2017) and is especially useful for fastidious microorganisms (Tenover et al., 2009), such as *Campylobacter* spp. (McGill et al., 2009).

3.4.5 Optical Microscopy-based Microcolony Detection Method (MCD)

Optical microscopy-based microcolony detection method (MCD) was developed for the rapid detection of antibiotic-resistant bacteria in blood (Kuchibiro et al., 2020).

Cultures were sub-cultured (1 μl) onto sheep blood agar and several other media including antibiotic-containing selective media for the detection of antibiotic-resistant bacteria. The selective medium used for Gram-positive cocci was CHROMagar MRSA (CHROMagar, Paris, France) whereas for Gram-negative bacilli ichromID ESBL (bioM_erieux) and CHROMagar mSuperCARBA medium (CHROMagar) were used. Microbes were incubated at 35^{0}C under aerobic conditions.

After 2–4 h of incubation, the selective media plates containing a known concentration of an antibiotic (with the lid on) were checked for microcolonies using an optical microscope set at 100x magnification using a 10x objective lens. This method takes approximately 3 h for detecting antibiotic-resistant bacterial growth and has been used to monitor methicillin resistance in staphylococci and the resistance of Gram-negative bacilli to third generation antibiotics namely cephalosporins and carbapenem. Thus, MCD is useful for the efficient and rapid detection of antibiotic-resistant bacterial colonies in media.

3.4.6 Conclusions on Phenotypic Methods

All the above phenotypic methods are simple to perform, MIC values can be estimated, pathogen identification is possible and more importantly the methods are validated. The disadvantages are that strain purification is a must, cultivation is required, growth may be slow (Leonard et al., 2018), may have some disagreement between standards and for some anti-microbials the cut-off values for susceptibility have not been established (Sundfjord et al., 2004; Kaprou et al., 2021). Another drawback is viable but non-cultivable bacteria cannot be cultivated in routine culture media. Further, diffusion-based methods are not useful for drugs that do not diffuse properly (Galhano et al., 2021).

3.5 AMR Screening through Metabolic Profiling of Bacteria

In contrast to the above culture-based methods for AST, automation of AST systems, using VITEK (bioMérieux), MicroScan WalkAway, PhenoTest BC Kit, BD phoenix, and Sensititre, reduce the time to 4.5–24 h (Maugeri et al., 2019). The VITEK 2 method involves the use of broth or any other liquid medium suitable for growth of the microbe since it measures the transmittance of an inoculum at regular intervals of time and decrease in transmittance corresponds to increasing turbidity (growth). Thus, by monitoring the growth over time it could be used to determine the MIC of antibiotics. MicroScan WalkAway also relies on monitoring growth of bacteria in the presence or absence of an antibiotic in growth medium containing a fluorogenic substrate, which releases a fluorescent signal with increase in bacterial growth. The fluorescence intensity correlates to the enzymatic activity of the bacteria (Delost et al., 2014). In contrast to these methods the PhenoTest BC Kit by Accelerate Diagnostics, can also identify and analyze antibiotic susceptibility (Pancholi et al., 2018). However, a major limitation is that these systems require bacterial isolates derived from an overnight culture, and thus the results cannot be obtained until the next day, which may be a disadvantage for a clinical sample (Avesar et al., 2017). Thus, there is a need for new technologies that can perform AST rapidly. In a recent review, Chen and Hong (2021) summarized recent rapid AST methods that take 30 min to a few hours based on the profiling of bacterial metabolism. These metabolism-based AST methods are based on detecting ATP levels, the copy number of nucleic acids, or macromolecule synthesis in bacteria. But the methods need to be automated to accelerate routine diagnosis in clinical applications of these rapid AST methods.

3.5.1 Quantification of Adenosine Triphosphate

Adenosine triphosphate (ATP) levels could serve as a quantitative method to monitor bacterial growth. ATP catalyses the oxidation of luciferin to adenyl oxyluciferin and releases photons (bioluminescence), that could be measured as a proxy for ATP levels with a luminometer. In *S. aureus,* ATP values showed a sharp drop in the methicillin susceptible strains and the results correlated with the disk diffusion and the MIC results obtained by the agar dilution method (McWalter 1984). However, the method failed in 13% of *S. aureus* tested with penicillin (Wheat et al., 1988; 1989). This was attributed to the fact that bacteria release ATP during growth and on cell lysis (which could occur due to antibiotic treatment) and thus there is a need to differentiate between ATP released due to cell growth and ATP released due to cell lysis for better quantitative susceptibility testing (Chen and Hong 2021). Heller et al., (2019) using these precautions recorded the ATP concentration/OD 600 nm for *B. subtilis* and *E. coli* in the absence and presence of antibiotics (levofloxacin, gentamicin, kanamycin) and observed a significant increase in the ATP/OD600 ratio between the antibiotic-treated and the control groups for susceptible strains, but not for the resistant strains when no increase in the ATP/OD600 ratio was detected. Ivancic et al., (2008) were the first to use the ATP bioluminescence method to determine whether the bacteria were resistant or susceptible in 104 urine samples. For accurate results, using blood, Matsui et al., (2019) eliminated ATP contributed by the corpuscles by removing the corpuscles by centrifugation and additionally by adding ATP-eliminating reagent to the supernatant bacterial suspension. The MIC value determined after 6 h tended to be higher than those of the conventional broth microdilution method. The advantages of ATP bioluminescence-based approach are: (a) the method is simple to perform, (b) rapid, (c) requires a low cost luminometer and (d) does not require experienced staff to perform the test. Therefore, this method has the potential for rapid AST in the clinic, but the results need to be validated using different species and antibiotics (Chen and Hong 2021).

3.6 AMR Screening of Ocular Samples by Molecular-based Methods

As opposed to phenotypic methods used for the detection of AMR in microbes the molecular-based method like PCR, isothermal amplification and DNA-microarrays offer several advantages such as: polymicrobial samples could be analyzed, AMR genes could be precisely characterized, multiplex targeting of AMR determinants could be done, ARG (genes conferring antimicrobial resistance) could be detected – ARGs could be detected directly in the ocular samples without the need for purifying the isolates. Detection of ARGs is rapid, and identification of ARGs is not dependent on phenotypic categories such as susceptibility, intermediate susceptibility and resistance and could be detected even in isolates in which susceptibility breakpoints have not been established and could identify introduced resistance factors (Sundfjord et al., 2004; Kaprou et al., 2021; Vasala et al., 2020). Despite the advantages, the molecular-based methods pose some problems such as (a) needs trained personal, (b) screens for resistance genes but not susceptibility, (c) incapable of defining MIC, (d) detects only known ARGs but not the presence of silent genes and pseudogenes, (e) mutations in primer binding sites may preclude PCR amplification, generating false-negative results, (f) PCR amplification of determinants in mixed flora samples may be affected, (g) limited number of targets, (h) regulatory mutations that affect gene expression are not detected, (i) is not cost effective and (j) do not correlate always with phenotypic results. But reduction in the cost and the availability of global databases on ARGs (Waseem et al., 2019) has made molecular-based methods the method of choice. A few of the molecular-based methods for detection of ARGs are discussed below.

3.6.1 PCR-based Methods

Among the various molecular-based methods, amplification of DNA by different PCR methods (Sundfjord et al., 2004; Seedy et al., 2017) like real-time (Mackay 2004), quantitative (Böckelmann et al., 2009), digital (Bogožalec Košir et al., 2020), and multiplex (Jousset et al., 2019) PCR have been used for the detection and identification of ARGs. PCR also has the advantage that it could amplify genes from both viable and viable but not cultivable and/dead, bacteria. Further, phenotypic methods often generate erroneous (e.g., false negative) interpretations that could be avoided by conventional PCR (Adzitey et al., 2013).

A recent study indicated that *dfrA*, a marker for phenotypic trimethoprim resistance, showed significant association between the presence of *dfrA* and trimethoprim resistance in *E. coli, Klebsiella* spp., *P. mirabilis* but not in *S. saprophyticus* isolated from urine samples (Somorin et al., 2022).

3.6.1.1 Multiplex PCR

Multiplex PCR is an improved method of the conventional PCR. In this method, several species-specific or gene-specific primers are used simultaneously to amplify, identify, and differentiate more than one type of microorganism or genes in a single run. A multiplex real-time PCR was used for characterization of ARGs in enterococci and staphylococci (Patel et al., 1997), *Staphylococcus aureus* and *S. epidermidis* (Martineau et al., 2000; Pèrez-Roth et al., 2001), *Neisseria gonorrhoeae* (Donà et al., 2016), in methicillin resistant *S. aureus* (MRSA) (Wang et al., 2019) and in extended-spectrum-beta-lactamase-producing *E. coli* (Woksepp et al., 2011) as well as *Enterococcus faecium, Staphylococcus aureus, Klebsiella pneumoniae, Acinetobacter baumannii, P. aeruginosa,* and *Enterobacter* (Woksepp et al., 2014). Freeman et al., (1999) demonstrated that 19 out of 25 *S. aureus* isolates obtained from milk, meat and other animal products displayed multiple resistance to penicillin, erythromycin, kanamycin and tetracycline and possessed the genes *bla*Z, *msr*A, *erm*B, *erm*C and *tet*K. AMR genes like *optr*A and *fex*A, which confer resistance to oxazolidinones and phenicol respectively were also present in 5 out of 10 *Enterococcus* isolates

(Domingues 2017). PCR could detect point mutations in AMR genes (Chahorm et al., 2018), identify quinolone resistance genes (*qnr*S1 and *qnr*B4) in 86 out of 142 β-lactamase-producing *E. coli* isolates (Seo and Lee 2019) and the *mgr*B gene, which confers resistance to colistin in commensal *Enterobacteriaceae* isolated from poultry (Pishnian et al., 2019). All these studies demonstrated the efficient and rapid use of PCR to detect AMR in bacteria.

3.6.1.2 PCR Combined with Restriction Fragment Length Polymorphism (PCR–RFLP)

In PCR combined with Restriction fragment length polymorphism (PCR-RFLP) method following PCR, the DNA amplicon is fragmented using restriction enzymes (endonucleases) to generate fragments of DNA, which could be electrophoresed and visualized by ethidium bromide staining. The generated fragments would be unique to the amplicon, which was restricted. PCR-RFLP identified mutations in *gyr*A associated with AMR in *Campylobacter* (Alonso et al., 2004) and in *Salmonella* (Gómez-Gómez et al., 2019), indicating that such point mutations may effectively contribute to AMR.

3.6.1.3 Real-Time Polymerase Chain Reaction

In real-time polymerase chain reaction (qPCR), amplification of DNA is quantified all through the amplification process (Bell et al., 2015) by the inclusion of fluorescently labelled primers that emit fluorescence signals (Yang et al., 2016). qPCR has a short reaction time, and the method could be used to determine the relative and absolute number of microorganisms in a sample (Bell et al., 2015) and also to identify ARGs in animal feces, soil, and surface water (Xu et al., 2019). qPCR has been used to detect antimicrobial-resistant bacteria like methicillin-resistant *Staphylococcus aureus* (Fang et al., 2003; Huletsky et al., 2004), *Helicobacter pylori* (Matsumura et al., 2001), *Enterococcus faecium* (Mohn et al., 2004), vancomycin-resistant enterococci (Palladino et al., 2003), *E. coli* (Randegger et al., 2001), *Enterococcus faecalis* and *E. faecium* (Woodford et al., 2002) and several other organisms (Sundfjord et al., 2004). This method has also been used to detect resistance genes like *tet*A and *tet*B genes in the slaughterhouse as well as in meat products (Guarddon et al., 2012). Waldeisen et al., (2011) observed that when *E. coli* was exposed to spectinomycin, chloramphenicol, and kanamycin the susceptible *E. coli* produced low bacterial loads.

qPCR was also used for the rapid identification of virulent pathogens in aqueous humor and/or vitreous fluid of 153 patients with postoperative endophthalmitis. qPCR assay targeted the *fem*A or *lyt*A gene in *S. aureus* and *S. pneumoniae* respectively and the *tuf* gene for detection and quantification of *S. epidermidis* (Kosacki et al., 2020). Studies have indicated that high-level expression of *femA* is essential for high-level MRSA (Li et al., 2012).

3.6.1.4 High Throughput Quantitative PCR

High throughput quantitative PCR (HT-qPCR) is comparatively fast, convenient and cost-effective and has been used to identify a large number of ARGs (Waseem et al.,

2019). In HT-qPCR, a 384-well reaction block is used to assay several hundreds of genes by qPCR combining SYBR green detection in a 5 µl reaction volume (Schmittgen et al., 2008). HT-qPCR has been used for detection of ARGs from various sample types (Waseem et al., 2020) like park soils (Wang et al., 2014a), from *Orientia tsutsugamushi* clinical isolates (Phuklia et al., 2018), broiler fecal samples (Che et al., 2019) and so forth and 100 environmental samples (Waseem et al., 2019). The analysis of the environmental samples indicated the presence of ARGs in the gut microbiome. It was proposed that in future HT-qPCR based resistance studies should focus on analysing linkages between resistance genes and bacterial communities to determine the health risk in various environmental samples (Waseem et al., 2019).

3.7 Isothermal Amplification Methods

Isothermal DNA amplification (IDA) facilitates exponential amplification of nucleic acid sequences at a constant temperature using strand displacing polymerases, rather than temperature changes and has been used for the detection of ARGs (Lee et al., 2019b). The main advantage is that IDA is carried out at a single temperature unlike PCR, thus eliminating the need for thermocycling. Further IDA is faster and more sensitive and offers greater specificity compared to PCR. In a recent study it was concluded that two of the IDA methods namely loop-mediated isothermal amplification (LAMP), and recombinase polymerase amplification (RPA) hold great potential for point-of-need diagnostics (Zou et al., 2020). Both are single step amplification methods and require minimum amount of DNA template (Zou et al., 2020). Further, in this method the LAMP amplicons can be detected with naked eye either through turbidity or color change (Srimongkol et al., 2020). On the other hand, isothermal methods are not suitable for multiplexing (Dhama et al., 2016) and may need several primers (Karami et al., 2011). Despite these shortcomings isothermal nucleic acid amplification technology (Morin et al., 2018) has been automated (Cantera et al., 2019). LAMP method has been used for detection of *msr*A, the macrolide–streptogramin type B resistance gene in *S. aureus* (Ross et al., 1996; Mu et al., 2016), ISAba1–blaOXA-51-like gene in *Acinetobacter ba*umanni (Mu et al., 2016) and for the detection of *Trichomonas vaginalis* (Reyes et al., 2014), *Mycoplasma genitalium* (Edwards et al., 2015), *Clostridium difficile* (Lloyd et al., 2015), *Caprine arthritis encephalitis virus* (Balbin et al., 2014) and so forth.

3.8 DNA Microarrays for Detecting AMR

In a DNA microarray 1000's of DNA sequences (probes) from a known organism are deposited and bound to a surface. At times the DNA sequences are synthesised on the surface and are used to quantify the relative concentration of nucleic acid sequences in an unknown mixture via hybridization (Bumgarner 2013). Normally an array is used to monitor gene expression levels (Figure 3.6). In this method, total RNA is extracted from the cells of interest and fluorescently labelled directly or transformed to a labelled cDNA or to a T7 RNA promoter tailed cDNA, which is further converted to cRNA prior to labelling. Labelling is normally accomplished by incorporating fluorescently labelled nucleotides during the synthesis of the cRNA or cDNA or by the incorporation of a biotin labelled nucleotide in the cRNA synthesis step. The labelled

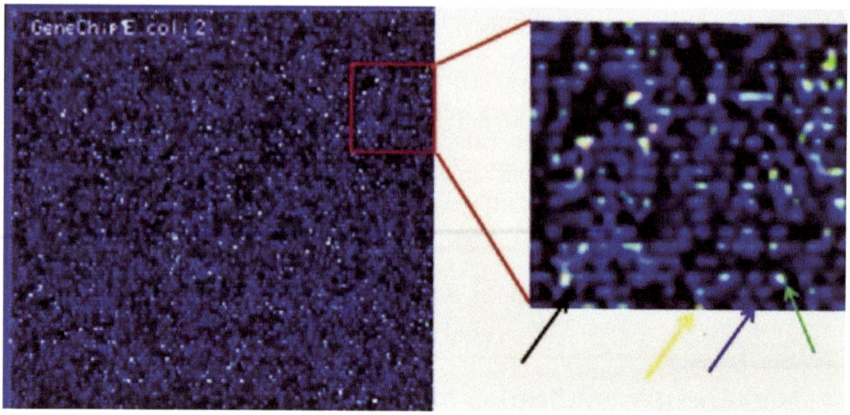

FIGURE 3.6 DNA microarray results following hybridisation of labelled ocular *E. coli* L-1216/2010 cDNA with *E. coli* Genome 2.0 gene chip Array. Each minute square in the gene chip contains a probe which when hybridised with labelled cDNA emits a fluorescence. When the cDNA does not hybridise with the probe the minute square is black in colour and denotes absence of fluorescence signal. When the cDNA hybridises with the probe, the colour of the minute square changes from black to another colour. The specific colours denote intensity of hybridisation which follows the order dark blue ->blue ->light Blue ->green ->yellow ->orange ->red->white and so forth. (Courtesy Konduri R., LVPEI, Hyderabad, India.)

cRNA or cDNA were then used to hybridize the probes on the microarray. The array is then washed and the signal is detected by measuring fluorescence at each spot with a scanning confocal microscope. In the case of biotin labelled samples, the array is stained post-hybridization with fluorescently labelled streptavidin and the intensity of the signal(s) corresponding to each spot on the array is taken as a measure of the expression level of the corresponding gene (Figure 3.6). A major limitation of a DNA array is that it can only detect sequences that the array was designed to detect and thus poses the problem of missing out on unknown genes (Bumgarner 2013).

For the detection of ARGs by DNA microarray, known DNA sequences of the ARG genes are immobilised on the microarray, which is then hybridised with the cDNA of an organism and the results are then analysed based on the colour of the spot that exhibited positive hybridisation. Thus, DNA microarray technologies offer a promising method for detection of AMR genes (Call et al., 2003; Wu et al., 2001). A multiplex asymmetric PCR-based DNA microarray assay was used for detection of the *mecA*, and *blaZ* (methicillin and penicillin resistance genes), *aac*(6_)-Ie-*aph*(2_) (aminoglycoside resistance), *ermA* and *ermC* genes (MLSB resistance), and the *msrA* gene (macrolide and streptogramin B resistance) in 178 clinically important *S. aureus* and 237 coagulase-negative *Staphylococci* (CoNS) and observed more than 90% correlation between antibiotic resistance and the genotypic analysis (Zhu et al., 2007). A few genes like the *mecA*-positive and 60 *aac*(6_)-Ie-*aph*(2_)-positive in CoNS isolates were found to be susceptible to the corresponding antibiotics based on disk diffusion assay. Other resistance genes reported in *S. aureus* include *aad*D (tobramycin), *tet* (tetracycline), *cat* (chloramphenicol), *van* B (vancomycin) (Vogt et al., 2014) and genes encoding enterotoxins (*sea*, *seb*, *sec*, *seh*, *sel* and *egc*) and so forth. (Morach et al., 2019). Kittler et al., (2019) detected *S. aureus* resistant to methicillin in chicken meat samples from different farms along with the genes for Gamma-hemolysin (*luk*F, *luk*S and *hlg*A), leucocidin (*luk*Y and *luk*X), and hemolysin (*hl*, *hla*, *hl*III and *hlb*). Vogt et al., (2014) also identified resistance genes to chloramphenicol (cmlA1-like), sulfonamides (sul), tetracycline (tet), and trimethoprim (dfrA) in *E. coli.*

In ocular isolates, DNA microarray was used to detect *K. pneumoniae*, *Streptococcus agalactiae* and *C. parapsilosis* in the vitreous of endophthalmitis patients (Sakai et al., 2012).

3.9 Detection of Antimicrobial Resistance Genes by Single-molecule Fluorescence Microscopy

Goyal et al., (2022) reported a simple, inexpensive, fast, reliable "cut and stretch" assay based on CRISPR/Cas9 excision and DNA combing to detect AMR genes on linearized circular bacterial plasmids using a fluorescence microscope. In this method, a clinical *K. pneumoniae* strain with a plasmid carrying the *bla*CTX-M gene was selected, and the plasmid was treated with Cas9 using either a gRNA targeting the *bla*CTX-M gene or a decoy gRNA (a gRNA targeting a human gene). After the Cas9 reaction, plasmids were stained and stretched on glass slides and as anticipated longer molecules were observed for the plasmid sample treated with Cas9 with *bla*CTX-M gRNA compared to the control with decoy gRNA. It was concluded that Cas9 targeted the *bla*CTX-M gene and linearized the circular plasmid and the change in length of the plasmid could be used to infer the presence of the gene of interest (Goyal et al., 2022). This single-molecule imaging-based assay in addition to detecting

AMR genes provides information on the number and size of plasmids. This method was successfully used to identify several β-lactamase-encoding genes in clinical samples (Goyal et al., 2022).

3.10 Whole Genome Sequencing and Metagenomics in AMR Diagnostics

Several genome sequencing methods have been employed for the identification and characterising of ARGs and these include pyrosequencing, whole genome sequencing (WGS), short and long read WGS and nanopore sequencing (Ronaghi et al., 1998; Kaprou et al., 2021). The advantages of these methods are: (a) there is no need for pure cultures, (b) non-culturable microorganisms also contribute to the DATA, (c) even long reads could be analysed, (d) could unravel genetic basis of AMR and (e) multiple AMR determinants could be identified (Kaprou et al., 2021). The disadvantages include (a) prohibitive equipment costs, (b) complex methodology, (c) need for trained personnel, (d) discrepancies with phenotypic tests (e) metagenome sequencing may identify new ARGs but the host is not easily identified from a mixed sample and (f) MIC is not defined (Kaprou et al., 2021). The application of the above sequencing and metagenomic methods is discussed below.

3.10.1 Whole Genome Sequencing

WGS is extremely reliable in the detection of ARGs but the limitation is that the identification is based on the accuracy of the whole genome sequence annotation. Nevertheless, WGS has been crucial in predicting or detecting AMR or ARGs in non-typhoidal *Salmonella* (McDermott et al., 2016; Matchawe et al., 2022), in *Streptococcus uberis* and *S. dysgalactiae* (Vélez et al., 2017), in *Campylobacter* (Zhao et al., 2016), in carbapenem-resistant *Klebsiella pneumoniae* (Fasciana et al., 2019) and so forth. WGS is thus a reliable indicator of resistance to several antibiotics (tetracycline, ciprofloxacin, nalidixic acid, erythromycin, gentamicin, azithromycin, clindamycin, telithromycin, and florfenicol) (Zhao et al., 2016; Argimón et al., 2020). Oniciuc et al., (2018), compiled a review on WGS for surveillance of AMR in food and food-related samples and observed that most studies were focused on pathogens like *Salmonella, Campylobacter spp., Listeria monocytogenes etc.*, with a view to identify resistant isolates, identify AMR genes, or to track the dissemination of AMR. It was also highlighted that WGS genotype and experimentally determined AMR phenotype were not always in agreement as observed in *S. enterica* (de Melo et al., 2021). In a recent study Matchawe et al., (2022) analysed 23 assembled genomes of non-typhoidal *Salmonella* (NTS) and reported that phenotypically, 22% (n = 5) of *Salmonella* isolates were streptomycin resistant, and 13% (n = 3) were MDR. But, genotypically, all the *Salmonella* isolates possessed MDR potential against several classes of antibiotics including carbapenems, third-generation cephalosporin and fluoroquinolone. A few also exhibited resistance potentials to polymyxin and 80% of isolates harboured "silent resistant genes" as reservoirs of drug resistance. In several studies WGS has also helped in establishing an association between ARG sequences and phenotypic resistance (MIC data) (Vélez et al., 2017; Konduri et al., 2020) and a strong correlation (99.2%) was observed between resistance genotypes and phenotypes (Zhao et al., 2016).

WGS of 10 ocular *E. coli* identified 22 ARGs (Konduri et al., 2020) (Table 3.1) and ARG genes correlated with resistance to a particular antibiotic whereas absence of a gene positively correlated with susceptibility to the specific antibiotic. Resistant genes for Minocycline and Sulphanilamide were not detected based on WGS implying that resistance to these two antibiotics is due to some unknown genes (Table 3.1). The study also concluded that expression of ARG was a must for the resistance phenotype. Resistance was also observed in the absence of the ARG implying that other undiscovered genes may be conferring resistance (Konduri et al., 2020). Another method that has been used to detect ARGs was nanopore sequencing (Judge et al., 2015) as in *Salmonella Typhi* (Ashton et al., 2015), in a multidrug-resistant enteroaggregative *E. coli* (Greig et al., 2018), in Shiga toxin-producing *E. coli* (González-Escalona et al., 2019) and in *Mannheimia haemolytica* (Lim et al., 2019).

3.10.2 Whole Metagenome Sequencing

Whole metagenome sequencing (WMS) allows for the identification of bacterial pathogens as well as ARGs (Schmidt et al., 2016) and could help to mitigate the risks of ARGs spreading in hospitals and environments (Kamathewatta et al., 2019) without the need for culture (Ferrari et al., 2017). In this approach, individual microorganisms are not required and uses the total DNA extracted from a community, which is

TABLE 3.1

Identification of Antimicrobial Resistance Genes in the Whole Genome Sequence of 10 Ocular *E. coli*[a] Isolates

S. No.	Class of Antibiotic	Resistant gene	L-1339/ 2013	L-2594/ 2017	L-494/ 2011	L-3003/ 2015	L-1010/ 2018	L-3137/ 2017	L-1149/ 2016	GMRV-476/2017	L-1534/ 2016	L-1216/ 2010	Count
1	**Aminoglycoside**	*aph(3'')-Ib*	-	+	-	-	-	-	-	-	+	-	2
		aph(6)-Id	-	+	-	-	-	-	-	-	+	-	2
		aac(6')-Ib-cr	+	+	-	+	-	-	-	+	-	-	4
		aadA5	+	+	+	-	+	-	-	-	-	-	4
		aac(3)-IIa	+	-	+	-	-	-	-	-	-	-	2
		aadA2	-	+	-	-	-	-	-	-	-	-	1
2	**Amphenicol**	*catB3*	+	+	+	+	-	-	-	-	-	-	4
		catA1	-	+	-	-	-	-	-	-	-	-	1
3	**Beta-lactam**	$bla_{CTX\text{-}M\text{-}15}$	+	+	+	+	+	+	+	+	-	-	8
		$bla_{NDM\text{-}5}$	-	+	-	-	-	-	-	-	-	-	1
		$bla_{TEM\text{-}1B}$	-	-	-	-	-	+	-	-	+	-	2
		$bla_{OXA\text{-}1}$	+	+	+	+	-	-	-	+	-	-	5
4	**Fluoroquinolone**	*aac(6')-Ib-cr*	+	+	-	+	-	-	-	+	-	-	4
5	**Lincosamide**	*mph(A)*	+	+	-	-	+	-	-	-	-	-	3
		emrB	-	-	-	+	-	-	-	-	-	-	1
		mdfA	-	+	+	+	+	+	+	+	+	+	9
6	**Sulfonamide**	*sul2*	-	+	-	-	-	-	-	-	+	-	2
		sul1	+	+	+	-	+	-	-	-	-	-	4
7	**Tetracycline**	*tet(B)*	-	+	-	-	-	-	-	-	-	-	1
		tet(A)	-	-	-	-	+	-	-	+	-	-	2

(continued)

TABLE 3.1 (Continued)

Identification of Antimicrobial Resistance Genes in the Whole Genome Sequence of 10 Ocular *E. coli*[a] Isolates

S. No.	Class of Antibiotic	Resistant gene	L-1339/ 2013	L-2594/ 2017	L-494/ 2011	L-3003/ 2015	L-1010/ 2018	L-3137/ 2017	L-1149/ 2016	GMRV-476/2017	L-1534/ 2016	L-1216/ 2010	Count
8	**Trimethoprim**	*dfrA17*	+	+	+	-	+	-	-	-	-	-	4
		dfrA12	-	+	-	-	-	-	-	-	-	-	1

Source: Modified from Konduri et al., 2020.

Notes: [a] Genes were identified from the whole genome sequence using ResFinder 3.2 (https://cge.cbs.dtu.dk/services/ResFinder/).
(+) and (-) indicate presence and absence of gene respectively.

sequenced to detect ARGs using public domain databases like SRST2, ShortBRED, PATRIC, SSTAR, KmerResistance, GROOT, DeepArgs, Resfinder, ARG-ANNOT, RGI, ARGs-OAP (v2), ARIBA, PointFinder and NCBI-AMRFinder (Rodrigues et al., 2018; Boolchandani et al., 2019). Several studies identified AMR and virulence genes in the intestine of cattle (Auffret et al., 2017; Singh et al., 2012; Thomas et al., 2017), and in the feces of poultry (Xiong et al., 2018) and observed that the resistance genes to tetracycline, erythromycin and aminoglycoside were common in fecal samples of pigs, chickens and humans (Ma et al., 2016). Nesme et al., (2014) used the metagenomic sequence method to detect ARGs in diverse environments (e.g., soil, oceans or human feces) and the data indicated that soil metagenomes had the most abundant and diverse pool of ARG including efflux pump genes and genes conferring resistance to vancomycin, tetracycline, or β-lactam antibiotics. Xiong et al., (2018) showed by metagenome analysis that tetracycline resistance genes increased in feces from broiler chicks treated with therapeutic doses of chlortetracycline. Kazimierczak et al., (2009) demonstrated that in pigs reared in an organic system, gene resistance to tetracyclines and doxycycline were detected in the bacterial population probably due to the presence of mobile genetic elements. Li et al., (2020) analysed commercially available chicken breast from USA and detected genera such as *Aeromonas, Enterobacter, Lactococcus* and *Pseudomonas* and observed 132 ARGs conferring resistance to 10 classes of antibiotics belonging to the classes of aminoglycosides and beta-lactams. Additionally in pigs, Lamendella et al., (2011) analyzed 637,722 sequencing reads of the intestinal microbiota and found that genes associated with AMR are homologous to carbohydrate metabolism genes.

3.10.3 WMS of Milk

Bacteria and AMR genes were predominant in milk filters compared to aliquots of bulk tank milk (Rubiola et al., 2020). *Acinetobacter* was the most abundant genus in filter aliquots, while *Pseudomonas* and *Ralstonia* were predominant in treated and untreated aliquots of milk samples respectively. *Bacillus*, *Serratia*, *Moraxella*, *Corynebacterium*, *Enterobacter*, and *Lactococcus* were abundant in both filter and milk aliquots (Rubiola et al., 2020).The relative proportion of AMR genes in the filter aliquots included genes resistant to multi-drug resistance (21–28%), rifampin (23–30%), aminoglycosides (14–24%), tetracyclines belonging to Tet39 group (9–15%), sulphonamides (3–5%), streptogramin (3–7%), β-lactams (3–5%) and multi-drug efflux pump genes belonging to SME group and *Mex*K, *Mex*B, and *Mex*F groups. Further at the lower AMR annotation levels and sequencing depths more AMR determinants were detected in milk filters and the determinants varied depending on the method used for the treatment of milk aliquots (Rubiola et al., 2020).

3.10.4 WMS of urine

Metagenomic nanopore sequencing (mNPS) technology was used to detect pathogens and also predict antibiotic susceptibility phenotypes in urine samples of 76 patients with acute urinary tract infections (Zhang et al., 2022). In the 43 culture positive samples *E. coli, K. pneumoniae, Enterococcus faecium, S. epidermidis, S. haemolyticus, Proteus mirabilis* and some unidentified bacteria were detected, and antibiotic resistance was recorded for 15 antibiotics including fluoroquinolone, penicillin, sulphonamide, cephalosporin, aminoglycoside, macrolide, nitrofurantoin, monobactam, tetracycline and so forth. A total of 73 ARGs across 31 samples were detected, and the largest number of ARGs mapped to *E. coli* and included efflux components and β-lactamase genes, that conferred multidrug resistance. The concordance between ARGs and clinical AST results was only 20.7%. The extended spectrum beta-lactamase phenotypes conferred resistance to cephalosporins due to blaCTX-M genes and fluoroquinolone resistance was due to *emr*H, and/or *gad*X. vanA gene, which is associated with vancomycin resistance was also detected in one sample. Other ARGs also conferred resistance to non-clinically relevant antibiotics (e.g., acridine_dye and rhodamine) (Zhang et al., 2022). Thus, metagenomics is an effective tool in the study and discovery of new resistance genes (Rodrigues et al., 2020).

3.11 Spectroscopic Methods in AMR Diagnostics

Several spectrophotometric methods (MALDI-TOF, FT-IR, AFM-IR and SERS) have also been used for AMR diagnostics. The methods are rapid, high throughput, automated, simple, low running cost, require small sample volume and could unravel the molecular basis of AMR. These methods have similar disadvantages. The equipment is expensive, needs purified strains, lack of databases of identified AMR biomarkers and lack portability and MIC is not defined.

3.11.1 MALDI-TOF Mass Spectrometry

Matrix-assisted laser desorption/ionization time-of-flight mass spectrometry (MALDI-TOF MS) has been used to characterize bacterial species (Oviaño and Bou 2018; Seng et al., 2009) based on the protein profile of the bacterium using whole bacterial cell extracts (Welker et al., 2019). To identify bacteria by MALDI-TOF MS, a bacterial colony is mixed with the MALDI matrix, dried within the matrix such that it crystalizes and then ionized with a laser beam producing protonated ions that are then analyzed based on their mass to charge (m/z) ratio. A "Peptide Mass Fingerprint" (PMF), a mass spectrum of peaks, is generated and compared to a database of microbial proteins of known microorganisms to identify the bacterium (Vrioni et al., 2018). MALDI-TOF MS has been used for the detection of AMR induced by bacterial enzymes like carbapenemases (Hoyos-Mallecot et al., 2014) and β-lactamases (Burckhardt et al., 2011; Oviaño et al., 2017a) or the AAC(6')-lb-cr enzyme (Oviaño et al., 2017b; 2018). Lange et al., (2014) developed the MALDI Biotyper antibiotic susceptibility test, which quantified the amounts of bacterial peptides using MALDI-TOF MS. The quantities of these peptides correlated with the number and growth of microorganisms and the method was applied to clinical isolates of *Klebsiella pneumoniae* and *K. oxytoca* in the presence of meropenem. This method was also evaluated for *Enterobacteriaceae* and Gram-negative bacteria in blood and were also tested against 4 non-β-lactam antibiotics (Jung et al., 2016). The susceptibility of bacteria were tested for gentamicin and cefotaxime by this method, with 5 mismatches for piperacillin-tazobactam within 4 h. Idelevich et al., (2018) developed a direct-on-target microdroplet growth assay and correctly categorized *K. pneumoniae* and *P. aeruginosa* isolates as susceptible or non-susceptible after 18 h of incubation. MALDI-TOF MS has also been used to identify and characterise proteins including ARGs (Justesen et al., 2018; Zürcher et al., 2012).

3.11.2 MALDI-TOF Mass Spectrometry fingerprinting

Zhu et al., (2018) developed an improved MALDI–TOF MS fingerprinting for intact bacteria using photo-reactive titanium dioxide (TiO2)-modified target plates. In this method the photo-reactivity of TiO2 would destroy the bacterial cell membranes and facilitate inner component desorption/ionization. Such experimental improvement could be used to detect resistance associated proteins in intact bacteria (Table 3.2). This methodology is reliable, rapid, accurate, convenient, cost-effective, and environmentally friendly (Vrioni et al., 2018). But the high cost, large size, careful maintenance, and requirement for an expert restrict its use as an AMR or AST testing platform (Welker et al., 2019).

TABLE 3.2

Applications of MALDI-TOF MS for Specific Antimicrobial Resistance (AMR) Detection in Bacteria

Organism	Antibiotic	Reference
E. coli	Polymyxins, Colistin	Dortet et al., 2018a.
E. coli, *Klebsiella pneumoniae*	Beta-lactams (ESBL-producing isolates)	Kawamoto et al., 2019.
Staphylococcus aureus, *S. intermedius,* *S. pseudintermedius*	Novobiocin, Polymyxin B, Acriflavine	Nisa et al., 2019.
S. aureus	Methicillin	Tang et al., 2019.
Candida auris	Echinocandins	Vatanshenassan et al., 2019.
K. pneumoniae	Carbapenems (carbapenemase-producing isolates)	Rocco et al., 2019.
Enterobacteriaceae	Carbapenems (carbapenemase-producing isolates)	Rocco et al., 2019.
Pseudomonas aeruginosa	Beta-lactams (MBL)	Li et al., 2019a.
Enterococcus faecium	Vancomycin	Parlapani et al., 2020.
Acinetobacter baumannii	Colistin	Dortet et al., 2018 b.
E. coli, *K. pneumoniae*	Cefotaxime, Meropenem, Ciprofloxacin	Axelsson et al., 2020.
S. aureus, *Enterococcus* spp., *E. coli,* *K. pneumoniae*	Oxacillin (methicillin), Vancomycin Ceftriaxone, Meropenem	Horseman et al., 2020.
Enterobacteriales	Imipenem/Relebactam	Oviaño et al., 2020.
S. aureus	Methicillin	Nix et al., 2020.

Source: Modified from Kaprou et al. 2021.

3.11.3 Mass Spectrometry and Stable Isotopes

MALDI-TOF MS combined with stable-isotope labelling could also be utilised to develop a method to determine the susceptibility of bacteria (Demirev et al., 2013; Sparbier et al., 2013; Jung et al., 2014). Bacteria are grown in media containing stable isotopes such as ^{13}C and ^{15}N so that the isotopes get incorporated into the biomolecules. As a result of this incorporation characteristic mass shifts are seen in the biomolecules which could be monitored to infer antibiotic susceptibility. Demirev et al., (2013) cultured *E. coli* in ^{13}C-labeled media with antibiotics and the ^{13}C-labeled labelled proteins when analysed exhibited a mass shift in the spectrum indicating that the bacteria were resistant. This method has been demonstrated in *E. coli* only but could be used for other bacteria. In contrast to ^{13}C-labeled culture media, Sparbier et al., (2013) used isotopically labeled amino acids ($^{13}C6$-$^{15}N2$ labeled lysine) to label methicillin-susceptible (MSSA) and methicillin resistant (MRSA) *S. aureus* strains in the presence of oxacillin and cefoxitin. As expected, mass shifts of peaks were observed in MRSA in the presence of antibiotics. In *P. aeruginosa* also similar studies were done using antibiotics with distinct modes of action like meropenem, tobramycin, and ciprofloxacin (Jung et al., 2014). The method is rapid and classifies bacteria as susceptible or non-susceptible, but MIC values or the intermediate criterion were not determined. Further, automation of procedures and methods that integrate identification and AST would expand the use and refine MALDI-TOF MS for clinical diagnosis (Chen and Hong 2021).

3.11.4 Fourier Transform Infrared (FTIR) Spectroscopy

Fourier transform infrared (FTIR) spectroscopy is a non-invasive, sensitive, rapid, low cost and simple method for detection and identification of pathogens. The output is the FTIR spectrum which represents the lipids, lipopolysaccharides, carbohydrates, proteins, and nucleic acids, in the sample (Vogt et al., 2019). The spectrum is unique for the given compound or organism under study and gives a unique 'fingerprint' (Naumann et al., 1991) (Figure 3.7) with absorbance peaks (FTIR wave numbers). The commonly seen peaks in the FTIR spectra of plants, bacteria, and cyanobacteria represent specific functional groups (Kenne and Merwe 2013) (Figure 3.7). Wenning and Scherer (2013) suggested almost a decade ago that FTIR spectroscopy could be used for species identification and strain typing from clinical sources (Coutinho et al., 2009), food pathogens (Kuhm et al., 2009), food contaminants (Kümmerle et al., 1998; Schmidt et al., 2012), probiotic cultures in food (Chen et al., 2009; Timmins et al., 1998) as well as in the environment (Behrendt et al., 2002; Fischer et al., 2006). FTIR has been used to differentiate between *Candida* species (Toubas et al., 2007), filamentous fungi (Fischer et al., 2006), enterococci (Kirschner et al., 2001) and *Mycobacterium* species (Rebuffo-Scheer et al., 2007).

Microbial FTIR spectra are identified by partial least square regression (PLSR) (Oust et al., 2004), partial least square discriminant analysis (PLSDA), artificial neural networks (ANN) (Udelhoven et al., 2000) and Pearson's product moment correlation coefficient (Helm et al., 1991). *Yersinia* species (Wenning and Scherer 2013), 92 species of lactic acid bacteria (Wenning et al., 2010) and *Burkholderia* species (Coutinho et al., 2009) were identified using FTIR and ANN and discriminant partial least squares analysis. These studies indicated that to identify unknown strains, the

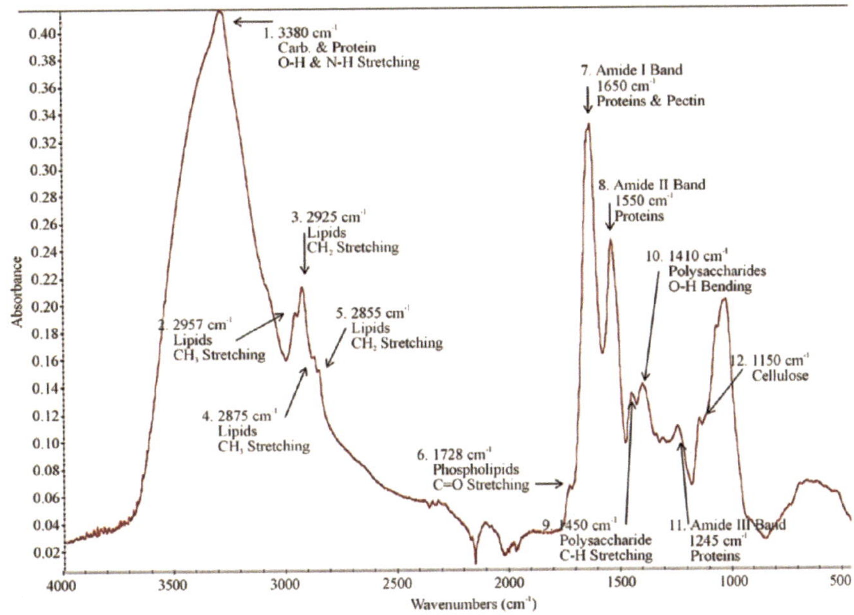

FIGURE 3.7 Absorbance peaks (FTIR wavenumbers) commonly seen in FTIR spectra of plants, bacteria, and cyanobacteria. Each peak represents a specific functional group from which they are derived. (Courtesy Kenne and Merwe 2013.)

FTIR dataset should include a number of reference strains as shown for *Actinobacteria* (Oberreuter et al., 2002), 73 yeasts species (Kümmerle et al., 1998), 21 *Pichia* species and *Issatchenkia* (Büchl et al., 2008) and *Listeria* species (Rebuffo et al., 2006) so as to overcome variance of phenotypes among strains, as in *Burkholderia* species (Coutinho et al., 2009).

FTIR has also been used for the detection and characterization of cancer cells (Erukhimovitch et al., 2007), cells infected with viruses (Salman et al., 2002), bacterial strain typing (Lechowicz et al., 2013), for tracking pharmacological substances injected to animals (Crupi et al., 2002) and to identify bacterial susceptibility to antibiotics (Sharaha et al., 2017; Wenning and Scherer 2013).

3.11.5 Fourier Transform Infrared microscopy

Initial studies indicated that Microspectrophotometry or Microspectroscopy which combines a standard microscope with the analytical tools of a spectrophotometer for material characterization could be used for identification of microorganisms. Several studies have also indicated that identification accuracy by conventional spectroscopy using dried suspensions of cells is superior to microspectroscopy (Oust et al., 2008; Rebuffo-Scheer et al., 2008). Several confounding factors like variation in incubation times, poor quality imprints of the colonies, and inability of the colonies to stick to the sample carrier were some of the problems (Wenning and Scherer 2013). With time application of Microspectrophotometry improved. A method was developed to analyze microbial IR spectra using pattern-recognition tools, to determine *E. coli* susceptibility to ceftazidime, gentamicin, nitrofurantoin, nalidixic acid, and ofloxacin (Sharaha et al., 2017). FTIR microscopy could detect structural molecular changes linked to AMR (Salman et al., 2017). The advantage of FTIR microspectroscopy over conventional FTIR spectroscopy is that it could be used for the analysis of restricted regions of the sample. Erukhimovitch et al., (2005) could distinguish between bacteria

(*Bacillus magaterium*, *E. coli* and *Pseudomonas stutzeri*) and fungi (*Penicillium* sp., *Memnoniella* sp. and *Fusarium* sp.) based on spectral differences. Kochan et al., (2019) identified changes in the chemical composition of *S. aureus* associated with vancomycin and daptomycin resistance.

3.11.6 Atomic Force Microscopy-infrared Spectroscopy (AFM-IR)

Atomic force microscopy-infrared spectroscopy (AFM-IR) is a single cell, nanoscale technique, which combines IR and scanning probe microscopy to map cell structures at the atomic scale. AFM-IR is a non-destructive technique (Kochan et al., 2020) and has been used to detect antimicrobial resistance- related changes such as thickening of the cell wall, as well as changes in the phospholipid content in relation to vancomycin resistance in *S. aureus* and daptomycin nonsusceptibility respectively (Kochan et al., 2019). However, this instrument is expensive, maintenance cost is prohibitive and requires a trained expert. Further databases with spectra would be required to differentiate susceptible and resistant strains. All these could be reasons as to why studies using AFM-IR for detection of AMR in microbes are limited (Kaprou et al., 2021).

3.11.7 Raman Spectroscopy

Raman spectroscopy is a molecule-specific fingerprinting technique providing quantitative information on molecules and their distribution in tissues and cells (Raman and Krishnan 1928; Kim 2015) and spectral changes provide information on changes in metabolism in biological samples. Raman spectroscopy has also been used to detect spectral changes of susceptible and resistant bacteria in the presence or absence of antibiotic (Kirchhoff et al., 2018; Schröder et al., 2015; 2017). Raman spectroscopy showed that ciprofloxacin induced concentration-dependent changes in *E. coli* strains after 90-min of exposure (Kirchhoff et al., 2018) with the 1458 cm^{-1} band increasing and the shoulder band at 1485 cm^{-1} decreasing with ciprofloxacin concentration. Germond et al., (2018) used Raman spectroscopy to detect antibiotic resistance in 11 strains of *E. coli* and additionally observed that spectral peak intensities correlated with the expression of antibiotic resistance genes. Using this approach 80% of the strains were correctly classified with respect to their susceptibility and MIC values were obtained in less than 2 h without the need for pre-cultivation. In a subsequent work vancomycin resistance in *enterococci* was determined within 3.5 h (Schröder et al., 2015) and two different Raman bands, around 1250 cm^{-1} and 1485 cm^{-1}, showed characteristic changes in 30 min after vancomycin addition. The sensitivity of this method was 87% and a specificity of 93% for *E. faecalis*, and 99% sensitivity and 77% specificity for *E. faecium*.

3.11.7.1 Neural Network and Raman Spectroscopy

A disadvantage of Raman spectroscopy-based AST methods is that spectral differences are easily masked by background noise (Jones et al., 2019). To overcome this, long integration time is needed to achieve high accuracy (Ho et al., 2019). Ho et al., (2019) used measurement times of 1 second and could distinguish MRSA and MSSA.

3.11.7.2 Combining Raman Spectroscopy with Isotope Labelling

Yet another method for antimicrobial susceptibility testing in bacteria is to combine Raman spectroscopy with stable isotope labelling. Tao et al., (2017) labelled bacterial cells with heavy water (D2O) so that the D+ from D2O gets incorporated to form C–D bonds in the macromolecules in the bacterial cells that could be detected by Raman shift at the C–D band from 2040 to 2300 cm−1. With the D2O-Raman method, *S. mutans* strains sensitive or resistant to fluoride-resistant could be discriminated in 0.5 h. In addition, the MIC based on metabolic activity was determined for *S. mutans* in the presence of two antiseptics (sodium fluoride and chlorhexidine) and one antibiotic (ampicillin) to quantify antibacterial efficacy via metabolism inhibition. However, the MIC-MA values determined by this method tend to be higher than conventional MIC values. Yang et al., (2019) extended Raman spectroscopy with D2O labelling method for rapid AST of UTI samples. The D2O-Raman method is time-consuming and low-throughput but with integration of deep learning, Raman spectroscopy could be less time consuming. With this potential, Raman spectroscopic assays could become an integral part of a novel diagnostic tool that provides MIC results within hours in the clinic.

3.11.7.3 Surface Enhanced Raman Spectroscopy (SERS)

Surface enhanced Raman spectroscopy (SERS) or surface-enhanced Raman scattering (SERS) is a variant of stimulated Raman spectroscopy (SRS) which is a weak process; about 1 out of 10^8 photons undergo Raman scattering spontaneously (Jones et al., 2019). SERS enhances Raman scattering by molecules adsorbed on rough metal surfaces or by nanostructures such as plasmonic-magnetic silica nanotubes. The enhancement factor can be as much as 10^{10} to 10^{11} (Blackie et al., 2007) and the technique could detect changes in single molecules (Le et al., 2006) that occur in bacteria because of antibiotic action (Wang et al., 2018; Galvan and Yu 2018). SERS is rapid, accurate, and ultrasensitive with respect to detection of resistant bacteria (Wang et al., 2018; Chen et al., 2019). However, the accuracy and sensitivity of SERS is further enhanced when coupled with microfluidics (MF-SERS) chip technology (Sackmann et al., 2014; Kaprou et al., 2019) and this was used for antibiotic susceptibility testing in *E. coli* (Chang et al., 2019) and could differentiate MSSA and MRSA (Lu et al., 2013; Leonard et al., 2018). A few of the limitations of MF-SERS are that the sample prior to analysis needs to be dried and this could lead to irreproducible results. The sample needs to be homogenous with respect to the presence of only one species and a comprehensive database of SERS spectra is needed (Bodelón et al., 2018).

3.11.7.4 SERS-based Detection of Antibiotic-resistant Bacteria in Food Sources and Water

The bacteria that could be detected from the environment and seawater include *Vibrio parahaemolyticus (*Xu et al.2013) which are known to be resistant to ampicillin, penicillin, and tetracycline and *Klebsiella pneumoniae* from water samples (Jones et al., 2017). But to enhance the sensitivity of SERS the method has been modified to include the use of antibodies as in the case of *S. typhimurium* (Khan et al., 2011), or aptamers in the place of monoclonal antibodies for the detection of both *S. typhimurium* and *S. aureus* (Zhang et al., 2015a, 2015b; Joshi et al., 2009). Further, the sensitivity of SERS assay could also be further improved when used in conjunction with principal component analysis (PCA), as in discriminating *E. coli* O157:H7 and *S. epidermidis* from other *Salmonella* strains (Galvan and Yu 2018). In some of these studies photothermal killing of the bacteria was possible (Khan et al., 2011; Poppe et al., 1998). Jones et al., (2017) developed a 3D SERS-based method for detection and photothermal inactivation of multidrug resistant bacteria from water samples like MRSA and multidrug resistant carbapenem-resistant *Enterobacteriaceae* (MDR-CRE).

3.11.7.5 SERS-based Detection of Antibiotic-resistant Bacteria in Urinary Tract Infection

Antibiotics are the first line of treatment for urinary tract infections (UTI) the world over and several instances of antibiotic-resistant bacteria in UTI have been reported (Andersson 2010). SERS has been used to detect 6 UTI isolates including *E. coli, Klebsiella oxytoca, K. pneumoniae, Citrobacter freundii, Enterococcus* spp., and *Proteus mirabillis* (Avci et al., 2015; Idress et al., 2010). In *E. coli* (Avci et al., 2015) the characteristic bands at 655, 730, and 1001 cm^{-1} were assigned to the outer membrane of the bacteria and were used to identify the pathogens. Cheong et al., (2017a) used a SERS-based drop coating method to rapidly identify quinolone-resistant *K. pneumoniae.*

3.11.7.6 SERS-based Detection and Differentiation of MRSA and Tuberculosis

MRSA was detected using a Graphene oxide- gold nanoparticle (GO-AuNP) system (Fan et al., 2013) using SERS. Lu et al., (2013) analysed 17400 SERS spectra from 58 *S. aureus* isolates for the early diagnosis of MSSA and MRSA. A similar approach using silver nanoparticles (AgNP) and SERS was developed by Mühlig et al., (2016) for the differentiation of *M. tuberculosis* and non-tuberculosis mycobacteria. The 3 strains of *M. tuberculosis* (Beijing 1934, Beijing 8394, and H37R*v*) were compared against *M. canettii, M. absessus*, and *M. szulgai.* (Lawn and Zumla 2011). Approximately 3000 SERS spectra were collected per isolate, to identify the isolate down to the strain level. Differences in the cell wall mycolic acid profiles enabled strain-level discrimination of the mycobacteria.

3.11.7.7 SERS-based Detection of Bacteria in Blood

Cheng et al., (2007; 2014) developed a modified SERS method in which the bacterial cells were separated from red blood cells (RBCs) based on differences in their dielectrophoretic properties (DEP) (Cheong et al., 2017b). This DEP based SERS was also accomplished in a microfluidic channel (Deng and Juang 2013) which could effectively separate bacterial cells from RBCs. However, detection of bacteria in experimental samples is limited in its application.

3.11.7.8 Antibiotic Susceptibility Testing with SERS

Liu et al., (2009) observed that the exposure of *S. aureus* to antibiotics that target the cell wall like oxacillin, ampicillin, vancomycin, and cefotaxime led to a significant decrease in the 730 cm^{-1} after ≈120 min whereas exposure of *S. aureus* to gentamicin and tetracycline took ≈9–12 h for significant spectral changes. The results clearly indicated that the cell wall targeting antibiotics rapidly altered the cell wall characteristics whereas gentamicin and tetracycline which target protein synthesis took a significantly longer time to affect the cell wall integrity and the cell wall signatures remained stable in the SERS spectra (Galvan and Yu 2018). This method was also used to differentiate live and dead bacterial suspensions after antibiotic exposure (Zhou et al., 2015) and to monitor SERS signals of *E. coli* treated with chloramphenicol, trimethoprim, polymyxin B, ampicillin, and formalin. These results also confirmed that polymyxin B and ampicillin, which aggressively degrade the bacterial cell wall led to SERS spectral changes within ≈5 min whereas chloramphenicol, trimethoprim and formalin, which do not target the cell wall, the SERS signals remained unchanged even after ≈2 h exposure, Additionally, antibiotic-resistant cells when exposed to polymyxin B and ampicillin the SERS signatures of wild-type bacteria diminished, whereas the SERS signals of antibiotic-resistant bacteria remained unaffected. Similar changes were also observed when *L. latis*, *A. baumannii*, MSSA and MRSA (Liu et al., 2009; Tien et al., 2016), and 12 UTI causing bacteria (including *E. coli* strains, *S. saprophyticus* strains, *K. pneumoniae* strains, and *E. faecalis* strain) were exposed to antibiotics (Premasiri et al., 2017).

3.11.7.9 Stimulated Raman Scattering

Raman-imaging, is a slow process in which 1 out of 10^8 photons undergo Raman scattering spontaneously (Jones et al., 2019; Stöckel et al., 2016). To overcome this two methods have been developed viz., Coherent anti-Stokes Raman Scattering (CARS) and Stimulated Raman Scattering (SRS) which improve the imaging sensitivity by 4–5 orders of magnitude (Evans and Xie 2008). SRS has been used for metabolic imaging in cells (Fu et al., 2014), tissues (Hu et al., 2015), model organisms (Wang et al., 2014b) and bacteria (Hong et al., 2018). SRS was also used for rapid phenotypic antimicrobial susceptibility testing in bacteria by imaging deuterated glucose metabolism in bacteria (Hong et al., 2018). Using vancomycin-susceptible and vancomycin-resistant enterococci, the uptake of deuterated glucose in a single *E. faecalis* living bacterium could be quantitatively monitored via hyperspectral SRS imaging since the susceptible and resistant bacteria responded differently, and the susceptibility to and MIC of antibiotics could be determined. This method is applicable to *E. coli, K. pneumoniae, and S. aureus* exposed to different antibiotics. In a subsequent study using *Pseudomonas aeruginosa* it was demonstrated that in D_2O containing medium, with 20 μg ml^{-1} gentamicin or cefotaxime the SRS image indicated that

C–D vibrational band, were significantly reduced in the presence of gentamicin, indicating inhibition of D_2O incorporation in *P. aeruginosa* by gentamicin whereas the C–D signals of *P. aeruginosa* in the presence of cefotaxime increased with time, indicating active D_2O incorporation in bacteria (Zhang et al., 2020). The response to antibiotic treatment was observed in as short as 10 min of incubation. MIC could also be determined. AST with D2O-SRS was further demonstrated for bacteria in urine and blood samples spiked with *E. coli* and *P. aeruginosa* respectively. In addition to rapidity, the main advantages of SRS-based AST are universality and direct measurement for clinical samples. A disadvantage is that the instrument of SRS is relatively expensive and bulky.

3.11.7.10 Phase-shift Reflectometric Interference Spectroscopic Measurement

Phase-shift reflectometric interference spectroscopic measurement (PRISM) was used as a platform for culture-free and label-free phenotypic AST (Leonard et al., 2017). This method consists of a biofunctionalized silicon micropillar arrays that serve as a solid–liquid interface for bacteria to colonize and a transducing element to monitor the PRISM of the bacteria when exposed to antibiotics, in real time. PRISM was used to monitor the responses of *E. coli* to the antibiotics gentamicin, ciprofloxacin, ceftriaxone, ampicillin and sulfamethoxazole–trimethoprim. The method allows for the determination of MICs and additionally could be employed to differentiate bacteriostatic and bactericidal antibiotics (Leonard et al., 2017).

3.12 Other Approaches in AMR Diagnosis

3.12.1 Colorimetric-based approach

Lee et al., (2019a) used a colorimetric assay on a microfluidic chamber for antibiotic susceptibility testing (AST) for vancomycin, gentamicin, and linezolid using two clinical methicillin-resistant *S. aureus* isolates. The integrated microfluidic system (IMS) could identify the optimal drug combination and the minimum effective dosage of the drug. Ma et al., (2020) described the identification and antimicrobial susceptibility of *Campylobacter* spp. (*C. jejuni, C. coli*, and *C. lari*) by using a a polymer-based microfluidic device. The microfluidic device consisted of an array of bacterial incubation chambers with chromogenic medium and antibiotics already loaded to monitor the growth of *Campylobacter* spp., visualized by color change due to chromogenic reactions. In both these studies a small bacterial sample was used, reduced reagent volume (<50 µl/well), and the method was automated and fast, and the result was a straight-forward colour change.

Xu et al., (2016) developed a polydimethylsiloxane (PDMS) microfluidic microchip with an array of cell culture chambers with a sample introduction channel. A paper substrate preloaded with chromogenic media and antimicrobial agents was embedded at the bottom of the chamber. This microchip was used for AST assays on multiple uro-pathogens. AST was determined as the lowest concentration of an antimicrobial agent that could inhibit the chromogenic reaction. Recently, He et al., (2020) reported the use of a laser-patterned paper-based device for detection and susceptibility testing of *E.coli*, based on observable colour change. Such integrated paper devices could be used for timely identification of bacterial infections at the point-of-care and for AST and thus contribute to tackling the global challenge of AMR.

3.12.2 pH-based Approaches

Monitoring small pH changes (using a pH-sensitive chitosan hydrogel) has been used for rapid antibiotic susceptibility testing (Tang et al., 2013) and for the detection of ARGs from *E. coli* and *K. pneumoniae* using a combination of Recombinase Polymerase Amplification (RPA) using just 100 CFUs (Hu et al., 2017).

3.12.3 Quartz-crystal Microbalance (QCM)-based Approaches

Quartz-crystal microbalance (QCM) is a nanogram-sensitive device with a piezoelectric sensor and the method is low cost, and rapid (within 10 min) (Reyes et al., 2017a, b). The method has been used for detection of AMR bacteria (Bragazzi et al., 2015). Reyes et al., (2017a, b) used a magnesium zinc oxide nanostructure-modified quartz crystal microbalance (MZOnano-QCM) biosensor for monitoring antimicrobial effects on *E. coli* and *Saccharomyces cerevisiae*, as well as for ARG detection.

3.12.4 Point-of-care Approaches

Point-of-care (POC) approaches offer many advantages such as rapid testing and identification directly from the sample thus overcoming the need to culture the organism. POC approaches have been developed for AMR diagnostics and phenotypic AST in UTI (Toosky et al., 2020). An added advantage is that the method could detect and quantify minimal loads of bacteria ranging from 50 to 105 CFU/ml using a combination of portable particle-counting device comprising a miniature confocal microscope with real-time data analysis. A disadvantage of this method is the negative effect of mixed cultures both on the specificity and the sensitivity of the results. As a modification of this method, Abram et al., (2020) reported a single step blood droplet digital PCR assay with a particle counter system capable of rapid and sensitive (10 CFU/ml) bacterial identification and AST simultaneously.

3.12.5 Multiplex approaches

The advantage of a multiplex approach is the ability to analyze several samples simultaneously and rapidly on a single chip for parallel testing of several antibiotics (Wistrand et al., 2020). In one such method a mix of bacterial isolates and agarose was loaded in an array of microchambers within the microfluidic chip and the growth rate was monitored with a custom-built dark-field microscope coupled to a motorized camera (taking snapshots every 10 min) followed by automated image analysis. The method takes about 5 h and the MIC values determined show 100% agreement with reference (broth microdilution) MIC values (Wistrand et al., 2020).

3.12.6 Single-cell or Single-molecule Approaches

Microfluidic agarose channels with immobilized bacteria have been used for rapid AST and MIC determination by microscopic monitoring (Choi et al., 2013) of time lapse images of the single bacterium cultured under various antibiotic concentrations. The method was rapid, and the turnaround time was less than 4 h (Choi et al., 2013). This rapid AST system was used for the determination UTIs caused by resistant bacteria (Baltekin et al., 2017). Li et al., (2019b) reported a versatile microfluidic system with real time visual detection (microscopy) which facilitated fast bacterial classification (3 min) based on their size and shape and phenotypic AST at the single-cell level. The turnaround time susceptibility testing based on the growth of the bacterium (single-cell level) in the presence of antibiotics, was 30 min. The system could also be applied for bacterial detection and complex (blood cultures, urine, whole blood) polymicrobial samples analysis (Li et al., 2019b).

3.13 Conclusions

Phenotypic methods developed for AMR diagnostics are useful for both AMR surveillance and detection of ARGs. But these methods are slow and require pure cultures and another drawback is that they cannot detect viable but non-cultivable bacteria. Methods based on detection of ATP levels, copy number of nucleic acids, or synthesis of macromolecules in bacteria are rapid but techniques are yet to be automated. In this context the molecular-based methods like PCR, whole genome sequencing, metagenome analysis and DNA-microarrays are rapid and could precisely characterise antimicrobial resistance genes. But these methods are not cost effective and may not correlate with phenotypic results. Spectrophotometric methods (MALDI-TOF, FT-IR, AFM-IR, and SERS) have also been used for AMR diagnostics. These methods are rapid, simple and running cost is low, but equipment is extremely expensive and also needs purified strains and large databases. Thus, the search has been for the most appropriate and convenient methods for AMR diagnostics, and this has led to the use of other spectroscopic methods like FTIR spectroscopy, FTIR microscopy, AFM-IR, Raman spectroscopy, PRISM and other recent approaches. Despite all these methods the phenotypic methods are considered the gold standard for AMR diagnostics and are routinely employed for AMR diagnostics in ocular samples.

REFERENCES

Abram, T.J.; Cherukury, H.; Ou, C.Y.; Vu, T.; Toledano, M.; Li, Y.; Grunwald, J.T.; Toosky, M.N.; Tifrea, D.F.; Slepenkin, A.; et al., Rapid bacterial detection and antibiotic susceptibility testing in whole blood using one-step, high throughput blood digital PCR. *Lab Chip* 2020, 20, 477–89.

Adzitey, F.; Huda, N.; Ali, G.R.R. Molecular techniques for detecting and typing of bacteria, advantages and application to foodborne pathogens isolated from ducks. *3 Biotech* 2013, 3, 97–107.

Alonso, R.; Mateo, E.; Girbau, C.; Churruca, E.; Martinez, I.; Fernández-Astorga, A. PCR-Restriction Fragment Length Polymorphism Assay for Detection of GyrA Mutations

Associated with Fluoroquinolone Resistance in Campylobacter Coli. *Antimicrob Agents Chemother* 2004, 48, 4886–8.

Álvarez-Molina, A.; de Toro, M.; Alexa, E.A.; Álvarez-Ordóñez, A. Applying Genomics to Track Antimicrobial Resistance in the Food Chain. *Ref. Modul. Food Sci.* 2020. doi: 10.1016/b978-0-08-100596-5.22700-5 , Corpus ID: 216498837.

Andersson D.I.; Hughes D. Antibiotic resistance and its cost: is it possible to reverse resistance? *Nat Rev Microbiol.* 2010 Apr, 8(4), 260–71. doi: 10.1038/nrmicro2319

Andrews, J.M. Determination of minimum inhibitory concentrations. *J. Antimicrob. Chemother.* 2001, 48, 5–16.

Argimón, S.; Masim, M.A.L.; Gayeta, J.M.; Lagrada, M.L.; Macaranas, P.K.V.; Cohen, V.; Limas, M.T.; Espiritu, H.O.; Palarca, J.C.; Chilam, J.; et al., Integratingwhole-genome sequencing within the National Antimicrobial Resistance Surveillance Program in the Philippines. *Nat Commun* 2020, 11, 2719.

Ashton, P.M.; Nair, S.; Dallman, T.; Rubino, S.; Rabsch,W.; Mwaigwisya, S.;Wain, J.; O'Grady, J. MinION nanopore sequencing identifies the position and structure of a bacterial antibiotic resistance island. *Nat Biotechnol* 2015, 33, 296–300.

Auffret, M.D.; Dewhurst, R.J.; Duthie, C.-A.; Rooke, J.A.; JohnWallace, R.; Freeman, T.C.; Stewart, R.;Watson, M.; Roehe, R. The Rumen Microbiome as a Reservoir of Antimicrobial Resistance and Pathogenicity Genes Is Directly Affected by Diet in Beef Cattle. *Microbiome* 2017, 5, 159.

Avci E.; Kaya N.S.; Ucankus G.; Culha M. Discrimination of urinary tract infection pathogens by means of their growth profiles using surface enhanced Raman scattering. *Anal Bioanal Chem* 2015, 407(27), 8233–41. doi: 10.1007/s00216-015-8950-5

Avesar, J.; Rosenfeld, D.; Truman-Rosentsvit, M.; Ben-Arye, T.; Geffen, Y.; Bercovici, M.; Levenberg, S. Rapid phenotypic antimicrobial susceptibility testing using nanoliter arrays. *Proc. Natl. Acad. Sci. USA* 2017, 114 (29), E5787–E5795. doi:10.1073/pnas.1703736114

Axelsson, C.; Rehnstam-Holm, A.-S.; Nilson, B. Rapid detection of antibiotic resistance in positive blood cultures by MALDI-TOF MS and an automated and optimized MBT-ASTRA protocol for Escherichia coli and Klebsiella pneumoniae. *Infect. Dis.* 2020, 52, 45–53.

Balbin M.M.; Belotindos L.P.; Abes N.S.; Mingala C.N. Caprine arthritis encephalitis virus detection in blood by loop-mediated isothermal amplification (LAMP) assay targeting the proviral gag region. *Diagn Microbiol Infect Dis.* 2014, 79(1), 37–42. doi: 10.1016/j.diagmicrobio.2013.12.012. Epub 2014 Jan 7. PMID: 24630755.

Baltekin, Ö.; Boucharin, A.; Tano, E.; Andersson, D.I.; Elf, J. Antibiotic susceptibility testing in less than 30 min using direct single-cell imaging. *Proc. Natl. Acad. Sci. USA* 2017, 114, 9170.

Behrendt U.; Ulrich A.; Schumann P.; Naumann D.; Suzuki K. Diversity of grass-associated Microbacteriaceae isolated from the phyllosphere and litter layer after mulching the sward; polyphasic characterization of Subtercola pratensis sp. nov., Curtobacterium herbarum sp. nov. and Plantibacter flavus gen. nov., sp. nov. *Int J Syst Evol Microbiol* 2002, 52, 1441–54.

Bell, J.A.; Weckstein, J.D.; Fecchio, A.; Tkach, V.V. A New Real-Time PCR Protocol for Detection of Avian Haemosporidians. *Parasites Vectors* 2015, 8, 383.

Blackie, Evan J.; Le Ru, Eric C.; Meyer, Matthias; Etchegoin, Pablo G. Surface Enhanced Raman Scattering Enhancement Factors: A Comprehensive Study. *J. Phys. Chem. C.* 2007, 111 (37), 13794–803. doi:10.1021/jp0687908

Böckelmann, U.; Dörries, H.-H.; Ayuso-Gabella, M.N.; de Salgot Marçay, M.; Tandoi, V.; Levantesi, C.; Masciopinto, C.; Van Houtte, E.; Szewzyk, U.; Wintgens, T.; et al., Quantitative PCR Monitoring of Antibiotic Resistance Genes and Bacterial Pathogens in Three European Artificial Groundwater Recharge Systems. *Appl. Environ. Microbiol.* 2009, 75, 154.

Bodelón, G.; Montes-García, V.; Pérez-Juste, J.; Pastoriza-Santos, I. Surface-Enhanced Raman Scattering Spectroscopy for Label- Free Analysis of P. aeruginosa Quorum Sensing. *Front. Cell. Infect. Microbiol.* 2018, 8, 143.

Bogožalec Košir, A.; Cvelbar, T.; Kammel, M.; Grunert, H.-P.; Zeichhardt, H.; Milavec, M. Digital PCR method for detection and quantification of specific antimicrobial drug-resistance mutations in human cytomegalovirus. *J. Virol. Methods* 2020, 281, 113864.

Boolchandani, M.; D'Souza, A.W.; Dantas, G. Sequencing-Based Methods and Resources to Study Antimicrobial Resistance. *Nat Rev. Genet.* 2019, 20, 356–70.

Bragazzi, N.L.; Amicizia, D.; Panatto, D.; Tramalloni, D.; Valle, I.; Gasparini, R. Chapter Six—Quartz-Crystal Microbalance (QCM) for Public Health: An Overview of Its Applications. In Advances in Protein Chemistry and Structural Biology; Donev, R., Ed.; Academic Press: Cambridge, MA, 2015; Volume 101, pp. 149–211.

Brinkac, L.; Voorhies A.; Gomez A.; Nelson K.E. The Threat of Antimicrobial Resistance on the Human Microbiome. *J. Microb Ecol.* 2017 November; 74(4), 1001–08. doi:10.1007/s00248-017-0985-z

Büchl, N.R.; Wenning, M.; Seiler, H.; Mietke-Hofmann, H.; Scherer, S. Reliable identification of closely related Issatchenkia and Pichia species using artificial neural network analysis of Fouriertransform infrared spectra. *Yeast* 2008, 25,787–98.

Bumgarner R. (2013). Overview of DNA microarrays: types, applications, and their future. *Current protocols in molecular biology, Chapter 22*, Unit–22.1. https://doi.org/10.1002/0471142727.mb2201s101

Burckhardt, I.; Zimmermann, S. Using matrix-assisted laser desorption ionization-time of flight mass spectrometry to detect carbapenem resistance within 1 to 2.5 hours. *J. Clin. Microbiol.* 2011, 49, 3321–4.

Call, D.R.; Bakko, M.K.; Krug, M.J.; Roberts, M.C. Identifying antimicrobial resistance genes with DNA microarrays. *Antimicrob Agents Chemother* 2003, 47, 3290–5.

Cansizoglu, M.F.; Tamer, Y.T.; Farid, M.; Koh, A.Y.; Toprak, E. Rapid ultrasensitive detection platform for antimicrobial susceptibility testing. *PLoS Biol.* 2019, 17, e3000291.

Cantera, J.L.; White, H.; Diaz, M.H.; Beall, S.G.; Winchell, J.M.; Lillis, L.; Kalnoky, M.; Gallarda, J.; Boyle, D.S. Assessment of eight nucleic acid amplification technologies for potential use to detect infectious agents in low-resource settings. *PLoS ONE* 2019, 14, e0215756.

Chahorm K., Prakitchaiwattana C. Application of Reverse Transcriptase-PCR-DGGE as a Rapid Method for Routine Determination of Vibrio Spp. in Foods. *Int. J. Food Microbiol.* 2018, 264, 46–52. doi: 10.1016/j.ijfoodmicro.2017.10.014

Chang, K.-W.; Cheng, H.-W.; Shiue, J.;Wang, J.-K.;Wang, Y.-L.; Huang, N.-T. Antibiotic Susceptibility Test with Surface-Enhanced Raman Scattering in a Microfluidic System. *Anal. Chem.* 2019, 91, 10988–95.

Che, J.; Lu, J.X.; Li, W.G.; Zhang, Y.F.; Zhao, X.F.; Yuan, M.; Bai, X.M.; Chen, X.; Li, J. A New High-throughput Real-time PCR Assay for the Screening of Multiple Antimicrobial Resistance Genes in Broiler Fecal Samples from China. *Biomed Environ Sci.* 2019 Dec, 32(12), 881–92. doi: 10.3967/bes2019.111. PMID: 31918793.

Chen, C.; Hong, W. Recent Development of Rapid Antimicrobial Susceptibility Testing Methods through Metabolic Profiling of Bacteria. *Antibiotics* 2021, 10, 311. https://doi.org/10.3390/ antibiotics10030311

Chen, G.; Kocaoglu-Vurma, N.A.; Harper, W.J.; Rodriguez-Saona, L.E. Application of infrared microspectroscopy and multivariate analysis for monitoring the effect of adjunct cultures during Swiss cheese ripening. *J Dairy Sci* 2009, 92,3575–3584

Chen, X.; Tang, M.; Liu, Y.; Huang, J.; Liu, Z.; Tian, H.; Zheng, Y.; de la Chapelle, M.L.; Zhang, Y.; Fu, W. Surface-enhanced Raman scattering method for the identification of methicillin-resistant Staphylococcus aureus using positively charged silver nanoparticles. *Microchim. Acta* 2019, 186, 102.

Cheng, I.F.; Chang, H.C.; Hou, D. Chang, H-C. An integrated dielectrophoretic chip for continuous bioparticle filtering, focusing, sorting, trapping, and detecting. *Biomicrofluidics*. 2007 Jun, 1(2): 021503. doi: 10.1063/1.2723669

Cheng, I.F.; Chen, T.Y.; Lu, R.J.; Wu, H.W. Rapid identification of bacteria utilizing amplified dielectrophoretic force-assisted nanoparticle-induced surface-enhanced Raman spectroscopy. *Nanoscale Res Lett*. 2014, 9(1), 324. doi:10.1186/1556-276X-9-324

Cheong, Y.; Kim, Y. J.; Kang, H.; Choi, S.; Lee, H. J. Rapid label-free identification of Klebsiella pneumoniae antibiotic resistant strains by the drop-coating deposition surface-enhanced Raman scattering method. *Spectrochim. Acta A Mol. Biomol. Spectrosc*. 2017a, 183, 53–9. ISSN 1386–1425. https://doi.org/10.1016/j.saa.2017.04.044

Cheong, Y.; Kim, Y. J.; Kang, H.; Choi, S.; Lee, H. J. Label-free identification of antibiotic resistant isolates of living Escherichia coli: Pilot study Microsc. *Res. Tech*. 2017b, 80, 177–82.

Choi, J.; Jung, Y.-G.; Kim, J.; Kim, S.; Jung, Y.; Na, H.; Kwon, S. Rapid antibiotic susceptibility testing by tracking single cell growth in a microfluidic agarose channel system. *Lab Chip* 2013, 13, 280–7.

Coutinho CP, Sa-Correia I, Lopes JA. Use of Fourier transform infrared spectroscopy and chemometrics to discriminate clinical isolates of bacteria of the Burkholderia cepacia complex from different species and ribopatterns. *Anal Bioanal Chem* 2009, 394:2161–71.

Crupi, V., D. Majolino, M.R. Mondello, P. Migliardo, V. Venuti. FT-IR spectroscopy: a powerful tool in pharmacology *J. Pharm. Biomed. Anal*. 2002, 29, 1149–52.

Delost, M. D. *Introduction to Diagnostic Microbiology for the Laboratory Sciences*. Jones & Bartlett Learning: Burlington, MA, 2014.

de Melo, A.N.F.; Monte, D.F.M.; de Souza Pedrosa, G.T.; Balkey, M.; Jin, Q.; Brown, E.; Allard, M.; de Oliveira, T.C.R.M.; Cao, G.; Magnani, M.; et al., Genomic Investigation of Antimicrobial Resistance Determinants and Virulence Factors in Salmonella Enterica Serovars Isolated from Contaminated Food and Human Stool Samples in Brazil. *Int. J. Food Microbiol*. 2021, 343, 109091.

Demirev, P.; Hagan, N.; Antoine, M.; Lin, J.; Feldman, A. Establishing drug resistance in microorganisms by mass spectrometry. *J. Am. Soc. Mass Spectrom*. 2013, 24, 1194–1201.

Deng, Y. L., Y. J. Juang Electrokinetic trapping and surface enhanced Raman scattering detection of biomolecules using optofluidic device integrated with a microneedles array *Biomicrofluidics* 2013, 7, 014111. doi: 10.1063/1.4793224

Dhama, K.; Karthik, K.; Chakraborty, S.; Tiwari, R.; Kapoor, S.; Kumar, A.; Thomas, P. Loop-mediated isothermal amplification of DNA (LAMP): A new diagnostic tool lights the world of diagnosis of animal and human pathogens: A review. *Pak. J. Biol. Sci*. 2016, 17, 151–66.

Domingues L., editor. *PCR: Methods and Protocols.* Humana Press: New York, NY, 2017.

Donà, V.; Kasraian, S.; Lupo, A.; Guilarte, Y.N.; Hauser, C.; Furrer, H.; Unemo, M.; Low, N.; Endimiani, A. Multiplex Real-Time PCR Assay with High-Resolution Melting Analysis for Characterization of Antimicrobial Resistance in Neisseria gonorrhoeae. *J. Clin. Microbiol.* 2016, 54, 2074–81.

Dortet, L.; Bonnin, R.A.; Pennisi, I.; Gauthier, L.; Jousset, A.B.; Dabos, L.; Furniss, R.C.D.; Mavridou, D.A.I.; Bogaerts, P.; Glupczynski, Y.; et al., Rapid detection and discrimination of chromosome- and MCR-plasmid-mediated resistance to polymyxins by MALDI-TOF MS in Escherichia coli: The MALDIxin test. *J Antimicrob Chemother* 2018a, 73, 3359–3367

Dortet, L.A.-O.; Potron, A.; Bonnin, R.A.-O.; Plesiat, P.; Naas, T.A.-O.; Filloux, A.A.-O.; Larrouy-Maumus, G. Rapid detection of colistin resistance in Acinetobacter baumannii using MALDI-TOF-based lipidomics on intact bacteria. *Sci. Rep.* 2018b, 8:16910. doi:10.1038/s41598-018-35041-y

Edwards, T.; Burke, P.; Smalley, H.B.; Gillies, L.; Longhurst, D.; Vipond, B.; Hobbs, G. Loop-mediated isothermal amplification (LAMP) for the rapid detection of Mycoplasma genitalium. *Diagn Microbiol Infect Dis* 2015, Sep, 83(1), 13–17. doi: 10.1016/j.diagmicrobio.2015.05.010

Erukhimovitch, V.; Pavlov, V.; Talyshinsky, M.; Souprun, Y.; Huleihel, M. FTIR microscopy as a method for identification of bacterial and fungal infections *J Pharm Biomed Anal* 2005, 37, 1105–08

Erukhimovitch, V.; Talyshinsky, M.; Souprun, Y.; Huleihel, M. Spectroscopic Characterization of Human and Mouse Primary Cells, Cell Lines and Malignant Cells. *Photochem. Photobiol.* 2007, 76, 446–451.

Evans, C.L.; Xie X.S. Coherent Anti-Stokes Raman Scattering Microscopy: Chemical Imaging for Biology and Medicine. *Annu. Rev. Anal. Chem.* 2008, 1, 883–909. doi: 10.1146/annurev.anchem.1.031207.112754

Fan, Z.; Kanchanapally, R; Ray, P.C. Hybrid Graphene Oxide Based Ultrasensitive SERS Probe for Label-Free Biosensing. *J Phys Chem Lett* 2013, 4 (21), 3813–18. doi:10.1021/jz4020597

Fang, H.; Hedin, G. Rapid screening and identification of methicillin-resistant *Staphylococcus aureus* from clinical samples by selective-broth and Real-Time PCR Assay. *J Clin Microbiol* 2003, 41, 2894–9.

Fasciana, T.; Gentile, B.; Aquilina, M.; Ciammaruconi, A.; Mascarella, C.; Anselmo, A.; Fortunato, A.; Fillo, S.; Petralito, G.; Lista, F.; et al., Co-existence of virulence factors and antibiotic resistance in new Klebsiella pneumoniae clones emerging in south of Italy. *BMC Infect. Dis.* 2019, 19, 928.

Ferrari, R.G.; Panzenhagen, P.H.N.; Conte-Junior, C.A. Phenotypic and Genotypic Eligible Methods for Salmonella Typhimurium Source Tracking. *Front. Microbiol.* 2017, 8, 2587.

Fischer G, Braun S, Thissen R, Dott W. FT-IR spectroscopy as a tool for rapid identification and intra-species characterization of airborne filamentous fungi. *J Microbiol Methods* 2006, 64, 63–77.

Freeman, W.M.; Walker, S.J.; Vrana, K.E. Quantitative RT-PCR: Pitfalls and Potential. *BioTechniques* 1999, 26, 112–25.

Friedman, N.D.; Temkin, E.; Carmeli, Y. The negative impact of antibiotic resistance. *Clin. Microbiol. Infect.* 2016, 22, 416–22.

Fu, D.; Yu, Y.; Folick, A.; Currie, E.; Farese, R.V.; Tsai, T.-H.; Xie, X.S.; Wang, M.C. In vivo metabolic fingerprinting of neutral lipids with hyperspectral stimulated Raman scattering microscopy. *J. Am. Chem. Soc.* 2014, *136*, 8820–8828.

Galhano, B.S.P.; Ferrari, R.G.; Panzenhagen, P.; de Jesus, A.C.S.; Conte-Junior, C.A. Antimicrobial Resistance Gene Detection Methods for Bacteria in Animal-Based Foods: A Brief Review of Highlights and Advantages. *Microorganisms* 2021, 9, 923. https://doi.org/10.3390/ microorganisms9050923

Galvan, D.D.; Yu, Q. Surface-Enhanced Raman Scattering for Rapid Detection and Characterization of Antibiotic-Resistant Bacteria. *Adv. Healthc. Mater.* 2018, 7, 1701335.

Germond, A.; Ichimura, T.; Horinouchi, T.; Fujita, H.; Furusawa, C.; Watanabe, T.M. Raman spectral signature reflects transcriptomic features of antibiotic resistance in Escherichia coli. *Commun. Biol.* 2018, 1, 85–85.

Gómez-Gómez, C.; Blanco-Picazo, P.; Brown-Jaque, M.; Quirós, P.; Rodríguez-Rubio, L.; Cerdà-Cuellar, M.; Muniesa, M. Infectious Phage Particles Packaging Antibiotic Resistance Genes Found in Meat Products and Chicken Feces. *Sci. Rep.* 2019, 9, 13281.

González-Escalona, N.A.O.; Allard, M.A.; Brown, E.W.; Sharma, S.; Hoffmann, M. Nanopore sequencing for fast determination of plasmids, phages, virulence markers, and antimicrobial resistance genes in Shiga toxin-producing Escherichia coli. *PLoS ONE* 2019, 14, e0220494. Published online 2019 July 30. doi: 10.1371/journal.pone.0220494

Goyal G., Ekedahl E., Nyblom M., Krog J., Fröbrant, E., Brander E.M., Sewunet T., Tangkoskul T., Giske C.G., Sandegren L., Thamlikitku V., Ambjörnsson T., Westerlund F. A simple cut and stretch assay to detect antimicrobial resistance genes on bacterial plasmids by single-molecule fluorescence microscopy. *Scientific Reports* 2022, 12, 9301. https://doi.org/10.1038/s41598-022-13315-w

Greig, D.R.; Dallman, T.J.; Hopkins, K.L.; Jenkins, C. MinION nanopore sequencing identifies the position and structure of bacterial antibiotic resistance determinants in a multidrug-resistant strain of enteroaggregative Escherichia coli. *Microb. Genom.* 2018, Oct, 4(10), e000213. Published online 2018 Sep 20. doi: 10.1099/mgen.0.000213

Guarddon, M.; Miranda, J.M.; Vázquez, B.I.; Cepeda, A.; Franco, C.M. Direct Quantification and Distribution of Tetracycline- Resistant Genes in Meat Samples by Real-Time Polymerase Chain Reaction. *J. Food Sci.* 2012, 77, M372–6.

Gy. Horváth, T. Bencsik, K. Ács, B. Kocsis. Sensitivity of ESBL-Producing Gram-Negative Bacteria to Essential Oils, Plant Extracts, and Their Isolated Compounds, Editor(s): Kateryna Kon, Mahendra Rai, *Antibiotic Resistance*, Academic Press, 2016, pp. 239–69, ISBN 9780128036426.

He, P.J.W.; Katis, I.N.; Kumar, A.J.U.; Bryant, C.A.; Keevil, C.W.; Somani, B.K.; Mahobia, N.; Eason, R.W.; Sones, C.L. Laserpatterned paper-based sensors for rapid point-of-care detection and antibiotic-resistance testing of bacterial infections. *Biosens. Bioelectron.* 2020, 152, 112008.

Heller, A.; Spence, D. A rapid method for post-antibiotic bacterial susceptibility testing. *PLoS ONE* 2019, 14, e0210534.

Helm D, Labischinski H, Naumann D. Elaboration of a procedure for identification of bacteria using Fourier-transform IR spectral libraries: a stepwise correlation approach. *J Microbiol Methods* 1991, 14, 127–42.

Ho, C.-S.; Jean, N.; Hogan, C.A.; Blackmon, L.; Jeffrey, S.S.; Holodniy, M.; Banaei, N.; Saleh, A.A.E.; Ermon, S.; Dionne, J. Rapid identification of pathogenic bacteria using Raman spectroscopy and deep learning. *Nat. Commun.* 2019, 10, 4927.

Hong, W.; Karanja, C.W.; Abutaleb, N.S.; Younis, W.; Zhang, X.; Seleem, M.N.; Cheng, J.-X. Antibiotic susceptibility determination within one cell cycle at single-bacterium level by stimulated Raman metabolic imaging. *Anal. Chem.* 2018, 90, 3737–3743.

Horseman, T.S.; Lustik, M.B.; Fong, K.S.K. Rapid qualitative antibiotic resistance characterization using VITEK MS. *Diagn. Microbiol. Infect. Dis.* 2020, 97, 115093.

Hoyos-Mallecot, Y.; Riazzo, C.; Miranda-Casas, C.; Rojo-Martín, M.D.; Gutiérrez-Fernández, J.; Navarro-Marí, J.M. Rapid detection and identification of strains carrying carbapenemases directly from positive blood cultures using MALDI-TOF MS. *J. Microbiol. Methods* 2014, 105, 98–101.

Hu, C.; Kalsi, S.; Zeimpekis, I.; Sun, K.; Ashburn, P.; Turner, C.; Sutton, J.M.; Morgan, H. Ultra-fast electronic detection of antimicrobial resistance genes using isothermal amplification and Thin Film Transistor sensors. *Biosens. Bioelectron.* 2017, 96, 281–7.

Hu, F.; Chen, Z.; Zhang, L.; Shen, Y.; Wei, L.; Min, W. Vibrational imaging of glucose uptake activity in live cells and tissues by stimulated Raman scattering. *Angew. Chem. Int. Ed.* 2015, 54, 9821–5.

Huletsky A, Giroux R, Rossbach V, Gagnon M, Vaillancourt M, Bernier M, Gagnon F, Truchon K, Bastien M, Picard FJ, van Belkum A, Ouellette M, Roy PH, Bergeron MG. New Real-Time PCR assay for rapid detection of methicillin-resistant *Staphylococcus aureus* directly from specimens containing a mixture of staphylococci. *J Clin Microbiol* 2004, 42, 1875–84.

Idelevich, E.; Sparbier, K.; Kostrzewa, M.; Becker, K. Rapid detection of antibiotic resistance by MALDI-TOF mass spectrometry using a novel direct-on-target microdroplet growth assay. *Clin. Microbiol. Infect.* 2018, 24, 738e43.

Idress, M., U. Mussarat, Y. Badshah, R. Qamar, H. Bokhari, Virulence factors profile of drug-resistant Escherichia coli isolates from urinary tract infections in Punjab, Pakistan. *Eur. J. Clin. Microbiol. Infect. Dis.* 2010, *29*, 1533–1537.

Ivancic, V.; Mastali, M.; Percy, N.; Gornbein, J.; Babbitt, J.; li, Y.; Landaw, E.; Bruckner, D.; Churchill, B.; Haake, D. Rapid antimicrobial susceptibility determination of uropathogens in clinical urine specimens by use of ATP bioluminescence. *J. Clin. Microbiol.* 2008, 46, 1213–19.

Jones, R.R.; Hooper, D.C.; Zhang, L.; Wolverson, D.; Valev, V.K. Raman techniques: Fundamentals and frontiers. *Nano Res. Lett.* 2019, 14, 231.

Jones, S., A. Pramanik, R. Kanchanapally, B. P. V. Nellore, S. Begum, C. Sweet, P. C. Rays, Multifunctional Three-Dimensional Chitosan/Gold Nanoparticle/Graphene Oxide Architecture for Separation, Label-Free SERS Identification of Pharmaceutical Contaminants, and Effective Killing of Superbugs ACS Sustain *Chem. Eng.* 2017, 5, 7175–71867. doi: 10.1021/acssuschemeng.7b01351

Joshi,R., H. Janagama, H. P. Dwivedi, T. Kumar, L. A. Jaykus,J. Schefers, S. Sreevatsan, Selection, characterization, and application of DNA aptamers for the capture and detection of *Salmonella enterica* serovars Mol. *Cell. Probes* 2009, 23, 20–28.

Jousset, A.B.; Bernabeu, S.; Bonnin, R.A.; Creton, E.; Cotellon, G.; Sauvadet, A.; Naas, T.; Dortet, L. Development and validation of a multiplex polymerase chain reaction assay for detection of the five families of plasmid-encoded colistin resistance. *Int. J. Antimicrob. Agents* 2019, 53, 302–09.

Judge, K.; Harris, S.R.; Reuter, S.; Parkhill, J.; Peacock, S.J. Early insights into the potential of the Oxford Nanopore MinION for the detection of antimicrobial resistance genes. *J. Antimicrob. Chemother.* 2015, 70, 2775–8.

Jung, J.; Eberl, T.; Sparbier, K.; Lange, C.; Kostrzewa, M.; Schubert, S.; Wieser, A. Rapid detection of antibiotic resistance based on mass spectrometry and stable isotopes. *Eur. J. Clin. Microbiol. Infect. Dis.* 2014, 33, 949–55.

Jung, J.S.; Hamacher, C.; Gross, B.; Sparbier, K.; Lange, C.; Kostrzewa, M.; Schubert, S. Evaluation of a semiquantitative matrix-assisted laser desorption ionization–time of flight mass spectrometry method for rapid antimicrobial susceptibility testing of positive blood cultures. *J. Clin. Microbiol.* 2016, *54*, 2820–24.

Justesen, U.S.; Acar, Z.; Sydenham, T.V.; Johansson, Å. Antimicrobial susceptibility testing of Bacteroides fragilis using the MALDI Biotyper antibiotic susceptibility test rapid assay (MBT-ASTRA). *Anaerobe* 2018, 54, 236–9.

Kamathewatta, K.I.; Bushell, R.N.; Young, N.D.; Stevenson, M.A.; Billman-Jacobe, H.; Browning, G.F.; Marenda, M.S. Exploration of antibiotic resistance risks in a veterinary teaching hospital with Oxford Nanopore long read sequencing. *PLoS ONE* 2019, 14, e0217600.

Kaprou, G.D.; Bergšpica, I.; Alexa, E.A.; Alvarez-Ordóñez, A.; Prieto, M. Rapid Methods for Antimicrobial Resistance Diagnostics. *Antibiotics* 2021, 10, 209. https://doi.org/10.3390/antibiotics10020209

Kaprou, G.D.; Papadopoulos, V.; Papageorgiou, D.P.; Kefala, I.; Papadakis, G.; Gizeli, E.; Chatzandroulis, S.; Kokkoris, G.; Tserepi, A. Ultrafast, low-power, PCB manufacturable, continuous-flow microdevice for DNA amplification. *Anal. Bioanal. Chem.* 2019, 411, 5297–307.

Karami, A.; Gill, P.; Motamedi, M.; Saghafinia, M. A review of the current isothermal amplification techniques: Applications, advantages and disadvantages. *J. Glob. Infect. Dis.* 2011, 3, 293–302.

Kawamoto, Y.; Kosai, K.; Yamakawa, H.; Kaku, N.; Uno, N.; Morinaga, Y.; Hasegawa, H.; Yanagihara, K. Detection of extended spectrum -lactamase (ESBL)-producing Enterobacteriaceae using the MALDI Biotyper Selective Testing of Antibiotic Resistance—Lactamase (MBT STAR-BL) assay. *J. Microbiol. Methods* 2019, 160, 154–6.

Kazimierczak, K.A.; Scott, K.P.; Kelly, D.; Aminov, R.I. Tetracycline Resistome of the Organic Pig Gut. *Appl. Environ. Microbiol.* 2009, 75, 1717–22.

Kenne, G. Deon van der Merwe Classification of Toxic Cyanobacterial Blooms by Fourier-Transform Infrared Technology (FTIR). *Adv. Microbiol.* 2013, 3(6A), Article ID:37921. doi:10.4236/aim.2013.36A001

Khan,S A., A. K. Singh, D. Senapati, Z. Fan, P. C. Ray, Bio-conjugated popcorn shaped gold nanoparticles for targeted photothermal killing of multiple drug resistant Salmonella DT104 *Chem. Commun.* 2011, 47, 9444

Kim, H.H. Endoscopic Raman spectroscopy for molecular fingerprinting of castric cancer: Principle to implementation. *Biomed. Res. Int.* 2015, 2015, 670121. doi: 10.1155/2015/670121. PMID: 26106612; PMCID: PMC4461730.

Kirchhoff, J.; Glaser, U.; Bohnert, J.A.; Pletz, M.W.; Popp, J.; Neugebauer, U. Simple ciprofloxacin resistance test and determination of minimal inhibitory concentration within 2 h using Raman spectroscopy. *Anal. Chem.* 2018, 90, 1811–18.

Kirschner C, Maquelin K, Pina P, Ngo Thi NA, Choo-Smith LP, Sockalingum GD, Sandt C, Ami D, Orsini F, Doglia SM, Allouch P, Mainfait M, Puppels GJ, Naumann D. Classification and identification of enterococci: a comparative phenotypic, genotypic, and vibrational spectroscopic study. *J Clin Microbiol* 2001, 39:1763–70.

Kittler, S.; Seinige, D.; Diana, M.; Müller, A.; Wendlandt, S.; Ehricht, R.; Monecke, S.; Kehrenberg, C. Characteristics of Methicillin- Resistant Staphylococcus Aureus from Broiler Farms in Germany Are Rather Lineage- than Source-Specific. *Poult. Sci.* 2019, 98, 6903–13.

Kochan, K.; Nethercott, C.; Perez-Guaita, D.; Jiang, J.-H.; Peleg, A.Y.; Wood, B.R.; Heraud, P. Detection of Antimicrobial Resistance-Related Changes in Biochemical Composition of Staphylococcus aureus by Means of Atomic Force Microscopy-Infrared Spectroscopy. *Anal. Chem.* 2019, 91, 15397–403.

Kochan, K.; Peleg, AY.; Heraud, P.; Wood, B.R. Atomic Force Microscopy Combined with Infrared Spectroscopy as a Tool to Probe Single Bacterium Chemistry. *Jove J Chem* 2020. doi: 10.3791/61728

Konduri R., Arunasri K, Reddy GS, Adicherla H, Sharma S, Shivaji S. Global gene expression in Escherichia coli, isolated from the diseased ocular surface of the human eye with a potential to form biofilm. *Gut Pathogens.* 2017, 9, 15.

Konduri R., Chinthala Reddy Sai Abhilash, Gumpili Sai Prashanthi, Shalem Raj Padakandla, Savitri Sharma and Sisinthy Shivaji Phylogenetic Grouping of Human Ocular *Escherichia coli* Based on Whole-Genome Sequence Analysis. *Microorganisms* 2020, 8(3), 422. https://doi.org/10.3390/microorganisms8030422

Konduri R., Sharma S and Shivaji S. Microbes of the human eye: Microbiome, antimicrobial resistance, and biofilm formation. *Experimental Eye Research* 2021, 108476, ISSN 0014-4835. https://doi.org/10.1016/j.exer.2021.108476

Kosacki J, Boisset S, Maurin M, Cornut P-L, Thuret G, Hubanova R, Vandenesch F, Carricajo A, Aptel F, Chiquet C. Friends Group Specific PCR and Quantitative Real-Time PCR in Ocular Samples from Acute and Delayed-Onset Postoperative Endophthalmitis . *Am J Ophthalmol.* 2020 Apr, 212, 34–42. doi: 10.1016/j.ajo.2019.11.026.2020

Kuchibiro T, Hirano A, Ogasawara S, Nakamura T. The microcolony detection method (MCD), a simple and rapid screening test for antimicrobial resistance bacteria on positive blood cultures. *Heliyon* 2020, 6, e05494

Kuhm AE, Suter D, Felleisen R, Rau J. Identification of Yersinia enterocolitica at the species and subspecies levels by Fourier transform infrared spectroscopy. *Appl Environ Microbiol* 2009, 75, 5809–13

Kümmerle M, Scherer S, Seiler H. Rapid and reliable identification of food-borne yeasts by Fourier-transform infrared spectroscopy. *Appl Environ Microbiol* 1998, 64, 2207–14.

Lamendella, R.; Domingo, J.W.S.; Ghosh, S.; Martinson, J.; Oerther, D.B. Comparative Fecal Metagenomics Unveils Unique Functional Capacity of the Swine Gut. *BMC Microbiol.* 2011, 11, 103.

Lange, C.; Schubert, S.; Jung, J.; Kostrzewa, M.; Sparbier, K. Quantitative matrixassisted laser desorption ionization-time of flight mass spectrometry for rapid resistance detection. *J. Clin. Microbiol.* 2014, 52, 4155e62.

Lawn, S.D., Zumla, A. (2011) Tuberculosis. *The Lancet*, 378, 57–72. http://dx.doi.org/10.1016/S0140-6736(10)62173-3

Lechowicz, L.; Urbaniak, M.; Adamus-Białek,W.; Kaca,W. The use of infrared spectroscopy and artificial neural networks for detection of uropathogenic Escherichia coli strains' susceptibility to cephalothin. *Acta Biochim. Pol.* 2013, 60, 713–18.

Lee, S.H.; Park, S.-M.; Kim, B.N.; Kwon, O.S.; Rho, W.-Y.; Jun, B.-H. Emerging ultrafast nucleic acid amplification technologies for next-generation molecular diagnostics. *Biosens. Bioelectron.* 2019b, 141, 111448. doi: 10.1016/j.bios.2019.111448

Lee, W.-B.; Chien, C.-C.; You, H.-L.; Kuo, F.-C.; Lee, M.S.; Lee, G.-B. An integrated microfluidic system for antimicrobial susceptibility testing with antibiotic combination. *Lab Chip* 2019a, 19, 2699–708.

Leonard H., Halachmi S, Ben-Dov N, Nativ O, Segal E. Unraveling Antimicrobial Susceptibility of Bacterial Networks on Micropillar Architectures Using Intrinsic Phase-Shift Spectroscopy. *ACS Nano.* 2017 Jun 27, 11(6), 6167–6177. doi: 10.1021/acsnano.7b02217. PMID: 28485961.

Leonard, H.; Colodner, R.; Halachmi, S.; Segal, E. Recent Advances in the Race to Design a Rapid Diagnostic Test for Antimicrobial Resistance. *Acs Sens.* 2018, 3, 2202–217.

Le Ru, Eric C.; Meyer, Matthias; Etchegoin, Pablo G. Proof of Single-Molecule Sensitivity in Surface Enhanced Raman Scattering (SERS) by Means of a Two-Analyte Technique. *J. Phys. Chem. B.* 2006, 110(4), 1944–8. doi:10.1021/jp054732v

Li, H.; Torab, P.; Mach, K.E.; Surrette, C.; England, M.R.; Craft, D.W.; Thomas, N.J.; Liao, J.C.; Puleo, C.;Wong, P.K. Adaptable microfluidic system for single-cell pathogen classification and antimicrobial susceptibility testing. *Proc. Natl. Acad. Sci. USA* 2019b, 116, 10270–9.

Li, J.; Hu,W.; Li, M.; Deng, S.; Huang, Q.; Lu,W. Evaluation of matrix-assisted laser desorption/ionization time-of-flight mass spectrometry for identifying VIM- and SPM-type metallo-_-lactamase-producing Pseudomonas aeruginosa clinical isolates. *Infect. Drug Resist.* 2019a, 12, 2781–8.

Li, S.; Mann, D.A.; Zhang, S.; Qi, Y.; Meinersmann, R.J.; Deng, X. Microbiome-Informed Food Safety and Quality: Longitudinal Consistency and Cross-Sectional Distinctiveness of Retail Chicken Breast Microbiomes. *mSystems* 2020, 5, e00589–20.

Li, X., Y. Xiong, X. Fan, P. Feng, H. Tang, T. Zhou, The role of femA regulating gene on methicillin-resistant Staphylococcus aureus clinical isolates, *Médecine et Maladies Infectieuses* 2012, 42(5), 218–25. https://doi.org/10.1016/j.medmal.2008.09.031

Lim, A.; Naidenov, B.; Bates, H.;Willyerd, K.; Snider, T.; Couger, M.B.; Chen, C.; Ramachandran, A. Nanopore ultra-long read sequencing technology for antimicrobial resistance detection in Mannheimia haemolytica. *J. Microbiol. Methods* 2019, 159, 138–47.

Liu, T-T, Y-H Lin, C-S Hung, T-J Liu, Y Chen, Y-C Huang, TH Tsai, H-H Wang, D-W Wang, J-K Wang, Y-L Wang, C-H Lin. A High Speed Detection Platform Based on Surface-Enhanced Raman Scattering for Monitoring Antibiotic-Induced Chemica lChanges in Bacteria Cell Wall. *PLoSONE* 2009, 4(5), e5470. doi:10.1371/journal.pone.0005470

Lloyd, A, Pasupuleti V, Thota P, Pant C, Rolston DD, Hernandez AV, Benites-Zapata VA, Fraser TG, Donskey CJ, Deshpande A. Accuracy of loop-mediated isothermal amplification for the diagnosis of Clostridium difficile infection: a systematic review. *Diagn Microbiol Infect Dis.* 2015 May, 82(1), 4–10. doi: 10.1016/j.diagmicrobio.2015.02.007. Epub 2015 Feb 23. PMID: 25752201.

Lu, X.; Samuelson, D.R.; Xu, Y.; Zhang, H.; Wang, S.; Rasco, B.A.; Xu, J.; Konkel, M.E. Detecting and tracking nosocomial methicillin-resistant Staphylococcus aureus using a microfluidic SERS biosensor. *Anal. Chem.* 2013, 85, 2320–27.

Ma, L.; Petersen, M.; Lu, X. Identification and Antimicrobial Susceptibility Testing of Campylobacter Using a Microfluidic Lab-on-a-Chip Device. *Appl. Environ. Microbiol.* 2020. Apr 17, 86(9), e00096–20. doi: 10.1128/AEM.00096-20

Ma, L.; Xia, Y.; Li, B.; Yang, Y.; Li, L.-G.; Tiedje, J.M.; Zhang, T. Metagenomic Assembly Reveals Hosts of Antibiotic Resistance Genes and the Shared Resistome in Pig, Chicken, and Human Feces. *Environ. Sci. Technol.* 2016, 50, 420–427.

Mackay, I.M. Real-time PCR in the microbiology laboratory. *Clin. Microbiol. Infect.* 2004, 10, 190–212.

Manente, R.; Santella, B.; Pagliano, P.; Santoro, E.; Casolaro, V.; Borrelli, A.; Capunzo, M.; Galdiero, M.; Franci, G.; Boccia, G. Prevalence and Antimicrobial Resistance of Causative Agents to Ocular Infections. *Antibiotics* 2022, 11, 463. https://doi.org/10.3390/ antibiotics11040463

Martineau F, Picard FJ, Lansac N et al., Correlation between resistance genotype determined by multiplex PCR assays and the antibiotic susceptibility patterns of *Staphylococcus aureus* and *Staphylococcus epidermidis. Antimicrob Agents Chemother* 2000, 44, 231–8.

Matchawe, C.; Machuka, E.M.; Kyallo, M.; Bonny, P.; Nkeunen, G.; Njaci, I.; Esemu, S.N.; Githae, D.; Juma, J.; Nfor, B.M.; et al., Detection of Antimicrobial Resistance, Pathogenicity, and Virulence Potentials of Non-Typhoidal Salmonella Isolates at the Yaounde Abattoir Using Whole-Genome Sequencing Technique. *Pathogens* 2022, 11, 502. https://doi.org/ 10.3390/pathogens11050502

Matsui, A.; Niimi, H.; Uchiho, Y.; Kawabe, S.; Noda, H.; Kitajima, I. A rapid ATP Bioluminescence-based test for detecting levofloxacin resistance starting from positive blood culture bottles. *Sci. Rep.* 2019, 9, 13565.

Matsumura, M, Hikiba Y, Ogura K et al., Rapid detection of mutations in the 23S rRNA gene of Helicobacter pylori that confers resistance to clarithromycin treatment to the bacterium. *J Clin Microbiol* 2001, 39:691–5.

Maugeri, G.; Lychko, I.; Sobral, R.; Roque, A.C. Identification and antibiotic-susceptibility profiling of infectious bacterial agents: A review of current and future trends. *J. Biotechnol.* 2019, 14, 1700750.

McDermott, P.F.; Tyson, G.H.; Kabera, C.; Chen, Y.; Li, C.; Folster, J.P.; Ayers, S.L.; Lam, C.; Tate, H.P.; Zhao, S. Whole-genome sequencing for detecting antimicrobial resistance in nontyphoidal Salmonella. *Antimicrob. Agents Chemother.* 2016, 60, 5515.

McGill, K.; Kelly, L.; Madden, R.H.; Moran, L.; Carroll, C.; O'Leary, A.; Moore, J.E.; McNamara, E.; O'Mahony, M.; Fanning, S.; Whyte P. Comparison of disc diffusion and epsilometer (E-test) testing techniques to determine antimicrobial susceptibiliy of Campylobacter isolates of food and human clinical origin. *J. Microbiol. Methods* 2009, 79, 238–41.

McWalter, P.W. Determination of susceptibility of Staphylococcus aureus to methicillin by luciferim-luciferase assay of bacterial adenosine triphosphate. *J. Appl. Bacteriol.* 1984, *56*, 145–50

Mohn SC, Ulvik A, Jureen R, Willems RJL, Top J, Leavis H, Harthug S, Langeland N. Duplex Real-Time PCR assay for rapid detection of ampicillin-resistant *Enterococcus faecium. Antimicrob Agents Chemother* 2004, 48, 556–60.

Morach, M.; Käppeli, N.; Hochreutener, M.; Johler, S.; Julmi, J.; Stephan, R.; Etter, D. Microarray Based Genetic Profiling of Staphylococcus Aureus Isolated from Abattoir Byproducts of Pork Origin. *PLoS ONE* 2019, 14, e222036.

Morin S, Bazarova N, Jacon P, Vella S. The Manufacturers' Perspective on World Health Organization Prequalification of In Vitro Diagnostics. *Clin Infect Dis.* 2018 Jan 6, 66(2), 301–05. doi: 10.1093/cid/cix719

Mu, X, Nakano R, Nakano A, Ubagai T, Kikuchi-Ueda T, Tansho-Nagakawa S, Kikuchi H, Kamoshida G, Endo S, Yano H, Ono Y. Loop-mediated isothermal amplification: Rapid and sensitive detection of the antibiotic resistance gene ISAba1-blaOXA-51-like in Acinetobacter baumannii. *J Microbiol Methods.* 2016 Feb, 121, 36–40. doi: 10.1016/j.mimet.2015.12.011

Muhlig, A., T Bocklitz, I. Labugger, S. Dees, S. Henk, E. Richter, S. Andres, M. Merker, S. Stockel, K. Weber, D. Cialla-May, J. Popp LOC-SERS: A Promising Closed System for the Identification of Mycobacteria, *Anal. Chem.* 2016, 88, 7998–8004.

Naumann, D., D. Helm, H. Labischinski, Microbiological characterizations by FT-IR spectroscopy. *Nature.* 1991 May 2, 351(6321), 81–2. doi: 10.1038/351081a0.

Nesme, J., Sébastien Cécillon, Tom O. Delmont, Jean-Michel Monier, Timothy M. Vogel, Pascal Simonet, Large-Scale Metagenomic-Based Study of Antibiotic Resistance in the Environment. *Curr Biol*, 2014, 24(10), 1096–100. https://doi.org/10.1016/j.cub.2014.03.036

Nisa, S.; Bercker, C.; Midwinter, A.C.; Bruce, I.; Graham, C.F.; Venter, P.; Bell, A.; French, N.P.; Benschop, J.; Bailey, K.M.; et al., Combining MALDI-TOF and genomics in the study of methicillin resistant and multidrug resistant Staphylococcus pseudintermedius in New Zealand. *Sci. Rep.* 2019, 9, 1271. https://doi.org/10.1038/s41598-018-37503-9

Nix, I.D.; Idelevich, E.A.; Storck, L.M.; Sparbier, K.; Drews, O.; Kostrzewa, M.; Becker, K. Detection of Methicillin Resistance in Staphylococcus aureus From Agar Cultures and Directly From Positive Blood Cultures Using MALDI-TOF Mass Spectrometry-Based Direct-on-Target Microdroplet Growth Assay. *Front. Microbiol.* 2020, 11, 232. doi: 10.3389/fmicb.2020.00232. PMID: 32117194; PMCID: PMC7033577.

Oberreuter, H, Seiler H, Scherer S. Identification of coryneform bacteria and related taxa by Fourier-transform infrared (FT-IR) spectroscopy. *Int J Syst Evol Microbiol* 2002, 52, 91–100

Oniciuc, E.; Likotrafiti, E.; Alvarez-Molina, A.; Prieto, M.; Santos, J.; Alvarez-Ordóñez, A. The Present and Future of Whole nGenome Sequencing (WGS) and Whole Metagenome Sequencing (WMS) for Surveillance of Antimicrobial Resistant Microorganisms and Antimicrobial Resistance Genes across the Food Chain. *Genes* 2018, 9, 268.

Oust, J.A.; Moretro, T.; Bertrand, D.; Kohler A. FT-IR microspectroscopy: a promising method for the rapid identification of Listeria species. *FEMS Microbiol Lett* 2008, 278, 164–70.

Oust, J.A.; Moretro, T.; Kirschner, C.; Narvhus, J.A.; Kohler, A. FT-IR spectroscopy for identification of closely related lactobacilli. *J Microbiol Methods* 2004, 59, 149–162.

Oviaño, M.; Bou, G. Matrix-assisted laser desorption ionization–time of flight mass spectrometry for the rapid detection of antimicrobial resistance mechanisms and beyond. *Clin. Microbiol. Rev.* 2018, 32, 927–37.

Oviaño, M.; Gato, E.; Bou, G. Rapid Detection of KPC-Producing Enterobacterales Susceptible to Imipenem/Relebactam by Using the MALDI-TOF MS MBT STAR-Carba IVD Assay. *Front. Microbiol.* 2020, 11, 328. doi: 10.3389/fmicb.2020.00328

Oviaño, M.; Gómara, M.; Barba, M.; Revillo, M.; Barbeyto, L.; Bou, G. Towards the early detection of β-lactamase-producing Enterobacteriaceae by MALDI-TOF MS analysis. *J. Antimicrob. Chemother.* 2017a, 72, 2259–62.

Oviaño, M.; Gómara, M.; Barba, M.; Sparbier, K.; Pascual, Á.; Bou, G. Quantitative and automated MALDI-TOF MS-based detection of the plasmid-mediated quinolone resistance determinant AAC(6′)-lb-cr in Enterobacteriaceae. *J. Antimicrob. Chemother.* 2017b, 72, 2952–4.

Palladino S, Kay ID, Flexman JP, Boehm I, Costa AMG, Lambert EJ, Christiansen KJ. Rapid detection of *vanA* and *vanB* genes directly from clinical specimens and enrichment broths by Real-Time multiplex PCR assay. *J Clin Microbiol* 2003, 41, 2383.

Pancholi, P.; Carroll, K. C.; Buchan, B. W.; Chan, R. C.; Dhiman, N.; Ford, B.; Granato, P. A.; Harrington, A. T.; Hernandez, D. R.; Humphries, R. M.; Jindra, M. R.; et al. Multicenter Evaluation of the Accelerate PhenoTest BC Kit for Rapid Identification and Phenotypic Antimicrobial Susceptibility Testing Using Morphokinetic Cellular Analysis. *J Clin Microbiol* 2018, 56(4), e01329–17. doi: 10.1128/JCM.01329-17. PMID: 29305546; PMCID: PMC5869823.

Parlapani, F.F.; Kyritsi, M.; Sakka, M.; Chatzinikolaou, K.; Donos, S.; Boziaris, I.S.; Hadjichristodoulou, C.; Athanassiou, C.G. Matrix-assisted laser desorption ionization–time of flight mass spectrometry reveals Enterococcus and Enterobacter spp. in major insect species involved in food security with resistance to common antibiotics. *J. Pest Sci.* 2020, 93, 159–70.

Patel R, Uhl JR, Kohner P, Hopkins MK, Cockerill III FR. Multiplex PCR detection of *vanA, vanB, vanC-1*, and *vanC2/3* genes in enterococci. *J Clin Microbiol* 1997, 35, 703–7.

Pèrez-Roth, E, Clavcric-Martìn F, Villar J, Mèndez-Àlvarez S. Multiplex PCR for simultaneous identification of *Staphylococcus aureus* and detection of methicillin and mupirocin resistance. *J Clin Microbiol* 2001, 39, 4037–41.

Phuklia, W.; Panyanivong, P.; Sengdetka, D.; Sonthayanon, P.; Newton, P.N.; Paris, D.H.; Day, N.P.J.; Dittrich, S. Novel highthroughput screening method using quantitative PCR to determine the antimicrobial susceptibility of Orientia tsutsugamushi clinical isolates. *J. Antimicrob. Chemother.* 2018, 74, 74–81.

Pishnian Z., Haeili M., Feizi A. Prevalence and Molecular Determinants of Colistin Resistance among Commensal Enterobacteriaceae Isolated from Poultry in Northwest of Iran. *Gut Pathog.* 2019, 11, 2. doi: 10.1186/s13099-019-0282-0

Poppe,C., N. Smart, R. Khakhria, W. Johnson, J. Spika, J. Prescott, (1998) Salmonella Typhimurium DT104: A Virulent and Drug-Resistant Pathogen. *Canadian Veterinary J*, 39, 559–65.

Premasiri, W. R. Y. Chen, P. M. Williamson, D. C. Bandarage, C. Pyles, L. D. Ziegler. Rapid urinary tract infection diagnostics by surface-enhanced Raman spectroscopy (SERS): identification and antibiotic susceptibilities. *Anal Bioanal Chem.* 2017 Apr, 409(11), 3043–54. doi: 10.1007/s00216-017-0244-7

Pujol, I, Capilla J, Fernández-Torres B, Ortoneda M, Guarro J. Use of the sensititre colorimetric microdilution panel for antifungal susceptibility testing of dermatophytes. *J Clin Microbiol.* 2002, 40(7), 2618–21. doi:10.1128/JCM.40.7.2618-2621.2002

Raman, C.V.; Krishnan, K.S. A new type of secondary radiation. *Nature* 1928, 121, 501–502.

Randegger, CC, Hachler H. Real-time PCR and melting curve analysis for reliable and rapid detection of SHV extended-spectrum b-lactamases. *Antimicrob Agents Chem* 2001, 45, 1730–6.

Rebuffo, CA, Schmitt J, Wenning M, von Stetten F, Scherer S. Reliable and rapid identification of Listeria monocytogenes and Listeria species by artificial neural network-based Fourier transform infrared spectroscopy. *Appl Environ Microbiol* 2006, 72, 994–1000.

Rebuffo-Scheer, CA, Dietrich J, Wenning M, Scherer S. Identification of five Listeria species based on infrared spectra (FTIR) using macrosamples is superior to a microsample approach. *Anal Bioanal Chem* 2008, 390, 1629–35.

Rebuffo-Scheer, CA, Kirschner C, Staemmler M, Naumann D. Rapid species and strain differentiation of non-tubercoulous mycobacteria by Fourier-transform infrared microspectroscopy. *J Microbiol Methods* 2007, 68, 282–90.

Reyes, JC, Solon JA, Rivera WL. Development of a loop-mediated isothermal amplification assay for detection of Trichomonas vaginalis. *Diagn Microbiol Infect Dis.* 2014 Jul, 79(3), 337–41. doi: 10.1016/j.diagmicrobio.2014.03.016

Reyes, P.I.; Yang, K.; Zheng, A.; Li, R.; Li, G.; Lu, Y.; Tsang, C.K.; Zheng, S.X.F. Magnesium Zinc Oxide Nanostructure-modified Quartz Crystal Microbalance for Dynamic Monitoring of Antibiotic Effects and Antimicrobial Resistance. *Procedia Technol.* 2017a, 27, 46–7.

Reyes, P.I.; Yang, K.; Zheng, A.; Li, R.; Li, G.; Lu, Y.; Tsang, C.K.; Zheng, S.X.F. Dynamic monitoring of antimicrobial resistance using magnesium zinc oxide nanostructure-modified quartz crystal microbalance. *Biosens. Bioelectron.* 2017b, 93, 189–97

Richter, S.S.; Karichu, J.; Otiso, J.; Van Heule, H.; Keller, G.; Cober, E.; Rojas, L.J.; Hujer, A.M.; Hujer, K.M.; Marshall, S.; et al., Evaluation of Sensititre Broth Microdilution Plate for determining the susceptibility of carbapenem-resistant Klebsiella pneumoniae to polymyxins. *Diagn. Microbiol. Infect. Dis.* 2018, 91, 89–92.

Rocco, V.G.; Intra, J.; Sarto, C.; Tiberti, N.; Savarino, C.; Brambilla, M.; Brambilla, P. Rapid Identification of Carbapenemaseproducing Klebsiella pneumoniae strains by Matrix-Assisted Laser Desorption/Ionization-Time of Flight using Vitek(®) *Mass Spectrometry System. Eurasian J. Med.* 2019, 51, 209–13.

Rodrigues, G.L.; Panzenhagen, P.; Ferrari, R.G.; dos Santos, A.; Paschoalin, V.M.F.; Conte-Junior, C.A. Frequency of Antimicrobial Resistance Genes in Salmonella From Brazil by in Silico Whole-Genome Sequencing Analysis: An Overview of the Last Four Decades. *Front. Microbiol.* 2020, 11, 1864.

Rodrigues, P.A.; Ferrari, R.G.; Conte-Junior, C.A. Application of Molecular Tools to Elucidate the Microbiota of Seafood. *J. Appl. Microbiol.* 2018, 124, 1347–65.

Ronaghi, M, Uhlen M, Nyren P. A sequencing method based on real-time pyrophosphate. *Science* 1998, 281, 363–5.

Ross, J I., E.Anne Eady, Jonathan H. Cove, Simon Baumberg, Minimal functional system required for expression of erythromycin resistance by msrA in Staphylococcus aureus RN4220. *Gene*, 1966, 183, 143–148, ISSN 0378-1119, https://doi.org/10.1016/S0378-1119(96)00541-0

Rubiola, S., Chiesa F., Dalmasso A., Di Ciccio P., Civera T. Detection of Antimicrobial Resistance Genes in the Milk Production Environment: Impact of Host DNA and Sequencing DepthFront. *Microbiol.* 2020, 11, 1983. doi: 10.3389/fmicb.2020.01983

Sackmann, E.K.; Fulton, A.L.; Beebe, D.J. The present and future role of microfluidics in biomedical research. *Nature* 2014, 507, 181–9.

Sakai, T, Kohzaki K, Watanabe A, Tsuneoka H, Shimadzu M. Use of DNA microarray analysis in diagnosis of bacterial and fungal endophthalmitis. *Clin Ophthalmol.* 2012, 6, 321–6. doi: 10.2147/OPTH.S29230

Salman, A., V. Erukhimovitch, M. Talyshinsky, M. Huleihil, M. Huleihel, FTIR spectroscopic method for detection of cells infected with herpes viruses. *Biopolymers*, 2002, 67(6), 406–12. doi:10.1002/bip.10171

Salman, A.; Sharaha, U.; Rodriguez-Diaz, E.; Shufan, E.; Riesenberg, K.; Bigio, I.J.; Huleihel, M. Detection of antibiotic resistant Escherichia Coli bacteria using infrared microscopy and advanced multivariate analysis. *Analyst* 2017, 142, 2136–44.

Schmidt, VS, Kaufmann V, Kulozik U, Scherer S, Wenning M. Microbial biodiversity, quality and shelf life of microfiltered and pasteurized extended shelf life (ESL) milk from Germany, *Austria and Switzerland. Int J Food Microbiol* 2012, 154:1–9.

Schmidt, K.; Mwaigwisya, S.; Crossman, L.C.; Doumith, M.; Munroe, D.; Pires, C.; Khan, A.M.;Woodford, N.; Saunders, N.J.; Wain, J.; et al., Identification of bacterial pathogens and antimicrobial resistance directly from clinical urines by nanopore-based metagenomic sequencing. *J. Antimicrob. Chemother.* 2016, 72, 104–14.

Schmittgen, TD, Lee EJ, Jiang J. High-throughput real-time PCR. *Methods Mol Biol.* 2008, 429, 89–98. doi: 10.1007/978-1-60327-040-3_7

Schröder, U.-C.; Beleites, C.; Assmann, C.; Glaser, U.; Hübner, U.; Pfister, W.; Neugebauer, U. Detection of vancomycin resistances in enterococci within 3 ½ hours. *Sci. Rep.* 2015, 5, 8217.

Schröder, U.-C.; Kirchhoff, J.; Hübner, U.; Mayer, G.; Glaser, U.; Henkel, T.; Pfister, W.; Fritzsche, W.; Popp, J.; Neugebauer, U. On-chip spectroscopic assessment of microbial susceptibility to antibiotics within 3.5 hours. *J. Biophotonics* 2017, 10, 1547–57.

Seedy, F.R.E.; Samy, A.A.; Salam, H.S.H.; Khairy, E.A.; Koraney, A.A. Polymerase chain reaction detection of genes responsible for multiple antibiotic resistance Staphylococcus aureus isolated from food of animal origin in Egypt. *Vet. World* 2017, 10, 1205–11.

Seng, P.; Drancourt, M.; Gouriet, F.; La Scola, B.; Fournier, P.-E.; Rolain, J.M.; Raoult, D. Ongoing revolution in bacteriology: Routine identification of bacteria by matrix-assisted laser desorption ionization time-offlight mass spectrometry. *Clin. Infect. Dis.* 2009, 49, 543–51.

Seo, K.W., Lee Y.J. Detection of Plasmid-Mediated Quinolone Resistance Genes in β-Lactamase-Producing Escherichia Coliisolates from Layer Hens. *Poult. Sci.* 2019, 98, 1480–7. doi: 10.3382/ps/pey545

Sharaha, U.; Rodriguez-Diaz, E.; Riesenberg, K.; Bigio, I.J.; Huleihel, M.; Salman, A. Using Infrared Spectroscopy and Multivariate Analysis to Detect Antibiotics' Resistant Escherichia coli Bacteria. *Anal. Chem.* 2017, 89, 8782–90.

Sharma, S. Antibiotic resistance in ocular bacterial pathogens *Indian Journal of Medical Microbiology* 2011, 29(3), 218–221.

Sharma, S. Diagnosis of infectious diseases of the eye. *Eye* (London, England) 2012, 26(2), 177–84. doi:10.1038/eye.2011.275

Shivaji, S. Rajagopalaboopathi Jayasudha, Sama Kalyana Chakravarthy, Chinthala Reddy SaiAbhilash, Gumpili Sai Prashanthi, Savitri Sharma, Prashant Garg, Somasheila I. Murthy,Alterations in the conjunctival surface bacterial microbiome in bacterial keratitis patients. *Experimental Eye Res*, 2021, 203, 108418. ISSN 0014-4835. https://doi.org/10.1016/j.exer.2020.108418

Singh, K.M.; Jakhesara, S.J.; Koringa, P.G.; Rank, D.N.; Joshi, C.G. Metagenomic Analysis of Virulence-Associated and Antibiotic Resistance Genes of Microbes in Rumen of Indian Buffalo (Bubalus Bubalis). *Gene* 2012, 507, 146–51.

Somorin Y M., N-J M. Weir, S H Pattison, M A. Crockard, C M. Hughes, M M. Tunney, D F. Gilpin. Antimicrobial resistance in urinary pathogens and culture-independent detection of trimethoprim resistance in urine from patients with urinary tract infection *BMC Microbiology* 2022, 22, 144 https://doi.org/10.1186/s12866-022-02551-9

Sparbier, K.; Lange, C.; Jung, J.; Wieser, A.; Schubert, S.; Kostrzewa, M. MALDI biotyper-based rapid resistance detectionby stable-isotope labeling. *J. Clin. Microbiol.* 2013, 51, 3741–8.

Srimongkol, G.; Ditmangklo, B.; Choopara, I.; Thaniyavarn, J.; Dean, D.; Kokpol, S.; Vilaivan, T.; Somboonna, N. Rapid colorimetric loop-mediated isothermal amplification for hypersensitive point-of-care Staphylococcus aureus enterotoxin A gene detection in milk and pork products. *Sci. Rep.* 2020, 10, 7768.

Stöckel, S.; Kirchhoff, J.; Neugebauer, U.; Rösch, P.; Popp, J. The application of Raman spectroscopy for the detection and identification of microorganisms. *J. Raman. Spectrosc.* 2016, 47, 89–109.

Sundfjord, A., Simonsen G. S., Haldorsen, B.C., Haaheim H., Hjelmevoll S., Littauer, P and Dahl, K. H., *Genetic methods for detection of antimicrobial resistance APMIS* 2004, 112, 815–37. ISSN 0903-4641.

Tang, W.; Ranganathan, N.; Shahrezaei, V.; Larrouy-Maumus, G. MALDI-TOF mass spectrometry on intact bacteria combined with a refined analysis framework allows accurate classification of MSSA and MRSA. *PLoS ONE* 2019, 14, e0218951.

Tang, Y.; Zhen, L.; Liu, J.;Wu, J. Rapid antibiotic susceptibility testing in a microfluidic pH sensor. *Anal. Chem.* 2013, 85, 2787–94.

Tao, Y.; Wang, Y.; Huang, S.; Zhu, P.; Huang, W.E.; Ling, J.; Xu, J. Metabolic-Activity-Based assessment of antimicrobial effects by D2O-labeled single-cell Raman microspectroscopy. *Anal. Chem.* 2017, 89, 4108–15.

Tenover, F.C. Antibiotic Susceptibility Testing. In *Encyclopedia of Microbiology*, 3rd ed.; Schaechter, M., Ed.; Academic Press: Oxford, UK, 2009; pp. 67–77.

Thomas, M.; Webb, M.; Ghimire, S.; Blair, A.; Olson, K.; Fenske, G.J.; Fonder, A.T.; Christopher-Hennings, J.; Brake, D.; Scaria, J. Metagenomic Characterization of the Effect of Feed Additives on the Gut Microbiome and Antibiotic Resistome of Feedlot Cattle. *Sci. Rep.* 2017, 7, 12257.

Thomas, R.K.; Melton, R.; Asbell, P.A. Antibiotic resistance among ocular pathogens: current trends from the ARMOR surveillance study (2009–2016) *Clini Optomet* 2019, 11, 15–26

Tien, N., H. C. Chen, S. L. Gau, T. H. Lin, H. S. Lin, B. J. You, P. C. Tsai, I. R. Chen, M. F. Tsai, I. K. Wang, C. J. Chen, C. T. Chang, Diagnosis of bacterial pathogens in the dialysate of peritoneal dialysis patients with peritonitis using surface-enhanced Raman spectroscopy *Clin. Chim. Acta* 2016, 461, 69–75.

Timmins, EM, Quain DE, Goodacre R. Differentiation of brewing yeast strains by pyrolysis mass spectrometry and Fourier transform infrared spectroscopy. *Yeast* 1998, 14, 885–93.

Toosky, M.N.; Grunwald, J.T.; Pala, D.; Shen, B.; Zhao,W.; D'Agostini, C.; Coghe, F.; Angioni, G.; Motolese, G.; Abram, T.J.; et al., A rapid, point-of-care antibiotic susceptibility test for urinary tract infections. *J. Med. Microbiol.* 2020, 69, 52–62.

Toubas D, Essendoubi M, Adt I, Pinon JM, Mainfait M, Sockalingum GD. FTIR spectroscopy in medical mycology: applications to the differentiation and typing of Candida. *Anal Bioanal Chem* 2007, 387, 1729–37.

Udelhoven, T, Naumann D, Schmitt J. Development of a hierarchical classification system with artificial neural networks and FTIR spectra for the identification of bacteria. *Appl Spectrosc* 2000, 54, 1471–9.

Vasala, A.; Hytönen, V.P.; Laitinen, O.H. Modern Tools for Rapid Diagnostics of Antimicrobial Resistance. *Front. Cell. Infect. Microbiol.* 2020, 10, 308. doi: 10.3389/fcimb.2020.00308.

Vatanshenassan, M.; Boekhout, T.; Meis, J.F.; Berman, J.; Chowdhary, A.; Ben-Ami, R.; Sparbier, K.; Kostrzewa, M. Candida auris Identification and Rapid Antifungal Susceptibility Testing Against Echinocandins by MALDI-TOF MS. *Front. Cell. Infect. Microbiol.* 2019, 9, 20.

Vélez, J.R.; Cameron, M.; Rodríguez-Lecompte, J.C.; Xia, F.; Heider, L.C.; Saab, M.; McClure, J.T.; Sánchez, J. Whole-Genome Sequence Analysis of Antimicrobial Resistance Genes in Streptococcus uberis and Streptococcus dysgalactiae Isolates from Canadian Dairy Herds. *Front. Vet. Sci.* 2017, 4, 63.

Vogt, D.; Overesch, G.; Endimiani, A.; Collaud, A.; Thomann, A.; Perreten, V. Occurrence and Genetic Characteristics of Third- Generation Cephalosporin-Resistant Escherichia Coli in Swiss Retail Meat. *Microb. Drug Resist.* 2014, 20, 485–94.

Vogt, S.; Löffler, K.; Dinkelacker, A.G.; Bader, B.; Autenrieth, I.B.; Peter, S.; Liese, J. Fourier-Transform Infrared (FTIR) Spectroscopy for Typing of Clinical Enterobacter cloacae Complex Isolates. *Front. Microbiol.* 2019, 10, 2582. doi: 10.3389/fmicb.2019.02582.

Vrioni, G.; Tsiamis, C.; Oikonomidis, G.; Theodoridou, K.; Kapsimali, V.; Tsakris, A. MALDI-TOF mass spectrometry technology for detecting biomarkers of antimicrobial resistance: Current achievements and future perspectives. *Ann. Transl. Med.* 2018, 6, 240.

Waldeisen, J.R.; Wang, T.; Mitra, D.; Lee, L.P. A real-time PCR antibiogram for drug-resistant sepsis. *PLoS ONE* 2011, 6, e28528.

Wang, F.-H.; Qiao, M.; Su, J.-Q.; Chen, Z.; Zhou, X.; Zhu, Y.-G. High Throughput Profiling of Antibiotic Resistance Genes in Urban Park Soils with Reclaimed Water Irrigation. *Environ. Sci. Technol.* 2014a, 48, 9079–85.

Wang, H.; Hecht, S.; Kline, D.; Leber, A.L. Staphylococcus aureus and methicillin resistance detection directly from pediatric samples using PCR assays with differential cycle threshold values for corroboration of methicillin resistance. *J. Microbiol. Methods* 2019, 159, 167–73.

Wang, K.; Li, S.; Petersen, M.; Wang, S.; Lu, X. Detection and Characterization of Antibiotic-Resistant Bacteria Using Surface-Enhanced Raman Spectroscopy. *Nanomaterials* 2018, 8, 762.

Wang, P.; Liu, B.; Zhang, D.; Belew, M.Y.; Tissenbaum, H.A.; Cheng, J.-X. Imaging lipid metabolism in live Caenorhabditis elegans using fingerprint vibrations. *Angew. Chem.* 2014b, 53, 11787–92

Waseem, H.; Jameel, S.; Ali, J.; Saleem Ur Rehman, H.; Tauseef, I.; Farooq, U.; Jamal, A.; Ali, M.I. Contributions and Challenges of High Throughput qPCR for determining Antimicrobial Resistance in the Environment: A Critical Review. *Molecules* 2019, 24, 163. doi:10.3390/molecules24010163

Waseem, H.; Saleem ur Rehman, H.; Ali, J.; Iqbal, M.J.; Ali, M.I. Chapter 14—Global trends in ARGs measured by HT-qPCR platforms. In *Antibiotics and Antimicrobial Resistance Genes in the Environment*; Hashmi, M.Z., Ed.; Elsevier: Amsterdam, The Netherlands, 2020; Volume 1, pp. 206–222.

Welker, M.; van Belkum, A. One System for All: Is Mass Spectrometry a Future Alternative for Conventional Antibiotic Susceptibility Testing? *Front. Microbiol.* 2019, 10, 2711.

Wenning, M.; Büchl, N.R.; Scherer, S. Species and strain identification of lactic acid bacteria using FTIR spectroscopy and artificial neural networks. *J Biophotonics* 2010, 3, 493–505.

Wenning, M.; Scherer, S. Identification of microorganisms by FTIR spectroscopy: perspectives and limitations of the method Appl Microbiol Biotechnol 2013, 97, 7111–20. doi: 10.1007/s00253-013-5087-3

Wheat, P.; Hastings, J.; Spencer, R. Rapid antibiotic susceptibility tests on Enterobacteriaceae by ATP bioluminescence. *J. Med. Microbiol.* 1988, 25, 95–99.

Wheat, P.; Spencer, R.; Hastings, J. A novel luminometer for rapid antimicrobial susceptibility tests on gram-positive cocci by ATP bioluminescence. *J. Med. Microbiol.* 1989, 29, 277–82.

Wiegand, I., Kai Hilpert; Robert E W Hancock Agar and broth dilution methods to determine the minimal inhibitory concentration (MIC) of antimicrobial substances. *Nat Protocols* 2008, 3(2), 163. doi:10.1038/nprot.2007.521

Wistrand-Yuen, P.; Malmberg, C.; Fatsis-Kavalopoulos, N.; Lübke, M.; Tängdén, T.; Kreuger, J. A Multiplex Fluidic Chip for Rapid Phenotypic Antibiotic Susceptibility Testing. *mBio* 2020, 11, e03109–19.

Woksepp, H.; Jernberg, C.; Tärnberg, M.; Ryberg, A.; Brolund, A.; Nordvall, M.; Olsson-Liljequist, B.; Wisell, K.T.; Monstein, H.-J.; Nilsson, L.E.; et al., High-resolution melting-curve analysis of ligation-mediated real-time PCR for rapid evaluation of an epidemiological outbreak of extended-spectrum-beta-lactamase-producing Escherichia coli. *J. Clin. Microbiol.* 2011, 49, 4032.

Woksepp, H.; Ryberg, A.; Billström, H.; Hällgren, A.; Nilsson, L.E.; Marklund, B.-I.; Olsson-Liljequist, B.; Schön, T. Evaluation of high-resolution melting curve analysis of ligation-mediated real-time PCR, a rapid method for epidemiological typing of ESKAPE (Enterococcus faecium, Staphylococcus aureus, Klebsiella pneumoniae, Acinetobacter baumannii, *Pseudomonas aeruginosa, and Enterobacter species) pathogens. J. Clin. Microbiol.* 2014, 52, 4339.

Woodford N, Tysall L, Auckland C et al., Detection of oxazolidinone-resistant *Enterococcus faecalis* and *Enterococcus faecium* strains by real-time PCR and PCR-restriction fragment length polymorphism analysis. *J Clin Microbiol* 2002, 40, 4298–300.

Wu, L, Thompson DK, Li G, Hurt RA, Tiedje JM, Zhou J. Development and evaluation of functional gene arrays for detection of selected genes in the environment. *Appl Environ Microbiol* 2001, 67, 5780–90.

Xiong, W.; Wang, Y.; Sun, Y.; Ma, L.; Zeng, Q.; Jiang, X.; Li, A.; Zeng, Z.; Zhang, T. Antibiotic-Mediated Changes in the Fecal Microbiome of Broiler Chickens Define the Incidence of Antibiotic Resistance Genes. *Microbiome* 2018, 6, 34.

Xu, B, Du Y, Lin J, Qi M, Shu B, Wen X, Liang G, Chen B, Liu D. Simultaneous Identification and Antimicrobial Susceptibility Testing of Multiple Uropathogens on a Microfluidic Chip with Paper-Supported Cell Culture Arrays. *Anal Chem.* 2016 Dec 6, 88(23), 11593–600. doi: 10.1021/acs.analchem.6b03052

Xu, J., J. W. Turner, M. Idso, S. V. Biryukov, L. Rognstad, H. Gong, V. L. Trainer, M. L. Wells, M. S. Strom, Q. Yu, In Situ Strain-Level Detection and Identification of Vibrio parahaemolyticus Using Surface-Enhanced Raman SpectroscopyAnal. *Chem.* 2013, 85, 2630–7.

Xu, L.; Chen, H.; Canales, M.; Ciric, L. Use of synthesized double-stranded gene fragments as qPCR standards for the quantification of antibiotic resistance genes. *J. Microbiol. Methods* 2019, 164, 105670.

Yang, K.; Li, H.-Z.; Zhu, X.; Su, J.-Q.; Ren, B.; Zhu, Y.-G.; Cui, L. Rapid antibiotic susceptibility testing of pathogenic bacteria using heavy-water-labeled single-cell Raman spectroscopy in clinical samples. *Anal. Chem.* 2019, 91, 6296–303.

Yang, X.; Noyes, N.R.; Doster, E.; Martin, J.N.; Linke, L.M.; Magnuson, R.J.; Yang, H.; Geornaras, I.;Woerner, D.R.; Jones, K.L.; et al., Use of Metagenomic Shotgun Sequencing Technology To Detect Foodborne Pathogens within the Microbiome of the Beef Production Chain. *Appl. Environ. Microbiol.* 2016, 82, 2433–43.

Zhang, H, Ma X, Liu Y, Duan N, Wu S, Wang Z, Xu B. Gold nanoparticles enhanced SERS aptasensor for the simultaneous detection of *Salmonella typhimurium* and *Staphylococcus aureus. Biosens Bioelectron* 2015a, 74, 872–7.

Zhang, L, Huang W, Zhang S, Li Q, Wang Y, Chen T, Jiang H, Kong D, Lv Q, Zheng Y, Ren Y, Liu P, Jiang Y and Chen Y. Rapid Detection of Bacterial Pathogens and Antimicrobial Resistance Genes in Clinical Urine Samples With Urinary Tract Infection by Metagenomic Nanopore Sequencing. *Front. Microbiol.* 2022, 13, 858777. doi: 10.3389/fmicb.2022.858777

Zhang, M, Hong W, Abutaleb NS, Li J, Dong PT, Zong C, Wang P, Seleem MN, Cheng JX. Rapid Determination of Antimicrobial Susceptibility by Stimulated Raman Scattering Imaging of D_2O Metabolic Incorporation in a Single Bacterium. *Adv Sci (Weinh).* 2020 Aug 16, 7(19), 2001452. doi: 10.1002/advs.202001452

Zhang, Q Q., G. G. Ying, C. G. Pan, Y. S. Liu, J. L. Zhao, Comprehensive Evaluation of Antibiotics Emission and Fate in the River Basins of China: Source Analysis,

Multimedia Modeling, and Linkage to Bacterial Resistance *Environ.Sci. Technol.* 2015b, 49, 6772.

Zhao, S.; Tyson, G.H.; Chen, Y.; Li, C.; Mukherjee, S.; Young, S.; Lam, C.; Folster, J.P.; Whichard, J.M.; McDermott, P.F. Whole- Genome Sequencing Analysis Accurately Predicts Antimicrobial Resistance Phenotypes in Campylobacter Spp. *Appl. Environ. Microbiol.* 2016, 82, 459.

Zhou, H. B., D. T. Yang, N. P. Ivleva, N. E. Mircescu, S. Schubert, R. Niessner, A. Wieser, C. Haisch, Label-Free in Situ Discrimination of Live and Dead Bacteria by Surface-Enhanced Raman Scattering *Anal. Chem.* 2015, 87, 6553–61 doi: 10.1021/acs.analchem.5b01271

Zhu, L.-X.; Zhang, Z.-W.; Wang, C.; Yang, H.-W.; Jiang, D.; Zhang, Q.; Mitchelson, K.; Cheng, J. Use of a DNA Microarray for Simultaneous Detection of Antibiotic Resistance Genes among Staphylococcal Clinical Isolates. *J. Clin. Microbiol.* 2007, 45, 3514.

Zhu, Y., Natalia Gasilova, Milica Jovi, Liang Qiao, Baohong Liu, Lysiane Tissìeres Lovey, Horst Picke, Hubert H. Girault. Detection of antimicrobial resistance-associated proteins by titanium dioxide-facilitated intact bacteria mass spectrometry†Chem. *Sci.*, 2018, 9, 2212

Zou, Y.; Mason, M.-G.; Botella, J.-R. Evaluation and improvement of isothermal amplification methods for point-of-need plant disease diagnostics. *PLoS ONE* 2020, 15, e0235216.

Zürcher, S.; Mooser, C.; Lüthi, A.U.; Mühlemann, K.; Barbani, M.T.; Mohacsi, P.; Garzoni, C.; Gorgievski-Hrisoho, M.; Schaller, A.; Flatz, L. Sensitive and rapid detection of ganciclovir resistance by PCR based MALDI-TOF analysis. *J. Clin. Virol.* 2012, 54, 359–63.

4

Antimicrobial Resistance in the Diseased Human Eye: Bacterial Keratitis

S. Shivaji

4.1 Introduction

Antimicrobial resistance (AMR) is a global phenomenon and impacts human health and economy. AMR is encountered in those regions of the human body that harbour or permit the colonisation of bacteria/ fungi like the skin (Perera and Hay 2005; Nielsen et al., 2021), nasal cavity (Karimi et al., 2017; Alharbi 2020), oral cavity, endodontic infections (Moraes et al. 2015; Brooks et al., 2022; Roberts 2002; Villedieu et al. 2003), dental plaque (Villedieu et al. 2003; Xu and Gunsolley 2014), blood (Kulauzov et al., 2008; Aish et al., 2022), urine/UTI (Nickel 2007; Haindongo et al., 2022) and so forth. The best studied is AMR in the human gut and in the gut of various animals (Lawther et al., 2022; Ma et al., 2021; 2022). Comparatively little is known about AMR in ocular surface, which is also known to have its own microbiome. This chapter highlights AMR on the ocular surface of the healthy eye and the eye effected with bacterial keratitis covering aspects such as surveillance of AMR, modes of transmission of AMR and antibiotic resistance in specific ocular bacteria.

4.1.1 AMR Surveillance

In the previous chapters man-made factors, microbial factors and genetic factors have been implicated in the acquisition of AMR (Blair et al., 2015; Walsh 2000). Intervention is needed to curb AMR and this is best achieved by AMR surveillance. The United States and many countries in Europe have national surveillance systems in place involving data sharing, dissemination of data and active engagement with policymakers, like the Latin American Antimicrobial Resistance Surveillance Network (ReLAVRA) created in 1996 by the World Health Organization and the ARMOR study in the United States (Asbell and Sanfilippo 2017; Thomas et al., 2019; Asbell et al., 2018a; 2018b; 2020a; 2020b; 2020c). In Europe the organisations include the Centre for Disease Prevention and Control (ECDC 2015), the Food Safety Authority and the Medicines Agency. Such efforts related to surveillance of AMR should be extended to involve all countries to develop a global response to AMR (Allcock et al., 2017).

DOI: 10.1201/9781003451105-4

4.1.2 Transmission of AMR

Spread of AMR and its occurrence in humans is closely associated with antimicrobials used in veterinary and their presence in the environment (Woolhouse et al., 2015; Finley et al., 2013). In 2015, surveillance data from across the European Union (ECDC 2015) showed positive associations between antimicrobial consumption in animals and resistance in bacteria from humans (ECDC 2015). For instance, methicillin-resistant *S. aureus* (MRSA) (Price et al., 2012) and *mec*C MRSA (Paterson et al., 2012; Harrison et al., 2013) represent a common antimicrobial-resistant lineage shared by humans and farm animals. But there was a lack of consistency in these results, with some studies unable to show that farm animals shared bacterial strains with humans (Mather et al., 2012). This lack of association could be influenced by several potentially confounding factors (Datta et al., 2014), thus highlighting the need to establish the link between AMR in humans and other animals.

4.1.2.1 Genetic and Epidemiological Approaches

Whole-genome sequencing was crucial in establishing the evolutionary origins of pathogens and the possible commonality in humans and animals. For instance, the transmission of MRSA lineage (ST5) may have occurred from human to poultry (Lowder et al., 2009), though the converse is also possible (Voss et al., 2005). For instance, cephalosporin-resistant *E. coli* are disseminated from livestock to humans (Machado et al., 2008; Doi et al., 2010) and through distinct plasmids (De Been et al., 2014). It was also observed that the multidrug-resistant genes in *Salmonella typhimurium* DT104 bacterium are different between animal and human populations (Mather et al., 2013). Thus, understanding the transmission of AMR among populations is complex.

4.2 Antibiotic Resistance in Ocular Bacteria in the Healthy Eye

The ocular bacteria *Staphylococcus aureus*, coagulase-negative staphylococci (CoNS), *S. epidermidis*, *Streptococcus pneumoniae, Haemophilus influenzae* and *Pseudomonas aeruginosa* colonize the surface of the eyes and eyelids (Chung et al., 2009; Tao et al., 2017). Tao et al. (2017) isolated 1,070 strains from normal conjunctiva and the predominant bacteria were characterised as *S. epidermidis* (39.7%), *S. pneumonia*e (4.5%), *S. aureus* (2.7%,) and *Moraxella catarrhalis* (2.2%). The AMR profiles differed between these ocular bacteria. CoNS resistance to several antibiotics was >50% and included penicillin (90.8%), azithromycin (85.8%), erythromycin (84.4%), oxacillin (76.8%), clindamycin (57.9%) and Trimethoprim-Sulfamethoxazole (63.4%). But CoNS was less resistant to quinolones (4–38.6%), rifampicin (3.5%), minocycline (1.0%), gentamycin (6.9%), and doxycycline (1.2%) (Figure 4.1) (Tao et al., 2017).

In contrast, *S. aureus* exhibited lower resistance to oxacillin (14.8%), co-trimoxazole, (13%), azithromycin (72.7%), erythromycin (72.2%), and quinolone (0–9.3%) and were not resistant to vancomycin, linezolid, rifampicin, or minocycline. *Streptococcus* spp. was sensitive to vancomycin, linezolid, and levofloxacin, but their resistance profile was >80% to erythromycin (98.8%), clindamycin (96.3%), and tetracycline

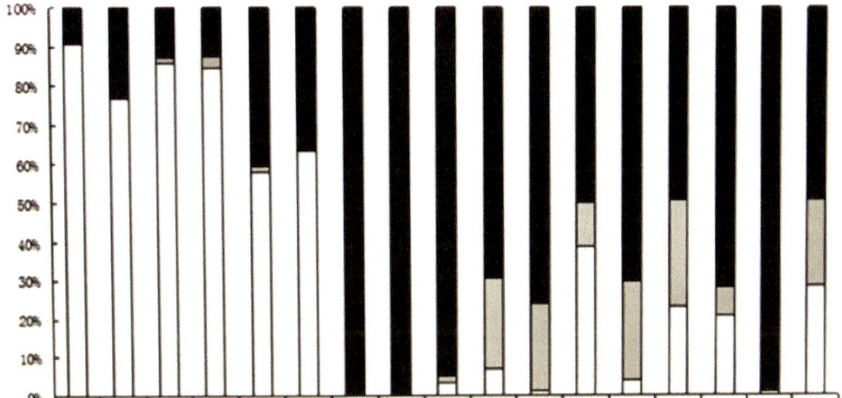

	PEN	OXA	AZM	ERY	CLI	SXT	VAN	LNZ	RIF	GEN	DOX	OFX	MXF	CIP	CHL	MNO	NOR
■ S%	9.2%	23.2%	12.7%	12.5%	40.4%	36.6%	99.8%	100.0%	94.8%	69.2%	76.1%	50.1%	70.1%	49.4%	71.9%	98.5%	49.5%
▨ I%	0.0%	0.0%	1.5%	3.2%	1.7%	0.0%	0.0%	0.0%	1.7%	23.9%	22.7%	11.3%	25.8%	27.5%	7.5%	0.5%	22.2%
□ R%	90.8%	76.8%	85.8%	84.4%	57.9%	63.4%	0.2%	0.0%	3.5%	6.9%	1.2%	38.6%	4.0%	23.1%	20.5%	1.0%	28.4%

FIGURE 4.1 Susceptibility of coagulase-negative *Staphylococcus* strains to different antibiotics namely penicillin (PEN), Oxacillin (OXA), Azithromycin (AZM), Erythromycin (ERY), Clindamycin (CLI), Trimethoprim-Sulfamethoxazole (SXT), Vancomycin (VAN), Linezolide (LNZ), Rifampicin (RIF), Gentamycin (GEN), Doxycycline (DOX), Ofloxacin (OFX), Moxifloxacin (MXF) and Ciprofloxacin (CIP). The percentage of CoNS showing susceptibility (S), Intermediate susceptibility (I) and Resistance (R) are indicated below the figure. (Adopted from Tao et al. (2017) an open-access article distributed under the terms of the Creative Commons Attribution License (CC BY).)

(82.7%) (Tao et al., 2017). In addition, conjunctival bacteria from refractive surgery patients were most susceptible to gemifloxacin, moxifloxacin, and gatifloxacin, whereas resistance to ofloxacin and levofloxacin increased in methicillin-resistant CoNS (Chung et al., 2009).

These conjunctival bacteria were also detected in contact lens storage cases and mobile phones (Waleed et al., 2021). It was suggested that the associated MDR genes in the resistant bacteria can be transferred to other pathogenic bacteria and subsequently to animals thus heralding the emergence of MDR pathogens (Sala et al., 2016). Ocular bacteria also cause ocular infections like bacterial keratitis (Stapleton and Carnt 2012), conjunctivitis (Comstock et al., 2010; Morris et al., 2011; Chen et al., 2018), and endophthalmitis (Schimel at al. 2013; Slean et al., 2017). Thus, the need to undertake antibiotic resistance surveillance in the diseased eye.

4.3 Antibiotic Resistance Surveillance in the Diseased Eye

Several classes of antibacterial drugs, including cephalosporins, β-lactams, fluoroquinolones and so forth, developed for systemic use have been effectively used to overcome ocular infections. But the modalities for use are slightly different compared to the systemic use of the drugs and these probably contribute to the emergence of antibiotic resistance in ocular bacteria. For instance, ocular infections are treated topically using high concentrations (several folds higher than the recommended systemic concentration) of the antibiotic, which is delivered directly to (or near to) the site of infection as eye drops. In addition, antibacterial agents are frequently used to augment treatment for intraocular infections. The use of antibiotics as a surgical prophylaxis several days before and after surgery could be another driving factor for the cause of antibiotic resistance in ophthalmology. It is also likely that increase in resistant bacteria in the eye is linked to the increase in resistant systemic pathogens. It is apparent that either of these above factors contribute to the emergence of AMR to antibiotics among ocular bacteria. Bacterial resistance is best studied when monitored over a period so that the transition from susceptibility to resistance becomes obvious. In a review Kowalski et al. (2003) highlighted the emerging resistance of bacterial pathogens associated with keratitis, conjunctivitis, endophthalmitis, and blepharitis to commonly used antibiotics. Further in a review, Tummanapalli and Willcox (2021), examined the data between January 2005–March 2020 on pathogens associated with ocular infection and concluded as follows: (a) bacterial infections of the eye is global and could result in vision loss, (b) ocular infections are caused by bacteria (70%), (c) bacterial AMR to fluoroquinolones and azoles has increased in the United States and Australia, (d) antimicrobial peptides could be the potential new drugs for ocular surface infections (Tummanapalli and Willcox, 2021). In this chapter AMR in ocular isolates is described with respect to bacterial keratitis.

4.4 Diagnosis of Adult Bacterial Keratitis

The most predominant bacteria that are the causative agents of keratitis are *Pseudomonas aeruginosa* and *Staphylococcus aureus* (Figure 4.2). Symptoms of bacterial keratitis are redness of the eye, pain, impaired vision due to blurring, vision loss, discharge, infiltrates in the cornea, presence of ulcers, photophobia, inflammation of anterior chamber, oedema of lid and conjunctiva and suppurative stromal infiltrates (Al-Mujaini et al., 2009; Egrilmez and Yildirim-Theveny 2020; Lin et al., 2019a; Tabbara and Chacra 2014). Typically, the infiltrates have indistinct edges and accompanied by oedema and infiltration by leukocytes. When *Pseudomonas* infection is the cause of keratitis, patients experience pain, rapid stromal thinning, descemetocele, yellow-green pus, ground-glass appearance, and loss of transparency in the stroma (Tabbara and Chacra 2014). But when the keratitis is caused by staphylococcal infection the ulcer associated with the cornea is usually round and circumscribed by an epithelial defect, with barely any oedema in the adjacent cornea (Tabbara and Chacra 2014). The ulcers caused by *Streptococcus pneumoniae* are serpiginous that infiltrate and span the cornea (Tabbara and Chacra 2014) (Figure 4.3). Between *S. aureus* and *S. epidermidis* the former presents with a greater degree of stromal infiltration (Al-Mujaini et al., 2009). Further *S. aureus* with recurrent bacterial keratitis causes scarring (Kaye et al., 2013).

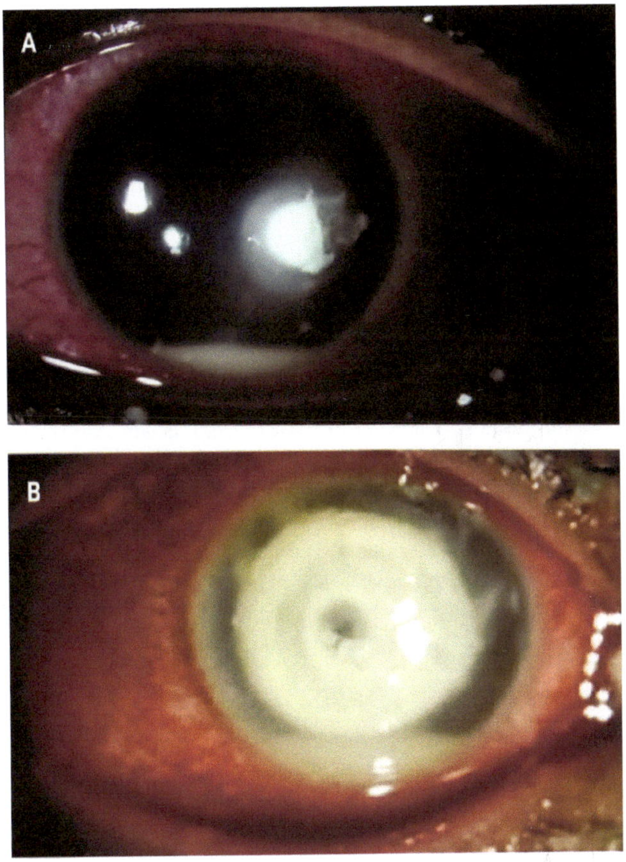

FIGURE 4.2 *Pseudomonas* keratitis showing early (top) and late stages of infection (bottom). (Courtesy Dr Bhupesh Bagga, L V Prasad Eye Institute, Hyderabad, India.)

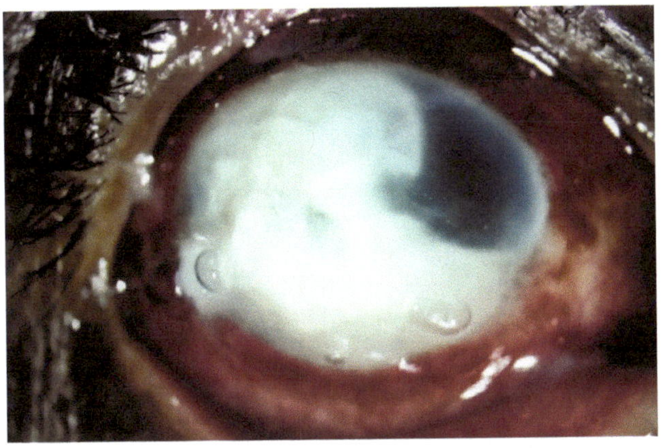

FIGURE 4.3 *Streptococcus* keratitis. (Courtesy Dr Bhupesh Bagga, L V Prasad Eye Institute, Hyderabad, India.)

4.4.1 Identification of the Causative Agent of Adult Bacterial Keratitis

Determination of a causative agent in microbial keratitis is based on the ability to culture and stain the microorganism. This often poses a problem with traditional microbiological culture methods correctly identifying only 4 out of 10 cases of ocular infections (Borroni et al., 2019). But in new approaches such as collecting corneal samples by the polytetrafluoroethylene (PTFE) impression membrane method the isolation rates were better and was also less invasive (Kaye et al., 2016). The most common stains used for the detection of bacteria are Gram-stain and Giemsa (Ung et al., 2019), whereas the media that is commonly utilized for culturing the bacteria include blood agar (sheep and horse) and chocolate agar. Both these methods have limitations and, over the years, it has been recognised that in clinically diagnosed cases of microbial keratitis (MK) the culture positivity rate is 50.3% and staining methods alone achieve diagnosis in only 27.3–61.6% of cases (Cariello et al., 2011; Leck et al., 2002). *In vivo* confocal microscopy (IVCM) has also been explored for definitive diagnosis (Hau et al., 2010) of *Acanthamoeba* cysts and hyphae within the cornea but was not suitable to visualize bacteria (Alzubaidi et al., 2016). The low efficiency in culture results could be due to prior antibiotic exposure, difficulty in cultivating organisms and the fastidious nature of *Streptococcus* spp. and *Propionibacterium* spp. (Ung et al., 2019). Molecular methods like PCR have been used to improve the efficiency of identification. Kim and co-workers (2016) using corneal ulcer specimens demonstrated that PCR was positive for 76% of bacterial cases and corneas (Kim et al., 2016). But next generation sequencing (NGS) determined the etiology in 16 infected corneas (Kim et al. 2016) and a putative culprit organism was identified. NGS is expensive and time consuming. According to Ung et al. (2019), challenges in the diagnosis of MK are yet to be overcome completely.

4.5 Prevalence, Incidence, and Risk factors of Bacterial Keratitis in Adults

4.5.1 Prevalence/Incidence of Bacterial Keratitis in Adults

In Southern Asia (mainly India) the percentage of bacterial keratitis (BK) has hovered around 50% (35.4 -51.9%), whereas in East Asia (China) it was <50% but in one study in Taiwan it was 81.1% (Table 4.1) (Ung et al., 2019). These findings on BK partially confirm the earlier findings of the Asia Cornea Society of Infectious Keratitis Study from 13 tertiary centres in India (38.3%), China (15.3%), Singapore (41.3%), South Korea (42.8%), Philippines (53.2%), Japan (47.8%), Thailand (50.5%) and Taiwan (61.8%) and observed that the per cent of BK ranged from a minimum of 15.3% in China to a maximum of 61.8% in Taiwan (Khor et al., 2018). In a recent study Ting et al. (2021a) reported that BK represents the most common type of infectious keratitis and was highest in Australia and Oceania (New Zealand and Australia) (>90–100%) (Cabrera-Aguas et al., 2019; Green et al., 2019a; b), in the UK (around 90%, 83–93%) (Ting et al., 2020 a;b; Tan et al., 2017), North America (88–92%) (Tam et al., 2017), South America (79–88%) (Cariello et al., 2011; Hernandez-Camarena et al., 2015) and Middle East (69.8–91.8%) (Politis et al., 2016). Studies in the UK also documented increase in *Moraxella* keratitis, often associated with longer corneal healing time (Ting et al., 2020a; 2020b; Tan et al., 2017). In England, increase in the incidence of

TABLE 4.1

Incidence of Bacterial Keratitis across the Globe

Location	Period	Bacterial keratitis (%)	Reference
Southern Asia, Tamil Nadu, India	2002–2013	35.4	Lalitha et al., 2016.
Southern Asia, Hyderabad, Pakistan	1991–2001	51.9	Gopinathan et al., 2009.
Southern Asia, Tamil Nadu, India	1999–2002	46.4	Bharathi et al., 2007.
East Asia, China	2009–2013	41.9	Lin et al., 2017.
East Asia, Taiwan	2003–2012	81.1	Hsiao et al., 2016.
East Asia, China	1999– 2004	16.2	Xie et al., 2006.
Europe, UK	2007–2019	83–93	Ting et al., 2020a; 2020b.
Europe, UK	2004–2015	90.6	Tan et al., 2017.
Europe, UK	1997–2003	85.4	Ibrahim et al., 2009.
North America, Mexico	2002–2011	88.3	Hernandez-Camarena et al., 2015.
North America, Toronto, Canada	2000–2010	91.8	Lichtinger et al., 2012.
North America, Miami, USA	1990–1998	91.1	Alexandrakis et al., 2000.
South America, Brazil	1975–2007	78.9	Cariello et al., 2011.
South America, Paraguay	1998–2001	51.0	Laspina et al., 2004.
Middle East and Africa, Israel	2002–2014	91.8	Politis et al., 2016.
Middle East and Africa, Iraq	2002–2005	69.8	Al-Shakarchi et al., 2015.
Middle East and Africa, Turkey	1990–2005	77.8	Yilmaz et al., 2006.
Australia and Oceania, New Zealand	2003–2007	98.3	Pandita and Murphy 2011.
Australia	2001–2003	90.2	Keay et al., 2006.
Australia	1991–1999	100	Mallari et al., 2001.

Source: Modified from Ung et al., 2019.

Gram-positive keratitis and a decrease in Gram-negative keratitis was observed over a period of ten years (Ting et al., 2020a; 2020b). In contrast, in Manchester Royal Eye Hospital a decreasing trend of Gram-positives was observed (Tan et al., 2017). But Tavassoli et al. (2019) evaluated BK patients for 11 years in the South-West of England and did not find a statistically significant trend in any of the microorganisms cultured. In San Francisco *S. aureus* BK was the most common (Peng et al., 2018) and MRSA incidence increased over 19 years. A four-year Pennsylvania study showed similar results (Siddiqui et al., 2015). Two studies from Sydney, Australia, observed that as in the UK studies, CoNS were the most discovered and MRSA occurrence was low (Watson et al., 2019; Cabrera-Aguas et al., 2019).

4.5.2 Risk Factors of Bacterial Keratitis in Adults

4.5.2.1 Exogenous Factors Associated with Bacterial Keratitis: Contact Lens Wear

The risk factors of BK have been categorised into 3 distinct categories namely exogenous local factors, endogenous local factors, and systemic risk factors (Egrilmez and Yildirim-Theveny 2020). The exogenous factors include use of contact lenses, topical anti-inflammatory drugs, topical antibiotics, glaucoma medications, post ocular trauma and post ocular surgery. Among the exogenous factors contact lens (CL) wear is a predisposing factor (29–64%) of MK in Portugal (Ferreira et al., 2018), France (Dethorey et al., 2013), the United States (Jeng et al., 2010), Singapore (Tong et al., 2019) and Australia (Khoo et al., 2020), compared to developing countries (0–18%) due to a lesser number of contact lens wearers (Khor et al., 2018). Patients are between 25 to 40 years old (Jeng et al., 2010; Tong et al., 2019; Stapleton 2020) and exhibited a female predominance of 57–69% (Jeng et al., 2010; Tong et al., 2019). Several other predisposing factors related to contact lens keratitis included the types of CL used (soft CL is associated with higher than rigid gas permeable CL), overnight wear, use of expired CL, types of CL solution used and CL prescribed/dispensed by non-ophthalmologists or non-opticians (Sauer et al., 2020; Carnt et al., 2018a; Stapleton et al., 2017). Tong et al. (2019) observed that *P. aeruginosa* was responsible for 63%–70% of the CL-related Mk.

4.5.2.2 Exogenous Factors Associated with Bacterial Keratitis: Ocular Trauma

BK due to trauma is predominant in individuals 30–55 years of age and occurs in farm-related workers (54–70%) and manual workers (11–17%) exposed to ocular trauma (Khor et al., 2018; Chidambaram et al., 2018; Mandour et al., 2016; Kumar et al., 2019) and is more common in males (74–78%) (Jeng et al., 2010; Tong et al., 2019) and correlated with a high male prevalence in countries such as South America (Cariello et al., 2011), Asia and Africa (Khor et al., 2018; Chidambaram et al., 2018). BK in general is low in the extremes of age (Jeng et al., 2010; Ferreira et al., 2018; Mandour et al., 2016). Trauma-related keratitis in European countries were caused by CoNS, *S. aureus*, *Streptococcus* spp., and *Corynebacterium* spp. (Ferreira et al., 2018). Ting et al. (2021a; 2021b) suggested that trauma of the cornea from nonvegetative matter followed by secondary opportunistic infection by ocular surface commensals could be responsible for the observed high rate of Gram-positive infection.

4.5.2.3 Exogenous Factors Associated with Bacterial keratitis: use of Ocular Drugs

Steroids are commonly used to manage intraocular and ocular surface inflammatory diseases like BK (Jones et al., 2017). In addition, topical steroids are used following surgery of ocular surface including transplantation of cornea (Di Zazzo et al., 2017). But topical steroids are also known to increase the risk of infectious keratitis (IK), particularly fungal keratitis (13%) and/or polymicrobial keratitis (Khoo et al., 2020; Ting et al., 2019). In India, 41% of *Acanthamoeba* keratitis (AK) cases are associated with use of topical steroids (Chidambaram et al., 2018; Carnt et al., 2018 a; b). BK was also evident following the use of nonsteroidal anti-inflammatory drugs (Guidera et al., 2001), glaucoma drugs (Chang et al., 2021) and other drugs (Hollander et al., 2004). Factors like traditional medicine and application of breast milk or vegetative products on the eye may influence the incidence of BK (Gautam et al., 2018).

4.5.2.4 Exogenous Factors Associated with Bacterial Keratitis: Post Ocular Surgery

Several ocular surgeries such as transplantation of cornea, cross-linking of cornea (CXL), pterygium, cataract surgeries and other ocular procedures (Cariello et al., 2011; Mandour et al., 2016; Dohse et al., 2020) are risk factors of BK (Mun et al., 2019; Das et al., 2020). The incidence of post operative keratitis varied with the ocular surgical procedure. Following post-keratoplasty the incidence of BK was 4% (Dohse et al., 2020) and in corneal grafting it was 22%. Keratitis rarely develops after refractive surgery (Randleman and Shah 2012) but in Brazil refractive surgery was the second commonest ocular surgery associated with BK (Cariello et al., 2011). Gram-positive bacteria and nontuberculous mycobacteria (NTM) cause post-refractive surgery keratitis. In addition, fungal and Acanthamoeba may also cause keratitis (Randleman and Shah 2012). BK by Gram-positives occurs following ocular surgeries like cataract and pterygium surgery most likely due to infections secondary to ocular surface commensals (Mandour et al., 2016; Dethorey et al., 2013).

4.5.2.5 Exogenous Factors Associated with Bacterial Keratitis: Corneal Crosslinking

Corneal collagen cross-linking (CXL) is used to cross-link the cornea following exposure of the corneal stroma to ultraviolet light after instillation of the cross-linking agent riboflavin (vitamin B2) to stabilize the melting of corneas. This procedure could successfully treat infectious ulcerative keratitis. Out of 524 patients who underwent CXL, only 7 cases developed postoperative infectious keratitis (Kato et al., 2021).

4.5.2.6 Endogenous Factors Associated with Bacterial Keratitis

The endogenous factors that cause BK include several eyelid disorders like lagophthalmos (Alhoutan and Alarfaj 2021), ectropion (Astori et al., 1998), bullous keratopathy (Luchs et al., 1997), cytomegalovirus (CMV) blepharitis (Long et al., 2022), dacryocystitis (Aasuri et al., 1999), vernal keratoconjunctivitis (Burrow et al., 2022), ocular pemphigoid (Khoo et al., 2019), neurotrophic keratopathy (NaPier et al., 2022), keratoplasty (Sony et al., 2002), corneal-crosslinking (Khoo et al., 2021), and keratorefractive surgery (Afsharpaiman et al., 2020).

4.5.2.6.1 Endogenous Factors Associated with Bacterial Keratitis: Lagophthalmos

Lagophthalmos is a rare ocular disease. In this disease the patient is unable to close the eyelids completely thus leading to corneal exposure and evaporation of the tear film. A patient with lagophthalmos and with keratitis may not be associated with the conventional risk factors like medication use, trauma, surgery, critical illness, or other ocular problems associated with it. Patients with bilateral lagophthalmos are treated with lubricants (Alhoutan and Alarfaj 2021).

4.5.2.6.2 Endogenous Factors Associated with Bacterial Keratitis: Cytomegalovirus

Cytomegalovirus (CMV) is known to cause blepharitis and keratitis. In a rare case of CMV an 84-year-old male (Long et al., 2022) patient presented with inflammation of the mucocutaneous junction of the eyelid. CMV presents as retinitis and rarely as endothelitis and keratitis.

4.5.2.6.3 Endogenous Factors Associated with Bacterial Keratitis: Ectropin

In ectropion the eyelid turns outward and this occurs in patients in the early postburn period and surgical intervention is needed. All the 5 patients of ectropin were noted to have exposure to keratitis (Astori et al., 1998).

4.5.2.6.4 Endogenous Factors Associated with Bacterial keratitis: Bullous Keratopathy

In bullous keratopathy the patients present with a blister-like swelling of the cornea and symptoms include photophobia, blurred vision, and feeling of a foreign body in the eye. A large series of 918 patients with bullous keratopathy were screened and ulcerative keratitis was reported in 4.7% of the patients (Luchs et al., 1997).

4.5.2.6.5 Endogenous Factors Associated with Bacterial Keratitis: Vernal Keratoconjunctivitis

Vernal keratoconjunctivitis (VKC) refers to an inflammatory process that involves both the conjunctiva and the superficial cornea and is a common bilateral atopic disease (Horner Tranta dots) having papillary, limbal, and mixed forms (Burrow et al., 2022). VKC is common in tropical and sub-tropical countries, which are warm and dry. In the United States the prevalence was around 0.29/10000 population, whereas in Europe it varied from 1.2 to 12.6 cases per 10000 population. In Africa it was more common varying from 2 to 37% (Alemayehu et al., 2019) and was predominant in males (Alemayehu et al., 2019; Nebbioso et al., 2015).

4.5.2.6.6 Endogenous Factors Associated with Bacterial Keratitis: Dacryocystitis

Dacryocystitis is an inflammation of the nasolacrimal sac caused by stagnation and accumulation of tears in the sac due to an obstruction in the nasolacrimal duct. Dacryocystitis occurs in 1 in 3884 live births, also in adults and predominantly in females (75%) (Chen et al., 2018). A retrospective analysis of 383 culture-proven bacterial keratitis patients between 1991 and 1995 revealed twelve patients with chronic dacryocystitis (Aasuri et al., 1999).

4.5.2.6.7 Endogenous Factors Associated with Bacterial Keratitis: Keratopathy

In neurotrophic keratitis (NK) the trigeminal corneal innervation is damaged, leading to breakdown of the epithelium and formation of corneal ulcers. NK is also quite common and herpetic keratitis, chemical burns, surgery of the cornea, diabetes, multiple sclerosis, and neurosurgical procedures can cause neurotrophic keratitis (NaPier et al., 2022). Keratopathy in which the cornea is damaged could occur due to prolonged exposure of the eye to the outside environment and thus could lead to ulceration, microbial keratitis, and permanent vision loss from scarring (Sony et al., 2002).

4.5.2.6.8 Endogenous Factors Associated with Bacterial Keratitis: Ocular Cicatricial Pemphigoid

Khoo et al. (2019) studied 171 patients and observed ocular cicatricial pemphigoid (OCP) in 2% of the patients. This is an autoimmune disease causing severe dry eye, scarring of the conjunctiva with inferior fornix shortening, entropion with trichiasis, corneal keratinization and ulcers, which may lead to permanent vision loss.

4.5.2.6.9 Endogenous Factors Associated with Bacterial Keratitis: Kerato-refractive Surgery

Kerato-refractive surgery is a surgical procedure in which the corneal epithelium of the patient is substituted with a donor's corneal disc (or lenticule) which is sutured to the Bowman's membrane. Keratitis prevalence after keratorefractive surgery is quite common and has been reported from the United States, Europe, and Asia (Afsharpaiman et al., 2020).

4.5.2.6.10 Endogenous Factors Associated with Bacterial Keratitis: Systemic Factors

Several systemic diseases are also known to cause BK-like diabetes mellitus (Lockwood et al., 2006), immunocompromised state (Hassan et al., 2010), alcoholism.

(Ormerod et al., 1988), connective tissue diseases (Young 2005), Stevens-Johnson syndrome (Tsai et al., 2022) and pemphigus (Takeshita 1996). The association between diabetes and infectious keratitis is significantly high (around 8–16%), (Chidambaram et al., 2018; Wang et al., 2018). Hyperglycaemia facilitates microbial growth especially *Pseudomonas* spp. and *Acinetobacter* spp. (Li et al., 2019), on the ocular surface and affect the homeostasis, corneal sensation, and wound healing, thereby increasing the risk of IK (Zhu et al., 2019) and could also lead to neuropathic keratopathy (Al-Aqaba et al., 2019). Infectious keratitis is also tenfold greater in individuals with HIV (238.1 vs. 27.6 per 100,000 population year) (Jeng et al., 2010). Renal disease common in diabetics is a recognised risk factor for infectious keratitis (Jan et al., 2018). Khoo et al. (2019) studied 171 patients with microbial keratitis and observed that blepharitis (79%), dry eye (25%) and Stevens–Johnson syndrome (4%) are common.

4.5.2.7 Socioeconomic Status Associated with Bacterial Keratitis

Several socioeconomic factors seem to influence BK, such as low socioeconomic status, poor ocular protection, low hygiene, and limited access to eye care (Khor et al., 2018; Chidambaram et al., 2018; Mandour et al., 2016; Gautam et al., 2018). For instance, three factors that appear to be associated with BK patients in Asia and Africa are illiteracy (~45–71%), living in rural areas (62–79%) and poor access to healthcare facilities (Mandour et al., 2016; Gautam et al., 2018; Kumar et al., 2011). In addition, workers involved in agricultural activities, individuals from rural areas, and illiterates had a higher risk of refractory keratitis (Mandour et al., 2016).

Microbial keratitis in general, has a negative impact on the quality of life (QOL) (Arunga et al., 2019) and is a huge economic burden on society. In 2010, the burden of treatment of MK was estimated as $175 million (Collier et al., 2014). Similar estimates for the rest of the world (India, Thailand, and China) (Khor et al., 2018) is not available, but complications of MK such as corneal perforations and scarring would be an additional burden.

4.6 Etiology of Bacterial Keratitis

The etiology of keratitis varies depending on the region and type of MK. BK is preponderant in Europe (91–93%) (Tan et al., 2017; Ting et al., 2020 a; b; Tavassoli et al., 2019; Khoo et al., 2020), North America (86–92%) (Alexandrakis et al., 2000; Hernandez-Camarena et al., 2015; Peng et al., 2018; Hsu et al., 2019), South America (79–88%) (Cariello et al., 2011; Hernandez-Camarena et al., 2015), Middle East (91.8%), Australia, and Oceania, (Mallari et al., 2001; Pandita and Murphy 2011; Watson et al., 2019; Cabrera-Aguas et al., 2019) (ranging from 85.4 to 98.3%). In these countries a higher proportion of Gram-positive bacteria were isolated (median 72.2%) compared to Gram-negative isolates (median 27.0%) in the MK patients (Table 4.2). Studies in the United States, China, Switzerland, South Korea, Spain, and Colombia also confirmed that Gram-positive organisms were in abundance (Mun et al., 2019). In contrast it was observed that between 2007–2016 in Taiwan Gram-negative bacteria were the most abundant (Liu et al., 2019).

Among the Gram-positive isolates, CoNS was the most commonly isolated organism (24–46%) (Ung et al., 2019; Kaye et al., 2013; Peng et al., 2018; Asbell et al., 2020c; Cariello et al., 2011; Hernandez-Camarena et al., 2015; Lin et al., 2017; 2019a; 2019b; Lalitha et al., 2015; Tavassoli et al., 2019; Zhang et al., 2017; Cabrera-Aguas et al., 2019; Laspina et al., 2004; Gopinathan et al., 2009; Politis et al., 2016; Tan et al., 2017; Hsu et al., 2019; Watson et al., 2019); the other Gram-positive isolates included *S. aureus* (Alexandrakis et al., 2000; Hernandez-Camarena et al., 2015; Laspina et al., 2004; Pandita and Murphy 2011; Tan et al., 2017; Watson et al., 2019; Ung et al., 2019; Cabrera-Aguas et al., 2019) followed by *S. epidermidis, Streptococcus* spp., *S. pneumoniae* (Bharathi et al., 2007; Gopinathan et al., 2009; Lichtinger et al., 2012; Pandita and Murphy 2011: Lalitha et al., 2016; Tan et al., 2017; Ung et al., 2019), *Corynebacterium* spp. (Watson et al., 2019; Cabrera-Aguas et al., 2019) and *Propionibacterium spp.* (Ung et al., 2019) (Fig. 4.2). The Gram-negative isolates included *Pseudomonas* spp., *P. aeruginosa* (Ung et al., 2019; Alexandrakis et al., 2000; Bharathi et al., 2007; Hernandez-Camarena et al., 2015; Ibrahim et al., 2009; Lalitha et al., 2016; Lin et al., 2017; Politis et al., 2016; Peng et al., 2018; Hsu et al., 2019; Watson et al., 2019; Cabrera-Aguas et al., 2019), *Escherichia spp.,* Klebsiella *spp., Serratia spp.* (Lalitha et al., 2016; Lichtinger et al., 2012; Konduri et.al. 2017), *Achromobacter xylosoxidans* and *Stenotrophomonas maltophilia* (Newman et al., 1984: Spierer et al., 2018). Actinomycetes included *Nocardia* spp. (Lalitha et al., 2016; Srinivasan et al., 2012a; 2012b) (Table 4.2). In the UK *Moraxella* keratitis (Ting et al., 2020a; Tan et al., 2017), *Nocardia* keratitis, (Srinivasan et al., 2012a; 2012b; Lalitha

TABLE 4.2

Bacteria and Protozoa Associated with Microbial Keratitis in Adult Humans

Microorganism	Region	Reference
Gram-positive bacteria: Coagulase negative staphylococci (CoNS), *Staphylococcus* spp., *S. epidermidis, S. aureus, Streptococcus* spp., *S. pneumoniae, Bacillus* spp. (24–46%), *Corynebacterium* spp., *Propionibacterium spp.*, *Nocardia* spp., non-tuberculous mycobacteria.	India	Lin et al., 2017; 2019 a; Lalitha et al., 2015;2016; Ung et al., 2019; Pandita and Murphy 2011; Bharathi et al., 2007; Gopinathan et al., 2009; Srinivasan et al., 2012a; 2012b.
	China	Zhang et al., 2017.
	Japan	Yamaguchi et al., 2011.
	UK	Tan et al., 2017; Tavassoli et al., 2019.
	Israel	Politis et al., 2016.
	Australia	Watson et al., 2019; Cabrera-Aguas et al., 2019.
	USA (Missouri, San Francisco, California	Hsu et al., 2019; Peng et al., 2018; Asbell et al., 2020c; Chandra et al., 2001.
	South America (South Florida, Brazil, Mexico)	Alexandrakis et al., 2000; Cariello et al., 2011; Hernandez-Camarena et al., 2015; Ung et al., 2019; Freitas et al., 2003.
	Canada	Lichtinger et al., 2012.
	Switzerland, South Korea, Spain, Colombia	Mun et al., 2019.
	Paraguay	Laspina et al., 2004.
Gram-negative bacteria: *Pseudomonas* spp., *P. aeruginosa*, *Escherichia spp., Klebsiella spp., Serratia spp.* *Achromobacter xylosoxidans, Stenotrophomonas maltophilia, Moraxella* spp.	India	Bharathi et al., 2007; Lalitha et al., 2016; Lin et al., 2017; Konduri et al., 2017; 2019.
	Further, India, Taiwan, and Malaysia	Das et al., 2019; Khor et al., 2020.
	Israel	Politis et al., 2016.

(continued)

TABLE 4.2 (Continued)
Bacteria and Protozoa Associated with Microbial Keratitis in Adult Humans

Microorganism	Region	Reference
	USA (San Francisco, California, Missouri).	Peng et al., 2018; Hsu et al., 2019.
	South America	Alexandrakis et al., 2000; Hernandez-Camarena et al., 2015.
	Canada	Lichtinger et al., 2012.
	UK	Ibrahim et al., 2009; Tan et al., 2017.
	Australia	Watson et al., 2019; Cabrera-Aguas et al., 2019.
Protozoa *Acanthamoeba* spp., *A. castellanii, A. polyphaga*	USA (Portland, San Francisco, California, Boston, Philadelphia	Panjwani et al., 2010; Shoff et al., 2008; Meier et al., 1998; Joslin et al., 2006; Mathers 2006; Sansanayudh et al., 2008; Thebpatiphat et al., 2007; Dart et al., 2008; 2009; Rayamajhee et al., 2021.
	Brazil	Cariello et al., 2011.
	UK	Carnt et al., 2018a; 2018b.
	India	Garg et al., 2017.

Source: Compiled by the author.

et al., 2012a; 2012b; Chu and Hu 2013) and nontuberculous mycobacteria (NTM) were the causative organisms of MK (Chu and Hu 2013). But these microorganisms are incapable of causing corneal infection when the corneal epithelium is intact unlike *Neisseria* spp., *Corynebacterium* spp., *Haemophilus* spp., and *Listeria* spp. Infections usually occur following trauma or conditions that disrupt the corneal epithelium (Tabbara and Chacra 2014).

4.7 Treatment of Bacterial Keratitis

4.7.1 Treatment of Bacterial Keratitis with Antibiotics

Several excellent reviews have dealt with the treatment of BK (Whitcher et al., 2001; Austin et al., 2017). Prior to specific medication, BK should be managed by various strategies that would contribute towards alleviating the disease by overcoming the predisposing risk factors and restoring the integrity of the corneal epithelium by protective and precautionary measures such as assuring proper contact lens use, discontinuing the use of deleterious drugs, eliminating explants or sutures, correcting ocular disorders (dacryocystitis, meibomian gland disease, and canaliculitis), overcoming dry eye, treating adjacent tissue infections, repairing thinned or perforated corneas, and managing systemic diseases (Egrilmez and Yildirim-Theveny 2020). It is recommended that treatment of keratitis should be initiated with a broad-spectrum drug guided by Gram-staining results (Hilliam et al., 2020) prior to microscopic inspection, staining and antibiogram. Antibiotic therapy is usually topical monotherapy (Egrilmez and Yildirim-Theveny 2020) and suggested even in extreme cases of keratitis involving the sclera along with intraocular infection as in gonococcal keratitis caused by *Neisseria gonorrhoeae* (Tabbara and Chacra 2014; Austin et al., 2017). In patients with central keratitis foci, and in severe cases, antibiotics are initially administered at 5–15 min intervals and subsequently administered at hourly intervals (Tabbara and Chacra 2014; Lin et al., 2019a; 2019b). Cycloplegic agents that provide comfort and prevent the formation of synechiae are also administered. Table 4.3 represents the specific antibiotic and the concentration recommended for BK by the American Academy of Ophthalmology (AAO) in 2018 (Lin et al., 2019a).

4.7.1.1 Treatment with Dual Antibiotic therapy

(a) Fluoroquinolones and Aminoglycosides

Fluoroquinolones namely ciprofloxacin, moxifloxacin, levofloxacin and ofloxacin are used as monotherapy for bacterial keratitis (ciprofloxacin 0.3%, ofloxacin 0.3%, and levofloxacin 1.5%) since this is as effective as combination therapy with fortified drops (Lin et al., 2019a) and no difference in treatment efficacy was observed between fortified cefazolin/tobramycin, moxifloxacin, and ofloxacin groups (Constantinou et al., 2007). Ciprofloxacin as an ointment or solution for the treatment of MRSA and *P. aeruginosa* is effective (O'Brien et al., 1988), but strains resistance to monotherapy using fluoroquinolones is a major concern (O'Brien et al., 1988; Hyndiuk et al., 1996). Srinivasan et al. (2012a; 2012b) also confirmed that, in the Steroids for Corneal Ulcers Trial (SCUT) study, patients on topical moxifloxacin therapy when

TABLE 4.3

Antibiotic Therapy Recommended by the AAO in 2018 for Bacteria Keratitis

Bacterium	Antibiotic (Topical concentration)
Gram-positive cocci	Cefazoline 50 mg/mL, vancomycin[a] 10–50 mg/mL, bacitracin[a] 10,000 IU, fluoroquinolones commercially available doses
Gram-negative rods	Tobramycin or gentamicin 9–14 mg/m, ceftazidime 50 mg/mL, fluoroquinolones[a] commercially available doses
Gram-negative cocci	Ceftriaxone 50 mg/mL, ceftazidime 50 mg/m, fluoroquinolones[a] commercially available doses
Gram-positive rods (Nontuberculous Mycobacteria)	Amikacin 20–40 mg/mL, clarithromycin 10 mg/mL, azithromycin 10 mg/mL, fluoroquinolones[a] commercially available doses
Gram-positive rods (Nocardia)	Sulfacetamide 100 mg/mL, amikacin 20/40 mg/mL, trimethoprim/sulfamethoxazole: trimethoprim 16 mg/mL, sulfamethoxazole 80 mg/mL
Undetermined or multiple microorganisms	Cefazoline or vancomycin 25–50 mg/mL, Tobramycin or Gentamicin 9–14 mg/mL, fluoroquinolones[a] commercially available doses.

Source: Modified from Egrilmez and Yildirim-Theveny, 2020.

Note: [a] Fluoroquinolones ciprofloxacin 0.3%, ofloxacin 0.3%, and levofloxacin 1.5% have been approved by the US Food and Drug department (Lin et al., 2019a).

administered adjunctive topical corticosteroid showed no improvement in 3-month vision in patients with bacterial corneal ulcers (Srinivasan et al., 2012a; 2012b). Besifloxacin also retained consistently low minimum inhibitory concentrations (Asbell et al., 2020a; 2020b; 2020c; Acar et al., 1993). Grandi et al. (2021), in a subsequent monocentric retrospective study between 1998–2017 in Turin, Italy, analysed 15,517 isolates of ocular infections, which included *Staphylococcus* spp., *Pseudomonas* spp., CoNS and *S. aureus* and reported that fluoroquinolones and chloramphenicol were the most effective towards bacterial ocular infections, followed by tetracycline, ampicillin, and aminoglycosides. In the UK, the Gram-positive were susceptible to cephalosporin (87–100%) whereas the Gram-negative bacteria were highly susceptible to aminoglycosides (97–100%) and fluoroquinolones (91–100%) (Ting et al., 2020a; Tan et al., 2017; Tavassoli et al., 2019), suggesting that fluoroquinolone monotherapy or cephalosporin-aminoglycoside dual therapy could safely be continued as the first line of treatment. In a recent 12-year Nottingham study, it was observed that susceptibility to aminoglycosides and fluoroquinolones was maintained and this antibiotic regimen could be continued (Ting et al., 2020a). Thus, the choice of a therapeutic agent in bacterial keratitis should depend on the resistance rate and additionally the penetration of the drug. For instance, the antibiotic resistance trend increased for both aminoglycosides and fluoroquinolones among Gram-positive pathogens, especially for *S. aureus* (Grandi et al., 2021). With respect to penetration rate, among the fluoroquinolones Lomefloxacin has the best penetration of corneal tissues and followed the order lomefloxacin 29–55 µg/g >norfloxacin 1.43–10.58> (Elena et al., 1997; McDermott et al., 1993), ciprofloxacin 8.82 (8.24) µg/g (Price et al., 1995). The concentration in the aqueous humour after topical instillation was 0.072 µg/ml for ciprofloxacin, 0.057 µg/ml for norfloxacin, 0.338 µg/ml for ofloxacin, and 2.2–7.6 µg/ml for lomefloxacin (McDermott et al., 1993).

(b) Treatment with Cefazolin and Gentamicin

The gold standard for the therapy of bacterial keratitis is topical treatment with fortified antibiotics (cefazolin and gentamicin). These topical antibiotics are applied frequently to achieve high concentration of antibiotics on the infected cornea to reduce the risk of AMR in ocular infections (Ting et al., 2021a; 2021b). An intrinsic problem with this combined therapy is that gentamicin concentration decreases on the ocular surface rapidly and persists only for 40 minutes on the conjunctiva when used as an ointment and only 10 minutes when used as drops (Lomholt et al., 2000). Furthermore, gentamicin has poor corneal penetration; and within 30–60 minutes the anterior chamber concentration is below limits of detection (<0.6 mg/l) and thus not very suitable for surgery, which needs to be completed within this period (Lomholt et al., 2000). Further fortified antibiotics are associated with local toxicity to the corneal epithelium (Cutarelli et al., 1991) and cause ocular irritation or intense conjunctival reactions.

(c) Treatment with Methicillin and Oxacillin

Methicillin and oxacillin are penicillin antibiotics that have also been used to treat eye infections. But S. aureus and CoNS, the main pathogens involved in eye infections (conjunctivitis and keratitis) in Naples, Italy, between 2017 and 2020 (Petrillo et al., 2021) showed a high resistance to methicillin (20%). This confirmed an earlier 10-year Mexico study that showed that a good proportion of ocular *S. aureus* (21–79%) and CoNS (48–71%) were resistant to oxacillin (or methicillin). *P. aeruginosa* and other Gram-negatives also exhibited high resistance against oxacillin (86% and 90%) and vancomycin (97% and 70%), with increase in resistance to ceftazidime observed (Hernandez-Camarena et al., 2015). In Taiwan, also, emergence of methicillin resistance was observed (Hsiao et al., 2016).

(d) Treatment with Other Dual Combination of Antibiotics

Several other studies have also indicated effective management of ocular microbial infections like BK using dual antibiotic therapy rather than monotherapy. For *S. aureus* the effective combinations include meropenem/ciprofloxacin (Kowalski et al., 2013), cefazolin/tobramycin (Elsahn et al., 2010), polymyxin B- trimethoprim/rifampin (Sueke et al., 2010) and balofloxacin/tigecycline (Jiang et al., 2014), which are potential alternatives to vancomycin. Polymyxin B–trimethoprim (PT) and rifampicin for the treatment of *S. aureus* and *Pseudomonas* corneal infections was more effective and exhibited a lower tendency to develop resistance (Chojnacki et al., 2018). Thymosin beta-4 (Tβ4), a naturally occurring protein, in combination with ciprofloxacin could fight off keratitis caused by *P. aeruginosa* and had minimal impact on host immune response (Carion et al., 2018). Thus, all these studies indicate that combination therapies are the preferred treatment for corneal infections.

4.7.2 Treatment of Bacterial Keratitis with Peptides

Antimicrobial peptides (AMPs) exhibit broad-spectrum killing of bacteria and cytotoxicity to host is low. Cornea expresses several AMPs like human β-defensins (hBD) (Mohammed et al., 2010) and α-defensins in the ocular surface secreted by neutrophils and lacrimal ductular epithelium (Kumar et al. 2006). In addition to hBD which is expressed by cornea several peptides with antimicrobial activity expressed by the liver like antimicrobial peptide-1 and -2 (LEAP-1 or 2) and LL-37/cathelicidin are also found in ocular surface epithelia (Paulsen et al. 2001). LL-37 expressed by corneal and conjunctival epithelial cells (Gordon et al. 2005) in combination with β-defensins protects the ocular surface from pathogens (Augustin et al. 2011). The S-100 proteins or calgranulins have also been reported in cornea, conjunctiva, nasolacrimal ducts, and lacrimal glands (Wilkinson et al. 2016). In fact, the ocular fluid is also associated with several other peptides like human neutrophil peptides (hNP-1 and -2), hBD-1, hBD-2, hBD- 3, lactoferrin, lysozyme, calprotectin, lipocalin and transferrin (Silva et al. 2013). Lysozyme, in tears could also hydrolyses bacterial cell wall (Mcdermott 2013). Studies have also indicated differential expression of AMPs in patients with bacterial keratitis (Sharma et al. 2018; Sharma et al. 2019); for instance, LL-37 was increased in patients with *S. pneumoniae* keratitis (Sharma et al. 2019), S100A8, S100A9, and hBD-1 was increased *in vitro* and in patients with *Corynebacterium pseudodiphtheriticum* infection (Roy et al. 2015) and decreased expression of hBD-9 was observed as in bacterial keratitis patients (Mohammed et al. 2010). Expression of hBD-3 and RNAse7 were also significantly increased in corneal epithelial cells in response to *A. castellanii* (Otri et al. 2010) and expression of β-defensins was decreased in patients with *A. flavus* corneal infections (Mallela et al. 2021). In contrast to AMPs, synthetic peptides have been tested against various bacterial pathogens of the eye. The peptides EC1-17KV, FK-16 and PA-13 have been screened against MDR *P. aeruginosa*. Pam-3, a palmitoleic acid–modified octapeptide fragment from hBD-1, was effective against ESKAPE pathogens (which includes six nosocomial pathogens, *Enterococcus faecium, S. aureus, K. pneumoniae, A. baumannii, P. aeruginosa,* and *Enterobacter* spp.), SAAP-148, was effective against ESKAPE pathogens as well as *E. cloacae, E. coli,* and *K. pneumoniae* (Mohammed et al. 2019). A hybrid peptide of cecropin A and melittin (a bee venom peptide) was also found to be effective against *P. aeruginosa* in a rabbit model of keratitis (Nos-Barbera et al. 1997). In 2021, Tummanapalli and Willcox, reported that the AMPs are broad-spectrum alternatives to antibiotics and show promise in treating keratitis.

4.7.3 Other Methods used for Treatment of Bacterial Keratitis

Several methods are also used to improve the delivery of the drug to the cornea. For instance, Xanthan gum vehicle improves the bactericidal effects of a combination of tobramycin and dexamethasone (Ustundag-Okur et al., 2014) and levofloxacin (Balzli et al., 2010) in treatment of *S. aureus* (Ustundag-Okur et al., 2014). Microemulsions of drugs is another novel method of ocular drug delivery (Bhattacharjee et al., 2019) to the corneal stroma. Liposomes also increase corneal penetration and stability of ocular drugs. Thus, liposomes containing the antibiotics can also be delivered to the eye (DelMonte and Kim 2011).

Corneal infections caused by *S. aureus, S. epidermidis, E. coli* and *P. aeruginosa* could be significantly reduced following Argon cold plasma treatment (Reitberger et al., 2018).

In another study, patients with bacterial keratitis responded to daily drops of topical steroids, including prednisolone acetate 1%, phenylephrine hydrochloride 0.12%, dexamethasone 0.1% and prednisolone sodium phosphate 0.5% (Green et al., 2019a; 2019b) and the patients had a 5.5-increased chance of better visual outcomes, but patients with *Nocardia* keratitis had poorer outcomes (Green et al., 2019a; 2019b).

The other novel therapeutics include corneal collagen cross-linking (CXL) (Price and Price 2016; Naranjo et al. 2019), plasma and phage therapy, photodynamic antimicrobial therapy (PDAT), and use of AMPs (Jadi et al., 2021). In CXL an ultraviolet-A (UVA) beam activates a photosensitive agent riboflavin, to generate free radicals, which induce the formation of chemical bonds between collagen fibrils of the stroma. CXL treatment is effective against pathogens such as *S. aureus* (Richoz et al., 2014), even when rose Bengal (Halili et al., 2016) and toluidine blue (Su et al., 2020) were used as the photosensitising agents. Further, CXL with topical antibiotics such as tobramycin have a synergistic effect (Kilic et al., 2018). Further, a topical CXL approach using sodium hydroxymethyl-glycinate as a cross-linking agent proved effective against MRSA and MSSA *in vitro* (Rapuano et al., 2018). But the outcomes of CXL for treatment of keratitis are uncertain, with corneal melt occurring after CXL (Davis et al., 2020).

Plasma and phage therapy are innovative treatment options for bacterial keratitis. In plasma therapy, a gas with antimicrobial properties and ability to generate reactive oxygen species is used. This method also facilitates wound healing and has anti-inflammatory properties (Reitberger et al., 2018). Argon-based plasma therapy in conjunction with antibiotics (Reitberger et al., 2018) could be used, but clinical validation is awaited.

In phage therapy viral bacteriophages are used to infect and kill bacteria (Jamal et al., 2019). Studies have shown that phage therapy is effective against *P. aeruginosa* keratitis in mice (Fukuda et al., 2012) and in a patient recovering from MRSA keratitis (Fadlallah et al., 2015).

Photodynamic therapy involves the use of three components: a photosensitizer, oxygen, and light. Thus, antimicrobial photodynamic therapy has an extraordinary potential for treatment of bacterial keratitis (De paiva et al., 2022).

4.8 Outcome of Therapy in Bacterial Keratitis

If not attended to, MK may cause blindness and the global estimates of unilateral blindness due to MK ranges from 1.5 to 2 million cases per year (Ung et al., 2019).

MK not only causes visual impairment but it is also associated with poor visual outcome (40–75% with acuity of <6/60) and causes ocular complications like corneal melting, perforation, and loss of eye (i.e., evisceration or enucleation) (Butler et al., 2005; Kunimoto et al., 2000) probably associated with higher rate of ocular co-morbidities and delay in presentation (Barua et al., 2017; Srivastava and Gill 2020). Ting et al. (2021a; 2021b) also indicated that MK was most prevalent during summer with *P. aeruginosa* being commonly isolated (Walkden et al., 2018).

In general, poor visual outcomes and predisposition to recurrence after bacterial keratitis was dependent on the characteristics of the corneal ulcer (size, locality, and depth), the risk factors, the bacteria involved and chronic diseases such as dry eye, eyelid deformities, bullous keratopathy, and rheumatoid disease (Schaefer et al., 2001; Miedziak et al., 1999). One study reported five patients with a poor clinical outcome. Three of the five had ulcers, two had central cornea involvement up to 60% or more of total corneal depth in all but one case. This last patient had undergone a pterygium excision that led to thinning of the cornea. The bacteria that were isolated from these patients were predominantly *Morganella morganii*, *S. pneumoniae*, *S. oralis*, *P. aeruginosa*, and *S. epidermidis*, but the numbers were small to conclude the prognosis related to specific bacterial patterns (Schaefer et al., 2001). In a recent review, Moe et al. (2022) studied all cases of *Acanthamoeba* keratitis between January 2006 to June 2011 in India and observed significant differences in the median time until re-epithelialization in *Acanthamoeba* keratitis (113 days) compared to 30 and 25 days for fungal and bacterial keratitis respectively. The median time until discontinuation of antimicrobials was also significantly longer in acanthamoeba keratitis compared to bacterial and fungal ulcers. In an earlier report it was observed that BK in UK presents with significant ocular morbidity with ocular surface diseases, contact lens wear, and systemic immunosuppression being the main risk factors. Further, older age, large infiltrate, and poor presenting vision were predictive of poor visual outcome and delayed corneal healing, highlighting the importance of prevention and early intervention for BK (Ting et al., 2021b).

4.9 Trends in Antimicrobial Resistance in Adult Bacterial Keratitis

The driving force of AMR in ocular infections is also multifactorial and includes injudicious use of antibiotics (Brown 2007), incorrect dosing regimen (Martinez et al., 2012), empirical method of prescribing antibiotics and community prevalence of drug resistance (Blomquist 2006). In 1988, Goodman and Gottsch (1988) reported methicillin-resistant *S. epidermidis* keratitis and ever since AMR in ocular infections was reported from several countries like USA (Asbell et al., 2020 a-c), China (Lin et al., 2019a) and India (Kaye et al., 2010). Studies have also addressed AMR patterns of specific bacteria involved in bacterial keratitis across several countries (Cherny et al., 2019).

4.9.1 Trends in Antimicrobial Resistance to Fluoroquinolones in Adult Bacterial Keratitis

In Table 4.4 an attempt has been made to capture the trends in AMR in ocular isolates from keratitis individuals to fluoroquinolones (Moxifloxacin, ciprofloxacin, levofloxacin and ofloxacin), penicillin antibiotics (methicillin and penicillin), cephalosporin antibiotics (cefuroxime, ceftazidime) and aminoglycoside antibiotics

TABLE 4.4

Trends in Resistance (Increase/Decrease) of Bacteria Associated with Microbial Keratitis in Humans

S. No.	Bacterium	Antibiotic	Country/ Year	Trend Increase / Decrease / No change	Reference
1	*Streptococcus* spp., *Staphylococcus* spp.	Moxifloxacin	USA- 2006–2014	Increase	Peng et al., 2018.
2	*S. aureus* strains, Methicillin-resistant CoNS (MRSA-CoNS)	Moxifloxacin	USA-8 years, ARMOR surveillance 2009–2016	No change	Thomas et al., 2019
3	MRSA, MSSA	Moxifloxaci, gatifloxacin	USA-20, 1993–2012	Increase	Chang et al., 2015
4	MRSA	Moxifloxacin	USA-20 years 1996–2015	Increase	Peng et al., 2018
5	MRSA, MRCNS	Methicillin	Mexico, 2002–2011	Increase	Hernandez-Camarena et al. (2015)
6	*P. aeruginosa*	Ceftazidime	Mexico, 2002–2011	Increase	Hernandez-Camarena et al. (2015).
7	*Streptococcus* spp.	Ofloxacin	UK, 2004–2015	Increase	Tan et al., 2017.
	S. aureus	Chloramphenicol,ciproflo-xacin, ofloxacin, cefuroxime, gentamycin, tobramycin, ceftazidime		Increase	
	Moraxella spp., *Pseudomonas* spp.			No change	
8	*S. aureus,* CoNS	Cipro-floxacin	UK- 10 years up to 2008	No change	Shalchi et al., 2011.
	Pseudomonas aeruginosa	Chloram-phenicol		Increase	

(*continued*)

TABLE 4.4 (Continued)

Trends in Resistance (Increase/Decrease) of Bacteria Associated with Microbial Keratitis in Humans

S. No.	Bacterium	Antibiotic	Country/ Year	Trend Increase / Decrease / No change	Reference
9	*S. aureus*, CoNS Oxacillin susceptible *S. aureus,* Oxacillin-resistant *S. aureus,* Oxacillin susceptible CoNS, Oxacillin-resistant CoNS, -*S. pneumoniae* CoNS and Oxacillin-resistant *S. aureus*	Oxacillin Vancomycin Gentamicin	USA-15 years, 1999–2013	No change No change Increase	Hsu et al., 2019.
10	*S. pneumoniae.*	Ofloxacin, cipro-floxacin, Moxi-floxacin, gatifloxacin, levofloxacin, chloram-phenicol, gentamicin, oxacillin, cefazolin	India-2002–2013	No change	Lalitha et al., 2017.
11	Viridans group *Streptococci*	Ofloxacin, Cipro-floxacin, moxi-floxacin, levofloxacin, gatifloxacin chloram-phenicol	India 2002–2013	No change	Lalitha et al., 2017.

12	Enterobacteriaceae: enteric Gram-negative rods	Cipro-floxacin, ofloxacin, gatifloxacin, levofloxacin Chloram-phenicol, gentamicin, tobramycin, amikacin, ceftazidime	India 2020–2013	Increase No change	Lalitha et al., 2017.
13	*Nocardia* spp.,	Cefaxzolin, vancomycin	India 2002–2013	No change	Lalitha et al., 2017.
14	*P. aeruginosa*	Ofloxacin, cipro-floxacin, gentamicin	India-2002–2013	No change	Lalitha et al., 2017.
15	*S. aureus,* CoNS	Cipro floxacin, levofloxacin	Iran-6 years period	Increase	Soleimani, et al., 2021.
16	CoNS, *S. aureus,* *P. aeruginosa*	Vancomyci, methicillin, ciprofloxacin. Ceftazidime, carbapenem gentamicin, tobramycin, amikacin, ciprofloxacin.	Korea, 2007–2016	No change No change	Mun et al., 2019.
17	*Staphylococcus* spp., *Streptococcus* spp., *Corynebacterium* spp. *Moraxella* spp., *Pseudomonas* spp.	cefazolin, gatifloxacin, moxi-floxacin, ofloxacin Cipro- floxacin, ceftazidime, tobramycin, gentamicin	Global, 2000–2020	Decrease No change	Zhang et al., 2022.

(continued)

TABLE 4.4 (Continued)

Trends in Resistance (Increase/Decrease) of Bacteria Associated with Microbial Keratitis in Humans

S. No.	Bacterium	Antibiotic	Country/ Year	Trend Increase / Decrease / No change	Reference
18	CoNS, *S. epidermidis,* other CoNS, Streptococci, *S. aureus* *Acinetobacter spp., Pseudomonas spp., Serratia spp., E. coli, Burkholderia*	Levo-floxacin, tobramycin, cefazolin, ceftazidime, cefuroxime chloram-phenicol Ofloxacin Cefazolin, chloram phenicol	China 2010 - 2018	Decrease Increase Increase Decrease	Lin et al., 2019a.

Source: Compiled by the author.

Note: The predominant bacteria that were screened included *Staphylococcus* spp., *Pseudomonas* spp., *Streptococcus* spp., *Moraxella* spp., *Serratia* spp., *Haemophilus* spp., *Enterobacter* spp., *Branhamella* spp., *Diphtheroid* spp., *Bacillus* spp., *Corynebacterium* spp., *Proteus* spp., *Escherichia coli., Citrobacter* spp., *Klebsiella* spp., *Acinetobacter* spp., *Nocardia* spp., *Propionibacterium* spp., *Alcaligenes* spp.

(gentamicin). From the year 1993–2020, AMR in ocular isolates in the United States, Mexico, Canada, UK, India, Iran, China, Korea, and so forth showed that with respect to fluoroquinolones (Table 4.4) 6 of the 12 studies indicated increase in resistance over time (Table 4.4) indicative of the emergence of AMR strains. This partially confirms the increasing rates of resistance to fluoroquinolones in ocular *P. aeruginosa* and *S. aureus* observed by Willcox (2011). Tuft and Matheson (2000) screened 1312 bacterial isolates over 16 years from keratitis patients and observed no change in the resistance to ofloxacin or to the combination of gentamicin and cefuroxime but resistance to chloramphenicol ($p=0.0019$) increased in Gram-negative bacteria. Decrease in AMR to fluoroquinolones was observed in just two studies (Lin et al., 2019a; Zhang et al., 2022) (Table 4.4) probably due to genetic modification of factors conferring resistance to antibiotics, thus making them more susceptible to antimicrobial drugs. Tan et al. (2017) showed that though *Streptococcus* spp. showed increase in resistance to ofloxacin other isolates like *S. aureus*, *Moraxella* spp., and *Pseudomonas* spp. showed no change in resistance to ciprofloxacin and ofloxacin (Table 4.3). A similar discrepancy was also observed by Lin et al. (2019a) who observed both decrease and increase in AMR to fluoroquinolones (Table 4.4).

4.9.2 Trends in Antimicrobial Resistance to Aminoglycosides in Adult Bacterial Keratitis

The aminoglycosides are broad-spectrum antibiotics prescribed for infections caused by Gram-negative pathogens. But when the aminoglycosides (gentamicin, tobramycin, amikacin) were screened against *S. aureus, Moraxella* spp. and *Pseudomonas* spp. (Tan et al., 2017); *S. pneumoniae, P. aeruginosa*, MSSA and Enterobacteriaceae (Laitha et al., 2017); CoNS, *S. aureus*, *P. aeruginosa* (Mun et al., 2019) and *Moraxella* spp., and *Pseudomonas spp.,* (Zhang et al., 2022) showed no change in AMR over time indicating that aminoglycosides have retained their antimicrobial activity (Table 4.4). In contradiction gentamicin resistance in ConS and oxacillin-resistance in *S. aureus* increased in just one study from the United States (Hsu et al., 2019).

4.9.3 Trends in Antimicrobial Resistance to Penicillin Antibiotics in Adult Bacterial Keratitis

The AMR of *S. aureus*, CoNS and *S. pneumoniae* to penicillin antibiotics (oxacillin, methicillin and penicillin) did not change with time (Hsu et al., 2019; Lalitha et al., 2017; Mun et al., 2019) (Table 4.4). But *S. aureus*, *Staphylococcus* spp., MRSA, methicillin-resistant coagulase-negative staphylococci (MRCNS) and CoNS showed increase in resistance to oxacillin, methicillin and (or) penicillin with time (Hernandez-Camarena et al., 2015; Lin et al., 2019a; Politis et al., 2016).

4.9.4 Trends in Antimicrobial Resistance to Glycopeptide Antibiotics in Adult Bacterial Keratitis

Resistance to vancomycin, a glycopeptide antibiotic, did not change (Hsu et al., 2019; Lalitha et al., 2017; Mun et al., 2019) irrespective of the ocular organism screened (Table 4.4). Resistance to chloramphenicol increased in *Pseudomonas* spp., *P. aeruginosa, S. aureus* and *Moraxella spp.* (Shalchi et al., 2011; Tan et al., 2017) whereas the resistance remained unchanged with time in *S. pneumoniae*, diptheroids, viridians group Streptococci, MSSA, CoNS and Enterobacteriaceae (Lalitha et al., 2017).

4.9.5 Trends in Antimicrobial Resistance to Cephalosporin Antibiotics in Adult Bacterial Keratitis

AMR trend in the cephalosporin antibiotics group (cefuroxime, ceftazidine, cefazolin and ceftazidime) has been very inconsistent with 5 studies showing no change in AMR over time (Mun et al., 2019; Tan et al., 2017; Lalitha et al., 2017; Afshari et al., 2008; Zhang et al., 2022), two showing a decrease in AMR (Lin et al., 2019 a; Zhang et al., 2022) and one study showing an increase in AMR (Hernandez-Camarena et al., 2015).

4.9.6 Trends in Antimicrobial Resistance to Chloramphenicol Antibiotic in Adult Bacterial Keratitis

Chloramphenicol is a broad-spectrum antibiotic that inhibits bacterial growth by inhibiting the production of proteins. Chloramphenicol susceptibility decreased in one study when screened against CoNS, *S. aureus, S. epidermidis,* Streptococci, *Acinetobacter* spp., *Pseudomonas spp., Serratia spp., E. coli,* and *Burkholderia* spp., implying the emergence of resistance (Lin et al., 2019a) (Table 4.4).

4.10 Multidrug Resistance to Antibiotics

Multidrug resistances (MDR) refer to resistance of a microorganism to a minimum of one antimicrobial agent in three or more different antimicrobial categories (Magiorakos et al., 2012). MDR is a commonly observed phenomenon in MRSA (76.2%) and MRCoNS (73.5%) (Thomas et al., 2019). Linezolid is effective against several bacteria, like vancomycin-resistant enterococci and MRSA (Elsahn et al., 2010) and MDR *P. aeruginosa* keratitis (MDR-PA). Budak et al. (2016) observed that a topical combination of linezolid/ceftazidime was effective in seven BK patients who discontinued vancomycin/ceftazidime topical combination therapy. In 23 MDR-PA from keratitis patients with decreased susceptibility (15%) to aminoglycosides, cephalosporins, and fluoroquinolones (43.38%) it was observed that all were susceptible to colistin and imipenem (Vazirani et al., 2015). In 15 eyes with MDR-PA keratitis Fernandes et al. (2016) detected susceptibility to imipenem in eight eyes, to colistin in four eyes, to neomycin in two eyes, and in one eye MDR-PA was not susceptible to any drug. MDR-PA were also susceptible to ciprofloxacin and 5% cefazolin. But Chatterjee and Agrawal (2016) achieved no visual improvement in any of the eight eyes of 12 cases of MDR-PA keratitis treated with fluoroquinolones alone. Thus, in MDR-PA, clinical success is less and anatomic integrity is affected when imipenem or colistin therapy is not administered within 48 h (Vazirani et al., 2015; Fernandes et al., 2016).

4.11 Conclusions

Bacterial keratitis if untreated would become a persistent burden in both developed and developing countries. A few important emerging points are (Schaefer et al., 2001):

(i) Prospective studies need to be more comprehensive and include all causative agents of keratitis (i.e., bacteria, fungi, protozoa and viruses).
(ii) Identification of the major risk factors of bacterial keratitis would help to design strategies to overcome bacterial keratitis.
(ii) Increase rate of AMR in ocular infections warrants the need for judicious use of antimicrobial agents and introduction of new antimicrobials for therapy.
(iii) Developing diagnostic tools based on NGS and artificial intelligence assisted platforms could provide guidance on the use of antimicrobial therapy to overcome AMR.
(iv) Contact lens wearers are more prone to bacterial keratitis and appropriate instructions on handling and care of the lenses is important.
(v) Patients with dry eye syndrome, acquired eyelid deformities, bullous keratopathy, and rheumatoid disease (Schaefer et al., 2001) should be forewarned about the development of corneal ulcers.
(vi) Isolation of bacteria and treatment with antibiotics represent decisive steps in management of bacterial keratitis.
(vii) The use of antibiotics with broad spectrum antibacterial activity, better tolerance, and good tissue penetration like lomefloxacin are recommended.
(vii) A double-blind study comparing the benefits of mono- and multi-drug treatment could help in management of bacterial keratitis.
(x) Several novel treatment modalities for treatment of keratitis have been developed but clinical validation is warranted.

REFERENCES

Aasuri MK, Reddy MK, Sharma S, Rao GN. Co-occurrence of pneumococcal keratitis and dacryocystitis. *Cornea.* 1999 May;18(3):273–6. doi: 10.1097/00003226-199905000-00005. PMID: 10336027.

Acar JF, O'brien TF, Goldstein FW et al., The epidemiology of bacterial resistance to quinolones. *Drugs* 1993; 45: 24–8.

Afshari N A, Ma J J K, Duncan S M, Pineda R, Starr C E, Decroos F C, Johnson C S, Adelman R A. Trends in Resistance to Ciprofloxacin, Cefazolin, and Gentamicin in the Treatment of Bacterial Keratitis . *J Ocul Pharmacol Ther* 2008 Apr;24(2):217–23. doi: 10.1089/jop.2007.0085.

Afsharpaiman S, Zare M, Yasemi M, Jamialahmadi T, Sahebkar A. The Prevalence of Infectious Keratitis after Keratorefractive Surgery: A Systematic Review and Meta-Analysis Study. *J Ophthalmol.* 2020 Jul 28;2020:6329321. doi: 10.1155/2020/6329321.

Aish K A E, Hanan Fayad, Rasha El Arkan, Naima Farwana, Najlaa Saquer, Antimicrobial resistance in bacteria isolated from blood cultures from babies with sepsis in neonatal intensive care units in the Gaza Strip: a cross-sectional study. *Lancet*, June 1, 2022; 399, S12. https://doi.org/10.1016/S0140-6736(22)01147-3

Al-Aqaba MA, Dhillon VK, Mohammed I, Said DG, Dua HS. Corneal nerves in health and disease. *Prog Retin Eye Res*. 2019; 73:100762.

Al-Mujaini, A.; Al-Kharusi, N.; Thakral, A.;Wali, U.P. Bacterial Keratitis-Perspective on Epidemiology, Clinico-Pathogenesis, Diagnosis and Treatment. *SQU Med. J.* 2009; 9: 184–95.

Al-Shakarchi F I and Marwa Aseel Ahmed Hussein and Abdulwahid B. Al-Shaibani. Profile of Microbial Keratitis at a Referral Center in Iraq, *J Al-Nahrain University-Science*, 2015,18: 141–7.

Alemayehu AM, Yibekal BT, Fekadu SA. Prevalence of vernal keratoconjunctivitis and its associated factors among children in Gambella town, southwest Ethiopia, June 2018. *PLoS One*. 2019; 14(4):e0215528.

Alexandrakis G, Alfonso EC, Miller D. Shifting trends in bacterial keratitis in South Florida and emerging resistance to fluoroquinolones. *Ophthalmology*. 2000; 107(8):1497–502

Alharbi,N S Screening of antibiotic-resistant staphylococci in the nasal cavity of patients and healthy individuals, *Saudi Journal of Biological Sciences*, 2020; 27(1): 100–05, ISSN 1319-562X, https://doi.org/10.1016/j.sjbs.2019.05.008.

Alhoutan K, Alarfaj K. Exposure Keratopathy: An Idiopathic Lagophthalmos Case Report. *Cureus*. 2021 Oct 21;13(10):e18945. doi: 10.7759/cureus.18945

Allcock S., E. H. Young, M. Holmes, D. Gurdasani, G. Dougan, M. S. Sandhu, L. Solomon, M. E. Török. Antimicrobial resistance in human populations: challenges and opportunities Global Health, *Epidemiology and Genomics* 2017; 2: e4. doi:10.1017/gheg.2017.4

Alzubaidi R, Sharif MS, Qahwaji R, et al., In vivo confocal microscopic corneal images in health and disease with an emphasis on extracting features and visual signatures for corneal diseases: a review study. *Br J Ophthalmol*. 2016;100(1):41–55

Arunga S, Wiafe G, Habtamu E, et al., The impact of microbial keratitis on quality of life in Uganda. *BMJ Open Ophthalmol*. 2019;4:e000351.

Asbell PA, DeCory HH. Antibiotic resistance among bacterial conjunctival pathogens collected in the Antibiotic Resistance Monitoring in Ocular Microorganisms (ARMOR) surveillance study. *PLoS One*. 2018a Oct 18;13(10):e0205814. doi: 10.1371/journal.pone.0205814

Asbell PA, Pandit RT, Sanfilippo CM. Antibiotic Resistance Rates by Geographic Region Among Ocular Pathogens Collected During the ARMOR Surveillance Study. *Ophthalmol Ther*. 2018b Dec;7(2):417–29. doi: 10.1007/s40123-018-0141-y

Asbell PA, Sanfilippo CM. Antibiotic Resistance Trends Among Ocular Pathogens in the US—Cumulative Results from the Antibiotic Resistance Monitoring in Ocular Microorganisms (ARMOR). *Surveil Study US Ophthal Rev*, 2017;10(1):35–8. doi: https://doi.org/10.17925/USOR.2017.10.01.35

Asbell PA, Sanfilippo CM, Sahm DF, DeCory HH. Trends in Antibiotic Resistance Among Ocular Microorganisms in the United States From 2009 to 2018 JAMA *Ophthalmol*. 2020a;138(5):439–450. doi:10.1001/jamaophthalmol.2020.0155

Asbell PA, Sanfilippo CM, DeCory HH. Trends in antibiotic resistance over time among staphylococci collected 2009–2019 in the ARMOR study. *IOVS*. 2020b;61:4885.

Asbell PA, Sanfilippo CM, Sahm DF, DeCory HH. Trends in Antibiotic Resistance Among Ocular Microorganisms in the United States From 2009 to 2018. *JAMA Ophthalmol*. 2020c;138:1–12.

Astori IP, Muller MJ, Pegg SP. Cicatricial, postburn ectropion and exposure keratitis. *Burns*. 1998 Feb;24(1):64–7. doi: 10.1016/s0305-4179(97)00076-4

Augustin DK, Heimer SR, Tam C, et al. Role of Defensins in Corneal Epithelial Barrier Function against *Pseudomonas aeruginosa* Traversal. *Infect. Immun.* 2011; 79: 595–605. doi:10.1128/iai.00854-10

Austin A, Lietman T, Rose-Nussbaumer J. Update on the Management of Infectious Keratitis. *Ophthalmology*. 2017;124(11):1678–89.

Balzli, C.L.; McCormick, C.C.; Caballero, A.R.; Tang, A.; O'Callaghan, R.J. The effectiveness of an improved combination therapy for experimental Staphylococcus aureus keratitis. *Adv. Ther.* 2010; 27: 933–40.

Barua K, Borah M, Deka C, Kakati R. Morbidity pattern and health-seeking behavior of elderly in urban slums: a crosssectional study in Assam, India. *J Fam Med Prim Care*. 2017;6:345–50.

Bharathi MJ, Ramakrishnan R, Meenakshi R, et al., Microbial keratitis in South India: influence of risk factors, climate, and geographical variation. *Ophthalmic Epidemiol.* 2007;14(2):61–9.

Bhattacharjee, A.; Das, P.J.; Adhikari, P.; Marbaniang, D.; Pal, P.; Ray, S.; Mazumder, B. Novel drug delivery systems for ocular therapy: With special reference to liposomal ocular delivery. *Eur. J. Ophthalmol.* 2019, 29, 113–26.

Blair JMA, Webber MA, Baylay AJ, Ogbolu DO, Piddcock LJV. Molecular mechanisms of antibiotic resistance. *Nature Reviews Microbiology* 2015; 13: 42–51.

Blomquist PH. Methicillin-resistant Staphylococcus aureus infections of the eye and orbit (an American Ophthalmological Society thesis). *Trans Am Ophthalmol Soc.* 2006;104:322–45.

Borroni, D.; Romano, V.; Kaye, S.B.; Somerville, T.; Napoli, L.; Fasolo, A.; Gallon, P.; Ponzin, D.; Esposito, A.; Ferrari, S. Metagenomics in ophthalmology: Current findings and future prospectives. *BMJ Open Ophthalmol.* 2019; 4: e000248.

Brooks L, Narvekar U, McDonald A, Mullany P. Prevalence of antibiotic resistance genes in the oral cavity and mobile genetic elements that disseminate antimicrobial resistance: A systematic review. *Mol Oral Microbiol.* 2022 Aug;37(4):133–53. doi: 10.1111/omi.12375

Brown L. Resistance to ocular antibiotics: an overview. *Clin Exp Optom.* 2007;90:258–62.

Budak B A, Baykara M, Kivanc SA, Yilmaz H, Cicek S. Comparing the ocular surface effects of topical vancomycin and linezolid for treating bacterial keratitis. *Cutan Ocul Toxicol.* 2016;35:126–30. doi:10.3109/15569527.2015.1058273

Burrow MK, Patel BC. Keratoconjunctivitis. 2022 May 24. In: *StatPearls* [Internet]. Treasure Island (FL): StatPearls Publishing; 2022 Jan. PMID: 31194419.

Butler TK, Spencer NA, Chan CC, Singh Gilhotra J, McClellan K. Infective keratitis in older patients: a 4 year review, 1998–2002. *Br J Ophthalmol.* 2005;89:591–6.

Cabrera-Aguas M, Khoo P, George CRR, Lahra MM, Watson S. Antimicrobial resistance trends in bacterial keratitis over 5 years in Sydney, Australia. *Clin Exp Ophthalmol.* 2019;48:183–91.

Cariello AJ, Passos RM, Yu MCZ, Hofling-Lima AL. Microbial keratitis at a referral center in Brazil. *Int Ophthalmol.* 2011;31(3):197–204.

Carion TW, Ebrahim AS, Kracht D, et al., Thymosin Beta-4 and ciprofloxacin adjunctive therapy improves Pseudomonas aeruginosa-induced keratitis. *Cells.* 2018;7(10):145.

Carnt N, Hoffman JJ, Verma S, et al., Acanthamoeba keratitis: confirmation of the UK outbreak and a prospective case-control study identifying contributing risk factors. *Br J Ophthalmol.* 2018a;102:1621–8.

Carnt N, Robaei D, Minassian DC, Dart JKG. Acanthamoeba keratitis in 194 patients: risk factors for bad outcomes and severe inflammatory complications. *Br J Ophthalmol.* 2018b;102:1431–5.

Chandra NS Torres MF Winthrop KL et al., Cluster of *Mycobacterium chelonae* keratitis cases following laser in-situ keratomileusis. *Am J Ophthalmol.* 2001; 132: 819–30

Chang HL, Kuo BI, Wu JH, Huang WL, Su CC, Chen WL. Anti-glaucoma agents-induced pseudodendritic keratitis presumed to be herpetic simplex keratitis: a clinical case series. *Sci Rep.* 2021 Nov 2;11(1):21443. doi: 10.1038/s41598-021-01073-0. PMID: 34728756; PMCID: PMC8563774.

Chang VS, Dhaliwal DK, Raju L, Kowalski RP. Antibiotic resistance in the treatment of Staphylococcus aureus keratitis: a 20-year review. *Cornea.* 2015; 34(6):698–703. https://doi.org/10.1097/ICO. 0000000000000431

Chatterjee S, Agrawal D. Multi-drug resistant Pseudomonas aeruginosa keratitis and its effective treatment with topical colistimethate. *Indian J Ophthalmol.* 2016; 64:153–7. doi:10.4103/0301-4738.17 9721

Chen FV, Chang TC, Cavuoto KM. Patient demographic and microbiology trends in bacterial conjunctivitis in children. *J AAPOS.* 2018; 22(1):66–7.

Cherny, CBS, Pratik Patel, OD, and Suzanne Sherman, ODHot Topics in Bacterial Keratitis Review of Cornea and contact lenses Published November 15, 2019.

Chidambaram JD, Venkatesh Prajna N, Srikanthi P, et al., Epidemiology, risk factors, and clinical outcomes in severe microbial keratitis in South India. *Ophthalmic Epidemiol.* 2018;25:297–305.

Chojnacki M, Philbrick A, Wucher B, et al., Development of a broad-spectrum antimicrobial combination for the treatment of Staphylococcus aureus and Pseudomonas aeruginosa corneal infections. *Antimicrob Agents Chemother.* 2018;63(1):e01929–18.

Chu HS, Hu FR. Non-tuberculous mycobacterial keratitis. *Clin Microbiol Infect.* 2013;19:221–6.

Chung JL, Seo KY, Yong DE, Mah FS, Kim TI, Kim EK, Kim JK. Antibiotic susceptibility of conjunctival bacterial isolates from refractive surgery patients. *Ophthalmol* 2009 Jun;116(6):1067–74. doi: 10.1016/j.ophtha.2008.12.064

Collier SA, Gronostaj MP, MacGurn AK, et al., Estimated burden of keratitis-United States, 2010. *MMWR Morb Mortal Wkly Rep.* 2014;63:1027–30.

Comstock TL, Paterno MR, Usner DW, Pichichero ME. Efficacy and safety of besifloxacin ophthalmic suspension 0.6% in children and adolescents with bacterial conjunctivitis: a post hoc, subgroup analysis of three randomized, double-masked, parallel-group, multicenter clinical trials. *Paediatr Drugs.* 2010;12(2):105–12.

Constantinou M, Daniell M, Snibson GR, Vu HT, Taylor HR. Clinical efficacy of moxifloxacin in the treatment of bacterial keratitis: a randomized clinical trial. *Ophthalmology.* 2007;114:1622–1629. doi:10.1016/j.ophtha.2006.12.011

Cutarelli PE, Lass JH, Lazarus HM, et al., Topical fluoroquinolones: antimicrobial activity and in vitro corneal epithelial toxicity. *Curr Eye Res* 1991;10:557–63.

Dart JK, Radford CF, Minassian D, et al., Risk factors for microbial keratitis with contemporary contact lenses: A case-control study. *Ophthalmol* 2008;115:10:1647–54.

Dart JK, Saw VP, Kilvington S. Acanthamoeba keratitis: diagnosis and treatment update 2009. *Am J Ophthalmol.* 2009;148:487–99.e2.

Das S, Garg P, Mullick R, Annavajjhala S. Keratitis following laser refractive surgery: Clinical spectrum, prevention and management. *Indian J Ophthalmol.* 2020 Dec; 68(12): 2813–18. doi: 10.4103/ijo.IJO_2479_20

Das S, Samantaray R, Mallick A, et al., Types of organisms and in-vitro susceptibility of bacterial isolates from patients with microbial keratitis: a trend analysis of 8 years. *Ind J Ophthalmol.* 2019; 67(1):49–53.

Datta R, Kleinman K, Rifas-Shiman S, Placzek H, Lankiewicz J, Platt R, Huang SS. Confounding by indication affects antimicrobial risk factors for methicillin-resistant Staphylococcus aureus but not vancomycin-resistant enterococci acquisition. *Antimicrobial Resistance Infect Control* 2014; 3: 19. doi: 10.1186/2047-2994-3-19. PMID: 24932407; PMCID: PMC4057914.

Davis, S.A.; Bovelle, R.; Han, G.; Kwagyan, J. Corneal collagen cross-linking for bacterial infectious keratitis. *Cochrane Database Syst. Rev.* 2020; 6: CD013001.

de Been M, Lanza VF, de Toro M, Scharringa J, Dohmen W, Du Y, Hu J, Lei Y, Li N, Tooming-Klunderud A, Heederik DJ, Fluit AC, Bonten MJ, Willems RJ, de la Cruz F, van Schaik W. Dissemination of cephalosporin resistance genes between Escherichia coli strains from farm animals and humans by specific plasmid lineages. *PLoS Genetics* 2014; 10: e1004776.

de Paiva ACM, Ferreira MDC, da Fonseca AS. Photodynamic therapy for treatment of bacterial keratitis. *Photodiagnosis Photodyn Ther.* 2022 Mar;37:102717. doi: 10.1016/j.pdpdt.2022.102717

DelMonte, D.W.; Kim, T. Anatomy and physiology of the cornea. *J. Cataract Refract. Surg.* 2011; 37: 588–98.

Dethorey G, Daruich A, Hay A, Renard G, Bourges JL. Severe bacterial keratitis referred to ophthalmology emergency departments: a retrospective study of 268 cases]. *J Fr Ophtalmol.* 2013;36:129–37.

Di Zazzo A, Kheirkhah A, Abud TB, Goyal S, Dana R. Management of high-risk corneal transplantation. *Surv Ophthalmol.* 2017; 62:816–27.

Dohse N, Wibbelsman TD, Rapuano SB, et al. Microbial keratitis and clinical outcomes following penetrating and endothelial keratoplasty. *Acta Ophthalmol.* 2020; 98:e895–e900 https://doi.org/10.1111/aos.14404

Doi Y, Paterson DL, Egea P, Pascual A, López-Cerero L, Navarro MD, Adams-Haduch JM, Qureshi ZA, Sidjabat HE, Rodríguez-Baño J. Extended-spectrum and CMY-type beta-lactamase-producing Escherichia coli in clinical samples and retail meat from Pittsburgh, USA and Seville, Spain. *Clin. Microbiol. Infect* 2010; 16: 33–8.

ECDC (European Centre for Disease Prevention and Control), E.E.F.S.A.a.E.E.M.A. ECDC/EFSA/EMA first joint report on the integrated analysis of the consumption of antimicrobial agents and occurrence of antimicrobial resistance in bacteria from humans and food-producing animals. *EFSA Journal* 2015; 13: 4006.

Egrilmez, S., and Yildirim-Theveny, Ş. (2020). Treatment-Resistant Bacterial Keratitis: Challenges and Solutions. *Clin. Opthamol.* Vol. 14, 287–97. doi:10.2147/opth.s181997

Elena PP, Jauch A. Ocular distribution of Lomefloxacin 0.3% after a single instillation in the infected eye of pigmented rabbits. *J Ocular Pharmacol Therapeutics* 1997; 16:551–8.

Elsahn AF, Yildiz EH, Jungkind DL, et al., In vitro susceptibility patterns of methicillin-resistant Staphylococcus aureus and coagulase-negative Staphylococcus corneal isolates to antibiotics. *Cornea.* 2010;29:1131–5. doi:10.1097/ICO.0b013e3181d2ce25

Fadlallah, A.; Chelala, E.; Legeais, J. Corneal Infection Therapy with Topical Bacteriophage Administration. *Open Ophthalmol. J.* 2015; 9: 167–8.

Fernandes M, Vira D, Medikonda R, Kumar N. Extensively and pan-drug resistant Pseudomonas aeruginosa keratitis: clinical features, risk factors, and outcome. *Graefes Arch Clin Exp Ophthalmol.* 2016;254:315–22. doi:10.1007/s00417-015-3208-7

Ferreira CS, Figueira L, Moreira-Gonçalves N, Moreira R, Torrão L, Falcão-Reis F. Clinical and Microbiological Profile of Bacterial Microbial Keratitis in a Portuguese Tertiary Referral Center-Where Are We in 2015? *Eye Contact Lens.* 2018;44:15–20.

Finley RL, Collignon P, Larsson DG, McEwen SA, Li XZ, Gaze WH, Reid-Smith R, Timinouni M, Graham DW, Topp E. The scourge of antibiotic resistance: the important role of the environment. *Clinical Infectious Diseases* 2013; 57: 704–10.

Freitas D Alvarenga L Sampaio J et al., An outbreak of *Mycobacterium chelonae* infection after LASIK. *Ophthalmology.* 2003; 110: 276–85

Fukuda, K.; Ishida, W.; Uchiyama, J.; et al., Pseudomonas aeruginosa keratitis in mice: Effects of topical bacteriophage KPP12 administration. *PLoS ONE* 2012; 7: e47742.

Garg P, Kalra P, Joseph J. Non-contact lens related Acanthamoeba keratitis. *Indian J Ophthalmol.* 2017; 65:1079–86.

Gautam V, Chaudhary A, Singh SK, Rai PG. Profile of Corneal Ulcer in a Month of harvesting Season in a Tertiary Level Eye Hospital of Eastern Nepal. *Nepal J Ophthalmol.* 2018;10:32–8.

Goodman DF, Gottsch JD. Methicillin-resistant *Staphylococcus epidermidis* keratitis treated with vancomycin. *Arch Ophthalmol* 1988;106(11):1570–1. doi:10.1001/archopht.1988.01060140738046

Gopinathan U, Sharma S, Garg P, Rao GN. Review of epidemiological features, microbiological diagnosis and treatment outcome of microbial keratitis: experience of over a decade. *Indian J Ophthalmol.* 2009;57(4):273–9.

Gordon, Y. J., Huang, L. C., Romanowski, E. G., Yates, K. A., Proske, R. J., and Mcdermott, A. M. (2005). Human Cathelicidin (LL-37), a Multifunctional Peptide, Is Expressed by Ocular Surface Epithelia and Has Potent Antibacterial and Antiviral Activity. *Curr. Eye Res.* 30, 385–94. doi:10.1080/02713680590934111

Grandi G, Bianco G, Boattini M, et al., Bacterial etiology and antimicrobial resistance trends in ocular infections: A 30-year study, Turin area, Italy. *European J Ophthalmol.* 2021;31(2):405–14. doi:10.1177/1120672119896419

Green M, Carnt N, Apel A, Stapleton F. Queensland Microbial Keratitis Database: 2005–2015. *Br J Ophthalmol.* 2019a;103:1481–6.

Green M, Hughes I, Hogden J, et al., High-dose steroid treatment of bacterial keratitis. *Cornea.* 2019b;38(2):135–40.

Guidera AC, Luchs JI, Udell IJ. Keratitis, ulceration, and perforation associated with topical nonsteroidal anti-inflammatory drugs. *Ophthalmology.* 2001 May;108(5):936–44. doi: 10.1016/s0161-6420(00)00538-8

Haindongo, E.H., Funtua, B., Singu, B. et al., Antimicrobial resistance among bacteria isolated from urinary tract infections in females in Namibia, 2016–2017. *Antimicrob Resist Infect Control* 2022;11: 33. https://doi.org/10.1186/s13756-022-01066-2

Halili, F.; Arboleda, A.; Durkee, H.; Taneja, M.; Miller, D.; Alawa, K.A.; Aguilar, M.C.; Amescua, G.; Flynn, H.W., Jr.; Parel, J.M. Rose Bengal- and Riboflavin-Mediated Photodynamic Therapy to Inhibit Methicillin-Resistant Staphylococcus aureus Keratitis Isolates. *Am. J. Ophthalmol.* 2016; 166: 194–202.

Harrison EM, Paterson GK, Holden MT, Larsen J, Stegger M, Larsen AR, Petersen A, Skov RL, Christensen JM, Bak Zeuthen A, Heltberg O, Harris SR, Zadoks RN, Parkhill J, Peacock SJ, Holmes MA. Whole genome sequencing identifies zoonotic transmission of MRSA isolates with the novel mecA homologue mecC. *EMBO Mol Med* 2013; 5: 509–15.

Hassan M, Papanikolaou T, Mariatos G, Hammad A, Hassan H. Candida albicans keratitis in an immunocompromised patient. *Clin Ophthalmol.* 2010;4:1211–15. https://doi.org/10.2147/OPTH.S7953

Hau SC, Dart JK, Vesaluoma M, et al., Diagnostic accuracy of microbial keratitis with in vivo scanning laser confocal microscopy. *Br J Ophthalmol.* 2010;94(8):982–7.

Hernandez-Camarena JC, Graue-Hernandez EO, Ortiz-Casas M, et al., Trends in microbiological and antibiotic sensitivity patterns in infectious keratitis: 10-year experience in Mexico City. *Cornea.* 2015;34(7):778–85.

Hilliam, Y.; Kaye, S.;Winstanley, C. Pseudomonas aeruginosa and microbial keratitis. *J. Med. Microbiol.* 2020, 69, 3–13.

Hollander, David A; Aldave, Anthony J. Drug-induced corneal complications. *Curr Opinion Ophthalmol:* December 2004; 15(6): 541–8 doi: 10.1097/01.icu.0000143688.45232.15

Hsiao CH, Sun CC, Yeh LK, et al., Shifting Trends in Bacterial Keratitis in Taiwan: a 10-Year Review in a Tertiary-Care Hospital. *Cornea.* 2016;35:313–7.

Hsu HY, Ernst B, Schmidt EJ, et al., Laboratory results, epidemiologic features, and outcome analyses of microbial keratitis: a 15-year review from St. *Louis. Am J Ophthalmol.* 2019; 198:54–62. doi: 10.1016/j.ajo.2018.09.032. PMID: 30308206; PMCID: PMC6349514.

Hyndiuk RA, Eiferman RA, Caldwell DR, et al., Comparison of cirpofloxacin ophthalmic solution 0.3% to fortified tobramycin-cefazolin in treating bacterial corneal ulcers. *Ophthalmol* 1996;103:1854–63

Ibrahim YW, Boase DL, Cree IA. Epidemiological characteristics, predisposing factors and microbiological profiles of infectious corneal ulcers: The Portsmouth corneal ulcer study. *Br J Ophthalmol.* 2009;93(10):1319–24.

Jadi PK, Sharma P, Bhogapurapu B and Roy S (2021) Alternative Therapeutic Interventions: Antimicrobial Peptides and Small Molecules to Treat Microbial Keratitis. *Front. Chem.* 9:694998. doi: 10.3389/fchem.2021.694998

Jamal, M.; Bukhari, S.; Andleeb, S.; Ali, M.; Raza, S.; Nawaz, M.A.; Hussain, T.; Rahman, S.U.; Shah, S.S. Bacteriophages: An overview of the control strategies against multiple bacterial infections in different fields. *J. Basic Microbiol.* 2019, 59, 123–33.

Jan RL, Tai MC, Weng SF, Chang C, Wang JJ, Chang YS. Risk of corneal ulcer in patients with end-stage renal disease: a retrospective large-scale cohort study. *Br J Ophthalmol.* 2018;102:868–72.

Jeng BH, Gritz DC, Kumar AB, et al., Epidemiology of ulcerative keratitis in Northern California. *Arch Ophthalmol.* 2010;128:1022–8.

Jiang, H.; Han, S.; Guo, C.; Liu, T.; Wu, X. In vitro and in vivo effectiveness evaluation of balofloxacin in experimental Staphylococcus aureus keratitis. *J. Ocul. Pharmacol. Ther.* 2014; 30: 482–8.

Jones L, Downie LE, Korb D, et al., TFOS DEWS II Management and Therapy Report. *Ocul Surf.* 2017;15:575–628.

Joslin CE, Tu EY, McMahon TT, Passaro DJ, Stayner LT, Sugar J. Epidemiological characteristics of a Chicago-area Acanthamoeba keratitis outbreak. *Am J Ophthalmol* 2006;142(2):212–17.

Karimi M, Esfahani BN, Halaji M, Mobasherizadeh S, Shahin M, Havaei SR, Shokri D, Havaei SA. Molecular characteristics and antibiotic resistance pattern of Staphylococcus aureus nasal carriage in tertiary care hospitals of Isfahan, *Iran. Infez Med.* 2017 Sep 1;25(3):234–40.

Kato N, Ide T, Kobashi H, Toda I. Infectious keratitis after corneal crosslinking for keratoconus caused by levofloxacin-resistant microorganisms. *BMC Ophthalmol.* 2021 Aug 31;21(1):317. doi: 10.1186/s12886-021-02081-4

Kaye R, Kaye A, Sueke H, et al., Recurrent bacterial keratitis. *Investig Ophthalmol Vis Sci.* 2013;54:4136–9.

Kaye S, Tuft S, Neal T, et al., Bacterial susceptibility to topical antimicrobials and clinical outcome in bacterial keratitis. *Investig Ophthalmol Vis Sci.* 2010;51:362–8.

Kaye, S.; Sueke, H.; Romano, V.; Chen, J.Y.; Carnt, N.; Tuft, S.; Neal, T. Impression membrane for the diagnosis of microbial keratitis. *Br. J. Ophthalmol.* 2016, 100, 607–10.

Keay L, Edwards K, Naduvilath T, Taylor HR, Snibson GR, Forde K, Stapleton F. Microbial keratitis predisposing factors and morbidity. *Ophthalmol.* 2006 Jan;113(1):109–16. doi: 10.1016/j.ophtha.2005.08.013

Khoo P, Cabrera-Aguas M, Robaei D, Lahra MM, Watson S. Microbial Keratitis and Ocular Surface Disease: A 5-Year Study of the Microbiology, Risk Factors and Clinical Outcomes in Sydney, Australia. *Curr Eye Res.* 2019 Nov;44(11):1195–202. doi: 10.1080/02713683.2019.1631852

Khoo P, Cabrera-Aguas MP, Nguyen V, Lahra MM, Watson SL. Microbial keratitis in Sydney, Australia: risk factors, patient outcomes, and seasonal variation. *Graefes Arch Clin Exp Ophthalmol.* 2020; 258(8):1745–55. doi: 10.1007/s00417-020-04681-0

Khoo, P; Cabrera-Aguas, M; Watson, S. L. Microbial Keratitis After Corneal Collagen Cross-Linking for Corneal Ectasia. *Asia-Pacific Journal of Ophthalmology*: July/August 2021; 10(4):355–9. doi: 10.1097/APO.0000000000000379

Khor HG, Cho I, Lee KRCK et al., Spectrum of microbial keratitis encountered in the tropics. *Eye Contact Lens* 2020; 46: 17–23.

Khor W-B, Prajna VN, Garg P, et al., The Asia Cornea Society Infectious Keratitis Study: A Prospective Multicenter Study of Infectious Keratitis In Asia. *Am J Ophthalmol.* 2018;195:161–70.

Kilic, B.B.; Altiors, D.D.; Demirbilek, M.; Ogus, E. Comparison between corneal cross-linking, topical antibiotic and combined therapy in experimental bacterial keratitis model. *Saudi J. Ophthalmol.* 2018, 32, 97–104.

Kim D, Song L, Breitwieser FP, Salzberg SL. Centrifuge: rapid and sensitive classification of metagenomic sequences. *Genome Res.* 2016;26(12):1721–9.

Konduri R, Arunasri K, Reddy GS, Adicherla H, Sharma S, Shivaji S. Global gene expression in Escherichia coli, isolated from the diseased ocular surface of the human eye with a potential to form biofilm. *Gut Pathogens.* 2017; 9: 15.

Konduri R, J Ramchiary, J S S Prakash, K Arunasri, S Sharma, S Shivaji. Gene Targets in Ocular Pathogenic *Escherichia coli* for Mitigation of Biofilm Formation to Overcome Antibiotic Resistance. *Front Microbiol.* 2019 Jun 21;10:1308. https://doi.org/10.3389/fmicb.2019.01308. PMID: 31293528; PMCID: PMC6598151.

Kowalski RP, Karenchak LM, Romanowski EG. Infectious disease: changing antibiotic susceptibility. *Ophthalmol. Clin. N. Am.* 2003;16:1–9. doi: 10.1016/s0896-1549(02)00061-5

Kowalski, R.P.; Kowalski, T.A.; Shanks, R.M.; Romanowski, E.G.; Karenchak, L.M.; Mah, F.S. In vitro comparison of combination and monotherapy for the empiric and optimal coverage of bacterial keratitis based on incidence of infection. *Cornea* 2013; 32: 830–4.

Kulauzov M, Medić D, Jovanović J, Mihajlović-Ukropina M, Stefan-Mikic S, Sevic S. (Antimicrobial resistance of the bacteria isolated from blood cultures in 2007). *Med Pregl.* 2008;61(1):21–6.

Kumar A, Khurana A, Sharma M, Chauhan L. Causative fungi and treatment outcome of dematiaceous fungal keratitis in North India. *Indian J Ophthalmol.* 2019;67:1048–53

Kumar A, Pandya S, Kavathia G, Antala S, Madan M, Javdekar T. Microbial keratitis in Gujarat, Western India: findings from 200 cases. *Pan Afr Med J.* 2011;10:48.

Kumar, A., Zhang, J., and Yu, F.-S. X. (2006). Toll-like Receptor 2-mediated mExpression of β-defensin-2 in Human Corneal Epithelial Cells. *Microbes Infect.* 8, 380–9. doi:10.1016/j.micinf.2005.07.006

Kunimoto DY, Sharma S, Garg P, Gopinathan U, Miller D, Rao GN. Corneal ulceration in the elderly in Hyderabad, south India. *Br J Ophthalmol.* 2000;84:54–9.

Lalitha P, Manoharan G, Karpagam R, et al., Trends in antibiotic resistance in bacterial keratitis isolates from South India. *Br J Ophthalmol.* 2016;101(2):108–13.

Lalitha P, Manoharan G, Karpagam R, Prajna N V, Srinivasan M, Mascarenhas J, Das M, Porco T C, Lietman T M, Cevallos V, Keenan, J D. Trends in antibiotic resistance in bacterial keratitis isolates from *South India Br J Ophthalmol.* 2017 February; 101(2): 108–13. doi:10.1136/bjophthalmol-2016-308487

Lalitha P, Prajna NV, Manoharan G, et al., Trends in bacterial and fungal keratitis in South India, 2002–2012. *Br J Ophthalmol.* 2015;99:192–4.

Lalitha P, Srinivasan M, Manikandan P, et al., Relationship of in vitro susceptibility to moxifloxacin and in vivo clinical outcome in bacterial keratitis. *Clin Infect Dis.* 2012a;54(10):1381–7. https://doi. org/10.1093/cid/cis189

Lalitha P, Srinivasan M, Rajaraman R, et al., Nocardia keratitis: clinical course and effect of corticosteroids. *Am J Ophthalmol.* 2012b;154:934–9.e1.

Laspina F, Samudio M, Cibils D, et al., Epidemiological characteristics of microbiological results on patients with infectious corneal ulcers: a 13-year survey in Paraguay. *Graefes Arch Clin Exp Ophthalmol.* 2004;242(3):204–9.

Lawther K, Santos FG, Oyama LB, Rubino F, Morrison S, Creevey CJ, McGrath JW and Huws SA (2022) Resistome Analysis of Global Livestock and Soil Microbiomes. *Front. Microbiol.* 13:897905. doi: 10.3389/fmicb.2022.897905

Leck A, Thomas P, Hagan M, et al., Aetiology of suppurative corneal ulcers in Ghana and south India, and epidemiology of fungal keratitis. *Br J Ophthalmol.* 2002;86(11):1211–5.

Li S, Yi G, Peng H, et al., How Ocular Surface Microbiota Debuts in Type 2 Diabetes Mellitus. *Front Cell Infect Microbiol.* 2019;9:202.

Lichtinger A, Yeung SN, Kim P, et al., Shifting trends in bacterial keratitis in Toronto: an 11-year review. *Ophthalmology.* 2012;119(9):1785–90.

Lin L, Duan F, Yang Y, Lou B, Liang L, Lin X. Nine-year analysis of isolated pathogens and antibiotic susceptibilities of microbial keratitis from a large referral eye center in southern China. *Infect Drug Resist.* 2019a;12:1295–302.

Lin L, Lan W, Lou B, et al., Genus distribution of bacteria and fungi associated with keratitis in a large eye center located in southern China. *Ophthalmic Epidemiol.* 2017;24(2):90–6.

Liu HY, Chu HS, Wang IJ, et al., Microbial keratitis in Taiwan: a 20-year update. *Am J Ophthalmol.* 2019b;205:74–81.

Lockwood, A., Hope-Ross, M., Chell, P. Neurotrophic keratopathy and diabetes mellitus. *Eye* 2006; 20; 837–839. https://doi.org/10.1038/sj.eye.6702053

Lomholt JA, Moller JK, Ehlers N. Prolonged persistence on the ocular surface of fortified gentamicin ointment as compared to fortified gentamicin eye drops. *Acta Ophthalmol Scand* 2000;78:34–6.

Long, Christopher P. M.D.*; Ozzello, Daniel J. M.D.†; Kikkawa, Don O. M.D., F.A.C.S.†; Liu, Catherine Y. M.D., Ph.D.†. Cytomegalovirus Blepharitis and Keratitis Masquerading as Eyelid Malignancy. *Ophthalmic Plast Reconstr Surg.* January/February 2022; 38(1): e34. doi: 10.1097/IOP.0000000000001992

Lowder BV, Guinane CM, Ben Zakour NL, Weinert LA, Conway-Morris A, Cartwright RA, Simpson AJ, Rambaut A, Nübel U, Fitzgerald JR. Recent human-to-poultry host jump, adaptation, and pandemic spread of Staphylococcus aureus. *Proceedings of the National Academy of Sciences of the United States of America* 2009; 106: 19545–50.

Luchs JI, Cohen EJ, Rapuano CJ, Laibson PR. Ulcerative keratitis in bullous keratopathy. *Ophthalmol.* 1997 May;104(5):816–22. doi: 10.1016/s0161-6420(97)30228-0. PMID: 9160028.

Ma T, McAllister TA, Guan L L. A review of the resistome within the digestive tract of livestock *J Animal Sci Biotechnol* 2021; 12:121. https://doi.org/10.1186/s40 104-021-00643-6

Ma T, Zaheer R, McAllister TA, Guo W, Li F, Tu Y, Diao Q, Guan LL. Expressions of resistome is linked to the key functions and stability of active rumenmicrobiome *Animal Microbiome* 2022; 4:38. https://doi.org/10.1186/s42523-022-00189-6

Machado E, Coque TM, Cantón R, Sousa JC, Peixe L. Antibiotic resistance integrons and extended-spectrum {beta}-lactamases among Enterobacteriaceae isolates recovered from chickens and swine in Portugal. *J Antimicrob Chemother.* 2008; 62: 296–302.

Magiorakos AP, Srinivasan A, Carey RB, et al., Multidrug-resistant, extensively drug-resistant and pandrug-resistant bacteria: an international expert proposal for interim standard definitions for acquired resistance. *Clin Microbiol Infect.* 2012;18:268–281. doi:10.1111/ j.1469-0691.2011.03570.x

Mallari PLT, McCarty DJ, Daniell M, Taylor H. Increased incidence of corneal perforation after topical fluoroquinolone treatment for microbial keratitis. *Am J Ophthalmol.* 2001;131(1):131–3.

Mallela, L. S., Sharma, P., Rao, T. S. R., Roy, S. Recombinant IL-22 Promotes protection in a Murine Model of Aspergillus flavus Keratitis and Mediates Host Immune Responses in Human Corneal Epithelial Cells. *Cell Microbiol* 2021; 24: e13367. doi:10.1111/cmi.13367

Mandour SS, Marey HM, Farahat HG. Resistant Microbial Keratitis in South Nile Delta, Egypt: influence of Regional Risk Factors. *Semin Ophthalmol.* 2016;31:473–8.

Martinez MN, Papich MG, Drusano GL. Dosing regimen matters:the importance of early intervention and rapid attainment of the pharmacokinetic/pharmacodynamic target. *Antimicrob Agents Chemother.* 2012;56:2795–805

Mather AE, Matthews L, Mellor DJ, Reeve R, Denwood MJ, Boerlin P, Reid-Smith RJ, Brown DJ, Coia JE, Browning LM, Haydon DT, Reid SW. An ecological approach to assessing the epidemiology of antimicrobial resistance in animal and human populations. *Proc Royal Society B: Biol Sci* 2012; 279: 1630–9.

Mather AE, Reid SW, Maskell DJ, Parkhill J, Fookes MC, Harris SR, Brown DJ, Coia JE, Mulvey MR, Gilmour MW, Petrovska L, de Pinna E, Kuroda M, Akiba M, Izumiya H, Connor TR, Suchard MA, Lemey P, Mellor DJ, Haydon DT, Thomson NR. Distinguishable epidemics of multidrug-resistant Salmonella Typhimurium DT104 in different hosts. *Science* 2013; 341: 1514–7.

Mathers W. Use of higher medication concentrations in the treatment of Acanthamoeba keratitis. *Arch Ophthalmol* 2006;124:6:923.

McDermott AM. Antimicrobial Compounds in Tears. *Exp. Eye Res.* 2013; 117: 53–61. doi:10.1016/j.exer.2013.07.014

McDermott ML, Tran TD, Cowden JW, et al., Corneal stromal penetration of topical ciprofloxacin in humans. *Ophthalmol* 1993;100:197–200.

Meier PA, Mathers WD, Sutphin JE, Folberg R, Hwang T, Wenzel RP. An epidemic of presumed Acanthamoeba keratitis that followed regional flooding. *Arch Ophthalmol* 1998;116:8:1090–4.

Miedziak AI, Miller MR, Rapuano CJ, et al., Risk factors in microbial keratitis leading to penetrating keratoplasty. *Ophthalmol* 1999;106:1166–71.

Moe CA, LalithaP, PrajnaNV, et al., Outcomesof amoebic,fungal,and bacterialkeratitis:A retrospective cohortstudy.*PLoSONE* 2022; 17(2):e0264021.

Mohammed I, Said DG, Nubile M, Mastropasqua L, Dua HS. Cathelicidin-Derived Synthetic Peptide Improves Therapeutic Potential of Vancomycin against Pseudomonas aeruginosa. *Front. Microbiol.* 2019; 10: 2190. doi:10.3389/fmicb.2019.02190

Mohammed I, Suleman H, Otri, et al., Localization and Gene Expression of Human β-Defensin 9 at the Human Ocular Surface Epithelium. *Invest. Ophthalmol. Vis. Sci.* 2010; 51: 4677–82. doi:10.1167/iovs.10-5334

Moraes, L. C., Só, M. V., Dal Pizzol Tda, S., Ferreira, M. B., Montagner, F. Distribution of genes related to antimicrobial resistance in different oral environments: A systematic review. *J Endodontics* 2015; 41(4): 434–41. https://doi.org/10.1016/j.joen.2014.12.018

Morris TW, Gearinger LS, Usner DW, et al., Integrated analysis of three bacterial conjunctivitis trials of besifloxacin ophthalmic suspension, 0.6%: microbiological eradication outcomes. *Clin Ophthalmol.* 2011;5:1359–67.

Mun Y, Kim MK, Oh JY. Ten-year analysis of microbiological profile and antibiotic sensitivity for bacterial keratitis in Korea. *PLoS ONE* 2019; 14(3): e0213103. https://doi.org/10.1371/journal.pone.0213103

NaPier E, Camacho M, McDevitt TF, Sweeney AR. Neurotrophic keratopathy: current challenges and future prospects. *Ann Med.* 2022 Dec;54(1):666–73. doi: 10.1080/07853890.2022.2045035

Naranjo, A., Arboleda, A., Martinez, J. D., et al., Rose Bengal Photodynamic Antimicrobial Therapy for Patients with Progressive Infectious Keratitis: A Pilot Clinical Study. *Am. J. Ophthalmol.* 2019; 208: 387–96. doi:10.1016/j.ajo.2019.08.027

Nebbioso M, Zicari AM, Celani C, Lollobrigida V, Grenga R, Duse M. Pathogenesis of Vernal Keratoconjunctivitis and Associated Factors. *Semin Ophthalmol.* 2015;30(5–6):340–4.

Newman P, Hider P, Waring G, et al., Corneal ulcer due to Achromobacter xylosoxidans. *Br J Ophthalmol.* 1984;68(7):472–4.

Nickel JC. Urinary Tract Infections and Resistant Bacteria: Highlights of a Symposium at the Combined Meeting of the *25th International Congress of Chemotherapy (ICC) and the 17th European Congress of Clinical Microbiology and Infectious Diseases (ECCMID)*, March 31–April 3, 2007, Munich, Germany. *Rev Urol.* 2007 Spring;9(2):78–80. PMID: 17592541; PMCID: PMC1892623.

Nielsen M C, Nan Wang2, and Sunny C. Jiang1,2 Acquisition of antibiotic resistance genes on human skin after swimming in the ocean. *Environ Res* 2021 Jun;197:110978. doi: 10.1016/j.envres.2021.110978

Nos-Barbera, S., Portoles, M., Morilla, A., Ubach, J., Andreu, D., and Paterson, C.A. (1997). Effect of Hybrid Peptides of Cecropin A and Melittin in an Experimental Model of Bacterial Keratitis. *Cornea* 16, 101–06. doi:10.1097/00003226-199701000-00017

O'Brien TP, Sawusah MR, Dick JD, et al., Topical ciprofloxacin treatment of pseudomonas keratitis in rabbits. *Arch Ophthalmol* 1988;106:1444–6.

Ormerod LD, Gomez DS, Schanzlin DJ, Smith RE. Chronic alcoholism and microbial keratitis. *Br J Ophthalmol.* 1988 Feb;72(2):155–9. doi: 10.1136/bjo.72.2.155. PMID: 3349017; PMCID: PMC1041391

Otri, A. M.,Mohammed, I., Abedin, A., et al., Antimicrobial Peptides Expression by Ocular Surface Cells in Response to Acanthamoeba Castellanii: an In Vitro Study. *Br. J. Ophthalmol.* 2010; 94: 1523–7. doi:10.1136/bjo.2009.178236

Pandita A, Murphy C. Microbial keratitis in Waikato, New Zealand. *Clin Exp Ophthalmol.* 2011;39(5):393–7.

Panjwani N. Pathogenesis of acanthamoeba keratitis. *Ocul Surf.* 2010 Apr;8(2):70–9. doi: 10.1016/s1542-0124(12)70071-x. PMID: 20427010; PMCID: PMC3072032.

Paterson GK, Larsen AR, Robb A, Edwards GE, Pennycott TW, Foster G, Mot D, Hermans K, Baert K, Peacock SJ, Parkhill J, Zadoks RN, Holmes MA. The newly described mecA homologue, mecALGA251, is present in methicillin-resistant Staphylococcus aureus isolates from a diverse range of host species. *J Antimicrob Chem* 2012; 67: 2809–13.

Paulsen, F. P., Pufe, T., Schaudig, U., et al. Detection of Natural Peptide Antibiotics in Human Nasolacrimal Ducts. *Invest. Ophthalmol. Vis. Sci.* 2001;42: 2157–63.

Peng MY, Cevallos V, McLeod SD, et al., Bacterial Keratitis: Isolated Organisms and Antibiotic Resistance Patterns in San Francisco. *Cornea.* 2018;37(1):84–7.

Perera G, Hay R. A guide to antibiotic resistance in bacterial skin infections. *J Eur Acad Dermatol Venereol.* 2005; Sep;19(5):531–45. doi: 10.1111/j.1468-3083.2005.01296.x. PMID: 16164705.

Petrillo, F.; Pignataro, D.; Di Lella, F.M.; Reibaldi, M.; Fallico, M.; Castellino, N.; Parisi, G.; Trotta, M.C.; D'Amico, M.; Santella, B.; Folliero, V.; Della Rocca, M.T.; Rinaldi, M.; Franci, G.; Avitabile, T.; Galdiero, M.; Boccia, G. Antimicrobial Susceptibility Patterns and Resistance Trends of *Staphylococcus aureus* and Coagulase-Negative Staphylococci Strains Isolated from Ocular Infections. *Antibiotics* 2021; 10: 527. https://doi.org/10.3390/antibiotics10050527

Politis M, Wajnsztajn D, Rosin B, et al., Trends of Bacterial Keratitis Culture Isolates in Jerusalem; a 13- Years Analysis. *PLoS One.* 2016;11(11):e0165223.

Price FW jr, Whitson WE, Collins KS, et al., Corneal tissue levels of topically applied ciprofloxacin. *Cornea* 1995;14: 152–6.

Price LB, Stegger M, Hasman H, Aziz M, Larsen J, Andersen PS, Pearson T, Waters AE, Foster JT, Schupp J, Gillece J, Driebe E, Liu CM, Springer B, Zdovc I, Battisti A, Franco A, Zmudzki J, Schwarz S, Butaye P, Jouy E, Pomba C, Porrero MC, Ruimy R, Smith TC, Robinson DA, Weese JS, Arriola CS, Yu F, Laurent F, Keim P, Skov R, Aarestrup FM. Staphylococcus aureus CC398: host adaptation and emergence of methicillin resistance in livestock. *mBio* 2012; 3: e00305–311.

Price, M. O., and Price, F. W., Jr. Corneal Cross-Linking in the Treatment of Corneal Ulcers. *Curr. Opin. Ophthalmol.* 2016;27: 250–5. doi:10.1097/icu.0000000000000248

Randleman JB, Shah RD. LASIK interface complications: etiology, management, and outcomes. *J Refract Surg.* 2012;28: 575–86.

Rapuano, P.B.; Scanameo, A.H.; Amponin, D.E.; Paulose, S.A.; Zyablitskaya, M.; Takaoka, A.; Suh, L.H.; Nagasaki, T.; Trokel, S.L.; nPaik, D.C. Antimicrobial Studies Using the Therapeutic Tissue Cross-Linking Agent, Sodium nHydroxymethylglycinate: Implication for Treating Infectious Keratitis. *Invest. Ophthalmol. Vis Sci.* 2018, 59, 332–7.

Rayamajhee, B., Mark DP Willcox, Fiona L Henriquez, Constantinos Petsoglou and Nicole Carnt *Acanthamoeba* keratitis: an increasingly common infectious disease of the cornea *Lancet Microbe*: August 1, 2021; 2: E345–6. https://doi.org/10.1016/S2666-5247(21)00093-8

Reitberger HH, Czugala M, Chow C, et al., Argon cold plasma—a novel tool to treat therapy-resistant corneal infections. *Am J Ophthalmol.* 2018;190:150–63.

Richoz, O.; Kling, S.; Hoogewoud, F.; Hammer, A.; Tabibian, D.; Francois, P.; Schrenzel, J.; Hafezi, F. Antibacterial efficacy of accelerated photoactivated chromophore for keratitis-corneal collagen cross-linking (PACK-CXL). *J. Refract. Surg*. 2014, 30, 850–4.

Roberts, M. C. (2002). Antibiotic toxicity, interactions and resistance development. *Periodontol* 2000; 28: 280–97. https://doi.org/10.1034/j. 1600-0757.2002.280112.x

Roy, S., Marla, S., and Praneetha, D. Recognition of Corynebacterium Pseudodiphtheriticum by Toll-like Receptors and Up-Regulation of Antimicrobial Peptides in Human Corneal Epithelial Cells. *Virulence* 2015;6: 716–21. doi:10.1080/21505594.2015.1066063

Sala A, Taddei S, Santospirito D, Sandri C, Magnone W, Cabassi CS. Antibiotic resistance in conjunctival and enteric bacterial flora in raptors housed in a zoological garden. *Vet Med Sci*. 2016 Jun 10;2(4):239–45. doi: 10.1002/vms3.38. PMID: 29067199; PMCID: PMC5645847.

Sansanayudh W, Cevallos V, Porco TC, Margolis TP, Lietman TM, Acharya NR. Fusarium and Acanthamoeba keratitis: Can a single center detect outbreaks? *Br J Ophthalmol* 2008;95:5:720–1.

Sauer A, Greth M, Letsch J, et al., Contact Lenses and Infectious Keratitis: from a Case-Control Study to a Computation of the Risk for Wearers. *Cornea*. 2020;39:769–74.

Schaefer F, Bruttin O, Zografos L et al., Bacterial keratitis: a prospective clinical and microbiological study. *Br J Ophthalmol* 2001; 85(7):842–7

Schimel AM, Miller D, Flynn HW Jr. Endophthalmitis isolates and antibiotic susceptibilities: a 10-year review of culture-proven cases. *Am J Ophthalmol*. 2013;156(1):e51:50–52.

Shalchi Z, Gurbaxani A, Baker M, Nash J. Antibiotic resistance in microbial keratitis: ten-year experience of corneal scrapes in the United Kingdom. *Ophthalmology*. 2011 Nov;118(11):2161–5. doi: 10.1016/j.ophtha.2011.04.021

Sharma, P., Guha, S., Garg, P., and Roy, S. Differential Expression of Antimicrobial Peptides in Corneal Infection and Regulation of Antimicrobial Peptides and Reactive Oxygen Species by Type III Secretion System of Pseudomonas aeruginosa. *Pathog. Dis*. 2018;76. doi:10.1093/femspd/fty001. PMID: 29325116.

Sharma, P., Sharma, N., Mishra, P., et al., Differential Expression of Antimicrobial Peptides in Streptococcus Pneumoniae Keratitis and STAT3-dependent Expression of LL-37 by Streptococcus Pneumoniae in Human Corneal Epithelial Cells. *Pathogens* 2019; 8: 31. doi:10.3390/pathogens8010031

Shoff ME, Joslin CE, Tu EY, Kubatko L, Fuerst PA. Efficacy of contact lens systems against recent clinical and tap water Acanthamoeba isolates. *Cornea* 2008;27:6:713–9.

Siddiqui, R.; Lakhundi, S.; Khan, N.A. Status of the effectiveness of contact lens solutions against keratitis-causing pathogens. *Cont. Lens Anterior Eye* 2015, 38, 34–8.

Silva, N. C., Sarmento, B., and Pintado, M. (2013). The Importance of Antimicrobial Peptides and Their Potential for Therapeutic Use in Ophthalmology. *Int. J. Antimicrob. Agents* 41, 5–10. doi:10.1016/ j.ijantimicag.2012.07.020

Slean GR, Shorstein NH, Liu L, Paschal JF, Winthrop KL, Herrinton LJ. Pathogens and antibiotic sensitivities in endophthalmitis. *Clin Exp Ophthalmol*. 2017;45(5):481–8.

Soleimani, M., Tabatabaei, S.A., Masoumi, A. et al., Infectious keratitis: trends in microbiological and antibiotic sensitivity patterns. *Eye* 2021; 35: 3110–15. https://doi.org/10.1038/s41433-020-01378-w

Sony P, Sharma N, Vajpayee RB, Ray M. Therapeutic keratoplasty for infectious keratitis: a review of the literature. *CLAO J*. 2002 Jul;28(3):111–8.

Spierer O, Miller D, O'brien TP. Comparative activity of antimicrobials against Pseudomonas aeruginosa, Achromobacter xylosoxidans and Stenotrophomonas maltophilia keratitis isolates. *Br J Ophthalmol*. 2018;102(5):708–12.

Srinivasan M, Mascarenhas J, Rajaraman R, et al., Corticosteroids for bacterial keratitis: the Steroids for Corneal Ulcers Trial (SCUT). *Arch Ophthalmol.* 2012a;130(2):143–50

Srinivasan M, Mascarenhas J, Rajaraman R, et al., The steroids for corneal ulcers trial: study design and baseline characteristics. *Arch Ophthalmol.* 2012b;130:151–7.

Srivastava S, Gill A. Untreated morbidity and treatment-seeking behaviour among the elderly in India: analysis based on National Sample Survey 2004 and 2014. *SSM Popul Health.* 2020;10: 100557.

Stapleton F, Carnt N. Contact lens-related microbial keratitis: how have epidemiology and genetics helped us with pathogenesis and prophylaxis. *Eye (Lond).* 2012;26(2):185–93.

Stapleton F, Naduvilath T, Keay L, et al., Risk factors and causative organisms in microbial keratitis in daily disposable contact lens wear. *PLoS ONE.* 2017;12:e0181343.

Stapleton F. Contact lens-related corneal infection in Australia. *Clin Exp Optom.* 2020;103:408–17.

Su, G.; Wei, Z.; Wang, L.; Shen, J.; Baudouin, C.; Labbé, A.; Liang, Q. Evaluation of Toluidine Blue-Mediated Photodynamic Therapy for Experimental Bacterial Keratitis in Rabbits. *Transl. Vis. Sci. Technol.* 2020; 9:13.

Sueke, H.; Kaye, S.B.; Neal, T.; Hall, A.; Tuft, S.; Parry, C.M. An in vitro investigation of synergy or antagonism between antimicrobial combinations against isolates from bacterial keratitis. *Invest. Ophthalmol. Vis. Sci.* 2010; 51: 4151–5.

Tabbara KF, Chacra BCT. Infectious keratitis. In: Tabbara KF, Abu El-Asrar AM, Khairallah M, editors. *Ocular Infections.* New York, NY: Springer Berlin Heidelberg; 2014:73–94.

Takeshita T. Bilateral herpes simplex virus keratitis in a patient with pemphigus vulgaris. *Clin Experimen Dermatol* 1996; 21(4):291–2. https://doi.org/10.1111/j.1365-2230.1996.tb00098.x

Tam ALC, Côté E, Saldanha M, Lichtinger A, Slomovic AR. Bacterial Keratitis in Toronto: a 16-Year Review of the Microorganisms Isolated and the Resistance Patterns Observed. *Cornea.* 2017;36:1528–34.

Tan S, Walkden A, Au L, et al., Twelve-year analysis of microbial keratitis trends at a UK tertiary hospital. *Eye.* 2017;31(8):1229–36.

Tao H, Wang J, Li L, Zhang H-Z, Chen M-P and Li L. Incidence and Antimicrobial Sensitivity Profiles of Normal Conjunctiva Bacterial Flora in the Central Area of China: A Hospital-Based Study. *Front. Physiol.* 2017; 8:363. doi: 10.3389/fphys.2017.00363

Tavassoli S, Nayar G, Darcy K, et al., An 11-year analysis of microbial keratitis in the South West of England using brain-heart infusion broth. *Eye (Lond)* 2019;33:1619–25.

Thebpatiphat N, Hammersmith KM, Rocha FN, Rapuano CJ, Ayres BD, Laibson PR, Eagle RC, Cohen EJ. Acanthamoeba keratitis: A parasite on the rise. *Cornea* 2007;26:6:701–6.

Thomas RK, Melton R, Asbell PA. Antibiotic resistance among ocular pathogens: current trends from the ARMOR surveillance study (2009–2016). *Clin Optom (Auckl).* 2019;11:15–26. doi:10.2147/OPTO.S189115

Ting DSJ, Ho CS, Cairns J, Elsahn A, et al., 12-year analysis of incidence, microbiological profiles and in vitro antimicrobial susceptibility of infectious keratitis: the Nottingham Infectious Keratitis Study. *Br J Ophthalmol.* 2020a; 105(3):328–333. 10.1136/bjophthalmol-2020-316128

Ting DSJ, McKenna M, Sadiq SN, et al., Arthrographis kalrae Keratitis Complicated by Endophthalmitis: a Case Report With Literature Review. *Eye Contact Lens*. 2020b; 46(6):e59–e65. 10.1097/ICL.0000000000000713. PMID: 32443014.

Ting DSJ, Ho CS, Deshmukh R, Said DG, Dua HS. Infectious keratitis: an update on epidemiology, causative microorganisms, risk factors, and antimicrobial resistance. *Eye (Lond)*. 2021a Apr;35(4):1084–1101. doi: 10.1038/s41433-020-01339-3

Ting DSJ, Cairns J, Gopal BP, Ho CS, Krstic L, Elsahn A, Lister M, Said DG and Dua HS. Risk Factors, Clinical Outcomes, and Prognostic Factors of Bacterial Keratitis: The Nottingham Infectious Keratitis Study. *Front. Med.* 2021b; 8:715118. doi: 10.3389/fmed.2021.715118

Ting DSJ, Said DG, Dua HS. Interface Haze After Descemet Stripping Automated Endothelial Keratoplasty. *JAMA Ophthalmol*. 2019;137:1201–2.

Tong W, Chen D, Chai C, Tan AM, Manotosh R. Disease patterns of microbial keratitis in Singapore: a retrospective case series. *Cont Lens Anterior Eye*. 2019;42:455–61.

Tsai TY, Adiyabazar D, Hsiao CH, Pan LY, Chen SY, Tsai YJ, Chen CB, Chung WH, Ma DH. Microbial Keratitis in Patients With Stevens-Johnson Syndrome and Toxic Epidermal Necrolysis: Experience From a Tertiary Centre in Taiwan. *Cornea*. 2022 May 13. doi: 10.1097/ICO.0000000000003063. Epub ahead of print. PMID: 35587449.

Tuft, S J and Melville Matheson In vitro antibiotic resistance in bacterial keratitis in London *Br J Ophthalmol* 2000;84:687–691.

Tummanapalli S S, Willcox MD. Antimicrobial resistance of ocular microbes and the role of antimicrobial peptides. *Clin Exp Optom*. 2021 Apr;104(3):295–307. doi: 10.1111/cxo.13125. Epub 2021 Mar 24. PMID: 32924208.

Ung L, PJM Bispo, SS Shanbhag, MS Gilmore, J Chodosh. The persistent dilemma of microbial keratitis: Global burden, diagnosis, and antimicrobial resistance *Survey Ophthalmol* May–June 2019; 64(3): 255–71.

Ustundag-Okur, N.; Gokce, E.H.; Egrilmez, S.; Ozer, O.; Ertan, G. Novel ofloxacin-loaded microemulsion formulations for ocular delivery. *J. Ocul. Pharmacol. Ther*. 2014; 30: 319–32.

Vazirani J, Wurity S, Ali MH. Multidrug-resistant Pseudomonas aeruginosa keratitis: risk factors, clinical characteristics, and outcomes. *Ophthalmology*. 2015;122(10):2110–4. doi:10.1016/j.ophtha.2015.06.007

Villedieu, A., Diaz-Torres, M. L., Hunt, N., McNab, R., Spratt, D. A., Wilson, M., Mullany, P. Prevalence of tetracycline resistance genes in oral bacteria. *Antimicrobial Agents and Chemother* 2003; 47(3): 878–82. https:// doi.org/10.1128/AAC.47.3.878-882.2003

Voss A, Loeffen F, Bakker J, Klaassen C, Wulf M. Methicillin-resistant Staphylococcus aureus in pig farming. *Emerg. Infect. Dis* 2005; 11: 1965–6. doi: 10.3201/eid1112.050428. PMID: 16485492; PMCID: PMC3367632.

Waleed, A M, Lua'i A-I, Wisam S, Sana J. Antimicrobial susceptibility of bacterial isolates from the conjunctiva, storage cases and mobile phones of university students using contact lenses. *Contact Lens Anterior Eye* 2021; 44(1): 62–6. ISSN 1367-0484. https://doi.org/10.1016/j.clae.2019.10.139

Walkden A, Fullwood C, Tan SZ, et al. Association Between Season, Temperature and Causative Organism in Microbial Keratitis in the UK. *Cornea*. 2018;37:1555–60.

Walsh C. Molecular mechanisms that confer antibacterial drug resistance. *Nature* 2000; 406: 775–81.

Wang B, Yang S, Zhai HL, et al., A comparative study of risk factors for corneal infection in diabetic and non-diabetic patients. *Int J Ophthalmol*. 2018; 11:43–7.

Watson, S.; Cabrera-Aguas, M.; Khoo, P.; Pratama, R.; Gatus, B.J.; Gulholm, T.; El-Nasser, J.; Lahra, M.M. Keratitis antimicrobial resistance surveillance program, Sydney, Australia: 2016 Annual Report. *Clin. Exp. Ophthalmol.* 2019; 47: 20–25.

Whitcher JP, Srinivasan M, Upadhyay MP. Corneal blindness: a global perspective. *Bull World Health Organ* 2001;79:214–21.

Wilkinson, A., Kawaguchi, N., Geczy, C., and Di Girolamo, N. S100A8 and S100A9 Proteins Are Expressed by Human Corneal Stromal Dendritic Cells. *Br. J. Ophthalmol.* 2016; 100: 1304–8. doi:10.1136/bjophthalmol-2016- 30882

Willcox MDP. Review of resistance of ocular isolates of Pseudomonas aeruginosa and staphylococci from keratitis to ciprofloxacin, gentamicin and cephalosporins. *Clin Exp Optom* 2011; 94: 161–8.

Woolhouse M, Ward M, van Bunnik B, Farrar J. Antimicrobial resistance in humans, livestock and the wider environment. *Philos Trans R Soc Lond B Biol Sci.* 2015 Jun 5;370(1670):20140083. doi: 10.1098/rstb.2014.0083

Xie L, Zhong W, Shi W, Sun S. Spectrum of fungal keratitis in north China. *Ophthalmology.* 2006;113(11):1943–8.

Xu, P., Gunsolley, J. Application of metagenomics in understanding oral health and disease. *Virulence* 2014; 5(3): 424–32. https://doi.org/10. 4161/viru.28532

Yamaguchi T, Bissen-Miyajima H, Hori-Komai Y, et al., Infectious keratitis outbreak after laser in situ keratomileusis at a single laser center in Japan. *J Cataract Refract Surg.* 2011; 37: 894–900.

Yilmaz S, Aras Saklamaz and Ahmet Maden Pseudomonas keratitis, *Ophthalmol* 2006; 113(5); 883–4.

Young S. Ocular involvement in connective tissue disorders. *Curr Allergy Asthma Rep.* 2005 Jul;5(4):323–6. doi: 10.1007/s11882-005-0076-y

Zhang Y, Wang ZQ, Sun XG. Etiological analysis and in vitro drug sensitivity of bacterial keratitis in northern China in the period of 2006–2015. *Zhonghua Yan Ke Za Zhi.* 2017;53:662–7.

Zhang Z, Cao K, Liu J, Wei Z, Xu X, Liang Q. Pathogens and Antibiotic Susceptibilities of Global Bacterial Keratitis: A Meta-Analysis. *Antibiotics (Basel)* 2022 Feb 12;11(2):238. doi: 10.3390/antibiotics11020238

Zhu L, Titone R, Robertson DM. The impact of hyperglycemia on the corneal epithelium: molecular mechanisms and insight. *Ocul Surf.* 2019;17:644–54.

5

Antimicrobial Resistance in the Diseased Human Eye: Fungal Keratitis

S. Shivaji

5.1 Introduction

Fungi are associated with various niches of the human body as in the skin (Theodore et al., 2014; Pai et al., 2018), gut (Pilmis et al., 2020), nasal cavity (Shin et al., 2007), oral cavity (Chandra et al., 2016; Niimi et al., 2010), dental plaque (Damasceno et al., 2017; Niimi et al., 2010), blood (Evertsson et al., 2000), urinary tract (Moro and David, 2021) and so forth and over time these fungi have acquired antimicrobial resistance (AMR). But comparatively little is known about AMR in the fungi that colonise the ocular surface, which is also known to have its own mycobiome. This chapter highlights methods for the detection and identification of fungi on the surface of the eye in both normal individuals and in individuals with fungal keratitis (FK), and enumerates their occurrence, diversity and AMR in the healthy human eye and in the diseased fungal keratitis eye.

5.2 Collection, Processing, and Identification of Ocular Fungi

Collection of ocular samples from the eye needs to be done carefully and precisely to avoid contaminating microbes from the skin and the environment. The sample should be collected without any transport/collection media and processed immediately (Shivaji, 2022). In case the sample needs to be transported, it should be noted that the slides, media, and samples (contact lenses, cases and solutions, intraocular lenses, corneal, and iris tissues, etc.) are transported in secure boxes and processed for cultivation of the microorganisms. A specific medium without charcoal is used for collection of corneal scrapings (McLeod et al., 2005). The collection device and procedure for recovery of the sample also varies depending on the type of keratitis (Table 5.1). Further, several microscopic methods are employed for the detection of the microorganisms in the ocular samples (Table 5.2).

DOI: 10.1201/9781003451105-5

TABLE 5.1

Collection of Samples for the Diagnosis of Various Ocular Infections

Ocular infection	Sample	Sample collection device and procedure
Keratitis	Corneal scraping	Kimura spatula, No. 15 surgical blade, bent needle
Deep seated stromal infiltrate in keratitis	Corneal biopsy	Lamellar biopsy
Non-healing keratitis requiring keratoplasty	Corneal buttons	Penetrating keratoplasty
Contact lens-associated keratitis	Contact lenses, lens cases, and lens solution	Aseptically removed, aseptically collected

Source: Modified from Sharma, 2012a and Shivaji 2022.

TABLE 5.2

Methods Used for the Diagnosis of Fungal Eye Infections in Conjunctival Swabs, Corneal Scrapings and Aqueous and Vitreous Fluid

Sample	Staining method	Microorganism
Conjunctival swabs/ scrapings	Gram, Giemsa, potassium hydroxide plus calcofluor white, direct/indirect immunofluorescence or immune-peroxidase staining.	Bacteria, fungi, parasites (Microsporidia) and Viral antigens.
Corneal scrapings	Gram, Giemsa, potassium hydroxide plus calcofluor white, Lactophenol cotton blue, Gomori methenamine silver, Ziehl-Neelsenstain, direct/ indirect immunofluorescence or immunperoxidase staining.	Bacteria, fungi, parasites (Acanthamoeba, Microsporidia) andviral antigens.
Aqueous/vitreous fluids/biopsy	Gram, Giemsa, calcofluor white, Gomori methenamine silver, Direct/indirect immunofluorescence or immune-peroxidase staining.	Bacteria, fungi and viral antigens.

Source: Modified from Sharma 2012a; 2012b and Shivaji 2022.

5.2.1 Detection of Fungi in Ocular Samples by Microscopy

A straight-forward method for the detection of fungi is to observe stained smears of clinical samples using a microscope (Shivaji, 2022). The most popular methods include:

(i) Detection of bacteria, fungi, and parasites in corneal scrapings by Gram staining (Sharma et al., 2002) (Figure 5.1a).

(ii) Giemsa staining to identify fungi including microsporidial spores in corneal scrapings from fungal keratitis patients (Zhang et al., 2010; Joseph et al., 2006).

(iii) Identification of fungi using potassium hydroxide plus calcofluor white stain (Chander et al., 1993) (Figure 5.1b) since this method detects 99.3%

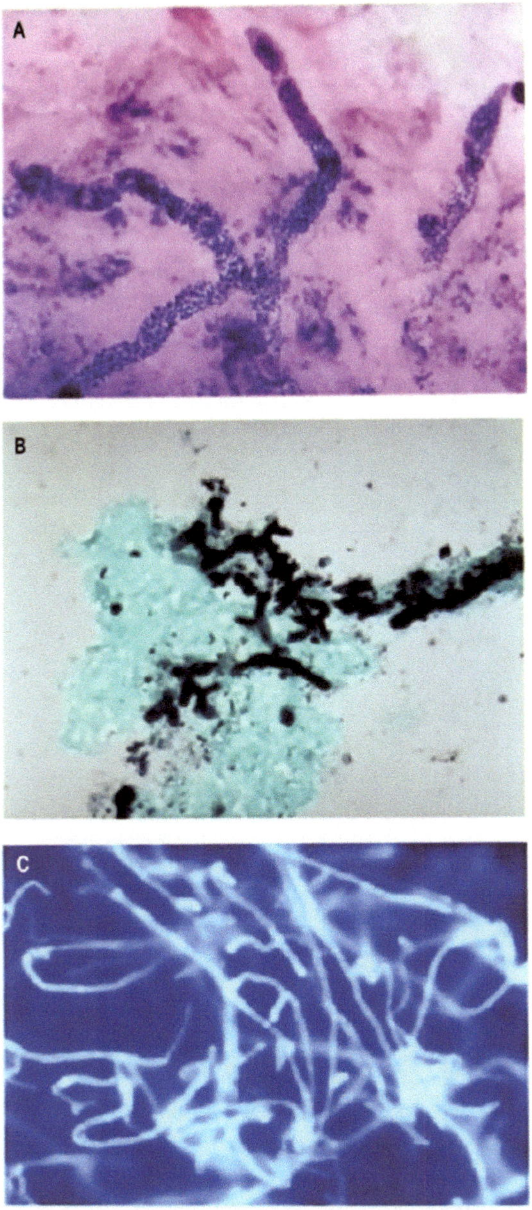

FIGURE 5.1 Detection of Fungi in Corneal Smears Following Gram staining 40X (a), Calcofluor white staining 40X (b) and Gomori's methenamine silver impregnation 40X (c). (Courtesy Venkatesh, LVPEI, Hyderabad.)

of the fungi, which is superior to results following Gram-staining (89.2%) (Bharathi et al., 2006).

(iv) Fungi could also be stained with lactophenol cotton blue using mounts of corneal ulcer scrapings (Thomas et al., 1991).

(v) Fungal filaments could also be detected using Gomori's methenamine silver impregnation technique (Vemuganti et al., 2002; Mukherjee et al., 2013) (Figure 5.1c).

(vi) Ziehl–Neelsen stain is also used to identify microsporidia (Joseph et al., 2006).

5.2.2 Detection of Fungi in Ocular Samples by Polymerase Chain Reaction

Uniplex and multiplex polymerase chain reaction (PCR) using primers specific to fungal phylogenetic markers targeting the 18S or 28S ribosomal RNA genes and ITS regions were used to detect fungal infections (Font et al., 2002; Embong et al., 2008; Ferrer and Alió, 2011; Sharma, 2012a; 2012b; Minkus et al., 2019; Shivaji et al., 2019; Sai Prashanthi et al., 2019). PCR results confirmed high concordance between PCR positivity and ocular mycotic infections caused by *Candida albicans*, *C. parasilopsis*, *Aspergillus flavus*, *Fusarium* spp., *Scedosporium apiospermum* and *Phoma* spp. (Font et al., 2002). The positivity of detecting fungal DNA was 40, 32 and 43% by 18S rRNA PCR, 28S rRNA PCR and ITS-PCR, respectively (Balne et al., 2012). More recently Next-generation sequencing (NGS), which is a high throughput sequencing method that sequences millions of small DNA fragments (150 to 500 bases) has been used to identify fungal isolates (Ma et al., 2019).

5.2.3 Detection of Fungi in Ocular Samples by Confocal Microscopy

Several studies have also used *in vivo* confocal microscopy (IVCM), for the diagnosis of *Acanthamoeba* keratitis (AK), filamentous fungal keratitis (FK) (Labbé et al., 2009; Vaddavalli et al., 2011), bacterial keratitis (BK) and HSV viral keratitis (VK) (Wang et al., 2019). The specificity of identifying AK and FK was 85.35 and 100% respectively. Further in FK, the accuracy of IVCM was comparable to the clinical observation (Wang et al., 2019). In addition, anterior segment Optical Coherence Tomography could be employed to diagnose Micro-cystic keratitis (Dembski et al., 2020). Hau et al. (2010) did caution that the accuracy of IVCM as a diagnosis tool for microbial keratitis is dependent on observer's experience.

5.2.4 Detection of Progression of Ocular Fungal Infections Using Animals as a Model System

Studies using experimental animals as models for FK should mimic the human ocular surface infection process by displaying clinical features comparable to the human counterpart and the results should be reproducible to evaluate trauma, infection, co-infection and should be convenient to isolate the fungi from the FK tissues (Kalkanci and Ozdek, 2011). Several rodents and mammals have been used as animal models because of the ease to breed, maintain and monitor responses to infection and additionally it is cost effective (Yuan and Wilhelmus, 2010).

5.2.4.1 Mouse as a Model

Both outbred (NIH Swiss) and inbred (BALB/C) mice are popular models for FK. Animals are immunosuppressed (methylprednisolone 100 mg/kg body weight or cyclophosphamide 180 mg/ kg body weight) prior to corneal inoculation with fungi (1 $\times 10^3$ CFU) on the scratched cornea of the anesthetised mice. Mice are sacrificed after the experiment and eyes removed and processed for histology for culturing of microbes (Wu et al., 2003 a) and for quantification of MIP2, KC, IL-1, and IL-6, the cytokines involved in inflammation (Mitchell et al., 2007; Zhong et al., 2009). Blood could also be collected for monitoring the immunity status, infection and so forth. *Fusarium* spp. grown as a biofilm on contact lenses has also been used as a model to study contact lens induced keratitis on injured corneas (Sun et al., 2010). The disadvantage of using mice is that they succumb to mortality even with ocular surface infections.

5.2.4.2 Rabbit as a Model

Rabbits have also been used to study contact lens (CL) induced keratitis (Velpandian et al., 2006,) and other ocular complication like keratomycosis (Ozturk et al., 2007) and endophthalmitis (Kocak et al., 2009). For these studies, rabbits were anaesthetised with ketamine (35 mg/kg) and xylazine (5 mg/kg) via., the intramuscular route and administered with corneal topical anaesthesia (0.2% novocaine) after which the corneal epithelium was removed, and the denuded cornea was inoculated with a fungal inoculum, protected using a CL and lids sutured to prevent CL extrusion (Goldblum et al., 2005). Keratomycosis was studied using Chinchilla rabbits injected with *Candida* cells into the corneal stroma and graded for conjunctival hyperaemia, corneal clouding, diameter of corneal neovascularisation and hypopyon (Ozturk et al., 2007).

5.2.4.3 Rat as a Model

Wistar or Lewis rats, due to their suitable body and eye size, immune response, and the ease of employing surgical procedures are also used as models for FK (Ren and Wu, 2010). For the induction of keratitis, a half thickness linear incision is made 2 mm from the centre of cornea with a syringe using a 30 G needle. Fungal inoculum is injected into the stroma.

5.2.4.4 Pig as a Model

The large size of the pig eye, and the ease in fitting soft contact lenses makes it a suitable and preferred animal model for contact lens associated keratitis. In addition to the normal pigs, the female Yucatan micropigs (about 25 Kg adult weight) were also used as a model for FK (He et al., 1992).

5.3 Fungi of the Normal Human Eye and AMR

5.3.1 Culturing Method for the Detection of Fungi from the Normal Eye

In the nineteenth century fungal signatures were discovered on the human eye (Von Graefe and Mittheilungen, 1857; Köllner, 1906) by culturing microorganisms on suitable media. But this method based on culture studies did not yield consistent results (Ozkan et al., 2017; Segal et al., 1977; Wu et al., 2003b). The frequency of fungal cultivability and detection based on positive PCR utilising DNA from swabs of conjunctivae and tissue scrapings from the cornea was not encouraging and a majority of the samples (> 80%) were negative for fungi by PCR and most conjunctival swabs yielded only one culture (Ozkan et al., 2017; Segal et al., 1977; Wu et al., 2003 a, b; Shivaji et al., 2019; Sai Prashanthi et al., 2019; Gopinathan et al., 2009; Sehgal et al., 1981; Wang et al., 2020; Williamson et al., 1968). In the ocular surface samples *Alternaria* spp., *Fusarium* spp., *Aspergillus niger*, *A. flavus*, *Curvularia* spp., *Penicillium* spp., *Helminthosporium* spp., *Candida albicans*, *C. guilliermondii*, *C. parapsilosis*, *Saccharomyces cerevisiae*, *Hormodendrum* spp., and *Rhodotorula rubra* were the common fungi (Mitsui and Hanbusa; 1955; Ainley and Smith, 1965; Williamson et al., 1968; Ando and Takatori, 1982; Sehgal et al., 1981; Karsten et al., 2012). Studies indicated that several confounding factors like the media used for monitoring the growth conditions and age of the individual (Williamson et al., 1968) influenced the success of cultivating fungi from ocular samples. Further, the fungi detected on the eye from an individual was not consistent and varied with time (one to four weeks), probably indicating that the ocular surface fungi are not residents but are acquired from the environment (Williamson et al., 1968). These fungi on the ocular surface vary in diversity across geographical regions with *Aspergillus flavus* and *A. niger*

TABLE 5.3

Cultivable Fungi of the Normal (Healthy) Human Eye as Detected by the Conventional Culturing Method

Fungi	Reference
Alternaria spp., *Aspergillus niger*, *A. flavus*, *Botrytis* spp., *Curvularia* spp., *Candida* spp., *Candida albicans*, *C. guilliermondii*, *C. parapsilosis*, *Cladosporium* spp., *Fusarium* spp., *Geotrichum* spp., *Gliocladium* spp., *Helminthosporium* spp., *Hormodendrum* spp., *Isaria* spp., *Mucor* spp., *Nigrospora* spp. *Papulospora* spp., *Penicillium* spp., *P. citrinum*, *Phialophora* spp., *Rhizoctonia* spp., *Rhizopus* spp., *Rhodotorula rubra*, *Rhodotorula* spp., *Saccharomyces* spp., *S. cerevisiae*, *Scopulariopsis* spp., *Stachybotrys* spp. and *Trichoderma* spp.	Ando and Takatori. 1982; Sehgal et al., 1981; Wang et al., 2020; Hammeke and Ellis, 1960; Ainley and Smith 1965; Mitsui and Hanabusa, 1955; Williamson et al., 1968.

Source: Modified from Shivaji et al., 2022; Shivaji 2022.

predominating in the tropics (Wang et al., 2020; Zeagans and Gelder 2014; Sharma, 2012b) whereas in Central Europe in addition to *Aspergillus* spp., *Penicillium* spp. and *Candida* spp. are dominant (Williamson et al., 1968). It has also been observed that though FK has a worldwide distribution it is predominantly reported from countries with a warm and dry climate than in temperate zones (Shukla et al., 2007; Brown et al., 2021). Table 5.3 is a comprehensive list of the viable fungi cultured from samples of the ocular surface.

5.3.2 Next Generation Sequencing (NGS) Method for the Detection of Ocular Fungi

5.3.2.1 Mycobiome of the Normal Human Eye

NGS is a culture-free method in which DNA from an ocular surface sample is used for identification of fungi. In this approach ITS2, a fungal phylogenetic marker, is amplified by PCR and sequenced. A major issue of this sensitive technique is to avoid contaminating DNA from used swabs, reagents used for DNA extraction or PCR, or from areas adjacent to the eye. Recently, NGS was used to analyse the fungal diversity from the conjunctival swabs of healthy eyes and fungal sequences were detected in 25 out of 34 conjunctival swab samples implying that the inability to detect fungi in all the samples is indicative of low fungal load in the conjunctival swabs of healthy eyes (Shivaji et al., 2019). More than 99% of the sequences were affiliated to 3 phyla. Ascomycota (47.74%) and Basidiomycota (26.87%) were the two dominant phyla in all the 25 eyes, whereas Zygomycota (0.08%) was detected only in 2 eyes (Figure 5.2a). At the genus level, 65 different genera were identified and the genera *Aspergillus* (21.05%), *Setosphaeria* (14.97%), *Malassezia* (7.37%), and *Hematonectria* (2.73%) were detected in the 25 eyes sampled whereas the genera *Emericella*, *Fusarium*, *Cladosporium*, and *Choiromyces* were present in 20 of the 25 eyes. The remaining 57 genera were very inconsistent in their occurrence and was shared between 1 to 17 eyes. Several sequences did not identify with any known fungus and were assigned either as unclassified (36.37%) or unidentified (11.11%) genera (Figure 5.2b; Table 5.4). The unclassified and unidentified taxa indicate the possibility of unknown fungal taxonomic groups associated with the eye and further reiterates the inadequacies of the existing fungal sequence databases. Studies on the diversity indices of the microbes associated with the eye indicated that alpha diversity of the fungal communities was similar in both the eyes (Shivaji et al., 2019). Further, the Beta diversity of the fungal communities was also independent of the eye sampled (left/right) and age of the individual but was dependent on sex of the individual sampled (Figure 5.3).

5.3.2.2 Ocular Pathogens on the Conjunctivae of Healthy Eye

Shivaji et al., (2019) identified several ocular pathogens affiliated to the genera *Acremonium*, *Candida*, *Rhodotorula*, *Aspergillus*, *Penicillium*, *Paecilomyces*, *Saccharomyces*, *Trichoderma*, *Malassezia*, *Fusarium*, *Cryptococcus*, *Ustilago*, *Pichia*, and *Arthrinium* (Shivaji et al., 2019) on the conjunctival surface of the normal healthy eye. Among these *Candida*, *Aspergillus*, *Malassezia*, and *Fusarium* were the most abundant genera (abundance, >0.50%) (Figure 5.2b).

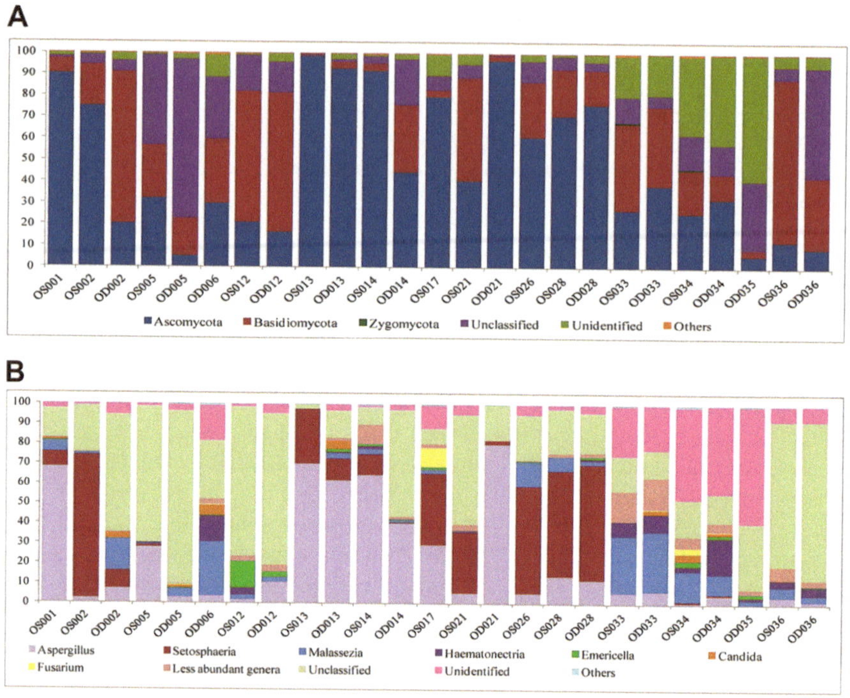

FIGURE 5.2 (a) Abundance of fungal phyla in the 25 conjunctival mycobiomes of healthy eyes. "Others" includes singletons, unassigned reads, and sparse OTUs (with <0.001% of total high-quality reads). (b) Abundance of the 7 dominant fungal genera (*Aspergillus* spp., *Setosphaeria* spp., *Malassezia* spp., *Haematonectria* spp., *Emericella* spp., *Candida* spp., and *Fusarium* spp.) with >0.50% abundance in the conjunctival mycobiomes of healthy eyes. All the other genera with <0.50% abundance were included as less abundant genera. Singletons, unassigned reads, and sparse OTUs (with <0.001% of total high-quality reads) are shown as "Others". (Modified from Shivaji et al., 2019.)

5.3.2.3 Comparison of Core and Putative Core Ocular Mycobiomes

According to Hamady and Knight (2009) and Turnbaugh et al., (2007) the 'core microbiome' includes those genera that are present in a specific microbiome in all the individuals analysed (Hamady and Knight 2009; Turnbaugh et al., 2007) whereas the "putative core microbiome" includes genera that were present in the microbiomes of >80% of the individuals. Further, microbiomes that varied depending on the environment and other human factors like lifestyle, and physiological differences, constituted the "variable microbiome" (Hamady and Knight 2009; Turnbaugh et al., 2007). In Indians, out of the 65 fungal genera reported from the ocular surface only four genera constituted the 'core' (Shivaji et al., 2019). In Chinese individuals, fungal diversity of the ocular surface was greater with 94 genera out of which 5 genera representing the 'core' and 3 genera represented the 'putative core' mycobiomes (Wang et al., 2020). Further, 11 genera were shared in the ocular mycobiomes between Indians and Chinese individuals (Wang et al., 2020; Shivaji et al., 2019). Several of these

TABLE 5.4

Abundance of Fungal Genera in the Ocular Surface Fungal Microbiomes from 25 Eyes (OD = 12 and OS = 13) of Healthy Normal Individuals

Serial Number	Genus	Abundance		Present Out of 25 Eyes Sampled
		Mean	Range	
1	*Aspergillus*	21.05	0.085–79.946	25
2	*Setosphaeria*	14.974	0.001–71.83	25
3	*Malassezia*	7.374	0.003–29.792	25
4	*Haematonectria*	2.735	0.0001–18.026	25
5	*Emericella*	1.146	0–13.176	24
6	*Fusarium*	0.52	0–9.372	21
7	*Cladosporium*	0.073	0–0.642	20
8	*Choiromyces*	0.003	0–0.028	20
9	*Candida*	0.892	0–5.494	17
10	*Neosartorya*	0.043	0–0.259	17
11	*Penicillium*	0.225	0–1.613	14
12	*Nectria*	0.003	0–0.016	14
13	*Pichia*	0.238	0–1.537	12
14	*Xeromyces*	0.388	0–8.069	12
15	*Leptosphaerulina*	0.244	0–4.271	9
16	*Arthrinium*	0.253	0–6.319	9
17	*Eurotium*	0.169	0–2.295	7
18	*Aureobasidium*	0.131	0–1.914	7
19	*Cryptococcus*	0.088	0–1.327	6
20	*Cercospora*	0.101	0–1.861	5
21	*Podospora*	0.019	0–0.226	5
22	*Saccharomyces*	0.069	0–0.833	5
23	*Colletotrichum*	0.106	0–2.656	5
24	*Hanseniaspora*	0.052	0–0.857	4
25	*Trichoderma*	0.04	0–0.309	4
26	*Phoma*	0.005	0–0.076	4
27	*Pestalotiopsis*	0.037	0–0.663	4
28	*Passalora*	0.08	0–1.998	4
29	*Acremonium*	0.048	0–0.635	3
30	*Clavispora*	0.01	0–0.211	3
31	*Khuskia*	0.002	0–0.024	3
32	*Myrothecium*	0.108	0–2.644	3
33	*Nigrospora*	0.01	0–0.176	3
34	*Alternaria*	0.119	0–2.966	3
35	*Ramichloridium*	0.013	0–0.326	3
36	*Metschnikowia*	0.004	0–0.061	2
37	*Rhizopus*	0.077	0–0.983	2
38	*Sagenomella*	0.024	0–0.564	2
39	*Wallemia*	0.002	0–0.036	2

(continued)

TABLE 5.4 (Continued)

Abundance of Fungal Genera in the Ocular Surface Fungal Microbiomes from 25 Eyes (OD = 12 and OS = 13) of Healthy Normal Individuals

Serial Number	Genus	Abundance		Present Out of 25 Eyes Sampled
		Mean	Range	
40	*Ceratocystis*	0.001	0–0.036	2
41	*Chrysosporium*	0.018	0–0.443	2
42	*Mycosphaerella*	0.032	0–0.791	2
43	*Paecilomyces*	0.018	0–0.44	2
44	*Pseudozyma*	0.003	0–0.065	2
45	*Stachybotrys*	0.007	0–0.18	2
46	*Ustilago*	0.002	0–0.046	2
47	*Arxula*	0.024	0–0.602	1
48	*Bullera*	0.022	0–0.551	1
49	*Calcarisporiella*	0.007	0–0.166	1
50	*Dactylella*	0.042	0–1.059	1
51	*Galactomyces*	0.006	0–0.141	1
52	*Hypoxylon*	0.015	0–0.38	1
53	*Ochrocladosporium*	0.003	0–0.082	1
54	*Phaeosphaeriopsis*	0.031	0–0.78	1
55	*Phanerochaete*	0.025	0–0.62	1
56	*Phialophora*	0.015	0–0.383	1
57	*Pyrenochaeta*	0.002	0–0.06	1
58	*Quambalaria*	0.159	0–3.963	1
59	*Rhodotorula*	0.015	0–0.369	1
60	*Simplicillium*	0.005	0–0.125	1
61	*Strelitziana*	0.014	0–0.337	1
62	*Sympodiomycopsis*	0.001	0–0.036	1
63	*Torulaspora*	0.06	0–1.492	1
64	*Villosiclava*	0.009	0–0.232	1
65	*Zygoascus*	0.026	0–0.644	1
66	Unclassified	36.374	2.099–87.168	25
67	Unidentified	11.109	0.007–58.637	25
68	Others[a]	0.482	0.187–0.985	25

Source: From Shivaji et al., 2019.

Note: [a] Include singletons, unassigned reads, and sparse OTUs (with <0.001% of total high-quality reads).

abundant genera (*Aspergillus, Malassezia, Setosphaeria, Hematonectria, Emericella, Fusarium, Cladosporium, Choiromyces, Rhodotorula, Davidiella, Alternaria, Meyerozyma* and *Petromyces*) are pathogenic (Wang et al., 2020; Shivaji et al., 2019). In addition, the genera *Fonsecaeapedrosoi, Lasiodiplodiatheobromae, Cylindrocarpon* spp., *Scedosporiumprolificans, Metarhizium anisopliae, Paecilomyces* and *Pythium*

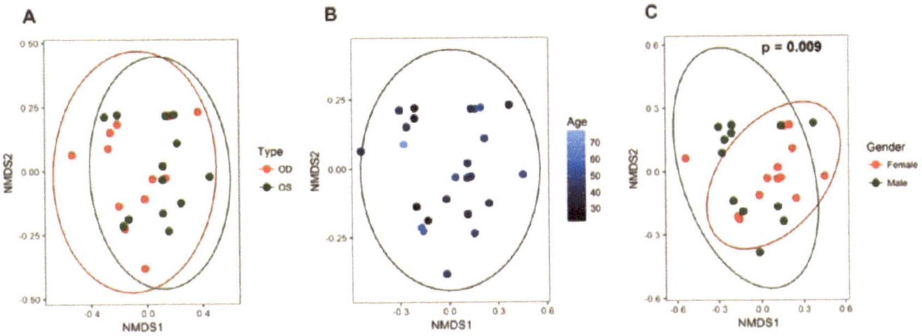

FIGURE 5.3 Comparison of the beta diversity in the ocular surface fungal microbiomes of the 25 eyes of healthy normal controls analysed using the NMDS plots and Bray-Curtis dissimilarity of OTUs. The fungal community structure was not dependent on the eye (left/right) that was sampled (PERMANOVA, $P = 0.533$) (a) nor was it influenced by age (b; PERMANOVA, $P = 0.859$) (b), but the fungal community appeared to vary significantly (PERMANOVA, $P = 0.009$) depending on the sex (c). (Modified from Shivaji et al., 2019.)

insidiosum were also detected. These are considered as rare fungal species causing mycotic keratitis (Maharana et al., 2016).

5.3.2.4 AMR in Fungi of the Normal Human Eye

Fungi of the genus *Aspergillus* was the only fungi detected in the conjunctival swabs of healthy controls indicating a paucity of fungi on the conjunctival swabs (6%) of healthy controls (HC-SW) (Table 5.5). In contrast, in the corneal scrapings of fungal keratitis patients (FK-CR) several more genera were identified and included *Aspergillus flavus, A. terreus, A. fumigatus, Fusarium spp., F. solani, Colletotrichum coccodes, Exserohilum rostratum, Curvularia lunata, Cladosporium* spp. and several unidentified septate or sparsely septate fungal filaments, aseptate fungal filaments and unidentified dematiaceous fungus (Table 5.5). Comparison of the studies by Karsten et al., (2012), Thomas and Kaliamurthy (2013) and Sai Prashanthi et al., (2019) indicated that 26 genera were common (Table 5.5). But none of these studies on the ocular surface fungi monitored the response of the fungi to antifungals.

5.4 Fungi of the Human Keratitis Eye and AMR

Keratitis is an ocular inflammatory disease in which the cornea becomes opaque and red and the patient presents with pain and has itching sensation of the eye. Keratitis, caused exclusively by fungi is referred to as fungal keratitis (FK) (Figure 5.4). FK if neglected, is known to cause several deleterious effects like scarring or perforation of the cornea, endophthalmitis, and may also cause blindness (Henry et al., 2012; Teweldemedhin et al., 2017). FK occurs due to corneal injury during farming activities when plant material accidentally penetrates the cornea. The predisposing factors for FK include immuno-compromised conditions as in HIV and AIDS patients, excessive use of broad-spectrum antibiotics, using corticosteroids indiscriminately and diseases

TABLE 5.5

Cultivable Fungi in Conjunctiva of Healthy Individuals (HC001-HC049), the Corneal Surface of Healthy Individuals (HCC1-HCC-10) and in Corneal Surface of Fungal Keratitis Patients (Fk01-FK37)

Sl. No.	Sample ID	Age (years)	Gender	Fungi identified from swab/ Corneal scraping
1	HC01	35	Male	*Aspergillus flavus*[a]
2	HC02	36	Female	*Aspergillus niger*[a]
3	HC03	28	Male	No Growth[a]
4	HC05	43	Male	No Growth[a]
5	HC06	77	Male	No Growth[a]
6	HC09	64	Female	No Growth[a]
7	HC12	59	Female	No Growth[a]
8	HC13	52	Female	No Growth[a]
9	HC14	54	Female	No Growth[a]
10	HC17	50	Female	No Growth[a]
11	HC21	31	Male	No Growth[a]
12	HC26	37	Female	No Growth[a]
13	HC28	41	Female	*Aspergillus flavus*[a]
14	HC33	24	Male	No Growth[a]
15	HC34	38	Male	No Growth[a]
16	HC35	37	Female	No Growth[a]
17	HC36	23	Male	No Growth[a]
18	HC37	24	Female	No Growth[a]
19	HC38	25	Female	No Growth[a]
20	HC40	41	Male	No Growth[a]
21	HC41	27	Male	No Growth[a]
22	HC43	28	Male	No Growth[a]
23	HC46	68	Male	No Growth[a]
24	HC48	61	Male	No Growth[a]
25	HC49	55	Male	No Growth[a]
26	HCC1	27	Female	Not done
27	HCC2	23	Male	Not done
28	HCC3	28	Female	Not done
29	HCC4	21	Female	Not done
30	HCC5	27	Male	Not done
31	HCC6	23	Male	Not done
32	HCC7	26	Male	Not done
33	HCC8	26	Female	Not done
34	HCC9	24	Male	Not done
35	HCC10	21	Male	Not done
36	FK01	41	Female	Fusarium species[b]
37	FK02	63	Female	Septate fungal filaments unidentified[b]
38	FK03	66	Male	*Colletotrichum coccodes*[b]
39	FK05	56	Male	Unidentified dematiaceous fungus[b]
40	FK06	41	Male	*Fusarium solani*[b]

TABLE 5.5 (Continued)

Cultivable Fungi in Conjunctiva of Healthy Individuals (HC001-HC049), the Corneal Surface of Healthy Individuals (HCC1-HCC-10) and in Corneal Surface of Fungal Keratitis Patients (Fk01-FK37)

Sl. No.	Sample ID	Age (years)	Gender	Fungi identified from swab/ Corneal scraping
41	FK07	54	Male	Aseptate fungal filaments unidentified[b]
42	FK09	80	Female	*Fusarium solani*[b]
43	FK10	50	Male	*Aspergillus flavus*[b]
44	FK11	60	Female	*Aspergillus flavus*[b]
45	FK12	29	Female	Fusarium species[b]
46	FK13	22	Male	Fusarium species[b]
47	FK14	45	Male	*Exserophilum rostratum*[b]
48	FK15	35	Male	*Aspergillus flavus*
49	FK16	65	Male	Unidentified dematiaceous fungus[b]
50	FK17	55	Female	*Aspergillus flavus*[b]
51	FK18	66	Female	*Aspergillus flavus*[b]
52	FK19	29	Male	Septate fungal filaments unidentified[b]
53	FK20	61	Female	*Aspergillus flavus*[b]
54	FK21	26	Male	Fusarium species[b]
55	FK22	66	Male	*Aspergillus terreus*[b]
56	FK23	18	Female	Fusarium species[b]
57	FK24	40	Female	Fusarium species[b]
58	FK25	56	Male	Cladosporium species[b]
59	FK26	47	Female	*Aspergillus flavus*[b]
60	FK27	24	Male	*Aspergillus flavus*[b]
61	FK28	31	Male	*Curvularia lunata*[b]
62	FK29	27	Male	*Aspergillus fumigatus*[b]
63	FK30	54	Male	Septate fungal filaments unidentified[b]
64	FK31	69	Female	*Curvularia lunata*[b]
65	FK32	59	Male	Septate fungal filaments unidentified[b]
66	FK33	54	Female	Sparsely septate fungal filaments unidentified[b]
67	FK34	68	Male	Unidentified dematiaceous fungus[b]
68	FK35	29	Female	Septate fungal filaments unidentified[b]
69	FK36	48	Female	*Fusarium solani*[b]
70	FK37	55	Male	Septate fungal filaments unidentified[b]

Source: Data from Sai Prashanthi et al., 2019.

Notes:
HC001-HC049, viable fungi from conjunctiva of healthy individuals;
HCC1-HCC-10, viable fungi from the corneal surface of healthy individuals;
Fk01-FK37, viable fungi from corneal surface of FK patients.
[a] Conjunctival swab
[b] Corneal scraping

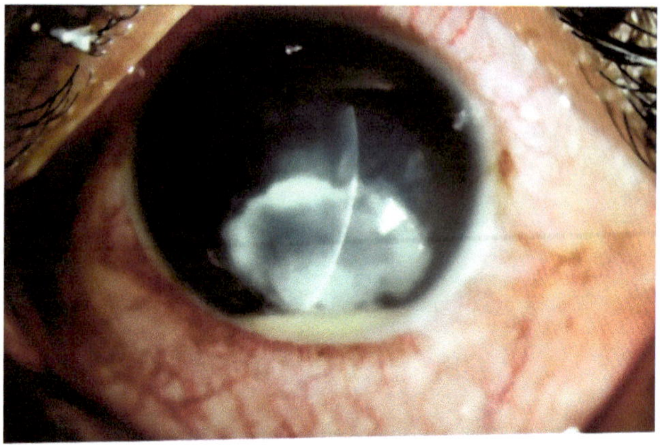

FIGURE 5.4 Fungal keratitis. (Courtesy Dr. Bhupesh Bagga, L V Prasad Eye Institute, Hyderabad, India.)

like diabetes. FK is also known as mycotic keratitis, keratomycosis, or hypopyon corneal ulcer (Foster, 1992; Shukla et al., 2007).

5.4.1 Culturing Method for the Detection of Fungi from the Human Keratitis Eye

Globally, the ocular disease of keratitis affects millions of people, and in India it is estimated that about 2 million people are annually affected (Gonzales et al., 1996: Gupta et al., 2013) and several fungi have been implicated in keratitis. Brown et al., (2021) estimated that the minimum annual incidence of FK is 1,051,787 cases with the highest rates in Asia and Africa. A few earlier studies indicated that the number of fungal genera in corneal infections varied from 56 (Shukla et al., 2007) to 92 genera (Karsten et al., 2012) whereas subsequently Thomas and Kaliamurthy (2013) reported 65 genera of fungi as associated with mycotic keratitis. Sai Prashanthi et al., (2019) confirmed that fungi in the conjunctival swabs of healthy controls were predominantly *Aspergillus flavus* and *A. niger*, whereas in the corneal scrapings of fungal keratitis individuals (FK-CR) apart from *A. flavus* several other species including *A. fumigatus*, *Fusarium* spp., *F. solani*, *Colletotrichum coccodes*, *Exserohilum rostratum*, *Cladosporium* spp., *Curvularia lunata*, were reported. In addition, unidentified septate and aseptate fungal filaments, and Dematceous fungus (Table 5.5) were identified (Sai Prashanthi et al., 2019). The above three studies by Karsten et al., (2012), Thomas and Kaliamurthy (2013) and Sai Prashanthi et al., (2019) shared 26 genera (*Acremonium*, *Candida*, *Aspergillus*, *Penicillium*, *Paecilomyces*, *Rhodotorula*, *Phoma*, *Cladosporium*, *Alternaria*, *Saccharomyces*, *Pyrenochaeta*, *Aureobasidium*, *Phialophora*, *Colletotrichum*, *Myrothecium*, *Stachybotrys*, *Chrysosporium*, *Malassezia*, *Fusarium*, *Torula*, *Trichoderma*, *Rhizopus*, *Cercospora*, *Cryptococcus*, *Ustilago* and *Phaeosphaeriopsis*). It is suggested that these 26 genera that are common in the three investigations in FK patients are potential opportunistic pathogens. Sai Prashanthi et al., (2019) based on the commonality in the diversity at the genera level in the conjunctival swabs and corneal scrapings from FK individuals, and heathy controls predicted that these genera are opportunistic pathogens, whereas the seven genera that were shared specifically between conjunctival swabs and corneal scrapings of FK individuals represented pathogens (Sai Prashanthi et al., 2019). Sai Prashanthi et al., (2019) also observed that a good proportion of the remaining genera detected (39 in number) were not reported earlier (Karsten et al., 2012; Thomas and Kaliamurthy 2013) despite the observation that it included fungal genera that cause keratitis, such as *Pichia*, *Nectria*, *Arthrinium*, *Pestalotiopsis*, *Khuskia*, and *Nigrospora*. In a recent review, Maharana et al., (2016) reported that *Fonsecaea pedrosoi*, *Lasiodiplodia theobromae*, *Cylindrocarpon* spp., *Scedosporium prolificans*, *Metarhizium anisopliae*, *Paecilomyces* spp., and *Pythium insidiosum* are rare in mycotic keratitis. The discrepancy in fungal diversity in the above studies may be attributed to variables like age, geographical origin, and the pathology in the individuals which could have been different (Williamson et al., 1968; Marchlewitz, 1966).

5.4.2 Worldwide Distribution of Fungi in Human Keratitis Patients

FK is predominant in regions that are warm and dry, as in the tropics (Asia and African subcontinent), then in temperate zones as in Europe, where mycotic keratitis is a mere 3% and is more common among agriculture workers than in any other occupation (Thomas, 1994; Brown et al., 2021). The minimal annual incidence of FK was highest in Asia and Africa (Brown et al., 2021). Houang et al., (2001) also observed that higher incidence of fungal keratitis was also dependent on the extent of urbanization. Between the warm countries subtle differences could be observed with respect to the most common cause of FK. For instance, in India, Nepal, Sri Lanka, and Bangladesh, *Aspergillus* spp., was the most common causal agent of keratomycoses followed by species of *Fusarium* spp. and *Candida* spp. (Chander and Sharma, 1994; Deshpande and Koppikar, 1999; Khanal et al., 2001; Dóczi et al., 2004). In Nepal, Rhizopus was also a major cause of mycotic keratitis (Chaudhary et al., 1999). A comprehensive study in China over a period of 15 years and involving 615 patients showed that *Fusarium* spp., was the most common pathogen followed by *Aspergillus* spp., *Penicillium* spp. and Curvularia spp. (Wang et al., 2000). While in African countries also *Fusarium* spp. was the most common pathogen but the other pathogens included *Aspergillus* spp. and *Curvularia* spp. (Carmichael et al., 1985; Gugani et al., 1976; Mselle 1999). In European countries, *Candida* spp., are the most common isolates of FK (Coster et al., 1981) while in America, *Fusarium* spp. dominated followed by *Candida* spp. (Harris et al., 1988; Liesegang and Forster, 1980; O'day, 1987; Rosa et al., 1994). WHO reported that mortality rate of keratitis is the highest (0.25 deaths per million people) in New Zealand. *C. parapsilosis* is the third leading cause of candidemia worldwide (Pappas et al., 2018; Pfaller et al., 2019).

5.4.3 Next Generation Sequencing Method for the Detection of Fungi from the Human Keratitis Eye

5.4.3.1 Mycobiome of the Keratitis Eye

Sai Prashanthi et al., (2019) analysed the mycobiomes from conjunctival swabs of healthy controls (HC-SW) and fungal keratitis patients (FK-SW), corneal scrapings from FK patients (FK-CR) and the corneal epithelium from patients undergoing photorefractive keratectomy (PRK). This study indicated that the Alpha diversity indices (Shannon diversity index, Simpson index, observed number of OTUs and Chao1 index) were significantly different between HC-SW and FK-CR mycobiomes and at the phylum level abundance of fungi affiliated to the phylum Ascomycota was significantly increased both in FK-SW (mean abundance, 69.99%) and in FK-CR (mean abundance, 88.36%) compared to HC-SW, whereas the abundance of Basidiomycota decreased in the keratitis individuals compared to HC-SW. The abundance of Zygomycota was low and was detected only in few eyes (Figure 5.5). Further at the genus level, majority (62%) of the sequences from HC-SW, FK-SW, and FK-CR were assigned to 111 different genera (Figure 5.6 a; b) very similar to the earlier published data between 1950–2012, which had identified 92 genera (Karsten et al., 2012). *Aspergillus* constituted the 'core' mycobiome whereas 11 genera namely *Aspergillus*, *Haematonectria*, *Emericella*, *Choiromyces*, *Setosphaeria*, *Candida*, *Malassezia*, *Penicillium*, *Fusarium*, *Cladosporium*, and *Cochliobolus* were present in majority of the eyes and constituted the 'variable microbiome' as per definition of Turnbaugh et al., (2007). Seven genera, namely *Malassezia*, *Haematonectria*, *Candida*, *Cladosporium*, *Fusarium*, *Choiromyces*, and *Cochliobolus* were equally abundant in HC-SW and FK-SW but differed in FK-CR. The remaining genera significantly varied in abundance (Figure 5.6 a; b) and HC-SW and FK-SW, HC-SW and FK-CR and FK-SW and FK-CR could be differentiated by 3, 33 and 27 genera respectively (Table 5.6). Heat map analysis also discriminated the mycobiomes in to three clusters indicating that the mycobiomes of HC-SW, FK-SW, and FK-CR are different (Sai Prashanthi et al., 2019) indicating dysbiosis in the ocular fungal microbiome of FK patients.

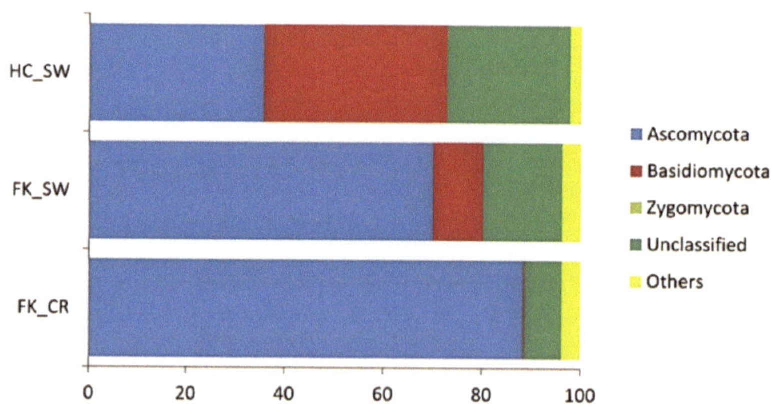

FIGURE 5.5 Abundance (%) of fungal phyla in the mycobiomes of conjunctivae of healthy controls (HC-SW, n=37), fungal keratitis patients (FK-SW, n=23) and corneal scrapings from fungal keratitis patients (FK-CR, n=29). (Modified from Sai Prashanthi et al., 2019.)

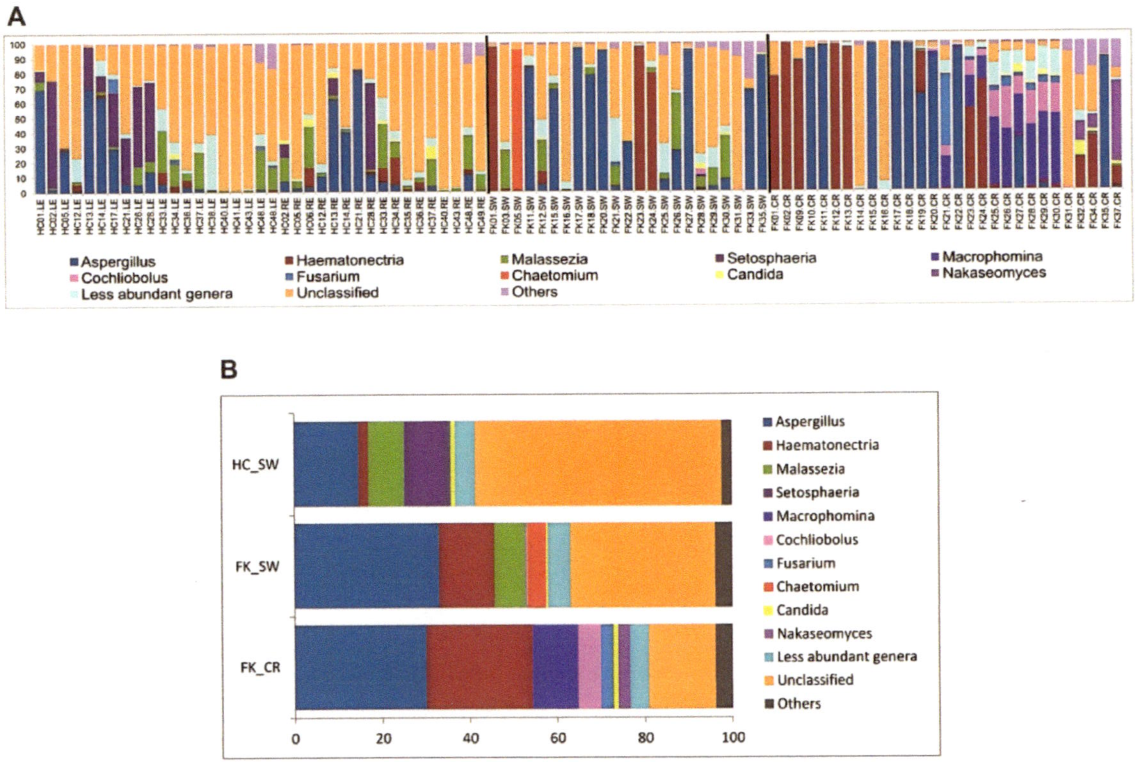

FIGURE 5.6 Abundance (%) (a) and average abundance of fungal genera (b) from mycobiomes of conjunctivae of healthy controls (HC-SW, n=37), conjunctivae of fungal keratitis patients (FK-SW, n=23) and corneal scrapings of fungal keratitis patients (FK-CR, n=29). (From Sai Prashanthi et al., 2019.)

TABLE 5.6

Significant Abundance Changes in Fungal Genera in the Mycobiomes of the Conjunctival Swabs of Healthy Controls (HC-SW), Conjunctival Swabs of Fungal Keratitis Patients (FK-SW) and Corneal Scrapings of Fungal Keratitis Patients (FK-CR)

		Median Abundance (%)		
S.No.	**Genera**	**HC-SW**	**FK-SW**	**BH-corrected *p* value**
1	*Setosphaeria*	0.0100	0	0.001
2	*Neosartorya*	0.0004	0	0.001
3	*Chaetomium*	0	0	0.001
		Median Abundance (%)		
S.No.	**Genera**	**HC-SW**	**FK-CR**	**BH-corrected *p* value**
1	*Malassezia*	2.7353	0.0137	0
2	*Emericella*	0.1851	0	0
3	*Cladosporium*	0.0208	0	0
4	*Fusarium*	0.0140	0.3567	0.015
5	*Setosphaeria*	0.0100	0	0.004
6	*Penicillium*	0.0022	0	0
7	*Choiromyces*	0.0007	0	0
8	*Cochliobolus*	0.0004	0.0101	0.004
9	*Neosartorya*	0.0004	0	0.004
10	*Arthrinium*	0	0	0.015
11	*Aureobasidium*	0	0	0.014
12	*Bipolaris*	0	0	0.015
13	*Botryosphaeria*	0	0.0005	0
14	*Cephaliophora*	0	0	0
15	*Cercospora*	0	0	0
16	*Clavispora*	0	0	0.041
17	*Curvularia*	0	0	0.014
18	*Diaporthe*	0	0	0
19	*Eurotium*	0	0	0
20	*Gibberella*	0	0	0
21	*Lasiodiplodia*	0	0	0.004
22	*Leptosphaerulina*	0	0	0
23	*Macrophomina*	0	0	0.004
24	*Myrothecium*	0	0	0.004
25	*Nakaseomyces*	0	0	0.030
26	*Neocosmospora*	0	0.0009	0.004
27	*Nigrospora*	0	0	0.041
28	*Phoma*	0	0	0.041
29	*Pleurostomophora*	0	0.0011	0
30	*Podospora*	0	0	0.004
31	*Retroconis*	0	0	0
32	*Rhizoctonia*	0	0	0
33	*Thielavia*	0	0	0.007

TABLE 5.6 (Continued)

Significant Abundance Changes in Fungal Genera in the Mycobiomes of the Conjunctival Swabs of Healthy Controls (HC-SW), Conjunctival Swabs of Fungal Keratitis Patients (FK-SW) and Corneal Scrapings of Fungal Keratitis Patients (FK-CR)

		Median Abundance (%)		
S.No.	Genera	FK-SW	FK-CR	BH-corrected *p* value
1	*Malassezia*	1.1787	0.0137	0
2	*Emericella*	0.0056	0	0
3	*Cladosporium*	0.0049	0	0
4	*Cochliobolus*	0.0007	0.0101	0.024
5	*Choiromyces*	0.0003	0	0
6	*Clavispora*	0.0002	0	0
7	*Botryosphaeria*	0	0.0005	0
8	*Cephaliophora*	0	0	0.006
9	*Cercospora*	0	0	0
10	*Chaetomium*	0	0	0.006
11	*Coniothyrium*	0	0	0.034
12	*Coriolopsis*	0	0	0.034
13	*Diaporthe*	0	0	0.006
14	*Eurotium*	0	0	0.020
15	*Flavodon*	0	0	0.020
16	*Gibberella*	0	0	0
17	*Lasiodiplodia*	0	0	0
18	*Leptosphaerulina*	0	0	0.006
19	*Macrophomina*	0	0	0.034
20	*Myrothecium*	0	0	0.006
21	*Peniophora*	0	0	0.020
22	*Phoma*	0	0	0.034
23	*Podospora*	0	0	0.034
24	*Retroconis*	0	0	0.006
25	*Rhizoctonia*	0	0	0.020
26	*Saccharomyces*	0	0	0.033
27	*Stagonosporopsis*	0	0	0.033

Source: Modified From Sai Prashanthi et al., 2019.

5.4.3.2 AMR in Fungi of Fungal Keratitis Patients

Antifungal resistance (AFR) is the ability of fungi to grow at drug concentrations that inhibit susceptible isolates and this could be measured using minimum inhibitory drug concentration (MIC). AFR is quite rare compared to resistance of bacteria to antibiotics. Nevertheless, AFR is of great clinical concern to clinicians who need to treat patients at high risk for invasive mycoses. Fisher et al., (2022) in a recent review highlighted that systemic fungal infection is an emerging crisis worldwide and is under-recognized compared to bacterial antimicrobial resistance. About the same time, Ben-Ami et al. (2021) reviewed the mechanism of resistance to antifungal drugs and observed that some fungal species are intrinsically resistant to certain drugs (e.g., fluconazole resistance in *C. krusei* and amphotericin B resistance in *C. lusitaniae*), whereas other fungi seem to acquire (secondary) the resistance. Intrinsic drug resistance is genetically determined and consequently fungal species are non- susceptible to drugs like antifungals, and the resistance is independent of prior exposure to the drugs (e.g., *Lomentospora*) (Lackner et al., 2014a; b; Walsh et al., 2004; Alastruey-Izquierdo et al., 2014). In contrast, acquired resistance develops because of exposure to drug that could be an antifungal agent or its structural analogue, and it may be either genetic or epigenetic (Beekman and Ene, 2020).

Antifungal drugs are classified into four classes based on their chemistry viz., azoles, echinocandins, polyenes, and flucytosine (Pfaller, 2012; Patil and Majumdar 2017). Of these the most widely used are the azoles due to the following attributes: exhibit broad-spectrum activity, are safe and their bioavailability. But the disadvantage of azoles is it exhibits a fungistatic activity, which facilitates target cells to overcome inhibitory concentrations of the drug (Shor and Perlin, 2015; Harrison et al., 2014) and thus leading to antifungal resistance or tolerance. The second class of antifungal drugs are the polyenes, which are nephrotoxic and thus not much preferred. The echinocandins, the third class of antifungals, are used in hospitals in the West but need to be administered intravenously (Pappas et al., 2016a, b; Chen et al., 2011a). The fourth class of antifungals, Flucytosine are often used in combination with other drugs, as resistance often appears rapidly with monotherapy (Vermes et al., 2000). Apart from these four classes of antifungals new drugs are needed to address multiple resistance in fungi like *C. auris*.

5.4.3.2.1 Fungal Resistance to Azoles: Fluconazole

Introduction Fluconazole is not an approved therapeutic drug but used in ocular fungal infections affecting both superficial and deep-seated mycosis respectively (Yee et al., 1997; Yilmaz and Maden, 2005; Tsai et al., 2014; Cheng et al., 2015) owing to its ability to permeate the ocular tissues (Gokhale, 2008) and attain therapeutic concentrations upon topical, oral, and systemic administration. Subconjunctival injection of fluconazole (2%) was more effective than topical or IV administration in managing fungal keratitis and fungal ulcers (Yilmaz and Maden, 2005). Fluconazole (0.2%) has also been effective in curing severe fungal keratitis in conjunction with lamellar keratectomy (You et al., 2015) when administered intra-stromally. Fluconazole is a preferred drug due to its favourable toxicity profile, good bioavailability, and potent antifungal activity against *Candida* spp., and *Aspergillus* spp. (Loeffler and Stevens 2003). But azole resistance in *Candida* spp. (Hachem et al., 2008; Wingard et al., 1991) and *Aspergillus fumigatus* deterred the use of the drug in fungal disease (Snelders et al., 2010; van der Linden et al., 2015).

Resistance Fungi develop resistance to fluconazole following prolonged exposure or due to an incomplete therapeutic regimen. But when fluconazole is administered along with other antifungals like amphotericin B or flucytosine as combination therapy it was found to be superior (Al-Hatmi et al., 2016; Mahdy et al., 2010) and resistance did not develop during combination therapy (Molloy et al., 2018). For the ocular delivery of fluconazole, hyalugel-integrated liposomes of fluconazole (0.9%) were preferred to monotherapy (fluconazole suspension) since the former approach increases uptake and permeation of fluconazole into the cornea by several fold (Moustafa et al., 2017). Silva et al., (2016) used fluconazole-loaded poly-lactideco-glycolide implants (25% w/w) for the treatment of *Candida* spp., endophthalmitis. Fetih (2016) used fluconazole-encapsulated vesicles incorporated into gels of poloxamer and chitosan and demonstrated antifungal activity against *C. albicans, Chrysosporium tropicum*, and *Penicillium chrysogenum*. Thus, fluconazole has been successfully used for the management of fungal keratitis and ulcers. Increase in fluconazole resistance from 0–50% has been recorded in South Africa (Chen et al., 2017; Jarvis et al., 2010). Further monotherapy with fluconazole is associated with significant (10% to 20%) relapse rate and the relapse isolates exhibited higher fluconazole MIC than incident isolates (Stone et al., 2019; Lee et al., 2012). In *Cryptococcus neoformans* resistance to fluconazole is associated with multiple ploidy changes, frequently involving chromosome 1, ERG11 point mutation, increased copy numbers of ERG11 and AFR1, an effluxtransporter (Stone et al., 2019).

5.4.3.2.2 Fungal Resistance to Azoles: Ketoconazole

Introduction Ketoconazole, like other azoles, inhibits ergosterol biosynthetic pathway in fungi (Van Tyle, 1984). Ketoconazole was introduced in the early 1980s and it was only after a decade that the first-generation triazoles, fluconazole and itraconazole were administered to cancer patients and to patients recruited for stem cell or solid organ transplantation. The side effects of prolonged oral ketoconazole therapy include decreased libido, oligospermia, menstrual changes and hepatoxicity (Müller et al., 2013). Further, it is active against only a few filamentous and yeast like fungal species (Sahu et al., 2014; Miller et al., 2016) and has been used to overcome deep-seated mycosis, due to its high bioavailability in the anterior chamber of the eye upon oral administration (Neoh et al., 2014). Ketoconazole is mostly used as an adjuvant in concomitant antifungal therapy. For better drug availability ketoconazole has been used as gels (ketoconazole-loaded proniosomal gels) (Abdelbary et al., 2017) and as nanoparticles of ketoconazole loaded into an alginate-chitosan gel (Ahmed and Aljaeid, 2017) or as ketoconazole-loaded polyethylene glycol-based solid lipid nanoparticles (Kakkar et al., 2015). Both the gels and nanoparticles were superior in delivering the drug at the ocular sites (Abdelbary et al., 2017) and the concentration in aqueous humor was significantly higher when the gel formulation was used indicating an improvement in the bioavailability. A standard dose for ketoconazole for ophthalmic fungal infections has not been indicated but ketoconazole eye drops exhibited effectiveness in treating *Aspergillus* spp. and *Fusarium* spp. corneal infections and both oral and/or topical application of ketoconazole were effective against cases of ocular infections due to *Phaeohyphomycetes* spp. (Thomas and Kaliamurthy, 2013). It was also observed that oral ketoconazole along with natamycin did not significantly affect the therapeutic outcomes (Torres et al., 1985; Rajaraman et al., 2015).

Resistance AFR to ketoconazole was studied *in vitro* using 130 ocular filamentous fungi and 50 ocular yeasts using ketoconazole (0.025–6.4 microg/mL) in doubling dilutions. Five yeasts (10%) and ten filamentous (7.6%) strains were resistant to ketoconazole and 10 (7.6%) *Aspergillus niger*, *A. terreus* and *C. krusei* were found to be resistant to ketoconazole (Therese et al., 2006). These results confirm earlier studies that indicated that ketoconazole was effective in treating mycoses of the eye but in some cases no effect was achieved heralding the emergence of AFR (Maĭchuk et al., 1990). Xu et al., (2021) observed that tolerance of *C. albicans* to ketoconazole depended on the composition of the growth medium used and the temperature of cultivation but these two variables had little effect on resistance. Deletion analysis demonstrated that tolerance to azoles was dependent on efflux pump gene *CDR1*. Further, Hsp90, calcineurin components and *CRZ1* which encodes a transcription factor were also required for azole tolerance but only partially. In addition*VMA11*, encoding a vacuolar ATPase subunit and concanamycin A, a V-ATPase inhibitor were required for azole tolerance (Xu et al., 2021). Thus, multiple factors and genes regulate tolerance to ketoconazole.

5.4.3.2.3 Fungal Resistance to Azoles: Voriconazole

Introduction Voriconazole is a new-generation azole antifungal agent with a good safety profile, few side effects, has high bioavailability and permeability across the ocular tissues (Hariprasad et al., 2004) and has multiple biological effects on fungi like inhibiting cell membrane ergosterol synthesis (Sanati et al., 1997). Voriconazole is fungistatic to yeasts but fungicidal to filamentous fungi (Manavathu et al., 1998). Voriconazole is active against all *Candida* species including *C. krusei* and *C. glabrata* that are inherently fluconazole-resistant, and effective against *C. albicans* that have acquired resistance to fluconazole (Manavathu et al., 1998; Barry and Brown 1996; Pfaller et al., 2001; Ruhnke et al., 1997). Voriconazole, *in vitro* is also active against other yeasts, including *Cryptococcus neoformans, Trichosporon beigelii*, and *Saccharomyces cerevisiae* (Hoban et al., 1999; Pfaller et al., 1999; Espinel-Ingroff, 1998) and filamentous fungi like *Aspergillus* spp., *Fusarium solani*, *F. moniliforme*, *F. oxysporum*, *Fusarium* spp. and other filamentous fungi (Pfaller et al., 1999; Klont et al., 2005; Al-Badriyehet al., 2010) when administered systemically, topically and/or by intraocular injection to keratitis patients. Voriconazole as a topical or oral administration along with concomitant administration with either amphotericin B or natamycin has shown effectiveness in healing FK caused by several fungi including *C. albicans*, *Alternaria* spp., *Fusarium* spp., and *Scopulariopsis* spp. (Bunya et al., 2007). Voriconazole monotherapy was also effective in treating *Aspergillus fumigatus* keratitis that was unresponsive to the standard natamycin eye drop therapy (Mehta et al., 2008) and in treating ocular keratitis caused by *Aspergillus* spp., and *Curvularia* spp. But was not effective against *Fusarium* keratitis, (Arora et al., 2011; Sharma et al., 2015a). Prajna et al., (2016a; b) found that oral voriconazole along with topical voriconazole (1%) or topical natamycin (5%) did not benefit the therapeutic outcomes in severe filamentous mycotic ulcers. Haddad and El-Mollayess (2012) used a combination of intracameral and intrastromal voriconazole to treat recalcitrant *Acremonium* fungal keratitis. In general, the corneal residence time of voriconazole is short thus requiring frequent administration thus leading to poor patient compliance (Suvarna et al., 2022). Attempts have also been made to improve ocular delivery of voriconazole by employing it as a microemulsion or nanoparticles. Microemulsion of voriconazole exhibited higher permeation, increased flux, and increased amount of voriconazole at the corneal surface in comparison to drug suspension (Kumar and

Sinha, 2014). Andrade et al., (2016) developed voriconazole-nanostructured lipid carriers as a prospective method. Additionally, solid lipid nanoparticles of voriconazole exhibited corneal permeation (Khare et al., 2016) and when these voriconazole loaded solid lipid nanoparticles were fabricated into *in situ* gelling system, they exhibited antifungal activity against *C. albicans*, without damaging the ocular surface making it a promising candidate for preclinical evaluations (Pandurangan et al., 2016).

Resistance Resistance of ocular fungi to fluconazole and ketoconazole are well documented as above but comparatively little is known on the resistance of ocular fungi to voriconazole. Studies have indicated that voriconazole has been successful when amphotericin B or fluconazole have been unsuccessful, even in cases of drug-resistant fungal keratitis (Bunya et al., 2007; Kramer et al., 2006). One study indicated that out of 244 fungal isolates from various clinical samples including *Aspergillus* spp., *Fusarium* spp., *Scedosporium apiospermum*, and *Rhizomucor pusillus*, all were susceptible except one *F. moniliforme* strain and *R. pusillus*, which were not suceptible to MIC_{90} ranges of ≤ 1 µg/ml (Linares et al., 2005). In 2019, resistance in *F. solani* in fungal keratitis patients was confirmed (Rosa et al., 2019). In one study it was observed that treatment with azoles before presentation and enrolment in MUTT I was not shown to increase voriconazole MICs (Prajna et al., 2016a; b). In a retrospective multicentre study, it was observed that voriconazole resistance, was associated with environmental CYP51A mutations, with significantly greater 42-day mortality (Clancy et al., 2005; Lestrade et al., 2018).

5.4.3.2.4 Fungal Resistance to Azoles: Itraconazole

Introduction Itraconazole, is an azole antifungal medication very effective against fungal infections including aspergillosis, blastomycsis, coccidioidomycosis, histoplasmosis, and paracoccidioidomycosis. Itraconazole was patented in 1978 and in 1992 it was approved for use in the United States and in some countries as an eye drop (Itral) for ocular fungal infections (Thakkar et al., 2019). It inhibits fungal growth by targeting the fungal membrane or by affecting their metabolism. It could be administered either orally or intravenously. The drug has several recognised side effects including nausea, diarrhoea, abdominal pain, rash, headache, liver problems, cardiac arrest, irregular heartbeat, dry mouth, convulsions, dizziness, nausea, vomiting, abdominal pain and anaphylaxis (Thakkar et al., 2019). Formulations of itraconazole for ocular drug delivery in fungal infections could be in the form of chitosan-based nanosuspension (Ahuja et al., 2013) or solid nanoparticles, which exhibited significantly higher corneal permeability (Mohanty et al., 2015). Kalavathy et al., (2005) reported that topical itraconazole could be used for treatment of fungal keratitis patients with *Fusarium* spp., *Aspergillus* spp., and *Curvularia* spp. as the causative fungal isolates. Lalitha et al. (2007) found that some or all the trizoles (itraconazole, voriconazole, and posaconazole), had the lowest MICs against *Aspergillus* spp. and *Fusarium* spp. and itraconazole was ineffective against *Fusarium* species (Lalitha et al., 2007) and fungal keratitis caused by *Trichophyton* species (Jin et al., 2014). Itraconazole was effective in patients suffering from severe fungal keratitis (Yao et al., 2003) when used as a single drug therapy or as a combination with fluconazole. Further, itraconazole when co-administered as a systemic or a topical drug was effective in successfully managing (over 77% success rate) corneal ulcers caused by *Aspergillus* spp., *Penicillium* spp., and *Fusarium* spp. (Agarwal et al., 2001). It is important that the *in vitro* studies be correlated with the clinical outcome.

Resistance In 1997, Snelders and colleagues (Snelders et al., 2010) reported the emergence of itraconazole-resistant *A. fumigatus* in the Netherlands and their prevalence increased to 6% in 2007. Subsequent azole-resistant *A. fumigatus* were also discovered across Europe and constituted 3.2% of clinical isolates (van der Linden et al., 2015). Several itraconazole resistant strains of *A. fumigatus* isolates had TR34/L98H genotype (Lockhart et al., 2011) in accordance with Snelders et al., (2010) who reported that 94% of azole-resistant isolates harboured the TR34/L98H genotype. Studies have also indicated that itraconazole-resistant mutant strains of *A. fumigatus* involved multiple mechanisms of resistance. These, mutant strains varied in their growth rates and the extent of itraconazole resistance and caused high mortality (88%) (van der Linden et al., 2011). In addition, the mutants in the presence of itraconazole accumulate ergosterol. Studies have indicated the involvement of *CYP51A* gene in itraconazole resistance since N22D and M220I mutations in *CYP51A* gene transformed the itraconazole susceptible strains to resistant strains. Most mutants constitutively expressed high levels of *Afumdr*3, *Afumdr*4, and *Atr*F which could also be mimicked with itraconazol treatment. Thus, over expression of efflux pumps and/or selection of drug target site mutations are at least partially responsible for itraconazole resistance (da Silva et al., 2004). Itraconazole is also known to have interactions with other drugs and as result decrease in azole concentration (Albengres et al., 1998) which is ineffective against some emerging pathogens like *Scedosporium* spp. and *Mucorales* spp. (Denning et al., 1997).

5.4.3.2.5 Fungal Resistance to Azoles: Posaconazole

Introduction Posaconazole is a broad spectrum triazole which is active *in vitro* against yeasts and filamentous fungi involved in mucormycosis (Herbrecht, 2004). Systemic and/or topical posaconazole was used for the treatment of a woman having *Fusarium* spp. keratitis which was resistant to amphotericin and natamycin (Sponsel et al., 2002). The ocular penetration of posaconazole was confirmed and a week later posaconazole (0.25 μg/ml) was recovered in the vitreous and 1.2 μg/ml in the plasma (Sponsel et al., 2002). Posaconazole has been used either as a standalone therapy or with other antifungals to treat deep pan-resistant *Fusarium* spp. keratitis and/or endophthalmitis (Tu et al., 2007a; b) and recalcitrant fungal keratitis, resistant to voriconazole, fluconazole, amphotericin B, and natamycin (Tu et al., 2007a; b; Sponsel et al., 2002; Herbrecht, 2004). The ocular bioavailability of posaconazole is extremely low and thus attempts were made to improve its bioavailability using micelles which proved successful (Durgun et al., 2022).

Resistance A large set of fungal strains (n = 300) including strains of *C. albicans*, *C. glabrata*, *Acremonium* spp., *Cladosporium* spp. and *Scedosporium apiospermum* were found to be either highly susceptible or less susceptible and posaconazole resistance was rare (Araujo et al., 2009). Posaconazole resistance in the eye is yet not reported.

5.4.3.2.6 Fungal Resistance to Azoles: Ravuconazole

Introduction Ravuconazole is a triazole antifungal agent that has broad-spectrum antifungal activity against several fungi like *Candida* spp., *Cryptococcus neoformans*, *Aspergillus* spp., and other fungal pathogens (Andes et al., 2003; Pfaller et al., 2002; Yamazumi et al., 2000). In 2004, Pfaller et al., (2004) observed that >98% (12567) of clinical isolates of *Candida* spp. from more than 200 medical centres worldwide were susceptible at ≤1 μg/ml and 229 (1.8%) were categorized as resistant, with MICs of ≥2 μg/ml. It was further observed that > 99% of the fluconazole-susceptible isolates were also susceptible to ravuconazole (≤1 μg/ml). Thus, the study demonstrated cross-resistance between fluconazole and ravuconazole in the above *Candida* spp. (Pfaller et al., 2004a).

Resistance Ergosterol biosynthesis inhibitors, such as posaconazole and ravuconazole, are also used as drugs for Chagas disease, caused by *Trypanosoma cruzi*. A clone of the *T. cruzi* Y strain was above 300-fold more tolerant to ravuconazole and showed reduced virulence. Ravuconazole cross-resistance to posaconazole was observed. Resistance has been attributed to a novel amino acid residue change, T297M, in CYP51 in *T. cruzi* (Caio et al., 2020). Ravuconazole resistance in the eye is yet not reported.

5.4.3.2.7 Fungal Resistance to Azoles: Miconazole

Introduction Miconazole is an imidazole that has been used for over 30 years for the treatment of superficial and cutaneous fungal diseases. Miconazole like other azoles inhibits ergosterol synthesis. In addition, miconazole inhibits peroxidases, which results in the accumulation of free radicals (peroxides) which eventually triggers cell death (Fothergill, 2006; Thevissen et al., 2007; Sud and Feingold, 1981). It was developed as a topical cream for treating superficial mycoses (Sawyer et al., 1975). It is not recommended for systemic administration due to cardiovascular and hepatotoxic side-effects (Coley and Crain, 1997; Barasch and Griffin, 2008). Topical administration as a 1% solution easily penetrates after epithelial scraping (Foster and Stefanyszyn, 1979; Mohan et al., 1989). It has also been used for subconjunctival administration (5–10 mg) to treat infections caused by *Candida* spp., *Fusarium* spp., *Curvularia* spp., and *Aspergillus* spp. (Foster, 1981; Foster and Stefanyszyn, 1979; Ishibashi and Kaufman, 1985). Fitzsimons and Peters (1986) in a study of 20 fungal keratitis patients observed that both topical and subconjunctival miconazole along with ketoconazole resulted in healing in 13 cases. *In vitro*, it was reported to be more effective than ketoconazole and itraconazole against *Aspergillus* spp., *C. albicans*, and others (Foster and Stefanyszyn, 1979; Uchida et al., 2006). Topical miconazole 1% exhibits minimal toxicity that is characterized by redness in the sclera of the eye due to inflammation in the eye and is an indication of an abrasion of the cornea. Systemic administration of miconazole has been successful in a few reported cases.

Resistance Fungal resistance to miconazole is unknown both in systemic and ocular use and its antifungal activity is maintained despite resistance seen with other azoles (Fothergill, 2006; Sawyer et al., 1975).

5.4.3.2.8 Fungal Resistance to Azoles: Econazole

Introduction Econazole, is an imidazole very similar to miconazole in molecular structure and used primarily in the treatment of fungi such as *Candida* spp., *Aspergillus* spp. *Paecilomyces* spp. and so forth. It is effective against superficial mycoses and has been used for systemic use (Heel et al., 1978) or as an ointment for prophylactic treatment (Mahashabde et al., 1987). Prajna et al., (2004), indicated that monotherapy with 5% natamycin was as effective against fungal keratitis as concurrent use of 5% natamycin and 2% econazole. Another reason that this drug is not widely used is that it has low water solubility thus limiting its incorporation into ophthalmic formulations. This difficulty was overcome recently by generating econazole-α-cyclodextrin inclusion complex embedded in a natural polysaccharides ion-sensitive hydrogel or in a hyaluronic acid hydrogel. Results suggested that econazole-α-cyclodextrin in hydrogels is suitable for topical fungal keratitis treatment.

Resistance Econazole has been used in the treatment of dermatomycoses caused by a variety of fungi. Yet, there are no reports on resistance in fungi to econazole.

5.4.3.3 Molecular Mechanisms of Resistance to Triazoles

Several strategies are in place in fungi to develop molecular resistance to azoles such as genetic modifications of the drug target, overexpression of efflux pumps, or modifications in metabolism (Pemán et al., 2009; Sanglard et al., 2009; Krishnasamy et al., 2018) (Table 5.7) or attributed to *cyp51A* and non-*cyp51A* mutations as in *Aspergillus* species (Sharma and Chowdhary, 2017). The predominant mechanisms of resistance to triazoles differ among fungal species.

5.4.3.3.1 Role of Efflux Pumps

Overexpression of efflux pumps is a major resistance mechanism to overcome the killing effect of azoles in pathogenic *Candida* spp., *C. albicans* and *C. glabrata* isolates (Schuetzer-Muehlbauer et al., 2003; Morschhäuser et al., 2007). In *C. albicans* and *C. auris* CDR1, a gene encoding for an ABC efflux pump (Sanglard et al., 1995) have been identified to be important for their role in azole resistance (Rybak et al., 2019). Deletion of CDR1 increased the susceptibility of resistant strains by several fold (from 64- to 128-fold) (Rybak et al., 2019). CDR2, which is similar to CDR1 and known as PDH1 was discovered in *C. glabrata* strains that were resistant to fluconazole. Both these drug transporters are inducible by azoles. An additional transporter, SNQ2, also causes resistance alongside CDR1 and CDR2 (Whaley and Rogers, 2016). Further, MDR1 another transporter is also associated with *C. albicans* resistance to azoles and human host defence mechanisms (Revie et al., 2018; Whaley and Rogers, 2016). QDR2, another transporter, is also associated with resistance to imidazoles (ketoconazole, miconazole, clotrimazole) in *C. glabrata* isolates (Whaley and Rogers, 2016). It has also been observed that in some *Candida* species, ABC efflux transporter mediated resistance is associated with extensive cross-resistance to the azole drugs itraconazole, voriconazole, and posaconazole (Sanguinetti et al., 2005). MFS transporters seem to have a minor role in *C. glabrata* azole resistance (Costa et al., 2016).

TABLE 5.7

Azole resistance mechanisms in *Candida* spp. and *C. auris*

Target	Effect	Resistance in *Candida* spp.	Resistance in *C. auris*
Azoles inhibit Lanosterol 14-alpha-demethylase which converts lanosterol to ergosterol	Weaker cell membrane, inhibits growth and sensitizes host immune system. Accumulates toxic intermediary compounds.	Overexpression of MSF an efflux pump gene due to mutation in Mrr1 transcription factor (Vandeputte et al., 2012). Overexpression of CDR genes encoding ABC efflux pumps due to mutation in Tac1 transcription factor (Vandeputte et al., 2012; Chabaane et al., 2019). Point mutation in gene encoding LD (*ERG11*). Upregulation of *ERG11* due to mutation in Upc2 transcription factor (Vandeputte et al., 2012). *ERG3* mutation to prevent toxic compound accumulation.	Presence of MFS efflux pumps genes (Sharma et al., 2016). Higher ABC efflux pumps activity compared to *C. glabrata* (Ben-Ami et al., 2017). Significant contribution of CDR1 to resistance (Chowdhary et al., 2018). ERG11 point mutations: (Y132F, S405F, and G464S), extended external loop (D446E, G448E, F449V, and G450E), and between the proximal surface and the heme (K143R) (Healey et al., 2018; Flowers et al., 2015). Presence of ERG3 (Rybak et al., 2019), but no mutation associated with increased resistance reported.

Source: Modified from Chaabane et al., 2019; Krishnasamy et al., 2018.

5.4.3.3.2 Mutations in ERG11Gene Family

In Candida species The enzyme Lanosterol 14-alpha-demethylase (LD), involved in the conversion of lanosterol to ergosterol, an important component of the cell membrane and which maintains membrane integrity (Sanglard et al., 1998) is the primary target of triazoles and alterations to this gene family in several species of *Candida* spp. results in triazole resistance. *C. albicans* strains that had acquired fluconazole resistance exhibited 3 types of genetic changes: (1) point mutations in the Erg11p coding sequence, which decreased its affinity to azoles (Marichal et al., 1999; Flowers et al., 2015), and reduced azole sensitivity in *Candida* spp. (Vandeputte et al., 2012); (2) upregulation in expression of Erg11p and (3) gain-of-function mutations in the transcription factors Tac1 and Mrr1 resulting in the upregulation of the multi drug efflux transporters Cdr1 and Cdr2 (Morschhäuser et al., 2007; Coste et al., 2019). In addition, Chowdhary et al. (2018) demonstrated that in *C. albicans* two mutations Y132F or K143R were responsible for triazole resistance (Chowdhary et al., 2018). Heterologous expression of these two mutations Y132F and K143R is known to increase MIC to azoles also in *S. cerevisiae* (Healey et al., 2018). In *C. parapsilosis* (Pappas et al., 2018; Pfaller et al., 2019) also resistance to triazoles is due to ERG11 mutations Y132F and K143R (Thomaz et al., 2018; Singh et al., 2019; Arastehfar et al., 2020). In addition, overexpression of ERG11 can result in azole resistance due to mutations in the UPC2 transcriptional activator (Hendrickson et al., 2019). ERG3-mediated resistance has also been implicated in resistance.

C. orthopsilosis isolates, which were fluconazole resistant also showed ERG11 mutation (Y132F mutation) (Arastehfar et al., 2019). Cross-resistance to other azoles was observed in >60% of fluconazole resistant *C. parapsilosis* isolates (Pfaller et al., 2004a). Whole genome sequencing of *C. parapsilosis* isolates revealed that fluconazole or voriconazole exposure induced resistance due to MRR1 mutations. Further the triazole, posaconazole induced cross-azole resistance and attenuated virulence owing to ERG3 mutations (Papp et al., 2020). *Candida orthopsilosis*, exhibits a higher MIC to itraconazole than *C. parapsilosis* sensu stricto (Vigezzi et al., 2019).

The mechanism of *C. krusei* resistance to fluconazole is a combination of decreased affinity of Erg11 to fluconazole, which is probably constitutive and inducible expression of the ABC-type efflux transporter Abc1 (Katiyar and Edlind, 2001; Guinea et al., 2006). Comparatively little is known about the resistance of *C. krusei* to fluconazole or other triazoles in eye infections.

C. auris is resistant to azoles and 90% of the isolates are resistant to fluconazole (Kathuria et al., 2015) except for isolates from east Asia (Sekizuka et al., 2019). Azole resistance may be attributed to ERG11 mutations (Y132F, F126T, and K143R) (Lockhart et al., 2016), modification of efflux transporters of both the ABC and MFS types (Chatterjee et al., 2015; Sharma et al., 2015) and ERG11 gene duplication (Bhattacharya et al., 2019). Among the triazoles posaconazole and isavuconazole are effective against *C. auris*, but 50% of isolates exhibit increase in MIC for voriconazole (Lockhart et al., 2016; Kathuria et al., 2015; Chowdhary et al., 2013a; 2014; Rudramurthy et al., 2017).

Candida (Meyerozyma) guillermondii also exhibits resistance to triazoles including fluconazole (Stavrou et al., 2020). In China, 164 isolates from various biological sources (blood, ascitic fluid, peritoneal fluid, pus) and from hospital material (catheter), revealed single and multiple point mutations in the ERG11 gene (Y132F,

W37C, P518R, P430Q, Q469K, and I303V) in non–wild-type isolates (Cheng et al., 2017). These were not ocular isolates.

In Aspergillus species Several *Aspergillus* species including *A. fumigatus, A. niger, A. flavus, and A. terreus* are associated with human diseases and are innately sensitive to the triazoles. However, the sibling or cryptic species that are morphologically indistinguishable from the *sensu stricto* species but could be differentiated by their DNA sequences, exhibit intrinsically high resistance to triazoles. The cryptic species, which exhibited azole resistance, include *A. calidoustus, A. pseudodeflectus, A. lentulus, A. felis, A. viridinutans, A. hiratsukae, A. tsurutae* and *Neosartory apseudofischeri* (Alastruey-Izquierdo et al., 2010; 2013; Balajee et al., 2009; Howard, 2011; 2014; Imbert et al., 2020; Talbot et al., 2019). Resistance to azoles in *A. fumigatus* is also mediated by the ERG11 gene family involving Cyp51A and Cyp51B (Warrilo et al., 2010). Acquired resistance of *A. fumigatus* to azoles is most frequently due to sequence changes in CYP51A gene and these changes could occur in the promoter or the coding sequence, or both, and only rarely with mutations in CYP51B (Snelders et al., 2010; Warrilo et al., 2010). Promoter tandem repeats in CYP51A upregulate the CYP51A gene but do not induce resistance (Snelders et al., 2011; 2015). In *A. fumigatus*, isolates from patients with no prior exposure to azoles, the CYP51 had promoter tandem repeats and specific substitutions (TR34/L98H and TR46/Y121F/T289A) (Snelders et al., 2010; Mellado et al., 2007; Hendrickson et al., 2019) and it was observed that the L98H substitution in CYP51 protein restricts the access of azoles to the heme molecule (Snelders et al., 2011) wheresas the Y121F substitution alters a hydrophobic pocket important for azole docking. The T289A substitution partially restores susceptibility to itraconazole (Snelders et al., 2015). Additional mutations, in positions G54 and M220 in *A. fumigatus* from patients with chronic pulmonary aspergillosis were detected (Tashiro et al., 2012; Singh et al., 2020). Coding sequence mutations decrease affinity of the catalytic domain of the enzyme to azole drugs (Snelders et al., 2010; 2011; 2015). Such mutations are yet to be described in ocular isolates.

5.4.3.3.3 Overexpression of ERG11

Overexpression of ERG11 is linked to azole resistance in *C. albicans* (Chowdhary et al., 2018). Azole-resistant *A. fumigatus* isolates also showed increased expression of CYP51 (Hagiwara et al., 2018). These azole-resistant *A. fumigatus* isolates also showed upregulation of efflux transporters, point mutations in HMG1 (encoding HMG CoA reductase), and in various genes involved in oxidative stress response (Hagiwara et al., 2018; Ballard et al., 2019). Both *C. albicans* and *A. fumigatus* import exogenous cholesterol and bypass the antifungal activity of sterol biosynthesis inhibitors (Xiong et al., 2021).

5.4.3.3.4 Mutations in Double-strand Break Repair and Mismatch Repair

Azole resistance in *C. albicans* is also associated with DNA repair pathways such as double-strand break repair (DSBR) and mismatch repair (MMR). Mutations in the DSBR gene RAD50 and MMR genes MSH2 and PMS1, result in fluconazole-resistant isolates. In *C. glabrata* MSH2 mutation show azole resistance and multi-drug resistance (Hendrickson et al., 2019; Healey et al., 2016).

5.4.3.3.5 Amino Acid Substitution Mutations and Other Mechanisms

Substitutions at Gly 54, Gly 138, Met 220, and Gly 448 are associated with acquired resistance in *A. fumigatus* isolates. Resistance could also be due to ABC and MSF transporters, HapE modification, Cyp51B overexpression, and biofilm production (Hendrickson et al., 2019).

5.4.3.3.6 Other Mechanisms of Resistance to Triazoles by Fungi

(i) *C. glabrata* also has the unique ability to tolerate the inhibition of the ergosterol biosynthesis pathway by using Aus1 to take up cholesterol from serum (Nakayama et al., 2000; 2007). In contrast cholesterol uptake is rapid compared with *C. albicans*, (Zavrel et al., 2013).

(ii) Uniquely, *Candida* and other fungi exhibit mitochondrial biogenesis in response to azole and other antifungals such as echinocandins (Defontaine et al., 1999; Shirazi and Kontoyiannis, 2013; Chamilos et al., 2006).

(iii) The upregulation of Cdr1, Mdr1, Tac1 and Mrr1, has been observed in some resistant *C. parapsilosis* isolates (Arastehfar et al., 2020; Grossman et al., 2015).

5.4.3.3.7 Changes in the Fitness in Triazole Resistant Fungi

Fungal fitness manifested as decreased growth rate and sporulation is a common phenomenon in triazole resistant fungi and has been attributed to genomic changes in fungi harbouring resistance-associated mutations (Fekete-Forgács et al., 2000; Graybill et al., 1998; Angiolella et al., 2008; Cowen et al., 2001; Sasse et al., 2012). In *Candida* spp. the azole resistant strains are normally associated with altered (reduced, unchanged or increased) fitness and virulence (Fekete-Forgács et al., 2000; Graybill et al., 1998; Angiolella et al., 2008). In *C. albicans* Tac1, Mrr1, and Upc2 mutations in drug resistant fungi was accompanied by a concomitant loss of fitness (Sasse et al., 2012). The fitness cost of such mutations was cumulative and was particularly robust when combining multidrug resistance regulator (MRR1) and transcriptional activator of CDR genes (TAC1) gain-of-function (GOF) mutations. Their findings suggest that overexpression of transporters Cdr1 and Cdr2 and major facilitator superfamily (MFS) transporters are disadvantageous, and rarely found together in fluconazole-resistant strains. These studies suggest that the dynamics of drug resistance are complicated because it involves multiple interactions between different resistance mechanisms and compensatory mutations that counterbalance fitness cost (Cowen et al., 2001). In contrast, azole resistance in *C. glabrata* is associated with enhanced virulence with fitness gain (Vale-Silva et al., 2015).

5.4.3.4 Resistance to Echinocandin Class of Antifungal Agents

Echinocandins are preferred for invasive candidiasis (Hendrickson et al., 2019) but resistance to echinocandins has been reported (Rocha et al., 2007; Hendrickson et al., 2019). The three drugs that are used in opthalomology under this class are caspofungin, micafungin, and anidulafungin.

5.4.3.4.1 Resistance to Echinocandin Class of Antifungal Agents: Caspofungin

Caspofungin, a lipopeptide is suitable for treating deep-seated fungal infections of the eye caused by *Candida* spp. and *Aspergillus* spp., but not very efficient in inhibiting *Fusarium* spp. (Goldblum et al., 2005; Shen et al., 2014). It effects fungal cell integrity by inhibiting β-(1,3)-D-glucan synthesis (Fesel and Zuccaro, 2016). Although caspofungin has shown favourable preclinical outcomes it is very expensive and has been used as an adjuvant or sometimes individually. The common mode of delivery of caspofungin (Cancidas®) in ocular mycoses is that it is either administered systemically or as eye drops or also administered intravitreally (deep-seatedmycoses) (Aydin et al., 2007). Therapeutic concentrations of caspofungin have been detected in aqueous humor (Spriet et al., 2009) indicating that it could be suitable in treating ocular mycoses. But it exhibits poor permeation in uninfected/uninflamed eyes (Neoh et al., 2011a). Studies have also demonstrated that topical caspofungin solution (1%) was effective in treating recurrent *Candida* keratitis and *Candida* keratitis that was refractory to voriconazole (Hurtado-Sarrió et al., 2010; Gregory et al., 2010). Further, topical caspofungin, along with intrastromal voriconazole and an independent caspofungin intrastromal therapy, was effective against *Alternaria* keratitis that did not respond optimally to natamycin therapy (Neoh et al., 2011b; Tu, 2008). Thus, caspofungin could be used off-label as a standalone drug or in combination with voriconazole to manage both refractory and recurrent fungal infections of the eye (Thakkar et al., 2019).

Resistance in Candida *species* Echinocandins are used for invasive candidiasis and aspergillosis (Perlin, 2015). In *C. glabrata*, 8% to 9% of blood stream isolates were found to be resistant to caspofungin and many also displayed cross-resistance to other echinocandins and triazoles (Pfaller, 2012; Beyda et al., 2014; Pham et al., 2014; Farmakiotis et al., 2014). *C. albicans* hyper-expressing fungal efflux transporters of ABC or MFS-type exhibited modest MIC changes to echinocandins (Niimi et al., 2006). Studies have shown that spontaneous *C. albicans* mutants were resistant to caspofungin (MICs were increased >70-fold), (Kurtz et al., 1996) and the mutant isolates required higher drug doses. Resistance of the mutants to caspofungin were mapped to the FKS1 and FKS2 genes in *C. glabrata*. Amino acid substitutions in the gene namely S645F, S645P, and S645Y in *C. albicans* (CaFks1 F641-P649), conferred decreased susceptibility to caspofungin (Park et al., 2005). Therefore, the mechanism for echinocandin resistance in *C. albicans*, is associated with mutations in Fks1, a large integral membrane protein, while mutations in both Fks1 and its paralog Fks2 are associated with resistance in *C. glabrata* (Douglas et al., 1997; 1994). Strains harbouring FKS mutations varied in their resistance depending on the region mutated. For instance, mutants in which the hot spot S645 locus of FKS is mutated the sensitivity to caspofungin is reduced by 1000-fold. In animal models also, these mutants show a poor response to caspofungin therapy (Arendrup et al., 2012; Cota et al., 2006; Wiederhold et al., 2011). Hot spot FKS mutations also showed increased resistance to anidulafungin and micafungin and in addition conferred cross-resistance to the available echinocandins.

Resistance to echinocandin in other *Candida* species including *C. krusei, C. glabrata, C. tropicalis, C. guillermondi, C. parapsilosis, and C. dubliniensis* also occurs through the FKS mutations (Perlin, 2015). The most prominent amino acid substitutions in

C. glabarata FKS include S629 in Fks1, S663 in Fks2, and P659 in Fks2 (Garcia-Effron et al., 2009; Wiederhold et al., 2011). In *A. fumigatus* echinocandin resistance was due to S645Y mutation in FKS (Rocha et al., 2007). But this resistance pattern was not obvious in clinical strains (Perlin, 2007). The risk factors associated with acquired echinocandin resistance in *Candida* spp. could be due to medical or surgical reasons (Fekkar et al., 2013), due to persistently low CD41 counts as in AIDS patients and in patients with recurrent esophagitis (Laverdière et al., 2006; Miller et al., 2006), due to retention of infected prosthetic material or a catheter (Garcia-Effron et al., 2008), due to endocarditis (Moudgal et al., 2005), due to complicated infection of the abdomen (Krogh-Madsen et al., 2006; Shields et al., 2014) or due to echinocandins therapy (Lewis et al., 2013; Fekkar et al., 2013). *C. glabrata* resistant to echinocandin may display cross-resistance to polyenes and azoles (Krogh-Madsen et al., 2006;). Heat shock protein (Hsp) 90 is also known to modulate echinocandin resistance mediated by calcineurin-dependent stress responses in *C. albicans*, *C. glabrata,* and *C. dubliniensis* (Singh et al., 2009; Singh-Babak et al., 2012; Chen et al., 2011b; Lamoth et al., 2014). Loss of function of Hsp90 decreased the resistance to echinocandins whereas Hsp90 expression enhanced the effect of echinocandin in a murine model of candidiasis. Deletion of Gcn5, a lysyl-acetyl transferase, which regulates adhesion, cell wall-mediated kinase signalling, hypersensitivity to oxidative stress, and regulation of the Fks1 glucan, decreases hypersensitivity to caspofungin (Shivarathri et al., 2019). Similarly, Efg1 and Cas5, which are transcription factors were shown to orchestrate the cell wall stress response of *C. albicans* to caspofungin (Xiong et al., 2021). FKS1 hot spot 1 mutations conferred resistant to echinocandins in *C. auris* (Chowdhary et al., 2014b; 2018; Kathuria et al., 2015; Biagi et al., 2019; Kordalewska et al., 2018). Further, Shivarathri and colleagues (2020) showed that in *C. auris*, the signalling pathways like the 2-component signal transduction and mitogen-activated protein kinase signalling regulate drug resistance to amphotericin B and caspofungin. For instance, deletion of the response regulator encoding SSK1 and the mitogen associated protein kinase HOG1 restored susceptibility to both antifungals (Shivarathri et al., 2020). Distinct mechanisms are associated with low-level echinocandin resistance or tolerance, which could be attributed to: (1) adaptive stress responses including paradoxical growth in the presence of supra-MIC drug concentration; (2) mutations in hot spot regions of *FKS1* and *FKS2* (Pfaller et al., 2016; Locke et al., 2016), and (3) intrinsic FKS mutations in *C. parapsilosis* and *C guilliermondii.*

Resistance in Aspergillus *species* In *Aspergillus* species resistance to echinocandin is attributed to either mutations in FKS1 or due to significant increase in the abundance of dihydrosphingosine and phytosphingosine. Caspofungin is known to induce mitochondrial-derived reactive oxygen species (ROS) and dampening of ROS levels eliminated caspofungin-induced resistance implying that build-up of ROS is associated with resistance to caspofungin. These findings confirm the observations of Chamilos and colleagues (2006), who implicated mitochondrial respiratory network in *C. parapsilosis* caspofungin non susceptibility. In innate resistant *Fusarium solani* and Mucorales also echinocandin resistance is largely FKS1 independent (Ibrahim et al., 2005)

Resistance in Cryptococcus neoformans *C. neoformans* is innately resistant to the echinocandins and was attributed to Cdc50, the β-subunit of the lipid translocase (flippase), which connects lipid flippase with cytoplasmic calcium homeostasis and signalling (Cao and Xue 2020; Cao et al., 2019). Fungal melanins also protect *C. neoformans* from the effect of echinocandins and amphotericin B, but this protection is not seen in *in vitro* (van Duin et al., 2002).

5.4.3.4.2 Resistance to Echinocandin Class of Antifungal Agents: Micafungin

Introduction Micafungin, is an echinocandin antifungal agent very similar to caspofungin, voriconazole and amphotericin B and other antifungal agents in the following aspects:

(a) micafungin like caspofungin exhibits activity against *Candida* spp. and *Aspergillus* spp. and a weak activity against *Fusarium* spp. (Diekema et al., 2003),
(b) micafungin is effective *in vitro* against many *Candida* isolates that are resistant to fluconazole and other triazoles (Messer et al.2006),
(c) intravitreal injection of micafungin is as effective as voriconazole and amphotericin B in treating *A. fumigatus* infections (Harrison et al., 2005) and
(d) pre-clinical topical micafungin administration in rabbits exhibited effects similar to topically administered natamycin in *Aspergillus* keratitis (Trujillo et al., 2004; Hiraoka et al., 2005).

In addition, intravitreal injection of micafungin was also effective in treating *Aspergillus* keratitis (Paris et al., 2004; Suzuki et al., 2008). Micafungin (Mycamine) is administered either systemically or topically for the management of keratitis and endophthalmitis. Clinically, micafungin concomitant therapy with voriconazole is effective against keratitis caused by *Beauveria bassiana* (when monotherapies with natamycin and voriconazole failed), against keratitis caused by *Wickerhamomyces anomalus* and endophthalmitis caused by *Trichosporon* species (Mitani et al., 2014; Kamoshita et al., 2014; Walia et al., 2009; Matsumoto et al., 2010). Topical and intravitreal administration of micafungin was effective and safe in the management of refractory *Pestalotiopsis clavispora* keratitis, which was refractory to voriconazole and natamycin-associated therapy (Monden et al., 2013). Further, micafungin administered concomitantly with fluconazole and voriconazole showed a better therapeutic outcome owing to a synergistic activity.

Resistance Echinocandins are the preferred drug for invasive candidiasis (Hendrickson et al., 2019). *C. glabrata* isolates are the most resistant species of *Candida* for the echinocandins and exhibits lowest resistance to micafungin (1.7%). Peak resistance rates for micafungin were 2.8% during 2006–2008 and lowest resistance rates were 1% during 2009–2011. Micafungin resistance was highest in North America compared to other regions of the world and the rates differed with the fungus screened. For *C. glabrata* the resistance was 2.8% versus 0–0.6% and for *C. tropicalis* 1.3% versus 0–0.5% respectively (Pfaller et al., 2019).

5.4.3.4.3 Resistance to Echinocandin Class of Antifungal Agents: Anidulafungin

Introduction Anidulafungin like the echinocandins caspofungin and micafungin, are effective against many *Candida* isolates including species resistant to fluconazole and other triazoles (Messer et al., 2006; Ghannoum et al., 2017; Arendrup et al., 2012). Shields et al., (2013) had also observed that anidulafungin and micafungin MIC breakpoints are superior to caspofungin for identifying FKS mutant *C. glabrata* strains and echinocandin resistance.

Resistance Resistance to anidulafungin was studied in *Candida tropicalis* and mutations in Fks1p were associated with anidulafungin resistance (Chew et al., 2019).

5.4.3.4.4 Mechanisms of Echinocandin Resistance in Candida *spp.*

Candida glabrata isolates resistant to echinocandin were found to possess a mutation in either FKS1 or FKS2 genes. The common mutant genotypes of FKS include S641F and S645P in FKS1 in *C. albicans* and S663P in FKS2 exclusively in *C. glabrata* (Perlin et al., 2017). Healey et al. (2016) demonstrated that 100% of the FKS1/2 mutants were resistant to echinocandins. Mutants with amino acid substitutions in *C. parapsilosis* (P649) and *C. guilliermondii* (M633/A634) (Perlin et al., 2017) were responsible for resistance. Echinocandin resistance are thought to be associated with prior echinocandin exposure (Shields et al., 2015; Beyda et al., 2014). It was also observed that echinocandin resistance due to FKS2 is reversible after administration of a calcineurin inhibitor such as tacrolimus indicating that it is calcineurin dependent (Revie et al., 2018). Additionally, Suwunnakorn et al. (2018) recently demonstrated that FKS2 and FKS3 in *C. albicans* can negatively regulate FKS1, thereby influencing echinocandin susceptibility, but this mechanism is yet to be demonstrated in clinical isolates (Suwunnakorn et al., 2018). Disruption in mismatch repair gene (MMR) MSH2 also results in multidrug-resistant strains of *C. glabrata* (Healey et al., 2016) including resistance to amphotericin B, fluconazole, voriconazole, caspofungin, and micafungin to various extents and the mutation may promote the emergence of resistance to antifungals through FKS mutations (Healey et al., 2016). But analysis of echinocandin-resistant isolates indicated that though a good majority of the resistant isolates had FKS2 mutation (81%), in a significant number of isolates (19%) the mutation was not detectable indicating that alternate mechanisms confer resistance. In this context it has been demonstrated that protein kinase C (PKC), protein phosphatase calcineurin, Hsp90, and Cas5 have all been implicated in increased resistance to echinocandin (Revie et al., 2018; Sharma et al., 2017). Even though multiple mechanisms contribute to echinocandin resistance, only mutants with FKS mutations are associated with worse outcomes for patients with candidemia (Beyda et al., 2014). It is hypothesized that the presence of FKS mutation may serve as a surrogate for echinocandin failure (Shields et al., 2015; Beyda et al., 2014).

5.4.3.4.5 Mechanisms of Echinocandin Resistance in Aspergillus *spp.*

Echinocandins are useful in the management of *Aspergillus* infections with azole resistance. But resistance mechanisms in *Aspergillus* spp., are less well described. In *A. fumigatus* resistance to echinocandins was either due to S6768Y or S678P substitutions in FKS1 (Hendrickson et al., 2019). In contradiction in a few isolates resistant to echinocandin, FKS mutations were not found, but these isolates exhibited increased FKS gene expression. Echinocandin resistance in *A. fumigatus* has also been attributed to increase in chitin synthesis, Δras mutations and Hsp90 mutations.

In *A. fumigatus* inhibition of Hsp90 and calcineurin has been shown to improve echinocandin susceptibility (Sharma and Chowdhary, 2017).

5.4.3.5 Resistance to Polyene Antifungal Agents

5.4.3.5.1 Resistance to polyene Antifungal Agents: Natamycin

Introduction Amphotericin B, nystatin, natamycin, rimocidin, filipin, candicin, hamycin, perimycin and dermostatin are all polyene antimycotics out of which only natamycin, and amphotericin B are used for ophthalmological purposes. Natamycin is a tetraene polyene antifungal drug effective against filamentous fungi. It inhibits fungi by binding to ergosterol (Prajna et al., 2016a; b; teWelscher et al., 2008) in the fungal cell membrane and thereby effecting the fungal cell integrity (teWelscher et al., 2008). Natamycin is an approved drug for treating keratitis and acts selectively against the filamentous fungal specie *Aspergillus* spp., *Fusarium* spp. and *Alternaria* spp. (Müller et al., 2013; Lalitha et al., 2007; 2008; Xuguang et al., 2006; Xu et al., 2010). Additionally, natamycin is more potent against *Fusarium* spp., in comparison to *Aspergillus* spp., which were isolated from eyes infected with keratitis (Wang et al., 2015). It was found that natamycin therapy was safer than the voriconazole therapy (Prajna et al., 2010) and effectiveness of natamycin in superficial and deep-seated mycosis, were higher than voriconazole (Pradhan et al., 2011). Natamycin is administered topically as eye drops (Patil et al., 2017) but due to their low retention and bioavailability at the surface of the eye repeated administration is required (Patil et al., 2018). Further it is effective against superficial fungal keratitis but not in deep-seated ocular keratomycosis (Patil et al., 2017) and it is administered along with other antifungal drugs such as amphotericin B, azoles, and echinocandins, either by the systemic or subconjunctival route (Qiu et al., 2015; Thakkar et al., 2019). Several strategies have been tried to improve the delivery of natamycin to the cornea such as using natamycin-loaded PEGylated nanolipid carriers (NT-PEG-NLC) (0.3%) which had higher trans-corneal natamycin delivery as in the iris–ciliary bodies and aqueous and vitreous humors during fungal keratitis therapy (Patil et al., 2018). Janga et al. (2018) used ion-sensitive *in situ* gels of natamycin bilosomes and showed 6 to 9-fold increase in the trans-corneal level of natamycin indicating the potential feasibility of these carriers in ocular fungal infections. Natamycin loaded -hydroxypropylbeta-cyclodextrin (NT-CD) complex *in situ* gels (NT-CD-IG) also improved corneal retention time and transcorneal permeation (Paradkar and Parmar, 2017; Patil et al., 2018; Zhang et al., 2018).

Resistance Natamycin is a preferred antifungal drug clinically, due to its favorable safety/toxicity profile. Over the years antifungal resistance to natamycin has been reported. In the Mycotic UlcerTreatment Trial I (MUTT I) it was observed that compared to topical voriconazole, natamycin was superior in the treatment of fungal keratitis (Prajna et al., 2016a; b) particularly those infected with *Fusarium* species (Streekstra et al., 2016). MUTT I data also showed that an increase in MIC by 2-fold (Prajna et al., 2016a; b) was associated with increased odds of perforation, increased 3-month scar size and reduced success in the treatment of corneal ulcers (Chowdhary et al., 2013a). Fungal strains have also been shown to develop resistance with prolonged exposure to natamycin *in vitro* (Streekstra et al., 2016). A subsequent analysis of 522 ocular samples from 3 randomized clinical trials including 651 patients with smear-positive fungal ulcers enrolled in Mycotic Ulcer Treatment Trial (MUTT) I (2010 to 2011) or II (2010 to 2015) or the Cross-Linking Assisted Infection Reduction (CLAIR) trial (2016 to 2018) in South India using *Fusarium* and *Aspergillus* indicated that susceptibility to natamycin and voriconazole over time showed a marginal increase of 1.02-fold per year for voriconazole and a 1.06-fold increase in natamycin resistance per year. Other filamentous fungi showed 1.07-fold increase in resistance to natamycin annually (Prajna et al., 2022). These results clearly suggested that the susceptibility trend to both natamycin and voriconazole was decreasing over the last decade in South India.

5.4.3.5.2 Resistance to Polyene Antifungal Agents: Amphotericin B

Introduction Amphotericin B is a broad-spectrum antifungal agent isolated from *Streptomyces nodosus* in the 1950s and approved by the FDA in the 1960s (O'Brien, 1999). The drug was introduced for use in 1958, and for more than two decades amphotericin B was a drug of choice for systemic fungal infections but lost favour due to accompanying adverse reactions, specifically nephrotoxicity and infusion reactions. Voriconazole replaced amphotericin B for treating invasive aspergillosis and other filamentous fungi (Herbrecht et al., 2002). It was initially used for systemic fungal infections and later recruited for ocular fungal infections (Müller et al., 2013). It is routinely used in the treatment of fungal infections of the eye like keratitis, and is effective against *Candida* spp., *Aspergillus* spp., *Fusarium* spp., *Penicillium* spp., *P. marneffi* and *Cryptococcus* spp. (Müller et al., 2013; Ito and Hooshmand, 2005; Segal and Walsh, 2006; Alastruey-Izquierdo et al., 2008; O'Brien, 1999). Amphotericin B is ineffective during the healing stages of corneal fungal infections when the integrity of the cornea is being restored (O'day et al., 1986). Amphotericin B lacks the ability to penetrate intact cornea and thus is used only when the cornea has been abraded. Therefore, it necessitates intravenous administrations of the drug at high dose of amphotericin B (to reach theocular sites). Reports have indicated that systemic administration of amphotericin B does not reach therapeutic levels in eye (Green et al., 1965; Goldblum et al., 2002; Kaur et al., 2008; O'Day et al., 1986; O'Day, 1987). Therefore, systemic administration is discouraged. It also has acute toxic effects such as aches, fever, chills, and rigors (Cornely et al., 2007). The other effects include Thrombophlebitis, Anaemia and CNS toxicity (Joseph and Sharma, 2016). Amphotericin-B also promotes immunopotentiation by reducing suppressor cells and concomitantly increasing pro-inflammatorycells (Shirley and Little, 1979a, b). The drug showed good tolerability and efficiency as eye drops and ointment (Hirose et al., 1997; Wood and Williford 1976) and intracorneal administration was better than subconjunctival administration. Various formulations were developed to improve the delivery of amphotericin such as amphotericin B-loaded chitosan-modified nanostructured lipid carriers (AmB-CH-NLC) (Fu et al., 2017) in rabbit eyes and amphotericin B-loaded lecithin/chitosan nanoparticles (Chhonker et al., 2015). In addition, nanosuspension and nanoparticles of amphotericin B exhibited activity against *F. solani* (Das and Suresh, 2011; Das et al., 2010) and improved ocular permeation. Amphotericin B microemulsion also showed high activity against *Candida* strains. But amphotericin B-loaded self-aggregating nanoparticles (AmB/PLA-g-CS) (0.15%) using amphiphilic poly-(lactic acid)-grafted-chitosan (PLA-g-CS) (Zhou et al., 2013) did not show antifungal activity against *C. albicans*. Yet another strategy used intracameral administration (0.15%) of amphotericin B rather than topical application (0.15%) (Shao et al., 2010; Yilmaz et al., 2007) and showed that intracameral amphotericin B was effective in treating refractory keratomycosis caused by *A. niger*, *Colletotrichum graminicola*, and *A. flavus* (Kaushik et al., 2001; Kermani and Aggarwal, 2000; Ritterband et al., 1997). A combination of intracameral and intrastromal injection of amphotericin B was effective in the treatment of severe FK, which was refractory to the conventional therapy involving fluconazole, natamycin, and itraconazole (Hu et al., 2016; Garcia-Valenzuela and Song, 2005). Amphotericin B and the echinocandins can treat patients with invasive aspergillosis, but each has its limitations. For instance, amphotericin B is not recommended as monotherapy for invasive aspergillosis (Patterson et al., 2016). Formulations of Amphotericin B and the echinocandins must be administered intravenously, which could also be a problem when therapy is prolonged. In patients with invasive fusariosis newer antifungals and formulations, including voriconazole and lipid amphotericin B may be effective. Lewis et al., (2005) studied the in vitro pharmacodynamics of amphotericin B, itraconazole, and voriconazole against various *Aspergillus*, *Fusarium*, and *Scedosporium* species.

Resistance Wiederhold (2017), in a review, listed several molds (*Scedosporium* species, *Lomentospora prolificans*, and *Fusarium* species) with reduced susceptibility or intrinsic resistance to antifungals. Not much is known about the molecular mechanism of resistance to amphotericin B by fungi. Amphotericin B treatment is known to increase expression of genes involved in iron homeostasis and ATP synthesis (Pang et al., 2017) and these genes are targets of transcription factors Aft1p and Zap1p. In fact aft1Δ and zap1Δ mutants were hypersensitive to amphotericin B and H_2O_2, suggesting that they are key regulators of the drug response. Mechanistically, amphotericin B and lactoferrin exhibit synergistic antifungal activity (Pang et al., 2017) affecting cell wall and membrane integrity and disrupting intracellular processes. Thus, Zap1p- and Aft1p-binding molecules along with antifungals could be used for synergistic effects (Pang et al., 2017). Amphotericin B resistance could also occur due to mutations in FLO8, the transcription factor and a putative transmembrane transporter (Escandon et al., 2019), due to mutations or altered expression of ergosterol biosynthetic genes in *S. cerevisiae* and *Candida* spp., (Young et al., 2003) and increased expression of HSp 70 and Hsp 90, scavenging of ROS and altered mitochondrial activity (Blum et al., 2013; Posch et al., 2018). In additionin *C. albicans* and *S. cerevisiae* the sphingolipid biosynthetic pathway may also be involved in conferring resistance to amphotericin B (Sharma et al., 2014). The ability to bind cholesterol in human cells, could be the reason for its side effects (Gallis et al., 1990) and replacement of ergosterol with other sterol intermediates like ergosta 5,7-dienol are associated with co-resistance to azoles and amphotericin B (Sanglard et al., 2003; Martel et al., 2010). *C. auris* is frequently non-susceptible to amphotericinB (Escandon et al., 2019).

5.4.3.5.3 Changes in the Fitness in Amphotericin B Resistant Fungi

Vincent and colleagues (2013) showed that several *C. albicans* strains including clinical and laboratory strains, showed increased vulnerability to oxidative stress, high temperature, and neutrophil phagocytosis and high levels of Hsp 90 were required for survival.

5.4.3.6 Resistance to Fluocytosine

Introduction Flucytosine is an inhibitor of nucleic acid synthesis, which could kill fungi. Monotherapy often leads to fungal resistance (Vermes et al., 2000). Thus, fluconazole is used in combination with amphotericin B to improve patient outcomes (Molloy et al., 2018). In immunodeficient HIV patients, *Cryptococcal meningitis* is the causative agent for more than 100,000 deaths per year. In such patients, treatment with amphotericin B plus flucytosine for 2 weeks was more effective than fluconazole monotherapy. This combination therapy caused lowest mortality after ten weeks of treatment (Molloy et al., 2018).

Resistance Unlike most antifungal agents flucytosine is activated after entry into the cell by a FUR1 gene encoded protein (Waldorf and Polak, 1983). In *Candida* spp. and *C. auris,* mutations in FUR1 were associated with flucytosine resistance (Vandeputte et al., 2012; Rhodes et al., 2018) and in the resistant *C. auris* strain the FUR1 gene had a F211I amino acid substitution. Such specific mutations or equivalent are yet to be detected in other *Candida* spp. Further studies would help to unravel the mechanisms conferring high resistance of the *C. auris* strains (Chaabane et al., 2019).

5.5 Tolerance and Hetero Resistance in Fungi

Variability in response, tolerance levels and resistance further complicate treatment of disease. For instance, *in vitro* MIC determination may not be the optimum dose to overcome the disease, since a considerable number of patients who die of candidemia are said to be infected with isolates that are susceptible to the drug by accepted MIC break points (Reboli et al., 2007; Lee et al., 2010). This discrepancy is ascribed to several factors such as factors associated with the host, disease, and drug and are thus unrelated to *in vitro* susceptibility. Additionally, the tolerance of the fungus to the drug, that is, the ability to grow at concentrations of the drug that are greater than the MIC could also contribute to the disease. The phenomenon of paradoxic growth, also known as the Eagle effect, has been reported in *Candida* spp. and *Aspergillus* spp., where the organisms are inhibited by low concentrations of an echinocandin, yet withstand much higher drug concentration. This type of tolerance is linked to stress pathways regulated by Hsp 90 and calcineurin and appears to be species-dependent (Kordalewska et al., 2018; Chamilos et al., 2007a). Yet, the mechanisms and clinical importance and relevance of paradoxic growth remains unclear (Delarze and Sanglard, 2015). Tolerance in each community of fungi could be heterogenous with some populations more tolerant. Tolerance could be ascribed to dormant (persister) cells (Berman and Krysan, 2020; Delarze and Sanglard, 2015) and ability to form biofilms (Nett and Andes, 2020). In addition, hetero-resistance, defined as significant cell-to-cell variation in drug response as seen in *C. glabrata* was hypothesized as a possible mechanism of tolerance. Using population analyses, hetero-resistance was observed in 57% of clinical isolates and shown to be a non-binary continuous phenotype. Studies have indicated that hetero-resistance was associated with increased efflux transporter activity and persistent visceral infection in a mouse model (Ben-Ami et al., 2016). In *C. glabrata*, a multi-drug resistant and a tolerant pathogenic fungus, several novel genes involved in tolerance to azoles and echinocandins have been reported (Vale-Silva et al., 2015; Schwarzmuller et al., 2014).

5.6 Implications of Antifungal Resistance

Mechanisms of resistance to antifungals have been reasonably studied in *Candida spp.* and *Aspergillus* spp. (Cowen et al., 2014) but our knowledge is restricted to non-clinical and systemic isolates only. Thus, there is a need to extend such studies to niche specific organisms like those associated with the eye and to emerging opportunistic yeasts *(Trichosporon* spp.) and opportunistic molds (*Fusarium* spp.). Further, the need to investigate the progression of resistance in high-risk patients with fungal infections (Morschhäuser, 2016), as in HIV/AIDS patients and other patients with systemic fungal diseases such as vulvovaginal candidiasis, chronic structural lung disease and so forth, and in immune-suppressed patients with pulmonary mycoses (Heo et al., 2017). It is likely that many clinical fungal strains have multiple mechanisms of resistance or uncharacterized mechanisms of resistance or tolerance (Heo et al., 2017; Ford et al., 2015). In the clinical scenario patients are exposed to repetitive cycles of antifungals, conducive for the acquisition of tolerance and resistance, and such practises should be curtailed. The use of animal models may not be relevant since these models do not simulate clinical scenarios, especially with respect to the emergence of antifungal resistance. For instance, fungal infections affect the extremes of age and especially immunosuppressed individuals and patients with comorbidities. In contrast the animals used are young, healthy, inbred and genetically less heterogeneous compared to genetic heterogeneity in humans. In addition, pharmacokinetic and pharmacodynamic properties of antifungals by resistant fungi (Lewis and Verweij, 2017) in infection models and a human vary, which could be due to the host's immune status and differences in anatomy and so forth. Further models, such as *Drosophila melanogaster*, *Galleria mellonella*, and *Caenorhabditis elegans* need to be authenticated for evaluating fungal resistance (Chamilos et al., 2007b). Studies have also indicated that certain antifungal resistance mechanisms predominate in patient populations, suggesting immunotherapeutic strategies for combating *Candida* resistance (Lewis et al., 2012). Furthermore, changes in the cell wall, evasion of the host immune response (Chamilos et al., 2010) and the immune-pharmacologic action of some antifungals, could enhance immune-mediated clearance and theoretically avert resistance (Ben-Ami et al., 2008).

5.7 Clinical Relevance of Antifungal Resistance: Unanswered Questions

In general, antifungal resistance is global (Pfaller et al., 2004b), and resistant fungal infections have been on the rise and need to be tackled worldwide using a multipronged attack involving generation of new antifungals and improvement in fungal diagnostics inclusive of both the cultivable method and culture-independent diagnostics to identify the resistant fungi.

In addition to PCR used for both early diagnosis and detection of resistance, MALD-TOF may potentially become a useful modality (Singhal et al., 2015).

The lack of concordance in *in vitro* fungal resistance (performed under conditions like high inoculum, aerobic growth, glucose-rich medium, planktonic pattern of growth) and *in vivo* fungal growth (Lamoth and Kontoyiannis, 2019) (involving low infecting inoculum, complex nutrient-poor substrates, biofilm formation, semi-anaerobic growth, and frequent sequential exposure of fungi to different antifungals) is a problem that is to be addressed.

The widespread use of combination therapy further complicates the translation of *in vitro* susceptibility data to the clinic (Lionakis et al., 2005; 2018). Further, a need to characterise clinical break points for all pathogens other than just *Candida* spp. and *Aspergillus* spp. (Lionakis et al., 2005; 2018; Lamoth and Kontoyiannis, 2019; Colombo et al., 2021; Espinel-Ingroff and Turnidge, 2016).

The implications of confounding factors such as underlying disease, comorbidities, drug–drug interactions, immunosuppression (Kontoyiannis and Lewis, 2014) should be accounted for.

There is a need for guidelines to be in place for the treatment of drug-resistant Candidiasis (Pappas et al., 2016 a; b) and Aspergillosis (Patterson et al., 2016), especially *C. auris*, (Lamoth and Kontoyiannis, 2018), *C. glabrata*, (Ostrosky-Zeichner, 2013) and azole-resistant *Aspergillus* spp. (Kontoyiannis and Lewis, 2002; Georgiadou and Kontoyiannis, 2012).

There is an urgent need to introduce new antifungal drugs with unique mechanisms of action (Osherov and Kontoyiannis, 2017) for instance drugs that target key virulence properties, such as invasion or adherence (Oever and Netea, 2014; Galloway-Peña and Kontoyiannis, 2020).

5.8 Conclusions on Antifungal Resistance

AFR is an area of emerging global public health concern and based on available data (Ben-Ami et al., 2021) the following conclusions could be made:

(i) several ocular fungi exhibit resistance to different classes of antifungal agents; but the mechanism of innate and acquired resistance are yet to be unveiled,
(ii) animal models need to be identified and developed to simulate the *in vivo* condition so that such models could be used to understand the molecular basis of resistance,
(iii) need to develop rapid and novel point-of-care assays to produce clinically meaningful end points,
(iv) need to develop both the conventional and the culture-free methods to identify the diseases causing fungi,
(v) undertake clinical trials on opportunistic mycoses caused by resistant fungi,
(vi) undertake effective measures involving government and non-government organisations to educate the public and frontline workers on curtailing and eradication of antifungal resistance by advocating judicious and proper use of the antifungals (Ben-Ami et al., 2021),
(vii) echinocandins are the preferred treatment option for invasive candidiasis,

(viii) amphotericin B–based treatment is preferred for immunosuppressed patients,

(ix) voriconazole, isavuconazole and posaconazole are treatment options for invasive pulmonary mold infections other than mucormycosis in immunosuppressed patients, and

(x) *C. glabrata* exhibits a high propensity to develop resistance to both azoles and echinocandins.

Nevertheless, though AFR to different antifungal classes in the eye is widely documented across the world and the causative fungi are identified, comparatively less is known about the molecular basis of fungal resistance in the eye.

5.9 Future Directions

Resistance of fungi on the ocular surface is of great concern since it impacts human health and economy. Thus, research needs to be an ongoing process to develop antifungals agents instead of using a given class of antifungals, which would eventually develop resistance. In fact, there is a need to think out of the box and try to discover and invent new antifungals beyond the conventional classes of antifungal drugs (polyene, azole, and echinocandin) to be prepared for the emergence of new fungal pathogens. Preclinical and clinical evaluations of antifungals should be undertaken on a war-footing so that a choice of antifungals is available depending on the causative agent. The strategy should also encompass optimum methods for drug delivery to the eye by developing carriers like nanoparticles, nano-emulsions, liposomes, gels and so forth that would enhance ocular retention and permeation when administered topically and be suitable for superficial fungal infections of the eye such as FK but also useful for treating deep-seated ocular fungal infections. Several preclinical studies also demonstrate the superiority of combination therapy versus monotherapy, and this aspect needs to be developed on a priority basis for the already available antifungals (Thakkar et al., 2019).

REFERENCES

Abdelbary GA, Amin MM, Zakaria MY. Ocular ketoconazole loadedproniosomal gels: Formulation, ex vivo corneal permeationand in vivo studies. *Drug Delivery.* 2017;24:309–19.

Agarwal P, Roy P, Das A, Banerjee A, Maity P, Banerjee A. Efficacy of topical and systemic itraconazole as a broadspectrum antifungal agent in mycotic corneal ulcer. A preliminary study. *Indian J Ophthalmol.* 2001;49:173–6.

Ahmed TA, Aljaeid BM. A potential in situ gel formulation loaded with novel fabricated poly(lactide-co-glycolide) nanoparticles for enhancing and sustaining the ophthalmic delivery of ketoconazole. *Int J Nanomedicine.* 2017;12:1863–75.

Ahuja M, Verma P, Bhatia M. Preparation and evaluation of chitosan–itraconazole co-precipitated nanosuspension for ocular delivery. *J Exp Nanosci.* 2013;10:209–21.

Ainley R, Smith B. Fungal flora of the conjunctival sac in healthy and diseased eyes. *Br J Ophthalmol.* 1965;49:505–515.

Alastruey-Izquierdo A, Alcazar-Fuoli L, Cuenca-Estrella M. Antifungal susceptibility profile of cryptic species of Aspergillus. *Mycopathologia.* 2014;178(5–6):427–33.

Alastruey-Izquierdo A, Cuenca-Estrella M, Monzón A, Mellado E, Rodríguez-Tudela J. Antifungal susceptibility profile of clinical Fusarium spp. isolates identified by molecular methods. *J AntimicrobChemother*. 2008;61:805–9

Alastruey-Izquierdo A, Cuesta I, Houbraken J, et al., In vitro activity of nine antifungal agents against clinical isolates of Aspergillus calidoustus. *Med Mycol*. 2010;48(1):97–102.

Alastruey-Izquierdo A, Mellado E, Pelaez T, et al., Population-based survey of filamentous fungi and antifungal resistance in Spain (FILPOP Study). *Antimicrob Agents Chemother.* 2013;57(7):3380–7.

Al-Badriyeh D, Neoh C, Stewart K, Kong D. Clinical utility of voriconazole eye drops in ophthalmic fungal keratitis. *Clin Ophthalmol*. 2010;4:391–405.

Albengres E, Louët H.Le, Tillement J.-P. "Systemic antifungal agents. Drug interactions of clinical significance." *Drug Safety*. 1998;18(2):83–97.

Al-Hatmi A, Meletiadis J, Curfs-Breuker I, Bonifaz A, Meis J, Hoog S. In vitro combinations of natamycin with voriconazole, itraconazole and micafungin against clinical Fusarium strains causing keratitis. *J AntimicrobChemother*. 2016;71:953–5.

Andes D, Marchillo K, Stamstad T, Conklin R. In vivo pharmacodynamics of a new triazole, ravuconazole, in a murine candidiasis model. *Antimicrob. Agents Chemother*. 2003;47:1193–1199.

Ando N, Takatori K. Fungal flora of the conjunctival sac. *Am J Ophthalmol*. 1982;94:67–74.

Andrade LM, Rocha KA, De Sa FA, Marreto RN, Lima EM, Gratieri T, et al., Voriconazole-loaded nanostructured lipid carriers for ocular drug delivery. *Cornea*. 2016;35:866–71.

Angiolella L, Stringaro AR, De Bernardis F, et al., Increase of virulence and itsphenotypic traits in drug-resistant strains of Candida albicans. *Antimicrob Agents Chemother.* 2008;52(3):927–36.

Arastehfar A, Daneshnia F, Hilmio_glu-Polat S. First report of candidemia clonal outbreak caused by emerging fluconazole-resistant candida parapsilosisisolates harboring Y132F and/or Y132F1K143R in Turkey. *Antimicrob Agents Chemother*. 2020;64(10):20.

Arastehfar A, Khodavaisy S, Daneshnia F, et al., Molecular identification, genotypic diversity, antifungal susceptibility, and clinical outcomes of infections caused by clinically underrated yeasts, Candida orthopsilosis, and Candida metapsilosis: an Iranian multicenter study (2014–2019). *Front Cell Infect Microbiol*. 2019;9:264.

Araujo R, Costa-de-Oliveira S, Coutinho I. *et al.,* Evaluating the resistance to posaconazole by E-test and CLSI broth microdilution methodologies of *Candida* spp. and pathogenic moulds. *Eur J Clin Microbiol Infect Dis*. 2009;28:1137–1140. https://doi.org/10.1007/s10096-009-0736-2

Arendrup MC, Perlin DS, Jensen RH. Differential in vivo activities of anidulafungin, caspofungin, and micafungin against Candida glabrata isolates with and without FKS resistance mutations. *Antimicrob Agents Chemother*. 2012;56(5): 2435–42.

Arora R, Gupta D, Goyal J, Kaur R. Voriconazole versus natamycin as primary treatment in fungal corneal ulcers. *Clin Exp Ophthalmol*. 2011;39:434–40.

Aydin S, Ertugrul B, Gultekin B, Uyar G, Kir E. Treatment of two postoperative endophthalmitis cases due to Aspergillus flavus and Scopulariopsis spp. with local and systemic antifungal therapy. *BMC Infect Dis*. 2007;7:87.

Balajee SA, Kano R, Baddley JW, et al., Molecular identification of Aspergillus species collected for the transplant-associated infection surveillance network. *J Clin Microbiol*. 2009;47(10):3138–41.

Ballard E, Weber J, Melchers WJG, et al., Recreation of in-host acquired single nucleotide polymorphisms by CRISPR-Cas9 reveals an uncharacterised gene playing a role in

Aspergillus fumigatus azole resistance via a non-cyp51A mediated resistance mechanism. *Fungal Genet Biol.* 2019;130:98–106.

Balne PK, Reddy AK, Kodiganti M et al., Evaluation of three PCR assays for the detection of fungi in patients with mycotic keratitis. *Br J Ophthalmol.* 2012;96:911–912. doi:10.1136/bjophthalmol-2011-300207

Barasch A, Griffin AV. Miconazole revisited: New evidence of antifungal efficacy from laboratory and clinical trials. *Future Microbiol.* 2008;3(3):265–269.

Barry AL, Brown SD. In vitro studies of two triazole antifungal agents (voriconazole [UK-104,496] and fluconazole) against Candida species. *Antimicrob Agents Chemother.* 1996;40:1948–9.

Beekman CN, Ene IV. Short-term evolution strategies for host adaptation and drug escape in human fungal pathogens. *PLoSPathog.* 2020;16(5):e1008519.

Ben-Ami R, Berman J, Novikov A, Bash E, Shachor-Meyouhas Y, Zakin S, et al., Multidrug-resistant Candida haemulonii and C. auris, Tel Aviv, Israel. *Emerg. Infect. Dis.* 2017;23:195–203. doi: 10.3201/eid2302.161486

Ben-Ami R, Dimitrios P, Kontoyiannis, MD. ScDb,* resistance to antifungal drugs. *Infect Dis Clin N Am.* 2021;35:279–311. https://doi.org/10.1016/j.idc.2021.03.003

Ben-Ami R, Lewis RE, Kontoyiannis DP. Immunocompromised hosts: Immunopharmacology of modern antifungals. *Clin Infect Dis.* 2008;47(2):226–35.

Ben-Ami R, Zimmerman O, Finn T, et al., Heteroresistance to fluconazole is a continuously distributed phenotype among Candida glabrata clinical strains associated with in vivo persistence. *mBio.* 2016;7(4):e00655–16.

Berman J, Krysan DJ. Drug resistance and tolerance in fungi. Nat Rev Microbiol. 2020;18:319–331. doi: 10.1038/s41579-019-0322-2

Beyda ND, John J, Kilic A, Alam MJ, Lasco TM, Garey KW. FKS mutant Candida glabrata: Risk factors and outcomes in patients with candidemia. *Clin Infect Dis.* 2014;59:819–25. https://doi.org/10. 1093/cid/ciu407

Bharathi MJ, Ramakrishnan R, Meenakshi R et al., Microbiological diagnosis of infective keratitis: Comparative evaluation of direct microscopy and culture results. *Br J Ophthalmol.* 2006;90:1271–1276. doi:10.1136/bjo.2006.096230

Bhattacharya S, Holowka T, Orner EP, et al., Gene duplication associated with increased fluconazole tolerance in Candida auris cells of advanced generational age. *Sci Rep.* 2019;9(1):5052.

Biagi MJ, Wiederhold NP, Gibas C, et al., Development of high-level echinocandin resistance in a patient with recurrent Candida auris Candidemia Secondary to Chronic Candiduria. *Open Forum Infect Dis.* 2019;6(7):ofz262.

Blum G, Hörtnagl C, Jukic E, et al., New insight into amphotericin B resistance in Aspergillus terreus. *Antimicrob Agents Chemother.* 2013;57(4):1583–8.

Brown, L., Leck A.K, Gichangi M, Burton M.J, Denning D.W. The global incidence and diagnosis of fungal keratitis. *Lancet Infect. Dis.* 2021;21(3). DOI: 10.1016/S1473-3099(20)30448-5

Bunya VY, Hammersmith KM, Rapuano CJ, Ayres BD, Cohen EJ. Topical and oral voriconazole in the treatment of fungal keratitis. *Am J Ophthalmol.* 2007;143:151–3.

Cao C, Wang Y, Husain S, et al., A Mechanosensitive channel governs lipid flippase-mediated echinocandin resistance in Cryptococcus neoformans. *mBio.* 2019;10(6):e01952–19.

Cao C, Xue C. More than flipping the lid: Cdc50 contributes to echinocandin resistance by regulating calcium homeostasis in Cryptococcus neoformans. *Microb Cell.* 2020;7(4):115–8.

Carmichael TR, Wolpert M, Koornhof HJ. Corneal ulceration at an urban African hospital., *Br J Ophthalmol.* 1985;69:920–6.

Chaabane F, Graf A, Jequier L, Coste AT. Review on antifungal resistance mechanisms in the Emerging Pathogen Candida auris. *Front. Microbiol.* 2019;10:2788. doi: 10.3389/fmicb.2019.02788

Chamilos G, Ganguly D, Lande R, et al., Generation of IL-23 producing dendritic cells (DCs) by airborne fungi regulates fungal pathogenicity via the induction of T(H)-17 responses. *PLoS One*. 2010;5(9):e12955.

Chamilos G, Lewis RE, Albert N, et al., Paradoxical effect of Echinocandins across Candida species in vitro: Evidence for echinocandin-specific and candida species-related differences. *Antimicrob Agents Chemother*. 2007a;51(6):2257–9.

Chamilos G, Lewis RE, Kontoyiannis DP. Inhibition of Candida parapsilosis mitochondrial respiratory pathways enhances susceptibility to caspofungin. *Antimicrob Agents Chemother*. 2006;50(2):744–7.

Chamilos G, Lionakis MS, Lewis RE, et al., Role of mini-host models in the study of medically important fungi. *Lancet Infect Dis*. 2007b;7(1):42–55.

Chander J, Chakrabarti A, Sharma A et al., Evaluation of Calcofluor staining in the diagnosis of fungal corneal ulcer. *Mycoses*. 1993;36:243–245. doi:10.1111/j.1439-0507.1993.tb00758.x

Chander J, Sharma A. Prevalence of fungal corneal ulcers in northern India. *Infection*. 1994;22:207–9.

Chandra J, Retuerto M, Mukherjee PK, Ghannoum M. The fungal biome of the oral cavity. *Methods Mol Biol*. 2016;1356:107–35. doi: 10.1007/978-1-4939-3052-4_9. PMID: 26519069

Chatterjee S, Alampalli SV, Nageshan RK, et al., Draft genome of a commonly misdiagnosed multidrug resistant pathogen Candida auris. *BMC Genomics*. 2015;16(1):686.

Chaudhary A, Singh TSK, Lalchandani S et al., Corneal ulceration and microbial keratitis in Pokhara. *Nepal., J Nep Med Assoc*. 1999;39:18–22.

Chen SC, Slavin MA, Sorrell TC. Echinocandin antifungal drugs in fungal infections: A comparison. *Drugs*. 2011a;71:11–41. doi: 10.2165/11585270-000000000-00000

Chen Y, Farrer RA, Giamberardino C, et al., Microevolution of serial clinical isolates of cryptococcus neoformans var. *grubii and C. gattii. mBio*. 2017;8(2):e00166-17.

Chen YL, Brand A, Morrison EL, et al., Calcineurin controls drug tolerance, hyphal growth, and virulence in Candida dubliniensis. *Eukaryot Cell*. 2011b;10(6):803–19.

Cheng JW, Liao K, Kudinha T, et al., Molecular epidemiology and azole resistance mechanism study of Candida guilliermondii from a Chinese surveillance system. *Sci Rep*. 2017;7(1):907.

Cheng S, Lin Y, Kuo C, Lai L. Cladosporium keratitis—a case report and literature review. *BMC Ophthalmol*. 2015;15:106.

Chew KL, Octavia S, Lin RTP, Yan GZ, Teo JWP. Delay in effective therapy in anidulafungin-resistant Candida tropicalis fungaemia: Potential for rapid prediction of antifungal resistance with whole-genome-sequencing. *J Glob Antimicrob Resist*. 2019 Mar;16:105–107. doi: 10.1016/j.jgar.2018.12.010. Epub 2018 Dec 21. PMID: 30583013

Chhonker YS, Prasad YD, Chandasana H, Vishvkarma A, Mitra K, Shukla PK, et al., Amphotericin-B entrapped lecithin/chitosan nanoparticles for prolonged ocular application. *Int J BiolMacromol*. 2015;72:1451–8.

Chowdhary A, Anil Kumar V, Sharma C, et al., Multidrug-resistant endemic clonal strain of Candida auris in India. *Eur J Clin Microbiol Infect Dis*. 2014;33(6): 919–26.

Chowdhary A, Kathuria S, Xu J, Meis JF. Emergence of azole-resistant aspergillus fumigatus strains due to agricultural azole use creates an increasing threat to human health. *PLoSPathog.* 2013a; 9(10):e1003633. doi:10.1371/journal.ppat. 1003633

Chowdhary A, Prakash A, Sharma C, Kordalewska M, Kumar A, Sarma S, et al., 2018. A multicentre study of antifungal susceptibility patterns among 350 Candida auris isolates (2009–17) in India: Role of the ERG11 and FKS1 genes in azole and echinocandin resistance. *J. Antimicrob. Chemother.* 73, 891–899. doi: 10.1093/jac/dkx480

Chowdhary A, Sharma C, Duggal S, et al., New clonal strain of Candida auris, Delhi, India. *Emerg Infect Dis.* 2013b;19(10):1670–3.

Clancy CJ, Yu VL, Morris AJ, et al., Fluconazole MIC and the fluconazole dose/ MIC ratio correlate with therapeutic response among patients with candidemia. *Antimicrob Agents Chemother.* 2005;49(8):3171–7.

Coley KC, Crain JL. Miconazole-induced fatal dysrhythmia. *Pharmacotherapy.* 1997;17(2):379–382.

Colombo AL, Agnelli C, Kontoyiannis DP. Knowledge gaps in candidaemia/invasive candidiasis in haematological cancer patients. *J AntimicrobChemother.* 2021;76(3):543–6.

Cornely OA, Maertens J, Bresnik M, et al., Liposomal amphotericin B as initial therapy for invasive mold infection: A randomized trial comparing a high-loading dose regimen with standard dosing (AmBiLoad trial). *Clin Infect Dis.* 2007;44(10):1289–1297.

Costa C, Ribeiro J, Miranda IM, et al., Clotrimazole drug resistance in Candida glabrata clinical isolates correlates with increased expression of the drug:H(1) Antiporters CgAqr1, CgTpo1_1, CgTpo3, and CgQdr2. *Front Microbiol.* 2016;7:526.

Coste AT. Review on Antifungal Resistance mechanisms in the emerging pathogen Candida auris. *Front. Microbiol.* 2019;10:2788.doi: 10.3389/fmicb.2019.02788

Coster DJ, Wilhelmus J, Peacock J, et al., Suppurative keratitis in London. The cornea in health and disease. *VIthCongr Euro Soc Ophthalmol.* 1981;40:395–8.

Cota J, Carden M, Graybill JR, et al., In vitro pharmacodynamics of anidulafungin and caspofungin against Candida glabrata isolates, including strains withdecreasedcaspofungin susceptibility. *Antimicrob Agents Chemother.* 2006;50(11):3926–8.

Cowen LE, Kohn LM, Anderson JB. Divergence in fitness and evolution of drug resistance in experimental populations of Candida albicans. *J Bacteriol.* 2001;183(10):2971–8.

Cowen LE, Sanglard D, Howard SJ, et al., Mechanisms of antifungal drug resistance. *Cold Spring HarbPerspect Med.* 2014;5(7):a019752.

Damasceno JL, Dos Santos RA, Barbosa AH, Casemiro LA, Pires RH, Martins CHG. Risk of fungal infection to dental patients. *Sci. World J.* 2017;2017:2982478. doi: 10.1155/2017/2982478. Epub 2017 Jun 14. PMID: 28695189; PMCID: PMC5488164

Das S, Suresh PK. Nanosuspension: A new vehicle for the improvement of the delivery of drugs to the ocular surface Applicati o n to amphotericin B. *Nanomedicine: Nanotechnology, Biology, and Medicine.* 2011;7:242–7.

Das S, Suresh PK, Desmukh R. Design of Eudragit RL 100 nanoparticles by nanoprecipitation method for ocular drug delivery. *Nanomedicine.* 2010;6:318–23.

da Silva Ferreira ME, Capellaro JL, dos Reis Marques E, Malavazi I, Perlin D, Park S, Anderson JB, Colombo AL, Arthington-Skaggs BA, Goldman MH, Goldman GH. In vitro evolution of itraconazole resistance in Aspergillus fumigatus involves multiple mechanisms of resistance. *Antimicrob Agents Chemother.* 2004 Nov;48(11):4405–13. doi: 10.1128/AAC.48.11.4405-4413.2004. PMID: 15504870; PMCID: PMC525395

Defontaine A, Bouchara JP, Declerk P, et al., In-vitro resistance to azoles associated with mitochondrial DNA deficiency in Candida glabrata. *J Med Microbiol.* 1999;48(7):663–70.

Delarze E, Sanglard D. Defining the frontiers between antifungal resistance, tolerance and the concept of persistence. *Drug Resist Updat.* 2015;23:12–9.

Dembski M, Nowinska A, Ulfik K et al., In vivo confocal microscopy and anterior segment optical coherence tomography analysis of the microcystic keratitis. *J Ophthalmol.* 2020;2020:8871904. doi:10.1155/2020/8871904

Denning DW, Venkateswarlu K, Oakley KL, et al. "Itraconazole resistance in *Aspergillus fumigatus*." *Antimicrob. Agents and Chemother.* 1997; 41(6):1364–1368.

Deshpande SD, Koppikar GV. A study of mycotic keratitis in Mumbai. *Ind J PatholMicrobiol.* 1999;42:81–87.

Diekema D, Messer S, Hollis R, Jones R, Pfaller M. Activities of caspofungin, itraconazole, posaconazole, ravuconazole, voriconazole, and amphotericin B against 448 recent clinical isolates of filamentous fungi. *J Clin Microbiol.* 2003;41:3623–6.

Dóczi1 T, Gyetvai2 L, Kredics3, Nagy1,2 E. Involvement of Fusarium spp. in fungal keratitis *Clin Microbiol Infect.* 2004;10:773–776.

Douglas CM, D'Ippolito JA, Shei GJ, et al., Identification of the FKS1 gene of Candida albicans as the essential target of 1,3-beta-D-glucan synthase inhibitors. *Antimicrob Agents Chemother.* 1997;41(11):2471–9.

Douglas CM, Foor F, Marrinan JA, et al., The Saccharomyces cerevisiae FKS1 (ETG1) gene encodes an integral membrane protein which is a subunit of 1,3- beta-D-glucan synthase. *Proc Natl Acad Sci U S A.* 1994;91(26):12907–11.

Durgun ME, Kahraman E, Hacıoğlu M, et al., Posaconazole micelles for ocular delivery: In vitro permeation, ocular irritation and antifungal activity studies. *Drug Deliv. and Transl. Res.* 2022;12:662–675. https://doi.org/10.1007/s13346-021-00974-x

Embong Z, Wan Hitam WH, Yean CY et al., Specific detection of fungal pathogens by 18S rRNA gene PCR in microbial keratitis. *BMC Ophthalmol.* 2008;8:7. doi:10.1186/1471-2415-8-7

Escandon P, Chow NA, Caceres DH, et al., Molecular epidemiology of Candida auris in Colombia reveals a highly related, countrywide colonization with regional patterns in Amphotericin B resistance. *Clin Infect Dis.* 2019;68(1):15–21.

Espinel-Ingroff A. In vitro activity of the new triazole voriconazole (UK-109,496) against opportunistic filamentous and dimorphic fungi and common and emerging yeast pathogens, *J Clin Microbiol.* 1998;36:198-202.

Espinel-Ingroff A, Turnidge J. The role of epidemiological cutoff values (ECVs/ECOFFs) in antifungal susceptibility testing and interpretation for uncommon yeasts and moulds. *Rev IberoamMicol.* 2016;33(2):63–75.

Evertsson U, Monstein HJ, Johansson AG. Detection and identification of fungi in blood using broad-range 28S rDNA PCR amplification and species-specific hybridisation. *APMIS.* 2000 May;108(5):385–92. doi: 10.1034/j.1600-0463.2000.d01-73.x. PMID: 10937778

Farmakiotis D, Tarrand JJ, Kontoyiannis DP. Drug-resistant Candida glabrata infection in cancer patients. *Emerg Infect Dis.* 2014;20(11):1833–40.

Fekete-Forgács K, Gyüre L, Lenkey B. Changes of virulence factors accompanying the phenomenon of induced fluconazole resistance in Candida albicans. *Mycoses.* 2000;43(7–8):273–9.

Fekkar A, Meyer I, Brossas JY, et al., Rapid emergence of echinocandin resistance during Candida kefyr fungemia treatment with caspofungin. *Antimicrob Agents Chemother.* 2013;57(5):2380–2.

Ferrer C, Alio JL. Evaluation of molecular diagnosis in fungal keratitis. Ten years of experience. *J Ophthalmic Inflamm Infect*. 2011;1:15–22. doi:10.1007/s12348-011-0019-9

Fesel P, Zuccaro A. β-Glucan: Crucial component of the fungal cell wall and elusive MAMP in plants. *Fungal Genet Biol*. 2016;90:53–60.

Fetih G. Fluconazole-loaded niosomal gels as a topical ocular drug delivery system for corneal fungal infections. *J Drug Deliv Sci Technol*. 2016;35:8–15.

Fisher MC, Alastruey-Izquierdo A, Berman J, Bicanic T, Bignell EM, Bowyer P, Bromley M, Brüggemann R, Garber G, Cornely OA, Gurr SJ, Harrison TS, Kuijper E, Rhodes J, Sheppard DC, Warris A, White PL, Xu J, Zwaan B, Verweij PE. Tackling the emerging threat of antifungal resistance to human health. *Nat. Rev. Microbiol.* 2022 Sep;20:557–571.

Fitzsimons R, Peters AL., Miconazole and ketoconazole as a satisfactory first-line treatment for keratomycosis. *Am J Ophthalmol*. 1986;101(5):605–608.

Flowers SA, Colon B, Whaley SG, et al., Contribution of clinically derived mutations in ERG11 to azole resistance in Candida albicans. *Antimicrob Agents Chemother*. 2015;59(1):450–60.

Font RL, Wu TG, Wilhelmus KR et al., Polymerase Chain Reaction (PCR) in the Diagnosis of Fungal Infections of the Eye. *Invest Ophthalmol Vis Sci*. 2002;43:1609–1609.

Ford CB, Funt JM, Abbey D, et al., The evolution of drug resistance in clinical isolates of Candida albicans. *Elife*. 2015;4:e00662.

Foster CS. Miconazole therapy for keratomycosis. *Am J Ophthalmol*. 1981;91(5):622–629.

Foster CS. Fungal keratitis. *Infect Dis Clin North Am*. 1992;6:851–7.

Foster CS, Stefanyszyn M. Intraocular penetration of miconazole in rabbits. *Arch Ophthalmol*. 1979;97(9):1703–1706.

Fothergill AW. Miconazole: A historical perspective. *Expert Rev Anti Infect Ther*. 2006 Apr;4(2):171–5. doi: 10.1586/14787210.4.2.171. PMID: 16597199

Franco CH, Warhurst DC, Bhattacharyya T, Au HYA, Le H, Giardini MA, Pascoalino BS, Torrecilhas AC, Romera LMD, Madeira RP, Schenkman S, Freitas-Junior LH, Chatelain E, Miles MA, Moraes CB. Novel structural CYP51 mutation in Trypanosoma cruzi associated with multidrug resistance to CYP51 inhibitors and reduced infectivity. *Int. J. Parasitol: Drugs and Drug Resistance* 13 June 2020. DOI: 10.1016/j.ijpddr.2020.06.001

Fu T, Yi J, Lv S, Zhang B. Ocular amphotericin B delivery by chitosan-modified nanostructured lipid carriers for fungal keratitis-targeted therapy. *J Liposome Res*. 2017;27:228–33.

Gallis HA, Drew RH, Pickard WW. Amphotericin B: 30 years of clinical experience. *Rev Infect Dis*. 1990;12(2):308–329.

Galloway-Peña JR, Kontoyiannis DP. The gut mycobiome: The overlooked constituent of clinical outcomes and treatment complications in patients with cancer and other immunosuppressive conditions. *PLoSPathog*. 2020;16(4):e1008353.

Garcia-Effron G, Kontoyiannis DP, Lewis RE, et al., Caspofungin-resistant Candida tropicalis strains causing breakthrough fungemia in patients at high risk for hematologic malignancies. *Antimicrob Agents Chemother*. 2008;52(11): 4181–3.

Garcia-Effron G, Lee S, Park S, et al., Effect of Candida glabrata FKS1 and FKS2 mutations on echinocandin sensitivity and kinetics of 1,3-beta-D-glucan synthase: Implication for the existing susceptibility breakpoint. *Antimicrob Agents Chemother*. 2009;53(9):3690–9.

Garcia-Valenzuela E, Song CD. Intracorneal injection of amphothericin B for recurrent fungal keratitis and endophthalmitis. *Arch Ophthalmol*. 2005;123:1721–3.

Georgiadou SP, Kontoyiannis DP. The impact of azole resistance on aspergillosis guidelines. *Ann N Y Acad Sci.* 2012;1272:15–22.

Ghannoum M, Larkin E, Long L. Susceptibilty of recent *Candida auris* isolates to the novel echinocandin CD101 and comparator antifungalagents. *27th European Congress on Clinical Microbiology and Infectious Diseases*; April 22–25, 2017; Vienna, Austria.

Gokhale N. Medical management approach to infectious keratitis. *Indian J Ophthalmol.* 2008;56:215–20.

Goldblum D, Frueh BE, Sarra GM et al., Topical caspofungin for treatment of keratitis caused by Candida albicans in a rabbit model. *Antimicrob Agents Chemother.* 2005;49:1359–1363. doi:10.1128/AAC.49.4.1359-1363.2005

Goldblum D, Rohrer K, Frueh B, Theurillat R, Thormann W, Zimmerli S. Ocular distribution of intravenously administered lipid formulations of amphotericin B in a rabbit model. *Antimicrob Agents Chemother.* 2002;46:3719–23.

Gonzales CA, Srinivasan M, Whitcher JP, Smolin G. Incidence of corneal ulceration in Madurai district, South India. *Ophthalmic Epidemiol.* 1996;3:159–166. doi: 10.3109/09286589609080122

Gopinathan U, Sharma S, Garg P et al., Review of epidemiological features, microbiological diagnosis and treatment outcome of microbial keratitis: Experience of over a decade. *Indian J Ophthalmol.* 2009;57:273–279. doi:10.4103/0301-4738.53051

Graybill JR, Montalbo E, Kirkpatrick WR, et al., Fluconazole versus Candida albicans: A complex relationship. *Antimicrob Agents Chemother.* 1998;42(11):2938–42.

Green WR, Bennett JE, Goos RD. Ocular penetration of amphotericin B: A report of laboratory studies and a case report of postsurgical Cephalosporium endophthalmitis. *Arch Ophthalmol.* 1965;73:769–775.

Gregory M, Macdonald E, Lockington D, Ramaesh K. Recurrent fungal keratitis following penetrating keratoplasty: An unusual source of infection. *Arch Ophthalmol.* 2010;128:1490–1.

Grossman NT, Pham CD, Cleveland AA, et al., Molecular mechanisms of fluconazole resistance in Candida parapsilosis isolates from a U.S. surveillance system. *Antimicrob Agents Chemother.* 2015;59(2):1030–7.

Gugani HC, Talwar RS, Njoku-Obi AN et al., Mycotic keratitis in Nigeria. A study of 21 cases. *Br J Ophthalmol.* 1976;60:607–13.

Guinea J, Sanchez-Somolinos M, Cuevas O, et al., Fluconazole resistance mechanisms in Candida krusei: The contribution of efflux-pumps. *Med Mycol.* 2006;44(6):575–8.

Gupta N, Tandon R, Gupta SK, Sreenivas V, Vashist P. Burden of corneal blindness in India. *Indian J. Community Med.* 2013;38:198–206. doi: 10.4103/0970-0218.120153

Hachem R, Hanna H, Kontoyiannis D, et al., The changing epidemiology of invasive candidiasis: Candida glabrata and Candida krusei as the leading causes of candidemia in hematologic malignancy. *Cancer.* 2008;112(11):2493–9.

Haddad RS, El-Mollayess GM. Combination of intracameral and intrastromal voriconazole in the treatment of recalcitrant Acremonium fungal keratitis. *Middle East Afr J Ophthalmol.* 2012 Apr-Jun;19(2):265–8. doi: 10.4103/0974-9233.95271. PMID: 22623874; PMCID: PMC3353683

Hagiwara D, Arai T, Takahashi H, et al., Non-cyp51A azole-resistant Aspergillus fumigatus isolates with mutation in HMG-CoA reductase. *Emerg Infect Dis.* 2018;24(10):1889–97.

Hamady M, Knight R. Microbial community profiling for human microbiome projects: Tools, techniques, and challenges. *Genome Res.* 2009;19(7):1141–1152.

Hammeke JC, Ellis PP. Mycotic flora of the conjunctiva. *Am J Ophthalmol.* 1960;49:1174–1178.

Hariprasad S, Mieler W, Holz E, Gao H, Kim J, Chi J, et al. Determination of vitreous, aqueous, and plasma concentration oforally administered voriconazole in humans. Arch Ophthalmol. 2004;122:42–7.

Harris DJ Jr, Stulting RD, Waring GO III et al., Late bacterial and fungal keratitis after corneal transplantation. Spectrum of pathogens, graft survival, and visual prognosis. *Ophthalmol.* 1988;95:1450–7.

Harrison BD, Hashemi J, Bibi M, Pulver R, Bavli D, Nahmias Y, Wellington M, Sapiro G, Berman J. A tetraploid intermediate precedes aneuploid formation in yeasts exposed to fluconazole. *PLoSBiol.* 2014;12:e1001815. doi: 10.1371/journal.pbio.1001815

Harrison J, Glickman R, Ballentine C, Trigo Y, Pena M, Kurian P, et al., Retinal function assessed by ERG before and after induction of ocular aspergillosis and treatment by the anti-fungal, micafungin, in rabbits. *Doc Ophthalmol.* 2005;110:37–55.

Hau SC, Dart JK, Vesaluoma M et al., Diagnostic accuracy of microbial keratitis with in vivo scanning laser confocal microscopy. *Br J Ophthalmol.* 2010;94:982–987. doi:10.1136/bjo.2009.175083

HeYG, McCulley JP, Alizadeh H et al., A pig model of Acanthamoeba keratitis: Transmission via contaminated contact lenses. *Invest Ophthalmol Vis Sci.* 1992;33:126–133.

Healey KR, Kordalewska M, Ortigosa CJ, Singh A, Berrío I, Chowdhary A, et al., Limited ERG11 mutations identified in isolates of Candida auris directly contribute to reduced azole susceptibility. *Antimicrob. Agents Chemother.* 2018;62, e01427–18. doi: 10.1128/AAC.01427-18

Healey KR, Zhao Y, PerezWB, Lockhart SR, Sobel JD, Farmakiotis D, et al., Prevalent mutator genotype identified in fungal pathogen Candida glabrata promotes multi-drug resistance. *Nat Commun.* 2016;7:11128. https://doi.org/10.1038/ncomms11128

Heel RC, Brogden RN, Speight TM, Avery GS. Econazole: A review of its antifungal activity and therapeutic efficacy. *Drugs.* 1978;16(3):177–201.

Hendrickson JA, Hu C, Aitken SL, Beyda N. Antifungal resistance: A concerning trend for the present and future current. *Infect. Dis. Rep.* 2019;21:47. https://doi.org/10.1007/s11908-019-0702-9

Henry CR, Flynn HW, Miller D, Forster RK, Alfonso EC. Infectious keratitis progressing to Endophthalmitis: A 15-year-study of microbiology, associated factors, and clinical outcomes. *Ophthalmol.* 2012;119:2443–2449. doi: 10.1016/j.ophtha.2012.06.030

Heo ST, Tatara AM, Jiménez-Ortigosa C, et al., Changes in in vitro susceptibility patterns of Aspergillus to Triazoles and correlation with aspergillosis outcome in a Tertiary Care Cancer Center, 1999–2015. *Clin Infect Dis.* 2017;65(2):216–25.

Herbrecht R, Denning DW, Patterson TF, et al., Voriconazole versus amphotericin B for primary therapy of invasive aspergillosis. *N Engl J Med.* 2002;347(6):408–15.

Herbrecht R. Posaconazole: A potent, extended-spectrum triazole anti-fungal for the treatment of serious fungal infections. *Int. J. Clin. Pract.* 2004;58(6):612–624.

HiraokaT, Wakabayashi T, Kaji Y, Nanbu P, Okamoto F, Kiuchi T, et al., Toxicological evaluation of micafungin ophthalmic solution in rabbit eyes. *J Ocul Pharmacol Ther.* 2005;21:149–56.

Hirose H, Terasaki H, Awaya S, Yasuma T. Treatment of fungal corneal ulcers with amphotericin B ointment. *Am J Ophthalmol.* 1997;124(6):836–838.

Hoban DJ, Zhanel GC, Karlowsky JA. In vitro susceptibilities of Candida and Cryptococcus neoformans isolates from blood cultures of neutropenic patients. *Antimicrob Agents Chemother.* 1999;43:1463–4.

Houang E, Lam D, Fan D et al., Microbial keratitis in Hong Kong: Relationship to climate, environment, and contact lens disinfection. *Trans R Soc Trop Med Hyg.* 2001;95:361–7.

Howard SJ. Multi-resistant aspergillosis due to cryptic species. *Mycopathologia.* 2014;178(5–6):435–9.

Howard SJ, Harrison E, Bowyer P, et al., Cryptic species and azole resistance in the Aspergillus niger complex. *Antimicrob Agents Chemother.* 2011;55(10):4802–9.

Hu J, Zhang J, Li Y, Han X, Zheng W,Yang J, et al., A combination of intrastromal and intracameral injections of amphotericin B in the treatment of severe fungal keratitis. *J Ophthalmol.* 2016;2016:7.

Hurtado-Sarrió M, Duch-Samper A, Cisneros-Lanuza A, Díiaz-Llopis M, Peman-Garcíia J, Vazquez-Polo A. Successful topical application of caspofungin in the treatment of fungal keratitis refractory to voriconazole. *Arch Ophthalmol.* 2010;128:941–2.

Ibrahim AS, Bowman JC, Avanessian V, et al., Caspofungin inhibits Rhizopus oryzae 1,3-beta-D-glucan synthase, lowers burden in brain measured by quantitative PCR, and improves survival at a low but not a high dose during murine disseminated zygomycosis. *Antimicrob Agents Chemother.* 2005;49(2):721–7.

Imbert S, Normand AC, Cassaing S, et al., Multicentric analysis of the species distribution and antifungal susceptibility of cryptic isolates from Aspergillus section Fumigati. *Antimicrob Agents Chemother.* 2020;64:20.

Ishibashi Y, Kaufman HE. The effects of subconjunctival miconazole in the treatment of experimental Candida keratitis in rabbits. *Arch Ophthalmol.* 1985;103(10):1570–1573.

Ito J, Hooshmand-Rad R. Treatment of Candida infections with amphotericin B lipid complex. *Clin Infect Dis.* 2005;40:S384–91.

Janga KY, Tatke A, Balguri SP, Lamichanne SP, Ibrahim MM, Maria DN, et al., Ion-sensitive in situ hydrogels of natamycin bilosomes for enhanced and prolonged ocular pharmacotherapy: In vitro permeability, cytotoxicity and in vivo evaluation. *Artif Cells NanomedBiotechnol.* 2018;23:1–12.

Jarvis JN, Meintjes G, Williams Z, et al., Symptomatic relapse of HIV-associated cryptococcal meningitis in South Africa: The role of inadequate secondary prophylaxis. *S Afr Med J* 2010;100(6):378–82.

Jin KW, Jeon HS, Hyon JY,Wee WR, Suh W, Shin YJ. A case of fungal keratitis and onychomycosis simultaneously infected by Trichophyton species. *BMC Ophthalmol.* 2014;14:90.

Joseph J Sharma S (2016) Antifungal Therapy in Eye Infections: New Drugs, New Trends. In: A. Basak et al., (eds.), *Recent Trends in Antifungal Agents and Antifungal Therapy.* DOI 10.1007/978-81-322-2782-3_9

Joseph J, Murthy S, Garg P et al., Use of different stains for microscopic evaluation of corneal scrapings for diagnosis of microsporidial keratitis. *J Clin Microbiol.* 2006;44:583–585 doi:10.1128/JCM.44.2.583-585.2006

Kakkar S, Karuppayil SM, Raut JS, Giansanti F, Papucci L, Schiavone N, et al., Lipid-polyethylene glycol based nano-ocular formulation of ketoconazole. *Int J Pharm.* 2015;495:276–89.

Kalavathy CM, Parmar P, Kaliamurthy J, Philip VR, Ramalingam MD, Jesudasan CA, Thomas PA. Comparison of topical itraconazole 1% with topical natamycin 5% for the treatment of filamentous fungal keratitis. *Cornea.* 2005 May;24(4):449–52. doi: 10.1097/01.ico.0000151539.92865.3e. PMID: 15829804

Kalkanci A, Ozdek S. Ocular fungal infections. *Curr Eye Res.* 2011;36:179–189. doi:10.3109/02713683.2010.533810

Kamoshita M, Matsumoto Y, Nishimura K, Katono Y, Murata M, Ozawa Y, et al., Wickerhamomycesanomalus fungal keratitis responds to topical treatment with antifungal micafungin. *J Infect Chemother*. 2014;21:141–3.

Karsten E, Watson SL, Foster LJ. Diversity of microbial species implicated in keratitis: A review. *Open Ophthalmol J*. 2012;6:110–124. doi:10.2174/1874364101206010110

Kathuria S, Singh PK, Sharma C, et al., Multidrug-resistant Candida auris Misidentified as Candida haemulonii: Characterization by matrix-assisted laser desorption ionization-time of flight mass spectrometry and DNA sequencing and its antifungal susceptibility profile variability by Vitek 2, CLSI Broth Microdilution, and Etest method. *J Clin Microbiol*. 2015;53(6):1823–30.

Katiyar SK, Edlind TD. Identification and expression of multidrug resistancerelated ABC transporter genes in Candida krusei. *Med Mycol*. 2001;39(1):109–16.

Kaur IP, Rana C, Singh H. Development of effective ocular preparations of antifungal agents. *J OculPharmacolTher*. 2008;24(5):481–493.

Kaushik S, Ram J, Brar GS, Jain AK, Chakraborti A, Gupta A. Intracameral amphotericin B: Initial experience in severe keratomycosis. *Cornea*. 2001;20:715–9.

Kermani NK, Aggarwal SP. Isolated post-operative Aspergillus niger endophthalmitis. *Eye (London, England)*. 2000;14(Pt 1):114–6.

Khanal B, Kaini KR, Deb M et al., Microbial keratitis in eastern Nepal., *Trop Doct*. 2001;31:168–9.

Khare A, Singh I, Pawar P, Grover K. Design and evaluation of voriconazole loaded solid lipid nanoparticles for ophthalmic application. *J Drug Deliv*. 2016;2016:6590361–1.

Klont R, Eggink C, Rijs A, Wesseling P, Verweij P. Successful treatment of Fusarium keratitis with cornea transplantation and topical and systemic voriconazole. *Clin Infect Dis*. 2005;40:e110–2.

Kocak N, Kaynak S, Kirdar S et al., Comparison of different antifungal treatment regimens for experimental Candida endophthalmitis in rabbit models. *Mikrobiyol Bul*. 2009;43:619–626.

Köllner H. III. Schimmelpilzerkrankung der Sklera. *Ophthalmologica*. 1906;16(5):441–447.

Kontoyiannis DP, Lewis RE. Antifungal drug resistance of pathogenic fungi. *Lancet* 2002;359(9312):1135–44.

Kontoyiannis DP, Lewis RE. Treatment principles for the management of mold infections. *Cold Spring HarbPerspect Med*. 2014;5(4):a019737.

Kordalewska M, Lee A, Park S, et al., Understanding echinocandin resistance in the emerging pathogen Candida auris. *Antimicrob Agents Chemother*. 2018;62:e00238-18.

Kramer M, Kramer MR, Blau H, Bishara J, Axer-Siegel R, Weinberger D. Intravitreal voriconazole for the treatment of endogenous *Aspergillus* endophthalmitis. *Ophthalmology*. 2006;113:1184–6.

Krishnasamy L, Krishnakumar S, Kumaramanickavel G, Saikumar C. Molecular mechanisms of antifungal drug resistance in Candida species. *J. Clin. Diagn. Res*. 2018;12:DE01–DE06. doi: 10.7860/JCDR/2018/36218.11961

Krogh-Madsen M, Arendrup MC, Heslet L, et al., Amphotericin B and caspofungin resistance in Candida glabrata isolates recovered from a critically ill patient. *Clin Infect Dis*. 2006;42(7):938–44.

Kumar R, Sinha VR. Preparation and optimization of voriconazole microemulsion for ocular delivery, colloids and surfaces B. *Biointerfaces*. 2014;117:82–8.

Kurtz MB, Abruzzo G, Flattery A, et al., Characterization of echinocandinresistant mutants of Candida albicans: Genetic, biochemical, and virulence studies. *Infect Immun*. 1996;64(8):3244–51.

Labbé A, Khammari C, Dupas B et al., Contribution of in vivo confocal microscopy to the diagnosis and management of infectious keratitis. *The Ocular Surface*. 2009;7:41–52 doi:10.1016/s1542-0124(12)70291-4

Lackner M, Hagen F, Meis JF, et al., Susceptibility and diversity in the therapy-refractory genus *scedosporium*. *Antimicrob Agents Chemother.* 2014a;58(10):5877–5885.

Lackner M, de Hoog GS, Yang L, et al., Proposed nomenclature for *Pseudallescheria*, *Scedosporium* and related genera. *Fungal Divers.* 2014b;67(1):1–10.

Lalitha P, Shapiro BL, Srinivasan M, Prajna NV, Acharya NR, Fothergil A.W, Ruiz J, Chidambaram D, Maxey KJ, Hong KC, Mcleod SD, Lietman TM. 2007. Antimicrobial susceptibility of *Fusarium*, *Aspergillus*, and other filamentous fungi isolated from keratitis. *Arch. Ophthalmol.* 125:789–793.

Lalitha P, Vijaykumar R, Prajna N, Fothergill A. In vitro natamycin susceptibility of ocular isolates of Fusarium and Aspergillus species: Comparison of commercially formulated natamycin eye drops to pharmaceutical-grade powder. *J Clin Microbiol.* 2008;46:3477–8.

Lamoth F, Juvvadi PR, Gehrke C, et al., Transcriptional activation of heat shock protein 90 mediated via a proximal promoter region as trigger of caspofungin resistance in Aspergillus fumigatus. *J Infect Dis.* 2014;209(3):473–81.

Lamoth F, Kontoyiannis DP. The Candida auris alert: Facts and perspectives. *J Infect Dis* 2018;217(4):516–20.

Lamoth F, Kontoyiannis DP. Therapeutic challenges of non-aspergillus invasive mold infections in immunosuppressed patients. *Antimicrob Agents Chemother.* 2019;63(11):e01244-19.

Laverdière M, Lalonde RG, Baril JG, et al., Progressive loss of echinocandin activity following prolonged use for treatment of Candida albicans oesophagitis. *J AntimicrobChemother.* 2006;57(4):705–8–60.

Lee CH, Chang TY, Liu JW, et al., Correlation of anti-fungal susceptibility with clinical outcomes in patients with cryptococcal meningitis. *BMC Infect Dis.* 2012;12:361.

Lee I, Morales KH, Zaoutis TE, et al., Clinical and economic outcomes of decreased fluconazole susceptibility in patients with Candida glabrata bloodstream infections. *Am J Infect Control.* 2010;38(9):740–5.

Lestrade PP, Bentvelsen RG, Schauwvlieghe A, et al., Voriconazole resistance and mortality in invasive aspergillosis: a multicenter retrospective cohort study. *Clin Infect Dis.* 2018;68:8.

Lewis JS 2nd, Wiederhold NP, Wickes BL, et al., Rapid emergence of echinocandin resistance in Candida glabrata resulting in clinical and microbiologic failure. *Antimicrob Agents Chemother.* 2013;57(9):4559–61.

Lewis RE, Verweij PE. Animal models for studying triazole resistance in aspergillus fumigatus. *J Infect Dis.* 2017;216(suppl_3):S466–73.

Lewis RE, Viale P, Kontoyiannis DP. The potential impact of antifungal drug resistance mechanisms on the host immune response to Candida. *Virulence.* 2012;3(4):368–76.

Lewis RE, Wiederhold NP, Klepser ME. In vitro pharmacodynamics of amphotericin B, itraconazole, and voriconazole against *Aspergillus*, Fusarium, and *Scedosporium*spp. *Antimicrob Agents Chemother.* 2005;49(3):945–951.

Liesegang TJ, Forster RK. Spectrum of microbial keratitis in south Florida. *Am J Ophthalmol.* 1980;90:38–47.

Linares MJ, Charriel G, Solís F, Rodriguez F, Ibarra A, Casal M. Susceptibility of filamentous fungi to voriconazole tested by two microdilution methods. *J Clin Microbiol.*

2005 Jan;43(1):250-3. doi: 10.1128/JCM.43.1.250-253.2005. PMID: 15634979; PMCID: PMC540146

Lionakis MS, Lewis RE, Chamilos G, et al., Aspergillus susceptibility testing inpatients with cancer and invasive aspergillosis: Difficulties in establishing correlation between in vitro susceptibility data and the outcome of initial amphotericinB therapy. *Pharmacotherapy.* 2005;25(9):1174–80.

Lionakis MS, Lewis RE, Kontoyiannis DP. Breakthrough invasive mold infections in the hematology patient: Current concepts and future directions. *Clin Infect Dis.* 2018;67(10):1621–30.

Locke JB, Almaguer AL, Zuill DE, Bartizal K. Characterization of in vitro resistance development to the novel echinocandin CD101 in Candida species. *Antimicrob Agents Chemother.* 2016;60(10):6100–6107.

Lockhart SR, Etienne KA, Vallabhaneni S, et al., Simultaneous emergence of multidrug-resistant Candida auris on 3 continents confirmed by whole genome sequencing and epidemiological analyses. *Clin Infect Dis.* 2016;64:134–40.

Lockhart SR, Frade JP, Etienne KA, et al., Azole resistance in Aspergillus fumigatus isolates from the ARTEMIS global surveillance study is primarily due to the TR/L98H mutation in the cyp51A gene. *Antimicrob Agents Chemother.* 2011;55(9):4465–8.

Loeffler J, Stevens D. Antifungal drug resistance. *Clin Infect Dis.* 2003;36:S31–41.

Ma L, Jakobiec FA, Dryja TP. A Review of Next-Generation Sequencing (NGS): Applications to the diagnosis of ocular infectious diseases. *Seminars in Ophthalmol.* 2019; 34:4, 223–231. DOI: 10.1080/08820538.2019.1620800

Maharana PK, Sharma N, Nagpal R, Jhanji V, Das S, Vajpayee RB. Recent advances in diagnosis and management of Mycotic Keratitis. *Indian J Ophthalmol.* 2016;64(5):346–357.

Mahashabde S, Nahata MC, Shrivastava U. A comparative study of anti-fungal drugs in mycotic corneal ulcer. *Indian J Ophthalmol.* 1987;35(5–6):149–152.

Mahdy R, Nada W, Wageh M. Topical amphotericin B and subconjunctival injection of fluconazole (combination therapy) versus topical amphotericin B (monotherapy) in treatment of keratomycosis. *J OculPharmacolTher.* 2010;26:281–5.

Maĭchuk IuF, Karimov MK, Lapshina NA. Ketokonazol v lechenii mikozov glaza [Ketoconazole in the treatment of ocular mycoses]. *Vestn Oftalmol.* 1990 Jan-Feb;106(1):44–6. Russian. PMID: 2378037

Manavathu EK, Cutright JL, Chandrasekar PH. Organism-dependant fungicidal activities of azoles. *Antimicrob Agents Chemother.* 1998;42: 3018–21.

Marchlewitz Ba MM. Die Pilzflora gesunder Augen. *Z Militärmed.* 1966;3:170–174.

Marichal P, Koymans L, Willemsens S, et al., Contribution of mutations in the cytochrome P450 14alpha-demethylase (Erg11p, Cyp51p) to azole resistance in Candida albicans. *Microbiology (Reading).* 1999;145(Pt 10):2701–13.

Martel CM, Parker JE, Bader O, et al., A clinical isolate of Candida albicans with mutations in ERG11 (encoding sterol 14alpha-demethylase) and ERG5 (encoding C22 desaturase) is cross resistant to azoles and amphotericin B. *Antimicrob Agents Chemother.* 2010;54(9):3578–83.

Matsumoto Y, Murat D, Kojima T, Shimazaki J, Tsubota K. The comparison of solitary topical micafungin or fluconazole application in the treatment of Candida fungal keratitis. *Br J Ophthalmol.* 2010;95:1406–9.

McLeod SD, Kumar A, Cevallos V et al., Reliability of transport medium in the laboratory evaluation of corneal ulcers. *Am J Ophthalmol.* 2005;140:1027–1031 doi:10.1016/j.ajo.2005.06.042

Mehta H, Mehta H, Garg P, Kodial H. Voriconazole for the treatment of refractory Aspergillus fumigatus keratitis. *Indian J Ophthalmol.* 2008;56:243–5.

Mellado E, Garcia-Effron G, Alcazar-Fuoli L, et al., A new Aspergillus fumigatus resistance mechanism conferring in vitro cross-resistance to azole antifungals involves a combination of cyp51A alterations. *Antimicrob Agents Chemother.* 2007;51(6):1897–904.

Messer SA, Diekema DJ, Boyken L, Tendolkar S, Hollis RJ, Pfaller MA. Activities of micafungin against 315 invasive clinical isolates of fluconazole-resistant *Candida* spp. *J Clin Microbiol.* 2006;44(2):324–326.

Miller CD, Lomaestro BW, Park S, et al., Progressive esophagitis caused by Candida albicans with reduced susceptibility to caspofungin. *Pharmacotherapy.* 2006;26(6):877–80.

Miller D, Galor A, Alfonso E, Fungal keratitis. In: Mannis M, Holland E (Eds.) *Cornea fundamentals, diagnosis, and management*, Elsevier Publishing China 2016.

Minkus CL, Bispo PJM, Papaliodis GN et al., Real-Time Multiplex PCR Analysis in Infectious Uveitis. *Semin Ophthalmol.* 2019;34:252–255. doi:10.1080/08820538.2019.1620803

Mitani A, Shiraishi A, Miyamoto H, Sunada A, Ueda A, Asari S, et al., Fungal keratitis caused by Beauveria bassiana: Drug and temperature sensitivity profiles: A case report. *BMC Research Notes.* 2014;7:677.

Mitchell BM, Wu TG, Chong EM et al., Expression of matrix metalloproteinases 2 and 9 in experimental corneal injury and fungal keratitis. *Cornea.* 2007;26:589–593. doi:10.1097/ICO.0b013e318033b504

Mitsui Y, Hanabusa J. Corneal infections after cortisone therapy. *Br J Ophthalmol.* 1955;39:244–250.

Mohan M, Panda A, Gupta SK. Management of human keratomycosis with miconazole. *Aust N Z J Ophthalmol.* 1989;17(3):295–297.

Mohanty B, Majumdar DK, Mishra SK, Panda AK, Patnaik S. Development and characterization of itraconazole-loaded solid lipid nanoparticles for ocular delivery. *Pharm Dev Technol.* 2015;20:458–64.

Molloy SF, Kanyama C, Heyderman RS, Loyse A, Kouanfack C, Chanda D, Mfinanga S, Temfack E, Lakhi S, Lesikari S, Chan AK, Stone N, Kalata N, Karunaharan N, Gaskell K, Peirse M, Ellis J, Chawinga C, Lontsi S, Ndong JG, Bright P, Lupiya D, Chen T, Bradley J, Adams J, van der Horst C, van Oosterhout JJ, Sini V, Mapoure YN, Mwaba P, Bicanic T, Lalloo DG, Wang D, Hosseinipour MC, Lortholary O, Jaffar S, Harrison TS, ACTA Trial Study Team . Antifungal combinations for treatment of cryptococcal meningitis in Africa. *N Engl J Med.* 2018;378:1004–1017. doi: 10.1056/NEJMoa1710922

Monden Y, Yamamoto S, Yamakawa R, Sunada A, Asari S, Makimura K, et al., First case of fungal keratitis caused by Pestalotiopsisclavispora. *Clin Ophthalmol.* 2013;7:2261–4.

Moro DD., Oluwole David Antimicrobial resistance among fungi from patients with urinary tract infections in Ojo, Lagos. *Nigeria GSC Biological and Pharmaceutical Sciences* March 2021. DOI: 10.30574/gscbps.2021.14.3.0082

Morschhäuser J, Barker KS, Liu TT, Blaß-Warmuth J, Homayouni R, Rogers PD. The transcription factor Mrr1p controls expression of the MDR1 efflux pump and mediates multidrug resistance in Candida albicans. *PLoSPathog.* 2007;3:e164. doi: 10.1371/journal.ppat.0030164

Morschhäuser J. The development of fluconazole resistance in Candida albicans- an example of microevolution of a fungal pathogen. *J Microbiol.* 2016;54(3):192–201.

Moudgal V, Little T, Boikov D, et al., Multiechinocandin- and multiazole-resistant Candida parapsilosis isolates serially obtained during therapy for prosthetic valve endocarditis. *Antimicrob Agents Chemother.* 2005;49(2):767–9.

Moustafa MA, Elnaggar YSR, El-Refaie WM, Abdallah OY. Hyalugel-integrated liposomes as a novel ocular nanosized delivery system of fluconazole with promising prolonged effect. *Int J Pharm*. 2017;534:14–24.

Mselle J. Fungal keratitis as an indicator of HIV infection in Africa. *Trop Doct*. 1999;29:133–5.

Mukherjee B, Chatterjee R, Biswas J. Reverse masquerade syndrome: fungal adnexal infection mimicking carcinoma in a HIV-positive patient. *Indian J Ophthalmol*. 2013;61:521–523. doi:10.4103/0301-4738.119454

Müller G, Kara-José N, de Castro R, Antifungals in eye infections: Drugs and routes of administration. *Rev Bras Oftalmol*. 2013;72:132–41.

Nakayama H, Izuta M, Nakayama N, et al., Depletion of the squalene synthase(ERG9) gene does not impair growth of Candida glabrata in mice. *Antimicrob Agents Chemother*. 2000;44(9):2411–8.

Nakayama H, Tanabe K, Bard M, et al., The Candida glabrata putative sterol transporter gene CgAUS1 protects cells against azoles in the presence of serum. *J AntimicrobChemother*. 2007;60(6):1264–72.

Neoh C, Daniell M, Chen S, Stewart K, Kong D. Clinical utility of caspofungin eye drops in fungal keratitis. *Int J Antimicrob Agents*. 2014;44:96–104.

Neoh C, Leung L, Misra A, Vajpayee R, Davies G, Fullinfaw R, et al., Penetration of topically administered 0.5-percent caspofungin eye drops into human aqueous humor. *Antimicrob Agents Chemother*. 2011a;55:1761–3.

Neoh C, Leung L, Vajpayee R, Stewart K, Kong D. Treatment of Alternaria keratitis with intrastromal and topical caspofunginincombination with intrastromal, topical, and oral voriconazole. *Ann Pharmacother*. 2011b;45:e24.

Nett JE, Andes DR. Contributions of the biofilm matrix to Candida pathogenesis. *J Fungi (Basel)*. 2020;6(1):21.

Niimi K, Maki K, Ikeda F, et al., Overexpression of Candida albicans CDR1, CDR2, or MDR1 does not produce significant changes in echinocandin susceptibility. *Antimicrob Agents Chemother*. 2006;50(4):1148–55.

Niimi M, Firth NA, Cannon RD. Antifungal drug resistance of oral fungi. *Odontology*. 2010 Feb;98(1):15–25. doi: 10.1007/s10266-009-0118-3. Epub 2010 Feb 16. PMID: 20155503

O'Brien TP. Therapy of ocular fungal infections. *Ophthalmol Clin NorthAm*. 1999;12:33–50.

O'Day D, Head W, Robinson R, Clanton J. Corneal penetration of topical amphotericin B and natamycin. *Curr Eye Res*. 1986;5:877–82.

O'Day DM. Selection of appropriate antifungal therapy. *Cornea*. 1987;6:238–45.

Oever JT, Netea MG. The bacteriome-mycobiome interaction and antifungal host defense. *Eur J Immunol*. 2014;44(11):3182–91.

Osherov N, Kontoyiannis DP. The anti-Aspergillus drug pipeline: Is the glass half full or empty? *Med Mycol*. 2017;55(1):118–24.

Ostrosky-Zeichner L. Candida glabrata and FKS mutations: Witnessing the emergence of the true multidrug-resistant Candida. *Clin Infect Dis*. 2013;56(12):1733–4.

Ozkan J, Nielsen S, Diez-Vives C et al., Temporal stability and composition of the ocular surface microbiome. *Sci Rep*. 2017;7:9880. doi:10.1038/s41598-017-10494-9

Ozturk F, Yavas GF, Kusbeci T et al., Efficacy of topical caspofungin in experimental fusarium keratitis. *Cornea*; 2007;26:726–728. doi:10.1097/ICO.0b013e3180553b9d

Pai V, Ganavalli A, Kikkeri NN. Antifungal resistance in dermatology. *Indian J Dermatol*. 2018:63:361-368.

Pandurangan DK, Bodagala P, Palanirajan VK, Govindaraj S. Formulation and evaluation of voriconazole ophthalmic solid lipid nanoparticles in situ gel. *Int J Pharm Investig*. 2016;6:56–62.

Pang CN, Lai YW, Campbell LT, et al., Transcriptome and network analyses in Saccharomyces cerevisiae reveal that amphotericin B and lactoferrin synergy disrupt metal homeostasis and stress response. *Sci Rep*. 2017;7:40232.

Papp C, Bohner F, Kocsis K, et al., Triazole evolution of Candida parapsilosis results in cross-resistance to other antifungal drugs, influences stress responses, and alters virulence in an antifungal drug-dependent manner. *mSphere*. 2020;5(5):20.

Pappas PG, Kauffman CA, Andes DR, Clancy CJ, Marr KA, Ostrosky-Zeichner L, Reboli AC, Schuster MG, Vazquez JA, Walsh TJ, Zaoutis TE, Sobel JD. 2016a. Clinical practice guideline for the management of candidiasis: 2016 update by the Infectious Diseases Society of America. *Clin Infect Dis*. 2016a;62:e1–e50. doi: 10.1093/cid/civ933

Pappas PG, Kauffman CA, Andes DR, et al., Executive summary: Clinical practice guideline for the management of candidiasis: 2016 update by the infectious diseases society of America. *Clin Infect Dis*. 2016b;62(4):409–17.

Pappas PG, Lionakis MS, Arendrup MC, et al., Invasive candidiasis. *Nat Rev Dis Primers* 2018;4:18026.

Paradkar M, Parmar M. Formulation development and evaluation of natamycin niosomal in-situ gel for ophthalmic drug delivery. *J Drug Delive Sci Technol*. 2017;39:113–22.

Paris G, Trujillo F,Woodward L, Trigo Y, Ballentine C, Najvar L, et al., Micafungin vs amphotericin B in the treatment of experimental aspergillosis endophthalmitis. *Invest Ophthalmol Vis Sci*. 2004;45:4014.

Park S, Kelly R, Kahn JN, et al., Specific substitutions in the echinocandin target Fks1p account for reduced susceptibility of rare laboratory and clinical Candida sp. isolates. *Antimicrob Agents Chemother*. 2005;49(8):3264–73.

Patil A, Lakhani P, Majumdar S. Current perspectives on natamycin in ocular fungal infections. *J Drug Deliv Sci Technol*. 2017;41:206–12.

Patil A, Lakhani P, Taskar P, Wu KW, Sweeney C, Avula B, et al., Formulation development, optimization, and in vitro-in vivo characterization of natamycin-loaded PEGylated nano-lipid carriers for ocular applications. *J Pharm Sci*. 2018;107:2160–71.

Patil A, Majumdar S. Echinocandins in ocular therapeutics. *J Ocul Pharmacol Ther*. 2017;33:340–52.

Patterson TF, Thompson GR 3rd, Denning DW, et al., Executive summary: Practice guidelines for the diagnosis and management of aspergillosis: 2016 update by the Infectious diseases society of America. *Clin Infect Dis*. 2016;63(4): 433–42.

Pemán J, Cantón E, Espinel-Ingroff A. Antifungal drug resistance mechanisms. *Expert Rev. Anti Infect. Ther.* 7, 453–460. doi: 10.1586/eri.09.18

Perlin DS. Resistance to echinocandin-class antifungal drugs. *Drug Resist Updat*. 2007;10(3):121–30.

Perlin DS. Echinocandin resistance in Candida. *Clin Infect Dis*. 2015;61(Suppl 6):S612–7.

Perlin DS, Rautemaa-Richardson R, Alastruey-Izquierdo A. The global problem of antifungal resistance: prevalence, mechanisms, and management. *Lancet Infect Dis*. 2017;17:e383–92. https://doi. org/10.1016/S1473-3099(17)30316-X

Pfaller MA. Antifungal drug resistance: Mechanisms, epidemiology, and consequences for treatment. Am J Med. 2012;125:S3–S13. doi: 10.1016/j.amjmed.2011.11.001

Pfaller MA, Diekema DJ, Jones RN, et al., International surveillance of bloodstream infections due to Candida species: Frequency of occurrence and in vitro susceptibilities to fluconazole, ravuconazole, and voriconazole of isolates collected from 1997 through 1999 in the SENTRY antimicrobial surveillance program. *J Clin Microbiol*. 2001;39:3254–9.

Pfaller MA, Messer SA, Boyken L, Rice C, Tendolkar S, Hollis RJ, Diekema DJ. Cross-resistance between fluconazole and ravuconazole and the use of fluconazole as a surrogate marker to predict susceptibility and resistance to ravuconazole among 12,796 clinical isolates of Candida spp. *J Clin Microbiol.* 2004a Jul;42(7):3137–41. doi: 10.1128/JCM.42.7.3137-3141.2004. PMID: 15243072; PMCID: PMC446250

Pfaller MA, Messer SA, Boyken L, et al., In vitro activities of voriconazole, posaconazole, and fluconazole against 4,169 clinical isolates of Candida spp. and Cryptococcus neoformans collected during 2001 and 2002 in the ARTEMIS global antifungal surveillance program. *Diagn Microbiol Infect Dis.* 2004b;48(3):201–5.

Pfaller, MA, Messer SA, Hollis RJ, Jones RN, Diekema DJ. In vitro activities of ravuconazole and voriconazole compared with those of four approved systemic antifungal agents against 6,970 clinical isolates of *Candida* spp. *Antimicrob. Agents Chemother.* 2002;46:1723–1727.

Pfaller MA, Messer SA, Rhomberg PR, Jones RN, CastanheiraM.n Activity of a long-acting echinocandin, CD101, determined using CLSI and EUCAST reference methods, against Candida and *Aspergillus* spp., including echinocandin- and azole-resistant isolates. *J AntimicrobChemother.* 2016;71(10):2868–2873.

Pfaller MA, Diekema DJ, Turnidge JD, Castanheira M, Jones RN. Twenty years of the SENTRY antifungal surveillance program: Results for Candida species from 1997-2016. *Open Forum Infect Dis.* 2019;6:S79–94. https://doi.org/10.1093/ofid/ofy358

Pfaller MA, Zhang J, Messer SA, et al., In vitro activities of voriconazole, fluconazole, and itraconazole against 566 clinical isolates of Cryptococcus neoformans from the United States and Africa. *Antimicrob Agents Chemother.* 1999;43:169–71.

Pham CD, Iqbal N, Bolden CB, Kuykendall RJ, Harrison LH, Farley MM, et al., Role of FKS mutations in Candida glabrata: MIC values, echinocandin resistance, and multidrug resistance. *Antimicrob Agents Chemother.* 2014;58:4690–6. https://doi.org/10.1128/AAC.03255-14

Pilmis B, Le Monnier A, Zahar JR. Gut Microbiota, antibiotic therapy and antimicrobial resistance: A narrative review. *Microorganisms.* 2020 Feb 17;8(2):269. doi: 10.3390/microorganisms8020269. PMID: 32079318; PMCID: PMC7074698

Posch W, Blatzer M, Wilflingseder D, et al., Aspergillus terreus: Novel lessons learned on amphotericin B resistance. *Med Mycol.* 2018;56(suppl_1):73–82.

Pradhan L, Sharma S, Nalamada S, Sahu S, Das S, Garg P. Natamycin in the treatment of keratomycosis: correlation of treatment outcome and in vitro susceptibility of fungal isolates. *Indian J Ophthalmol.* 2011;59:512–4.

Prajna N, Mascarenhas J, Krishnan T, Reddy P, Prajna L, Srinivasan M, et al. Comparison of natamycin and voriconazolefor the treatment of fungal keratitis. *Arch Ophthalmol.* 128(2010):672–8.

Prajna, NV, Lalitha P, Krishnan T, Rajaraman R, Radnakrishnan N, Srinivasan M, Devi L, Das M, Liu Z, Zegans ME, Acharya NR, Porco TC, Lietman TM, Rose-Nussbaumer J. Patterns of antifungal resistance in adult patients with fungal keratitisin South India a post hoc analysis of 3 randomized clinical trials. JAMA *Ophthalmol.* 2022;140(2):179–184. doi:10.1001/jamaophthalmol.2021.5765

Prajna NV, Nirmalan PK, Mahalakshmi R, Lalitha P, Srinivasan M. Concurrent use of 5% natamycin and 2% econazole for the management of fungal keratitis. *Cornea.* 2004;23(8):793–796.

Prajna NV, Lalitha P, Rajaraman R, et al. Mycotic ulcer treatment trial group. Changing azole resistance: A secondary analysis of the MUTT I Randomized Clinical Trial., *JAMA Ophthalmol.* 2016a;134(6):693–696. doi:10.1001/jamaophthalmol.2016. 0530

Prajna NV, Krishnan T, Rajaraman R, et al. Mycotic ulcer treatment trial II group. Effect of oral voriconazole on fungal keratitis in the Mycotic Ulcer Treatment Trial II (MUTT II): a randomized clinical trial., *JAMA Ophthalmol.* 2016b;134(12):1365–1372. doi:10.1001/jamaophthalmol.2016.4096

Qiu S, Zhao G, Lin J,Wang X, Hu L, Du Z, et al., Natamycin in the treatment of fungal keratitis: A systematic review and meta-analysis. *Int J Ophthalmol.* 2015;8:597–602.

Rajaraman R, Bhat P, Vaidee V, Maskibail S, Raghavan A, Sivasubramaniam S, et al., Topical 5% natamycin with oral ketoconazole in filamentous fungal keratitis: A randomized controlled trial., *Asia-Pacific J Ophthalmol (Philadelphia, Pa).* 2015;4:146– 50.

Reboli AC, Rotstein C, Pappas PG, et al., Anidulafungin versus fluconazole for invasive candidiasis. *N Engl J Med.* 2007;356(24):2472–82.

Ren M, Wu X . Evaluation of three different methods to establish animal models of Acanthamoeba keratitis. *Yonsei Med J.* 2010;51:121–127 doi:10.3349/ymj.2010.51.1.121

Revie NM, Iyer KR, Robbins N, Cowen LE. Antifungal drug resistance: Evolution, mechanisms and impact. *CurrOpinMicrobiol.* 2018;45:70–6. https://doi.org/10.1016/j.mib.2018.02.005

Rhodes J, Abdolrasouli A, Farrer RA, Cuomo CA, Aanensen DM, Armstrong-James D, et al. Genomic epidemiology of the UK outbreak of the emerging human fungal pathogen Candida auris. *Emerg.Microbes Infect.* 2018;7:43. doi: 10.1101/201343

Ritterband DC, Shah M, Seedor JA. Colletotrichumgraminicola: A new corneal pathogen. *Cornea.* 1997;16:362–4.

Rocha EMF, Garcia-Effron G, Park S, Perlin DS. 2007. A Ser678Pro substitution in Fks1p confers resistance to echinocandin drugs in *Aspergillus fumigatus. Antimicrob Agents Chemother.* 51(11):4174–4176.

Rosa PD, Sheid K, Locatelli C, Marinho D, Goldani L. Fusarium solani keratitis: Role of antifungal susceptibility testing and identification to the species level for proper management. *Braz J Infect Dis.* 2019 May-Jun;23(3):197–199. doi: 10.1016/j.bjid.2019.05.002. Epub 2019 May 23. PMID: 31129063; PMCID: PMC9428218

Rosa RH, Miller D, Alfonso EC. The changing spectrum of fungal keratitis in south Florida. *Ophthalmol.* 1994;101:1005–13.

Rudramurthy SM, Chakrabarti A, Paul RA, et al., Candida auris candidaemia in Indian ICUs: analysis of risk factors. *J AntimicrobChemother.* 2017;72:1794–801.

RuhnkeM, Schmidt-Westhausen A, Trautmann M. In vitro activities of voriconazole (UK-109,496) against fluconazole-susceptible and -resistant Candida albicans isolates from oral cavities of patients with human immunodeficiency virus infection. *Antimicrob Agents Chemother.* 1997;41:575–7.

Rybak JM, Doorley LA, Nishimoto AT, Barker KS, Palmer GE, Rogers PD. Abrogation of triazole resistance upon deletion of CDR1 in a clinical isolate of Candida auris. *Antimicrob. Agents Chemother.* 2019;63:e00057–19. doi: 10.1128/AAC.00057-19

Sahu R, Singh B, Saraf S, Kaithwas G, Kishor K. Photochemical toxicity of drugs intended for ocular use. *Arhiv za higijenuradatoksikologiju.* 2014;65:157–67.

Sai Prashanthi G, Jayasudha R, Chakravarthy SK, Padakandla SR, SaiAbhilash CR, Sharma S, Bagga B, Murthy SI, Garg P, Shivaji S. Alterations in the ocular surface fungal microbiome in fungal keratitis patients. *Microorganisms.* 2019 Sep 2;7(9):309. doi: 10.3390/microorganisms7090309. PMID: 31480776; PMCID: PMC6780152

Sanati H, Belanger P, Fratti B, Ghannoum M. A new triazole, voriconazole (UK-109,496), blocks sterol biosynthesis in Candida albicans and Candida krusei. *Antimicrob Agents Chemother.* 1997;41:2492–6.

Sanglard D, Coste A, Ferrari S. Antifungal drug resistance mechanisms in fungal pathogens from the perspective of transcriptional gene regulation. *FEMS Yeast Res.* 2009;9, 1029–1050. doi: 10.1111/j.1567-1364.2009.00578.x

Sanglard D, Ischer F, Koymans L, Bille J. Amino acid substitutions in the cytochrome P-450 lanosterol 14-demethylase (CYP51a1) from azoleresistant Candida albicans clinical isolates contribute to resistance to azole antifungal agents. *Antimicrob. Agents Chemother.* 1998;42:241–253.

Sanglard D, Ischer F, Parkinson T, et al., Candida albicans mutations in the ergosterol biosynthetic pathway and resistance to several antifungal agents. *Antimicrob Agents Chemother.* 2003;47(8):2404–12.

Sanglard D, Kuchler K, Ischer F, Pagani JL, Monod M, Bille J. 1995. Mechanisms of resistance to azole antifungal agents in Candida albicans isolates from AIDS patients involve specific multidrug transporters. *Antimicrob. Agents Chemother.* 39:2378–2386. doi: 10.1128/AAC.39.11.2378

Sanguinetti M, Posteraro B, Fiori B, et al., Mechanisms of azole resistance in clinical isolates of Candida glabrata collected during a hospital survey of antifungal resistance. *Antimicrob Agents Chemother.* 2005;49(2):668–79.

Sasse C, Dunkel N, Schafer T, et al., The stepwise acquisition of fluconazole resistance mutations causes a gradual loss of fitness in Candida albicans. *Mol Microbiol.* 2012;86(3):539–56.

Sawyer PR, Brogden RN, Pinder RM, Speight TM, Avery GS. Miconazole: A review of its antifungal activity and therapeutic efficacy. *Drugs.* 1975;9(6):406–423.

Schuetzer-Muehlbauer M, Willinger B, Egner R, Ecker G, Kuchler K. Reversal of antifungal resistance mediated by ABC efflux pumps from Candida albicans functionally expressed in yeast. *Int. J. Antimicrob. Agents.* 2003;22, 291–300. doi: 10.1016/S0924-8579(03)00213-9

Schwarzmuller T, Ma B, Hiller E, et al., Systematic phenotyping of a large-scale Candida glabrata deletion collection reveals novel antifungal tolerance genes. *PLoSPathog* 2014;10(6):e1004211.

Segal B, Walsh T. Amphotericin B is still the drug of choice for invasive aspergillosis. *Am J Respir Crit Care Med.* 2006;174:102.

Segal E, Romano A, Eylan E et al., Fungal flora of the normal conjunctival sac. *Mykosen.* 1977;20:9–14. doi:10.1111/j.1439-0507.1977.tb01475.x

Sehgal SC, Dhawan S, Chhiber S et al., Frequency and significance of fungal isolations from conjunctival sac and their role in ocular infections. *Mycopathologia.* 1981;73:17–19. doi:10.1007/BF00443007

Sekizuka T, Iguchi S, Umeyama T, et al., Clade II Candida auris possess genomic structural variations related to an ancestral strain. *PLoS One* 2019;14(10):e0223433.

Shao Y, Yu Y, Pei C, Tan Y, Zhou Q, Yi J, et al., Therapeutic efficacy of intracameral amphotericin B injection for 60 patients with keratomycosis. *Int J Ophthalmol.* 2010;3:257–60.

Sharma C, Chowdhary A. Molecular bases of antifungal resistance in filamentous fungi. *Int. J. Antimicrob. Agents.* 2017;50:607–16. http://dx.doi.org/doi: 10.1016/j.ijantimicag.2017.06.018

Sharma S. Diagnosis of fungal keratitis: Current options. *Expert Opin Med Diagn.* 2012a;6:449–455. doi:10.1517/17530059.2012.679656

Sharma S. Diagnosis of infectious diseases of the eye. *Eye (Lond).* 2012b;26:177–184 doi:10.1038/eye.2011.275

Sharma S, Kunimoto DY, Gopinathan U et al., Evaluation of corneal scraping smear examination methods in the diagnosis of bacterial and fungal keratitis: a survey of eight years of laboratory experience. *Cornea*. 2002;21:643–647. doi:10.1097/00003226-200210000-00002

Sharma S, Alfatah M, Bari VK, et al., Sphingolipid biosynthetic pathway genes FEN1 and SUR4 modulate amphotericin B resistance. *Antimicrob Agents Chemother*. 2014;58(4):2409–14.

Sharma S, Das S, Virdi A, Fernandes M, Sahu SK, Kumar Koday N, et al., Re-appraisal of topical 1% voriconazole and 5% natamycin in the treatment of fungal keratitis in a randomised trial., *Br J Ophthalmol*. 2015a;99:1190.

Sharma C, Kumar N, Meis JF, et al., Draft genome sequence of a fluconazoleresistant Candida auris strain from a Candidemia patient in India. *Genome Announc*. 2015b;3(4):e00722-15.

Sharma C, Kumar N, Pandey R, Meis J.F, Chowdhary A. Whole genome sequencing of emerging multidrug resistant Candida auris isolates in India demonstrates low genetic variation. *New Microbes New Infect*. 2016;13:77–82. doi: 10.1016/j.nmni.2016.07.003

Shen Y, Liang C, Wang C, Lin K, Hsu M, Yuen H, et al., Pharmacokinetics and safety of intravitreal Caspofungin. *Antimicrob Agents Chemother*. 2014;58:7234–9.

Shields RK, Nguyen MH, Clancy CJ. Clinical perspectives on echinocandin resistance among Candida species. *CurrOpin Infect Dis*. 2015;28:514–22. https://doi.org/10.1097/QCO. 0000000000000215

Shields RK, Nguyen MH, Press EG, Updike CL, Clancy CJ. Anidulafungin and micafungin MIC breakpoints are superior to that of caspofungin for identifying FKS mutant *Candida glabrata* strains and Echinocandin resistance. *Antimicrob Agents Chemother*. 2013;57(12):6361–6365.

Shields RK, Nguyen MH, Press EG, et al., Abdominal candidiasis is a hidden reservoir of echinocandin resistance. *Antimicrob Agents Chemother*. 2014;58(12):7601–5.

Shin SH, Ye MK, Lee YH. Fungus culture of the nasal secretion of chronic rhinosinusitis patients: seasonal variations in Daegu, Korea. *Am J Rhinol*. 2007 Sep-Oct;21(5):556–9. doi: 10.2500/ajr.2007.21.3065. PMID: 17999789

Shirazi F, Kontoyiannis DP. Mitochondrial respiratory pathways inhibition in Rhizopus oryzae potentiates activity of posaconazole and itraconazole via apoptosis. *PLoS One* 2013;8(5):e63393.

Shirley SF, Little JR. Immunopotentiating effects of amphotericin B. I. Enhanced contact sensitivity in mice. *J Immunol*. 1979a;123(6):2878–2882.

Shirley SF, Little JR. Immunopotentiating effects of amphotericin B. II. Enhanced in vitro proliferative responses of murine lymphocytes. *J Immunol*. 1979b;123(6):2883–2889.

Shivaji S. 2022. *Human Ocular Microbiome: Bacteria, Fungi and Viruses in the Human Eye* 15 May 2022, Springer Verlag, Singapore. ISBN13 9789811917530

Shivaji S, Jayasudha R, Prashanthi GS, Arunasri K, Das T. Fungi of the human eye: Culture to mycobiome. *Exp Eye Res*. 2022 Apr;217:108968. doi: 10.1016/j.exer.2022.108968. Epub 2022 Feb 1. PMID: 35120870

Shivaji S, Jayasudha R, Sai Prashanthi G, Kalyana Chakravarthy S, Sharma S. The human ocular surface fungal microbiome. *Invest Ophthalmol Vis Sci*. 2019;60(1):451–459.doi: 10.1167/iovs.18-26076

Shivarathri R, Jenull S, Stoiber A, et al., The two-component response regulator Ssk1 and the mitogen-activated protein kinase Hog1 control antifungal drug resistance and cell wall architecture of Candida auris. *mSphere*. 2020;5(5):e00973-20.

Shivarathri R, Tscherner M, Zwolanek F, et al., The fungal histone acetyl transferase Gcn5 controls virulence of the human pathogen Candida albicans through multiple pathways. *Sci Rep.* 2019;9(1):9445.

Shor E, Perlin DS. Coping with stress and the emergence of multidrug resistance in fungi. PLoSPathog. 2015;11:e1004668. doi: 10.1371/journal.ppat.1004668

Shukla PK, Kumar M, Keshava G.B.S. . Mycotic keratitis: An overview of diagnosis and therapy. *Mycoses* 2008, 51:183–199. doi:10.1111/j.1439-0507.2007.01480.x

Silva GR, Almeida APR, Fernandes-Cunha GM, Castro BFM, Vieira LC, Fulgêncio GO, et al., Safety and in vivo release of fluconazole-loaded implants in rabbits' eyes. *J Drug Deliv Sci Technol.* 2016;35:323–6.

Singh A, Sharma B, Mahto KK. High-frequency direct detection of triazole resistance in Aspergillus fumigatus from patients with chronic pulmonary fungal diseases in India. *J Fungi (Basel).* 2020;6(2):67.

Singh A, Singh PK, de Groot T, et al., Emergence of clonal fluconazole-resistant Candida parapsilosis clinical isolates in a multicentre laboratory-based surveillance study in India. *J AntimicrobChemother.* 2019;74(5):1260–8.

Singh SD, Robbins N, Zaas AK, et al., Hsp90 governs echinocandin resistance in the pathogenic yeast Candida albicans via calcineurin. *PLoSPathog.* 2009;5(7):e1000532.

Singh-Babak SD, Babak T, Diezmann S, et al., Global analysis of the evolution and mechanism of echinocandin resistance in Candida glabrata. *PLoSPathog.* 2012;8(5):e1002718.

Singhal N, Kumar M, Kanaujia PK, et al., MALDI-TOF mass spectrometry: An emerging technology for microbial identification and diagnosis. *Front Microbiol.* 2015;6:791.

Snelders E, Camps SM, Karawajczyk A, et al., Genotype-phenotype complexity of the TR46/Y121F/T289A cyp51A azole resistance mechanism in Aspergillus fumigatus. *Fungal Genet Biol.* 2015;82:129–35.

Snelders E, Karawajczyk A, Schaftenaar G, et al., Azole resistance profile of amino acid changes in Aspergillus fumigatus CYP51A based on protein homology modeling. *Antimicrob Agents Chemother.* 2010;54(6):2425–30.

Snelders E, Karawajczyk A, Verhoeven RJ, et al., The structure-function relationship of the Aspergillus fumigatuscyp51A L98H conversion by site-directed mutagenesis: the mechanism of L98H azole resistance. *Fungal Genet Biol.* 2011;48(11):1062–70.

Sponsel WE, Graybill JR, Nevarez HL, Dang D. Ocular and systemic posaconazole(SCH-56592) treatment of invasive Fusarium solani keratitis and endophthalmitis. *Br J Ophthalmol.* 2002 Jul;86(7):829–30. doi: 10.1136/bjo.86.7.829-a. PMID: 12084760; PMCID: PMC1771206

Spriet I, Delaere L, Lagrou K, Peetermans W, Maertens J,Willems L. Intraocular penetration of voriconazole and caspofungin in a patient with fungal endophthalmitis. *J AntimicrobChemother.* 2009;64:877–8.

Stavrou AA, Perez-Hansen A, Lackner M, et al., Elevated minimum inhibitory concentrations to antifungal drugs prevail in 14 rare species of candidemiacausing Saccharomycotina yeasts. *Med Mycol.* 2020;58:8.

Stone NR, Rhodes J, Fisher MC, et al., Dynamic ploidy changes drive fluconazole resistance in human cryptococcal meningitis. *J Clin Invest.* 2019;129:15.

Streekstra H, Verkennis AE, Jacobs R, Dekker A, Stark J, Dijksterhuis J. Fungal strains and the development of tolerance against natamycin. *Int JFood Microbiol.* 2016;238:15–22. doi:10.1016/ j.ijfoodmicro.2016.08.006

Sud IJ, Feingold DS. Mechanisms of action of the antimycotic imidazoles. *J Invest Dermatol.* 1981;76(6):438–441.

Sun Y, Chandra J, Mukherjee P, et al., A murine model of contact lens-associated fusarium keratitis. *Invest Ophthalmol Vis Sci.* 2010;51:1511–1516. doi:10.1167/iovs.09-4237

Suvarna P, Chaudhari P, Birangal S, Mallela LS, Roy S, Koteshwara A, Aranjani JM, Lewis SA. Voriconazole–Cyclodextrin Supramolecular ternary complex- loaded ocular films for management of fungal Keratitis. *Mol. Pharmaceutics.* 2022;19:258–273.

Suwunnakorn S, Wakabayashi H, Kordalewska M, Perlin DS, Rustchenko E. FKS2 and FKS3 genes of opportunistic human pathogen Candida albicans influence Echinocandin susceptibility. *Antimicrob Agents Chemother.* 2018;62:e02299–17. https://doi.org/10.1128/aac.02299-17

Suzuki T, Uno T, Chen G, Ohashi Y. Ocular distribution of intravenously administered micafungin in rabbits. *J Infect Chemother.* 2008;14:204–2017.

Talbot JJ, Frisvad JC, Meis JF. cyp51A mutations, extrolite profiles, and antifungal susceptibility in clinical and environmental isolates of the Aspergillus viridinutans Species Complex. *Antimicrob Agents Chemother.* 2019;63(11):19.

Tashiro M, Izumikawa K, Hirano K, et al., Correlation between triazole treatment history and susceptibility in clinically isolated Aspergillus fumigatus. *Antimicrob Agents Chemother.* 2012;56(9):4870–5.

Teweldemedhin M, Gebreyesus H, Atsbaha AH, Asgedom S.W, Saravanan M. Bacterial profile of ocular infections: A systematic review. *BMC Ophthalmol.* 2017;17:212. doi: 10.1186/s12886-017-0612-2

teWelscher YM, ten Napel HH, Balague MM, et al., Natamycin blocks fungal growth by binding specifically to ergosterol without permeabilizing the membrane. *J Biol Chem.* 2008;283(10):6393–6401. doi:10.1074/jbc.M707821200

Thakkar R, Patil A, Mehraj T, Dudhipala N, Majumdar S. Updates in ocular antifungal pharmacotherapy: Formulation and clinical perspectives. *Curr. Fungal Infect. Rep.* 2019;13:45–58.

Therese KL, Bagyalakshmi R, Madhavan HN, Deepa P. In-vitro susceptibility testing by agar dilution method to determine the minimum inhibitory concentrations of amphotericin B, fluconazole and ketoconazole against ocular fungal isolates. *Indian J Med Microbiol.* 2006 Oct;24(4):273–9. doi: 10.4103/0255-0857.29386. PMID: 17185846

Thevissen K, Ayscough KR, Aerts AM, Du W, De Brucker K, Meert EM, Ausma J, Borgers M, Cammue BP, François IE. Miconazole induces changes in actin cytoskeleton prior to reactive oxygen species induction in yeast. *J Biol Chem.* 2007;282(30):21592–21597.

Thomas PA. Mycotic keratitis- An underestimated mycosis. *J Med Vet Mycol.* 1994;32:235–56.

Thomas PA, Kaliamurthy J. Mycotic keratitis: Epidemiology, diagnosis and management. *Clin Microbiol Infect.* 2013;19(3):210–220.

Thomas PA, Kuriakose T, Kirupashanker MP, et al., Use of lactophenol cotton blue mounts of corneal scrapings as an aid to the diagnosis of mycotic keratitis. *Diagnostic Microbiology and Infectious Disease.* 1991;14:219–224. doi:10.1016/0732-8893(91)90035-e

Thomaz DY, de Almeida JN Jr, Lima GME, et al., An azole-resistant candida parapsilosis outbreak: Clonal persistence in the intensive care unit of a Brazilian teaching hospital., *Front Microbiol.* 2018;9:2997.

Torres M, Mohamed J, Cavazos-Adame H, Martinez L. Topical ketoconazole for fungal keratitis. *Am J Ophthalmol.* 1985;100:293–8.

Trujillo F, Paris G, Woodward L, Graybill J, Pena M, Najvar L, et al., Comparison of two antifungal agents, natamycin and micafungin, for fungal keratitis. *Invest Ophthalmol Vis Sci*. 2004;45:113.

Tsai S, Lin Y, Hsu H, Chen Y. Subconjunctival injection of fluconazole in the treatment of fungal Alternaria keratitis. *Ocul Immunol Inflamm*. 2014;24:103–6.

Tu E. Alternaria keratitis: Clinical presentation and resolution with topical fluconazole or intrastromal voriconazole and topical caspofungin. *Cornea*. 2008;28:116–9.

Tu EY, McCartney DL, Beatty RF, Springer KL, Levy J, Edward D. Successful treatment of resistant ocular fusariosis with posaconazole (SCH-56592). *Am J Ophthalmol*. 2007a Feb;143(2):222–227. doi: 10.1016/j.ajo.2006.10.048. Epub 2006 Dec 8. PMID: 17258521

Turnbaugh PJ, Ley RE, Hamady M, Fraser-Liggett CM, Knight R, Gordon JI. The human microbiome project. *Nature*. 2007; 449(7164): 804–810.

Uchida K, Abe S, Yamaguchi H. The postantifungal effect (PAFE) of itraconazole, in comparison with those of miconazole and fluconazole, on Candida species. *Microbiol Immunol*. 2006;50(9):679–685.

Vaddavalli PK, Garg P, Sharma S, et al., Role of confocal microscopy in the diagnosis of fungal and acanthamoeba keratitis. *Ophthalmol*. 2011;118:29–35. doi:10.1016/j.ophtha.2010.05.018

Vale-Silva LA, Sanglard D. Tipping the balance both ways: Drug resistance and virulence in Candida glabrata. *FEMS Yeast Res*. 2015;15(4):fov025.

Vandeputte P, Ferrari S, Coste AT. Antifungal resistance and new strategies to control fungal infections. *Int. J. Microbiol*. 2012;2012:713687. doi: 10.1155/2012/713687

van der Linden JW, Arendrup MC, Warris A, et al., Prospective multicenter international surveillance of azole resistance in Aspergillus fumigatus. *Emerg Infect Dis*. 2015;21(6):1041–4.

van der Linden JW, Snelders E, Kampinga GA, et al., Clinical implications of azole resistance in Aspergillus fumigatus, The Netherlands, 2007–2009. *Emerg Infect Dis*. 2011;17(10):1846–54.

van Duin D, Casadevall A, Nosanchuk JD. Melanization of Cryptococcus neoformans and Histoplasma capsulatum reduces their susceptibilities to amphotericin B and caspofungin. *Antimicrob Agents Chemother*. 2002;46(11):3394–400.

Van Tyle JH. Ketoconazole. Mechanism of action, spectrum of activity, pharmacokinetics, drug interactions, adverse reactions and therapeutic use. *Pharmacotherapy*. 1984;4:343–373. doi: 10.1002/j.1875-9114.1984.tb03398.x

Velpandian T, Narayanan K, Nag TC et al., Retinal toxicity of intravitreally injected plain and liposome formulation of fluconazole in rabbit eye. *Indian J Ophthalmol*. 2006;54:237–240. doi:10.4103/0301-4738.27947

Vemuganti GK, Naidu C, Gopinathan U. Rapid detection of fungal filaments in corneal scrapings by microwave heating-assisted Grocott's methenamine silver staining. *Indian J Ophthalmol*. 2002;50:326–328.

Vermes A, Guchelaar HJ, Dankert J. Flucytosine: A review of its pharmacology, clinical indications, pharmacokinetics, toxicity and drug interactions. *J Antimicrob Chemother*. 2000;46:171–179. doi: 10.1093/jac/46.2.171

VigezziC, Icely PA, Dudiuk C, et al., Frequency, virulence factors and antifungal susceptibility of Candida parapsilosis species complex isolated from patients with candidemia in the central region of Argentina. *J Mycol Med*. 2019;29(4):285–91.

Vincent BM, Lancaster AK, Scherz-Shouval R, et al., Fitness trade-offs restrict the evolution of resistance to amphotericin B. *PLoSBiol.* 2013;11(10):e1001692.

Von Graefe A, Mittheilungen K. Verschwärung der Sclera nach einer Schieloperation. *Arch Ophthal.,* 1857;3:409–411.

Waldorf AR, Polak A. Mechanisms of action of 5-fluorocytosine. *Antimicrob. Agents Chemother.* 1983;23:79–85. doi: 10.1128/AAC.23.1.79

Walia H, Tucci V, Greene J, Tordilla-Wadia J, Kelty P, Walia S. A case of endogenous trichosporon endophthalmitis treated with micafungin and voriconazole. *J Global Infect Dis.* 2009;1:71–4.

Walsh TJ, Groll A, Hiemenz J, Fleming R, Roilides E, Anaissie E. Infections due to emerging and uncommon medically important fungal pathogens. *Clin Microbiol Infect.* 2004;10(Suppl 1):48–66.

Wang L, Wang L, Han L, Yin W. Study of pathogens of fungal keratitis and the sensitivity of pathogenic fungi to therapeutic agents with the disk diffusion method. *Curr Eye Res.* 2015;40: 1095–101.

Wang L, Zhang Y, Wang Y, Wang G, Lu J, Deng J. Spectrum of mycotic keratitis in China. *Zhonghua Yan Ke Za Zhi.* 2000;36:138–40.

Wang Y, Chen H, Xia T et al., Characterization of fungal microbiota on normal ocular surface of humans. *Clin Microbiol Infect.* 2020;26:123 e129–123 e113. doi:10.1016/j.cmi.2019.05.011

Wang YE, Tepelus TC, Vickers LA et al., Role of in vivo confocal microscopy in the diagnosis of infectious keratitis. *Int Ophthalmol.* 2019;39:2865–2874. doi:10.1007/s10792-019-01134-4

Warrilow AG, Melo N, Martel CM, et al., Expression, purification, and characterization of Aspergillus fumigatus sterol 14-alpha demethylase (CYP51) isoenzymes A and B. *Antimicrob Agents Chemother.* 2010;54(10):4225–34.

Whaley SG, Rogers PD. Azole Resistance in Candida glabrata. *Curr Infect Dis Rep.* 2016;18:19–21. https://doi.org/10.1007/s11908- 016-

White TC, Findley K, Dawson TL Jr, Scheynius A, Boekhout T, Cuomo CA, Xu J, Saunders CW. Fungi on the Skin: Dermatophytes and Malassezia cold spring. *Harb Perspect Med.* 2014; doi: 10.1101/cshperspect.a019802.

Wiederhold NP Antifungal resistance: Current trends and future strategies to combat. *Infect and Drug Resist.* 2017;10:249–259.

Wiederhold NP, Najvar LK, Bocanegra RA, et al., Caspofungin dose escalation for invasive candidiasis due to resistant Candida albicans. *Antimicrob Agents Chemother.* 2011;55(7):3254–60.

Williamson J, Gordon AM, Wood R et al., Fungal flora of the conjunctival sac in health and disease. Influence of topical and systemic steroids. *Br J Ophthalmol.* 1968;52:127–137 doi:10.1136/bjo.52.2.127

Wingard JR, Merz WG, Rinaldi MG, et al., Increase in Candida kruseiinfection among patients with bone marrow transplantation and neutropenia treated prophylactically with fluconazole. *N Engl J Med.* 1991;325(18):1274–7.

Wood TO, Williford W. Treatment of keratomycosis with amphotericin B 0.15%. *Am J Ophthalmol.* 1976;81(6):847–849.

Wu TG, Wilhelmus KR, Mitchell BM. Experimental keratomycosis in a mouse model. *Invest Ophthalmol Vis Sci.* 2003a;44:210–216. doi:10.1167/iovs.02-0446

Wu T, Mitchell B, Carothers T et al., Molecular analysis of the pediatric ocular surface for fungi. *Curr Eye Res.* 2003b;26:33–36. doi:10.1076/ceyr.26.1.33.14253

Xiong K, Su C, Sun Q, et al., Efg1 and Cas5 orchestrate cell wall damage response to Caspofungin in Candida albicans. *Antimicrob Agents Chemother.* 2021;65(2):e01584-20.

Xu Y, Lu H, Zhu S, Li WQ, Jiang YY, Berman J, Yang F. Multifactorial mechanisms of tolerance to Ketoconazole in Candida albicans. *Microbiol Spectr.* 2021 Sep 3;9(1):e0032121. doi: 10.1128/Spectrum.00321-21. Epub 2021 Jun 23. PMID: 34160280; PMCID: PMC8552639

Xu Y, Pang G, Zhao D, Gao C, Zhou L, Sun S, et al., In vitro activity of thimerosal against ocular pathogenic Fungi. *Antimicrob Agents Chemother.* 2010;54:536–9.

Xuguang S, Zhixin W, Zhiqun W, Shiyun L, Ran L. Ocular fungal isolates and antifungal susceptibility in northern China. *Am J Ophthalmol.* 2006;143:131–3.

Yamazumi T, Pfaller M.A, Messer S.A, Houston A, Hollis R.J, Jones R.N.. In vitro activities of ravuconazole (BMS-207147) against 541 clinical isolates of *Cryptococcus neoformans. Antimicrob. Agents Chemother.* 2000;44:2883−2886.

Yao YF, Zhang YM, Zhou P, Zhang B, Qiu WY, Tseng SCG. Therapeutic penetrating keratoplasty in severe fungal keratitis using cryopreserved donor corneas. *Br J Ophthalmol.* 2003;87: 543–7.

Yee R, Cheng C, Meenakshi S, Ludden T, Wallace J, Rinaldi M. Ocular penetration and pharmacokinetics of topical fluconazole. *Cornea.* 1997;16:64–71.

Yilmaz S, Maden A. Severe fungal keratitis treated with subconjunctival fluconazole. *Am J Ophthalmol.* 2005;140:454–8.

Yilmaz S, Ture M, Maden A. Efficacy of intracameral amphotericin B injection in the management of refractory keratomycosis and endophthalmitis. *Cornea.* 2007;26:398–402.

You X, Li J, Li S, Shi W. Effects of lamellar keratectomy and intrastromal injection of 0.2% fluconazole on fungal keratitis. *J Ophthalmol.* 2015;2015:10.

Young LY, Hull CM, Heitman J. Disruption of ergosterol biosynthesis confers resistance to amphotericin B in Candida lusitaniae. *Antimicrob Agents Chemother.* 2003;47(9):2717–24.

Yuan X, Wilhelmus KR. Toll-like receptors involved in the pathogenesis of experimental Candida albicans keratitis. *Invest Ophthalmol Vis Sci.* 2010;51:2094-2100. doi:10.1167/iovs.09-4330

Zavrel M, Hoot SJ, White TC. Comparison of sterol import under aerobic and anaerobic conditions in three fungal species, Candida albicans, Candida glabrata, and Saccharomyces cerevisiae. *Eukaryot Cell.* 2013;12(5):725–38.

Zeagans ME, Van Gelder RN. Considerations in understanding the ocular surface microbiome. *Am J Ophthalmol.* 2014;158:420-422 doi:10.1016/j.ajo.2014.06.014

Zhang J, Xie Y, Zhou T, Li J, He J, Xia H, et al., Development of natamycin-hydroxypropyl-beta-Cyclodextrin inclusion complex, ion-triggered in situ gel for sustained ocular delivery: in vitro, ex vivo evaluation and ocular pharmacokinetics study. *Invest Ophthalmol Vis Sci.* 2018;59:2676.

Zhang W, Yang H, Jiang L et al., Use of potassium hydroxide, Giemsa and calcofluor white staining techniques in the microscopic evaluation of corneal scrapings for diagnosis of fungal keratitis. *J Int Med Res.* 2010;38:1961-1967. doi:10.1177/147323001003800609

Zhong W, Yin H, Xie L. Expression and potential role of major inflammatory cytokines in experimental keratomycosis. *Mol Vis.* 2009;15:1303-1311.

Zhou W, Wang Y, Jian J, Song S. Self-aggregated nanoparticles based on amphiphilic poly(lactic acid)-grafted-chitosan copolymer for ocular delivery of amphotericin B. *Int J Nanomedicine.* 2013;8:3715–28.

6

Antimicrobial Resistance in the Diseased Human Eye: Viral Keratitis

S. Shivaji

6.1 Introduction

Keratitis is an ocular disease manifested by partial or total loss of vision. Apart from bacteria and fungi, less common are keratitis caused by viruses. It is estimated that the prevalence of infectious keratitis ranged from 0.148% to 0.192% which included viral 0.065% to 0.11%, bacterial 0.068% to 0.075%, and fungal 0.007% to 0.015% in two different studies (Song et al., 2014; Cao et al., 2014). Several diverse viruses are known as etiological agents of viral keratitis and the dominant viral etiological agent is the herpes simplex virus (HSV) (Carmichael, 2012; Farooq and Shukla, 2012; Hjalgrim et al., 2007; Wharton, 1996). The other viruses that cause keratitis include cytomegalovirus (CMV), varicella-zoster virus (VZV), and Epstein-Barr virus (EBV). This chapter briefly deals with the prevalence, diagnosis and pathobiology of viral keratitis and presents the resistance pattern of the viruses to treatment protocols.

6.2 Herpes Simplex Virus

6.2.1 Prevalence of HSV

HSV is a common etiological agent of corneal infection. It is an enveloped double-stranded DNA virus and exists as type 1 (HSV-1) and type 2 (HSV-2). Of the two types HSV-1 causes ocular infections while HSV-2 infects the genitalia (Whitley et al., 1998; Ritterband and Friedberg, 1998). HSV-1 infects about two-thirds of the population (65%) across the globe (Liesegang, 2001b; Kaufman, 2000). It is estimated that the incidence of keratitis due to HSV (HSK) is about 1.5 million (Farooq and Shukla, 2012), with 40,000 new cases causing monocular visual impairment or blindness each year across the world (Cabrera-Aguas et al., 2022a; 2022b). In developed countries HSK causes opacification of the stroma (Cabrera-Aguas et al., 2022a; 2022b). Extensive prevalence of HSV-1 is due to direct contact with saliva, tears, and so forth, of infected individuals. The recent seroprevalence data indicated a linear decrease in HSV-1 over a period of 16 years (1999–2000 to 2015–2016), which varied depending on the ethnicity

DOI: 10.1201/9781003451105-6

of the individuals with a maximum of 82.0% to 71.7% for Mexican-Americans to a minimum of 52.4% to 36.9% for non-Hispanic white, and in non-Hispanic black the decrease was 68.4% to 58.8% (Koganti et al., 2019; 2022b; Kaufman, 2000). In adults a large proportion of HSV-1 antibodies could be recovered from the ganglia (Baringer and Swoveland, 1973; Stevens et al., 1988). HSV in primary infection usually occurs within the first few years of life and induces cutaneous lesions and illness accompanied with severe conjunctivitis, keratoconjunctivitis, and periocular and eyelid skin vesicles and corneal lesions (Ritterband and Friedberg, 1998).

6.2.2 HSV-1 Infection Pattern in Children and Adults

Infection pattern of HSV-1 in children and adults is different. Children normally present with ocular inflammation and experience more recurrences, and complications attributed to their stronger immunity (Cabrera-Aguas et al., 2022a; 2022b; White and Chodosh, 2014). In children HSV-1 is ulcerative and causes painful stomatitis, corneal opacification, irregular astigmatism, amblyopia, overt fever, anorexia and edema of oral mucosa (White and Chodosh, 2014). Further, bilateral ocular HSV disease is common in children during primary infections rather than in recurrent infections, with 3.4% to 26% primary infections and a recurrence rate of 45 to 50% within the first year compared to 1.3% to 12% primary and 18% recurrence respectively in adults (White and Chodosh, 2014).

In the United States, HSV-1 causes pharyngitis and tonsillitis. The virus after initial infection moves to the trigeminal ganglia (TG), and persists for a prolonged period, and in most cases lifelong in a latent form (Koujah, et al., 2019a; 2022b) prior to getting reactivated (Wald and Corey, 2007). HSV-1 also causes encephalitis but occurs only in a few individuals (0.0001–0.0005% of the population), which is highly fatal causing mortality in 70% of the affected individuals (Wald and Corey. 2007; McQuillan et al., 2018). Further, it has also been observed that individuals with atopic disease were more prone to ocular HSV disease (2 to 4.8–fold higher) compared to people free of atopy (Cabrera-aguas et al., 2022b). The other predisposing factors that influence susceptibility of the cornea include medications (latanoprost and corticosteroids), trauma due to surgery of the eye and inflammation (White and Chodosh, 2014).

6.2.3 HSV-1 and Ocular Disease

In the United States over 500,000 people are infected with HSV-1 and approximately 8% of these individuals develop ocular diseases (Farooq and Shukla. 2012; Holland et al., 1992; Reynaud et al., 2017) like conjunctivitis, iridocyclitis, retinal necrosis and keratitis (Farooq and Shukla. 2012; Liesegang et al., 1989). This is in difference to the observations made in 1985, which indicated that ~15% of individuals infected with HSV-1 develop blepharoconjunctivitis and dendritic ulcers (Darougar et al., 1985). Further, recurrence rates over a ten-year period increased from 9.6% to 67% (Farooq and Shukla. 2012; Liesegang, 2001; Shuster et al., 1981). In a considerable proportion of the HSK patients (approximately 40%) 2–5 relapses are observed (Wishart et al., 1987). Table 6.1 summarises the prevalence, incidence, seroprevalence, recurrence and incidence of resistance.

TABLE 6.1

The Epidemiology of Herpes Simplex Keratitis

Statistics	Prevalence and percentages	References
Prevalence of HSK in the United States	150 in 100,000, and 20K new cases annually	Liesegang et al., 1989.
Annual incidence in Western countries	10–30 per 100,000	Liesegang et al., 1989; Young et al., 2010.
People (%) with HSV-1 antibodies by middle age	More than 90%	Liesegang, 2001.
HSV-1 infections (%) that are asymptomatic	More than 90%	Liesegang, 2001.
Recurrence (%) of HSV-1 at 1, 2, 5, and 10 years	9.6%, 22.9%, 40.0% and 67.0% respectively.	Farooq and Shukla. 2012; Liesegang. 2001; Shuster et al., 1981.
Global lifetime risk of developing HSK	1%	Reynaud et al., 2017.
Patients with 2–5 relapses of HSK over their lifetime vs 6–15 relapses	40% vs 11%.	Wishart et al., 1987.
Patients (%) who develop severe complications	15%	Darougar et al., 1985.
Incidence (%) of acyclovir - resistant strains among immunocompetent patients	0–0.6%	Christophers et al., 1998.
Incidence (%) of acyclovir - resistant strains among immunocompromised patients	3–6% with 14% for bone marrow transplant recipients	Christophers et al., 1998.

Source: Modified from Koganti et al., 2021.

6.2.3.1 HSV-1 Infection Process in Ocular Disease

The infection of corneal epithelial cells by HSV-1 starts with the replication of the virus in the epithelial cells of the cornea, formation of syncytia (Pertel and Spear, 1996), lysis of the syncytia and release of HSV-1 virions, which infect and form lesions in the various layers of the cornea (Toriyama et al., 2014). During active epithelial keratitis caused by HSV-1 upregulation of several toll-like receptors (TLRs) TLR-4, 7, 8, and 9 occurs (Jin et al., 2007). Using specific knock-out mice it was demonstrated that TLR-2 and TLR-4 are involved in keratitis with TLR-2 knock outs showing inhibition in keratitis lesions whereas in TLR-4 knock out mice keratitis lesions were more rapid (Sarangi et al., 2007). Antagonists of TLR-9 were effective in preventing exacerbation of disease thus confirming their involvement in keratitis. Studies have confirmed that TLR-9 has a significant role with respect to the clearance of HSV-1 and promotes inflammation leading to corneal destruction (Sarangi et al., 2007; Takeda et al., 2011).

HSV-1 is also trophic for dendritic cells (DCs) in the basement membrane of the cornea (Albers et al., 1989) where HSV-1 exists in a latent state (Mott et al., 2008). Conversely when the activation of DCs was prevented or hindered it elevated viral titres (Bryant-Hudson and Carr, 2012). Innate immune system in epithelial cells also activates antiviral type 1 interferons (Egan et al., 2013) for the clearance of HSV-1 (Cheng et al., 2000). Following replication of HSV-1 macrophages, neutrophils are activated (Kodukula et al., 1999) and these activated cells secrete interferons and cytokines (Kodukula et al., 1999). Interferons and cytokines along with interferon-γ and granzymes A and B secreted by activated natural killer (NK) cells induce apoptosis of HSV-infected epithelial cells (Grubor-Bauk et al., 2008). These immune cells activated at various time points of HSV-1 infection (Liu et al., 1996) protect against HSV-1 infection, reduce the severity of virus-induced epithelial lesions, and prevent viral encephalitis (Sciammas et al., 1997). It has also been observed that CD8+ T cells inhibit reactivation but do not affect the latently infected neurons (Liu et al., 2000). Furthermore, during HSV-1 infection, the B cells are involved in the antigen presentation and also secrete cytokines (Deshpande et al., 2000). These inflammatory reactions damage the cornea by secreting diverse cytokines including interferons (IFN-α, IFN-β, and IFN-γ), interleukins (IL-1α, IL-1β, IL-6, IL-8, IL-12, and IL-17), tumor growth factor (TGF) and tumor necrosis factor (TNF-α) (Cathcart et al., 2011; Remeijer et al., 2004). HSV-1 during the earlier infection and latency ensures suppression of the immune response (Burgess and Mohr, 2018; Matundan and Ghiasi, 2019). But, in later stages of HSV-1 infection the immune response is robust thus leading to inflammation of the tissue and occurrence of ocular infections such as blepharitis, keratitis, and keratoconjunctivitis (Lobo et al., 2019).

6.2.3.2 HSV-1 Pathobiology

The major cause for the transmission of the virus is attributed to contact between a seronegative individual and a seropositive patient actively shedding the virus (Whitley et al., 1998). The virus attaches to the host surfaces and travels to the trigeminal ganglia (TG) (Whitley et al., 1998), the latent site of infection where the virus persists. Psychological stress, fatigue, immunosuppression, and UV exposure in the host (Jones 2003; Perng and Jones, 2010) reactivates HSV-1 in its latent state following, which the activated virus travels through TG and ophthalmic nerves and reaches the eye and causes chronic inflammation and corneal damage (Koganti et al., 2019) (Figure 6.1).

6.2.3.3 Herpes Simplex Epithelial Keratitis and Therapy

Epithelial keratitis is the most predominant type of HSV keratitis comprising about 60% of ocular HSV keratitis cases (Kaufman, 2000; Lobo et al., 2019). It is mild, self-limiting and resolves in two to three weeks (Labib and Chigbu, 2022). Epithelial keratitis first appears as granular spots, which progress to form punctate lesions, and eventually forms dendritic lesions (Azher et al., 2017) with branches and terminal bulbs containing live viruses (Lobo et al., 2019; Shoji et al., 2016; Kaufman, 2000). The dendritic lesions coalesce and form an ulcer with flat edges that can change conformation as infection progresses (Liesegang et al., 1989; Lobo et al., 2019). Immunofluorescence assays or PCR are used to confirm that HSV-1 is the cause of the

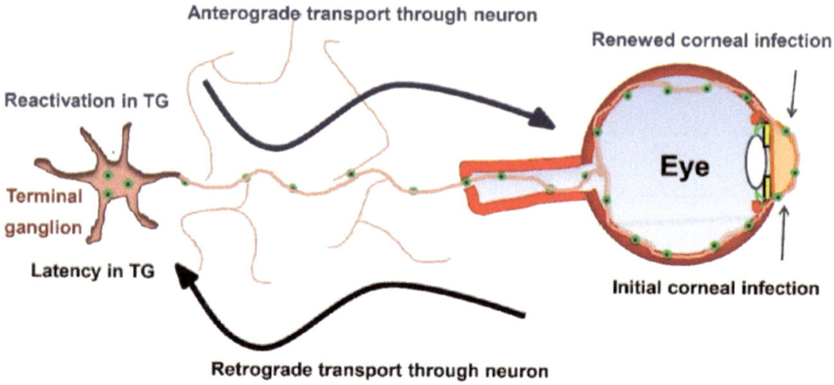

FIGURE 6.1 HSV-1 life cycle during ocular infection. The virus attaches to the host surfaces and travels to the trigeminal ganglia the latent site of infection where the virus persists still it gets reactivated after which the virus travels through trigeminal ganglia and ophthalmic nerves and reaches the eye and causes chronic inflammation and corneal damage. (Compiled by the author.)

symptoms observed (El-Aal et al., 2006). Diagnosis of epithelial keratitis is through slit lamp bio-microscopy and confocal microscopy (Kobayashi et al., 2015). The typical, dendritic lesion with end bulbs could be visualised using Lissamine green and rose Bengal staining (Azher et al., 2017).

Therapy

Idoxuridine (Stoxil), is used for the treatment of epithelial herpes keratitis (Kaufman, 1962), which is supplanted with trifluridine (Viroptic) (Kaufman and Heidelberger, 1964). Initially the treatment involves six to eight times of topical trifluridine a day tapering the treatment as the ulcer heals. The other drug used is the potent nucleoside analog acyclovir (ACV) (Elion et al., 1977; Wilhelmus, 2015; Koganti et al., 2019; Hovding, 1989), but the disadvantage is that ACV as an ointment blurs vision. In a recent study, encapsulated acyclovir was used for sustained release. Valacyclovir, is recommended for children and adults with arthritis since these individuals have difficulty in using eye drops (Kaufman, 2000). Viral resistance to acyclovir is a possibility with topical use (Hung et al., 1984). Debridement in conjunction with antiviral use have also been used to treat and prevent recurrences (Wilhelmus, 2015; Thygeson, 1976; Liesegang, 1989). Amniotic membranes in conjunction with antivirals, have also been used (Cheng et al., 2017).

6.2.3.4 HSV Stromal Keratitis

Keratitis by HSV-1 could specifically infect the stroma to cause stromal keratitis, which is encountered in $\geq$ 20–48% of ocular HSV keratitis cases (Remeijer et al., 2004). In HSV-1 stromal keratitis patients the vision loss is due to scarring and neovascularization of the cornea (Lobo et al., 2019). Stromal keratitis is attributed to the toxic effects of the virus and the host's immunological response and increased risk of recurrences leads to risk of vision loss (Lobo et al., 2019). Often stromal HSK is accompanied by stromal edema and stromal opacities in the anterior chamber and in the acute phase haze may become obvious. When stromal keratitis presents with inflammation accompanied with corneal vascularization, it is referred to as interstitial keratitis. Neovascularisation ranges from a single frond of vessels in one quadrant of cornea to all quadrants of the cornea. In such cases, angiography guided fine needle diathermy of afferent feeder vessels (Spiteri et al., 2015) and anti-VEGF agent bevacizumab treatment have been used (Koenig et al., 2012). These vessels can leak lipid, resulting in significant scarring. Some patients with HSV infection develop endothelitis (corneal stromal oedema without stromal infiltration), which could lead to endothelial cell loss with subsequent corneal oedema (Vannas et al., 1983). Patients with HSV keratitis may also develop Iridocyclitis (inflammation of the iris and ciliary body) and trabeculitis (due to acute intraocular pressure rise and secondary glaucoma) (Kaufman et al., 1971; Ritterband and Friedberg, 1998). Stromal HSK is of two types, disciform and necrotizing stromal keratitis (Shoji et al., 2016).

6.2.3.4.1 HSV Disciform Stromal Keratitis

Disciform stromal keratitis occurs as circular, focal stromal edema along with keratic precipitates and Descemet membrane folds (Lobo et al., 2019; Shoji et al., 2016). Disciform keratitis is a cell-mediated immune reaction (Kaufman, 2000) and the observed edema is like the corneal edema caused due to degeneration of the endothelium as in Fuchs' endothelial dystrophy or that observed due to damage of the endothelium after cataract surgery or the swelling that occurs after ocular trauma (Kaufman, 2000).

Therapy for Disciform Stromal Keratitis Corticosteroids and trifluridine drops are given concurrently to reverse the edema and inflammation (Kaufman et al., 1963). Typically, disciform keratitis resolves following corticosteroid-antiviral combination within 7 to 10 days. Recurrence of edema occurs in up to 20% of patients and in such patients the corticosteroid-antiviral therapy is resumed. Viral resistance to the above drugs in disciform keratitis has not been reported.

6.2.3.4.2 HSV Necrotizing Stromal Keratitis

Necrotizing stromal keratitis is a rare manifestation of HSK, which primarily occurs in patients who have had recurrent attacks of herpetic keratitis. The virus invades the stroma and may or may not cause ulceration (Labib and Chigbu, 2022; Ritterband and Friedberg, 1998); but the virus causes dense white infiltrates and edema of the stroma (Heiligenhaus et al., 2003). Furthermore, the stromal tissue could develop an abscess-like opacity and exhibit eponymous necrosis of the stromal tissue (Toriyama et al., 2014). In necrotizing stromal keratitis, there is an increased risk of stromal melting and epithelial ulcers may also be present (Lobo et al., 2019). This condition is probably due to invading virus and cell-mediated immune disease and considered the most severe and prolonged type of keratitis. The severe inflammation may lead to rapid corneal thinning and subsequently perforation.

Therapy for necrotizing stromal keratitis Oral acyclovir is not effective against necrotising stromal keratitis (Herpetic Eye Disease Study Group, 1996; 1997), but oral ACV prevents recurrences (Herpetic Eye Disease Study Group, 1998) and long-term use is recommended (Moyes et al., 1994). When stromal herpes results in scarring of cornea, penetrating keratoplasty is performed (Kaufman, 2000). Viral resistance to ACV in necrotising stromal keratitis has not been reported.

6.2.3.4.3 Diagnosis of Stromal HSV Keratitis

Clinical examination, multimodal imaging utilizing fluorescein and indocyanine green, enzyme-linked immunosorbent assay (ELISA) and viral culture of the virus are used for diagnosis of stromal HSV keratitis. The timing of detection and accuracy of identification of the virus are critical to initiate treatment and prevent the possibility of vision loss (Azher et al., 2017).

In the context of stromal HSV keratitis the 'The Herpetic Eye Disease Study' (HEDS), involving multiple institutes (Louisiana State University Medical Center in New Orleans, the Cullen Eye Institute of Baylor College of Medicine, and the Francis I. Proctor Foundation of the University of California in San Francisco and many private physicians) the outcomes of treatment of patients with herpetic eye diseases is significant and the salient findings include:

(a) oral acyclovir is not effective against herpetic stromal disease (Barron et al., 1994),
(b) corticosteroids are beneficial in the treatment of stromal disease (Wilhelmus et al., 1994),
(c) oral acyclovir prevents recurrences in patients with established herpes (Herpetic Eye Disease Study Group, 1998),
(d) oral acyclovir also reduces recurrences of other viruses like orofacial herpes (Moyes et al., 1994; HEDS, 1998; Karsten et al., 2012).

HEDS apart from providing guidance on the efficacy of topical antivirals for both epithelial and stromal herpes keratitis also confirm that a combination therapy involving antivirals and corticosteroids would help to overcome symptoms of stromal herpes keratitis (Kaufman, 2000).

6.2.3.5 HSV Meta-herpetic keratitis

Metaherpetic keratitis is a severe form of HSV keratitis caused by recurrent attacks of HSV, superficial bullae, stromal involvement, endotheliitis, and anterior uveitis caused by HSV. Marked neovascularization is observed in metaherpetic keratitis. The damage observed is attributed to both the immune and inflammatory mechanisms following HSV-1 infection of the cornea (Liesegang, 1999; Wilhelmus et al., 1987). Epithelial detachment of the corneal epithelium and ulceration could also occur in the absence of viral multiplication (Kaufman, 1964). Diagnosis is based on patient's history, microscopic observation of deep ulcers, inflammation of the stroma, peripheral corneal neovascularization and absence of growth of bacteria or fungi from the corneal scrapings from the affected individuals.

Therapy for Meta-herpetic Keratitis Topical steroid therapy is beneficial when combined with protective anti-viral coverage (Schwab, 1988). Loteprednol etabonate, valacyclovir or acyclovir were used for treatment (Goldblum et al., 2008; Perry and Faulds, 1996). Fibronectin eye drops from patients own blood plasma is effective in treating non-healing corneal epithelial defects after HSV-1 keratitis (Nishida et al., 1985). Early onset of systemic antiviral therapy in conjunction with topical steroids, MMPs inhibitors, vitamin C, fibronectin eye drops and preservative free lubricants are recommended for treatment. A lubricating ointment or a soft contact lens are recommended to avoid friction from the eye lids (Kaufman, 2000). Viral resistance to the above drugs in meta-herpetic keratitis has not been reported.

6.2.3.6 Recurrence of Herpetic disease

Recurrences of ocular herpetic disease is more severe causing corneal scarring (Shuster et al., 1981) and encompasses a variety of clinical manifestations including infectious epithelial keratitis, neurotrophic keratopathy, immune stromal keratitis, necrotising stromal keratitis and endotheliitis and induces acellular immunological response (Ritterband and Friedberg, 1998). Sunlight, trauma, emotional stress, menstruation, and other infectious diseases have been found to activate a recurrence of herpetic disease. Ocular HSV recurrence rates were 36% at 5 years and 63% at 20 years after a primary episode. After a second episode between 70% to 80% had another recurrence within 10 years (Liesegang et al., 1989). Low-dose oral acyclovir reduces recurrences of ocular herpes by almost half (Group HEDS, 1998). This low-dose oral acyclovir treatment is advisable since the stromal disease could impair vision.

6.2.3.7 HSV Endothelial Keratitis

Patients with endothelial keratitis or Endotheliitis present with inflammation of the endothelium and may show stromal opacity caused by endothelial dysfunction of endothelium rather than stromal inflammation. Endotheliitis is characterised by keratic precipitates, corneal edema, and anterior chamber inflammation and classified as linear, sectoral, disciform, and diffuse depending on the arrangement of keratic precipitates (Alfawaz, 2013). HSK endotheliitis can be distinguished from anterior uveitis since patients present with iritis and edema of the stroma (Carrillo-Arroyo et al., 2012).

Therapy for Endothelial keratitis Antiviral and anti-inflammatory treatment are effective for endothelial keratitis but a significant decrease in endothelial cell density (10.3% per year) compared with healthy fellow eyes was observed (Hillenaar et al., 2009). Viral resistance to the drugs used in endothelial keratitis has not been reported.

6.2.3.8 HSV Neurotrophic Keratitis

About 27% of all neurotrophic keratitis are caused by Herpes infections (Bonini et al., 2000). In neurotrophic keratitis (NK) HSV damages the nerves, which innervate the cornea. Patients do not complain of pain but present with dry eyes and photophobia (Sacchetti and Lambiase, 2014). Individual with NK may also present with superficial punctate keratopathy and edema of the cornea. In the second stage, oval or circular epithelial defects with edema of the stroma are common (Versura et al., 2018). Changes in the third stage include corneal ulcers, scarring, or perforations of the eye (Sacchetti and Lambiase, 2014; Versura et al., 2018).

Therapy for neurotrophic keratitis Pharmacological treatments for NK are not available. But preservative-free artificial tears and topical steroids have been used to control ocular inflammation in NK but with caution (Bonini et al., 2003; Petroutsos et al., 1982). In the event of stromal melting, topical collagenase inhibitors, such as N-acetylcysteine, and systemic administration of tetracycline or medroxyprogesterone may be considered (Hossain, 2012; Perry and Golub, 1985). Topical antibiotic eye drops have also been used to prevent infection in eyes with NK. Additionally corneal or scleral contact lenses could be used to promote corneal epithelial healing (Grey et al., 2012). Surgical treatments such as partial or total tarsorrhaphy, botulinum A toxin injection of the eyelid elevator muscle (Khokhar et al., 2005), amniotic membrane transplantation (AMT) for ocular surface reconstruction, TachoSil (Nycomed Pharma, Zurich, Switzerland) an absorbable fibrin sealant patch, application of a soft bandage contact lens or AMT and conjunctival flap or lamellar/penetrating keratoplasty (Golubović and Parunović, 1990) are some of the methods used to overcome NK. Viral resistance to the drugs used in neurotrophic keratitis has not been reported.

6.2.3.9 New Drug Therapies for Treatment of HSV Keratitis

The drugs in use to treat HSV keratitis are restricted in their efficacy and in a few cases antiviral resistance have been reported (Labib and Chigbu, 2022). For instance, the drugs already in use do not kill the virus in its latent form, and do not treat neurotrophic keratitis. It is in this context that several new drugs have been tried and include protelivir and amenamevir (helicase-primase inhibitors, which inhibit viral DNA synthesis), SC93305 and BX795 (inhibitors of viral replication) (Zinser et al., 2018), BX795 (a TANK-binding kinase 1 inhibitor, which blocks HSV-1 in infected cells) (Jaishankar et al., 2018), aganirsen (an antisense oligonucleotide) (Cursiefen, et al., 2014), CRISPR-Cas9 (effective in blocking viral replication) (Van Diemen et al., 2016), a peptide-acyclovir combination (Bősze, et al., 2020), lipid mediator resolvin D1 and microRNAs particularly Mirl 55, (Rajasagi et al., 2017), 6-thioguanine (6-TG), which was effective in acyclovir resistant strains (Chen et al., 2021) and Harringtonine, a natural alkaloid (Liu et al., 2021), which was effective against HSV-1 strains resistant to acyclovir (Liu et al., 2021). In 2021, Zhang et al., found C3 botulinum toxin substrate 1 could be used against HSV-1 (Zhang et al., 2021). Several proteins have also been used against HSV keratitis and include VEGF, an Src kinase inhibitor molecule and Robo4, also involved in VEGF signaling (Giménez et al., 2015).

Recently, to overcome variations in the quality and consistency of care for HSV keratitis, a range of clinical- and technology-based innovations were proposed by Chodosh and Ung (2020). Emphasis was to provide an immediate link between the anatomical localization of disease (corneal epithelial, stromal, or endothelial) and the appropriate treatment, and use electronic medical record system for evidence-based treatment algorithms. Population-wide disease registries would facilitate identification of implementation-targets for quality improvement, education, and research. These innovations would reduce the human and economic burdens of this highly morbid, and often blinding disease.

6.3 Cytomegalovirus (CMV)

6.3.1 Incidence and Prevalence of CMV

CMV is a Herpes virus that causes congenital infections worldwide (Manicklal et al., 2013). CMV from an infected mother is transmitted to the child about 30–40% of the times (Carlson et al., 2010) across the world and babies born CMV positive ranged from 0.3–2.4% in developed countries whereas in the United States it is reported to be 1% (Bonalumi et al., 2011). CMV is also the most common virus detected in intrauterine infection and it is fatal, causing approximately 20–30% deaths mainly attributed to vascular coagulation, liver dysfunction, and bacterial infections (Pass et al., 2006). In addition, 5% to 15% of CMV positive live infants are associated with several manifestations including hearing loss, delayed psychomotor development, and visual impairment (Bonalumi et al., 2011) and a good proportion (40%) develop deafness, optic atrophy, microcephaly, encephalitis, and mental retardation (Ahlfors et al., 1986; Carmichael, 2012). These infections can also manifest as pneumonia, mononucleosis, and meningitis (Beam et al., 2018; Rafailidis et al., 2007).

6.3.2 CMV Pathobiology

CMV in infected individuals may exist in the blood in the latent state and transmission occurs through saliva, breast-feeding, blood transfusion and sexual intercourse, (Alfawaz. 2013; Hayes et al., 1972). In certain countries transmission is very severe and seroprevalence was observed up to 100% (Cannon et al., 2011). CMV is latent in CD34+ myeloid progenitor cells and in the pulmonary, corneal, retinal, and salivary tissues (Reeves and Sinclair, 2008). Asymptomatic viremia, CMV syndrome and tissue-invasive disease was observed in immunocompromised and immunosuppressed hosts due to type 1 immunity (Reeves and Sinclair, 2008).

6.3.3 Characteristics of CMV Keratitis

6.3.3.1 CMV Endotheliitis

CMV is an emerging etiological agent of corneal endotheliitis (Oka et al., 2015) and the other ocular pathologies include anterior uveitis, iris atrophy with iritis and retinitis (Inoue, et al., 1998; Koizumi et al., 2006; Wilhelmus, et al., 1996). Studies have indicated that idiopathic corneal endotheliitis were positive for CMV DNA in 24.1% of cases with corneal edema and the copy number of CMV ranged from 6.3×10^4 to 3.6×10^6 per mL implying a propensity of CMV towards corneal endotheliitis compared to epithelial or stromal keratitis. CMV endotheliitis in Japan was more frequent (80.2%) in middle-aged men (mean age 66.9 ± 10.9 years) (Koizumi et al., 2015). Both CMV and HSK endotheliitis are similar and show coin-shaped keratic precipitates, corneal edema, anterior chamber inflammation (Alfawaz. 2013), increased intraocular pressure, loss of cells from the endothelium, and anterior uveitis (Chee et al., 2008). CMV endotheliitis based on the distribution of keratic precipates have been categorised into linear, sectoral, disciform, and diffuse. In linear CMV endolthelitis, keratic precipitates are arranged in a linear fashion and edema is visible whereas in sectoral type, keratic precipitates are spread throughout the swollen cornea (Alfawaz, 2013; Koizumi et al., 2008). The edema and keratic precipitates are visible in the disciform variant but the edema is circular in the central or paracentral cornea (Alfawaz, 2013). In the diffuse endotheliitis, as the term indicates edema and small keratic precipitates are distributed across the cornea (Alfawaz. 2013). At times the keratic precipitates are pigmented (Tan and Tan. 2019; Wang, et al., 2010). Confocal microscopy of CMV endotheliitis revealed a unique shape the "owl's eye", which appeared as a reflective region in the nucleus surrounded by a ring of low reflection (Shiraishi et al., 2007).

6.3.3.2 CMV Keratitis

CMV is involved in epithelial and stromal keratitis and endotheliitis (Alfawaz, 2013). and the virus is trophic to endothelial, smooth muscle and myeloid cells. Keratitis, due to CMV, normally occurs post keratoplasty of immunocompetent patients (Kandori et al., 2010; Tan and Tan. 2019; Wang, et al., 2010). In one case, CMV infected keratocytes and lead to the formation of a corneal button that had to be excised. Endotheliitis due to CMV can occur post laser in-situ keratomileusis (LASIK) surgery (Tan et al., 2016).

Therapy and Resistance To protect the cornea from permanent damage from CMV early intervention is a must. In patients with AIDS, oral and topical antiviral treatments were not always effective since the patients developed opacity, branching and epitheliopathy without ulcers (Wilhelmus, et al., 1996) indicative of resistance. Later stromal keratouveitis could also develop in these patients (Wilhelmus, et al., 1996; Inoue et al., 1998). Resistance to drugs has been reported for CMV retinitis. Resistance to ganciclovir results from mutations in the CMV genome at the UL97 locus (Chou et al., 1995a; 1995b) and in the polymerase gene (Sullivan et al., 1993). Foscarnet, a pyrophosphate analogue, inhibits CMV replication by acting directly on viral DNA polymerase. Resistance to foscarnet results from mutation in the polymerase gene (Sullivan and Coen, 1991). Thus, CMV exhibits resistance against antiviral drugs including gangciclovir and foscarnet.

6.4 Varicella-zoster Virus

6.4.1 Incidence and Prevalence of VZV

Varicella-zoster virus (VZV) is an herpes virus and also known as human herpesvirus 3 (HHV-3) or Human alpha-herpesvirus *3*. VZV is dormant in the sensory nerve ganglia after primary infection and causes chicken pox. The virus after causing primary infection lies dormant in neurons and several years after post-infection from chickenpox infection, VZV gets reactivated and causes shingles (Nagel and Gilden, 2007; Kaufman, 2000). Approximately 4 million cases of VZV occur each year in childhood and about 100 children succumb annually (Wharton, 1996).

6.4.2 Management of VZV

VZV cases following vaccination decreased substantially (~90%) from 1995 to 2000 among children (Seward et al., 2002) and by 2010 VZV were 97% lower than in 1995. Despite the live attenuated vaccine reducing the incidence of VZV infections, the virus after a latency period gets reactivated and causes diseases such as chickenpox, shingles and sometimes stroke (Yawn and Gilden. 2013). VZV infection also manifests as keratitis (Khodabande, 2009), which could be overcome using antiviral and corticosteroid therapy (Khodabande, 2009). The mean incidence of VZV was estimated as 4–4.5 per 1,000 person-years (Yawn and Gilden. 2013) and the incidence increased from 3.2 in 1996–1997 to 4.1 in 2000–2001 perv 1000 individuals (Yawn and Gilden. 2013). A good proportion of immunocompromised individuals were effected (< 10%) and about 3% required hospitalization (Yawn and Gilden. 2013).

6.4.3 Pathobiology of VZV

VZV transmission occurs due to direct contact with infectious lesions and through respiratory aerosol or droplet inhalation (Gershon and Gershon, 2013). The upper respiratory epithelium is commonly infected and symptoms are obvious after 10–21 days (Zerboni et al., 2014). Many VZV proteins (IE62, IE63, ORF47, and ORF66) inhibit the host immune system and thus promote infection (Zerboni et al., 2014). VZV is latent in TG and divides after reaching the corneal tissues (Laing et al., 2018). VZV also possesses tropism for T cells (in tonsils) and dendritic cells (Zerboni et al., 2014).

6.4.4 VZV Keratitis

6.4.4.1 Incidence and Prevalence of VZV Keratitis

Reactivation of VZV in the ophthalmic branch of the trigeminal nerve causes Herpes zoster ophthalmicus (HZO) (Liesegang, 1985; 2008; Kong et al., 2020; Thompson et al., 2021; Miyakoshi et al., 2012). HZO constitutes 25% of HZV infections (Shaikh and Ta, 2002). Keratitis is a common presentation of HZO and includes stromal (6.6%), epithelial (5.0%) and endothelial (1.2%) keratitis (Szeto et al., 2017). In the United States 200,000 new cases of HZO are reported each year. HZO can progress into chronic or recurrent disease and the rates of recurrence increases from 8.0% in the first year to 25.0% after 5 years (Tran et al., 2016). In the early stages of infection HZK presents as dendriform epithelial keratitis and later HZK causes an inflammatory immune response, leading to stromal keratitis or endotheliitis and finally result in neurotrophic keratitis (NK) (Cohen et al., 2015).

6.4.4.2 Characteristics of VZV Keratitis

VZV can affect any layer of the cornea and causes keratitis in 6 to 10% of cases of HZO and the patients experience loss of vision due to corneal scarring or haze following epithelial and/or stromal VZV keratitis (Li, 2018; Cabrera-Aguas et al., 2022a). In HZO, 50–72% patients show ocular involvement (Pavan-Langston, 1995) due to denervation of the eye, reduced corneal sensation and decrease in tear pro-duction (Kaufman, 2000). A fraction of these HZO patients with ocular involvement develop keratitis (Shaikh and Ta, 2002) and symptoms include pain, excessive production of tears, and a foreign-body sensation (Miyakoshi et al., 2012). When VZV infects the corneal epithelium, it manifests as pseudo-dendrites or punctate keratitis (Kaufman, 2000). Corneal lesions are grey, elevated, pleomorphic (Hu et al., 2010a) and form linear mucous plaques on the corneal surface. Inflammation is mild and corneal ulcers are visible with anterior corneal stromal infiltration with opacity in the infected eye (Miyakoshi et al., 2012) without any change in intraocular pressure. In the stroma, VZV keratitis manifests as disciform distributions (Silverstein et al., 1997) with coin-shaped keratitis in the superficial stroma and as sclero-keratitis in patients with VZV DNA (Denier et al., 2020). In children also VZV keratitis manifests as coin-shaped keratitis and some develop deep diffuse keratitis in the stroma (Denier et al., 2020). VZV keratitis relapse occurs in majority of the children with a median follow-up time of 31 months (Denier et al., 2020). Dry eye signs and symptoms appear as sequelae of HZV. VZV may affect the eyelids. Eye ointment with lubrication properties need to be instilled to restore normal function of the lids. VZV may invade the iris, and cause segmental iris atrophy (Kaufman, 2000).

6.4.4.3 Clinical Diagnosis of VZV Keratitis

VZV can cause either stromal keratitis (VZV-SK) or endotheliitis (VZV-E). VZV-SK typically occurs after about 10 days and is characterized by stromal inflammation and can lead to neovascularization and scarring. VZV-SK manifests as anterior stromal keratitis and is often accompanied by nummular infiltrates in the stroma just below the epithelium—these infiltrates are an immune-mediated response to viral antigens (Liesegaang, 1985). In the absence of treatment, stromal inflammation may result in neovascularization and/or corneal scarring. Deep stromal keratitis can manifest several months later (Liesegaang, 1985). VZV-E presents clinically with corneal edema, keratic precipitates, and mild anterior chamber inflammation, and can lead to significant loss of endothelial cells (Reijo et al., 1983). Often, it coexists with uveitis (Moshirfar, 2019). Chronologically, Descemet folds often appear between 4 days. Inflammation also can lead to chronic edema, endothelial decompensation, and an associated increase in intraocular pressure. VZV-SK and VZV-E are typically preceded by development of punctate epitheliopathy and formation of pseudodendrites within 10 days (Liesegaang, 1985).

6.4.4.4 Therapy for VZV Keratitis

Chronic corticosteroid therapy is usually required. Iritis and the corticosteroid when used for therapy can cause cataracts, secondary glaucoma, and prolonged morbidity and decreased vision (Kaufman, 2000). Therapy includes oral antivirals (Valtrex, valacyclovir, famvir and famciclovir) within 72 h of the infection (Degreef et al., 1994). The antivirals famciclovir and acyclovir reduce ocular recurrences if treatment is initiated within 72 h (Degreef et al., 1994). Antiviral agents alone or in combination with topical corticosteroids were also effective for treating pseudodendritic keratitis (Hu et al., 2010b). In cases of zoster infection synechiae with chronic inflammation may be overcome by dilation of the pupil and anti-glaucoma therapy to decrease the elevated intraocular pressure(Kaufman, 2000). Viral resistance to the antivirals used in VZV keratitis has not been reported.

6.5 Epstein-Barr virus

6.5.1 Incidence and Prevalence

Epstein-Barr virus (EBV) or human herpesvirus 4, is found all over the world and persists as a lifelong infection. Children (3-4 years) and adolescents are seropositive in developing countries and in developed countries respectively (de-Thé et al., 1975; Hjalgrim et al., 2007). The global prevalence of EBV is ≥90% (Smatti et al., 2018) and increased seropositivity appears to be both age and education dependent in Taiwan (Chen, et al., 2015). In tropical countries sexually active women and smokers were associated with EBV infections (Balfour et al., 2013). A vaccine for the prevention of EBV is not available (Cohen, 2018) even though EBV causes upper respiratory infections in children and mononucleosis in young adults (Nowalk and Green, 2016). Unlike many other viruses EBV is associated with certain lymphomas and carcinomas. It has also been reported that patients undergoing organ transplants are prone to the virus (Nowalk and Green, 2016). Very rarely, EBV has been implicated in retinitis and keratitis (Sato et al., 2018).

6.5.2 Pathobiology of Epstein-Barr virus

Transmission of EBV is via., contact with saliva (Nowalk and Green, 2016) and due to intimate contact with individuals seropositive for EBV and are actively shedding the virus. EBV causes infectious mononucleosis, which persists asymptomatically all through the lifetime in adults (Smatti et al., 2018). EBV is recognised as a high-risk factor for Hodgkin's lymphoma and certain carcinomas (Jarrett, 2003; Sathiyamoorthy et al., 2016). EBV can infect many cell types including B-cells, T-cells, epithelial cells, and myocytes, and the virus is known to remain latent in B-cells (Sathiyamoorthy et al., 2016). EBV transforms B-cells causing different types of lymphomas like Burkitt and Hodgkin's lymphomas (Küppers. 2003). Normally EBV does not cause symptomatic pathologies but it is associated with disorders related to lymphoproliferation in transplant/immunocompromised patients (Niller et al., 2008). B-cell receptor (BCR) activation is a stimulus for reactivation of EBV post latency (Lv et al., 2017). The EBV lifecycle is regulated by proteins involved in RNA processing changes induced upon BCR activation (Lv et al., 2017), psychological stress, and weakening host immunity (Kerr, 2019). Luteolin a flavonoid inhibits EBV reactivation by inhibiting the activity of immediate-early gene promoters (Wu et al., 2016).

6.5.3 Characteristics of Epstein-Barr Virus Keratitis

EBV keratitis are normally independent of mononucleosis (Tanner, 1954) but a few reports do indicate that patients with infectious mononucleosis exhibit nummular keratitis or interstitial keratitis (Pinnolis et al., 1980). EBV is reported in young children (9 years old) and in young adults (Wong et al., 1987). EBV keratitis manifests as dendritic lesions of the epithelium (Matoba, and Jones, 1987; Tanner, 1954) or as deeper infiltrates in the stroma. Patients with EBV stromal keratitis complain of blurred vision, photophobia, corneal irritation, and multifocal or ring-shaped granular anterior stromal opacities and at times soft, pleomorphic infiltrates in the peripheral cornea with little or complete absence of neovascularization (Matoba, et al., 1986). EBV keratitis when severe could cause necrosis and neovascularization of the cornea (Park, et al., 2014). EBV infection is also associated with autoimmune or malignant disease in immune-suppressed patients and in patients with prolonged EBV primary infection, thus necessitating early diagnosis and treatment (Niller and Bauer, 2017). Studies have indicated that pathways like the PI3K/AKT and ERK pathways (Park et al., 2014), and activation of TRIF/TRAF or RIP-1 signalling (Park et al., 2015) are involved in EBV infections. Chronic EBV disease can result in anterior uveitis and vitritis as well as macular edema and optic nerve swelling (Wong et al., 1987).

6.5.4 Clinical Diagnosis of EBV Keratitis

Individuals infected with EBV show mononucleosis. Matoba et al. (1986) observed that patients with clinical and serological features suggestive of EBV stromal keratitis had granular anterior stromal opacities, which were discrete, demarcated, multifocal, pleomorphic or ring-shaped. Further, some patients had soft, blotchy, multifocal infiltrates, predominantly all through the peripheral cornea, and also showed minimal neovascularization. Chodosh, (1994) showed that EBV stromal keratitis was associated with three principal forms, which included:

- Type I: multifocal subepithelial infiltrates like adenoviral keratitis,
- Type II: multifocal, blotchy, pleomorphic infiltrates with active inflammation or granular ring-shaped opacities (inactive form) in anterior to mid-stroma, and
- Type III: multifocal deep or full thickness peripheral infiltrates, with or without vascularization.

6.5.5 Therapy for EBV Keratitis

EBV keratitis resolves rapidly with the introduction of topical corticosteroids and in many cases, inflammation is resolved without antiviral therapy. Thus, immunological processes were involved rather than viral replication (Matoba et al., 1986). Viral resistance to the drugs used in EBV keratitis has not been reported.

6.6 Adenovirus Keratitis

6.6.1 Incidence and Prevalence of Adenovirus

Adenovirus causes a spectrum of disease including systemic (such as a severe pharyngitis or gastroenteritis) and ocular disease like keratoconjunctivitis, epidemic keratoconjunctivitis (EKC), pharyngoconjunctival fever (PCF) and a focal epithelial keratitis. Several millions are infected with viral conjunctivitis each year (Kaneko et al., 2011), with a substantial percentage having corneal involvement in adults (20 to 40, years of age) and with PCF, in children (Jhanji et al., 2015). There is no sex predilection (Fujimoto et al., 2012) and the most prone are patients in close contact with hospitals (ophthalmic units), schools, nursing homes, and workplaces (Kaneko et al., 2011).

6.6.2 Pathobiology of Adenovirus

Adenoviral transmission is primarily through contact with respiratory droplets or ocular secretions (Harnett and Newnham, 1981). Adenoviral keratitis is painful and vision is compromised (Omari and Main, 2018). In epithelial keratitis the cornea may be swollen due to iritis accompanied with severe pain and develop subepithelial transient infiltrates that appear as semi-transparent snowballs in the stroma of the cornea, which reduces vision. When the subepithelial infiltrates persist up to 3 months to years, it causes discomfort and vision is compromised. Approximately 25% of cases of adenoviral infections occur only in one eye (Kaufman, 2000). In EKC the virus replicates on the conjunctival epithelial surface, spreads into the corneal epithelium and underlying stroma and may spread to the other eye. Adenovirus triggers an innate immune cellular response (involving neutrophils, macrophages/ monocytes, and natural killer cells) that destroys the infected corneal cells or alternatively produce inflammatory cytokines that damage the infected cells (Liu and Muruve, 2003). When the response is delayed adenoviral infection damages the corneal epithelium/stroma via. Th1 helper cells, which activate macrophage-mediated phagocytosis via the release of IFN-gamma (Liu and Muruve, 2003). Thus, immunomodulators may play a positive role in treatment.

6.6.3 Characteristics of Adenoviral Keratitis

Adenoviruses are nonenveloped, double-stranded DNA and Adenoviruses 8, 19, and 37 are implicated in epidemic keratoconjunctivitis (EKC), a very contagious disease (Mader et al., 1992). Adenovirus spreads from physician to other patients, between health care professionals and in ophthalmic units through contaminated instruments like tonometer and also by finger-to eye transmission by healthcare workers (Mader et al., 1992). Thus, adequate hand washing and cleansing of ophthalmic instruments is important to control the spread (Ritterband and Friedberg, 1998). EKC effects the cornea but the corneal involvement occurs 3–4 days after infection and it is usually mild and transient. EKC manifests as diffuse, fine epithelial keratitis that could be stained with fluorescein and persists for 2–3 weeks. EKC develops a central ulceration with irregular borders and grey-white dots, which persist for 1 to 2 weeks. About two weeks after the onset, subepithelial infiltrates appear, which remain for weeks to years. These infiltrates resolve spontaneously. In rare cases, disciform keratitis or anterior uveitis can occur (González-López et al., 2013). Superinfection by *Streptococcus pyogenes* could occur, especially in paediatric patients and could lead to amblyopia (Kuo et al., 2008).

6.6.4 Clinical Features of Adenoviral Keratitis

Patients with adenoviral infections complain of redness in one eye, tearing, irritation, and photophobia and the contralateral eye may get infected 1–2 days later, but is usually less severe (Bowling, 2016). Symptoms of adenoviral infection varies depending on the disease caused.

In EKC, patients manifest with similar symptoms, but photophobia is usually stronger and, additionally, the patients experience foreign body sensation with keratitis occurring in 80% of the patients (Bowling, 2016). EKC may cause severe corneal pathology with three full layer corneal epithelial detachments (Lee et al., 2015).

In contrast to EKC, in chronic keratoconjunctivitis, symblepharon or pseudomembrane formation may occur. In addition, opacification of the cornea, decrease in visual acuity, distortions of the corneal surface, astigmatism and higher refractive errors are common symptoms (Aydin Kurna et al., 2015). In chronic adenoviral keratitis, patients complain of photophobia and in addition decreased vision, halo, and glare, due to the recurrent corneal damage (Aydin Kurna et al., 2015). EKC typically starts with epithelial microcysts and may progress to focal punctate epithelial keratopathy (Bialasiewicz, 2017). In more severe cases, subepithelial/anterior stromal infiltrates and grey epithelial infiltrates develop after around 2–3 weeks. Opacities observed are a result of viral replication, and due to inflammatory response to the virus (Mader et al., 1992).

In pharyngoconjunctival fever (PCF) patients suffer from fever, pharyngitis, acute follicular conjunctivitis and regional lymphoid hyperplasia, experience photophobia and keratitis occurs only 30% of the time (Bowling, 2016).

6.6.5 Diagnosis of Adenoviral Keratitis

One definitive method for diagnosing adenoviral keratitis is to culture the virus. This is sensitive and specific with nearly 100% sensitivity (Bowling, 2016). But this straightforward approach is not routinely followed since it is expensive, requires a specific media for transport, the method is slow and takes weeks to obtain cultures. Many a times the corneal infiltrates may have resolved even before the culture results are available (Mader et al., 1992). More recent alternatives like immunoassays with high sensitivity and specificity are promising (Jhanji et al., 2015). Immunoassay detection of the virus is cheap, simple, rapid and can detect 53 adenovirus serotypes in a short period of time (<10 minutes) (Jhanji et al., 2015). PCR is a promising alternative to detect the virus and is more sensitive than immunoassay (Kuo et al., 2016; Elnifro et al., 1999). But PCR is time consuming and requires dedicated facilities. Sequencing techniques and serology could also be used for adenoviral infection. In serology adenovirus-IgM or increasing IgG are detected (Bowling, 2016). But the serology method is not preferred since antibodies in the serum may have returned to normal level. Giemsa stain of a conjunctival specimen can also be useful (Weiss, 1994) and the specimen will contain more mononuclear cells (Bowling, 2016) as compared to multinucleate giant cells (also known as Tzanck cells) in herpetic keratitis. This technique due to reasons of cost and a longer time is undertaken when the clinical picture is similar to a herpetic infection.

6.6.6 Management of Ocular Adenovirus Infections

Hygienic practices are of primary importance to minimize adenovirus infections and these include:

(a) judicious washing of hands, avoiding touching and rubbing of eyes, avoiding sharing of towels,
(b) avoidance of contact with hospital staff, patients, optical instruments (tonometer, eye-drop dispensers, and optical aids like lenses) (Bowling, 2016),
(c) patients with EKC should be treated separately (Meyer-Rusenberg et al., 2011),
(d) disinfection should be carried out only with virucidal agents such as povidone-iodine and sodium hypochlorite and 70% isopropyl alcohol swabs (Meyer-Rusenberg et al., 2011),
(e) care providers should wear gloves and disinfect their hands, the equipment, and surfaces after examining the patient (Meyer-Rusenberg et al., 2011) and
(f) follow-up of patients is avoided unless necessary after the acute phase (Meyer-Rusenberg et al., 2011).

6.6.7 Therapy for Ocular Adenovirus Infections

Therapy is not required in the acute phase of mild disease but corticosteroid treatment with either fluorometholone or other steroid drops are given in case of pain. Treatment may also include warm or cold compresses, lubrication, topical vasoconstrictors, cycloplegic drops and artificial tears (Ritterband and Friedberg, 1998; Omari and Mian, 2018). Some standard therapies that are used are listed below.

6.6.7.1 Anti-inflammatory Agents

Anti-inflammatory agents, including topical steroids, are used but their use is limited due to ocular and systemic complications. Ocular complications include elevated intraocular pressure and cataract. In a clinical trial, Kovalyuk et al. (2017), used a combination of topical povidone-iodine and dexamethasone to treat corneal infiltrates and punctate keratitis and concomitantly reduction in the redness, discharge, and pseudomembranes in the patients was observed (Kovalyuk et al., 2017). In this combination povidone-iodine kills the extracellular adenoviral particles (Jhanji et al., 2015) and the steroid helps to overcome the associated inflammatory adverse events. At times subepithelial infiltrates are resistant to topical steroids (Hillenkamp et al., 2001; 2002) and in such cases cyclopsorine, a calcinuerin inhibitor is effective (Pflugfelder and de Paiva, 2017). A retrospective study indicated that topical cyclosporine or topical steroids were more effective and caused less severe disease than when treated with artificial tears (Asena et al., 2017).

6.6.7.2 Antivirals

Ganciclovir, an antiviral nucleoside analog of guanine (Kinchington et al., 2005), trifluridine and vidarabine are completely or partially ineffective in treating ocular adenoviral diseases (Jhanji et al., 2015; Kinchington et al., 2005) both *in vitro* and in clinical trials (Little et al., 1968). But Cidofovir, a cytidine analog and an inhibitor of adenoviral DNA polymerase (Kinchington et al., 2002) is a potent antiviral drug *in vitro* and in animal models (Kinchington et al., 2005; Gordon et al., 1992) and showed significant reduction in the number of subepithelial infiltrates of adenoviral particles (Kinchington et al., 2005). But topical cidofovir was associated with lacrimal canalicular obstruction (Kinchington et al., 2005) and caused resistance to the drug (Kinchington et al., 2005). Trifluridine (Carmine et al., 1982) and cidofovir (Gordon et al., 1992) inhibit adeno-virus infection, but this result has not yet been replicated in humans. Another antiviral that has been tried is Ribavirin, an inhibitor of inosine monophosphate dehydrogenase, which could be used for treating adenoviral infections (Kinchington et al., 2005). But Ribavirin did not qualify for treatment of adenoviral ocular diseases since it exhibits limited activity against Adenovirus serotypes 8, 19, and 37, which commonly cause keratitis (Kinchington et al., 2002; 2005). Additionally, Ribavirin does not clear adenovirus (Mann et al., 1998). Another drug that inhibited ocular adenoviral infection is 2, 3-dideoxycytidine (ddC) (Romanowski et al., 2009). But supporting preclinical and clinical trials are needed before the use of this drug for therapy.

6.6.7.3 Immunomodulators

Reports indicate that some interferons (IFN), like IFNb when used topically prevented the formation of corneal infiltrates (Romano et al., 1981; Wilhelmus et al., 1987). Further, IFNg is effective in conjunctivitis (Mistchenko and Falcoff 1987), but not for treating keratitis. IFNa has no apparent therapeutic effects (Sundmacher et al., 1982). Omari and Mian (2018) suggested that interferons may have some use for prophylaxis of corneal infiltrates (Rossa and Sundmacher 1991). Resistance to some of the drugs used for adenoviral keratitis has been studied.

6.6.7.4 Surgery

Phototherapeutic keratectomy (PTK) with concurrent mitomycin C is a popular surgical option for adenoviral keratitis. The surgery reduces corneal opacities, and improves visual acuity and vision (Yamazaki et al., 2011). In some patients, excimer laser ablation have been used to remove chronic infiltrates and restore a clear visual axis. Otherwise, surgery is mostly reserved for persistent membranes or symblepharon formation. Occasionally following adenoviral infection mydriasis enlargement of pupils even in bright light is required (Omari and Main, 2018).

6.7 Resistance to Acyclovir

Among the various drugs used for treatment of keratitis caused by different viruses the widely used is acyclovir (ACV), which is not effective in the case of disciform and necrotising keratitis but effective against epithelial keratitis, stromal disease, meta herpetic keratitis and VZV keratitis. ACV is also known to reduce or inhibit ocular recurrences of the disease. In the later cases the chances of resistance to acyclovir does exist. The prevalence of acyclovir-resistance strains of HSV in immunocompetent patients is approximately 0.1–3% (Piret and Boivin, 2011; Pottage and Kessler, 1995). Development of recurrent resistance HSK has been attributed to long-term prophylactic usage of ACV (Ogura et al., 2017; Toriyama et al., 2014). Piret and Boivin (2011) confirmed acyclovir-resistance using a plaque reduction assay with an EC50 value > 9 μM and qPCR was also used to confirm the presence of the virus (Virók et al., 2017). Normally when resistance is developed to ACV other nucleoside analogs like Foscarnet could be used. Foscarnet is effective against acute retinal necrosis caused by ACV-resistant strain of HSV (Ogura et al., 2017). Additionally, topical trifluoro-thymidine is effective in treating necrotizing keratitis due to ACV-resistant strains (Toriyama et al., 2014). Other analogs to ACV include topical or intravenous administrations of cidofoir or imiquimoid (Martinez et al., 2006; Ronkainen et al., 2018). Acyclovir and trifluridine are not recommended on a regular basis because they are nephrotoxic and cause ocular disorders (Carmine et al., 1982; Yildiz et al., 2013).

Glucocorticoids like prednisolone, dexamethasone and loteprednol are also prescribed for the treatment of keratitis. Interestingly these glucocorticoids do not target the virus but nevertheless they are used because they reduce inflammation and pain (Austin et al., 2017). Glucocorticoids may also cause blurred vision and occasionally foreign body sensation in the eye. Thus, there is a need for developing novel drugs to overcome viral keratitis.

6.7.1 Other Effects of Acyclovir

In 1994, the Herpetic Eye Disease Study (HEDS, 1998) demonstrated that corticosteroid therapy treatment reduced stromal kerato-uveitis by 68% (Wilhelmus et al., 1994) and reduced stromal keratitis symptoms from 72 to 26 days (Austin et al., 2017). It was also observed in the HEDS-Epithelial Keratitis trial oral acyclovir (400 mg 5 times per day for three weeks) did not confer additional benefits compared to topical trifluridine in preventing HSK (Barron et al., 1994) but a yearlong daily ACV administration reduced the relapse rates for ocular and orofacial HSV diseases (Wilhelmus et al., 1998). Thus, prophylactic use of acyclovir and corticosteroids became the norm for HSK treatment.

6.8 Other Ocular Viral Infections

HHV-7, Coxsackievirus, Rhabdovirus, (Inoue et al., 2010; Jhanji, et al., 2015; Lim et al., 2014; Madhavan et al., 2002), human T-lymphotropic virus-1 (HTLV-1), mumps, Rubeola (measles), vaccinia, and variola viruses are known to cause keratitis

(Bhaskaram et al., 1986; Shibata et al., 2016) (Figure 6.2). These viruses vary in their effects with respect to the extent to which they cause infectious keratitis as judged by precipitates, opacity, and lesions.

Rubeola virus cause measles in children but it also causes severe photophobia, conjunctivitis, and some iritis in the Western world. In contrast in Africa, Rubeola causes dense corneal opacities and measles, which could contribute to blindness (Kaufman, 2000). Preventive vaccination is available for measles but there is no post-infection therapy.

Rubella is also called German measles and is accompanied by follicular conjunctivitis, which is mild and usually appears 2 to 3 days before the rash and lasts only a few days. Congenital rubella is caused by maternal infection and is associated with the classic triad of congenital birth defects, deafness, and cataracts. Edema of the cornea is prominent and it is cloudy centrally at birth. Maternal vaccination could prevent Rubella, but the disease does not respond to chemotherapy after the occurrence of infection (Ratnam et al., 1996).

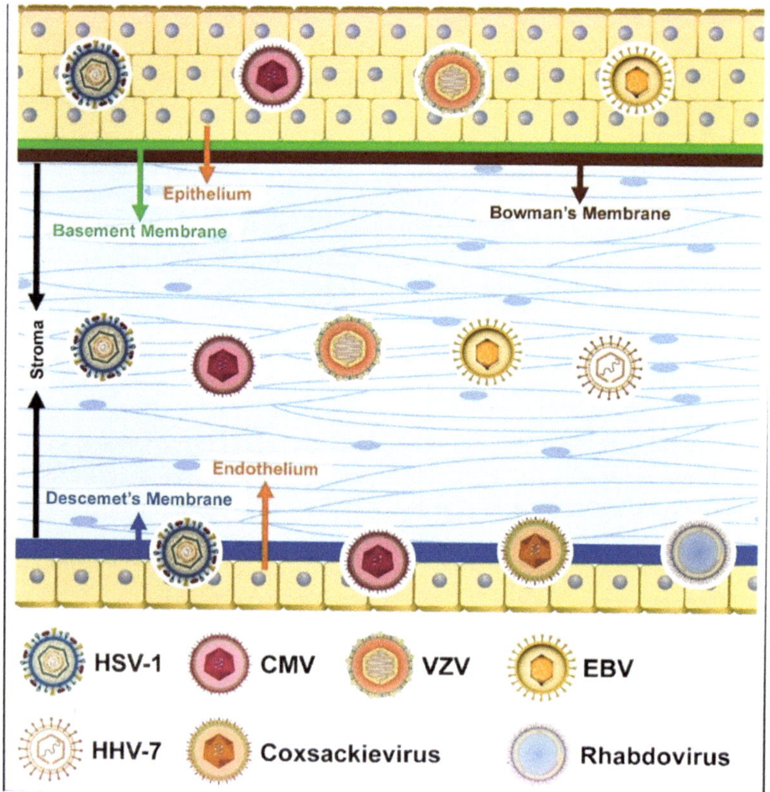

FIGURE 6.2 Viral keratitis caused by several different viruses infecting different layers of the cornea. Figure based on data from Gueudry et al., 2013; Holland and Schwartz. 1999; Inoue, et al., 2010; Lim et al., 2014; Madhavan et al., 2002; Matoba, et al., 1986; Oka et al., 2015. (Modified from Koganti et al., 2021.)

Mumps are associated with ocular symptoms and patients present with acute conjunctivitis with mucoid discharge, diffuse episcleritis and, rarely, sclero-keratitis. Iritis and interstitial keratitis may occur. Keratitis resolves in 2 to 3 weeks with no treatment (Kaufman, 2000). Only some of these above viruses cause keratitis and reports on whether these viruses are resistant to drugs when used is not available.

6.9 Clinical Diagnosis

Clinical features that would help in diagnosis has been described under each virus and the respective diseases. This section only summarises briefly the salient clinical diagnostic features of viral keratitis. Visualisation of dendritic lesions is a classic symptom of HSK (Azher et al., 2017; Darougar et al., 1985) and coin-shaped keratic precipitaes (KPs) is characteristic of CMV endotheliitis or VZV/EBV stromal keratitis (Alfawaz, 2013). Other tests such as the Snellen Acuity Test, the penlight exam and fluorescein staining, which assesses damage to the corneal epithelium are routine tests in the diagnosis of keratitis (Wolffsohn et al., 2017).

Subsequently, a positive PCR, during the onset of infection or during reactivation, for the viral genomes from the anterior chamber could confirm the diagnosis (Reinhard, 2009; Alfawaz, 2013; Kandori et al., 2010; 2022b; Jd et al., 2006). Detection and quantification of viral antibodies by enzyme linked immunosorbent assay (ELISA), (Alfawaz, 2013; Wang, et al., 2016) and the Goldmann-Witmer coefficient (GWC) of antibody production (GWC > 3 is considered positive) (Ys et al., 2010) are also important techniques to confirm the diagnosis (Westeneng et al., 2007). But the results should be interpreted with caution since healthy eyes are also positive for IgG antibodies of some viruses like EBV. Presence of EBV IgM antibodies confirm primary EBV keratitis (Reinhard, 2009). In one study of infectious keratitis, only 43% of infected specimens were positive with both PCR and GWC (Jd et al., 2006). In contrast, immunoblotting, GWC, and PCR had a sensitivity of 81% when used as individual tests but increased to 97% when all three tests were used together (Fekkar et al., 2008). PCR of corneal scrapings can also be used to confirm the initial diagnosis (Austin et al., 2017; El-Aal et al., 2006; Inoue, et al., 2010; Kandori et al., 2010).

6.10 Treatment of Viral Keratitis

Treatment modalities have already been discussed with respect to each type of viral keratitis and the use of acyclovir for treatment in the previous sections. Therefore, this part will only mention the other treatment therapies used.

6.10.1 Use of Aptamers for Treatment of Herpes Simplex Virus Keratitis

Aptamers are short oligonucleotide sequences, which could be used for anti-viral activity (Zhu et al., 2015). For instance, RNA aptamers, which target the viral glycoprotein D (gD) have been used to inhibit the entry of the virus in *in vitro* (Gopinath et al., 2012). Synthetic DNA aptamers have been used to inhibit HSV-1 in corneal models (Yadavalli et al., 2017). But the efficacy of aptamers in keratitis models has yet to be demonstrated (Yadavalli et al., 2017). Humanized monoclonal antibodies have also not been tested for their efficacy in keratitis models (Bauer et al., 2017; Krawczyk et al., 2013).

6.10.2 Use of Cationic Peptides for Treatment of Herpes Simplex Virus Keratitis

Positively charged linear cationic peptides are the biggest class of antimicrobial peptides (AMP) (Vishnepolsky and Pirtskhalava, 2014). Such cationic peptides are effective in treating ocular HSV-1 infections since these peptides bind to the cell surface of HSV and interfere with viral attachment. These peptides demonstrated positive outcomes using *in vitro, ex vivo*, and *in vivo* corneal models (Jaishankar et al., 2016; Park et al., 2013; Tiwari et al., 2011). Cationic peptides, like retrocyclin-2 and TAT-Cd, reduced the severity of symptoms in a keratitis murine model (Jose et al., 2013). However, their efficacy deteriorates when applied 24 hours post-infection or longer (Jose et al., 2013). Thus, their role in treating patients with developed HSK is questionable.

6.10.3 CRISPR/Cas9 Technology for Viral Keratitis

CRISPR/Cas9 technology for the removal of latent viral genomes from the host, could be used for preventing recurring keratitis episodes (Koujah et al., 2019a). In 2014, CRISPR/Cas9 was used to remove latent EBV genomes from patient-derived cells (van Diemen et al., 2016). In addition, human Herpes virus (HHV) encode microRNAs (miRNA), which facilitate host immune evasion and anti-viral defence. This could be overcome using CRISPR/Cas9 to mutate miRNAs. *In vitro* studies using a latently infected gastric carcinoma cell line NSU-719 have highlighted promising potential of CRISPR/Cas9-mediated editing of EBV genome by targeting miRNAs BART5, BART6, and BART15. Further development of this technology could lead to similar results in murine models and eventually humans. Thus, CRISPR/Cas9-mediated editing of EBV genome would reduce the likelihood of recurrence (Koganti et al., 2021)

6.10.4 Use of PDK1 Inhibitors of Viral Protein Synthesis

BX795 a viral translation inhibitor has been utilized (Clark et al., 2009) to inhibit viral protein synthesis (Jaishankar et al., 2018; Yadavalli et al., 2020). This drug is effective when administered either therapeutically or prophylactically and could be extremely useful in patients with resistance to acyclovir or its derivatives (Koganti et al., 2021).

6.10.5 Surgical Interventions for HSK

Necrotizing HSK or neurotrophic herpes keratitis (Tuli et al., 2018) require surgical interventions like corneal gluing to seal perforations in the cornea (< 3 mm) using a cyanoacrylate-based adhesive (Rana and Savant, 2013). Further, amniotic membrane transplants (AMTs) are used for healing neurotrophic ulcers (Chen et al., 2000) and for overcoming the thinning of the cornea by increasing corneal thickness (Nubile et al., 2011). Corneal transplantation or keratoplasty to manage cases of herpes keratitis or corneal scarring are surgical interventions to improve visual acuity in keratitis patients (Serna-Ojeda et al., 2017; Wang, et al., 2011). Keratoplasty is quite often rejected (Lyall et al., 2012). Vascularization of the cornea is common in HSK and increases the chances of rejection and recurrent infections in the transplanted cornea (Kuffova et al., 2016). Oral acyclovir is often given after keratoplasty to reduce recurrent infections (Kanclerz and Alio, 2020), but acyclovir usage for lamellar keratoplasties is less clear (Kanclerz and Alio. 2020). Anterior lamellar transplants have better outcomes than penetrating keratoplasty operations, but they are inferior when patients have deep scarring of the cornea or endothelial involvement (Tuli et al., 2018). One of the major complications of HSK is that it could damage the corneal nerves causing neurotrophic keratopathy and patients present with a decreased corneal sensation, blink reflex and tear production (Tsatsos et al., 2016). Further, the corneal epithelium breakdown should be treated early otherwise it could cause corneal scarring, thinning, vascularisation, perforation, or secondary corneal infection (Tsatsos et al., 2016). Several treatments like ocular lubricants, bandage contact lens, tarsorrhaphy, botulinum toxin-induced ptosis, growth factors and autologous plasma and so forth could stimulate epithelial growth and prevent further disruption of the ocular surface (Tsatsos et al., 2016; Cabrera-Aguas et al., 2022a; 2022b). Corneal mucous plaques or delayed pseudo-dendrites may also occur months or year later (Cabrera-Aguas et al. 2022a; 2022b).

6.11 Ocular Infection of Molluscum Contagiosum Virus

Molluscum contagiosum virus effects the eyelids, conjunctiva and rarely the cornea (Charles and Friedberg, 1992). Punctate keratopathy, subepithelial infiltrates and a corneal pannus resembling trachoma may be seen. The disease is self-limiting and managed by surgical treatments including resection, curettage, cryotherapy and carbon dioxide laser (Ritterband and Friedberg, 1998).

6.12 Conclusions

Several viruses like HSV, CMV, VZV, and EBV are etiological agents of infectious keratitis. The most studied are HSV and CMV keratitis. These viral etiological agents of keratitis have clinical manifestations unique to each virus. Treatment modalities vary depending on the virus involved, and viral resistance to therapy is limited and has been reported only for acyclovir in epithelial keratitis. In contrast viral resistance to any of the drugs used in necrotising stromal keratitis, disciform keratitis and meta-herpetic keratitis has not been reported. This indeed is a positive sign for therapy implicating that the above drugs could still be used. But it is also possible that studies related to susceptibility and resistance are yet to be conducted with respect to many of the ocular diseases and the concerned drugs. The challenges in treatment of ocular viral diseases are manifold and includes:

(a) need to develop rapid and accurate methods for diagnosis and therapy,
(b) devise optimum treatment protocols to avoid recurrence of HSV infection of the anterior segment,
(c) need to understand the molecular basis of drug resistance and
(d) continue research in identifying new antiviral drugs.

REFERENCES

Ahlfors K, Ivarsson SA, Bjerre I. Microcephaly and congenital cytomegalovirus infection: a combined prospective and retrospective study of a Swedish infant population. *Pediatrics* 1986;78; 1058–1063. [PubMed: 3024098]

Albers I, Kirchner H, Domke-Opitz I. Resistance of human blood monocytes to infection with herpes simplex virus. *Virology* 1989;169; 466–469. [PubMed: 2539701]

Alfawaz A. Cytomegalovirus-related corneal endotheliitis: a review article. *Saudi J Ophthalmol.* 2013;27; 47–49.

Asena L, Singar Ozdemir E, Burcu A, et al. Comparison of clinical outcome with different treatment regimens in acute adenoviral keratoconjunctivitis. *Eye (Lond.)* 2017; 31: 781–787.

Austin A, Lietman T, Rose-Nussbaumer J. Update on the management of infectious keratitis. *Ophthalmology* 2017;124: 1678–1689.

Aydin Kurna S, Altun A, Oflaz A, Karatay Arsan A. Evaluation of the impact of persistent subepithelial corneal infiltrations on the visual performance and corneal optical quality after epidemic keratoconjunctivitis. *Acta Ophthalmol.* 2015; 93: 377–382.

Azher TN, Yin X, Tajfirouz D, Huang AJ, Stuart PM. Herpes simplex keratitis: challenges in diagnosis and clinical management. *Clin Ophthalmol.* 2017;11; 185–191.

Balfour HH, Odumade OA, Schmeling DO, Mullan BD, Ed JA, Knight JA, Vezina HE, Thomas W, Hogquist KA. Behavioral, virologic, and immunologic factors associated with acquisition and severity of primary Epstein-Barr virus infection in university students. *J Infect Dis.* 2013;207; 80–88.

Baringer JR, Swoveland P. Recovery of herpes-simplex virus from human trigeminal ganglions. *N Engl J Med.* 1973;288: 648±650.

Barron BA, Gee L, Hauck WW, Kurinij N, Dawson CR, Jones DB, Wilhelmus KR, Kaufman HE, Sugar J, Hyndiuk RA. Herpetic eye disease study. A controlled trial of oral acyclovir for herpes simplex stromal keratitis. *Ophthalmology* 1994;101; 1871–1882. [PubMed: 7997323]

Bauer D, Alt M, Dirks M, Buch A, Heilingloh CS, Dittmer U, Giebel B, Görgens A, Palapys V, Kasper M, Eis-Hübinger AM, Sodeik B, Heiligenhaus A, Roggendorf M, Krawczyk A. A therapeutic antiviral antibody inhibits the anterograde directed neuron-to-cell spread of herpes simplex virus and protects against ocular disease. *Front Microbiol.* 2017;8; 2115. doi: 10.3389/fmicb.2017.02115. PMID: 29163407; PMCID: PMC5671610.

Beam E, Germer JJ, Lahr B, Yao JDC, Limper AH, Binnicker MJ, Razonable RR. Cytomegalovirus (CMV) DNA quantification in bronchoalveolar lavage fluid of immunocompromised patients with CMV pneumonia. *Clin Transplant.* 2018; 32(1). doi: 10.1111/ctr.13149. PMID: 29112278.

Bhaskaram P, Mathur R, Rao V, Madhusudan J, Radhakrishna KV, Raghuramulu N, Reddy V. Pathogenesis of corneal lesions in measles. *Hum Nutr Clin Nutr.* 1986;40; 197–204. [PubMed: 3721904]

Bialasiewicz A. Adenoviral keratoconjunctivitis. *Sultan Qaboos Univ Med J.* 2007; 7: 15–23.

Bonalumi S, Trapanese A, Santamaria A, D'Emidio L, Mobili L. Cytomegalovirus infection in pregnancy: review of the literature. *J Pren Med.* 2011;5: 1–8. [PubMed: 22439067]

Bonini S, Lambiase A, Rama P, Caprioglio G, Aloe L. Topical treatment with nerve growth factor for neurotrophic keratitis. *Ophthalmology* 2000;107: 1347–1352.

Bonini S, Rama P, Olzi D, Lambiase A. Neurotrophic keratitis. *Eye (Lond.)* 2003;17: 989–995.

Bősze S, Zsila F, Biri-Kovács B, Szeder, B, Majer Z, Hudecz F, Uray K. Tailoring uptake efficacy of HSV-1 gD derived carrier peptides. *Biomolecules* 2020;10: 721.

Bowling B. Kanski's Clinical Ophthalmology: A Systematic Approach. 8th ed. London, UK: Elsevier; 2016.

Bryant-Hudson KM, Carr DJJ. PD-L1-expressing dendritic cells contribute to viral resistance during acute HSV-1 infection. *Clin Dev Immunol.* 2012; 924619.

Burgess HM, Mohr I. Defining the role of stress granules in innate imsune Suppression by the herpes simplex virus 1 endoribonuclease VHS. *J Virol.* 2018;92:e00829-18. doi: 10.1128/JVI.00829-18. PMID: 29793959; PMCID: PMC6052315.

Cabrera-Aguas M, Khoo P, McCluskey P, et al. Viral Ocular Infections, in: Rezaei N (Ed.), Encyclopedia of Infection and Immunity. Oxford: Elsevier; 2022a. pp. 219–233.

Cabrera-Aguas M, Khoo P, Watson SL. Infectious keratitis: a review. *Clin Exp Ophthalmol.* 2022b July;50(5): 543–562. doi: 10.1111/ceo.14113. Epub 2022 Jun 3. PMID: 35610943; PMCID: PMC9542356.

Cannon MJ, Hyde TB, Schmid DS. Review of cytomegalovirus shedding in bodily fluids and relevance to congenital cytomegalovirus infection. *Rev Med Virol.* 2011;21; 240–255.

Cao J, Yang Y, Yang W, Wu R, Xiao X, Yuan J, Xing Y, Tan X. Prevalence of infectious keratitis in Central China. *BMC Ophthalmol.* 2014;14: 43. [PubMed: 24690368]

Carlson A, Norwitz ER, Stiller RJ. Cytomegalovirus infection in pregnancy: should all women be screened? *Rev Obstet Gynecol.* 2010 Fall;3(4):172–179. PMID: 21364849; PMCID: PMC3046747.

Carmichael A. Cytomegalovirus and the eye. *Eye (Lond.)* 2014;26: 237–240.

Carmine AA, Brogden RN, Heel RC, Speight TM, Avery GS. Trifluridine: a review of its antiviral activity and therapeutic use in the topical treatment of viral eye infections. *Drugs* 1982;23; 329–353.

Carrillo-Arroyo I, Gutiérrez-Díaz E, Mencía-Gutiérrez E, Gómez-Pérez P, Montero-Rodríguez M. Herpetic endotheliitis and trabeculitis with delayed corneal involvement. *Arch Soc Esp Oftalmol.* 2012;87; 47–49.

Cathcart HM, Zheng M, Covar JJ, Liu Y, Podolsky R, Atherton SS. Interferon-gamma, macrophages, and virus spread after HSV-1 injection. *Invest Ophthalmol Vis Sci.* 2011;52, 3984–3993.

Charles N, Friedberg D. Epibulbar molluscum contagiosum in acquired immune deficiency syndrome. *Ophthalmology* 1992;99: 1123–1126.

Chee S, Bacsal K, Jap A, Se-Thoe S, Cheng CL, Tan BH. Clinical features of cytomegalovirus anterior uveitis in immunocompetent patients. *Am J Ophthalmol.* 2008;145; 834–840. [PubMed: 18255045]

Chen C, Huang KA, Shen J, Tsao K, Huang Y. A large-scale seroprevalence of Epstein-Barr virus in Taiwan. *PLoS One* 2015;10; e0115836.

Chen D, Liu Y, Zhang F, You Q, Ma W, Wu J, Wu Z. 6-Thioguanine inhibits herpes simplex virus 1 infection of eyes. *Microbiol Spectr.* 2021;9: e0064621.

Chen HJ, Pires RT, Tseng SC. Amniotic membrane transplantation for severe neurotrophic corneal ulcers. *Br J Ophthalmol.* 2000;84: 826–833.

Cheng AMS, Tseng SCG. Self-retained amniotic membrane combined with antiviral therapy for herpetic epithelial keratitis. *Cornea* 2017;36: 1383–1386.
Cheng H, Tumpey TM, Staats HF, van Rooijen N, Oakes JE, Lausch RN. Role of macrophages in restricting herpes simplex virus type 1 growth after ocular infection. *Invest Ophthalmol Vis Sci.* 2000;41: 1402–1409. [PubMed: 10798656]
Chodosh J. Epstein-Barr virus stromal keratitis. *Ophthalmol Clin North Am.* 1994;7: 549–556.
Chodosh J, Ung L. Adoption of Innovation in Herpes Simplex Virus Keratitis Cornea _ Supplement to Volume 39, Number 11, November 2020, S7–S18.
Chou S, Erice A, Jordan M et al. Analysis of the UL97 phosphotransferase coding sequence in clinical cytomegalovirus isolates and identification of mutations conferring ganciclovir resistance. *J Infect Dis.* 1995a;171: 576–583.
Chou S, Guentzel S, Michaels K, Miner R, Drew W. Frequency of UL97 phosphotransferase mutations related to ganciclovir resistance in clinical cytomegalovirus isolates. *J Infect Dis.* 1995b;172: 239–242.
Christophers J, Clayton J, Craske J, Ward R, Collins P, Trowbridge M, Darby G. Survey of resistance of herpes simplex virus to acyclovir in northwest England. *Antimicrob Agents Chemother.* 1998;42: 868–872.
Clark K, Plater L, Peggie M, Cohen P. Use of the pharmacological inhibitor BX795 to study the regulation and physiological roles of TBK1 and IκB kinase. *J Biol Chem.* 2009;284: 14136–14146.
Cohen EJ. Management and prevention of herpes zoster ocular disease. *Cornea* 2015;34(Suppl. 10): S3–8.
Cohen JI. Vaccine development for Epstein-Barr virus. *Adv Exp Med Biol.* 2018;1045: 477–493.
Cursiefen C, Viaud E, Bock F, Geudelin B, Ferry A, Kadlecová P, Lévy M, al Mahmood S, Colin S, Thorin E et al. Aganirsen antisense oligonucleotide eye drops inhibit keratitis-induced corneal neovascularization and reduce need for transplantation: the I-CAN study. *Ophthalmology* 2014;121: 1683–1692.
Darougar S, Wishart MS, Viswalingam ND. Epidemiological and clinical features of primary herpes simplex virus ocular infection. *Br J Ophthalmol.* 1985;69: 2–6.
Degreef H, the Famciclovir Herpes Zoster Clinical Study Group. Famciclovir, a new oral antiherpes drug: results of the first controlled clinical study demonstrat-ing its eficacy and safety in the treatment of uncomplicated herpes zoster in immunocompetent patients. *Int J Antimicrob Agents* 1994;4: 241±246.
Denier M, Gabison E, Sahyoun M, Labetoulle M, Dureau P, Cochereau I, Doan S. Stromal keratitis After varicella in children. *Cornea* 2020;39: 680–684.
Deshpande SP, Kumaraguru U, Rouse BT. Dual role of B cells in mediating innate and acquired immunity to herpes simplex virus infections. *Cell Immunol.* 2000;202: 79–87. [PubMed: 10896767]
De Groot-Mijnes JD, Rothova A, Van Loon AM, Schuller M, Ten Dam-Van Loon NH, De Boer JH, Schuurman R, Weersink AJ. 2006. Polymerase chain reaction and Goldmann-Witmer coefficient analysis are complimentary for the diagnosis of infectious uveitis. *Am J Ophthalmol.* 2006;141: 313–318. [PubMed: 16458686]
de-Thé G, Day NE, Geser A, Lavoué MF, Ho JH, Simons MJ, Sohier R, Tukei P, Vonka V, Zavadova H. Sero-epidemiology of the Epstein-Barr virus: preliminary analysis of an international study – a review. *IARC Sci Publ.* 1975;1971 (11 Pt 2):3–16. PMID: 191375. 3–16.
Egan KP, Wu S, Wigdahl B, Jennings SR. Immunological control of herpes simplex virus infections. *J Neurovirol.* 2013;19: 328–345. [PubMed: 23943467]

El-Aal AMA, El Sayed M, Mohammed E, Ahmed M, Fathy M. Evaluation of herpes simplex detection in corneal scrapings by three molecular methods. *Curr Microbiol.* 2006;52: 379–382.

Elion GB, Furman PA, Fyfe JA, de Miranda P, Beauchamp L, Schae€er HJ. Selectivity of action of an antiherpetic agent, 9-(2-hydroxyethoxy- methyl)guanine. *Proc Natl Acad Sci USA* 1977;74: 5716± 5720.

Elnifro EM, Cooper RJ, Klapper PE, et al. Diagnosis of viral and chlamydial keratoconjunctivitis: which laboratory test? *Br J Ophthalmol.* 1999; 83: 622–627.

Farooq AV, Shukla D. Herpes simplex epithelial and stromal keratitis: an epidemiologic update. *Surv Ophthalmol.* 2012;57: 448–462. [PubMed: 22542912]

Fekkar A, Bodaghi B, Touafek F, Le Hoang P, Mazier D, Paris L. Comparison of immunoblotting, calculation of the Goldmann-Witmer coefficient, and real-time PCR using aqueous humor samples for diagnosis of ocular toxoplasmosis. *J Clin Microbiol.* 2008;46: 1965–1967. [PubMed: 18400917]

Fujimoto T, Matsushima Y, Shimizu H, et al. A molecular epidemiologic study of human adenovirus type 8 isolates causing epidemic keratoconjunctivitis in Kawasaki City Japan in 2011. *Jpn J Infect Dis.* 2012; 65: 260–263

Gershon AA, Gershon MD. Pathogenesis and current approaches to control of Varicella-Zoster virus infections. *Clin Microbiol Rev.* 2013;26: 728–743.

Giménez F, Mulik S, Veiga-Parga T, Bhela S, Rouse BT. Robo 4 counteracts angiogenesis in herpetic stromal keratitis. *PLoS One* 2015;10; e0141925.

Goldblum D, Bachmann C, Tappeiner C, et al. Comparison of oral antiviral therapy with valacyclovir or acyclovir after penetrating keratoplasty for herpetic keratitis. *Br J Ophthalmol.* 2008;92: 1201–1205.

Golubović S, Parunović A. Cyanoacrylate glue in the treatment of corneal ulcerations. *Fortschr Ophthalmol.* 1990;87: 378–381.

González-López JJ, Morcillo-Laiz R, Muñoz-Negrete FJ. Adenoviral keratoconjunctivitis: an update. *Arch Soc Esp Oftalmol.* 2013 Mar;88(3): 108–115.

Gopinath SCB, Hayashi K, Kumar PKR. Aptamer that binds to the gD protein of herpes simplex virus 1 and efficiently inhibits viral entry. *J Virol.* 2012;86: 6732–6744.

Gordon YJ, Romanowski E, Araullo-Cruz T, De Clercq E. Pretreatment with topical 0.1% (S)-1-(3-hydroxy-2-phosphonylmethoxypropyl)cytosine inhibits adenovirus type 5 replication in the New Zealand rabbit ocular model. *Cornea* 1992; 11: 529–533.

Grey F, Carley F, Biswas S, Tromans C. Scleral contact lens management of bilateral exposure and neurotrophic keratopathy. *Cont Lens Anterior Eye.* 2012;35: 288–291.

Group HEDS. Acyclovir for the prevention of recurrent herpes simplex virus eye disease. *Herpetic Eye Disease Study Group. N Engl J Med.* 1998;339(5): 300–6.

Grubor-Bauk B, Arthur JL, Mayrhofer G. Importance of NKT cells in resistance to herpes simplex virus, fate of virus-infected neurons, and level of latency in mice. *J Virol.* 2008;82: 11073–11083. [PubMed: 18614643]

Gueudry J, Boutolleau D, Gueudin M, Burrel S, Miri A, Bodaghi B, Muraine M. Acyclovir-resistant varicella-zoster virus keratitis in an immunocompetent patient. *J Clin Virol.* 2013;58: 318–320.

Harnett GB, Newnham WA. Isolation of adenovirus type 19 from the male and female genital tracts. *Br J Vener Dis.* 1981; 57: 55–57.

Hayes K, Danks DM, Gibas H, Jack I. Cytomegalovirus in human milk. *N Engl J Med.* 1972;287: 177–178.

Heiligenhaus A, Li H, Hernandez Galindo EE, Koch JM, Steuhl K, Meller D. Management of acute ulcerative and necrotising herpes simplex and zoster keratitis with amniotic membrane transplantation. *Br J Ophthalmol.* 2003;87: 1215–1219.

Herpetic Eye Disease Study Group. A controlled trial of oral acyclovir for iridocyclitis caused by herpes sim- plex virus. *Arch Ophthalmol.* 1996;114: 1065±1072.

Herpetic Eye Disease Study Group. A controlled trial of oral acyclovir for the prevention of stromal keratitis or iritis in patients with herpes simplex virus epithelial keratitis. The epithelial keratitis trial. *Arch Ophthalmol.* 1997;115: 703±712.

Herpetic Eye Disease Study Group. Acyclovir for the prevention of recurrent herpes simplex virus eye disease. *New Engl J Med.* 1998;339: 300±306.

Hillenaar T, Weenen C, Wubbels RJ, Remeijer L. Endothelial involvement in herpes simplex virus keratitis: an in vivo confocal microscopy study. *Ophthalmology.* 2009 Nov;116(11): 2077-86.e1-2.

Hillenkamp J, Reinhard T, Ross RS, et al. The effects of cidofovir 1% with and without cyclosporin a 1% as a topical treatment of acute adenoviral keratoconjunctivitis: a controlled clinical pilot study. *Ophthalmology* 2002; 109: 845–850.

Hillenkamp J, Reinhard T, Ross RS, et al. Topical treatment of acute adenoviral keratoconjunctivitis with 0.2% cidofovir and 1% cyclosporine: a controlled clinical pilot study. *Arch Ophthalmol.* 2001; 119: 1487–1491.

Hjalgrim H, Friborg J, Melbye M. The Epidemiology of EBV and its Association with Malignant Disease, in: Arvin A, Campadelli-Fiume G, Mocarski E, Moore PS, Roizman B, Whitley R, Yamanishi K (Eds.), *Human Herpesviruses: Biology, Therapy, and Immunoprophylaxis.* Cambridge: Cambridge University Press; 2007. PMID: 21348071

Holland EJ, Mahanti RL, Belongia EA, Mizener MW, Goodman JL, Andres CW, Osterholm MT. Ocular involvement in an outbreak of herpes gladiatorum. *Am J Ophthalmol.* 1992;114: 680–684.

Holland EJ, Schwartz GS. Classification of herpes simplex virus keratitis. *Cornea* 1999; 18: 144–154.

Hossain P. The corneal melting point. *Eye (Lond)* 2012;26: 1029–1030.

Hovding G. A comparison between acyclovir and trifluorothymidine ophthalmic ointment in the treatment of epithelial dendritic keratitis: a double blind, randomized parallel group trial. *Acta Ophthalmol (Copenh)* 1989;67: 51±54.

Hu AYH, Strauss EC, Holland GN, Chan MF, Yu F, Margolis TP. Late varicella-zoster virus dendriform keratitis in patients with histories of herpes zoster ophthalmicus. *Am J Ophthalmol.* 2010a;149: 214–220.e3.

Hung SO, Patterson A, Rees PJ. Pharmacokinetics of oral acyclovir (Zovirax) in the eye. *Br J Ophthalmol.* 1984;68: 192±195.

Inoue T, Hayashi K, Omoto T, Kosaki R, Higaki S, Shimomura Y. Corneal infiltration and CMV retinitis in a patient with AIDS. *Cornea* 1998;17: 441–442.

Inoue T, Kandori M, Takamatsu F, Hori Y, Maeda N. Corneal endotheliitis with quantitative polymerase chain reaction positive for human herpesvirus 7. *Arch Ophthalmol.* 2010;128: 502–503.

Jaishankar D, Buhrman JS, Valyi-Nagy T, Gemeinhart RA, Shukla D. Extended release of an anti–heparan sulfate peptide from a contact lens suppresses corneal herpes simplex virus-1 infection. *Invest Ophthalmol Vis Sci.* 2016;57: 169–180.

Jaishankar D, Yakoub AM, Yadavalli T, Agelidis A, Thakkar N, Hadigal S, Ames J, Shukla D. An off-target effect of BX795 blocks herpes simplex virus type 1 infection of the eye. *Sci Transl Med.* 2018;10(10): eaan5861.

Jarrett RF. Risk factors for Hodgkin's lymphoma by EBV status and significance of detection of EBV genomes in serum of patients with EBV-associated Hodgkin's lymphoma. *Leuk Lymphoma* 2003;44 Suppl 3: 27.

Jhanji V, Chan TCY, Li EYM, Agarwal K, Vajpayee Rasik B. Adenoviral keratoconjunctivitis. *Surv Ophthalmol.* 2015;60: 435–443.

Jin X, Qin Q, Chen W, Qu J. Expression of toll-like receptors in the healthy and herpes simplex virus-infected cornea. *Cornea* 2007;26: 847–852.

Jones C. Herpes simplex virus type 1 and bovine herpesvirus 1 latency. *Clin Microbiol Rev.* 2003;16: 79–95.

Jose GG, Larsen IV, Gauger J, Carballo E, Stern R, Brummel R, Brandt CR. A cationic peptide, TAT-Cd°, inhibits herpes simplex virus type 1 ocular infection in vivo. *Invest Ophthalmol Vis Sci.* 2013;54: 1070–1079.

Kanclerz P, Alio JL. Ocular surgery after herpes simplex and herpes zoster keratitis. *Int Ophthalmol.* 2020;40: 3599–3612.

Kandori M, Inoue T, Takamatsu F, Kojima Y, Hori Y, Maeda N, Tano Y. Prevalence and features of keratitis with quantitative polymerase chain reaction positive for cytomegalovirus. *Ophthalmology* 2010;117: 216–222.

Kaneko H, Suzutani T, Aoki K, et al. Epidemiological and virological features of epidemic keratoconjunctivitis due to new human adenovirus type 54 in Japan. *Br J Ophthalmol.* 2011; 95: 32–36

Karsten E, Watson SL, Foster LJR. Diversity of microbial species implicated in keratitis: a review. *Open Ophthalmol J.* 2012; 6: 110–124.

Kaufman HE. Clinical cure of herpes simplex keratitis by 5-iodo-2'-deoxyuridine. *Proc Soc Exp Biol Med.* 1962;109: 251±252.

Kaufman HE. Treatment of viral diseases of the cornea and external eye. *Prog Ret Eye Res.* 2000; 19 (1): 69–85.

Kaufman HE, Heidelberger C. Therapeutic anti-viral action of 5-trifluoromethyl-2'-deoxyuridine. *Science* 1964;145: 585±586.

Kaufman HE, Kanai A, Ellison ED. Herpetic iritis: demonstration of virus in the anterior chamber by fluorescent antibody techniques and electron microscopy. *Am J Ophthalmol.* 1971; 71: 465–469.

Kaufman HE, Martola E-L, Dohlman CH. Herpes simplex treatment with IDU and corticoster- oids. *Arch Ophthalmol.* 1963;69: 468±472.

Kerr JR. Epstein-Barr virus (EBV) reactivation and therapeutic inhibitors. *J Clin Pathol.* 2019;72: 651–658.

Khodabande A. Varicella endotheliitis: a case report. *Eur J Ophthalmol.* 2009;19: 1076–1078.

Khokhar S, Natung T, Sony P, et al. Amniotic membrane transplantation in refractory neurotrophic corneal ulcers: a randomized, controlled clinical trial. *Cornea* 2005; 24: 654–660.

Kinchington PR, Araullo-Cruz T, Vergnes JP, et al. Sequence changes in the human adenovirus type 5 DNA polymerase associated with resistance to the broad spectrum antiviral cidofovir. *Antiviral Res.* 2002; 56: 73–84.

Kinchington PR, Romanowski EG, Jerold Gordon Y. Prospects for adenovirus antivirals. *J Antimicrob Chemother.* 2005; 55: 424–429.

Kobayashi A, Yokogawa H, Mori N, Sugiyama K. Mapping of dendritic lesions in patients with herpes simplex keratitis using in vivo confocal microscopy. *Clin Ophthalmol.* 2015;9: 1771–1777.

Kodukula P, Liu T, Rooijen NV, Jager MJ, Hendricks RL. Macrophage control of herpes simplex virus type 1 replication in the peripheral nervous system. *J Immunol.* 1999;162: 2895–2905.

Koenig Y, Bock F, Kruse FE, Stock K, Cursiefen C. Angioregressive pretreatment of mature corneal blood vessels before keratoplasty: fine-needle vessel coagulation combined with anti-VEGFs. *Cornea* 2012: 31: 887–892.

Koganti R, Yadavalli T, Shukla D. Current and emerging therapies for ocular herpes simplex virus type-1 infections. *Microorganisms* 2019;7 :429. doi: 10.3390/microorganisms7100429. PMID: 31658632; PMCID: PMC6843252.

Koganti R, Yadavalli T, Naqvi RA, Shukla D, Naqvi AR. Pathobiology and treatment of viral keratitis. *Exp Eye Res.* 2021 April; 205: 108483. doi:10.1016/j.exer.2021.10848

Koizumi N, Inatomi T, Suzuki T, Shiraishi A, Ohashi Y, Kandori M, Miyazaki D, Inoue Y, Soma T, Nishida K, Takase H, Sugita S, Mochizuki M, Kinoshita S. Clinical features and management of cytomegalovirus corneal endotheliitis: analysis of 106 cases from the Japan corneal endotheliitis study. *Br J Ophthalmol.* 2015;99: 54–58.

Koizumi N, Suzuki T, Uno T, Chihara H, Shiraishi A, Hara Y, Inatomi T, Sotozono C, Kawasaki S, Yamasaki K, Mochida C, Ohashi Y, Kinoshita S. Cytomegalovirus as an etiologic factor in corneal endotheliitis. *Ophthalmology* 2008;115: 292–297.e3.

Koizumi N, Yamasaki K, Kawasaki S, Sotozono C, Inatomi T, Mochida C, Kinoshita S. Cytomegalovirus in aqueous humor from an eye with corneal endotheliitis. *Am J Ophthalmol.* 2006;141: 564–565. [PubMed: 16490509]

Kong CL, Thompson RR, Porco TC, et al. Incidence rate of herpes zoster ophthalmicus: a retrospective cohort study from 1994 through 2018. *Ophthalmology.* 2020;127: 324–30.

Koujah L, Shukla D, Naqvi AR. CRISPR-Cas based targeting of host and viral genes as an antiviral strategy. *Semin Cell Dev Biol.* 2019a;96: 53–64.

Koujah L, Suryawanshi RK, Shukla D. Pathological processes activated by herpes simplex virus-1 (HSV-1) infection in the cornea. *Cell Mol Sci.* 2019b;76: 405–419. [PubMed: 30327839]

Kovalyuk N, Kaiserman I, Mimouni M, et al. Treatment of adenoviral keratoconjunctivitis with a combination of povidone-iodine 1.0% and dexamethasone 0. 1% drops: a clinical prospective controlled randomized study. *Acta Ophthalmol.* 2017; 95:e686–e692

Krawczyk A, Arndt MAE, Grosse-Hovest L, Weichert W, Giebel B, Dittmer U, Hengel H, Jäger D, Schneweis KE, Eis-Hübinger AM, Roggendorf M, Krauss J. Overcoming drug-resistant herpes simplex virus (HSV) infection by a humanized antibody. *PNAS* 2013;110: 6760–6765.

Kuffova L, Knickelbein JE, Yu T, Medina C, Amescua G, Rowe AM, Hendricks RL, Forrester JV. High-risk corneal graft rejection in the setting of previous corneal herpes simplex virus (HSV)-1 infection. *Invest Ophthalmol Vis Sci.* 2016;57: 1578–1587.

Kuo IC, Espinosa C, Forman M, Valsamakis A. A polymerase chain reaction-based algorithm to detect and prevent transmission of adenoviral conjunctivitis in hospital employees. *Am J Ophthalmol.* 2016; 163: 38–44.

Kuo SC, Shen SC, Chang SW, Huang SC, Hsiao CH. Corneal superinfection in acute viral conjunctivitis in young children. *J Pediatr Ophthalmol Strabismus.* 2008 Nov-Dec;45(6): 374–376.

Küppers R. B cells under influence: transformation of B cells by Epstein-Barr virus. *Nat Rev Immunol.* 2003;3: 801–812. [PubMed: 14523386]

Labib BA, Chigbu DI. Clinical management of herpes simplex virus keratitis. *Diagnostics* 2022; 12: 2368. https://doi.org/ 10.3390/diagnostics12102368

Laing KJ, Ouwendijk WJD, Koelle DM, Verjans Georges MGM. Immunobiology of Varicella-Zoster Virus infection. *J Infect Dis.* 2018;218, S68–S74. [PubMed: 30247598]

Lee YC, Chen N, Huang IT, et al. Human adenovirus type 8 epidemic keratoconjunctivitis with large corneal epithelial full-layer detachment: an endemic outbreak with uncommon manifestations. *Clin Ophthalmol.* 2015; 9: 953–957.

Li JY. Herpes zoster ophthalmicus: acute keratitis. *Curr Opin Ophthalmol.* 2018; 29(4): 328- 333.

Liesegang TJ. Classification of herpes simplex virus keratitis and anterior uveitis. *Cornea* 1999;18: 127–143. http://dx.doi.org/10.1097/00003226-199903000-00001 [

Liesegang TJ. Corneal complications from herpes zoster ophthalmicus. *Ophthalmology.* 1985;92(3): 316–324

Liesegang TJ. Epidemiology of ocular herpes simplex. Natural history in Rochester, Minn, 1950 through 1982. *Arch Ophthalmol.* 1989; 107: 1160–1165.

Liesegang, T.J. (1999) Classification of Herpes Simplex Virus Keratitis and Anterior Uveitis. *Cornea,* 18, 127-143. http://dx.doi.org/10.1097/00003226-199903000-00001

Liesegang TJ. Herpes simplex virus epidemiology and ocular importance. *Cornea* 2001;20: 1–13. [PubMed: 11188989]

Liesegang TJ. Herpes zoster ophthalmicus natural history, risk factors, clinical presentation, and morbidity. *Ophthalmology*. 2008;115(Suppl. 2):S3–12.

Liesegang TJ, Melton LJ III, Daly PJ, Ilstrup DM. Epidemiology of ocular herpes simplex: incidence in Rochester, Minn, 1950 Through 1982. *Arch Ophthalmol.* 1989;107: 1155–1159. [PubMed: 278798]

Lim DH, Kim J, Lee JH, Chung E, Chung T, 2014. A case of corneal endothelial dysfunction due to coxsackievirus A24 corneal endotheliitis after cataract surgery. *Cornea* 33: 533–535. [PubMed: 24608251]

Little JM, Lorenzetti DW, Brown DC, et al. Studies of adenovirus type 3 infection treated with methisazone and trifluorothymidine. *Proc Soc Exp Biol Med.* 1968; 127: 1028–1032.

Liu Q, Muruve DA. Molecular basis of the inflammatory response to adenovirus vectors. *Gene Ther.* 2003; 10: 935–940.

Liu T, Khanna KM, Chen X, Fink DJ, Hendricks RL. Cd8+ T Cells Can Block Herpes Simplex Virus Type 1 (HSV-1) Reactivation from Latency in Sensory Neurons. *J Exp Med.* 2000;191: 1459–1466. [PubMed: 10790421]

Liu T, Tang Q, Hendricks RL. Inflammatory infiltration of the trigeminal ganglion after herpes simplex virus type 1 corneal infection. *J Virol.* 1996; 70: 264–271. [PubMed: 8523535]

Liu Y, You Q, Zhang F, Chen D, Huang Z, Wu Z. Harringtonine inhibits herpes simplex virus type 1 infection by reducing herpes virus entry mediator expression. *Front Microbiol.* 2021; 12: 722748.

Lobo A-M, Agelidis AM, Shukla D. Pathogenesis of herpes simplex keratitis: the host cell response and ocular surface sequelae to infection and inflammation. *Ocular Surface* 2019;17: 40–49.

Lv D, Zhong J, Zhang K, Pandey A, Li R. Understanding Epstein-Barr virus life cycle with proteomics: a temporal analysis of ubiquitination during virus reactivation. *Omics (Larchmont, N.Y.)* 2017;21: 27–37.

Lyall DAM, Tarafdar S, Gilhooly MJ, Roberts F, Ramaesh K. Long term visual outcomes, graft survival and complications of deep anterior lamellar keratoplasty in patients with herpes simplex related corneal scarring. *Br J Ophthalmol.* 2012 Sep;96(9): 1200–1203.

Mader TH, Stulting RD. Viral keratitis. *Infect Dis Clin North Am.* 1992; 6: 831–849.

Madhavan HN, Goldsmith CS, Rao SK, Fogla R, Malathi J, Priya K. Isolation of a vesicular virus belonging to the family rhabdoviridae from the aqueous humor of a patient with bilateral corneal endotheliitis. *Cornea* 2002;21: 333–335.

Manicklal S, Emery VC, Lazzarotto T, Boppana SB, Gupta RK. The "silent" global burden of congenital cytomegalovirus. *Clin Microbiol Rev.* 2013;26: 86–102.

Mann D, Moreb J, Smith S, Gian V. Failure of intravenous ribavirin in the treatment of invasive adenovirus infection following allogeneic bone marrow transplantation: a case report. *J Infect.* 1998; 36: 227–228.

Martinez V, Molina J, Scieux C, Ribaud P, Morfin F. Topical imiquimod for recurrent acyclovir-resistant HSV infection. *Am J Med.* 2006;119: 9. [PubMed: 16483863]

Matoba AY, Jones DB. Corneal subepithelial infiltrates associated with systemic Epstein-Barr viral infection. *Ophthalmology* 1987;94: 1669–1671.

Matoba AY, Wilhelmus KR, Jones DB. Epstein-Barr viral stromal keratitis. *Ophthalmology* 1986;93: 746–751.

Matundan H, Ghiasi H. Herpes simplex virus 1 ICP22 suppresses CD80 expression by murine dendritic cells. *J Virol.* 2019;93:e01803–18. doi: 10.1128/JVI.01803-18.

McQuillan G, Kruszon-Moran D, Flagg EW, Paulose-Ram R, 2018. Prevalence of herpes simplex virus type 1 and type 2 in persons aged 14–49: United States, 2015–2016. *NCHS Data Brief,* 1–8.

Meyer-Rusenberg B, Loderstadt U, Richard G, et al. Epidemic keratoconjunctivitis: the current situation and recommendations for prevention and treatment. *Dtsch Arztebl Int.* 2011; 108: 475–480.

Mistchenko AS, Falcoff R. Recombinant human interferon-gamma inhibits adenovirus multiplication in vitro. *J Gen Virol.* 1987; 68(Pt 3): 941–944.

Miyakoshi A, Takemoto M, Shiraki K, Hayashi A. Varicella-Zoster virus keratitis with asymptomatic conjunctival viral shedding in the contralateral eye. *Case Rep Ophthalmol.* 2012;3: 343–348. [PubMed: 23162462]

Moshirfar M. A review of corneal endotheliitis and endotheliopathy: differential diagnosis, evaluation, and treatment. *Ophthalmol Ther.* 2019;8(2): 195–213.

Mott KR, Underhill D, Wechsler SL, Ghiasi H. Lymphoid-related CD11c+ CD8alpha+ dendritic cells are involved in enhancing herpes simplex virus type 1 latency. *J Virol.* 2008;82: 9870–9879.

Moyes AL, Sugar A, Musch DC, Barnes RD. Antiviral therapy after penetrating keratoplasty for herpes simplex keratitis. *Arch Ophthalmol.* 1994;112: 601±607

Nagel MA, Gilden DH. The protean neurologic manifestations of varicella-zoster virus infection. *Cleveland Clin J Med.* July 2007;74(7): 489–494, 496, 498–499.

Niller HH, Bauer G. Epstein-Barr virus: clinical diagnostics. *Methods Mol Biol.* 2017;1532: 33–55. doi: 10.1007/978-1-4939-6655-4_2. PMID: 27873265.

Niller HH, Wolf H, Minarovits J. Regulation and dysregulation of Epstein-Barr virus latency: implications for the development of autoimmune diseases. *Autoimmunity* 2008;41: 298–328. [PubMed: 18432410]

Nishida T, Nakagawa S, Manabe R. Clinical evaluation of fibronectin eye drops on epithelial disorders after herpetic keratitis. *Ophthalmology* 1985;92: 213–216.

Nowalk A, Green M. Epstein-Barr virus. *Microbiol Spectr.* 2016. doi: 10.1128/microbiolspec.DMIH2-0011-2015 Corpus ID: 6969443.

Nubile M, Dua HS, Lanzini M, Ciancaglini M, Calienno R, Said DG, Pocobelli A, Mastropasqua R, Carpineto P. In vivo analysis of stromal integration of multilayer amniotic membrane transplantation in corneal ulcers. *Am J Ophthalmol.* 2011;151: 809–822.e1.

Ogura H, Fukae J, Kimura S, Aoki M, Nabeshima K, Tsuboi Y. Acyclovir resistant acute herpes simplex encephalitis associated with acute retinal necrosis: a case report and review of the literature. *Rinsho Shinkeigaku* 2017;57: 230–233.

Oka N, Suzuki T, Inoue T, Kobayashi T, Ohashi Y. Polymorphisms in cytomegalovirus genotype in immunocompetent patients with corneal endotheliitis or iridocyclitis. *J Med Virol.* 2015;87: 1441–1445.

Omari AA and Mian SI. Adenoviral keratitis: a review of the epidemiology, pathophysiology, clinical features, diagnosis, and management *Curr Opin Opthamol.* July 2018;29(4), 265.

Park GB, Hur DY, Kim YS, Lee H, Yang JW, Kim D. TLR3/TRIF signalling pathway regulates IL-32 and IFN-β secretion through activation of RIP-1 and TRAF in the human cornea. *J Cell Mol Med.* 2015;19: 1042–1054.

Park GB, Kim D, Kim YS, Kim S, Lee H, Yang JW, Hur DY. The Epstein-Barr virus causes epithelial–mesenchymal transition in human corneal epithelial cells via Syk/Src and Akt/Erk signaling pathways. *Invest Ophthalmol Vis Sci.* 2014;55: 1770–1779.

Park PJ, Antoine TE, Farooq AV, Valyi-Nagy T, Shukla D. An investigative peptide-acyclovir combination to control herpes simplex virus type 1 ocular infection. *Invest Ophthalmol Vis Sci.* 2013;54: 6373–6381.

Pass RF, Fowler KB, Boppana SB, Britt WJ, Stagno S. Congenital cytomegalovirus infection following first trimester maternal infection: symptoms at birth and outcome. *J Clin Virol.* 2006;35: 216–220.

Pavan-Langston D. Herpes zoster ophthalmicus. *Neurology* 1995;45: 50.

Perng G, Jones C. Towards an understanding of the herpes simplex virus type 1 latency-reactivation cycle. *Interdiscip Perspect Infect Dis.* 2010;2010: 262415.

Perry CM, Faulds D. Valacyclovir: a review of its antiviral activity, pharmacokinetic properties and therapeutic efficacy in herpes virus infections. *Drugs* 1996;52: 754–772.

Perry HD, Golub LM. Systemic tetracyclines in the treatment of non-infected corneal ulcers: a case report and proposed new mechanism of action. *Ann Ophthalmol.* 1985;17: 742–744.

Pertel PE, Spear PG. Modified entry and syncytium formation by herpes simplex virus type 1 mutants selected for resistance to heparin inhibition. *Virology* 1996;226: 22–33.

Petroutsos G, Guimaraes R, Giraud JP, Pouliquen Y. Corticosteroids, and corneal epithelial wound healing. *Br J Ophthalmol.* 1982;66: 705–708.

Pflugfelder SC, de Paiva CS. The pathophysiology of dry eye disease: what we know and future directions for research. *Ophthalmology* 2017; 124(11S):S4–S13.

Pinnolis M, McCulley JP, Urman JD. Nummular keratitis associated with infectious mononucleosis. *Am J Ophthalmol.* 1980;89: 791–794.

Piret J, Boivin G. Resistance of herpes simplex viruses to nucleoside analogues: mechanisms, prevalence, and management. *Antimicrob Agents Chemother.* 2011;55: 459–472.

Pottage JC, Kessler HA. Herpes simplex virus resistance to acyclovir: clinical relevance. *Infect Agents Dis.* 1995;4: 115–124. [PubMed: 8548189]

Rafailidis PI, Kapaskelis A, Falagas ME. Cytomegalovirus meningitis in an immunocompetent patient. *Med Sci Monit.* 2007;13, CS107–109. [PubMed: 17767122]

Rajasagi NK, Bhela S, Varanasi SK, Rouse BT. Frontline science: aspirin-triggered resolvin D1 controls herpes simplex virus-induced corneal immunopathology. *J Leukoc Biol.* 2017;102: 1159–1171.

Rana M, Savant V. A brief review of techniques used to seal corneal perforation using cyanoacrylate tissue adhesive. *Cont Lens Anterior Eye* 2013;36: 156–158.

Ratnam S, West R, Gadag V, Burris J. Measles immunization strategy: measles antibody re-sponse following MMR II vaccination of children at one year of age. *Can J Public Health* 1996;87: 97±100.

Reeves M, Sinclair J. Aspects of human cytomegalovirus latency and reactivation. *Curr Top Microbiol Immunol.* 2008;325: 297–313.

Reijo A, Antti V, Jukka M. Endothelial cell loss in herpes zoster keratouveitis. *Br J Ophthalmol.* 1983;67(11): 751–754.

Reinhard T. Suspected Epstein-Barr Virus (EBV) Disease of the Anterior Segment, in: *Color Atlas of Herpetic Eye Diseases: A Practical Guide to Clinical Management,* ed. Rainer Sundmacher, Springer Berlin, Heidelberg, pp. 163–165; 2009.

Remeijer L, Osterhaus A, Verjans G. Human herpes simplex virus keratitis: the pathogenesis revisited. *Ocul Immunol Inflamm.* 2004;12: 255–285.

Reynaud C, Rousseau A, Kaswin G, M'garrech M, Barreau E, Labetoulle M. Persistent impairment of quality of life in patients with herpes simplex keratitis. *Ophthalmology* 2017;124: 160–169.

Ritterband DC, Dorothy N. Friedberg virus infections of the eye. *Rev. Med Virol.* 1998;8: 187–201.

Romano A, Ladizensky E, Guarari-Rotman D, Revel M. Clinical effect of human-fibroblast-derived (Beta) interferon in treatment of adeno-virus epidemic keratoconjunctivitis and its complications. *Tex Rep Biol Med.* 1981; 41: 559–565.

Romanowski EG, Yates KA, Gordon YJ. The in vitro and in vivo evaluation of ddC as a topical antiviral for ocular adenovirus infections. *Invest Ophthalmol Vis Sci.* 2009; 50: 5295–5299.

Ronkainen S, Ronkainen S, Rothenberger M, Rothenberger M. Herpes vegetans: an unusual and acyclovir-resistant form of HSV. *J Gen Intern Med.* 2018;33: 393.

Rossa V, Sundmacher R. Local prevention with interferon of 'epidemic' conjunctivitis caused by a currently unidentifiable virus. *Klin Monbl Augenheilkd* 1991; 199: 192–194.

Sacchetti M, Lambiase A. Diagnosis and management of neurotrophic keratitis. *Clin Ophthalmol.* 2014;8: 571–579.

Sarangi PP, Kim B, Kurt-Jones E, Rouse BT. Innate recognition network driving herpes simplex virus-induced corneal immunopathology: role of the toll pathway in early inflammatory events in stromal keratitis. *J Virol.* 2007;81: 11128–11138.

Sathiyamoorthy K, Hu YX, Möhl BS, Chen J, Longnecker R, Jardetzky TS. Structural basis for Epstein–Barr virus host cell tropism mediated by gp42 and gHgL entry glycoproteins. *Nature Communications* 2016;7: 13557.

Sato T, Kitamura R, Kaburaki T, Takeuchi M. Retinitis associated with double infection of Epstein-Barr virus and varicella-zoster virus: a case report. *Medicine (Baltimore)* 2018;97, e11663.

Schwab IR. Oral Acyclovir in the management of herpes simplex ocular infections. *Ophthalmology* 1988;95: 423–430.

Sciammas R, Kodukula P, Tang Q, Hendricks RL, Bluestone JA. T cell receptor–γ/δ cells protect mice from herpes simplex virus type 1–induced lethal encephalitis. *J Exp Med.* 1997;185: 1969–1975.

Serna-Ojeda JC, Loya-Garcia D, Navas A, Lichtinger A, Ramirez-Miranda A, Graue-Hernandez EO. Long-term outcomes of pediatric penetrating keratoplasty for herpes simplex virus keratitis. *Am J Ophthalmol.* 2017;173: 139–144.

Seward JF, Watson BM, Peterson CL, Mascola L, Pelosi JW, Zhang JX, Maupin TJ, Goldman GS, Tabony LJ, Brodovicz KG, Jumaan AO, Wharton M. Varicella disease after introduction of varicella caccine in the United States, 1995–2000. *JAMA* 2002;287: 606–611.

Shaikh S, Ta CN. Evaluation and management of herpes zoster ophthalmicus. *Am Fam Physician* 2002;66: 1723–1730. [PubMed: 12449270]

Shibata T, Takahashi Y, Okamoto A, Sasaki H, Kitagawa K. An adult case of fulminant mumps keratitis with positive viral RNA in aqueous humor detected by RT-PCR. *Cornea* 2016;35: 569–571. [PubMed: 26863498]

Shiraishi A, Hara Y, Takahashi M, Oka N, Yamaguchi M, Suzuki T, Uno T, Ohashi Y. Demonstration of "owl's eye" morphology by confocal microscopy in a patient with presumed cytomegalovirus corneal endotheliitis. *Am J Ophthalmol.* 2007;143: 715–717.

Shoji J, Sakimoto T, Inada N, Kamei Y, Matsubara M, Takamura E, Sawa M. A diagnostic method for herpes simplex keratitis by simultaneous measurement of viral DNA and virus-specific secretory IgA in tears: an evaluation. *Jpn J Ophthalmol.* 2016;60: 294–301.

Shuster JJ, Kaufman HE, Nesburn AB. Statistical analysis of the rate of recurrence of herpesvirus ocular epithelial disease. *Am J Ophthalmol.* 1981;91: 328–331.

Silverstein BE, Chandler D, Neger R, Margolis TP. Disciform keratitis: a case of herpes zoster sine herpete. *Am J Ophthalmol.* 1997;123: 254–255.

Smatti MK, Al-Sadeq DW, Ali NH, Pintus G, Abou-Saleh H, Nasrallah GK. Epstein–Barr virus epidemiology, serology, and genetic variability of LMP-1 oncogene among healthy population: an update. *Front Oncol.* 2018;8: 211.

Song X, Xie L, Tan X, Wang Z, Yang Y, Yuan Y, Deng Y, Fu S, Xu J, Sun X, Sheng X, Wang Q. A multi-center, cross-sectional study on the burden of infectious keratitis in China. *PloS One* 2014;9, e113843.

Spiteri N, Romano V, Zheng Y, Yadav S, Dwivedi R, Chen J, Ahmad S, Willoughby CE, Kaye SB. Corneal angiography for guiding and evaluating fine-needle diathermy treatment of corneal neovascularization. *Ophthalmology* 2015;122: 1079–1084.

Stevens JG, Haarr L, Porter DD, Cook ML, Wagner EK Prominence of the herpes simplex virus latency-associated transcript in trigeminal ganglia from seropositive humans. *J Infect Dis.* 1988;158: 117±123.

Sullivan V, Biron K, Talarico C et al. A point mutation in the human cytomegalovirus DNA polymerase gene confers resistance to ganciclovir and phosphonylmethoxyalkyl derivatives. *Antimicrob Agents Chemother.* 1993;37: 19–25.

Sullivan V, Coen D. Isolation of foscarnet-resistant human cytomegalovirus patterns of resistance and sensitivity to other antiviral drugs. *J Infect Dis.* 1991;164: 781–784.

Sundmacher R, Wigand R, Cantell K. The value of exogenous interferon in adenovirus keratoconjunctivitis. Preliminary results. *Graefes Arch Clin Exp Ophthalmol.* 1982; 218: 139–140.

Szeto SK, Chan TC, Wong RL, et al. Prevalence of ocular manifestations and visual outcomes in patients with herpes zoster ophthalmicus. *Cornea.* 2017;36: 338–42.

Takeda S, Miyazaki D, Sasaki S, Yamamoto Y, Terasaka Y, Yakura K, Yamagami S, Ebihara N, Inoue Y. Roles played by toll-like receptor-9 in corneal endothelial cells after herpes simplex virus type 1 infection. *Invest Ophthalmol Vis Sci.* 2011;52: 6729–6736.

Tan T, Cheung CMG, Mehta JS. Activation of cytomegalovirus corneal endotheliitis following laser in situ keratomileusis. *BMJ Case Rep.* 2016 Nov 29:2016:bcr2016216774. doi: 10.1136/bcr-2016-216774.

Tan T, Tan DTH. Cytomegalovirus corneal endotheliitis after descemet membrane endothelial keratoplasty. *Cornea* 2019;38: 413–418.

Tanner OR. Ocular manifestations of infectious mononucleosis. *AMA Arch Ophthalmol.* 1954;51: 229–241. [PubMed: 13123609]

Thompson RR, Kong CL, Porco TC, et al. Herpes zoster and post-herpetic neuralgia: changing incidence rates from 1994 to 2018 in the United States. *Clin Infect Dis*. 2021;73:e3210–7.

Thygeson P. Historical observations on herpetic keratitis. *Surv Ophthalmol.* 1976;21: 82–90.

Tiwari V, Liu J, Valyi-Nagy T, Shukla D. Anti-heparan sulfate peptides that block herpes simplex virus infection in vivo. *J Biol Chem.* 2011;286, 25406–25415.

Toriyama K, Inoue T, Suzuki T, Kobayashi T, Ohashi Y. Necrotizing keratitis caused by acyclovir-resistant herpes simplex virus. *Case Rep Ophthalmol.* 2014;5: 325–328.

Tran KD, Falcone MM, Choi DS, et al. Epidemiology of herpes zoster ophthalmicus: recurrence and chronicity. *Ophthalmology*. 2016;123: 469–75 .

Tsatsos M, MacGregor C, Athanasiadis I, Moschos MM, Hossain P, Anderson D. Herpes simplex virus keratitis: an update of the pathogenesis and current treatment with oral and topical antiviral agents. *Clin Experiment Ophthalmol.* 2016; 44(9): 824- 837.

Tuli S, Gray M, Shah A. Surgical management of herpetic keratitis. *Curr Opin Ophthalmol.* 2018;29: 347–354.

van Diemen FR, Kruse EM, Hooykaas MJ, et al. CRISPR/Cas9-mediated genome editing of herpesviruses limits productive and latent infections. *PLoS Pathogens.* 2016 Jun;12(6):e1005701. DOI: 10.1371/journal.ppat.1005701.

Vannas A, Ahonen R, Makitie J. Corneal endothelium in herpetic keratouveitis. *Arch Ophthalmol.* 1983;101: 913–915.

Versura P, Giannaccare G, Pellegrini M, Sebastiani S, Campos EC. Neurotrophic keratitis: current challenges and future prospects. *Eye Brain* 2018;10: 37–45.

Virók DP, Eszik I, Mosolygó T, Önder K, Endrész V, Burián K. A direct quantitative PCR-based measurement of herpes simplex virus susceptibility to antiviral drugs and neutralizing antibodies. *J Virol Methods* 2017 April;242: 46–52.

Vishnepolsky B, Pirtskhalava M. Prediction of linear cationic antimicrobial peptides based on characteristics responsible for their interaction with the membranes. *J Chem Inf Model.* 2014 May 27;54(5): 1512–1523. doi: 10.1021/ci4007003. Epub 2014 Apr 29. PMID: 24730612; PMCID: PMC4038373.

Wald A, Corey L. Persistence in the Population: Epidemiology, Transmission, in: Arvin A, Campadelli-Fiume G, Mocarski E, Moore PS, Roizman B, Whitley R, Yamanishi K (Eds.), *Human Herpesviruses: Biology, Therapy, and Immunoprophylaxis.* Cambridge; Cambridge University Press; 2007. PMID: 21348071.

Wang J, Xie L, Song X, Zhao J. Trends in the indications for penetrating keratoplasty in Shandong, 2005–2010. *Int J Ophthalmol.* 2011;4: 492–497. [PubMed: 22553709]

Wang S, Tsai I-L, Lin H, Kuo L, Tsai C, Liou S. Recurrent cytomegalovirus corneal endotheliitis after penetrating keratoplasty. *Eur J Ophthalmol.* 2010;20: 457–459. [PubMed: 20037910]

Wang ZJ, Zhou M, Cao WJ, Ji J, Bi YW, Huang X, Xu GZ. Evaluation of the Goldmann-Witmer coefficient in the immunological diagnosis of ocular toxocariasis. *Acta Trop.* 2016;158: 20–23.

Weiss A. Acute conjunctivitis in childhood. *Curr Probl Pediatr.* 1994; 24: 4–11.

Westeneng AC, Rothova A, de Boer JH, de Groot-Mijnes JDF. Infectious uveitis in immunocompromised patients and the diagnostic value of polymerase chain reaction and Goldmann-Witmer coefficient in aqueous analysis. *Am J Ophthalmol.* 2007;144: 781–785.

Wharton M. The epidemiology of Varicella-Zoster virus. *Infect Dis Clin North Am.* 1996;10: 571–581.

White ML, Chodosh J. *Herpes Simplex Virus Keratitis: A Treatment Guideline San Francisco: Hoskins Center for Quality Eye Care.* American Academy of Ophthalmology; 2014.

Whitley RJ, Kimberlin DW, Roizman B. Herpes simplex viruses. *Clin Infect Dis.* 1998;26: 541–555.

Wilhelmus KR. Antiviral treatment and other therapeutic interventions for herpes simplex virus epithelial keratitis. *Cochrane Database Syst Rev.* 2015;1, CD002898.

Wilhelmus KR, Dunkel EC, Herson J. Topical human fibroblast interferon for acute adenoviral conjunctivitis. *Graefes Arch Clin Exp Ophthalmol.* 1987;225: 461–464

Wilhelmus KR, Font RL, Lehmann RP, Cernoch PL. Cytomegalovirus keratitis in acquired immunodeficiency syndrome. *Arch Ophthalmol.* 1996;114: 869–872.

Wilhelmus KR, Gee L, Hauck WW, Kurinij N, Dawson CR, Jones DB, Barron BA, Kaufman HE, Sugar J, Hyndiuk RA, Laibson PR, Stulting RD, Asbell PA. Herpetic eye disease study: a controlled trial of topical corticosteroids for herpes simplex stromal keratitis. *Ophthalmology* 1994;101: 1883–1896.

Wilhelmus K, Beck R, Moke P, Dawson C, Barron B, Jones D, Kaufman H, Kurinji N, Stulting D, Sugar J, Cohen E, Hyndiuk R, Asbell P. Acyclovir for the prevention of recurrent herpes simplex virus eye disease. Herpetic Eye Disease Study Group. *N Engl J Med.* 1998;339: 300–306.

Wishart MS, Darougar S, Viswalingam ND. Recurrent herpes simplex virus ocular infection: epidemiological and clinical features. *Br J Ophthalmol.* 1987;71: 669–672.

Wolffsohn JS, Arita R, Chalmers R, Djalilian A, Dogru M, Dumbleton K, Gupta PK, Karpecki P, Lazreg S, Pult H, Sullivan BD, Tomlinson A, Tong L, Villani E, Yoon KC, Jones L, Craig JP. TFOS DEWS II diagnostic methodology report. *Ocul Surf.* 2017;15: 539–574.

Wong KW, D'Amico DJ, Hedges TR, Soong HK, Schooley RT, Kenyon KR. Ocular involvement associated with chronic Epstein-Barr virus disease. *Arch Ophthalmol.* 1987;105 : 88–792.

Wu C, Fang C, Hsu H, Chen Y, Chou S, Huang S, Cheng Y, Lin S, Chang Y, Tsai C, Chen J. Luteolin inhibits Epstein-Barr virus lytic reactivation by repressing the promoter activities of immediate-early genes. *Antiviral Res.* 2016;132: 99–110.

Yadavalli T, Agelidis A, Jaishankar D, Mangano K, Thakkar N, Penmetcha K, Shukla D. Targeting herpes simplex virus-1 gD by a DNA aptamer can be an effective new strategy to curb viral infection. *Mol Ther Nucleic Acids* 2017;9: 365–378. [PubMed: 29246315]

Yadavalli T, Suryawanshi R, Ali M, Iqbal A, Koganti R, Ames J, Aakalu VK, Shukla D. Prior inhibition of AKT phosphorylation by BX795 can define a safer strategy to prevent herpes simplex virus-1 infection of the eye. *Ocul Surf.* 2020; 18(2): 221–230.

Yamazaki ES, Ferraz CA, Hazarbassanov RM, et al. Phototherapeutic keratectomy for the treatment of corneal opacities after epidemic keratoconjunctivitis. *Am J Ophthalmol.* 2011; 151: 35–43; e31.

Yawn BP, Gilden D. The global epidemiology of herpes zoster. *Neurology* 2013;81: 928–930.

Yildiz C, Ozsurekci Y, Gucer S, Cengiz A, Topaloglu R. Acute kidney injury due to acyclovir. *CEN Case Rep.* 2013;2: 38–40.

Young RC, Hodge DO, Liesegang TJ, Baratz KH. Incidence, recurrence, and outcomes of herpes simplex virus eye disease in Olmsted County, Minnesota, 1976–2007: the effect of oral antiviral prophylaxis. *Arch Ophthalmol.* 2010;128: 1178–1183. [PubMed: 20837803]

Ys H, Cr S, Sh C, Cc L, Cl L, Kj C, Kk L, Tl C, Ch H. The validity of clinical feature profiles for cytomegaloviral anterior segment infection. *Graefes Arch Clin Exp Ophthalmol.* 2010;249: 103–110.

Zerboni L, Sen N, Oliver SL, Arvin AM. Molecular mechanisms of varicella zoster virus pathogenesis. *Nat Rev Microbiol.* 2014;12: 197–210.

Zhang F, Liu Y, You Q, Yang E, Liu B, Wang H, Xu S, Nawaz W, Chen D, Wu Z. NSC23766 and Ehop016 suppress herpes simplex virus-1 replication by inhibiting Rac1 activity. *Biol Pharm Bull.* 2021; 44: 1263–1271.

Zhu Q, Liu G, Kai M. DNA aptamers in the diagnosis and treatment of human diseases. *Molecules* 2015;20: 20979–20997. [PubMed: 26610462]

Zinser E, Krawczyk A, Mühl-Zürbes P, Aufderhorst U, Draßner C, Stich L, Zaja M, Strobl S, Steinkasserer A, Heilingloh CS. A new promising candidate to overcome drug resistant herpes simplex virus infections. *Antivir Res.* 2018;149: 202–210.

7

Antimicrobial Resistance in the Diseased Human Eye: Acanthamoeba *Keratitis*

Bhupesh Bagga and S. Shivaji

7.1 Introduction

Acanthamoeba is a eukaryote which is a common inhabitant of the environment, distributed worldwide and found in soil, air, fresh water, sewage, seawater, swimming pools, hospitals, contact lens cases and so forth. In human beings *Acanthamoeba* are isolated from skin, nasal cavities, oral cavity, and intestines. In addition, it has also been isolated from plants and other mammals (De Jonckheere, 1991). *Acanthamoeba* are bacterivores (Marciano-Cabral and Cabral, 2003*)* feeding on bacteria, and several bacteria also survive as endosymbionts (Amann et al., 1997; Horn and Wagner, 2004; Newsome et al., 1998; Fritsche et al., 1999; Horn et al., 2002). The bacterial endosymbionts in acanthamoeba include *Legionella pneumophila*, *Pseudomonas aeruginosa*, *Escherichia coli* and *Staphylococcus aureus* (Huws et al., 2008; *Marciano-Cabral and Cabral,* 2003). These endosymbionts increase disease severity in *Acanthamoeba* keratitis (AK) (Karsenti et al., 2022).

Acanthamoeba spp. were divided into three distinct groups based on cyst size and shape as groups (I, II, and III) (Page, 1967). Subsequently, *Acanthamoeba* strains have been categorised into 12 groups, designated T1-T12. T4 includes *A. castellanii*, *A. divionensis*, *A. lugdunensis*, *A. mauritaniensis*, *A. paradivionensis*, *A. polyphaga*, *A. rhysodes* and *A. triangularis* which are pathogens (Kong, 2009). At the molecular level *Acanthamoeba* species have been categorised based on mitochondrial DNA restriction fragment length polymorphism (RFLP) (Bogler et al., 1983; Gautom et al., 1994), sequencing of small subunit rRNAs (Johnson et al., 1990) and nuclear rRNA gene sequences (18S rDNA) (Gast et al., 1996). Riboprinting was used by Chung et al. (1998) for subgenus classification of *Acanthamoeba.* Comparison of all the above approaches revealed major inconsistencies and to overcome this problem the genus *Acanthamoeba* is being revised based on sequence comparisons (Booton et al., 2001; Byers et al., 2001).

7.2 Diseases Caused by *Acanthamoeba*

Acanthamoeba is a pathogen causing granulomatous amoebic encephalitis (GAE), amoebic keratitis (AK), and skin infections (Jones et al., 1975; Martinez et al., 1977;

DOI: 10.1201/9781003451105-7

Nagington et al., 1974). The two diseases that have been extensively studied include AK and GAE (Culbertson, 1961; Jones et al., 1975; Nagington et al., 1974). In AIDS and immunocompromised patients *Acanthamoeba* spp. causes cutaneous lesions and sinusitis in immunocompromised individuals (Dunand et al., 1997; Helton et al., 1993; Martinez and Janitschke, 1985; May et al., 1992; Torno, et al., 2000).

7.2.1 Granulomatous Amoebic Encephalitis (GAE)

GAE is a fatal disease killing about 95% of those infected with amoebae (Baig et al., 2013). GAE is an amoebic infection of the central nervous system (CNS) causing headache, seizures, and mental-status abnormalities (Marciano-Cabral and Cabral, 2003), in children and adults associated with immunodeficiency (Ringsted et al., 1976; Sighal et al., 2001), diabetes, malignancies, malnutrition, systemic lupus erythematosus, and alcoholism (Marciano-Cabral and Cabral, 2003). Predisposing factors include alcohol consumption, drug abuse, steroid treatment, cancer chemotherapy and radiotherapy for cancer and organ transplantation (Anderlini et al., 1994; Martinez and Janitschke 1985; Sell et al., 1997). *Acanthamoeba* enters the human body through ruptures in the skin or by inhalation (Marciano-Cabral and Cabral, 2003), finds its way into the circulatory system and reaches the CNS, crosses the blood–brain barrier and causes neuronal damage that can be fatal within days. Edema is observed in the cerebral hemispheres and several multifocal lesions are obvious in various parts of the brain like the midbrain, brain stem, corpus callosum, and cerebellum and inflammatory exudates are seen over the cortex (Marciano-Cabral and Cabral., 2003).

7.2.1.1 Diagnosis of GAE

Visualisation of *Acanthamoeba* in the CSF of patients is normally accomplished by microscopy of wet mounts of CSF or smears of CSF sediment (Callicott, 1968; Cleland et al., 1982; Lalitha et al., 1985) fixed and stained with Giemsa-Wright stain. *Acanthamoeba* trophozoites are stained with haematoxylin and eosin (Newsome et al., 1992) and the trophozoites could be visualised with a prominent nucleolus and contractile and cytoplasmic vacuoles. GAE is normally characterised by pleocytosis with lymphocytes and polymorphonuclear (PMN) leukocytes and low levels of glucose with high protein levels (Newsome et al., 1992). Tomography and magnetic resonance imaging does not provide a definitive diagnosis of GAE (Martinez and Visvesvara, 1997). *In vitro* cultivation of *Acanthamoeba* has also been attempted using non nutrient agar enriched with *E. coli* (Martinez and Visvesvara, 1991). In addition, *Acanthamoeba* spp. could also be observed as trophozoites and cysts in frozen or haematoxylin and eosin-stained sections of brain or cutaneous lesion biopsy material (Martinez and Visvesvara, 1997; Rosenberg and Morgan, 2001; Teknos et al., 2000). Immunofluorescent or immunoperoxidase cytochemical staining of cryostat sections as well as transmission electron microscopy of infected tissues, were more sensitive in identifying *Acanthamoeba* (Ma et al., 1990; Stevens et al., 1977).

7.2.1.2 Treatment of GAE

Treatment of *Acanthamoeba* infection involved the use of amphotericin B, rifampicin, trimethoprim, sulfamethoxazole, ketoconazole, fluconazole or albendazole (Baig, 2014).

7.2.2 *Acanthamoeba* Infections in Patients with AIDS

In 1998, for the first time, *Acanthamoeba* infection was detected in a patient with AIDS (Gonzalez et al., 1986) and the frequency of detection increased in the subsequent years (Bonilla et al., 1999; Chandrasekar, et al., 1997; Gordon et al., 1992; May et al., 1992; Rivera and Padhya, 2002; Teknos et al., 2000; Torno et al., 2000; Wiley et al., 1987). The majority of patients with *Acanthamoeba* died within a month of neurological symptoms (Calore et al., 1997; Gordon et al., 1992; Seijo et al., 2000; Wiley et al., 1987). In HIV individuals *Acanthamoeba* infections included sinusitis, otitis (Rivera, and Padhya, 2002; Teknos et al., 2000) and cutaneous lesions (Bonilla et al., 1999; Casper et al., 1999; Gonzalez et al., 1986; Teknos et al., 2000) and in a few cases leukocytoclastic vasculitis, osteomyelitis, and endophthalmitis have been reported (Helton et al., 1993; Selby et al., 1998).

7.2.2.1 Treatment of Acanthamoeba *Infections in AIDS Patients*

In an HIV positive patient with cutaneous lesions and sinusitis due to Acanthamoeba infection treatment with the drug 5-fluorocytosine was successful (Helton et al., 1993). Combination therapy was effective in treating *Acanthamoeba* infection in immunocompetent children using a combination of ketoconazole and rifampin added to trimethoprim-sulfamethoxazole (Singhal et al., 2001). Further, in a renal transplant patient with a disseminated *Acanthamoeba* infection, a 4-week combination therapy including IV pentamidine isethionate, topical chlorhexidine gluconate, and 2% ketoconazole cream was successful (Slater et al., 1994). Surgical procedures have also been used to overcome *Acanthamoeba* infection. In 2000, Seijo et al. (2000) demonstrated that CNS lesions due to *Acanthamoeba* infection could be removed by surgery and post-surgery the patient is treated with fluconazole and sulfadiazine. Alternatively, Teknos et al. (2000) observed that surgical debridement of nasal and paranasal sinuses and post-surgery long-term therapy with 5-fluorocytosine for CNS infection and for renal transplant patients was effective. At about the same time, Rivera and Padhya (2002) observed that rhinosinusitis in an AIDS patient could be surgically removed and then subjected to multidrug therapy with pentamidine, amphotericin B, flucytosine, rifampin, itraconazole, and chlorhexidine. But, unfortunately, the results using multidrug regimens with disseminated *Acanthamoeba* infections have yielded inconsistent results. Rosenberg and Morgan (2001) successfully treated *Acanthamoeba* infection in an HIV-positive patient with recurrent sinusitis and cutaneous lesions, using a combination of itraconazole, azithromycin, 5-fluorocytosine and rifampin for 7 weeks (Rosenberg and Morgan 2001); the treatment was also successful in treating a lung transplant recipient without CNS (Oliva et al., 1999). Amphotericin B, rifampin, and 5-fluorocytosine when used as a combination therapy did show improvement but eventually it was ineffective and the patient succumbed to Gram-negative septicemia. Even when the patients were treated with a combination of amphotericin B, rifampin, and 5-fluorocytosine, followed by itraconazole, rifampin, and 5-fluorocytosine, after recurrence of disease the patient died (Hunt et al., 1995.; Dunand et al., 1997; Levine et al., 2001; Migueles and Kumar, 1998; Murakawa et al., 1995).

7.2.3 Cutaneous Acanthameobiasis

Cutaneous infections caused by *Acanthamoeba* generally manifest as hard erythematous nodules or skin ulcers (Murakawa et al., 1995; Tan et al., 1993; Torno et al., 2000) both in HIV-infected and non-HIV-infected patients (Casper et al., 1999; Duluol et al., 1996), in immunosuppressed patients (van Hamme et al., 2001), or in individuals with immunological diseases (Martinez, 1982; Slater et al., 1994). In cutaneous form of acanthamoebiasis, papulonodules that drain purulent material and develop into nonhealing ulcers are seen (May et al., 1992; Rosenberg and Morgan, 2001). Cutaneous Acathamoebiasis cause high mortality rate ranging from 73% in individuals without CNS involvement while the mortality is 100% in patients with CNS disease (Torno et al., 2000). The cutaneous lesions are generally necrotic and are surrounded by inflammatory cells, vasculitis, trophozoite and cysts (Friedland et al., 1992).

7.2.3.1 Treatment of Cutaneous Acanthameobiasis

In an AIDS patient with cutaneous and sinus lesions, 5-fluorocytosine therapy was reported successful (Helton et al., 1993) whereas in a lung transplant patient, intravenous pentamidine and itraconazole along with topical ketoconazole was effective (Van Hamm et al., 2001). Cutaneous acanthamoebiasis could also be treated with itraconazole, pentamidine, 5-fluocytosine, and topical chlorhexidine gluconate and ketoconazole cream (Helton et al., 1993). Patients administered combination treatments showed initial improvement, but eventually most of them died (Dunand et al., 1997; Hunt et al., 1995).

7.2.4 *Acanthamoeba* Keratitis

Acanthamoeba castellanii, A. polyphaga, A. hatchetti, A. culbertsoni, A. rhysodes, A. griffini, A. quina, and *A. lugdunensis*, cause *Acanthamoeba* keratitis (AK) (Bacon et al., 1993; Moore et al., 1985; 1987; Claerhout and Kestelyn, 1999) characterised by corneal ulcers which could lead to blindness (Lorenzo-Morales et al., 2015). It commonly occurs in individuals who are immunocompetent (Niederkorn et al., 1999a; 1999b). An epidemic of AK occurred in 1980 and was predominant among contact lens wearers due to the use of contaminated lenses (Seal and Hay, 1993; 1994).

Additionally, it was observed that the soft contact lens use was the largest risk factor (Scruggs et al., 2022) attributed to poor contact lens hygiene practices like swimming in a stock tank, rubbing eyes after handling frogs, using topical anaesthesia, and/or corneal damage with organic material. Sleeping with contact lenses (34.7%) was the most reported risk factor (Scruggs et al., 2022). Mascarenhas et al. (2014) analysed 115 patients with AK in India, from 2006–2011 and observed that patients with AK were younger with a longer duration of symptoms with a ring infiltrate and the disease was confined to the corneal epithelium. Additionally, use of home-made saline solutions, and corneal abrasions (Auran et al., 1987; Doren et al., 1991; Illingworth and Cook. 1998; Niederkorn, et al., 1999) also contributed to the increase in AK.

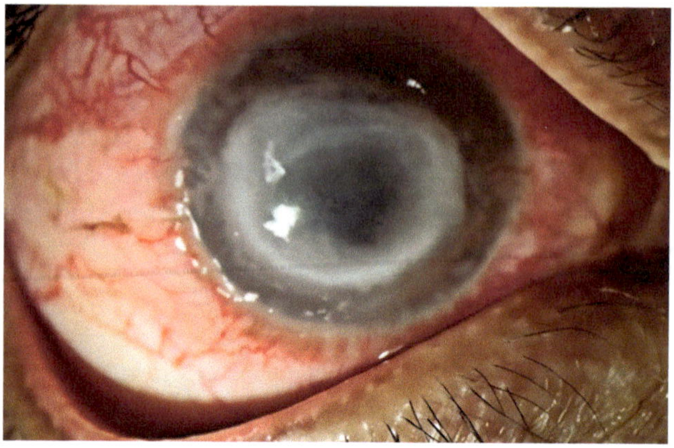

FIGURE 7.1 *Acanthamoeba* keratitis. (Courtesy Dr Bhupesh Bagga, L V. Prasad Eye Institute, Hyderabad, India.)

7.2.4.1 Clinical Features of Acanthamoeba Keratitis

Symptoms of AK include redness, watering, photophobia, and lid edema as in any other form of keratitis. Classical presentation of *Acanthamoeba* keratitis include ring infiltrates (Figure 7.1), which can progress further to loss of epithelium with thinning of stroma and perforations (Clarke and Niederkorn, 2006). The disease can also involve episcleritis and scleritis.

The histopathologic picture of AK varies (Kremer et al., 1994; Mietz and Font, 1997). Initially, amoebae infect the corneal epithelium, then invade and damage the stroma and cause mild to severe inflammation. A ring-like stromal infiltrate, of infiltrating inflammatory cells such as neutrophils is seen in AK patients. Trophozoites infiltrate the nerves of the cornea causing neuritis and necrosis (Garner, 1993; Mathers et al., 1987). In rare circumstances, *Acanthamoeba* from the cornea could spread to the retina, causing chorioretinitis (Johns et al., 1988; Moshari et al., 2001). In severe cases *Acanthamoeba* may not be susceptible to the drug and surgical therapy thus necessitating enucleation of the eye (Dougherty et al., 1994).

7.2.4.2 Diagnosis of Acanthamoeba keratitis

AK can be difficult to diagnose (Sherman, 2022) and treat due to close mimicker with other more common diseases such as HSV and fungal keratitis. But since *Acanthamoeba* contamination of contact lenses and cleaning solutions are the source of infection decontaminating these with hydrogen peroxide (Shoff et al., 2008; Johnston et al., 2009) is recommended whereas commercial contact lens washing solution products (Wright et al., 1985) are not effective. Around 85% of all AK cases in the USA are contact lens wearers due to improper storing or handling of the lenses (Sherman, 2022). Therefore, it makes sense that Pinna (2002) recommended the use of "1-day" disposable contact lenses to avoid keratitis. Microwave irradiation could also kill *Acanthamoeba* spp. (Hiti et al., 2001). Corneal scrapes or corneal biopsy specimens are used for the identification of cysts or trophozoites of *Acanthamoeba* (Hiti et al., 2001; Clarke and Niederkorn, 2006) but swabs obtained either from the cornea or conjunctiva are not suitable (Wright et al., 1981). But, *Acanthamoeba* could be cultured from contact lenses, lens storage cases, lens-cleaning solutions (Armstrong, 2000; Illingworth et al., 1995; Moore and McCulley. 1989; Newman et al., 1997) and from approximately 70–73% of clinical specimens (Penland and Wilhelmus. 1997; Tay-Kearney et al., 1993). Indirect immunofluorescent-antibody assay and several cytological staining methods like Calcofluor white+ were used to identify trophozoites and cysts in paraffin-embedded tissues (Silvany et al., 1987). Staining of corneal scrapings or CSF with Acridine orange has been recommended for rapid histological diagnosis of AK or GAE (Hahn et al., 1998). PCR for detection of *Acanthamoeba* has been demonstrated for the specific detection of *Acanthamoeba* (Lai et al., 1994; Dykova, et al., 1999; Schroeder et al., 2001). More recently real-time quantitative PCR for *Acanthamoeba* detection was recommended over gel-based electrophoresis method (Kandori et al., 2010b: Itahashi et al., 2011; Ikeda et al., 2012). In a review comparing direct smear analysis, culture, and PCR for *Acanthamoeba* ribosomal DNA indicated that the detection of *Acanthamoeba* was best when contaminated contact lenses and casings were used and highest diagnostic sensitivities were reported for direct smear analysis, culture, and PCR (Boggild et al., 2009).

H1 nuclear magnetic resonance (NMR) spectroscopy has also been used to identify *Acanthamoeba* (Hauber et al., 2011). Further, *in vivo* corneal tandem scanning confocal microscopy (TSCM), is rapid, non-invasive, and equally sensitive to either culture or biopsy-based methods for identification of *Acanthamoeba* (Parmar et al., 2006; Hau et al., 2010; Vaddavalli et al., 2011; Tu et al., 2008) but it also depends on the skill and experience of the user.

More recently, fluorescence in situ hybridization was used to enhance the detection of *Acanthamoeba* (Stothard et al., 1999). A variety of molecular methods such as DNA restriction fragment length polymorphism (RFLPs) of the complete nuclear 18S rRNA gene sequences or the conserved region of 18S rRNA gene (Kim et al., 1996; Chung et al., 1998; Szenasi et al., 1998), variation in complete mitochondrial 16S rRNA (Kong and Chung, 1996) or the mitochondrial genome sequence (Byers et al., 1990; Gautom, et al., 1994; Kong et al., 1995; Yagita and Endo. 1990; Dykova et al., 1999; Lehmann et al., 1998; Stothard et al., 1998), and randomly amplified polymorphic DNA (RAPD) analysis (Alves et al., 2000) have been used for identification of *Acanthamoeba*.

Immunodiagnostic probes like antibody fragments specific to *Acanthamoeba* spp. have also been used to detect *Acanthamoeba* by immunofluorescence or flow cytometry (Khan et al., 2000).

Tandem scanning confocal microscopy and scanning confocal microscopy have also been employed for the detection of *Acanthamoeba* (Mathers et al., 1996; Pfister et al., 1996; Sherman, 2022). Page and Mathers, (2013) compared 372 AK cases diagnosed by confocal microscopy and corneal scrapings that were analysed in Oregon, from 1999 to 2011, and recorded an average of 31 new cases of AK per year over the twelve-year period and observed that the sensitivities were 97.2% and 87.8%, respectively (Scruggs et al., 2022).

7.2.4.3 Treatment of Acanthamoeba *keratitis*

Earlier diagnosis of AK (Claerhout and Kestelyn, 1999; Sherman, 2022) is very important to treat effectively with medications (Brooks et al., 1994; Holland et al.,

1987; Lindquist et al., 1988; Ishibashi et al., 1990). When *Acanthamoeba* infection penetrates deeper into the cornea the treatment is prolonged (up to 1 year or longer) (Bacon et al., 1993) since many therapeutic agents are not effective in penetrating tissues beneath the cornea (Dougherty et al., 1994; Ficker et al., 1990; Horne et al., 1994; Murdoch et al., 1998). Binder (1989) suggested cryotherapy for patients who were refractory to medicinal and surgical treatment. The target of AK therapy is to destroy cysts and trophozoites of *Acanthamoeba* and rapidly overcome inflammation (Skryabina et al., 2019). Several therapeutic agents have been used for the successful treatment of AK (Scruggs et al., 2022). A few of the specific treatments are listed below:

(i) 0.1% propamidine isethionate (Brolene) topically with 0.15% dibromopropamidine early in infection (Bacon et al., 1993; Maudgal, 1989; Moore and McCulley. 1989; Seal et al., 1996; Wright et al., 1985; Skryabina et al., 2019),
(ii) topical administration of chlorhexidine and propamidine for extended periods (Seal et al., 1996),
(iii) 0.1% hexamidine for patients with superficial AK (Brasseur et al., 1994),
(iv) 0.006% chlorhexidine solution alone (Kosrirukvongs, et al., 1999) for AK treatment.,
(v) imidazoles (miconazole, itraconazole, and ketoconazole) have been used with limited success (Berger et al., 1990; D'Aversa et al., 1995; Horne et al., 1994; Ishibashi, et al.1990; Lindquist et al., 1988; Moore et al., 1985; Wright et al., 1985; Skryabina et al., 2019; Bang et al., 2010; Tu et al., 2010),
(vi) Polyhexamethylene biguanide (PHMB), a disinfectant with both amoebicidal and cytocidal activity against *Acanthamoeba* strains (Seal et al., 1995) has been employed (Larkin et al., 1992; Clarke and Niederkorn, 2006),
(vii) a combination therapy including chlorhexidine digluconate (a bisbiguanide) with PHMB (a polymeric biguanide) or with aromatic diamidines (hexamidine, pentamidine, and propamidine isethionate) or PHMB and hexamidine with debridement has also been reported for the treatment of AK (Duguid, et al., 1997; Gray et al., 1994; Hay and Seal, 1996; Illingworth and Cook. 1998; Lindquist, 1998; Murdoch et al., 1998; Radford et al., 2002; Larkin et al., 1992; Seal et al., 1995; Wright et al., 1985; Clarke and Niederkorn, 2006; Skryabina et al., 2019),
(viii) eight different alkylphosphocholines have been used against *Acanthamoeba* spp. (Hiti et al., 2002; Skryabina et al., 2019),
(ix) corticosteroids have been used in conjunction with therapeutic agents for the treatment of AK (Auran et al., 1987; Scruggs et al., 2022),
(x) recently, Pitavastatin loaded nanoparticles were designed for the management of *Acanthamoeba* Keratitis. Amoebicidal activity of the nanoparticles was able to eliminate *A. castellanii*, *A. polyphaga, A. griffini* and *A. quina* by chromatin condensation, autophagic vacuoles and mitochondria dysfunction (Sifaoui et al., 2022),
(xi) photorefractive surgery has also been used in the setting of AK. Kandori et al. (2010a; 2010b) observed that AK treated with standard topical drugs sometimes developed large corneal abscesses and this could be removed using laser phototherapeutic keratectomy (PTK),

(xii) surgery is undertaken when the cornea is permanently scarred or in cases when AK is refractory to medical treatment (Clarke and Niederkorn, 2006). Awwad et al. (2005) and Nguyen et al. (2007) reported penetrating keratoplasties (PK) as a surgical approach for AK. But, if AK patients are treated within 6 weeks of presentation, PK may not be required (Illingworth et al., 1995; Skryabina et al., 2019). PK is more successful if performed after resolution of inflammation (Bacon et al., 1993). Szentmary et al. (2012), reported improved outcomes of keratoplasty after three months of keratitis inactivity, suggesting that surgery should be performed later in the clinical course, if possible,

(xiii) Cremona et al. (2002) and Por et al. (2009) reported that deep lamellar keratectomy (DLK) with a conjunctival flap was effective for controlling infection and relieving pain in AK patients. Parthasarathy and Tan (2007) also reported a case of DLK for treatment of refractive AK and

(xiv) corneal crosslinking (CXL) is currently being proposed as one of the possible treatment options before keratoplasty in cases of severe nonresponsive infectious keratitis (Kumar and Lloyd, 2002). Garduno-Vieyra et al. (2011) and Khan et al. (2011) observed significant improvement in administering collagen cross-linking to AK patients. A recent meta-analysis (Ting et al., 2019) showed that in bacterial or fungal keratitis, photoactivated chromophores for keratitis-corneal cross-linking (PACK-CXL) may reduce the time to corneal healing. In this method, the microbicidal effect of PACK-CXL is attributed to ultraviolet A (UVA) induced DNA damage and reactive oxygen species release (Pettersson et al., 2019), while relief from pain may be due to the suppression of inflammation and nociception by subepithelial nerves (Pettersson et al., 2019; Varacalli et al., 2021).

In a recent study, Richardson et al. (2022) compared the anti-amoebic susceptibility patterns between *Acanthamoeba* isolates from India and those of the United States using 5 anti-amoebic agents for all isolates and found no statistically significant differences in anti-amoebic susceptibility suggesting that differential outcomes in *Acanthamoeba* keratitis between the 2 locations is not due to differences in anti-amoebic susceptibility.

7.3 Resistance to Drugs in *Acanthamoeba* Keratitis

Resistance to antimicrobial agents used in the treatment of AK is attributed to the physical barrier of the cyst wall (Murdoch et al., 1998; Turner et al., 2000) (Table 7.1). Alkyl phosphocholines are now being screened as amoebicidal agents in *Acanthamoeba* and results indicated that hexadecyl phosphocholine may be useful for treatment of GAE as well as AK since Alkyl phosphocholines have the ability to cross the blood-brain barrier (Hiti et al., 2002). Recent studies suggested that sIgA, complement, neutrophils, and macrophages may influence resistance to infection (Marciano-Cabral and Guy Cabral, 2003).

TABLE 7.1

Therapy and Outcomes with Respect to *Acanthamoeba* Keratitis

Patient No.	Therapy[a,b]	Outcome	Reference
1	Propamidine isethionate, Neosporin	Toxicity	Wright et al. (1985), Skryabina et al. (2019), Kenyon and Binotti (2020) and Sherman (2022).
2	PHMB[b]	Successful	Larkin et al. (1992) and Sherman, (2022).
3	1% clotrimazole eye drops, Dibromopropamidine	Recurrence	Dougherty et al. (1994) and Kenyon and Binotti (2020).
	Oral fluconazole, clotrimazole, PHMB	Recurrence Progression to enucleation	Dougherty et al. (1994), Skryabina et al. (2019) and Kenyon and Binotti (2020).
4	Propamidine isethionate, neosporin	Treatment failure	Gray et al. (1994) and Kenyon and Binotti, (2020).
	Miconazole, PNA, itraconazole	Toxicity	
	PHMB	Resolution of disease	
5	Miconazole, metronidazole, PNA, neomycin, ketoconazole	Residual infiltrate	Horne et al. (1994) and Kenyon and Binotti (2020).
6	Miconazole, metronidazole, propamidine, ketoconazole	Recurrence	Horne et al. (1994) and Kenyon and Binotti (2020).
7	Hexamidine	Well tolerated cure	Brasseur et al. (1994) and Sherman (2022).
8	PHMB and propamidine isethionate; steroids only after clinical improvement	Cure, long-term treatment necessary	Illingworth et al. (1995).
9	Chlorhexidine and propamidine	Effective if used for the long term	Seal et al. (1996), Kenyon and Binotti (2020) and Sherman (2022).

10	Itraconazole, miconazole	Unresponsive	Murdoch et al. (1998) and Skryabina et al. (2019).
	Penetrating keratoplasty	Successful	Skryabina et al. 2019.
11	Propamidine, neosporin, itraconazole, miconazole	Corneal perforation, keratoplasty	Murdoch et al. (1998) and Sherman (2022).
12	Propamidine, PHMB	Resolution of disease	Murdoch et al. (1998) and Skryabina et al. (2019).
13	PHMB, propamidine, PNA, chlorhexidine	Propamidine toxicity	Murdoch et al. (1998) and Skryabina et al. (2019).
14	Chlorhexidine, PHMB, propamidine isethionate	Treatment of choice	Lindquist et al. (1988).
15	Corneal crosslinking	Nonresponsive infectious keratitis	Skryabina et al. (2019) and Sherman (2022).

Source: Adopted from Marciano-Cabral F. and Cabral, G., 2003.

Notes:

[a] PNA, prednisolone acetate.

[b] PHMB, polyhexamethylene biguanide is used as an experimental drug.

7.3.1 Role of the Immune System in *Acanthamoeba* Keratitis

Immune response to *Acanthamoeba* has been investigated in several animals. In pigs immunized with an aqueous extract of *Acanthamoeba* intramuscularly, subconjunctivally, or by both routes yielded conflicting reports of protection when challenged with Acanthamoeba-laden "soft contact lenses". Furthermore, prior ocular infection with *Acanthamoeba* did not provide protection from reinfection. The response to immunization in hamsters was also poor (Van Klink et al., 1997). It was also observed that drug-depletion of macrophages exacerbated severity of infection in the Chinese hamster and prolonged the disease (Van Klink et al., 1996). McCulley et al. (1995) reported that stromal necrosis in *Acanthamoeba* lesions is mediated by proteases released by neutrophils whereas in the cornea of the Chinese hamster, resolution of disease occurs with production of macrophage inflammatory protein 2 (MIP-2) (Hurt et al., 2001). Walochnik et al. (2001) found that compared to uninfected individuals sera from AK patients showed weak immunoreactivity to the pathogenic strain. In contrast, oral immunization with *Acanthamoeba* antigens induced corneal resistance to *Acanthamoeba* infection in two animal models, the hamster and pig models of AK (Van Klink et al., 1997; Leher et al., 1998a; 1988b) and the protection positively correlated with production of IgA antibodies. *Acanthamoeba*-specific monoclonal IgA antibodies also protected against AK in hamster (Leher et al., 1999). It was observed that amoeba-specific tear IgA were significantly lower in patients with AK but serum immunoglobulin levels were similar between healthy subjects and patients with AK (Alizadeh et al., 2001). The available data suggested that both innate and acquired immune responses are involved in AK infection. The immunoglobulins (IgA) may inhibit the attachment of *Acanthamoebae* to corneal cells, while neutrophils and macrophages destroy *Acanthamoebae*.

7.3.2 Virulence Factors of *Acanthamoeba* spp.

Virulence in *Acanthamoeaba* (Walochnik, et al., 2000) is dependent on adherence of *Acanthamoeba* trophozoites to the target cells followed by damage and invasion of tissue to establish the infection. In the first step of infection the *Acanthamoeba* (*A. polyphaga*, *A. castellanii*, and *A. culbertsoni*) attach to mannose-containing glycoproteins on the corneal epithelium (Kennett et al., 1999) through a 136-kDa mannose-binding protein on the surface of *Acanthamoeba* (Yang et al., 1997) thus mediating the adherence between the parasite and the cornea (Panjwani et al., 1992). *Acanthamoeba* binds preferentially to collagen IV, laminin, and fibronectin on the host cells (Gordon et al., 1993). Following adhesion, amoeba invades and destroys human epithelial cells, stromal keratocytes, and stromal cell homogenates by phagocytosis (Moore et al., 1991; Stopak et al., 1991) and by the release of cytolytic substances specific to the target cells (Badenoch et al., 1995; Larkin et al., 1991; Stopak et al., 1991). Enzymes secreted by *Acanthamoeba* (Alfieri et al., 2000) like serine proteinases, cysteine proteinases, metalloproteinases and elastases play a role in contact-independent injury to cells and facilitate the invasion of host tissue (Hadas and Mazur, 1993; Mitro et al., 1994; Cho et al., 2000; Ferrante and Bates, 1988; He et al., 1990; Mitra et al., 1995). Secretion of these proteinases, phospholipases and sphingomyelinases correlated positively with invasive properties of *Acanthamoeba* (Van Klink et al.,

1992; Cursons and Brown, 1978; Cursons et al., 1978; Elson et al., 1970; Mishra et al., 1986; Victoria et al., 1975a; 1975b; Visvesvara and Balamuth, 1975).

7.3.3 Cytopathic Effects of *Acanthamoeba* spp.

Adherence to host cells is a prerequisite for amoeba-induced cytolysis of host cells (Cao et al., 1998), which is referred to as cyto-pathetic effects (CPE) of the amoeba. As a consequence of the CPE the host cells die (Jahnes et al., 1957; Leher et al., 1998b; Marciano-Cabral et al., 2000; Pettit et al.1996). Badenoch et al. (1988) observed an 86% correlation between virulence of trophozoites *in vivo* and CPE *in vitro*. *A. castellanii* ATCC 30868 from human corneal tissue produced CPE on a variety of cell types including primary pig epithelium, human corneal endothelium, murine neuroblastoma cells, human ocular melanomas, and murine fibrosarcomas (Cursons and Brown, 1978; Cursons et al., 1978; Niszl et al., 1998). Lagmay et al. (2000) also reported that amoebae isolated from AK patients and from an environmental isolate produced CPE on C6 rat glial cells. This may be similar to the amebostomes (i.e., food cups) produced by *A. lenticulata* and an unidentified species of *Acanthamoeba*, which ingested Vero cells (Diaz et al., 1991). Pettit et al. (1996) also demonstrated that trophozoites of *Acanthamoeba* induced CPE on cultured rat B103 nerve cells and ingested the cells through the formation of food cups. Studies have indicated that CPE is distinct from apoptosis observed in the lysis of tumor cells by *A. castellanii* (Alizadeh et al., 1994) and calcium channels and cytoskeletal elements were involved in the CPE of ocular melanoma cells by *A. castellanii* (Taylor et al., 1995). Several *Acanthamoeba* species including *A. culbertsoni*, *A. castellanii*, *A. polyphaga*, and *A. aastronyxis* caused cytolysis through a contact-dependent process involving digipodia from the surface of the amoeba to the membrane of the target cell (Bacon et al., 1993). In addition, during CPE amoebae release ADP into the culture medium, resulting in a biphasic rise in $[Ca]_i$, which culminates in apoptosis (Mattana et al., 1997; 2001). Cell death induced by *Acanthamoeba* spp. may also occur by necrosis (Pettit et al., 1996).

7.4 Animal Models of *Acanthamoeba* Keratitis

Appropriate animal models for studying AK are yet to be identified since many animals are not susceptible to infection. A few of the models that have been used include:

(a) A rabbit model: animals were administered subconjunctival or topical steroids and subsequently inoculated intrastromally with *A. castellanii.* Rabbits exhibited necrosis and an inflammatory response. Treatment with steroids did not enhance the severity of AK or the rate of isolation of amoebae (Font et al. (1982),
(b) A rat model: AK could be induced in rats by inoculating amoebae into the corneal stroma or when the amoebae were co-injected with *Corynebacterium* (Badenoch et al. 1990; 1991),
(c) A pig model: Pigs have been used to study AK especially with respect to using "contact lenses" laden with amoebae to mimic human exposure. But this was

not acceptable since, unlike in human infection with *Acanthamoeba*, in pig the infection spontaneously resolved in 8 to 10 weeks (He et al., 1992) and

(d) A hamster model: In this model "contact lenses" loaded with *Acanthamoeba* were applied to abraded corneal surfaces of the hamster. The infection sustained for 5 to 7 days, closely resembling the human infection. The corneal disease was not produced when "contact lenses" with amebae were applied to intact cornea (Van Klink et al., 1993). This model differs in that the disease is acute and self-limiting.

7.5 *Acanthamoeba*–bacterium Interactions

Acanthamoeba spp. are predators of bacteria preferring "edible" bacteria like *E. coli* K-12, *Klebsiella aerogenes* and cynanobacteria over the "non-edible" *Bacillus subtilis* or *Serratia marcescens* (Weekers et al., 1993; Wright et al., 1981; Rodriguez-Zaragoza, 1994). Several different types of bacteria have been identified in *Acanthamoeba* isolates, including chlamydia-like and rickettsial-related organisms (Aman et al., 1997; Birtles et al., 1997; Fritsche et al., 1993; 1999). Human corneal specimens also contained intracellular endosymbionts, which were not prone to culture (Fritsche et al., 1993) and were tentatively assigned to *Candidatus Caedibacter acanthamoebae*, *Candidatus paracaedibacter*, and *Candidatus Paracaedibacter symbiosus.* The presence of *Rickettsiae* like bacteria suggested that *Acanthamoeba* may be involved in the transmission of rickettsia in human beings (Horn et al., 1999). Additionally, chlamydia-like endosymbionts of the protozoan may be involved in respiratory disease (Birtles et al., 1997). In fact studies have indicated that the endosymbiont Legionella *pneumophila,* the causative agent of atypical pneumonia, is associated with *Acanthamoeba* in humans suggesting that *Acanthamoeba* are the source of legionellosis (Rowbotham, 1980; Harf 1994; Harf and Monteil. 1988; Abu Kwaik et al., 1998a; 1998b; Hagele et al., 1998). Gao and Kwaik (2000), confirmed that *L. pneumophila* kill amebae by necrosis. *Legionella*-like amoebic pathogens (LLAP) were first described in 1956 by Drozanski (1956) and subsequently a number of LLAPs 1 through 12 (Adeleke et al., 1996; Rowbotham, 1993) were described and suggested to play a role in community-acquired pneumonia (Marrie et al., 2001).

Apart from *Legionella* and *Legionella*-like bacteria several other species including *Candidatus parachlamydia acanthamoebae*, *Candidatus Odyssella thessalonicensis*, *Rickettsia* spp., *P. aeruginosa*, and *Comamonas acidovoran* (Aman et al., 1997; Birtles et al., 1997; Fritsche et al., 1993; Michel et al., 1995; Michel and Hauroder, 1997), *Mycobacterium avium*, *M. leprae*, *Burkholderia cepacia*, *B. pseudomallei*, *Simkania negevensis*, *E. coli* O157:H7, *Helicobacter pylori*, *Chlamydia pnuemoniae*, and at least three species of *Listeria* are endosymbionts in *Acanthamoeba* (Barker et al., 1999; Essig et al., 1997; Jadin, 1975; Krishna prasad and Gupta, 1978; Landers et al., 2000; Ly et al., 1990; Winiecka-Krusnell et al., 2002). In addition, *Vibrio cholerae* strains (Thom et al., 1992), *Coxiella burnetti* (La Scola and Raoult, 2001) *Salmonella enterica* serovar *Typhimurium*, (Gaze et al., 2001), *Helicobacter pylori* (Badenoch et al., 1988), *P. aeruginosa* (Badenoch et al., 1990) could survive in acanthamoeba. The yeast *Cryptococcus neoformans* also replicates in *Acanthamoeba*

(Steenbergen et al., 2001). Thus, several bacteria, and one yeast, can proliferate within *Acanthamoeba.*

7.6 Effect of Antimicrobials on the Endosymbiont Bacteria inside *Acanthamoeba*

Bacteria residing in *Acanthamoeba* exhibit increased resistance to biocides and increased bacterial virulence (Barker et al., 1995; King et al., 1988). For example, *L. pneumophila* and *M. avium* in *A. polyphaga* and *A. castellanii* respectively were more protected from antimicrobials. *L. pneumophila* endosymbionts recovered from *Acanthamoeba*, were about 1,000-fold more resistant to rifampin or ciprofloxacin compared to in vitro cultivated *L. pneumophila* (Barker et al., 1993). In fact, Miltner and Bermudez (2000) observed that *M. avium* in *A. castellanii* was not susceptible to rifabutin, azithromycin, and clarithromycin. But when *M. avium* infected U937 macrophage-like cells the bacterium was susceptible to the same set of antibiotics. *Acanthamoeba* spp. are not susceptible to chlorination and confer protection against chlorine to the bacteria residing within like *S. Typhimurium*, *Yersinia enterocolitica*, *Shigella sonnei*, and *Campylobacter jejuni* (King et al., 1988). *Acanthamoeba* organisms even in the cyst stage protect the endosymbiont bacteria (Rowbotham, 1986; Turner et al., 2000; Kilvington and Price, 1990). Thus, it is possible that *Acanthamoeba* serve as bacterial reservoirs for human infection for different types of bacteria like *Legionella*, *Pseudomonas* spp., from various habitats such as eye wash stations (Paszko-Kolva et al., 1991) and contaminated drinking water system in a hospital (Michel et al., 1995; Walochnik et al., 1999). Studies have also indicted that coinfection with both the bacterium and protozoa was perquisite for keratitis. For instance, rat corneas infected with both Corynebacterium and *Acanthamoeba* caused suppurative keratitis whereas infection with either the bacterium or just the potozoan did not (Larkin and Easty 1990). Further, *Acanthamoeba* with bacterial endosymbionts cause CPE on cultured human embryonic fibroblast cells more quickly than those devoid of endosymbionts (Fritsche et al., 1998). Thus, *Acanthamoeba* spp. act as cofactors in human infection and also are reservoirs for bacterial pathogens.

7.7 Conclusions

Acanthamoeba spp. are distributed all over the world and cause GAE, AK, and cutaneous and sinus lesions. AK is predominant in immunodeficient individuals and in individuals with compromised immune systems, including those with AIDS but also occurs in immunocompetent individuals. Successful therapy of *Acanthamoeba* infections is dependent on rapid diagnosis which is hindered by factors such as the lack of familiarity of most clinicians with the disease and limited therapeutic options. Apart from routine *in vitro* culture of amoebae diagnosis of the parasite is also carried out on tissue sections or cytological preparations including newer molecular methods. *Acanthamoeba* spp. causes disease both by cell contact-dependent and -independent processes. *Acanthamoeba* spp. sequester a variety of bacteria and act as a reservoir of pathogens. The observed increase in the incidence of *Acanthamoeba* infections may be due to better diagnosis and the fact that there are more immunocompromised individuals like HIV-infected or patients undergoing cancer chemotherapy or immunosuppressive therapy.

REFERENCES

Abu Kwaik, Y., L. Y. Gao, B. J. Stone, and O. S. Harb. 1998a. Invasion of mammalian and protozoan cells by Legionella pneumophila. *Bull. Inst. Pasteur.* 96:237–247.

Abu Kwaik, Y., L. Y. Gao, B. J. Stone, C. Venkataraman, and O. S. Harb. 1998b. Invasion of protozoa by Legionella pneumophila and its role in bacterial ecology and pathogenesis. *Appl. Environ. Microbiol.* 64:3127–3133.

Adeleke, A., J. Pruckler, R. Benson, T. Rowbotham, M. Halablab, and B. Fields. 1996. Legionella-like amebal pathogens—phylogenetic status and possible role in respiratory disease. *Emerg. Infect. Dis.* 2:225–230.

Alfieri, S. C., C. E. Correia, S. A. Motegi, and E. M. Pral. 2000. Proteinase activities in total extracts and in medium conditioned by Acanthamoeba polyphaga trophozoites. *J. Parasitol.* 86:220–227.

Alizadeh, H., S. Apte, M. S. El Agha, L. Li, M. Hurt, K. Howard, H. D. Cavanagh, J. P. McCulley, and J. Y. Niederkorn. 2001. Tear IgA and serum IgG antibodies against Acanthamoeba in patients with Acanthamoeba keratitis. *Cornea* 20:622–627.

Alizadeh, H., M. S. Pidherney, J. P. McCulley, and J. Y. Niederkorn. 1994. Apoptosis as a mechanism of cytolysis of tumor cells by a pathogenic free-living amoeba. *Infect. Immun.* 62:1298–1303.

Alves, J. M., C. X. Gusmao, M. M. Teixeira, D. Freitas, A. S. Foronda, and H. T. Affonso. 2000. Random amplified polymorphic DNA profiles as a tool for the characterization of Brazilian keratitis isolates of the genus Acanthamoeba. *Braz. J. Med. Biol. Res.* 33:19–26.

Amann, R., N. Springer, W. Schönhuber, W. Ludwig, E. N. Schmid, K. D. Müller, and R. Michel. Obligate intracellular bacterial parasites of acanthamoebae related to Chlamydia spp. *Appl. Environ. Microbiol.* 1997, 63, 115–121.

Anderlini, P., D. Przepiorka, M. Luna, L. Langford, M. Andreeff, D. Claxton, and A. B. Deisseroth. 1994. Acanthamoeba meningoencephalitis after bone marrow transplantation. *Bone Marrow Transplant.* 14:459–461.

Armstrong, M. 2000. The pathogenesis of human Acanthamoeba infection. *Infect. Dis. Rev.* 2:65–73.

Auran, J. D., M. B. Starr, and F. A. Jakobiec. 1987. Acanthamoeba keratitis. A review of the literature. *Cornea.* 6:2–26.

Awwad, S T., D. N. Parmar, M. Heilman, R. W. Bowman, J. P. McCulley, and H. D. Cavanagh. 2005. Results of penetrating keratoplasty for visual rehabilitation after Acanthamoeba keratitis. *Am. J. Ophthalmol.* 140 (6). 1080.e2–1084.e2

Bacon, A. S., D. G. Frazer, J. K. Dart, M. Matheson, L. A. Ficker, and P.Wright. 1993. A review of 72 consecutive cases of Acanthamoeba keratitis,1984–1992. *Eye* 7:719–725.

Badenoch, P. R., M. Adams, and D. J. Coster. 1995. Corneal virulence, cytopathic effect on human keratocytes and genetic characterization of Acanthamoeba. *Int. J. Parasitol.* 25:229–239.

Badenoch, P. R., T. R. Grimmond, J. Cadwgan, S. E. Deayton, and M. S. L. Essery. 1988. Nasal carriage of free-living amoebae. *Microb. Ecol. Health Dis.* 1:209–211.

Badenoch, P. R., A. M. Johnson, P. E. Christy, and D. J. Coster. 1990. Pathogenicity of Acanthamoeba and a Corynebacterium in the rat cornea. *Arch. Ophthalmol.* 108:107–112.

Badenoch, P. R., A. M. Johnson, P. E. Christy, and D. J. Coster. 1991. A model of Acanthamoeba keratitis in the rat. *Rev. Infect. Dis.* 13 (Suppl. 5):S445.

Baig, A. M. (December 2014). Granulomatous amoebic encephalitis: Ghost response of an immunocompromised host? *J. Med. Microbiol.* 63 (Pt 12):1763–1766.

Baig, A. M, Iqbal J., Khan N. A. (August 2013). In vitro efficacies of clinically available drugs against growth and viability of an Acanthamoeba castellanii keratitis isolate belonging to the T4 genotype. *Antimicrob. Agents Chemother.* 57 (8):3561–3567.

Bang, S., E. Edell, A. O. Eghrari, and J. D. Gottsch. 2010. Treatment with voriconazole in 3 eyes with resistant Acanthamoeba keratitis. *Am. J. Ophthalmol.* 149 (1):66–69.

Barker, J., P. A. Lambert, and M. R. Brown. 1993. Influence of intraamoebic and other growth conditions on the surface properties of Legionella pneumophila. *Infect. Immun.* 61:3503–3510.

Barker, J., H. Scaife, and M. R. Brown. 1995. Intraphagocytic growth induces an antibiotic-resistant phenotype of Legionella pneumophila. *Antimicrob. Agents Chemother.* 39:2684–2688.

Barker, J., T. J. Humphrey, and M. W. Brown. 1999. Survival of Escherichia coli O157 in a soil protozoan: Implications for disease. *FEMS Microbiol. Lett.* 173:291–295.

Berger, S. T., B. J. Mondino, R. H. Hoft, P. B. Donzis, G. N. Holland, M. K. Farley, and J. E. Levenson. 1990. Successful medical management of Acanthamoeba keratitis. *Am. J. Ophthalmol.* 110:395–403.

Binder, P. S. 1989. Cryotherapy for medically unresponsive acanthamoeba keratitis. *Cornea* 8:106–114.

Birtles, R. J., T. J. Rowbotham, C. Storey, T. J. Marrie, and D. Raoult. 1997. Chlamydia-like obligate parasite of free-living amoebae. *Lancet.* 349: 925–926

Bogler, S. A., C. D. Zarley, L. L. Burianek, P. A. Fuerst, and T. J. Byers. 1983. Interstrain mitochondrial DNA polymorphism detected in Acanthamoeba by restriction endonuclease analysis. *Mol. Biochem. Parasitol.* 8:145–163.

Boggild, A. K., Martin DS, Lee TY, Yu B, Low DE.. (May 2009). Laboratory diagnosis of Amoebic Keratitis: Comparison of four diagnostic methods for different types of clinical specimens. *J. Clin. Microbiol.* 47(5):1314–1318. doi: 10.1128/JCM.00173-09. Epub 2009 Mar 25. PMID: 19321727; PMCID: PMC2681827.

Bonilla, H. F., A. Whitehurst, and C. A. Kauffman. 1999. Acanthamoeba sinusitis and disseminated infection in a patient with AIDS. *Infect. Med.* 16:397–400.

Booton, G. C., D. R. Ledee, M. H. Awwad, S. Sharma, I. Nizsl, M. M. Markus, P. A. Fuerst, and T. J. Byers. 2001. Acanthamoeba mitochondrial 16S rDNA sequences: Inferred phylogeny and support of nuclear ribosomal 18S rDNA gene sequence types. In S. Billot-Bonef, P. A. Cabanes, F. Marciano-Cabral, P. Pernin, and E. Pringuez (ed.), *IXth International Meeting on the Biology and Pathogenicity of Free-Living Amoebae Proceedings*. John Libbey Eurotext, Paris, France, p. 227–234.

Brasseur, G., L. Favennec, D. Perrine, J. P. Chenu, and P. Brasseur. 1994. Successful treatment of Acanthamoeba keratitis by hexamidine. *Cornea.* 13:459–462.

Brooks, J. G., Jr., D. J. Coster, and P. R. Badenoch. 1994. Acanthamoeba keratitis. Resolution after epithelial debridement. *Cornea.* 13:186–189.

Byers, T. J., G. C. Booton, D. R. Stothard, R. J. Gast, D. R. Ledee, J. M. Schroeder, M. H. Awwad, and P. A. Fuerst. 2001. Studies of the phylogeny, systematics and pathogenicity of Acanthamoeba using ribosomal and transfer RNA gene sequences. In

S. Billot-Bonef, P. A. Cabanes, F. Marciano-Cabral, P. Pernin, and E. Pringuez (ed.), *IXth International Meeting on the Biology and Pathogenicity of Free-Living Amoebae Proceedings*. John Libbey Eurotext, Paris, France, p. 211–218.

Byers, T. J., E. R. Hugo, and V. J. Stewart. 1990. Genes of Acanthamoeba: DNA, RNA and protein sequences (a review). *J. Protozool.* 37:17S–25S.

Cao, Z., D. M. Jefferson, and N. Panjwani. 1998. Role of carbohydratemediated adherence in cytopathogenic mechanisms of Acanthamoeba. *J. Biol. Chem.* 273:15838–1584

Callicott, J. H., Jr. 1968. Amebic meningoencephalitis due to free-living amebas of the Hartmannella (Acanthamoeba)-Naegleria group. *Am. J. Clin. Pathol.* 49:84–91.

Calore, E. E., M. J. Cavaliere, and N. M. Calore. 1997. Cerebral amebiasis in the acquired immunodeficiency syndrome. *Acta Neurol. Belg.* 97:248–250.

Casper, T., D. Basset, C. Leclercq, J. Fabre, N. Peyron-Raison, and J. Reynes. 1999. Disseminated acanthamoeba infection in a patient with AIDS: Response to 5-fluorocytosine therapy. *Clin. Infect. Dis.* 29:944–945.

Chandrasekar, P. H., P. S. Nandi, M. R. Fairfax, and L. R. Crane. 1997. Cutaneous infections due to Acanthamoeba in patients with acquired immunodeficiency syndrome. *Arch. Intern. Med.* 157:569–572.

Cho, J. H., B. K. Na, T. S. Kim, and C. Y. Song. 2000. Purification and characterization of an extracellular serine proteinase from Acanthamoeba castellanii. *IUBMB Life.* 50:209–214.

Chung, D. I., H. S. Yu, M. Y. Hwang, T. H. Kim, T. O. Kim, H. C. Yun, and H. H. Kong. 1998. Subgenus classification of Acanthamoeba by riboprinting. *Korean J. Parasitol.* 36:69–80.

Claerhout, I., and P. H Krestelyn. 1999. Acantahamoeba keratitis: Review Bull. *Soc. belge Ophtalmol.* 274:71–82.

Clarke, D. W., and J. Y. Niederkorn. 2006. The immunobiology of Acanthamoeba keratitis. *Microbes and Infection*, 8(5):1400–1405. ISSN 1286-4579 https://doi.org/10.1016/j.micinf.2005.12.009

Cleland, P. G., R. V. Lawande, G. Onyemelukwe, and H. C. Whittle. 1982. Chronic amebic meningoencephalitis. *Arch. Neurol.* 39:56–57.

Cremona, G., M. A. Carrasco, A. Tytiun, and M. J. Cosentino. 2002. Treatment of advanced acanthamoeba keratitis with deep lamellar keratectomy and conjunctival flap. *Cornea* 21:705–708.

Culbertson, C. G. 1961. Pathogenic Acanthamoeba (Hartmannella). *Am. J. Clin. Pathol.* 35:195–202.

Cursons, R. T., and T. J. Brown. 1978. Use of cell cultures as an indicator of pathogenicity of free-living amoebae. *J. Clin. Pathol.* 31:1–11.

Cursons, R. T., T. J. Brown, and E. A. Keys. 1978. Virulence of pathogenic free-living amebae. *J. Parasitol.* 64:744–745.

D'Aversa, G., G. A. Stern, and W. T. Driebe, Jr. 1995. Diagnosis and successful medical treatment of Acanthamoeba keratitis. *Arch. Ophthalmol.* 113:1120–1123.

Diaz, J., A. Osuna, M. J. Rosales, J. Cifuentes, and C. Mascaro. 1991. Sucker-like structures in two strains of Acanthamoeba: Scanning electron microscopy study. *Int. J. Parasitol.* 21:365–367.

Dougherty, P. J., P. S. Binder, B. J. Mondino, and B. J. Glasgow. 1994. Acanthamoeba sclerokeratitis. *Am. J. Ophthalmol.* 117:475–479.

Duguid, I. G., J. K. Dart, N. Morlet, B. D. Allan, M. Matheson, L. Ficker, and S. Tuft. 1997. Outcome of acanthamoeba keratitis treated with polyhexamethyl biguanide and propamidine. *Ophthalmol.* 104:1587–1592.

De Jonckheere, J. F. 1991. Ecology of Acanthamoeba. *Rev. Infect. Dis.* 13 (Suppl 5):S385–387.

Doren, G. S., E. J. Cohen, S. E. Higgins, I. J. Udell, R. C. Eagle, Jr., J. J. Arentsen, and P. R. Laibson. 1991. Management of contact lens associated Acanthamoeba keratitis. *CLAO J.* 17:120–125.

Drozanski, W. 1956. Fatal bacterial infection in soil amoebae. *Acta Microbiol. Pol.* 5:315–317.

Duluol, A. M., M. F. Teilhac, and J. L. Poirot. 1996. Cutaneous lesions due to Acanthamoeba sp. in patients with AIDS. *J. Eukaryot. Microbiol.* 43:130–131.

Dunand, V. A., S. M. Hammer, R. Rossi, M. Poulin, M. A. Albrecht, J. P. Doweiko, P. C. DeGirolami, E. Coakley, E. Piessens, and C. A. Wanke. 1997. Parasitic sinusitis and otitis in patients infected with human immunodeficiency virus: Report of five cases and review. *Clin. Infect. Dis.* 25:267–272.

Dykova, I., J. Lom, J. M. Schroeder-Diedrich, G. C. Booton, and T. J. Byers. 1999. Acanthamoeba strains isolated from organs of freshwater fishes. *J. Parasitol.* 85:1106–1113.

Elson, C., R. P. Geyer, and R. S. Chang. 1970. A lecithinase from the amoeba Hartmannella rhysodes. *J. Protozool.* 17:440–445.

Essig, A., M. Heinemann, U. Simnacher, and R. Marre. 1997. Infection of Acanthamoeba castellanii by Chlamydia pneumoniae. *Appl. Environ. Microbiol.* 63:1396–1399.

Ferrante, A., and E. J. Bates. 1988. Elastase in the pathogenic free-living amoebae Naegleria and Acanthamoeba spp. *Infect. Immun.* 56:3320–3321.

Ficker, L., D. Seal, D. Warhurst, and P. Wright. 1990. Acanthamoeba keratitis—resistance to medical therapy. *Eye* 4:835–838.

Font, R. L., M. J. Tapert, N. M. Robinson, G. Visvesvara, D. Murphy, and M. S. Osato. 1982. An animal model of Acanthamoeba keratitis. Further studies with emphasis on the early destruction of trophozoites. *Investig. Ophthalmol. Visual Sci.* 22:S163.

Friedland, L. R., S. A. Raphael, E. S. Deutsch, J. Johal, L. J. Martyn, G. S. Visvesvara, and H. W. Lischner. 1992. Disseminated Acanthamoeba infection in a child with symptomatic human immunodeficiency virus infection. *Pediatr. Infect. Dis. J.* 11:404–407.

Fritsche, T. R., M. Horn, S. Seyedirashti, R. K. Gautom, K. H. Schleifer, M. Wagner. 1999. In situ detection of novel bacterial endosymbionts of Acanthamoeba spp. phylogenetically related to members of the order Rickettsiales. *Appl. Environ. Microbiol.* 65:206–212.

Fritsche, T. R., R. K. Gautom, S. Seyedirashti, D. L. Bergeron, and T. D. Lindquist. 1993. Occurrence of bacterial endosymbionts in Acanthamoeba spp. isolated from corneal and environmental specimens and contact lenses. *J. Clin. Microbiol.* 31:1122–1126.

Fritsche, T. R., D. Sobek, and R. K. Gautom. 1998. Enhancement of in vitro cytopathogenicity by Acanthamoeba spp. following acquisition of bacterial endosymbionts. *FEMS Microbiol. Lett.* 166:231–236.

Gao, L. Y., and Y. A. Kwaik. 2000. The mechanism of killing and exiting the protozoan host Acanthamoeba polyphaga by Legionella pneumophila. *Environ. Microbiol.* 2:79–90.

Garduno-Vieyra, L., C. R. Gonzalez-Sanchez, and S. E. Hernandez-da Mota. 2011. Ultraviolet—a light and riboflavin therapy for acanthamoeba keratitis: A case report. *Ophthalmol.* 2 (2):291–295.

Gast, R. J., D. R. Ledee, P. A. Fuerst, and T. J. Byers. 1996. Subgenus systematics of Acanthamoeba: four nuclear 18S rDNA sequence types. *J. Eukaryot. Microbiol.* 43:498–504.

Gautom, R. K., S. Lory, S. Seyedirashti, D. L. Bergeron, and T. R. Fritsche. 1994. Mitochondrial DNA fingerprinting of Acanthamoeba spp. Isolated from clinical and environmental sources. *J. Clin. Microbiol.* 32:1070–1073.

Gaze, W. H., N. Burroughs, M. P. Gallagher, and E. M. H. Wellington. 2001. Interactions between Salmonella typhimurium and Acanthamoeba polyphaga: A new mode of intracellular growth within contractile vacuoles. In S. Billot-Bonef, P. A. Cabanes, F. Marciano-Cabral, P. Pernin, and E. Pringuez (ed.), *IXth International Meeting on the Biology and Pathogenicity of Free-Living Amoebae Proceedings*. John Libbey Eurotext, Paris, France, p. 155–157.

Gonzalez, M. M., E. Gould, G. Dickinson, A. J. Martinez, G. Visvesvara, T. J. Cleary, and G. T. Hensley. 1986. Acquired immunodeficiency syndrome associated with Acanthamoeba infection and other opportunistic organisms. *Arch. Pathol. Lab. Med.* 110:749–751.

Gordon, V. R., E. K. Asem, M. H. Vodkin, and G. L. McLaughlin. 1993. Acanthamoeba binds to extracellular matrix proteins in vitro. *Investig. Ophthalmol. Visual Sci.* 34:658–662.

Gordon, S. M., J. P. Steinberg, M. H. DuPuis, P. E. Kozarsky, J. F. Nickerson, and G. S. Visvesvara. 1992. Culture isolation of Acanthamoeba species and leptomyxid amebas from patients with amebic meningoencephalitis, including two patients with AIDS. *Clin. Infect. Dis.* 15:1024–1030.

Gray, T. B., K. A. Gross, R. T. Cursons, and J. F. Shewan. 1994. Acanthamoeba keratitis: A sobering case and a promising new treatment. *Aust. N. Z. J. Ophthalmol.* 22:73–76.

Hadas, E., and T. Mazur. 1993. Proteolytic enzymes of pathogenic and non-pathogenic strains of Acanthamoeba spp. *Trop. Med. Parasitol.* 44:197–200.

Hagele, S., J. Hacker, and B. C. Brand. 1998. Legionella pneumophila kills human phagocytes but not protozoan host cells by inducing apoptotic cell death. *FEMS Microbiol. Lett.* 169:51–58.

Hahn, T. W., T. P. O'Brien, W. J. Sah, and J. H. Kim. 1998. Acridine orange staining for rapid diagnosis of Acanthamoeba keratitis. *Jpn. J. Ophthalmol.* 42:108–114.

Harf, C. 1994. Free-living amoeba: Interactions with environmental pathogenic bacteria. *Endocytobiosis Cell Res.* 10:167–183.

Harf, C., and H. Monteil. 1988. Interaction between free-living amoebae and Legionella in the environment. *Water Sci. Technol.* 20:235–239.

Hau, S. C., J. K. G. Dart, M. Vesaluoma et al., 2010. Diagnostic accuracy of microbial keratitis with in vivo scanning laser confocal microscopy. *Br. J. Ophthalmol.* 94 (8):982–987.

Hauber, S., H. Parkes, R. Siddiqui, and N. A. Khan. 2011. The use of high-resolution 1H nuclear magnetic resonance (NMR) spectroscopy in the clinical diagnosis of Acanthamoeba. *Parasitol. Res.* 109 (6):1661–1669.

Hay, J., and D. V. Seal. 1996. Acanthamoeba keratitis: Risk factors and outcome. *Br. J. Ophthalmol.* 80:773.

He, Y. G., J. P. McCulley, H. Alizadeh, M. Pidherney, J. Mellon, J. E. Ubelaker, G. L. Stewart, R. E. Silvany, and J. Y. Niederkorn. 1992. A pig model of Acanthamoeba keratitis: Transmission via contaminated contact lenses. *Investig. Ophthalmol. Visual Sci.* 33:126–133.

He, Y. G., J. Y. Niederkorn, J. P. McCulley, G. L. Stewart, D. R. Meyer, R. Silvany, and J. Dougherty. 1990. In vivo and in vitro collagenolytic activity of Acanthamoeba castellanii. *Investig. Ophthalmol. Visual Sci.* 31:2235–2240.

Helton, J., M. Loveless, and C. R. White, Jr. 1993. Cutaneous acanthamoeba infection associated with leukocytoclastic vasculitis in an AIDS patient. *Am. J. Dermatopathol.* 15:146–149.

Hiti, K., J. Walochnik, C. Faschinger, E. M. Haller-Schober, and H. Aspock. 2001. Microwave treatment of contact lens cases contaminated with acanthamoeba. *Cornea.* 20:467–470.

Hiti, K., J. Walochnik, E. M. Haller-Schober, C. Faschinger, and H. Aspock. 2002. Viability of Acanthamoeba after exposure to a multipurpose disinfecting contact lens solution and two hydrogen peroxide systems. *Br. J. Ophthalmol.* 86:144–146.

Holland, G. N., and P. B. Donzis. 1987. Rapid resolution of early Acanthamoeba keratitis after epithelial debridement. *Am. J. Ophthalmol.* 104: 87–89.

Horn, M., T. R. Fritsche, R. K. Gautom, K. H. Schleifer, and M. Wagner. 1999. Novel bacterial endosymbionts of Acanthamoeba spp. related to the Paramecium caudatum symbiont Caedibacter caryophilus. *Environ. Microbiol.* 1:357–367.

Horn, M., T. R. Fritsche, T. Linner, R. K. Gautom, M. D. Harzenetter, M. Wagner. Obligate bacterial endosymbionts of Acanthamoeba spp. related to the - Proteobacteria: Proposal of "Candidatus Procabacter acanthamoebae" gen. *nov., sp. nov. Int. J. Syst. Evol. Microbiol.* 2002, 52, 599–605.

Horn, M., M. Wagner. Bacterial endosymbionts of free-living amoebae. *J. Eukaryot. Microbiol.* 2004, 51, 509–514.

Horne, D. D., M. E. Frizell, L. Ingham, R. G. Jans, S. M. Gubash, C. M. Anand, and M. A. Athar. 1994. Acanthamoeba keratitis: An emerging clinical problem. *CMAJ.* 150:923–925.

Hunt, S. J., S. L. Reed, W. C. Mathews, and B. Torian. 1995. Cutaneous Acanthamoeba infection in the acquired immunodeficiency syndrome: Response to multidrug therapy. *Cutis.* 56:285–287.

Hurt, M., S. Apte, H. Leher, K. Howard, J. Niederkorn, and H. Alizadeh. 2001. Exacerbation of Acanthamoeba keratitis in animals treated with antimacrophage inflammatory protein 2 or antineutrophil antibodies. *Infect. Immun.* 69:2988–2995.

Huws, S. A, R. J. Morley, M. V. Jones, M. R. Brown, A. W. Smith. (May 2008). Interactions of some common pathogenic bacteria with Acanthamoeba polyphaga. *FEMS Microbiol. Lett.* 282 (2): 258–265.

Garner, A. 1993. Pathogenesis of acanthamoebic keratitis: Hypothesis based on a histological analysis of 30 cases. *Br. J. Ophthalmol.* 77:366–370.

Ikeda, Y., D. Miyazaki, K. Yakura et al. (June 2012). Assessment of realtime polymerase chain reaction detection of Acanthamoeba and prognosis determinants of Acanthamoeba keratitis. *Ophthalmol.* 119 (6):1111–1119. doi: 10.1016/j.ophtha.2011.12.023. Epub 2012 Mar 3. PMID: 22381810

Illingworth, C. D., and S. D. Cook. 1998. Acanthamoeba keratitis. *Surv. Ophthalmol.* 42:493–508.

Illingworth, C. D., S. D. Cook, C. H. Karabatsas, and D. L. Easty. 1995. Acanthamoeba keratitis: Risk factors and outcome. *Br. J. Ophthalmol.* 79:1078–1082.

Ishibashi, Y., Y. Matsumoto, T. Kabata, R. Watanabe, S. Hommura, K. Yasuraoka, and K. Ishii. 1990. Oral itraconazole and topical miconazole with debridement for Acanthamoeba keratitis. *Am. J. Ophthalmol.* 109:121–126.

Itahashi, M., S. Higaki, M. Fukuda, H. Mishima, and Y. Shimomura. 2011. Utility of realtime polymerase chain reaction in diagnosing and treating Acanthamoeba keratitis. *Cornea.* 30 (11):1233–1237.

Jahnes, B., H. M. Fullmer, and C. P. Li. 1957. Free-living amoebae as contaminants in monkey kidney tissue cultures. *Proc. Soc. Exp. Biol. Med.* 96:484–488.

Jadin, J. B. 1975. Free-living pathogenic amoebae. *Wiad. Parazytol.* 21:493–498. (In Polish.)

Johns, K. J., D. M. O'Day, and S. S. Feman. 1988. Chorioretinitis in the contralateral eye of a patient with Acanthamoeba keratitis. *Ophthalmol.* 95:635–639.

Johnson, A. M., R. Fielke, P. E. Christy, B. Robinson, and P. R. Baverstock. 1990. Small subunit ribosomal RNA evolution in the genus Acanthamoeba. *J. Gen. Microbiol.* 136:1689–1698.

Johnston, S. P., R. Sriram, Y. Qvarnstrom, S. Roy, J. Verani, J. Yoder, S. Lorick, J. Roberts, M. J. Beach, G. Visvesvara. (July 2009). Resistance of Acanthamoeba cysts to disinfection in multiple contact lens solutions. *J. Clin. Microbiol.* 47 (7):2040–2045. doi:10.1128/JCM.00575-09

Jones, B. R., G. S. Visvesvara, and N. M. Robinson. 1975. Acanthamoeba polyphaga keratitis and Acanthamoeba uveitis associated with fatal meningoencephalitis. *Trans. Ophthalmol. Soc. U. K.* 95:221–232.

Kandori, M., T. Inoue, F. Takamatsu et al., 2010a. Two cases of Acanthamoeba keratitis diagnosed only by real-time polymerase chain reaction. *Cornea.* 29 (2):228–231.

Kandori, M., T. Inoue, M. Shimabukuro et al., 2010b. Four cases of Acanthamoeba keratitis treated with phototherapeutic keratectomy. *Cornea.* 29 (10):1199–1202.

Karsenti, N., A. Purssell, R. Lau, F. Ralevski, S. Bhasker, H. Raheel, A. K.Boggild. 2022. Surveillance of Amoebic Keratitis-Causing Acanthamoebae for potential bacterial Endosymbionts in Ontario, Canada. *Pathogens.* 11:661.

Kennett, M. J., R. R. Hook, Jr., C. L. Franklin, and L. K. Riley. 1999. Acanthamoeba castellanii: Characterization of an adhesin molecule. *Exp. Parasitol.* 92:161–169.

Kenyon, K. R., and W. Binotti. Acanthamoeba: The Keratitis that Won't Quit! *Rev of Ophthalmol.* Published 9 April 2020

Khan, N. A., J. Greenman, K. P. Topping, V. C. Hough, G. S. Temple, and T. A. Paget. 2000. Isolation of Acanthamoeba-specific antibodies from a bacteriophage display library. *J. Clin. Microbiol.* 38:2374–2237.

Khan, Y. A., R. T. Kashiwabuchi, S. A. Martins et al., 2011. Riboflavin and ultraviolet light a therapy as an adjuvant treatment for medically refractive Acanthamoeba keratitis: Report of 3 cases. *Ophthalmol.* 118 (2):324–331.

Kilvington, S., and J. Price. 1990. Survival of Legionella pneumophila within cysts of Acanthamoeba polyphaga following chlorine exposure. *J. Appl. Bacteriol.* 68:519–525.

Kim, Y. H., M. S. Ock, H. C. Yun, M. Y. Hwang, H. S. Yu, H. H. Kong, and D. I. Chung. 1996. Close relatedness of Acanthamoeba pustulosa with Acanthamoeba palestinensis based on isoenzyme profiles and rDNA PCR-RFLP patterns. *Korean J. Parasitol.* 34:259–266.

King, C. H., E. B. Shotts, Jr., R. E. Wooley, and K. G. Porter. 1988. Survival of coliforms and bacterial pathogens within protozoa during chlorination. *Appl. Environ. Microbiol.* 54:3023–3033.

Kong, H. H. (October 2009). Molecular phylogeny of acanthamoeba. *The Korean J. Parasitol.* 47 (Suppl):S21–8.

Kong, H. H., and D. I. Chung. 1996. PCR and RFLP variation of conserved region of small subunit ribosomal DNA among Acanthamoeba isolates assigned to either A. castellanii or A. polyphaga. *Korean J. Parasitol.* 34:127–134.

Kong, H. H., J. H. Park, and D. I. Chung. 1995. Interstrain polymorphisms of isoenzyme profiles and mitochondrial DNA fingerprints among seven strains assigned to Acanthamoeba polyphaga. *Korean J. Parasitol.* 33:331–340. 231.

Kosrirukvongs, P., D. Wanachiwanawin, and G. S. Visvesvara. 1999. Treatment of acanthamoeba keratitis with chlorhexidine. *Ophthalmol.* 106:798–802.

Kremer, I., E. J. Cohen, R. C. Eagle, Jr., I. Udell, and P. R. Laibson. 1994. Histopathologic evaluation of stromal inflammation in Acanthamoeba keratitis. *CLAO J.* 20:45–48.

Krishna Prasad, B. N., and S. K. Gupta. 1978. Preliminary report on the engulfment and retention of Mycobacteria by trophozoites of exenically grown Acanthamoeba castellanii Douglas, 1930. *Curr. Sci.* 47:245–247.

Kumar, R., and D. Lloyd. 2002. Recent advances in the treatment of Acanthamoeba keratitis. *Clin Infect Dis.* 35 (4):434–441. doi: 10.1086/341487

Lagmay, J. P., R. R. Matias, F. F. Natividad, and G. L. Enriquez. 2000. Cytopathogenicity of Acanthamoeba isolates on rat glial C6 cell line. *Southeast Asian J Trop Med Public Health* 30 :670–677.

Lai, S., M. Asgari, and H. R. Henney, Jr. 1994. Non-radioactive DNA probe and polymerase chain reaction procedures for the specific detection of Acanthamoeba. *Mol. Cell Probes.* 8:81–89.

Lalitha, M. K., V. Anandi, A. Srivastava, K. Thomas, A. M. Cherian, and S. M. Chandi. 1985. Isolation of Acanthamoeba culbertsoni from a patient with meningitis. *J. Clin. Microbiol.* 21:666–667.

Landers, P., K. G. Kerr, T. J. Rowbotham, J. L. Tipper, P. M. Keig, E. Ingham, and M. Denton. 2000. Survival and growth of Burkholderia cepacia within the free-living amoeba Acanthamoeba polyphaga. *Eur. J. Clin. Microbiol. Infect. Dis.* 19:121–123.

Larkin, D. F., and D. L. Easty. 1990. Experimental Acanthamoeba keratitis. *I. Preliminary findings. Br. J. Ophthalmol.* 74:551–555.

Larkin, D. F., M. Berry, and D. L. Easty. 1991. In vitro corneal pathogenicity of Acanthamoeba. *Eye* 5:560–568.

Larkin, D. F., S. Kilvington, and J. K. Dart. 1992. Treatment of Acanthamoeba keratitis with polyhexamethylene biguanide. *Ophthalmol.* 99:185–191.

La Scola, B., and D. Raoult. 2001. Survival of Coxiella burnetii within free-living amoeba Acanthamoeba castellanii. *Clin. Microbiol. Infect.* 7 :75–79.

Leher, H., K. Kinoshita, H. Alizadeh, F. L. Zaragoza, Y. He, and J. Niederkorn. 1998a. Impact of oral immunization with Acanthamoeba antigens on parasite adhesion and corneal infection. *Investig. Ophthalmol. Visual Sci.* 39:2337–2343.

Leher, H., R. Silvany, H. Alizadeh, J. Huang, and J. Y. Niederkorn. 1998b. Mannose induces the release of cytopathic factors from Acanthamoeba castellanii. *Infect. Immun.* 66:5–10.

Leher, H., F. Zaragoza, S. Taherzadeh, H. Alizadeh, and J. Y. Niederkorn. 1999. Monoclonal IgA antibodies protect against Acanthamoeba keratitis. *Exp. Eye Res.* 69:75–84.

Lehmann, O. J., S. M. Green, N. Morlet, S. Kilvington, M. F. Keys, M. M. Matheson, J. K. Dart, J. I. McGill, and P. J. Watt. 1998. Polymerase chain reaction analysis of corneal epithelial and tear samples in the diagnosis of Acanthamoeba keratitis. *Investig. Ophthalmol. Visual Sci.* 39:1261–1265.

Levine, S., A. E. Goldstein, M. Dahdouh, P. Blank, C. Hoffman, and C. A. Gropper. 2001. Cutaneous Acanthamoeba in a patient with AIDS: A case study with a review of new therapy. *Cutis* 67:377–380.

Lindquist, T. D. 1998. Treatment of Acanthamoeba keratitis. *Cornea* 17:11–16.

Lindquist, T. D., N. A. Sher, and D. J. Doughman. 1988. Clinical signs and medical therapy of early Acanthamoeba keratitis. *Arch. Ophthalmol.* 106:73–77.

Lorenzo-Morales, J., N. A. Khan, and J. Walochnik. 2015. An update on Acanthamoeba keratitis: Diagnosis, pathogenesis and treatment. *Parasite.* 22:10. doi:10.1051/parasite/2015010

Ly, T. M. C., and H. E. Muller. 1990. Ingested Listeria monocytogenes survive and multiply in protozoa. *J. Med. Microbiol.* 33:51–54.

Ma, P., G. S. Visvesvara, A. J. Martinez, F. H. Theodore, P. M. Daggett, and T. K. Sawyer. 1990. Naegleria and Acanthamoeba infections: Review. *Rev. Infect. Dis.* 12:490–513.

Marciano-Cabral, F., and G. Cabral. (April 2003). Acanthamoeba spp. as agents of disease in humans. *Clin. Microbiol. Rev.* 16 (2):273–307.

Marciano-Cabral, F., R. Puffenbarger, and G. A. Cabral. 2000. The increasing importance of Acanthamoeba infections. *J. Eukaryot. Microbiol.* 47:29–36.

Marrie, T. J., D. Raoult, B. La Scola, R. J. Birtles, and E. de Carolis. 2001. Legionella-like and other amoebal pathogens as agents of community-acquired pneumonia. *Emerg. Infect. Dis.* 7:1026–1029.

Martinez, A. J. 1982. Acanthamoebiasis and immunosuppression. *Case report. J. Neuropathol. Exp. Neurol.* 41:548–557.

Martinez, A. J., and K. Janitschke. 1985. Acanthamoeba, an opportunistic microorganism: A review. *Infection.* 13:251–256.

Martinez, A. J., C. Sotelo-Avila, J. Garcia-Tamayo, J. T. Moron, E. Willaert, and W. P. Stamm. 1977. Meningoencephalitis due to Acanthamoeba sp. Pathogenesis and clinico-pathological study. *Acta Neuropathol. (Berlin).* 37:183–191.

Martinez, A. J., and G. S. Visvesvara. 1991. Laboratory diagnosis of pathogenic free-living amoebas: Naegleria, Acanthamoeba, and Leptomyxida. *Clin. Lab Med.* 11:861–872.

Martinez, A. J., and G. S. Visvesvara. 1997. Free-living, amphizoic and opportunistic amebas. *Brain Pathol.* 7:583–598.

Mascarenhas, J., N. Prajna Lalitha, Venkatesh Prajna, Muthiah Srinivasan, Manoranjan Das, Sean S. D'Silva, C. E. Oldenburg, D. S. Borkar, E. J. Esterberg, T. M. Lietman, and J. D. Keenan. (January 2014). Acanthamoeba, fungal, and bacterial keratitis: A comparison of risk factors and clinical features. *Am J Ophthalmol.* 157 (1):56–62. doi: 10.1016/j.ajo.2013.08.032.

Mathers, W., G. Stevens, Jr., M. Rodrigues, C. C. Chan, J. Gold, G. S. Visvesvara, M. A. Lemp, and L. E. Zimmerman. 1987. Immunopathology and electron microscopy of Acanthamoeba keratitis. *Am. J. Ophthalmol.* 103:626–663.

Mathers, W. D., J. E. Sutphin, R. Folberg, P. A. Meier, R. P. Wenzel, and R. G. Elgin. 1996. Outbreak of keratitis presumed to be caused by Acanthamoeba. *Am. J. Ophthalmol.* 121:129–142.

Mattana, A., F. Bennardini, S. Usai, P. L. Fiori, F. Franconi, and P. Cappuccinelli. 1997. Acanthamoeba castellanii metabolites increase the intracellular calcium level and cause cytotoxicity in wish cells. *Microb. Pathog.* 23:85–93.

Mattana, A., M. G. Tozzi, M. Costa, G. Delogu, P. L. Fiori, and P. Cappuccinelli. 2001. By releasing ADP, Acanthamoeba castellanii causes an increase in the cytosolic free calcium concentration and apoptosis in wish cells. *Infect. Immun.* 69:4134–4140.

Maudgal, P. C. 1989. Acanthamoeba keratitis: Report of three cases. *Bull. Soc. Belge Ophtalmol.* 231:135–148.

May, L. P., G. S. Sidhu, and M. R. Buchness. 1992. Diagnosis of Acanthamoeba infection by cutaneous manifestations in a man seropositive to HIV. *J. Am. Acad. Dermatol.* 26:352–355.

McCulley, J. P., H. Alizadeh, and J. Y. Niederkorn. 1995. Acanthamoeba keratitis. *CLAO J.* 21:73–76.

Michel, R., H. Burghardt, and H. Bergmann. 1995. Acanthamoeba, naturally intracellularly infected with Pseudomonas aeruginosa, after their isolation from a microbiologically contaminated drinking water system in a hospital. *Zentbl. Hyg. Umweltmed.* 196:532–544.

Michel, R., and B. Hauroder. 1997. Isolation of an Acanthamoeba strain with intracellular Burkholderia pickettii infection. *Zentbl. Bakteriol.* 285: 541–557.

Mietz, H., and R. L. Font. 1997. Acanthamoeba keratitis with granulomatous reaction involving the stroma and anterior chamber. *Arch. Ophthalmol.* 115:259–263.

Migueles, S., and P. Kumar. 1998. Primary cutaneous acanthamoeba infection in a patient with AIDS. *Clin. Infect. Dis.* 27:1547–1548.

Miltner, E. C., and L. E. Bermudez. 2000. Mycobacterium avium grown in Acanthamoeba castellanii is protected from the effects of antimicrobials. *Antimicrob. Agents Chemother.* 44:1990–1994.

Mishra, S. K., S. C. Maitra, N. K. Garg, and A. M. Kidwai. 1986. Enzyme and chemical composition of plasma membranes of virulent and avirulent strains of Acanthamoeba culbertsoni. *Indian J. Biochem. Biophys.* 23:5–8.

Mitra, M. M., H. Alizadeh, R. D. Gerard, and J. Y. Niederkorn. 1995. Characterization of a plasminogen activator produced by Acanthamoeba castellanii. *Mol. Biochem. Parasitol.* 73:157–164.

Mitro, K., A. Bhagavathiammai, O. M. Zhou, G. Bobbett, J. H. McKerrow, R. Chokshi, B. Chokshi, and E. R. James. 1994. Partial characterization of the proteolytic secretions of Acanthamoeba polyphaga. *Exp. Parasitol.* 78: 377–385.

Moore, M. B., and J. P. McCulley. 1989. Acanthamoeba keratitis associated with contact lenses: Six consecutive cases of successful management. *Br. J. Ophthalmol.* 73:271–275.

Moore, M. B., J. P. McCulley, M. Luckenbach, H. Gelender, C. Newton, M. B. McDonald, and G. S. Visvesvara. 1985. Acanthamoeba keratitis associated with soft contact lenses. *Am. J. Ophthalmol.* 100:396–403.

Moore, M. B., J. P. McCulley, C. Newton, L. M. Cobo, G. N. Foulks, D. M. O'Day, K. J. Johns, W. T. Driebe, L. A. Wilson, R. J. Epstein, et al. 1987. Acanthamoeba keratitis. A growing problem in soft and hard contact lens wearers. *Ophthalmol.* 94:1654–1661.

Moore, M. B., J. E. Ubelaker, J. H. Martin, R. Silvany, J. M. Dougherty, D. R. Meyer, and J. P. McCulley. 1991. In vitro penetration of human corneal epithelium by Acanthamoeba castellanii: A scanning and transmission electron microscopy study. *Cornea* 10:291–298.

Moshari, A., I. W. McLean, M. T. Dodds, R. E. Damiano, and P. L. McEvoy. 2001. Chorioretinitis after keratitis caused by Acanthamoeba: Case report and review of the literature. *Ophthalmol.* 108:2232–2236.

Murakawa, G. J., T. McCalmont, J. Altman, G. H. Telang, M. D. Hoffman, G. R. Kantor, and T. G. Berger. 1995. Disseminated acanthamebiasis in patients with AIDS. A report of five cases and a review of the literature. *Arch. Dermatol.* 131:1291–1296.

Murdoch, D., T. B. Gray, R. Cursons, and D. Parr. 1998. Acanthamoeba keratitis in New Zealand, including two cases with in vivo resistance to polyhexamethylene biguanide. *Aust. N. Z. J. Ophthalmol.* 26:231–223

Nagington, J., P. G. Watson, T. J. Playfair, J. McGill, B. R. Jones, and A. D. Steele. 1974. Amoebic infection of the eye. *Lancet.* ii:1537–1540.

Newman, W., J. Hay, B. Browne, and D. V. Seal. 1997. Acanthamoeba as a "transient" in the corneal scrape of a poorly compliant soft contact lens wearer with peripheral keratitis. *Eye* 11:937–939.

Newsome, A. L., F. T. Curtis, C. G. Culbertson, and S. D. Allen. 1992. Identification of Acanthamoeba in bronchoalveolar lavage specimens. *Diagn. Cytopathol.* 8:231–234.

Newsome, A. L., T. M. Scott, R. F. Benson, B. S.Fields. 1998. Isolation of an Amoeba naturally harboring a distinctive Legionella species. *Appl. Environ. Microbiol.* 64:1688–1693.

Nguyen, T. H., R. W. Weisenthal, G. J. Florakis, J. J. Reidy, R. N. Gaster, and D. Tom. 2010. Penetrating keratoplasty in active Acanthamoeba keratitis. *Cornea.* 29 (9):1000–1004.

Niederkorn, J. Y., H. Alizadeh, H. F. Leher, and J. P. McCulley. 1999a. The immunobiology of Acanthamoeba keratitis. *Springer Semin. Immunopathol.* 21:147–160.

Niederkorn, J. Y., H. Alizadeh, H. Leher, and J. P. McCulley. 1999b. The pathogenesis of Acanthamoeba keratitis. *Microbes Infect.* 1:437–443.

Niszl, I. A., R. B. Veale, and M. B. Markus. 1998. Cytopathogenicity of clinical and environmental Acanthamoeba isolates for two mammalian cell lines. *J. Parasitol.* 84:961–967.

Oliva, S., M. Jantz, R. Tiernan, D. L. Cook, and M. A. Judson. 1999. Successful treatment of widely disseminated acanthamoebiasis. *South. Med. J.* 92:55–57.

Page, F. C. 1967. Re-definition of the genus Acanthamoeba with descriptions of three species. *J. Protozool.* 14:709–724.

Page, M A., and W. D. Mathers. 2013. Acanthamoeba Keratitis: A 12-Year experience covering a wide spectrum of presentations, diagnoses, and outcomes. *J. Ophthalmol.* :6. Article ID 670242. http://dx.doi.org/10.1155/2013/670242

Panjwani, N., Z. Zhao, J. Baum, M. Pereira, and T. Zaidi. 1992. Acanthamoebae bind to glycolipids of rabbit corneal epithelium. *Infect. Immun.* 60:3460–3463.

Parmar, D. N., S. T. Awwad, W. M. Petroll, R. W. Bowman, J. P. McCulley, and H. D. Cavanagh. 2006. Tandem scanning confocal corneal microscopy in the diagnosis of suspected Acanthamoeba keratitis. *Ophthalmolo.* 113 (4):538–547.

Parthasarathy, A., and D. T. H. Tan. 2007. Deep lamellar keratoplasty for Acanthamoeba keratitis. *Cornea.* 26 (8):1021–1023.

Paszko-Kolva, C., H. Yamamoto, M. Shahamat, T. K. Sawyer, G. Morris, and R. R. Colwell. 1991. Isolation of amoebae and Pseudomonas and Legionella spp. from eyewash stations. *Appl. Environ. Microbiol.* 57:163–167.

Penland, R. L., and K. R. Wilhelmus. 1997. Comparison of axenic and monoxenic media for isolation of Acanthamoeba. *J. Clin. Microbiol.* 35:915–922.

Pettersson, M. N., N. Lagali, J. Mortensen, V. Jofré, P. Fagerholm. 2019. High fluence PACK-CXL as adjuvant treatment for advanced Acanthamoeba keratitis. *Am. J. Ophthalmol. Case Rep.* 15:100499.

Pettit, D. A., J. Williamson, G. A. Cabral, and F. Marciano-Cabral. 1996. In vitro destruction of nerve cell cultures by Acanthamoeba spp.: A transmission and scanning electron microscopy study. *J. Parasitol.* 82:769–777.

Pfister, D. R., J. D. Cameron, J. H. Krachmer, and E. J. Holland. 1996. Confocal microscopy findings of Acanthamoeba keratitis. *Am. J. Ophthalmol.* 121:119–128.

Pinna, A. 2002. Acanthamoeba and disinfecting contact lens solutions. *Br. J. Ophthalmol.* 86:1461–1462.

Por, Y. M,. J. S. Mehta, J. L. L. Chua et al., 2009. Acanthamoeba 6 Journal of Ophthalmology keratitis associated with contact lens wear in Singapore. *Am. J. Ophthalmol.* 148 (1):7.e2– 12.e2.

Radford, C. F., D. C. Minassian, and J. K. Dart. 2002. Acanthamoeba keratitis in England and Wales: incidence, outcome, and risk factors. *Br. J. Ophthalmol.* 86:536–542.

Richardson, Q. R., L. Prajna, S. Elakkiya, F. Sulthana Kamal, M. Talbott, N. Venkatesh Prajna, R. Rajaraman, V. Cevallos, G. D. Seitzman, T. M. Lietman, and J. D. Keenan. *Antiamoebic susceptibility in Acanthamoeba Keratitis: Comparison of isolates from South India and Northern California Cornea* 2022 Aug 25. doi: 10.1097/ICO.0000000000003060

Ringsted, J., B. V. Jager, D. Suk, and G. S. Visvesvara. 1976. Probable acanthamoeba meningoencephalitis in a Korean child. *Am. J. Clin. Pathol.* 66:723–730.

Rivera, M. A., and T. A. Padhya. 2002. Acanthamoeba: A rare primary cause of rhinosinusitis. *Laryngoscope.* 112:1201–1203.

Rodriguez-Zaragoza, S. 1994. Ecology of free-living amoebae. *Crit Rev. Microbiol.* 20:225–241.

Rosenberg, A. S., and M. B. Morgan. 2001. Disseminated acanthamoebiasis presenting as lobular panniculitis with necrotizing vasculitis in a patient with AIDS. *J. Cutan. Pathol.* 28:307–313.

Rowbotham, T. J. 1980. Preliminary report on the pathogenicity of Legionella pneumophila for freshwater and soil amoebae. *J. Clin. Pathol.* 33:1179–1183.

Rowbotham, T. J. 1986. Current views on the relationships between amoebae, legionellae and man. *Isr. J. Med. Sci.* 22:678–689.

Rowbotham, T. J. 1993. Legionella-like amoebal pathogens. In J. M. Barbaree, R. F. Breiman, and A. P. Dufour (ed.), *Legionella—current status and emerging perspectives.* American Society for Microbiology, Washington, D.C., p. 137–140.

Schroeder, J. M., G. C. Booton, J. Hay, I. A. Niszl, D. V. Seal, M. B. Markus, P. A. Fuerst, and T. J. Byers. 2001. Use of subgenic 18S ribosomal DNA PCR and sequencing for genus and genotype identification of acanthamoebae from humans with keratitis and from sewage sludge. *J. Clin. Microbiol.* 39:1903–1911.

Scruggs B. A., Tyler S. Quist a, M. Bridget Zimmerman c, Jorge L. Salinas d, Mark A. Greiner. 2022. Risk factors, management, and outcomes of Acanthamoeba keratitis: A retrospective analysis of 110 cases. *Am. J. Ophthalmol.Case Rep.* 25: 101372.

Seal, D. V., and J. Hay. 1993. The microbiologist as a contact lens wearer. *J. Med. Microbiol.* 39:1–2.

Seal, D. V., and J. Hay. 1994. Acanthamoeba keratitis. *Br. Med. J.* 309:1019.

Seal, D. V., J. Hay, and C. M. Kirkness. 1995. Chlorhexidine or polyhexamethylene biguanide for acanthamoeba keratitis. *Lancet* 345:136.

Seal, D., J. Hay, C. Kirkness, A. Morrell, A. Booth, A. Tullo, A. Ridgway, and M. Armstrong. 1996. Successful medical therapy of Acanthamoeba keratitis with topical chlorhexidine and propamidine. *Eye* 10:413–421.

Seijo, M. M., G. Gonzalez-Mediero, P. Santiago, D. L. Rodriguez, J. Diz, C. Conde, and G. S. Visvesvara. 2000. Granulomatous amebic encephalitis in a patient with AIDS: Isolation of Acanthamoeba sp. group II from brain tissue and successful treatment with sulfadiazine and fluconazole. *J. Clin. Microbiol.* 38:3892–3895.

Selby, D. M., R. S. Chandra, T. A. Rakusan, B. Loechelt, B. M. Markle, and G. S. Visvesvara. 1998. Amebic osteomyelitis in a child with acquired immunodeficiency syndrome: A case report. *Pediatr. Pathol. Lab Med.* 18:89–95.

Sell, J. J., F. W. Rupp, and W. W. Orrison, Jr. 1997. Granulomatous amebic encephalitis caused by Acanthamoeba. *Neuroradiol.* 39:434–436.

Sherman, S. *Attacking Acanthamoeba Keratitis Review Cornea Contact lens* June 15, 2022.

Shoff, M. E., C. E. Joslin, E. Y. Tu, L. Kubatko, and P. A. Fuerst. (July 2008). "Efficacy of contact lens systems against recent clinical and tap water Acanthamoeba isolates". *Cornea.* 27 (6):713 9. doi:10.1097/QAI.0b013e31815e7251

Sifaoui, I., P. Díaz-Rodríguez, R. L Rodríguez-Exposito, M. Reyes-Batlle, A. L′opez-Arencibia, L. Salazar Villatoro, I. Castelan-Ramírez, M. Omãna-Molina, A. Oliva, J. E. Piñero. 2022. Jacob Lorenzo-Morales Pitavastatin loaded nanoparticles: A suitable ophthalmic treatment for Acanthamoeba Keratitis inducing cell death and autophagy in *Acanthamoeba polyphaga. Eur J Pharm Biopharm.* 180:11–22.

Silvany, R. E., M. W. Luckenbach, and M. B. Moore. 1987. The rapid detection of Acanthamoeba in paraffin-embedded sections of corneal tissue with calcofluor white. *Arch. Ophthalmol.* 105:1366–1367.

Singhal, T., A. Bajpai, V. Kalra, S. K. Kabra, J. C. Samantaray, G. Satpathy, and A. K. Gupta. 2001. Successful treatment of Acanthamoeba meningitis with combination oral antimicrobials. *Pediatr. Infect. Dis. J.* 20:623–627.

Skryabina Y.V., Y. S. Astakhov, Y. S. Konenkova, T. S. Varganova, V. P. Petukhov, K. V. Nokhrina, K. O. Dnestryanskaya. 2019. Acanthamoeba keratitis. *Review of literature. Case Rep Opthalmol J* 12 (1):59–71. https://doi.org/10.17816/OV2019 159-71

Slater, C. A., J. Z. Sickel, G. S. Visvesvara, R. C. Pabico, and A. A. Gaspari. 1994. Brief report: Successful treatment of disseminated Acanthamoeba infection in an immunocompromised patient. *N. Engl. J. Med.* 331:85–87.

Steenbergen, J. N., H. A. Shuman, and A. Casadevall. 2001. Cryptococcus neoformans interactions with amoebae suggest an explanation for its virulence and intracellular pathogenic strategy in macrophages. *Proc. Natl. Acad. Sci. USA.* 98:15245–15250.

Stevens, A. R., T. Kilpatrick, E. Willaert, and A. Capron. 1977. Serologic analyses of cell-surface antigens of Acanthamoeba spp. with plasma membrane antisera. *J. Protozool.* 24:316–324.

Stopak, S. S., M. I. Roat, R. C. Nauheim, P. W. Turgeon, G. Sossi, R. P. Kowalski, and R. A. Thoft. 1991. Growth of Acanthamoeba on human corneal epithelial cells and keratocytes in vitro. *Investig. Ophthalmol. Visual Sci.* 32:354–359.

Stothard, D. R., J. Hay, J. M. Schroeder-Diedrich, D. V. Seal, and T. J. Byers. 1999. Fluorescent oligonucleotide probes for clinical and environmental detection of Acanthamoeba and the T4 18S rRNA gene sequence type. *J. Clin. Microbiol.* 37:2687–2693.

Stothard, D. R., J. M. Schroeder-Diedrich, M. H. Awwad, R. J. Gast, D. R. Ledee, S. Rodriguez-Zaragoza, C. L. Dean, P. A. Fuerst, and T. J. Byers. 1998. The evolutionary history of the genus Acanthamoeba and the identification of eight new 18S rRNA gene sequence types. *J. Eukaryot. Microbiol.* 45:45–54.

Szenasi, Z., T. Endo, K. Yagita, and E. Nagy. 1998. Isolation, identification and increasing importance of "free-living" amoebae causing human disease. *J. Med. Microbiol.* 47:5–16.

Szentmary, N., S. Goebels, P. Matoula, F. Schirra, and B. Seitz. 2012. Die Akanthamöbenkeratitis—ein seltenes und oft spät diagnostiziertes Chamäleon. *Klin. Monatsbl Augenheilkd.* 229 (5):521–528.

Tan, B., C. M. Weldon-Linne, D. P. Rhone, C. L. Penning, and G. S. Visvesvara. 1993. Acanthamoeba infection presenting as skin lesions in patients with the acquired immunodeficiency syndrome. *Arch. Pathol. Lab Med.* 117:1043–1046.

Tay-Kearney, M. L., C. N. McGhee, G. J. Crawford, and K. Trown. 1993. Acanthamoeba keratitis. A masquerade of presentation in six cases. *Aust. N. Z. J. Ophthalmol.* 21:237–245.
Taylor, W. M., M. S. Pidherney, H. Alizadeh, and J. Y. Niederkorn. 1995. In vitro characterization of Acanthamoeba castellanii cytopathic effect. *J. Parasitol.* 81:603–609.
Teknos, T. N., M. D. Poulin, A. M. Laruentano, and K. K. Li. 2000. Acanthamoeba rhinosinusitis: characterization, diagnosis, and treatment. *Am. J. Rhinol.* 14:387–391.
Thom, S., D. Warhurst, and B. S. Drasar. 1992. Association of Vibrio cholerae with fresh water amoebae. *J. Med. Microbiol.* 36:303–306.
Ting, D. S. J., C. Henein, D. G. Said, H. S. Dua. 2019. Photoactivated chromophore for infectious keratitis—Corneal cross-linking (PACK-CXL): A systematic review and meta-analysis. *Ocul. Surf.* 17:624–634.
Torno, M. S., Jr., R. Babapour, A. Gurevitch, and M. D. Witt. 2000. Cutaneous acanthamoebiasis in AIDS. *J. Am. Acad. Dermatol.* 42:351–354.
Tu, E. Y., C. E. Joslin, and M. E. Shoff. 2010. Successful treatment of chronic stromal Acanthamoeba keratitis with oral voriconazole monotherapy. *Cornea.* 29 (9):1066–1068.
Tu, E. Y., C. E. Joslin, J. Sugar, G. C. Booton, M. E. Shoff, and P. A. Fuerst. 2008. The relative value of confocal microscopy and superficial corneal scrapings in the diagnosis of Acanthamoeba keratitis. *Cornea.* 27 (7):764–772.
Turner, N. A., J. Harris, A. D. Russell, and D. Lloyd. 2000. Microbial differentiation and changes in susceptibility to antimicrobial agents. *J. Appl. Microbiol.* 89:751–759
Vaddavalli, P. K., P. Garg, S. Sharma, V. S. Sangwan, G. N. Rao, and R. Thomas. 2011. Role of confocal microscopy in the diagnosis of fungal and Acanthamoeba keratitis. *Ophthalmol.* 118 (1):29–35.
Van Hamme, C., M. Dumont, M. Delos, and J. M. Lachapelle. 2001. Acanthamibiase cutanee chez un transplante pulmonaire. *Ann. Dermatol. Venereol.* 128:1237–1240.
Van Klink, F., H. Alizadeh, Y. He, J. A. Mellon, R. E. Silvany, J. P. McCulley, and J. Y. Niederkorn. 1993. The role of contact lenses, trauma, and Langerhans cells in a Chinese hamster model of Acanthamoeba keratitis. *Investig. Ophthalmol. Visual Sci.* 34:1937–1944.
Van Klink, F., H. Alizadeh, G. L. Stewart, M. S. Pidherney, R. E. Silvany, Y. He, J. P. McCulley, and J. Y. Niederkorn. 1992. Characterization and pathogenic potential of a soil isolate and an ocular isolate of Acanthamoeba castellanii in relation to Acanthamoeba keratitis. *Curr. Eye Res.* 11:1207–1220.
Van Klink, F., H. Leher, M. J. Jager, H. Alizadeh, W. Taylor, and J. Y. Niederkorn. 1997. Systemic immune response to Acanthamoeba keratitis in the Chinese hamster. *Ocul. Immunol. Inflamm.* 5:235–244.
Van Klink, F., W. M. Taylor, H. Alizadeh, M. J. Jager, N. van Rooijen, and J. Y. Niederkorn. 1996. The role of macrophages in Acanthamoeba keratitis. *Investig. Ophthalmol. Visual Sci.* 37:1271–1281.
Varacalli, G., A. Di Zazzo, T. Mori, T. H. Dohlman, S. Spelta, M. Coassin, and S. Bonini. 2021. Challenges in Acanthamoeba Keratitis: A Review. *J. Clin. Med.* 10 (5):942. https://doi.org/10.3390/jcm10050942
Victoria, E. J., and E. D. Korn. 1975a. Enzymes of phospholipid metabolism in the plasma membrane of Acanthamoeba castellanii. *J. Lipid Res.* 16:54–60.
Victoria, E. J., and E. D. Korn. 1975b. Plasma membrane and soluble lysophospholipases of Acanthamoeba castellanii. *Arch. Biochem. Biophys.* 171:255–258.
Visvesvara, G. S., and W. Balamuth. 1975. Comparative studies on related free-living and pathogenic amebae with special reference to Acanthamoeba. *J. Protozool.* 22:245–256.

Walochnik, J., E. Haller-Schober, H. Kolli, O. Picher, A. Obwaller, and H. Aspock. 2000. Discrimination between clinically relevant and nonrelevant Acanthamoeba strains isolated from contact lens- wearing keratitis patients in Austria. *J. Clin. Microbiol.* 38:3932–3936.

Walochnik, J., A. Hassl, K. Simon, G. Benyr, and H. Aspock. 1999. Isolation and identification by partial sequencing of the 18S ribosomal gene of freeliving amoebae from necrotic tissue of Basilliscus plumifrons (Sauria: Iguanidae). *Parasitol. Res.* 85:601–603.

Walochnik, J., A. Obwaller, E. M. Haller-Schober, and H. Aspock. 2001. Anti-Acanthamoeba IgG, IgM, and IgA immunoreactivities in correlation to strain pathogenicity. *Parasitol. Res.* 87:651–656.

Weekers, P. H., P. L. E. Bodelier, J. P. Wijen, and G. D. Vogels. 1993. Effects of grazing by the free-living soil amoebae Acanthamoeba castellanii, Acanthamoeba polyphaga, and Hartmannella vermiformis on various bacteria. *Appl. Environ. Microbiol.* 59:2317–2319.

Wiley, C. A., R. E. Safrin, C. E. Davis, P. W. Lampert, A. I. Braude, A. J. Martinez, and G. S. Visvesvara. 1987. Acanthamoeba meningoencephalitis in a patient with AIDS. *J. Infect. Dis.* 155:130–133.

Winiecka-Krusnell, J., K. Wreiber, A. von Euler, L. Engstrand, and E. Linder. 2002. Free-living amoebae promote growth and survival of Helicobacter pylori. *Scand. J. Infect. Dis.* 34:253–256.

Wright, J. L., K. Redhead, and H. Maudsley. 1981. Acanthamoeba castellanii, a predator of cyanobacteria. *J. Gen. Microbiol.* 125:293–300.

Wright, P., D. Warhurst, and B. R. Jones. 1985. Acanthamoeba keratitis successfully treated medically. *Br. J. Ophthalmol.* 69:778–782.

Yagita, K., and T. Endo. 1990. Restriction enzyme analysis of mitochondrial DNA of Acanthamoeba strains in Japan. *J. Protozool.* 37:570–575.

Yang, Z., Z. Cao, and N. Panjwani. 1997. Pathogenesis of Acanthamoeba keratitis: Carbohydrate-mediated host-parasite interactions. *Infect. Immun.* 65:439–445.

8

Antimicrobial Resistance in the Diseased Human Eye: Dry Eye Disease

Arunasri Anakala Kotakonda

8.1 Introduction

Dry eye disease (DED) is defined as a multifactorial disease of the ocular surface, and the ocular symptoms include tear film instability, hyperosmolarity, ocular surface inflammation and damage, and neurosensory abnormalities (Craig et al., 2017). Based on the clinical symptoms DED may be broadly categorised into aqueous deficient and evaporative DED (Craig et al., 2017) (Figure 8.1). According to estimates the prevalence of DED ranged from 9.5–90% (Shanti et al., 2020) across different parts of the world including Europe, United States of America, Asia and India (Wan we al., 2019; Titiyal et al., 2018) (Table 8.1). A hospital-based study indicated that, in the Indian population between 21 to 40 years, DED is more prevalent in males (65.3%) than females (34.7%) (Titiyal et al., 2018). The above study also cautioned about the preponderance of DED in the urban population compared to rural (Titiyal et al., 2018). On a similar note, a study conducted in Raipur, India, during 2019–2020, reported high DED symptoms in males (58.7%) compared to females (41.3%) (Samrat et al., 2021). In a population-based study conducted in West Bank towns in Palestine in the age group of 18–90 years indicated that DED was prevalent in 64% of the population and more prevalent in females (52.7%) and those who were elderly (Shanti et al., 2020). The discrepancy in the estimates of DED prevalence could be due to the influence of geographical location, ethnicity, the diagnostic approaches followed in the absence of a standard diagnostic criteria. A feature common to all these studies on DED is that, in all these countries, DED prevalence showed an increasing trend (Courtin et al., 2016; Dana et al., 2019; Siffel et al., 2020; Stapleton et al., 2017).

Patients with DED normally complained about mild to severe itching, burning, irritation, eye-fatigue, and ocular inflammation—may damage the cornea and conjunctiva and lead to vision loss (Craig et al., 2017). In the Indian population the most common symptom of DED was red eyes (2.8%), burning sensation (1.8%), foreign body sensation (1.7%), dry eyes (1.2%), gummy eyes (1.2%), and crusts on eyelashes (0.8%) (Samrat et al., 2021). Several risk factors are associated with DED and these include for the female sex, the use of digital display, smoking and air-conditioned environments (Samrat et al., 2021; Tityal et al., 2018). According to the "Tear film and ocular surface society" (TFOS), in addition to the above risk factors other consistent risk factors for DED include age, race, computer use, contact lens wearing,

DOI: 10.1201/9781003451105-8

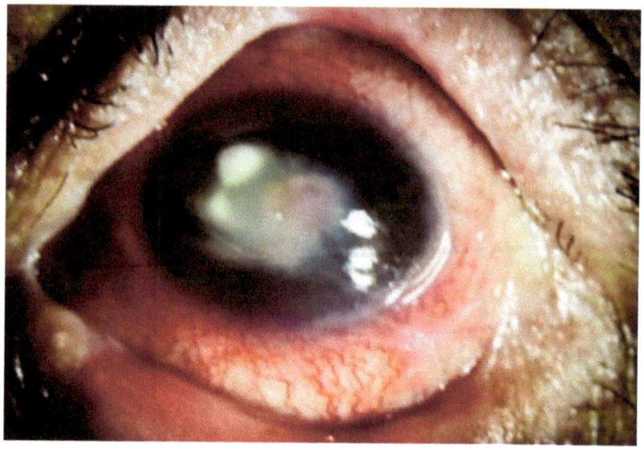

FIGURE 8.1 Ocular surface modifications in a Stevens-Johnson syndrome patient with dry eye disease clearly showing ocular inflammation as evidenced by damage to the cornea and conjunctiva and red eyes. (Courtesy Dr Syan Basu, L V Prasad Eye Institute, Hyderabad, India.)

TABLE 8.1

Prevalence of Dry Eye Disease in Different Countries

Sl. No	Country	Number of studies	Survey years	Total prevalence (%)	Female prevalence (%)	Male prevalence (%)
1	Brazil	1	NA	12.8	14.4	9.8
2	China	5	2001–2007	21–75.6	23.3 - 52.9	17.2 - 52.1
3	France	2	2006–2013	21.9–34.4	27.2–39.5	13.6–25.8
4	Ghana	1	NA	44.3	49.8	41.6
5	India	1	2014–2016	32	39.3	29.1
6	Iran	1	2009	18.3	20	15.7
7	Japan	3	2008–2016	11.2–21.6	13.3–24.4	8.3–21
8	Jordan	1	2015	59	60.4	57.2
9	Korea	2	2008–2012	16.16–30.3	22.4–34.7	11.5–25.6
10	Mexico	2	2013–2014	41.1–70.4	45.3–74.8	35–63.7
11	Saudi Arabia	1	2016	6.5	4.9	8.2
12	Singapore	2	2004–2010	12.3–18.4	14.8–21.8	9.0–12.5
13	Spain	1	2005	18.4	21.8	12.5
14	Thailand	1	NA	14.2	16.6	10.6
15	UK	2	2012–2013	9.6–15.5	9.6	NA
16	USA	3	2004–2013	4.3–14.5	8.4–17.9	4.3–10.5

Source: Adapted from Wan et al. (2019).

estrogen replacement therapy, stem cell transplantation, androgen deficiency, inflammatory diseases such as meibomian gland dysfunction, connective tissue disease, Sjogren syndrome, pollution, arid environment, low humidity, sick building syndrome and use of specific medications such as antihistamines, antidepressants, anxiolytics, and isotretinoin. Other probable health associated risk factors include diabetes, rosacea, viral infection, thyroid disease, psychiatric conditions, pterygium, low fatty acid intake, refractive surgery, allergic conjunctivitis, and use of medications such as anticholinergic, diuretics, β-blockers and anxiety (Craig et al., 2017; Shanthi et al., 2020). However, the involvement as DED potential risk factors of ethnicity, menopause, acne, sarcoidosis, smoking, alcohol, pregnancy, demodex infestation, botulinum toxin injection, multivitamins and oral contraceptives were all inconclusive (Craig et al., 2017).

8.2 Diagnosis of Dry Eye Disease

In dry eye disease all or part of the ocular surface comprising cornea, conjunctiva, lacrimal glands and meibomian glands may be affected (Zhang et al., 2017b). The clinical diagnosis is challenging because of the ambiguity in the pathophysiology of the disease. However, many a times a greater degree of discordance between the symptoms and clinical signs are observed. Based on the report of the "Tear film and ocular surface society" (TFOS) meeting at the second "Dry Eye Workshop" (DEWS II) it was suggested that DED should be confirmed in patients by evaluating through triage questions and excluding the non-DED possibilities (Wolffsohn et al., 2017). The general diagnosis for DED includes symptom evaluation using questionnaires and clinical tests that evaluate tear film break up time, tear film osmolarity and ocular surface staining (Begley et al., 2002; Abetz et al., 2011; Savini et al., 2008). Further, hyper evaporative and aqueous deficient DED must be differentiated by the meibomian gland and tear lipid evaluation and by ascertaining tear volume assessment (Tomlinson et al., 2006). Apart from the above conventional tests, new molecular marker-based tests for diagnosis such as MMP-9 and use of lipid layer interferometry may be useful (Bang et al., 2020; Arita et al., 2016). Further, testing positive for the tear film homeostasis loss biomarkers, such as lacritin and lipocalin, would constitute a DED diagnosis (Dartt and Willcox 2013). In general, the optimal diagnosis of DED involves a combination of several tests and disease symptoms.

8.3 Etiology of Dry Eye Disease

The etiology of DED is as yet unclear, but DED may be caused or exacerbated by several risk factors enumerated above which include environmental and health-related factors (Clayton, 2018). Microbial etiology in DED is associated mostly with the blepharitis of the eye and Meibomian gland dysfunction (Jackson 2008; Zhang et al., 2017b). Studies in India and the UK have indicated that DED samples were culture positive at 59% and 97% respectively (Venugopal et al., 2016; Graham et al., 2007). In the conjunctival samples coagulase-negative *Staphylococcus* was the predominant microorganism in both the countries (Venugopal et al., 2016; Graham et al., 2007). In fact, several other bacteria such as *S. aureus*, *Streptococcus* spp.,

TABLE 8.2

Microbiome Associated with Different Categories of Dry Eye Diseases

Dry eye disease categories	Predominant bacteria	Reference
Meibomian gland dysfunction	*Rubrobacter*, *Novibacillus*, *Campylobacter*, *Geobacillus*, *Sphingomonas*, *Corynebacterium*, *Sphingobium*, *Pedobacter*, *Fictibacillus*, and *Enterococcus*	Graham et al., 2007; Watters et al., 2017; Zhao et al., 2020
Blepharitis	Coagulase negative *Staphylococci*, *S. aureus*, *Streptococcus* spp., *Bacillus subtilis*, *Rhodococcus* spp., *P. aeruginosa* and *Hemophilus influenza*	Narayana et al., 2013
Stevens-Jhonson syndrome	*Corynebacterium*, *Streptococcus*, *Serratia*, *Staphylococcus*, *E. coli*, *Proteus mirabilis* and *Haemophilus*.	Frizon et al., 2014; Kittipibul et al., 2020
Sjogren's syndrome	*Corynebacterium* and *Pelomonas*	Qi et al., 2021
Aqueous deficient dry eye	*Acinetobacter, Staphylococcus, Bacillus, Corynebacterium, and Clostridium_sensu_stricto_1*	Song et al., 2022

Source: Compiled by the author.

Bacillus subtilis, *Rhodococcus* spp., *P. aeruginosa*, and *Hemophilus influenzae* were reported in DED associated with blepharitis (Narayanan et al., 2013). Recently, with the establishment of ocular surface microbiome in healthy individuals, the link between the microbiome changes and the dry eye disease have been studied (Table 8.2). In 2007 Graham and colleagues studied the microbiome of the ocular surface of individuals and reported distinct changes in the microbiome composition in individuals with dry eye compared to healthy individuals. Using the conventional cultivable method and 16S rDNA approach, it was observed that coagulase negative *Saphylococci* increased in bacterial count (CFU/swab) in DED patients (Graham et al., 2007) and *S. epidermidis* was present in 100% of DED samples (Graham et al., 2007). Venugopal et al. (2016) observed *Corynebacterium* spp., *Streptococcus* spp., and *Staphylococcus* spp. common in Stevens-Johnson Syndrome (SJS) patients. Frizon et al. (2014) reported the detection of *Staphylococcus* spp. followed by *Corynebacterium* spp. in SJS patients. Several atypical species were also found on the ocular surface of SJS patients, like *Serratia* non-liquefaciens, *E. coli, Proteus mirabilis* and *Haemophilus* spp. Kittipibul et al. (2020) also reported the presence of atypical bacteria in SJS, including *Streptococcus agalactiae* and *P. mirabilis*. In the closed eyes of DED five taxa, namely *Methylobacteriaceae, Pseudomonas, Bradyrhizobium* and *Allobaculum* were abundant compared to the control samples (Willis et al., 2020). Dry eye disease with and without autoimmune disorder may be distinguished by the abundance of *Corynebacterium* spp. and *Pelomonas* spp. (Qi et al., 2021). Significant potential pathogenic bacterial strains such as coagulase- negative staphylococci, *Pseudomonas acnes*, coryneform bacteria, and *S. aureus* were cultured from the meibomian gland dysfunction (MGD) patients (Zhao et al., 2020 a, 2020b). Based on the above studies it is apparent that the microbiome plays an important role in the DED etiology and several bacterial pathogens were associated with dry eye disease.

8.4 Treatment of Dry Eye Disease

Dry eye disease is a substantial burden to the health care system as well as to the patient's economy. The treatment of the disease is simple and is based on clinical signs and symptoms. Untreated severe dry eyes may lead to corneal ulceration and conjunctival scarring. Although the advancements in the understanding of the disease helped in endorsing new drugs for treating DED, many drugs are in the pipeline for their approval (O'Neil et al., 2019). The treatment of DED would comprise advocating use of artificial tear drops, tear retention treatment and stimulation of tear secretion or anti-inflammatory drugs (Yavuz et al., 2012). Artificial tear drops would help in improving the ocular lubrication and these drug formulations contain sodium hyaluronate, polyvinyl alcohol, hydroxypropyl methyl cellulose, carboxymethyl cellulose, and polyethylene glycol. Anti-inflammatory drugs such as corticosteroids are the most common drugs used for treating DED. Corticosteroids reduce the symptoms associated with moderate to severe dry eye and clinical signs are no longer obvious (Colligris et al., 2014). However, the prolonged use of steroids may encourage bacterial and fungal infections. Therefore, the use of non-steroidal anti-inflammatory drugs is recommended. Antibiotics are required to treat the lid margin bacterial burden in blepharitis and MGD patients. Treatment requires topical medications such as fusidic acid and fluroquinolones along with administration of oral antibiotics like cloxacillin and rifampin for recalcitrant disease (Jackson 2008).

8.5 AMR in Dry Eye Disease

Dry eye disease is treated mostly with anti-inflammatory drugs. Antibiotics are less frequently used in DED compared to other ocular diseases and the antibiotics are mostly used in treating the severe DED. Both topical and oral antibiotics may be used to ensure relief from the symptoms. Topical antibiotics such as tobramycin, azithromycin, and oral antibiotics such as doxycycline and minocycline have been used to treat blepharitis (Narayanan et al., 2013). The topical medications may alter the normal flora associated with the ocular surface and the eye lids and thus reduce the inflammation of the eye (Hori et al., 2008; Williamson et al., 1968; Yan et al., 2011). A case report of a methicillin-resistant *S. aureus* keratitis in a dry eye patient with a therapeutic contact lens highlights the prevalence of antibiotic resistance bacteria (Koh et al., 2012). Hori et al (2008) observed no significant difference in the bacterial isolation rates from conjunctiva of normal and dry eye patients but observed the presence of fluoroquinolone-resistant bacterial strains in DED individuals (Hori et al., 2008).

8.6 Trends in AMR in Dry Eye Disease

Reports on antimicrobial resistance in dry eye disease are lacking. A study conducted in a tertiary eye care centre in Spain has indicated that CoNS and *S. aureus* were the predominant bacteria that were isolated from the eye lid margins of blepharitis patients and were resistant to erythromycin and rifampicin through the years from 2012 to 2018 (de Paula et al., 2020). Large scale studies and updates on the antimicrobial resistance trends are important for the ophthalmologists to deliver the appropriate treatment to the patients.

8.7 Multidrug Resistance in Dry Eye Disease

Resistance to antibiotics was observed in the bacterial isolates obtained from blepharitis and keratitis patients with DED who were treated with antibiotics such as fluoroquinolone. Fluoroquinolones are a risk factor for colonization and infection of methicillin-resistant *S. aureus, Pseudomonas aeruginosa*, and multidrug resistant bacteria (Nseir et al., 2005). Therefore, the dosage and duration of treatment with fluoroquinolones should be conducted with utmost caution to reduce the emergence of multidrug resistant bacteria (Nseir et al., 2005). Studies should be conducted to understand the prevalence of multidrug resistance in dry eye disease.

8.8 Conclusions

Microbial resistance is less researched in dry eye disease patients. With the growing abundance of antimicrobial resistance strains in different ophthalmic diseases, a proper assessment of the antimicrobial resistance in the isolates from DED patients is required. From the sporadic reports available, it is apparent that the use of antibiotics showed changes in the ocular surface microbiome and identified fluoroquinolone resistant bacteria in dry eye disease patients. The antimicrobial resistance burden in dry eye disease is yet to be studied comprehensively.

REFERENCES

Abetz L, Rajagopalan K, Mertzanis P, et al. Impact of Dry Eye on Everyday Life (IDEEL) Study Group. Development and validation of the impact of dry eye on everyday life (IDEEL) questionnaire, a patient-reported outcomes (PRO) measure for the assessment of the burden of dry eye on patients. *Health Qual. Life Outcomes* 2011;9:111.

Arita R, Morishige N, Fujii T, Fukuoka S, Chung JL, Seo KY, Itoh K. Tear interferometric patterns reflect clinical tear dynamics in dry eye patients. *Invest. Ophthalmol. Vis. Sci.* 2016;57(8): 3928–3934. doi: https://doi.org/10.1167/iovs.16-19788

Bang SP, Son MJ, Kim H, et al. In vitro validation of the tear matrix metalloproteinase 9 in-situ immunoassay. *Sci. Rep.* 2020;10: 15126. https://doi.org/10.1038/s41598-020-71977-w

Begley C, Caffery B, Chalmers R, et al. Use of the Dry Eye Questionnaire to measure symptoms of ocular irritation in patients with aqueous tear deficiency. *Cornea* 2002;21: 664–70.

Clayton JA. Dry Eye. *N. Engl. J. Med.* 2018;378: 2212–2223.doi:10.1056/NEJMra1407936

Colligris B, Crooke A, Huete-Toral F, Pintor J. An update on dry eye disease molecular treatment: advances in drug pipelines. *Expert Opin. Pharmacother.* 2014;15(10):1371–1390. doi:10.1517/14656566.2014.914492

Courtin R, Pereira B, Naughton G, Chamoux A, Chiambaretta F, Lanhers C, et al. Prevalence of dry eye disease in visual display terminal workers: a systematic review and meta-analysis. *BMJ Open* 2016;6: e009675–13. doi:10.1136/bmjopen-2015-009675

Craig JP, et al. TFOS DEWS II Report Executive Summary, *Ocul. Surf.* 2017. http://dx.doi.org/10.1016/ j.jtos.2017.08.003

Dana R, Meunier J, Markowitz JT, et al. Patient-reported burden of dry eye disease in the United States: results of an online cross-sectional survey. *Am. J. Ophthalmol.* 2020;S0002-9394: 30156-2.

Dartt DA, Willcox MD. Complexity of the tear film: importance in homeostasis and dysfunction during disease. *Exp. Eye Res.* 2013 Dec;117: 1–3. doi: 10.1016/j.exer.2013.10.008. PMID: 24280033; PMCID: PMC4225770.

de Paula A, Oliva G, Barraquer RI, de la Paz MF. Prevalence and antibiotic susceptibility of bacteria isolated in patients affected with blepharitis in a tertiary eye centre in Spain. *Eur. J. Ophthalmol.* 2020 Sep;30(5): 991–997. doi: 10.1177/1120672119854985. Epub 2019 Jun 24. PMID: 31232091.

Frizon, L, et al. Evaluation of conjunctival bacterial flora in patients with Stevens-Johnson syndrome. *Clinics* 2014;69: 168–172.

Graham JE, Moore JE, Jiru X, Moore JE, Goodall EA, Dooley JSG, Hayes VEA, Dartt DA, Downes CS, Moore TCB. Ocular pathogen or commensal: a PCR-based study of surface bacterial flora in normal and dry eyes. *Investig. Ophthalmol. Vis. Sci.* 2007;48: 5616–5623. doi: 10.1167/iovs.07-0588

Hori Y, Maeda N, Sakamoto M, et al. Bacteriologic profile of the conjunctiva in the patients with dry eye. *Am. J. Ophthalmol.* 2008;146:729–734.

Jackson WB. Blepharitis: current strategies for diagnosis and management. *Can. J. Ophthalmol.* 2008;43: 170–179. doi: 10.3129/i08-016

Kittipibul T, Puangsricharern V, Chatsuwan T. Comparison of the ocular microbiome between chronic Stevens-Johnson syndrome patients and healthy subjects. *Sci. Rep.* 2020;10: 4353.

Koh S, Maeda N, Soma T, et al. Development of methicillin-resistant *Staphylococcus aureus* keratitis in a dry eye patient with a therapeutic contact lens. *Eye Contact Lens* 2012;38: 200–202.

Narayanan S, Redfern RL, Miller WL, Nichols KK, McDermott AM. Dry eye disease and microbial keratitis: is there a connection? *Ocul. Surf.* 2013 Apr;11(2):75–92.doi:10.1016/j.jtos.2012.12.002.Epub2013Jan29.PMID:23583043; PMCID: PMC3637884.

Nseir S, Ader F, Marquette CH, Durocher A. Impact de l'utilisation des fluoroquinolones sur l'émergence des bactéries multirésistantes [Impact of fluoroquinolone use on multidrug-resistant bacteria emergence]. *Pathol Biol (Paris)* 2005 Oct-Nov;53(8–9): 470–475. French. doi: 10.1016/j.patbio.2005.07.008. Epub 2005 Sep 19. PMID: 16176863.

O'Neil EC, Henderson M, Massaro-Giordano M, Bunya VY. Advances in dry eye disease treatment. *Curr. Opin. Ophthalmol.* 2019 May;30(3): 166–178. doi: 10.1097/ICU.0000000000000569. PMID: 30883442; PMCID: PMC6986373.

Qi Y, Wan Y, Li T, Zhang M, Song Y, Hu Y, Sun Y, Li L. Comparison of the ocular microbiomes of dry eye patients with and without autoimmune disease. *Front. Cell Infect. Microbiol.* 2021 Sep 22;11: 716867. doi: 10.3389/fcimb.2021.716867. PMID: 34631599; PMCID: PMC8493086.

Samrat, C, Agrawal D, Sanowar G, Kandoi R. Prevalence of symptoms of dry eye disease in an urban Indian population. *Indian J. Ophthalmol.* May 2021;69(5): 1061–1066 doi: 10.4103/ijo.IJO_1796_20.

Savini G, Prabhawasat P, Kojima T, Grueterich M, Espana E, Goto E. The challenge of dry eye diagnosis. *Clin. Ophthalmol.* 2008 Mar;2(1): 31–55. doi: 10.2147/opth.s1496. PMID: 19668387; PMCID: PMC2698717.

Shanti Y, Shehada R, Bakkar MM, et al. Prevalence and associated risk factors of dry eye disease in 16 northern West bank towns in Palestine: a cross-sectional study. *BMC Ophthalmol.* 2020;20: 26. https://doi.org/10.1186/s12886-019-1290-z

Siffel C, Hennies N, Joseph C, et al. Burden of dry eye disease in Germany: a retrospective observational study using German claims data. *Acta Ophthalmol.* 2020;98: e504–12.

Song H, Xiao K, Chen Z, Long Q. Analysis of conjunctival sac microbiome in dry eye patients with and without Sjögren's Syndrome. *Front. Med. (Lausanne).* 2022 Mar 8;9: 841112. doi: 10.3389/fmed.2022.841112. PMID: 35350577; PMCID: PMC8957797.

Stapleton F, Alves M, Bunya VY, et al. TFOS DEWS II epidemiology report. *Ocul. Surf.* 2017;15: 334–365.

Titiyal JS, Falera RC, Kaur M, Sharma V, Sharma N. Prevalence and risk factors of dry eye disease in North India: Ocular surface disease index-based cross-sectional hospital study. *Indian J. Ophthalmol.*, 2018;66(2): 207–211.

Tomlinson SK, Ramaesh K, et al. Tear film osmolarity: determination of a referent for dry eye diagnosis. *Invest. Ophthalmol. Vis. Sci.*, 2006;47: 4309–4315.

Venugopal R, Satpathy G, Sangwan S, Kapil A, Aron N, Agarwal T, Pushker N, Sharma N. Conjunctival microbial flora in ocular Stevens-Johnson syndrome sequelae patients at a tertiary eye care center. *Cornea* 2016 Aug;35(8): 1117–1121. doi: 10.1097/ICO.0000000000000857.

Wan Y, Zhang M, Li X. The global prevalence of dry eye disease and its association with economy: a systematic review. *Res. Square*; 2019. doi: 10.21203/rs.2.10986/v1.

Watters GA, Turnbull PR, Swift S, Petty A, Craig JP. Ocular surface microbiome in meibomian gland dysfunction. *Clin. Exp. Ophthalmol.* 2017;45: 105–111. doi: 10.1111/ceo.12810.

Williamson J, Gordon AM, Wood R, et al. Fungal flora of the conjunctival sac in health and disease. Influence of topical and systemic steroids. *Br. J. Ophthalmol.* 1968;52: 127–37.

Willis KA, Postnikoff CK, Freeman A, et al. The closed eye harbours a unique microbiome in dry eye disease. *Sci. Rep.* 2020; 10(1): 12035.

Wolffsohn JS, Arita R, Chalmers R, Djalilian A, Dogru M, Dumbleton K, Gupta PK, Karpecki P, Lazreg S, Pult H, et al. TFOS DEWS II Diagnostic Methodology report. *Ocul. Surf.* 2017;15: 539–574. doi: 10.1016/j.jtos.2017.05.001

Yan Z, Young AL, Hua H, Xu Y. Multiple oral Candida infections in patients with Sjogren's syndrome – prevalence and clinical and drug susceptibility profiles. *J. Rheumatol.* 2011;38: 2428–2431.

Yavuz B, Bozdağ Pehlivan S, Unlü N. An overview on dry eye treatment: approaches for cyclosporin a delivery. *ScientificWorldJournal* 2012;2012: 194848. doi: 10.1100/2012/194848. Epub 2012 Apr 24. PMID: 22619624; PMCID: PMC3349326.

Zhang SD, He JN, Niu TT, Chan CY, Ren CY, Liu SS, Qu Y, Chong KL, Wang HL, Tao J, et al. Bacteriological profile of ocular surface flora in meibomian gland dysfunction. *Ocul. Surf.* 2017a;15: 242–247.

Zhang X, M VJ, Qu Y, He X, Ou S, Bu J, Jia C, Wang J, Wu H, Liu Z, Li W. Dry eye management: targeting the ocular surface microenvironment. *Int. J. Mol. Sci.* 2017b Jun 29;18(7): 1398. doi: 10.3390/ijms18071398. PMID: 28661456; PMCID: PMC5535891.

Zhao F, Zhang D, Ge C, Zhang L, Reinach PS, Tian X, Tao C, Zhao Z, Zhao C, Fu W, Zeng C, Chen W. Metagenomic profiling of ocular surface microbiome changes in Meibomian gland dysfunction. *Invest. Opthalmol. Vis. Sci.* 2020a;61(8); 22. doi: 10.1167/iovs.61.8.22. PMID: 32673387; PMCID: PMC7425691.

9

Antimicrobial Resistance in the Diseased Human Eye: Endophthalmitis

Joveeta Joseph

9.1 Introduction

Microorganisms breaching the ocular bulbus acquire access into the otherwise sterile interior of the eye and can replicate unimpeded, leading to infection and inflammation. The inflammation of the intraocular layers resulting from the seeding of microorganisms is known as endophthalmitis (Durand 2013). It is one of the most dreaded complications and potentially vision-threatening diseases. It may be categorized by etiology (infectious versus non-infectious), or the organism(s) involved (parasites, bacteria, and fungi). Furthermore, on the basis of mode of entry of microorganism, endophthalmitis is also categorised into "endogenous" and "exogenous", and based on the predisposing factor, exogenous endophthalmitis is further divided into "postoperative" and "post-traumatic" episodes (Safneck 2012). The clinical presentation of endophthalmitis include ocular pain, vitritis leading to vitreous opacification, decreased vision, corneal edema, hypoyon, insidious inflammation, photophobia, and redness (Figure 9.1). The resulting infection might spread to the adjacent soft tissues of the orbit, sclera and the tenon capsule leading to pan-ophthalmitis (Dave et al., 2019) and phthisis bulbi. In pan-ophthalmitis, the affected eye usually needs to undergo evisceration or enucleation. The evisceration rate of pan-ophthalmitis ranges from 14.3% to 23.2% (Lu et al., 2016). The reported incidence of endophthalmitis differs depending on various contributing factors, which include the period of analysis, geographical factors, prevalence of predisposing factors, rural versus urban settings, sample size of reported cases and outpatient versus inpatient populations (Safneck 2012).

Endophthalmitis is associated with several organisms comprising Gram-negative bacteria, Gram-positive bacteria and fungi with bacteria isolated with the greatest frequency (Teweldemedhin et al., 2017). While, in Western countries, the commonest etiology of bacteria causing endophthalmitis is Gram-positive bacteria (Gentile et al., 2014); in India, Gram-negative bacteria play an equivalent role in the disease pathogenesis (Pathengay et al., 2011). Two studies have revealed a higher proportion of Gram-negative isolates in culture-proven endophthalmitis, one reporting 41.7% of postoperative cases (Anand et al., 2000) while the other showed 30.1% of overall cases (Pathengay et al., 2011). *Pseudomonas* was the predominant Gram-negative organism in both studies, accounting for 25.9% and 13% of all isolates respectively and being

DOI: 10.1201/9781003451105-9

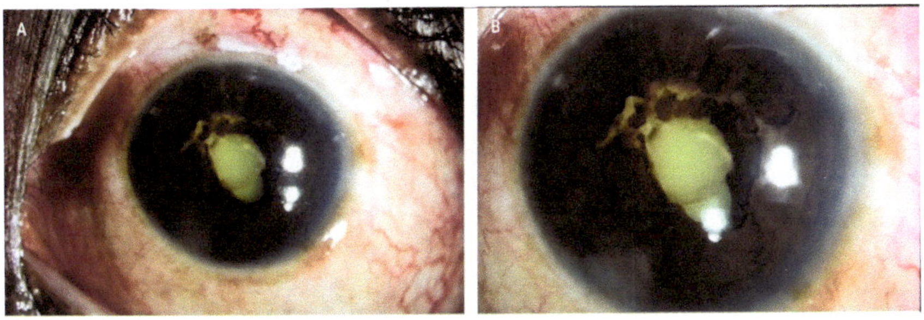

FIGURE 9.1 A patient with infectious endophthalmitis developed after post-cataract surgery with severe vitritis leading to vitreous opacification, edema of the cornea and redness, seen at L.V. Prasad Eye Institute, Hyderabad, India. (Courtesy Joveeta Joseph, L.V. Prasad Eye Institute, Hyderabad, India.)

the single most frequent organism in the earlier study (Anand et al., 2000; Pathengay et al., 2011). The other common Gram-negative organisms causing endophthalmitis included *Proteus* spp., and *Klebsiella* spp., whereas Gram-positive organisms included *Streptococcus* spp., *Staphylococcus* spp., and *Bacillus* spp. (Jambulingam et al., 2010; Anand et al., 2000; Teweldemedhin et al., 2017). The complexity and severity of the disease are due to the complex combination of bacterial load, rapid multiplication, migration potential, virulence factor expression, and levels of toxin production (Packer et al., 2011). The invading pathogen can in turn affect the retina leading to irreversible vision loss (Durand 2013). Early identification of infection, biopsy and culture of the vitreous fluid and appropriate selection of antibiotics is the key for effective treatment of infectious endophthalmitis. The advocated management of bacterial endophthalmitis involves direct injection of antibiotics into the vitreous cavity (Baum 1982; Vahey 1999). It is believed that early vitrectomy may be favourable as it provided earlier visualization of the retina, removes the debridement of bacteria, eliminates inflammatory cells from the vitreous cavity and assists better diffusion of antibiotics (Endophthalmitis Vitrectomy Study, 1995). Endophthalmitis is always associated with inflammation, and ocular inflammatory response is essential for the clearance of pathogens during infection. On the contrary, with the eye being an immune privilege organ, the persistent inflammatory response can prompt collateral damage to delicate neurologic tissues. Hence along with antibiotics, corticosteroids/dexamethasone is frequently administrated as an adjunct therapy. Nevertheless, the aid of corticosteroid administration and the use of intravitreal steroids to manage bacterial endophthalmitis has been controversial and contradictory. While few studies showed that intravitreal dexamethasone/antibiotic combinations are effective (De Kaspar et al., 2008; Liu et al., 2000), few other studies reported its inefficacy (Wiskur e al., 2008; Kufoy et al., 1990; Kocak et al., 2009) in reducing inflammation during experimental bacterial endophthalmitis. This controversial result further warrants the discovery of newer non-immunosuppressive anti-inflammatory therapeutics. Since antibiotics are the mainstay of treatment for bacterial endophthalmitis, antibiotic resistance constitutes an immense threat in the management of the endophthalmitis and restoration of vision.

9.2 Multidrug-Resistant Endophthalmitis: A Therapeutic Challenge

Over the last few decades, indiscriminate use of antibiotics globally has led to the rise of multidrug-resistant organisms or "superbugs" (Fair et al., 2014), which necessitates constant vigilance in order that the successes achieved so far will not be lost. Multidrug resistance among ocular pathogens is an emerging concern and is increasing, as is worldwide systemic antibiotic resistance (Relhan et al., 2016). This ever-increasing resistance to antibiotics is particularly catastrophic as immediate empirical treatment is administered in cases of sight-threatening infections like endophthalmitis (Wakley et al., 2014). The changing trend in the antibiotic susceptibility of the causative organism has already been reported by several earlier studies (Benz et al., 2004; Anand et al., 2000). In an earlier study at The LV Prasad Eye Institute (LVPEI) Hyderabad, India, which looked at the trends amongst multidrug-resistant bacteria (Pathengay et al., 2011), and observed that Gram-negative bacteria (78.6%) was more common compared to Gram-positive bacteria (21.4%). Resistance to ceftazidime and amikacin was seen in 45% and 54.5% of the Gram-negative bacteria respectively while 33.3% were resistant to both ceftazidime and amikacin.

9.2.1 Endophthalmitis caused by *Pseudomonas* spp.

Endophthalmitis caused by *Pseudomonas* sp. is characterized by a rapid fulminant course, poor visual and anatomical outcomes, and severe ocular tissue destruction even when the organism is sensitive to intraocular antibiotics (Eifrig et al., 2003; Pollack 1984). The study reported that out of 28 eyes, 18 eyes infected with *P. aeruginosa* (PA) either developed phthisis bulbi or underwent evisceration. Moreover, the remaining nine patients couldn't achieve a final visual acuity of 5/200 or better. Drug resistance to *Pseudomonas* further complicates the management and worsens prognosis (Pathengay et al., 2011). *Pseudomonas* was found to be the most common isolate in 72.73% of multidrug-resistant (MDR) bacterial endophthalmitis cases (Pathengay et al., 2011). Another study showed that 27% of the endophthalmitis cases were MDR, which increased from 10.3% to 50% in a 6 year span (Pan et al., 2020). A study by Bawankar et al. (2019), reported that despite early vitrectomy and prompt administration of intravitreal antibiotics, the outcome was poor in about 50% of patients who had poor visual outcomes and 46% of patients underwent evisceration or phthisis. Another study by Parchand et al. (2020), revealed that the overall outcome was poor in MDR-PA induced endophthalmitis patients (85.5%), of which 37% of patients needed evisceration, 30.6% of the eyes developed phthisis bulbi while 14.5% of the patients had better than 20/200 visual acuity.

9.2.2 Endophthalmitis Caused by *Staphylococcus aureus*

S. aureus aided by a large repertoire of virulence factors, along with metabolic flexibility, facilitates its persistence within the host and represents a huge public health burden (Brandt et al., 2018). Over and above that, when it becomes resistant to an entire class of antibiotics called β-lactams, management becomes challenging. Methicillin-resistant *S. aureus* (*MRSA*) has become a more frequent cause of ocular infection (Adebayo et al., 2011; Asbell et al., 2008). The proportion of *MRSA* in endophthalmitis is said to have increased to 38% in 2008 from around 26% in 2006 (Chuang et al., 2012). In a study conducted by the United States surveillance program called the Antibiotic Resistance Monitoring in Ocular Microorganisms (ARMOR) (Haas W et al., 2011), that tracks resistance profiles from 207 presumed endophthalmitis isolates (69 aqueous and 138 vitreous) obtained from 2009 to 2016, found that 60 of the 119 (50%) coagulase-negative *staphylococci* and 15 of the 40 (38%) *S. aureus* isolates were methicillin-resistant. Basically, once they're methicillin-resistant, many of them are multi-drug resistant, meaning if you switch classes, you're likely to hit another resistance pattern. Deramo et al. (2008) observed that 18% of the culture-positive cases accounted for MRSA infection and was associated with a poor visual outcome. A study at Bascom Palmer Eye Institute reported that in 59% of MSSA, a final visual acuity of 20/400 or better was achieved while only 36% was achieved in patients infected with MRSA. Furthermore, MRSA endophthalmitis was associated with a higher rate of vitrectomy and no visual improvement, despite treatment, possibly reflecting a more severe clinical presentation (Major et al., 2010). MRSA-associated endophthalmitis frequently leads to a poor visual outcome, including loss of all vision and potential loss of the eye (Winokur et al., 2007; Flynn et al., 2008). To halt the global spread of drug resistance, it is essential that novel antimicrobial therapeutics that target resistant bacteria be discovered. Furthermore, in this battle of drug resistance, these pathogens have become accustomed to the milieu of the internal structures of the eye and have now acquired resistance to the innate immune system that usually helps in preventing endophthalmitis. Also, since the current therapeutic modalities are often unsuccessful in mitigating inflammatory tissue destruction during drug-resistant endophthalmitis, newer molecules or adjuvants are needed to combat these infections.

9.3 Impact of Multidrug-Resistance in Endophthalmitis

Recent literature has begun to elucidate, that the clinical course and severity of bacterial endophthalmitis is not only associated with the strain and virulence of the infecting bacterium but also, on the immune response of the host (Kernt et al., 2010). It is now becoming increasingly evident that bacterial instigation of the host's immune response notably aid in the inflammatory process and intraocular damage observed during this potentially blinding infection (Miller FC et al., 2019). There is a high amount of data available, pertaining to the difference in the virulence factors and pathogenicity of non-MDR and MDR bacteria (Tzouvelekis et al., 2012; De breji et al., 2012; Xiong et al., 2015). To investigate whether there is a similar difference in the phenotypic characteristics and possible virulence factors of antibiotic-resistant and antibiotic-susceptible ocular isolates, a study using 46 clinical isolates of *P. aeruginosa* showed that there is increased production of phospholipase (91.6%), protease (62.5%), pyoverdine (58.3%), and pyocyanin (30.4%) in the case of MDR-PA isolate, and they formed stronger biofilms along with exo-A (30.4%) production (Naik et al., 2021b). An important, but often-overlooked factor in the clearance of infection and emergence of drug resistance is the host's immune response. Lately, research has begun to illustrate the underlying molecular and cellular events that contribute to the impairment that occurs in intraocular infection. Vaccinations against individual toxins produced by more virulent pathogens might not prevent the damage caused by the multitude of other toxins expressed by these strains, nor reduce the level of inflammation associated with the elaboration of these toxins. Infection with more virulent strains of bacteria might result in greater inflammation, and the resultant bystander damage could surpass the benefits of the innate immune response in the eye (Miller et al., 2019). The specific effects of multidrug-resistant bacterial toxins on the cells and tissues present in the posterior segment have not been explored. These immunologically privileged areas induce intraocular inflammation just by sensing the mere presence of certain pathogens. Considering the restricted efficacy of antibiotics in the case of multidrug-resistance, bacterial clearance warrants an efficacious immune response.

Understanding how MDR pathogens instigate and modulate innate immune signalling, and whether this response is unrestrained or insubstantial leading to collateral damage may guide future immunomodulatory therapy. While a great deal of effort has gone into understanding the role of bacterial virulence factors in disease initiation and propagation, it is becoming increasingly clear that the host response to infection plays a major role in causing cellular damage associated with endophthalmitis, which influences the outcome by either constraining the inflammation or spreading the successive retinal damage. Unfortunately, the crucial role of the host immune response in the visual function damage is still not properly understood.

9.4 Differential Host Immune Response in Infections with Resistant Strains

Despite recognizing the importance of the role of the host's immune defences, considerable studies have emphasized only the dynamics of the pathogen and pharmacokinetics (PK) of the particular antibiotic, without evaluating the role of host immune response (Lipsitch et al., 1997; Drusano 2004; De Ryke et al., 2006; Ambrose et al., 2007). With the increase in virulence, there might be a difference in the host recognition patterns as well as the initiation of this immune response. A study by Wang et al. (2016) showed that drug-sensitive and resistant strains cause a different immune response in the pathogenesis of MDR-tuberculosis (MDR-TB), where about 50 proteins and 43 mi-RNAs, were differentially expressed in samples from drug-resistant patients compared to drug-sensitive patients. A similar study on MDR-TB shows modulated activation of CD8+ T Cells (Geffner et al., 2014) in resistant strains. Basingnaa et al. (2018) determined the differences in plasma concentrations of "IFN- ϒ and TNF-α and IL-10" in MDR-Tuberculosis and Drug-Susceptible (DS) Tuberculosis patients. Levels of "IL-10, TNF- α and IFN-ϒ" cytokines were significantly higher in MDR-TB patients compared to DS-TB patients. A study by Yang et al. (2018) outlined that, in macaques, infection by MDR-TB strain resulted in higher bacterial burdens followed by extensive immune responses in the airways. Tam et al. (2018) also concluded that, in an experimental pneumonia model, antibiotic resistance in *Pseudomonas aeruginosa* is associated with distinct activation patterns of the innate immune system. Another earlier study showed that extensively drug-resistant (XDR)-TB patients had a substantial increase in the frequency of "CD4+CD25+FoxP3+ Tregs" compared with DS-TB

(Davids et al., 2018). Additionally, Geffner (2014) reported that MDR-TB instigated a modulated CD8+ T cell activation. Thus, as evident by the literature, multidrug-resistant pathogens induces diverse immune responses which necessitates a thorough understanding of the immunologic responses of this immune-privileged organ aiding in infection control, thereby preventing the associated damage. A better understanding of the complex and often redundant role of these pathways and immune mediators may identify host molecules as potential therapeutic targets. As more information becomes available with respect to the MDR bacteria-induced host immune response, several steps in the evolution of infection may emerge as new therapeutic opportunities.

9.4.1 Host Immune Response in Microglial and Retinal Cells Infected with Resistant Strains

The increasing incidence of MDR-PA and MRSA along with the rapid fulminant course and poor visual and anatomical outcomes despite prompt treatment, poses a significant challenge in the treatment of endophthalmitis (Eifrig et al., 2003; Marangon et al., 2004). Our *in-vitro* analysis (Naik et al., 2020; Naik and Joseph 2021) reveals a robust inflammatory response due to MDR and MRSA infection in human microglial and retinal cells versus antibiotic susceptible pathogens, during infection, also manifesting that these cells play a crucial role in the pathogenesis of endophthalmitis. Our study revealed that *P. aeruginosa* and *S. aureus* strains, which were isolated from culture-proven endophthalmitis patients induced an innate response depending on their antibiotic susceptibility profile. The divergent immune response seen during infection with resistant strains was characterized by a modulated expression of TLRs, various inflammatory cytokines and chemokines and MMPs. More importantly, we observed that MDR-PA induced a higher inflammatory response compared to sensitive PA, which coincides with its fitness rate and increased survival. Further validation in RPE cells, demonstrated that the differential innate responses exists, but was delayed compared to expression in CHME-3 cells.

A study of TLR expression patterns in cultured retinal cells revealed exacerbated expression levels of TLR-4, 5 and 9 in MDR-PA and TLR-1, 2, *6* and *9* in MRSA infected cells. Earlier studies have shown that cell membrane lipoproteins, nucleic acids, LPS and flagellin are ligands for TLR-4, *5* and *9* in *P. aeruginosa* and *TLR1, 2,* and *6* recognize the *S. aureus* cell wall components, such as lipopolysaccharide (LPS) and peptidoglycan (PGN) and induce a robust transcriptional activation leading to inflammatory responses (McIsaac et al., 2012; Zarember et al., 2002). The expression of *TLR-9* upon sensing unmethylated CpG-motifs of *P. aeruginosa* leading to cytokine induction has already been reported by Greene et al. (2005) and Duggan et al. (2011). Additionally, the recruited macrophages and neutrophils express TLR-2, which rapidly responds to *S. aureus* infection and further stimulates the production of cytokine and phagocytosis (Fournier; 2013). Another earlier study (Gomez-Zorrilla et al., 2014) correlated the level of cytokine and bacterial clearance during intraperitoneal infections and discovered that MDR isolates induced a stronger immune response. These data correlate with the results in this study which shows an exacerbated immune response with MDR strains in both microglia and retinal pigment epithelium (RPE) cells. Studies have shown that intratracheal administration of recombinant IL-1β causes accumulation of lung neutrophils and lung injury in hamsters (Patton et. 1995).

Additionally, endogenous IL-1β is reported to protect neutrophils from apoptosis (Bossu et al., 2001) which has the potential for a greater inflammatory response. Both MDR-PA and MRSA evoked an accelerated expression of IL-1β. Furthermore, it has been suggested that in presence of microbial components, inflammasomes like NLRP3 via activation of caspase-1 can tune the production and release of IL-1β (Hirao et al., 2009). Our study did demonstrate the participation of caspase-1 activity in MRSA infected cells, which was comparatively upregulated compared to MSSA (Naik and Joseph 2021). This finding was similar to a report by Rosenzweig et al. (2012), which showed that the inflammasome complex (caspase-1/ NLRP3) contributed to IL-1β synthesis in the eye during endotoxin-induced uveitis. This observed upregulation of NLRP3 and Caspase-1 activity in MRSA infected cell lines could be speculated that methicillin-resistance in *S. aureus* may stimulate NLRP3 leading to the up-regulation of caspase-1 activity as well as an accelerated IL-1β expression. IFN-γ acts as a crucial regulatory cytokine, which is proficient in coordinating both host resistance and plays a pivotal role in controlling *staphylococcal* colonization (Culshaw et al., 2005; Satorres et al., 2009). A report by Kolonitsiou et al. (2019) showed that the MRSA strain could induce a significantly lower IFN-γ production, which they linked with its ability to escape the host immune response and promote chronic colonization. The same observation was made in the present study, thereby explaining its potential to evade the host immune response. Another study by Cole et al. (2016) found that in experimental endocarditis animals challenged by MRSA, the elevated release of *IL-10* and *IL-8* was observed compared with MSSA and the opposite observation was made for TNF-α in the serum. The authors concluded that there is a distinctive pattern of stimulation by MSSA and MRSA leading to shorter survival. Leech et al. (2017) have reported that MRSA evoked a rapid IL-10 production, which corroborated with our study. Another report by Lamas et al. (2018) demonstrated that IL-10 production reduces the clearance of *Salmonella*, which results in increased mortality in the host. Cytokines like IL-1β and TNF-α could upregulate the expression of nitric oxide, which causes induction of MMP genes, leading to excessive proteolysis and tissue damage (Vatansever et al., 2013; Kim et al., 2005). Thus, we speculated that the induced expression of MMP-9 by MDR-PA and MRSA could contribute to enhanced pathogenesis. These results corroborate with McClellan et al. (2006), who reported a six-fold overexpression of MMP-9 in response to *P. aeruginosa* infection in the cornea. Our study also suggests that in case of MRSA infection, excessive expression of MMP-9 and MMP-2 could be involved, at least in part, in tissue destruction as observed in an earlier similar study in MRSA-infected lung cancer cells (An et al., 2016). Thus, excessive expression of these metalloproteinases induced by MRSA strains could accelerate the destruction of the retinal tissue architecture. To summarize, by employing two *in vitro* model systems, it is demonstrated that a non-cell specific inflammatory "cytokine storm" is associated with MDR-PA and MRSA-induced infection. These data highlight the association between antibiotic resistance and host immune response during endophthalmitis and suggest newer frontiers in supporting targeted immune enhancement for better clinical outcomes.

The initial innate response instigated by the host holds an important role in the clearance of pathogen (Giese et al., 2003); nonetheless, its impairment can lead to collateral damage to sensitive non-regenerative retinal cells (Callegan et al., 2002). Therefore, a fine-tuned response that includes recruiting activated immune cells that allows efficient bacterial clearance with minimal host-mediated damage is warranted. This can be achieved by using immunomodulation therapies, which permits fine-tuning the intraocular inflammatory response and though the present study is preliminary, it gives us clues to newer immunomodulatory targets. It was also proposed that based on the type of pathogen or the insult, the resulting inflammatory response and the sequential inflammatory pathways can be delineated.

9.5 Validation in Endophthalmitis Patients with Resistant *P. aeruginosa / S. aureus*

Much of what is known about the role of the inflammatory response has been determined in severe pneumonia (Fernández-Serrano et al., 2003; Endeman et al., 2011; Menéndez et al., 2012) and sepsis (Kellum et al., 2007; Muenzer et al., 2010) only. Nevertheless, inflammatory responses in MDR-PA and MRSA induced endophthalmitis in comparison to S-PA and MSSA induced endophthalmitis respectively remains unexplored. Understanding the immune factors in the human vitreous fluid would lead us to establish a possible correlation between the development of multidrug resistance and the potential of pathogens to instigate a varied immune response and the signalling sequelae. Focusing on the pathways associated with inflammation would be a promising immune-modulatory strategy to limit inflammation and protect vital retinal functions. Naik et al. (2021a), sought to validate the divergent immune response exhibited by the immune mediators as well as to explore the probable association with phosphorylated signalling intermediates in MDR-PA and MRSA endophthalmitis in human vitreous samples. As the visual outcome of patients is the only measure of the endophthalmitis outcome, the expressions of these inflammatory mediators were further studied for the probable association.

In their study Naik et al. (2021a) demonstrated an aggravated inflammatory response, characterised by elevated IL-8, IL-6, IL-10, IL-1β, and TNF-α levels in MDR-PA endophthalmitis and higher IL-10, IL-6, IL-8 and GM-CSF levels in MRSA and MSSE infected vitreous compared to its susceptible counterparts (unpublished data), suggesting a probable link between resistance to antibiotics and inflammation-induced irreversible retinal damage (Manson et al., 2010; Taylor et al., 2018). These findings agree with findings in diseases like MDR-Tuberculosis (MDR-TB), bacteraemia and acute organ dysfunction. Basingnaa and co-workers, (2018) reported significantly elevated levels of IL-1β in MDR-TB. Additionally, they also observed higher levels of TNF- α and IL-10 in MDR-TB compared to DS-TB cases. Previous studies have reported an elevated level of IL-10 in MDR-PA induced sepsis compared to non-MDR *P. aeruginosa* as well as in mice models of pneumonia (Gómez-Zorrilla et al., 2017; Tam et al., 2018). IL-10 also appears to play a crucial role in downregulating the immune response and limiting injury of tissue in response to MTB infection, but an elevated expression of IL-10 may incite a negative impact on the ability to manage the infection (Romero-Adrian et al., 2015). Several other studies assessing the role of IL-6 concluded that its expression is linked with the sensitivity to the pathogen and disease severity (Breen et al., 1990; Havlir et al., 2001; Wong et a., 1992; Yoshimoto et al., 1992; Havell et al., 1990). Moreover, Gomez-Zorrilla et al. (2014) also observed upregulation of IL-6 levels in MDR-PA bacteraemia. McNicholas et al. (2018) compared cytokine levels between MSSA and MRSA bloodstream infections and reported a higher level of IL-6 in the MRSA infected group. A study by Verdrengh et al. (1998) suggested that while granulocyte has a protective role during *S. aureus* infection, upregulation of granulocyte-macrophage production had no added protection against tissue injury. However, we found elevated levels of GM-CSF in the vitreous fluid of MRSA/MSSE infected individuals (Naik and Joseph 2021). Studies (Naik et al., 2020; Naik and Joseph 2021) also reported a significantly higher level of IL-8 in vitreous of S-PA *and* MRSA/MRSE infected patients which is a known stimulator of neutrophils and PMNs (Tacke et al., 2019). This occurrence may, in part, explain the poor outcome in MDR-PA and MRSA patients, regardless of rapid initiation of treatment with second-line intravitreal antibiotics. Interestingly, the inflammatory response is much stronger in MDR-PA infected vitreous samples compared to the MRSA/MRSE infected samples.

Our earlier study (Naik et al., 2021a) also demonstrated significant differences in the phosphorylation levels of NF-κB, Akt, JNK, STAT3 and p70 S6 kinase between S-PA and MDR-PA study group though no significant difference was observed in MRSA/MRSE versus MSSA infected vitreous samples. Previous work reported that JNK1/2 is activated by TNF-α in cultured human chondrocytes (Malemud et al., 2012) and this could explain the increased phosphorylation level of JNK1/2 in *S-PA* patients in our study. Another study reported that STAT3 phosphorylation impacts the activation of NF-κB and assists the inflammatory response in *S. aureus* (Rajamani et al., 2016) and *Bacillus* sp. (Mursalin et al., 2020a; b) endophthalmitis. While the role of GM-CSF and IL-6 is well studied in activating STAT5 and STAT3 respectively, (Lehtonen et al., 2002, Wang et al., 2013) our studies (Naik et al., 2020; Naik and Joseph, 2021) reported a distinctive expression of IL-6 and STAT3, although there was no statistical significance observed in case of GM-CSF and STAT5. Nonetheless, both our studies underline the role of cytokines and the immune signalling pathways to be involved in both initiations, severity and extent of MDR-PA infection in retinal cells. It remains an important finding that a significant difference was observed in our patient vitreous fluid despite large variations in clinical and demographic factors. Given the growing body of evidence linking cytokines to hyperinflammation, it's not astonishing that all these inflammatory markers could be used as diagnostic markers to help detect MDR-PA and MRSA infections. Furthermore, the investigation did unequivocally validate our proposed hypothesis in *in-vitro* conditions.

9.5.1 Comparative Whole Transcriptome Analysis of Eyes Infected with Antibiotic-resistant and Susceptible *P. aeruginosa and S. aureus* in a Murine Model

Most research focus almost completely on the pathogen's virulence and the antibiotic's mechanism of action, without specifically taking host immune response into consideration, only certain studies, imply that host-pathogen interactions influence the extremity and outcome of an infectious disease (Mursalin et al., 2020 a). A better understanding of the host responses to multidrug resistant pathogens during endophthalmitis, in comparison to its antibiotic susceptible counterpart will assist the development of adjunctive therapies and help alleviate the poor visual outcome associated with MDR-PA and MRSA endophthalmitis. While a number of animal models have proven invaluable in unravelling the fundamental mechanisms of endophthalmitis pathogenesis, Naik et al. (2022a) developed an experimental mice model of endophthalmitis, while focussing on the disease pathology of MDR-PA and MRSA and classified the modulated molecular pathways and master genes that were divergently regulated during drug-resistant endophthalmitis by whole transcriptome analysis. The microarray results of MDR-PA infected eyes (Naik et al., 2022a), revealed a significant increase of GFAP, which was also confirmed by histopathological investigation. However, in case of MRSA (Naik et al., 2022b), although the ocular sections revealed increased GFAP intensity, the difference was not appreciable and statistically significant compared to MSSA, and the transcriptomic data was also in agreement with it. A hallmark of these studies is the extent of retinal deterioration, suggesting a key role of antibiotic resistance in disease progression. The differentially expressed cytokines unravelled in this study (Naik et al., 2022a) was like our findings in human vitreous samples and human retinal cells. Our findings further strengthen our hypothesis of a cytokine storm that occurs in drug resistant endophthalmitis (Naik et al., 2021a). Apart from cytokines, our comparative transcriptomics data in MDR-PA endophthalmitis, outlined the involvement of a broad range of chemokines in inducing a modulated host immune response, including "CXCL (2,5,9,12,13,14,16), CCL (6,24,12,3,9,4,7,19), and CCR (2,5)". In *MRSA* transcriptomic profiling, IL-10, IL-15, IL-1 receptor, IL-11, Interferon receptors, GM-CSF and M-CSF and chemokines (CXCL-1, CCL-11) were significantly upregulated (Naik et al., 2021a). Although various studies have linked these chemokines to inflammation (Gerard et al., 2001), nothing has been written on their relationship with MDR-PA and MRSA infection. Upregulation of CCL-2 and 3 and CXCL-2 in *Bacillus* endophthalmitis, according to Coburn et al. (2018), leads to significant retinal injury. They also suggested that CCL3 interacts with CCR5 to recruit and activate polymorphonuclear leukocytes, and that blocking CCR5 could minimise infiltration of neutrophils into the vitreous and hence reduce retinal injury. Rajmani et al. (2016) conducted a study on *S. aureus* endophthalmitis and reported a marked upregulation of several cytokines compared to uninfected controls, however in the study by Naik et al. (2022b), only differential upregulation of IL-10 and IL-6 was observed compared to MSSA. This suggests that unlike MDR-PA, MRSA elicits a subordinate cytokine and chemokine response. An earlier investigation in atopic dermatitis, found that IL-1β and IFN-γ produce a considerable growth boost on *S. aureus* strains in a concentration-dependent manner, and these findings are consistent with our findings in MRSA endophthalmitis (Di Domenico et al., 2018). In addition, MDR-PA study also unveiled the involvement of many caspases, as well as NLRP3, as differentially upregulated in comparison to S-PA endophthalmitis (Naik et al., 2022a). Toxins released by *P. aeruginosa* (exoT, exoS, and exoU) have been demonstrated to activate the NLRP3 inflammasome (Sutterwala et al., 2007), resulting in tissue damage. While, inflammasomes are manifested to play a critical role in the pathogenesis of autoimmune disorders, inflammatory diseases, and malignancies (Lee et al., 2013; Dinarello, 2010), it has been postulated that during infection, the inflammasome is assembled via IL-1β, which further draws inflammatory mediators (Khan et al., 2007; McClellan et al., 2006, Van den Steen et al., 2000), causing collateral damage to the host tissue in the process of managing the pathogenic effect, resulting in partial or irreversible blindness (Yerramothu et al., 2018). The above-mentioned process along with the release of cytokines and caspase-1 activation leads to pyroptosis. During infection, early elevation of IL-1β could be suggestive of playing an important role in signalling, besides leukocyte transport. Hence, targeted therapy that would block the production of IL-1β and the sequential inflammasomes activation in the eye, may aid in the improvement of clinical outcomes. According to a previous study, suppression of caspase-1 as an adjunct therapy combined with ciprofloxacin minimized the degree of intraocular inflammation in a mouse model of *PA* keratitis, (Thakur et al., 2004). MMP dysregulation also has been linked to disease characterised by unrestrained proteolysis of the connective tissue (Wong et al., 2003) and ECM deterioration. As a result, it's not surprising that, in the case of MDR-PA endophthalmitis, retinal deterioration was greater due to elevated expression of MMPs, resulting in breakage of blood-retinal-barrier. In addition, during MDR-PA infection, certain "intercellular tight junction proteins, such as Claudin (9 and 7), TGFBR2 and gap junction proteins", were incessantly elevated (Naik et al., 2022a). Tight-junction dysregulation causes altered barrier function, resulting in alterations in cytokines such as "IFN-α, IFN-gamma, IL-6 and IL-1β," which are also observed in other inflammatory disorders like multiple sclerosis and malignancies (Harhaj et al., 2004; Cereijido et al., 2007). Again, the infiltration of inflammatory mediators is mediated by adhesion molecules, expressed on endothelial cell surfaces and increased expression of these adhesion molecules have been earlier shown in some ocular diseases (Chen et al., 2022; Giese et al., 2000). Similarly, in the whole transcriptome analysis of MRSA endophthalmitis, many tight junction and gap junction proteins were disproportionately elevated compared to *MSSA* induced endophthalmitis (Naik et al., 2022b). Dejana (1996) found that leukocyte attachment to endothelial cells is usually followed by extravasation, which regulates leukocyte movement through the clefts of endothelial cells. Elevated leukocyte migration in MRSA endophthalmitis could be the result of the aforesaid observation. Although both MRSA and MSSA infected eyes exhibited increased leukocyte and neutrophil influx, there was no statistically significant change (Naik et al., 2022b).

Innate immunity includes the complement system as well. Complement activation produces anaphylatoxins, which cause inflammation, attract phagocytes chemotactically to the infection site, enhances opsonization leading to bacterial lysis, and causes vascular permeability. While each of these activities aids in the clearing of bacteria, when left uncontrolled, the same mechanisms can cause substantial damage to the host tissue (Tosi 2005). In one study, C3a, C5a, and C5b-C9, as well as other components of the complement system have enhanced expression (Chen at al., 2015). Complements can cause severe tissue remodelling in the case of prolonged injury and over-activation of the inflammatory reaction, as previously observed (Chen at al., 2015). Modulation of the complement system is a hot topic in inflammation-related tissue damage research (Galeas-Pena et al., 2019), therefore it's crucial to distinguish between the favourable and harmful effects of complement cascade activation in MDR Gram-negative induced endophthalmitis. Conversely, in a murine model of *S. aureus* endophthalmitis, Engelbert et al. (2005) found that the absence of complement system had no effect on the disease prognosis.

In terms of master regulators of pathways, it was observed that there were several involved in MDR-PA induced endophthalmitis including CARD6, TRAF6, TRAF1, TRADD, Rgs7, STAT3, CREB/ATF bZIP transcription factor, NF-κβ, Tank, CASP8 and FADD-like apoptosis regulator etc. Albeit, in MRSA very few regulators were found to be involved. Upregulation of CREB, is known to regulate the gene transcription of IL-2, IL-6, IL-10, and TNF-α, as well as inhibition of NF-κβ activation and

Acknowledgments: Hyderabad Eye Research Foundation, Hyderabad, India.

Funding: DST-SERB to Dr. Joveeta Joseph (File Number: CRG/2019/004386) India and Hyderabad Eye Research Foundation (HERF).

REFERENCES

Adebayo A, Parikh JG, McCormick SA, Shah MK, Huerto RS, Yu G, et al., Shifting trends in in vitro antibiotic susceptibilities for common bacterial conjunctival isolates in the last decade at the New York Eye and Ear Infirmary. *Graefes Arch Clin Exp Ophthalmol*. 2011 Jan;249(1):111–9. doi: 10.1007/s00417-010-1426-6.

Ambrose PG, Bhavnani SM, Rubino CM, Louie A, Gumbo T, Forrest A, et al., Pharmacokinetics-pharmacodynamics of antimicrobial therapy: it's not just for mice anymore. *Clin Infect Dis*. 2007 Jan 1;44(1):79–86. doi: 10.1086/510079.

An J, Li Z, Dong Y, Ren J, Guo K. Methicillin-resistant Staphylococcus Aureus infection exacerbates NSCLC cell metastasis by up-regulating TLR4/MyD88 pathway. Cell Mol Biol (Noisy-le-grand). 2016 Jul 31;62(8):1–7.

Anand AR, Therese KL, Madhavan HN. Spectrum of aetiological agents of postoperative endophthalmitis and antibiotic susceptibility of bacterial isolates. *Indian J Ophthalmol*. 2000 Jun;48(2):123–8.

Asbell PA, Sahm DF, Shaw M, Draghi DC, Brown NP. Increasing prevalence of methicillin resistance in serious ocular infections caused by Staphylococcus aureus in the United States: 2000 to 2005. *J Cataract Refract Surg*. 2008 May;34(5):814–8. doi: 10.1016/j.jcrs.2008.01.016.

Astrakhantseva IV, Efimov GA, Drutskaya MS, Kruglov AA, Nedospasov SA. Modern anti-cytokine therapy of autoimmune diseases. *Biochemistry (Mosc)*. 2014 Dec;79(12):1308–21. doi: 10.1134/S0006297914120049.

Bambouskova M, Gorvel L, Lampropoulou V, Sergushichev A, Loginicheva E, Johnson K, et al., Electrophilic properties of itaconate and derivatives regulate the IκBζ-ATF3 inflammatory axis. *Nature*. 2018 Apr;556(7702):501–504. doi: 10.1038/s41586-018-0052-z.

Barman S, Dhanda G, Naik P, Mukherjee R, Jolly L, Joseph J, et al., Multi-functional small molecules with temporal charge-switchability tackle infection and inflammation. *Adv Ther*. 2022; 5:2100234.

Basingnaa A, Antwi-Baffour S, Nkansah DO, Afutu E, Owusu E. Plasma levels of cytokines (IL-10, IFN-γ and TNF-α) in multidrug resistant tuberculosis and drug responsive tuberculosis patients in Ghana. *Diseases*. 2018 Dec 23;7(1):2. doi: 10.3390/diseases7010002.

Baum J, Peyman GA, Barza M. Intravitreal administration of antibiotic in the treatment of bacterial endophthalmitis. *III. Consensus. Surv Ophthalmol*. 1982 Jan-Feb;26(4):204–6. doi: 10.1016/0039-6257(82)90080-7.

Bawankar P, Bhattacharjee H, Barman M, Soibam R, Deka H, Chandra Kuri G, et al., Outbreak of multidrug-resistant pseudomonas aeruginosa endophthalmitis due to contaminated Trypan Blue solution. *J Ophthalmic Vis Res*. 2019 Jul 18;14(3):257–266. doi: 10.18502/jovr.v14i3.4781.

Bazzoni F, Beutler B. The tumor necrosis factor ligand and receptor families. *N Engl J Med*. 1996 Jun 27;334(26):1717–25. doi: 10.1056/NEJM199606273342607.

Benz MS, Scott IU, Flynn HW Jr, Unonius N, Miller D. Endophthalmitis isolates and antibiotic sensitivities: a 6-year review of culture-proven cases. *Am J Ophthalmol. 2004* Jan;137(1):38–42. doi: 10.1016/s0002-9394(03)00896-1.

Bossù P, del Grosso E, Cesaroni MP, Maurizi G, Balestro N, Stoppacciaro A, et al., Balance between autocrine interleukin-1beta and caspases defines life versus death of polymorphonuclear cells. *Eur Cytokine Netw.* 2001 Mar;12(1):177–86.

Brandt SL, Putnam NE, Cassat JE, Serezani CH. Innate immunity to Staphylococcus aureus: evolving paradigms in soft tissue and invasive infections. *J Immunol.* 2018 Jun 15;200(12):3871–3880. doi: 10.4049/jimmunol.1701574.

Bravo-Santano N, Ellis JK, Mateos LM, Calle Y, Keun HC, Behrends V, et al., Intracellular Staphylococcus aureus modulates host central carbon metabolism to activate autophagy. *mSphere.* 2018 Aug 8;3(4):e00374–18. doi: 10.1128/mSphere.00374-18.

Callegan MC, Engelbert M, Parke DW 2nd, Jett BD, Gilmore MS. Bacterial endophthalmitis: epidemiology, therapeutics, and bacterium-host interactions. *Clin Microbiol Rev.* 2002 Jan;15(1):111–24. doi: 10.1128/CMR.15.1.111-124.2002.

Cereijido M, Contreras RG, Flores-Benítez D, Flores-Maldonado C, Larre I, Ruiz A, et al., New diseases derived or associated with the tight junction. *Arch Med Res.* 2007 Jul;38(5):465–78. doi: 10.1016/j.arcmed.2007.02.003.

Chen M, Xu H. Parainflammation, chronic inflammation, and age-related macular degeneration. *J Leukoc Biol.* 2015 Nov;98(5):713–25. doi: 10.1189/jlb.3RI0615-239R.

Chen YG, Zhang C, Chiang SK, Wu T, Tso MO. Increased nuclear factor-kappa B p65 immunoreactivity following retinal ischemia and reperfusion injury in mice. *J Neurosci Res.* 2003 Apr 1;72(1):125–31. doi: 10.1002/jnr.10548.

Chen YH, Lightman S, Eskandarpour M, Calder VL. Adhesion molecule targeted therapy for non-infectious uveitis. *Int J Mol Sci.* 2022 Jan 3;23(1):503. doi: 10.3390/ijms23010503.

Chuang CC, Hsiao CH, Tan HY, Ma DH, Lin KK, Chang CJ, et al., Staphylococcus aureus ocular infection: methicillin-resistance, clinical features, and antibiotic susceptibilities. *PLoS One.* 2012;8(8):e42437. doi: 10.1371/journal.pone.0042437.

Coburn PS, Miller FC, LaGrow AL, Parkunan SM, Blake Randall C, Staats RL, et al., TLR4 modulates inflammatory gene targets in the retina during Bacillus cereus endophthalmitis. *BMC Ophthalmol.* 2018 Apr 16;18(1):96. doi: 10.1186/s12886-018-0764-8.

Cole LE, Zhang J, Kesselly A, Anosova NG, Lam H, Kleanthous H, et al., Limitations of murine models for assessment of antibody-mediated therapies or vaccine candidates against staphylococcus epidermidis bloodstream infection. *Infect Immun.* 2016 Mar 24;84(4):1143–49. doi: 10.1128/IAI.01472-15.

Cordes T, Wallace M, Michelucci A, Divakaruni AS, Sapcariu SC, Sousa C,et.al. Immunoresponsive gene 1 and itaconate inhibit succinate dehydrogenase to modulate intracellular succinate levels. *J Biol Chem.* 2016 Jul 1;291(27):14274–84. doi: 10.1074/jbc.M115.685792.

Culshaw S, Leung BP, Gracie JA, Campbell CC, Thomson D, Gemmell C, et al., Prior elevation of IL-18 promotes rapid early IFN-gamma production during staphylococcal infection. *Eur J Immunol.* 2005 May;35(5):1438–44. doi: 10.1002/eji.200425661.

Dave TV, Dave VP, Sharma S, Karolia R, Joseph J, Pathengay A, et al., Infectious endophthalmitis leading to evisceration: spectrum of bacterial and fungal pathogens and antibacterial susceptibility profile. *J Ophthalmic Inflamm Infect.* 2019 May 16;9(1):9. doi: 10.1186/s12348-019-0174-y.

Davids M, Pooran AS, Pietersen E, Wainwright HC, Binder A, Warren R, et al., Regulatory T cells subvert mycobacterial containment in patients failing extensively drug-resistant tuberculosis treatment. *Am J Respir Crit Care Med.* 2018 Jul 1;198(1):104–16. doi: 10.1164/rccm.201707-1441OC.

de Breij A, Eveillard M, Dijkshoorn L, van den Broek PJ, Nibbering PH, Joly-Guillou ML. Differences in Acinetobacter baumannii strains and host innate immune response determine morbidity and mortality in experimental pneumonia. *PLoS One*. 2012;7(2):e30673. doi: 10.1371/journal.pone.0030673.

De Kaspar HM, Ta CN, Engelbert M, Mette M, Thiel M, Kampik A. Effects of intravitreal corticosteroid in the treatment of Staphylococcus aureus-induced experimental endophthalmitis. *Retina*. 2008 Feb;28(2):326–32. doi: 10.1097/IAE.0b013e3181237cf8.

Dejana E. Endothelial adherens junctions: implications in the control of vascular permeability and angiogenesis. *J Clin Invest*. 1997 Dec 1;100(11 Suppl):S7-10.

Deramo VA, Lai JC, Winokur J, Luchs J, Udell IJ. Visual outcome and bacterial sensitivity after methicillin-resistant Staphylococcus aureus-associated acute endophthalmitis. *Am J Ophthalmol*. 2008 Mar;145(3):413–417. doi: 10.1016/j.ajo.2007.10.020.

DeRyke CA, Lee SY, Kuti JL, Nicolau DP. Optimising dosing strategies of antibacterials utilising pharmacodynamic principles: impact on the development of resistance. *Drugs*. 2006;66(1):1–14. doi: 10.2165/00003495-200666010-00001.

Di Domenico EG, Cavallo I, Bordignon V, Prignano G, Sperduti I, Gurtner A, et al., Inflammatory cytokines and biofilm production sustain Staphylococcus aureus outgrowth and persistence: a pivotal interplay in the pathogenesis of Atopic Dermatitis. *Sci Rep*. 2018 Jun 28;8(1):9573. doi: 10.1038/s41598-018-27421-1.

Dinarello CA. Why not treat human cancer with interleukin-1 blockade? *Cancer Metastasis Rev*. 2010 Jun;29(2):317–29. doi: 10.1007/s10555-010-9229-0.

Drusano GL. Antimicrobial pharmacodynamics: critical interactions of 'bug and drug'. *Nat Rev Microbiol*. 2004 Apr;2(4):289–300. doi: 10.1038/nrmicro862.

Duggan JM, You D, Cleaver JO, Larson DT, Garza RJ, Guzmán Pruneda FA, et al., Synergistic interactions of TLR2/6 and TLR9 induce a high level of resistance to lung infection in mice. *J Immunol*. 2011 May 15;186(10):5916–26. doi: 10.4049/jimmunol.1002122.

Durand ML. Endophthalmitis. *Clin Microbiol Infect*. 2013 Mar;19(3):227–34. doi: 10.1111/1469-0691.12118.

Eifrig CW, Scott IU, Flynn HW Jr, Miller D. Endophthalmitis caused by Pseudomonas aeruginosa. *Ophthalmology*. 2003 Sep;110(9):1714–7. doi: 10.1016/S0161-6420(03)00572-4.

Endeman H, Meijvis SC, Rijkers GT, van Velzen-Blad H, van Moorsel CH, Grutters JC, et al., Systemic cytokine response in patients with community-acquired pneumonia. *Eur Respir J*. 2011 Jun;37(6):1431–8. doi: 10.1183/09031936.00074410.

Endophthalmitis Vitrectomy Study. A randomized trial of immediate vitrectomy and of intravenous antibiotics for the treatment of postoperative bacterial endophthalmitis. *Arch Ophthalmol*. 1995 Dec;113(12):1479–96.

Engelbert M, Gilmore MS. Fas ligand but not complement is critical for control of experimental Staphylococcus aureus Endophthalmitis. *Invest Ophthalmol Vis Sci*. 2005 Jul;46(7):2479–86. doi: 10.1167/iovs.04-1139.

Fair RJ, Tor Y. Antibiotics and bacterial resistance in the 21st century. *Perspect Medicin Chem*. 2014 Aug 28;6:25–64. doi: 10.4137/PMC.S14459.

Fernández-Serrano S, Dorca J, Coromines M, Carratalà J, Gudiol F, Manresa F. Molecular inflammatory responses measured in blood of patients with severe community-acquired pneumonia. *Clin Diagn Lab Immunol*. 2003 Sep;10(5):813–20. doi: 10.1128/cdli.10.5.813-820.2003.

Flynn HW Jr, Scott IU, Brod RD, Han DP. Current management of endophthalmitis. *Int OphthalmolClin*.2004Fall;44(4):115–37.doi:10.1097/00004397-200404440-00010.

Fournier B. The function of TLR2 during staphylococcal diseases. *Front Cell Infect Microbiol.* 2013 Jan 4;2:167. doi: 10.3389/fcimb.2012.00167.

Galeas-Pena M, McLaughlin N, Pociask D. The role of the innate immune system on pulmonary infections. *Biol Chem.* 2019 Mar 26;400(4):443–456. doi: 10.1515/hsz-2018-0304.

Geffner L, Basile JI, Yokobori N, Kviatcovsky D, Sabio y García C, Ritacco V, et al., Mycobacterium tuberculosis multidrug resistant strain M induces an altered activation of cytotoxic CD8+ T cells. *PLoS One.* 2014 May 16;9(5):e97837. doi: 10.1371/journal.pone.0097837.

Gentile RC, Shukla S, Shah M, Ritterband DC, Engelbert M, Davis A, Hu DN. Microbiological spectrum and antibiotic sensitivity in endophthalmitis: a 25-year review. *Ophthalmology.* 2014 Aug;121(8):1634–42.

Gerard C, Rollins BJ. Chemokines and disease. *Nat Immunol.* 2001 Feb;2(2):108–15. doi: 10.1038/84209.

Giese MJ, Rayner SA, Fardin B, Sumner HL, Rozengurt N, Mondino BJ, et al., Mitigation of neutrophil infiltration in a rat model of early Staphylococcus aureus endophthalmitis. *Invest Ophthalmol Vis Sci.* 2003 Jul;44(7):3077–82. doi: 10.1167/iovs.02-1250.

Gómez-Zorrilla S, Camoez M, Tubau F, Periche E, Cañizares R, Dominguez MA, et al., Antibiotic pressure is a major risk factor for rectal colonization by multidrug-resistant Pseudomonas aeruginosa in critically ill patients. *Antimicrob Agents Chemother.* 2014 Oct;58(10):5863–70. doi: 10.1128/AAC.03419-14.

Greene CM, Carroll TP, Smith SG, Taggart CC, Devaney J, Griffin S, et al., TLR-induced inflammation in cystic fibrosis and non-cystic fibrosis airway epithelial cells. *J Immunol.* 2005 Feb 1;174(3):1638–46. doi: 10.4049/jimmunol.174.3.1638.

Haas W, Pillar CM, Torres M, Morris TW, Sahm DF. Monitoring antibiotic resistance in ocular microorganisms: results from the Antibiotic Resistance Monitoring in Ocular micRorganisms (ARMOR) 2009 surveillance study. *Am J Ophthalmol.* 2011 Oct;152(4):567–574.e3. doi: 10.1016/j.ajo.2011.03.010.

Harhaj NS, Antonetti DA. Regulation of tight junctions and loss of barrier function in pathophysiology. *Int J Biochem Cell Biol.* 2004 Jul;36(7):1206–37. doi: 10.1016/j.biocel.2003.08.007.

Havell EA, Sehgal PB. Tumor necrosis factor-independent IL-6 production during murine listeriosis. *J Immunol.* 1991 Jan 15;146(2):756–61.

Havlir DV, Torriani FJ, Schrier RD, Huang JY, Lederman MM, Chervenak KA, et al., Serum interleukin-6 (IL-6), IL-10, tumor necrosis factor (TNF) alpha, soluble type II TNF receptor, and transforming growth factor beta levels in human immunodeficiency virus type 1-infected individuals with Mycobacterium avium complex disease. *J Clin Microbiol.* 2001 Jan;39(1):298–303. doi: 10.1128/JCM.39.1.298-303.2001.

Hirao K, Yumoto H, Takahashi K, Mukai K, Nakanishi T, Matsuo T. Roles of TLR2, TLR4, NOD2, and NOD1 in pulp fibroblasts. *J Dent Res.* 2009 Aug;88(8):762–7. doi: 10.1177/0022034509341779.

Jambulingam M, Parameswaran SK, Lysa S, Selvaraj M, Madhavan HN. A study on the incidence, microbiological analysis and investigations on the source of infection of postoperative infectious endophthalmitis in a tertiary care ophthalmic hospital: an 8-year study. *Indian J Ophthalmol.* 2010 Jul-Aug;58(4):297–302. doi: 10.4103/0301-4738.64132.

Kellum JA, Kong L, Fink MP, Weissfeld LA, Yealy DM, Pinsky MR, et al., Understanding the inflammatory cytokine response in pneumonia and sepsis: results of the Genetic and Inflammatory Markers of Sepsis (GenIMS) Study. *Arch Intern Med.* 2007 Aug 13–27;167(15):1655–63. doi: 10.1001/archinte.167.15.1655.

Kernt M, Kampik A. Endophthalmitis: pathogenesis, clinical presentation, management, and perspectives. *Clin Ophthalmol.* 2010 Mar 24;4:121–35. doi: 10.2147/opth.s6461.

Khan S, Cole N, Hume EB, Garthwaite L, Conibear TC, Miles DH, et al., The role of CXC chemokine receptor 2 in Pseudomonas aeruginosa corneal infection. *J Leukoc Biol.* 2007 Jan;81(1):315–8. doi: 10.1189/jlb.0506344.

Kim YS, Kim SS, Cho JJ, Choi DH, Hwang O, Shin DH, et al., Matrix metalloproteinase-3: a novel signaling proteinase from apoptotic neuronal cells that activates microglia. *J Neurosci.* 2005 Apr 6;25(14):3701–11. doi: 10.1523/JNEUROSCI.4346-04.2005.

Koçak N, Kaynak S, Kirdar S, Irmak O, Bahar IH. Tavşanlarda oluşturulan deneysel Candida endoftalmisinde farkli tedavi seçeneklerinin karşilaştirilmasi [Comparison of different antifungal treatment regimens for experimental Candida endophthalmitis in rabbit models]. *Mikrobiyol Bul.* 2009 Oct;43(4):619–26. Turkish.

Kolonitsiou F, Papadimitriou-Olivgeris M, Spiliopoulou A, Drougka E, Jelastopulu E, Anastassiou ED, Spiliopoulou I. Methicillin-resistant *Staphylococcus aureus* ST80 induce lower cytokine production by monocytes as compared to other sequence types. *Front Microbiol.* 2019 Jan 9;9:3310. doi: 10.3389/fmicb.2018.03310. PMID: 30687287; PMCID: PMC6333658.

Kufoy EA, Fox K, Fox A, Parks C, Pakalnis VA. Modulation of the blood-aqueous barrier by gram positive and gram negative bacterial cell wall components in the rat and rabbit. *Exp Eye Res.* 1990 Feb;50(2):189–95. doi: 10.1016/0014-4835(90)90230-r.

Lamas A, Miranda JM, Regal P, Vázquez B, Franco CM, Cepeda A. A comprehensive review of non-enterica subspecies of Salmonella enterica. *Microbiol Res.* 2018 Jan;206:60–73. doi: 10.1016/j.micres.2017.09.010.

Lampropoulou V, Sergushichev A, Bambouskova M, Nair S, Vincent EE, Loginicheva E, et al., Itaconate links inhibition of succinate dehydrogenase with macrophage metabolic remodeling and regulation of inflammation. *Cell Metab.* 2016 Jul 12;24(1):158–66. doi: 10.1016/j.cmet.2016.06.004.

Lee HM, Kim JJ, Kim HJ, Shong M, Ku BJ, Jo EK. Upregulated NLRP3 inflammasome activation in patients with type 2 diabetes. *Diabetes.* 2013 Jan;62(1):194–204. doi: 10.2337/db12-0420.

Leech JM, Lacey KA, Mulcahy ME, Medina E, McLoughlin RM. IL-10 plays opposing roles during Staphylococcus aureus systemic and localized infections. *J Immunol.* 2017 Mar 15;198(6):2352–65. doi: 10.4049/jimmunol.1601018.

Lehtonen A, Matikainen S, Miettinen M, Julkunen I. Granulocyte-macrophage colony-stimulating factor (GM-CSF)-induced STAT5 activation and target-gene expression during human monocyte/macrophage differentiation. *J Leukoc Biol.* 2002 Mar;71(3):511–9.

Lipsitch M, Levin BR. The population dynamics of antimicrobial chemotherapy. *Antimicrob Agents Chemother.* 1997 Feb;41(2):363–73. doi: 10.1128/AAC.41.2.363.

Liu SM, Way T, Rodrigues M, Steidl SM. Effects of intravitreal corticosteroids in the treatment of Bacillus cereus endophthalmitis. *Arch Ophthalmol.* 2000 Jun;118(6):803–6. doi: 10.1001/archopht.118.6.803.

Lu X, Ng DS, Zheng K, Peng K, Jin C, Xia H, et al., Risk factors for endophthalmitis requiring evisceration or enucleation. *Sci Rep.* 2016 Jun 15;6:28100. doi: 10.1038/srep28100.

Major JC Jr, Engelbert M, Flynn HW Jr, Miller D, Smiddy WE, Davis JL. Staphylococcus aureus endophthalmitis: antibiotic susceptibilities, methicillin resistance, and clinical outcomes. *Am J Ophthalmol.* 2010 Feb;149(2):278–83.e1. doi: 10.1016/j.ajo.2009.08.023.

Malemud CJ, Sun Y, Pearlman E, Ginley NM, Awadallah A, Wisler BA, et al., Monosodium urate and tumor necrosis factor-α increase apoptosis in human chondrocyte cultures. *Rheumatology (Sunnyvale).* 2012 Dec 26;2:113. doi: 10.4172/2161-1149.1000113.

Manson JM, Hancock LE, Gilmore MS. Mechanism of chromosomal transfer of Enterococcus faecalis pathogenicity island, capsule, antimicrobial resistance, and other traits. *Proc Natl Acad Sci USA*. 2010 Jul 6;107(27):12269–74. doi: 10.1073/pnas.1000139107.

Marangon FB, Miller D, Muallem MS, Romano AC, Alfonso EC. Ciprofloxacin and levofloxacin resistance among methicillin-sensitive Staphylococcus aureus isolates from keratitis and conjunctivitis. *Am J Ophthalmol.* 2004 Mar;137(3):453–8. doi: 10.1016/j.ajo.2003.10.026.

McClellan SA, Huang X, Barrett RP, Lighvani S, Zhang Y, Richiert D, et al., Matrix metalloproteinase-9 amplifies the immune response to Pseudomonas aeruginosa corneal infection. *Invest Ophthalmol Vis Sci.* 2006 Jan;47(1):256–64. doi: 10.1167/iovs.05-1050.

McIsaac SM, Stadnyk AW, Lin TJ. Toll-like receptors in the host defense against Pseudomonas aeruginosa respiratory infection and cystic fibrosis. *J Leukoc Biol.* 2012 Nov;92(5):977–85. doi: 10.1189/jlb.0811410.

McNicholas S, Talento AF, O'Gorman J, Hannan MM, Lynch M, Greene CM, et al., Cytokine responses to Staphylococcus aureus bloodstream infection differ between patient cohorts that have different clinical courses of infection. *BMC Infect Dis. 2014* Nov 15;14:580. doi: 10.1186/s12879-014-0580-6.

Menéndez R, Sahuquillo-Arce JM, Reyes S, Martínez R, Polverino E, Cillóniz C, et al., Cytokine activation patterns and biomarkers are influenced by microorganisms in community-acquired pneumonia. Chest. 2012 Jun;141(6):1537–1545. doi: 10.1378/chest.11-1446.

Miller FC, Coburn PS, Huzzatul MM, LaGrow AL, Livingston E, Callegan MC. Targets of immunomodulation in bacterial endophthalmitis. *Prog Retin Eye Res.* 2019 Nov;73:100763. doi: 10.1016/j.preteyeres.2019.05.004.

Mills EL, Ryan DG, Prag HA, Dikovskaya D, Menon D, Zaslona Z, et al., Itaconate is an anti-inflammatory metabolite that activates Nrf2 via alkylation of KEAP1. *Nature.* 2018 Apr 5;556(7699):113–117. doi: 10.1038/nature25986.

Muenzer JT, Davis CG, Chang K, Schmidt RE, Dunne WM, Coopersmith CM, et al., Characterization and modulation of the immunosuppressive phase of sepsis. *Infect Immun.* 2010 Apr;78(4):1582–92. doi: 10.1128/IAI.01213-09.

Mursalin MH, Coburn PS, Livingston E, Miller FC, Astley R, Flores-Mireles AL, et al., Bacillus S-layer-mediated innate interactions during endophthalmitis. Front Immunol. 2020a, Feb 12;11:215. doi: 10.3389/fimmu.2020.00215.

Mursalin MH, Coburn PS, Miller FC, Livingston ET, Astley R, Callegan MC. Innate immune interference attenuates inflammation in bacillus endophthalmitis. *Invest Ophthalmol Vis Sci.* 2020b Nov 2;61(13):17. doi: 10.1167/iovs.61.13.17.

Naik P, Singh S, Vishwakarma S, Kaur I, Dave VP, Kumar A, et al., Multidrug-resistant *pseudomonas aeruginosa* evokes differential inflammatory responses in human microglial and retinal pigment epithelial cells. *Microorganisms.* 2020 May 14;8(5):735. doi: 10.3390/microorganisms8050735.

Naik P, Joseph J. Difference in host immune response to methicillin-sensitive and resistant Staphylococcus aureus (MSSA and MRSA) endophthalmitis. *Ocul Immunol Inflamm.* 2021 9:1–11. doi: 10.1080/09273948.2020.1859551.

Naik P, Singh S, Rudraprasad D, Dave VP, Kumar A, Joseph J. Multidrug-resistant pseudomonas aeruginosa triggers differential inflammatory response in patients with endophthalmitis. *Transl Vis Sci Technol.* 2021a;10(9):26. doi: 10.1167/tvst.10.9.26.

Naik P, Pandey S, Gagan S, Biswas S, Joseph J. Virulence factors in multidrug (MDR) and Pan-drug resistant (XDR) Pseudomonas aeruginosa: a cross-sectional study of isolates recovered from ocular infections in a high-incidence setting in southern India. *J Ophthalmic Inflamm Infect.* 2021b Sep 28;11(1):36. doi: 10.1186/s12348-021-00268-w. PMID: 34585284.

Naik P, Pandey S, Naik MN, Mishra DK, Boyenpally SR, Joseph J. Transcriptomic and histological analysis of exacerbated immune response in multidrug-resistant Pseudomonas aeruginosa in a murine model of endophthalmitis. Front Immunol. 2022a Jan 3;12:789023. doi: 10.3389/fimmu.2021.789023.

Naik P, Naik MN, Mishra DK, Joseph J. Methicillin resistance in Staphylococcus aureus modulates the transcriptome and disease pathology in a murine model of endophthalmitis. *Exp Eye Res.* 2022b Mar 4:109016. doi: 10.1016/j.exer.2022.109016. Epub ahead of print. PMID: 35257714.

Packer M, Chang DF, Dewey SH, Little BC, Mamalis N, Oetting TA, et al., ASCRS cataract clinical committee. Prevention, diagnosis, and management of acute postoperative bacterial endophthalmitis. *J Cataract Refract Surg.* 2011 Sep;37(9):1699–714. doi: 10.1016/j.jcrs.2011.06.018.

Pan U, Jain A, Gubert J, Kumari B, Sindal MD. Antibiotic sensitivity trends of pseudomonas endophthalmitis in a tertiary eye care center in South India: a 12-year retrospective study. *Indian J Ophthalmol.* 2020 Apr;68(4):627–631.

Parchand SM, Agrawal D, Chatterjee S, Gangwe A, Mishra M, Agrawal D. Post-cataract surgery cluster endophthalmitis due to multidrug-resistant Pseudomonas aeruginosa: a retrospective cohort study of six clusters. *Indian J Ophthalmol.* 2020 Jul;68(7):1424–31. doi: 10.4103/ijo.IJO_1612_19.

Parkunan SM, Randall CB, Coburn PS, Astley RA, Staats RL, Callegan MC. Unexpected roles for toll-like receptor 4 and TRIF in intraocular infection with gram-positive bacteria. *Infect Immun.* 2015 Oct;83(10):3926–36. doi: 10.1128/IAI.00502-15.

Pathengay A, Moreker MR, Puthussery R, Ambatipudi S, Jalali S, Majji AB, et al., Clinical and microbiologic review of culture-proven endophthalmitis caused by multidrug-resistant bacteria in patients seen at a tertiary eye care center in southern India. *Retina.* 2011 Oct;31(9):1806–11. doi: 10.1097/IAE.0b013e31820f4b9d.

Patton LM, Saggart BS, Ahmed NK, Leff JA, Repine JE. Interleukin-1 beta-induced neutrophil recruitment and acute lung injury in hamsters. *Inflammation.* 1995 Feb;19(1):23–9. doi: 10.1007/BF01534377.

Pollack M. The virulence of Pseudomonas aeruginosa. *Rev Infect Dis.* 1984 Sep-Oct;6 Suppl 3:S617–26. doi: 10.1093/clinids/6.supplement_3.s617.

Rajamani D, Singh PK, Rottmann BG, Singh N, Bhasin MK, Kumar A. Temporal retinal transcriptome and systems biology analysis identifies key pathways and hub genes in Staphylococcus aureus endophthalmitis. *Sci Rep.* 2016 Feb 11;6:21502. doi: 10.1038/srep21502.

Ramadan RT, Ramirez R, Novosad BD, Callegan MC. Acute inflammation and loss of retinal architecture and function during experimental Bacillus endophthalmitis. *Curr Eye Res.* 2006 Nov;31(11):955–65. doi: 10.1080/02713680600976925.

Relhan N, Albini TA, Pathengay A, Kuriyan AE, Miller D, Flynn HW. Endophthalmitis caused by Gram-positive organisms with reduced vancomycin susceptibility: literature review and options for treatment. *Br J Ophthalmol.* 2016 Apr;100(4):446–52. doi: 10.1136/bjophthalmol-2015-307722.

Rider P, Carmi Y, Cohen I. Biologics for targeting inflammatory cytokines, clinical uses, and limitations. *Int J Cell Biol.* 2016;2016:9259646. doi: 10.1155/2016/9259646.

Romero-Adrian TB, Leal-Montiel J, Fernández G, Valecillo A. Role of cytokines and other factors involved in the Mycobacterium tuberculosis infection. *World J Immunol.* 2015; 5(1): 16–50

Rosenzweig HL, Woods A, Clowers JS, Planck SR, Rosenbaum JT. The NLRP3 inflammasome is active but not essential in endotoxin-induced uveitis. *Inflamm Res.* 2012 Mar;61(3):225–31.

Safneck JR. Endophthalmitis: a review of recent trends. *Saudi J Ophthalmol.* 2012 Apr;26(2):181–9. doi: 10.1016/j.sjopt.2012.02.011.

Satorres SE, Alcaráz LE, Cargnelutti E, Di Genaro MS. IFN-gamma plays a detrimental role in murine defense against nasal colonization of Staphylococcus aureus. *Immunol Lett.* 2009 Apr 27;123(2):185–8. doi: 10.1016/j.imlet.2009.03.003.

Singh S, Singh PK, Jha A, Naik P, Joseph J, Giri S, et al., Integrative metabolomics and transcriptomics identifies itaconate as an adjunct therapy to treat ocular bacterial infection. *Cell Rep Med.* 2021 May 18;2(5):100277. doi: 10.1016/j.xcrm.2021.100277.

Sutterwala FS, Mijares LA, Li L, Ogura Y, Kazmierczak BI, Flavell RA. Immune recognition of Pseudomonas aeruginosa mediated by the IPAF/NLRC4 inflammasome. *J Exp Med.* 2007 Dec 24;204(13):3235–45. doi: 10.1084/jem.20071239.

Tacke R, Sun J, Uchiyama S, Polovina A, Nguyen D, Nizet V. Protection against lethal multidrug-resistant bacterial infections using macrophage cell therapy. *Infect Microbes Dis.* 2019;1(2):61–69. doi: 10.1097/im9.0000000000000012.

Tam VH, Pérez C, Ledesma KR, Lewis RE. Transcriptional profiles of pulmonary innate immune responses to isogenic antibiotic-susceptible and multidrug-resistant Pseudomonas aeruginosa. *Microbiol Immunol.* 2018 Apr;62(4):291–294. doi: 10.1111/1348-0421.12581.

Taylor AW, Ng TF. Negative regulators that mediate ocular immune privilege. *J Leukoc Biol.* 2018 Feb 12. doi: 10.1002/JLB.3MIR0817-337R.

Teweldemedhin M, Gebreyesus H, Atsbaha AH, Asgedom SW, Saravanan M. Bacterial profile of ocular infections: a systematic review. *BMC Ophthalmol.* 2017 Nov 25;17(1):212. doi: 10.1186/s12886-017-0612-2.

Thakur A, Barrett RP, Hobden JA, Hazlett LD. Caspase-1 inhibitor reduces severity of pseudomonas aeruginosa keratitis in mice. *Invest Ophthalmol Vis Sci.* 2004 Sep;45(9):3177–84. doi: 10.1167/iovs.04-0041.

Tosi MF. Innate immune responses to infection. *J Allergy Clin Immunol.* 2005 Aug;116(2):241–9; quiz 250. doi: 10.1016/j.jaci.2005.05.036.

Tzouvelekis LS, Markogiannakis A, Psichogiou M, Tassios PT, Daikos GL. Carbapenemases in Klebsiella pneumoniae and other Enterobacteriaceae: an evolving crisis of global dimensions. *Clin Microbiol Rev.* 2012 Oct;25(4):682–707. doi: 10.1128/CMR.05035-11.

Vahey JB, Flynn HW Jr. Results in the management of Bacillus endophthalmitis. *Ophthalmic Surg.* 1991 Nov;22(11):681–6.

Van den Steen PE, Proost P, Wuyts A, Van Damme J, Opdenakker G. Neutrophil gelatinase B potentiates interleukin-8 tenfold by aminoterminal processing, whereas it degrades CTAP-III, PF-4, and GRO-alpha and leaves RANTES and MCP-2 intact. *Blood.* 2000 Oct 15;96(8):2673–81.

Vatansever F, de Melo WC, Avci P, Vecchio D, Sadasivam M, Gupta A, et al., Antimicrobial strategies centered around reactive oxygen species—bactericidal antibiotics, photodynamic therapy, and beyond. *FEMS Microbiol Rev.* 2013 Nov;37(6):955–89. doi: 10.1111/1574-6976.12026.

Verdrengh M, Tarkowski A. Granulocyte-macrophage colony-stimulating factor in Staphylococcus aureus-induced arthritis. *Infect Immun.* 1998 Feb;66(2):853–5. doi: 10.1128/IAI.66.2.853-855.1998.

Wakely L, Sheard R. Recent advances in endophthalmitis management. *R Coll Ophthalmol Focus.* 2014;2:5–6.

Wang C, Liu CM, Wei LL, Shi LY, Pan ZF, Mao LG, et al., A group of novel serum diagnostic biomarkers for multidrug-resistant tuberculosis by iTRAQ-2D LC-MS/MS and solexa sequencing. *Int J Biol Sci.* 2016 Jan 1;12(2):246–56. doi: 10.7150/ijbs.13805.

Wang Y, van Boxel-Dezaire AH, Cheon H, Yang J, Stark GR. STAT3 activation in response to IL-6 is prolonged by the binding of IL-6 receptor to EGF receptor. *Proc Natl Acad Sci USA.* 2013 Oct 15;110(42):16975–80. doi: 10.1073/pnas.1315862110.

Winokur J, Deramo VA, Lai JC, Luchs J, Ailani H, Udell IJ. Visual outcome and bacterial sensitivity after MRSA-associated acute endophthalmitis. *Invest Ophthalmol Vis Sci.* 2007;48(13):693.

Wiskur BJ, Robinson ML, Farrand AJ, Novosad BD, Callegan MC. Toward improving therapeutic regimens for Bacillus endophthalmitis. *Invest Ophthalmol Vis Sci.* 2008 Apr;49(4):1480–7. doi: 10.1167/iovs.07-1303.

Wong SY, Beaman MH, Abrams JS, Remington JS. Kinetics of IFN-g and IL-6 secretion in murine toxoplasmosis [abstract 330]. In: Proceedings of the 32d Interscience Conference on Antimicrobial Agents and Chemotherapy (Anaheim, CA). Washington, DC: American Society for Microbiology, 1992.

Wong TT, Mead AL, Khaw PT. Matrix metalloproteinase inhibition modulates postoperative scarring after experimental glaucoma filtration surgery. *Invest Ophthalmol Vis Sci.* 2003 Mar;44(3):1097–103. doi: 10.1167/iovs.02-0366.

Xiong H, Carter RA, Leiner IM, Tang YW, Chen L, Kreiswirth BN, et al., Distinct contributions of neutrophils and CCR2+ monocytes to pulmonary clearance of different Klebsiella pneumoniae strains. *Infect Immun.* 2015 Sep;83(9):3418–27. doi: 10.1128/IAI.00678-15.

Xu F, Kang Y, Zhang H, Piao Z, Yin H, Diao R, et al., Akt1-mediated regulation of macrophage polarization in a murine model of Staphylococcus aureus pulmonary infection. *J Infect Dis.* 2013 Aug 1;208(3):528–38. doi: 10.1093/infdis/jit177.

Yang E, Yang R, Guo M, Huang D, Wang W, Zhang Z, et al., Multidrug-resistant tuberculosis (MDR-TB) strain infection in macaques results in high bacilli burdens in airways, driving broad innate/adaptive immune responses. *Emerg Microbes Infect.* 2018 Dec 12;7(1):207. doi: 10.1038/s41426-018-0213-z.

Yerramothu P, Vijay AK, Willcox MDP. Inflammasomes, the eye and anti-inflammasome therapy. *Eye (Lond).* 2018 Mar;32(3):491–505. doi: 10.1038/eye.2017.241.

Yoshimoto T, Nakanishi K, Hirose S, Hiroishi K, Okamura H, Takemoto Y, et al., High serum IL-6 level reflects susceptible status of the host to endotoxin and IL-1/tumor necrosis factor. *J Immunol.* 1992 Jun 1;148(11):3596–603.

Zarember KA, Godowski PJ. Tissue expression of human Toll-like receptors and differential regulation of Toll-like receptor mRNAs in leukocytes in response to microbes, their products, and cytokines. *J Immunol.* 2002 Jan 15;168(2):554–61. doi: 10.4049/jimmunol.168.2

10

Antimicrobial Resistance in the Diseased Human Eye: Scleritis

S. Shivaji

10.1 Introduction

Scleritis is a rare, sight-threatening ocular inflammatory disease (McCluskey and Wakefield, 1996; Okhravi et al., 2005; Murthy et al., 2020) characterized by inflammation of the sclera associated with engorgement of the conjunctival, superficial, and deep episcleral vessels. Scleral inflammatory disorders are classified as episcleritis when the inflammation is superficial, and as scleritis when associated with deeper and more destructive inflammation (Jabs et al., 2000; Goldstein et al., 2014). Multiple ocular complications of scleritis have been identified and include scleral and corneal thinning, perforation, keratitis, anterior uveitis, glaucoma, cataract, and exudative retinal detachment (Watson and Hayreh, 1976). Scleritis is associated with several life-threatening systemic autoimmune diseases in 30–50% cases (Bin Ismail et al., 2017). The most common systemic diseases with which scleritis is associated are rheumatoid arthritis (RA), granulomatosis with polyangiitis (formerly Wegener's granulomatosis), seronegative spondyloarthropathies, relapsing polychondritis, and systemic lupus erythematosus (SLE) (Smith et al., 2007; Jabs et al., 2000; Lin et al., 2008). Scleritis may also be associated with infectious causes, previous ocular surgery, exposure to drugs, or malignancies (Smith et al., 2007; Pavesio and Meier, 2001).

10.2 Classification of Scleritis

Watson and Hayreh (1976) classified scleritis as either anterior or posterior types. Of the two types, anterior scleritis is the most common, in which the front of the sclera is inflamed. In posterior scleritis the back of the sclera is inflamed and so it is difficult to visualise. Posterior scleritis is less common compared to anterior scleritis, but it causes serious eye problems, such as detachment of the retina and or glaucoma. The ora serrata is an arbitrary boundary between the anterior and posterior types of scleritis (Watson and Hayreh, 1976; Benson, 1988). One study reported a rare case of anterior scleritis in association with posterior scleritis (Dastidar et al., 2022). In 1991, scleritis was graded based on the presence or absence of eight common clinical signs, such as area of inflammation, tenderness, nodules, scleral necrosis, corneal involvement, anterior chamber cells, vitreous cells, and retinal detachment (McCluskey and Wakefield, 1991). Ahn

DOI: 10.1201/9781003451105-10

et al. (2010) and Murthy et al. (2020) classified anterior scleritis as diffuse, nodular, and necrotizing scleritis with or without inflammation (Figure 10.1–10.4). Further, in a Korean population it was observed that nodular scleritis was the most common and 21.1% had connective tissue diseases, 14.5% had infectious scleritis, and 43.4% had previous ocular surgery, mostly pterygium excision (Ahn et al., 2010). In India, in patients with infectious scleritis examined from March 2005 to December 2009, it was observed that surgery was the major risk factor (58.5%) and it was more common after vitreoretinal surgery, cataract surgery, pterygium surgery, corneoscleral tear repair and

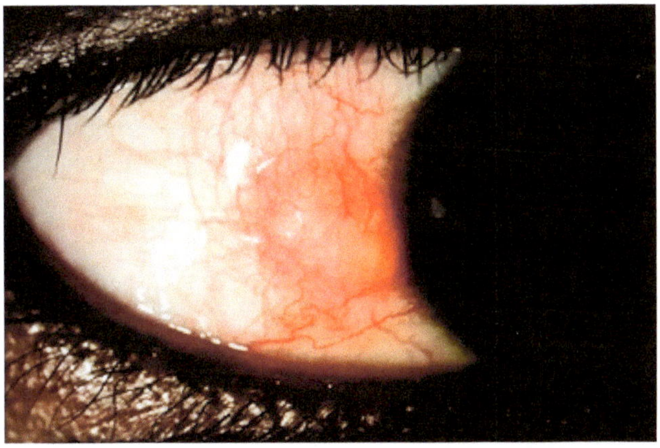

FIGURE 10.1 Nodular anterior scleritis with a scleral nodule. (Adapted from Promelle et al. (2021). This is an open access article distributed under the terms and conditions of the Creative Commons Attribution (CC BY) license (https://creativecommons.org/licenses/by/4.0/).)

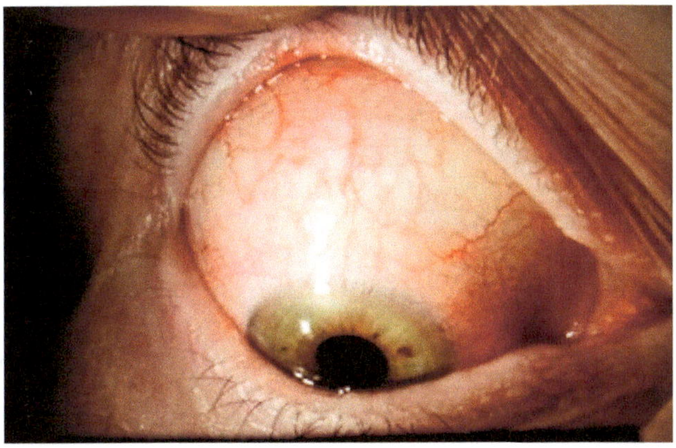

FIGURE 10.2 Diffuse anterior scleritis showing widespread inflammation of sclera. (Adapted from Promelle et al. (2021). This is an open access article distributed under the terms and conditions of the Creative Commons Attribution (CC BY) license (https://creativecommons.org/licenses/by/4.0/).)

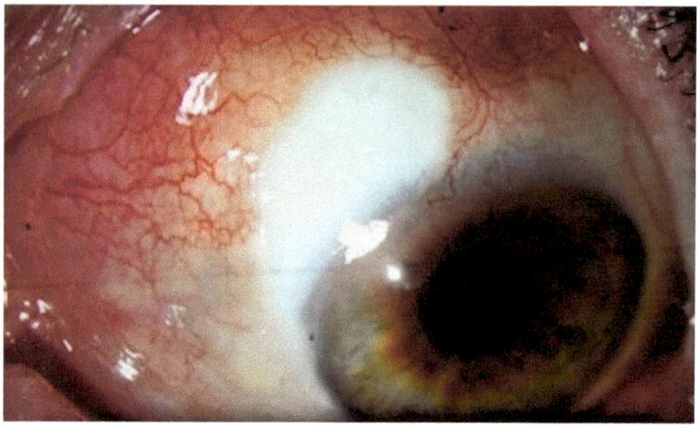

FIGURE 10.3 Necrotizing scleritis with inflammation showing diffuse scleritis with an area of avascular, necrotic sclera. (Adapted from Promelle et al. (2021). This is an open access article distributed under the terms and conditions of the Creative Commons Attribution (CC BY) license (https://creativecommons.org/licenses/by/4.0/).)

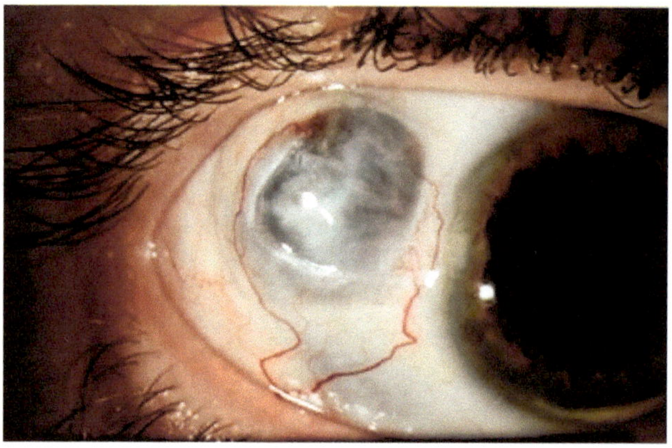

FIGURE 10.4 Necrotizing scleritis without inflammation, or scleromalaciaperforans. (Adapted from Promelle et al. (2021).This is an open access article distributed under the terms and conditions of the Creative Commons Attribution (CC BY) license (https://creativecommons.org/licenses/by/4.0/).)

scleral buckle surgery (Reddy et al., 2015). In India, diffuse anterior scleritis constituted 41% of the cases and the common systemic diseases it was associated with was rheumatoid arthritis (18%). Necrotizing scleritis was at 3.5 times higher risk of developing ocular complications, and patients received immunosuppressive treatment (74%) along with corticosteroids. Subsequently, after the classification of scleritis in 2010 by Ahn et al. (2010), Sen et al. (2011) reclassified scleritis based on standardized digital images of the sclera showing variation in the degrees of inflammation. The images were graded on a scale of 0 to 4+ in two sessions without reference images in session 1 and in session 2 with standard images according to the proposed grading system. This system for assessing scleritis demonstrated a good interobserver agreement.

10.3 Prevalence of Scleritis

The incidence of infectious scleritis is about 17% in Indian scleritis patients (Reddy et al., 2015), whereas in various other countries it ranges from 5–10% (Sainz de la Maza et al., 1994). Thus, the incidence of infectious etiology in the West is significantly less and probably indicates that the prevalence depends on the geographical location (Honik et al., 2013; Ho et al., 2014; Homayounfar et al., 2013). In Hawai, the overall incidence rates of scleritis were 4.1 cases per 100 000 person-years and women were overrepresented among scleritis patients, and Blacks had the highest incidence of scleritis compared to Pacific Islanders (Homayounfar et al., 2013). In Singapore, scleritis was categorised by the "Ocular Autoimmune Systemic Inflammatory Infectious Study" (OASIS), which indicated that most patients (62.5%) had diffuse anterior scleritis followed by nodular anterior scleritis (20.8%), posterior scleritis (10.8%) and necrotizing anterior scleritis (5.8%) (Bin Ismail et al., 2017). Patnaik et al. (2020) studied anterior nodular scleritis in a retrospective study in India for about a decade between 2007 and 2018 and the demographic parameters indicated that the mean age at presentation was 46.8 ± 13.1 years, females were more affected (70.7%) and only 14% of the patients had bilateral involvement. The predominant symptoms were redness (92.6%) and ocular pain (69.1%) and almost one-fifth of the patients (22%) were associated with a systemic disease and rheumatoid arthritis (5%) was the most common autoimmune disease. In India, Tuberculosis was associated with nodular scleritis. In 2021, Dutta Majumder et al. (2021) analysed 354 non-infectious scleritis patients in the Asia-Pacific region. The Indian cohort of scleritis patients differed in demographic parameters and the patients were younger by age, mostly seen in males and only fewer bilateral cases were seen compared with patients from Singapore and Australia (Lane et al., 2019). These findings contrasted with that observed in the West, where scleritis typically affects the middle-aged (Lane et al., 2019; Kumar et al., 2018). Further, necrotizing scleritis is common in older patients (Keino et al., 2010), posterior scleritis in younger patients (Dutta Majumder et al., 2019 a-b; Dutta Majumder, 2019c) as young as a 7-month-old baby (van der Maesen et al., 2007). Bilaterality of scleritis was detected in 50% of the cases and followed the order most frequent in diffuse anterior scleritis> necrotizing anterior scleritis> posterior scleritis> nodular anterior scleritis (Lane et al., 2019; González-López et al., 2016).

10.4 Non-infectious Scleritis

10.4.1 Etiology of Non-Infectious Scleritis

Non-infectious scleritis is commonly associated with systemic autoimmune disorders and systemic vasculitis (Dutta Majumder et al., 2021). In fact, scleritis is a common manifestation of a systemic immune-mediated disease and 15–59% of patients may present with scleritis (Okhravi et al., 2005; Wakefield et al., 2013; Sainz de la Maza et al., 1995; 2012a-c). In a recent clinical case report and review it was reported that necrotizing scleritis with ocular manifestation of scleritis is the first sign of rheumatologic disease (Moreno et al., 2023). In addition to immune-mediated diseases commonly associated with scleritis, several miscellaneous factors such as medication could cause medication-induced scleritis (Table 10.1). Among the autoimmune causes

TABLE 10.1

Diseases Associated with Scleritis

A. Non-infectious diseases
1. Associated with connective tissue diseases and other inflammatory conditions in which vasculitis may occur
Rheumatoid arthritis
Systemic lupus erythematosus
Relapsing polychondritis
HLA-B27 + without spondyloarthropathy
Ankylosing spondylitis
Reactive arthritis
Psoriatic arthritis
Arthritis and inflammatory bowel disease
Antiphospholipid syndrome
2. Associated with vasculitis diseases
Granulomatosis with polyangiitis
Allergic granulomatous angiitis (Churg-Strauss syndrome)
Polyarteritis nodosa
IgA vasculitis
Takayasu's arteritis
Giant cell arteritis
Behçet's disease
Cogan's syndrome
Hypocomplementeric urticarial vasculitis
3. Medication-induced scleritis
Biphosphonates: pamidronate, alendronate, risedronate, zoledronate, ibandronate,
Anticonvulsants: topiramate
Biologics: etanercept
Anticancer immunotherapy: nivolumab/cabiralizumab
4. Scleritis due to Miscellaneous Causes
Atopy
Rosacea
Gout
Foreign body granuloma
Chemical injury
B. Infectious diseases
Bacterial scleritis
Fungal scleritis
Viral scleritis
Protozoan scleritis

Source: Modified from Murthy et al., 2020.

the most common was rheumatoid arthritis (7.5%), followed by granulomatosis with polyangiitis (4.2%) (GPA) (Sainz de la Maza et al., 1995), relapsing polychondritis (3.3%), Sjogren's syndrome (3.3%), and psoriatic arthropathy (1.7%) (Jan et al., 2022; Dutta Majumder et al., 2021). Scleritis in GPA is often necrotizing and could lead to permanent visual loss (Sainz de la Maza et al., 1995). Among the different types of scleritis and their association with an autoimmune condition it was observed that necrotizing anterior scleritis showed 85.7% association with an autoimmune condition, followed by posterior scleritis (30.8%), and diffuse scleritis (25.3%). A very small proportion of nodular scleritis patients (2 out of 25, 8.0%) had an associated autoimmune condition (Bin Ismail et al., 2017).

Scleritis in spondylo-arthropathies or in systemic lupus erythematosus (SLE) is benign and self-limiting, whereas scleritis in RA and relapsing polychondritis (RP) can range from mild to severe (Sainz de la Maza et al., 1995). When Scleritis is associated with SLE it manifests as nodular, diffuse, and posterior scleritis but rarely necrotizing (Pavesio, 2017). In a few cases scleritis may be associated with more than one autoimmune disease. For instance, in RP-associated scleritis, 30% of scleritis patients were also associated with systemic vasculitis or other autoimmune diseases (Sainz de le Maza et al., 2016). GPA tends to progress rapidly, whereas RA has an insidious nature of progression (Cocho et al., 2016). Apart from the above autoimmune disorders with which scleritis is associated it may rarely be associated with autoimmune conditions such as inflammatory bowel disease, sarcoidosis, disease, Cogan's syndrome, Takayasu disease and polyarteritis nodosa, carcinoma of lung and lymphoma (Sainz de la Maza et al., 2012b). Scleritis could also be drug-induced like bisphosphonate-associated scleritis, which may affect one or both eyes and usually resolves upon discontinuation of the drug (Abtahi et al., 2012). Various other drugs such as topiramate (Abtahi et al., 2012), procainamide (Turgeon et al., 1989), etanercept (Taban et al., 2006), and a combination of nivolumab and cabiralizumab (Gonzales et al., 2018) are known to cause scleritis. Surgically induced necrotizing scleritis (SINS), has been reported following conjunctival manipulation, episcleral vessel cautery, pterygium, or extra capsular cataract extraction surgery, trabeculectomy (Aktas et al., 2019), retinal detachment surgery (Madanagopalan et al., 2019), strabismus surgery (Ram, 2018), and intravitreal injection (Ryu et al., 2017). Several of these surgery-induced cases (up to 90%) may have an undiagnosed systemic autoimmune condition (Murthy et al., 2020). No definite genetic predisposition has been identified in patients with scleritis (Dutta Majumder et al., 2021). But studies have indicated that HLA-B27 is associated with both anterior and posterior scleritis (Yang et al., 2018). Sainz de la Maza et al. (2012a) observed that in a retrospective case series of 500 scleritis patients, HLA-B27-associated ocular inflammation without spondyloarthropathy was the second most common cause of the systemic disease after RA (Sainz de la Maza et al., 2012a). Coincidentally, RA, SLE, and antinuclear cytoplasmic antibody (ANCA)-associated vasculitis have also been found to be associated with HLA-DR15 (Vergouwen et al., 2020). This is a significant observation considering that HLA-DR15 has been reported as a predisposing factor for corneal ulceration (Watson and Young, 2004). Additionally, 2 other genes, cytotoxic T lymphocyte-associated antigen-4 and protein tyrosine phosphatase nonreceptor type 22, have been found to confer genetic susceptibility to scleritis (Li et al., 2019).

10.4.2 Clinical Features of Non-infectious Scleritis

10.4.2.1 Anterior Scleritis

Pain is a characteristic feature of scleritis and radiates beyond the eye to the rest of the face to include forehead, brow, ear, face, or jaw. Pain is normally triggered by touch and movement such as accidental touching of the eye or temple and even by movements of the eye. Further, the triggered pain could be severe enough to disrupt the sleep of the patient. Thus, patients with necrotizing scleritis normally experience extreme discomfort though very rarely patients with scleromalaciaperforans, a rare non-inflammatory form of anterior necrotizing scleritis, almost lack pain symptoms (Murthy et al., 2020). In anterior scleritis patients, large, firm, immobile, nodules (nodule) may be present behind the limbus. These nodules are tender to palpation and pus points are not seen. The sclera and conjunctiva are white, avascular with surrounding scleral edema and congestion and the uveal tissue overlying the sclera is thin and translucent. Due to a thin sclera a bulging staphyloma may develop. In children nodular scleritis is the most common subtype of anterior scleritis (Dutta Majumder et al., 2019b) followed by diffuse anterior scleritis with involvement of the peripheral cornea (Sainz de la Maza et al., 2012b).

10.4.2.2 Posterior Scleritis

Posterior scleritis, which comprises up to 20% of scleritis cases, is normally difficult to diagnose (McCluskey et al., 1999). It is common in children (Dutta Majumder et al., 2019b) with male predominance, less associated with systemic diseases, and exhibit frequent disc swelling. Concurrent anterior scleritis is common in patients with posterior scleritis and ranges from 59.6% (McCluskey et al., 1999) to 67.7% (Sainz de la Maza et al., 2012b). Posterior scleritis is often bilateral and idiopathic and has similarities with other inflammatory conditions thus making diagnosis a challenge (González-López et al., 2016). Pain is common in patients with posterior scleritis and the pain could be periocular as in 64% and headache as in 13% of the patients (Lavric et al., 2016). Reduced vision is also common in patients with posterior scleritis (Tanaka et al., 2018). The other features include optic nerve edema, compression of the optic nerve and manifest as optic disc edema, optic perineuritis (Cheung and Chee, 2012), serous retinal detachment, macular edema, annular ciliochoroidal detachment, internal ophthalmoplegia, and choroidal thickening, mass, and folds (Benson, 1988), occlusion of the blood vessels (Kamei, 2012) and secondary angle-closure glaucoma (Ugurbas et al., 2012; Jain et al., 2004). Sen et al. (2011) developed a quantitative scoring system, based on clinical signs of scleritis, to grade the severity of scleritis and to predict the response to systemic immunosuppressive therapy. Giant nodular scleritis is a challenging presentation of posterior scleritis (Agrawal et al., 2016a, b). In contrast to all these studies, McCluskey et al. (1999), observed that 17.1% of the patients with posterior scleritis were asymptomatic, except that they presented with reduced vision.

10.4.3 Treatment for Scleritis

Considering that scleritis manifests as non-infectious and infectious types and could be associated with systemic diseases the treatment is not straight forward and is best that the treatment modalities for each type is detailed separately. In a review Promelle et al. (2021) presented an update on treatment perspectives and suggested that oral non-steroidal anti-inflammatory drugs (NSAIDs) and/or topical steroids could be used for episcleritis which is benign. But scleritis which is associated with rheumatoid arthritis needs to be treated with NSAIDs, steroids and disease-modifying anti-rheumatic drugs (DMARDs). Further in refractory cases, scleritis treatment should include anti-TNF biologic agents (infliximab and adalimumab). Several other reviews have highlighted the treatment modalities for non-infectious (Abdel-Aty et al., 2022; Beardsley et al., 2013; Sainz de la Maza et al., 2012c; Stem et al., 2017; Sota et al., 2021; Dutta Majumder et al., 2021) and infectious scleritis (Murthy et al., 2020).

10.4.3.1 Treatment of Non-infectious Scleritis

10.4.3.3.1 Steroids

Topical prednisolone acetate 1%, often used for mild cases of anterior scleritis with intraocular inflammation was abandoned with time since topical application is prone to high failure rate and majority of the patients require systemic therapy. Injection of triamcinolone, another steroid, in the periocular region demonstrated efficacy and safety in both non-necrotizing and non-infectious scleritis (Sen et al., 2005) and in one study 89.7% of eyes had complete resolution following a single injection (Sohn et al., 2011). But this approach increased the risk of progressive scleral necrosis and perforation and elevated the intraocular pressure (Sohn et al., 2011) and was thus discontinued (Watson, 1980). Triamcinolone is thus used only if the inflammation is non-necrotizing and the patients did not exhibit elevated intra-ocular pressure.

10.4.3.3.2 Oral Non-steroidal Anti-inflammatory Drugs (NSAIDS)

NSAIDS (Abdel-Aty et al., 2022), such as flurbiprofen (Agrawal et al., 2016a) or indomethacin (Sainz de la Maza et al., 2012c) are used to manage non-necrotizing, and non-infectious scleritis with low degree of inflammation. But when the ocular inflammation is refractory to NSAIDs use of systemic steroids or immunomodulatory agents is the way forward (Jabs et al., 2001; Abdel-Aty et al., 2022; Sainz de la Maza et al., 2012 c; Beardsley et al., 2013). In a recent study immunosuppressive treatments with methotrexate, cyclosporin, and adalimumab was favourable in patients with recalcitrant non-infectious posterior scleritis (Pan et al., 2022).

10.4.3.3.3 Corticosteroids

Corticosteroid with the addition of immunosuppressive agents is normally used for the treatment of non-infectious scleritis. In patients intolerant to oral steroids intravenous methylprednisolone was effective in the treatment of scleritis (McCluskey and Wakefield, 1987). Co-administering NSAIDs and corticosteroids increases the incidence of GI ulceration and thus needs to be avoided (Piper et al., 1991). Dutta Majumder et al. (2021) used pulse intravenous corticosteroid therapy when rapid control of inflammation was required. Further, rapid resolution of scleritis was also possible using subconjunctival injections of depot corticosteroid in patients with non-infectious, non-necrotizing anterior scleritis (Johnson and Chu, 2010; Sohn et al., 2011). This method of delivery of corticosteroid reduced the need for systemic therapy (Roufas et al., 2010). But complications like scleral melt or necrosis may occur and this led to discontinuity in this mode of treatment (Zamir et al., 2002).

10.4.3.3.4 Antimetabolites

Methotrexate, an antimetabolite was effective in curing 38.6% of patients with scleritis when used along with systemic corticosteroid treatment (Beardsley et al., 2013; Gangaputra et al., 2019). Other studies indicated that methotrexate alone was effective in resolution of inflammation in 61.9% patients (Sands et al., 2018) and brought about full or partial response in 79% of patients (Kaplan-Messas et al., 2003). Mycophenolate mofetil (MMF) another antimetabolite was also effective in treating scleritis, but the response varied from adequate treatment response in 49% of patients (Daniel et al., 2010), resolution of inflammation in 25.5% to 55% (Sobrin et al., 2008) and decrease in the dose of corticosteroids in some (Kolomeyer et al., 2012) and successful treatment of scleritis in one patient (Yalçindağ et al., 2008). Mycophenolate also proved to be a steroid sparing agent (82%–100%) (Hwang et al., 2011; Choudhary et al., 2006) but was not effective as an adjunct to cyclosporine or methotrexate (Sen et al., 2003). Though most studies with MMF treatment of scleritis were positive, there were a few studies which were not effective and showed lower rates of resolution of inflammation. Apart from being less effective in overcoming inflammation, it increased the possibility of relapse and increased the possibility of recurrence in other autoimmune disorders like lupus nephritis and immunoglobulin A (IgA) nephropathy, GI complications, bone marrow suppression, and elevated liver enzymes (Sobrin et al., 2008; Burgos et al., 2009; Chandrakantan et al., 2005; Galor et al., 2008).

Azathioprine (AZA) is yet another antimetabolite which has been used to overcome different types of scleritis specifically necrotizing scleritis, posterior scleritis, and vasculitis-associated scleritis (Beardsley et al., 2013; Sainz de la Maza et al., 2016; Kumar et al., 2018). But AZA does cause a flu-like response in some patients about one month after starting and therefore it is discontinued (Galor et al., 2008).

10.4.3.3.5 Calcineurin Inhibitors

Calcineurin inhibitors like cyclosporine have been used to treat scleritis associated with granulomatosis with polyangiitis (Foster et al., 2016; Wakefield and McCluskey, 1989). Cyclosporine inhibitors could reduce corticosteroid dose and was effective in patients with idiopathic and necrotizing scleritis (McCarthy et al., 1992; Gumus et al., 2009). Tacrolimus, another calcineurin inhibitor, has been successfully used in the treatment of scleritis in both systemic and topical forms (Young et al., 2005; Miyazaki et al., 2008) and in combination with sirolimus, another calcineurin inhibitor to treat necrotizing scleritis (Bhatt et al., 2015). Tacrolimus treatment is effective in treating non-infectious, non-necrotizing anterior scleritis patients refractory to topical steroids. The treated patients showed reduced clinical signs and experience less ocular pain (Yazu et al., 2021). Further, topical tacrolimus ointment could completely resolve scleral inflammation and reduce steroid use in patients with necrotizing and nodular scleritis (Lee et al., 2013).

10.4.3.3.6 Antibiotics

Dapsone (4,4-diaminediphenyl sulfone), is an anti-inflammatory antibiotic (Kirtschig et al., 2003) that has been used in the treatment of relapsing polychondritis-associated scleritis (Yoo et al., 2011) and Sweet syndrome-(acute febrile neutrophilic dermatosis) associated nodular scleritis as adjunctive therapy (Wong et al., 2007). But Dapsone was not effective and failed to control inflammation in 6 of 8 patients with necrotizing scleritis (Hoang-Xaun et al., 1990). Further it is toxic and causes methemoglobinemia and hemolytic anemia in patients with G6PDH deficiency (Naik et al., 2008).

10.4.3.3.7 Alkylating Agents

Alkylating agents, like cyclophosphamide, are effective in treating scleritis patients. Cyclophosphamide treatment was effective in reducing the dose of steroids in 30%-60.5% of patients with scleritis within 12 months (Pujari et al., 2010). In another study, combination therapy with cyclophosphamide and methotrexate was effective in treating 16 patients with corticosteroid-refractory necrotizing scleritis or peripheral ulcerative keratitis associated with rheumatoid arthritis (Galor and Thorne, 2007). But the disadvantage of using cyclophosphamide is that the drug exerts side effects like leukopenia (18.1%), cystitis (7.7%), anaemia (3.3%) and infection (2.3%). It was also observed that one patient following treatment with cyclophosphamide died from pneumocystis pneumonia (Pujari et al., 2010). Further, cyclophosphamide increases the rate of bladder cancer (Radis et al., 1995).

Chlorambucil is also an alkylating agent used in the treatment of scleritis. But this drug causes several side effects which include secondary malignancy, amenorrhea in women, testicular dysfunction in men, alopecia, herpes zoster reactivation, and at times thrombocytopenia (Goldstein et al., 2002).

10.4.3.3.8 Biological Agents

Several biological agents like inhibitors of tumor necrosis factor-α (TNF-α) viz., monoclonal antibodies adalimumab, infliximab, and certolizumab, the soluble TNF receptor fusion protein etanercept, and the CD20 inhibitor rituximab have been studied for the management of scleritis (De Fidelix et al., 2015; 2016).

Infliximab Infliximab, is a TNF-α inhibitor, used in the treatment of scleritis. In 18 studies encompassing 47 patients (14 idiopathic, 14 with rheumatoid arthritis, and 19 with other inflammatory diseases), infliximab therapy, improved either visual acuity or inflammation by about 96%, reduced or helped to discontinue corticosteroids in 64% and in 36% infliximab was discontinued (De Fidelix et al., 2015; Abdel-Aty et al., 2022).

Adalimumab Adalimumab is another TNF-α inhibitor, which has been approved for treating different types of uveitis in children (non-infectious intermediate, posterior, and panuveitis) and was also effective in treating scleritis in two scleritis patients, one associated with psoriasis and the other with rheumatoid arthritis (Huynh et al., 2008; Restrepo et al., 2010). Treatment with infliximab, adalimumab, or both achieved resolution in 15 of 17 scleritis patients (88%) (Ragam et al., 2014). In one study treatment with adalimumab brought about resolution of symptoms in 47% of the patients with scleritis (Durrani et al., 2012) but subsequent studies indicated favourable resolution as in a case series of two patients with necrotizing scleritis (Lawuyi et al., 2016) or in a case of Takayasu arteritis–associated scleritis (Akhtar et al., 2015). Further, in a prospective study on 43 patients afflicted with uveitis or scleritis, treatment with adalimumab or infliximab resolved symptoms in 91% of the patients, induced an adverse event in 54% of patients and, in 9%, therapy was discontinued due to a serious adverse event (Sharma et al., 2021).

Rituximab Several studies report the efficacy of rituximab treatment in vasculitis-associated scleritis and necrotizing scleritis. In addition, in patients with granulomatosis with polyangiitis-associated scleritis, quiescence or resolution of ocular symptoms were observed on rituximab treatment (Cao et al., 2016; Ng et al., 2021).

Abatacept Abatacept was effective in the resolution of ocular inflammation in two of three rheumatoid arthritis patients, who had active diffuse anterior scleritis in four eyes (Fabiani et al., 2020). Abatacept was also effective in achieving resolution in other ocular inflammatory conditions such as peripheral ulcerative keratitis (Dominguez-Casas et al., 2020) and idiopathic arthritis-related uveitis in children, who were refractory to TNF-α inhibitors (Birolo et al., 2016).

Campath Campath is a monoclonal antibody against CD52, which was first used to treat uveitis. In addition, in a series of ten patients with recalcitrant ocular inflammation due to various ocular etiologies, a sustained response was achieved in two scleritis patients (Beardsley et al., 2013; Dick et al., 2000).

Tofacitinib Tofacitinib is a Janus kinase pathway inhibitor effective against refractory scleritis (Paley et al., 2018), including cases of necrotizing scleritis (Pyare et al., 2020). Paley et al. (2018) observed that two cases of recalcitrant uveitis and scleritis, which were not cured by multiple immunomodulatory regimes were responsive to tofacitinib. In addition, Pyare et al. (2020) also observed that tofacitinib was effective in treating **a** necrotizing scleritis patient.

Golimumab Golimumab is a human monoclonal antibody, which is used as an immunosuppressive medication for treatment of adults with rheumatoid arthritis, psoriatic arthritis, or ankylosing spondylitis (Sota et al., 2021). More recently it has been used in treating non-infectious scleritis (Amer et al., 2020) in a woman affected by psoriatic arthritis and associated with bilateral progressive necrotizing posterior scleritis and associated with retinal vasculitis. Golimumab also controlled scleral inflammation in a patient affected with necrotizing scleritis and having advanced glaucoma (Dutta-Majumder et al., 2019b).

Certolizumab, Etanercept, Tocilizumab, Anakinra, Gevokizumab Certolizumab, a PEGylated TNF-α inhibitor, was also effective in one case of scleritis (Smith et al., 2001). But Etanercept, a soluble TNF receptor fusion protein when evaluated for scleritis treatment presented mixed results. For instance, etanercept treatment resulted in clinical improvement in 33% of patients with arthritis associated scleritis (Smith et al., 2001), whereas in some patients it was very effective with minimal side effects (Hernandez-Illas et al., 2004). But, the disadvantage is that etanercept is known to cause uveitis (Taban et al., 2006; Gaujoux-Viala et al., 2012) and therefore it is not recommended.

Other less commonly used biologics like tocilizumab, a monoclonal antibody that targets the interleukin-6 receptor and Anakinra an inhibitor of the interleukin-1 receptor are effective in treating scleritis, and the patients exhibited complete quiescence of symptoms (Poelman et al., 2020; Bottin et al., 2018; Sota et al., 2021). But, paradoxically, a few reports indicate worsening inflammatory eye disease upon discontinuation of etanercept (Gaujoux-Viala et al., 2012). Although many different types of drugs have helped in the management of non-infectious scleritis, currently standard management guidelines are still lacking (Ingaralingam et al., 2020). Gevokizumab is an anti-IL-1β monoclonal antibody effective in the control of refractory RA-associated scleritis (Botsios et al., 2007; Sota et al., 2021) and subcutaneous injections of Gevokizumab (60 mg) effectively treated both non-infectious non-necrotizing scleritis with no serious adverse events (Knickelbein et al., 2016).

10.5 Infectious Scleritis

10.5.1 Etiology of Infectious Scleritis

The sclera is impermeable to infection due to its resilient, avascular, and compact structure and the fact that it is covered by the overlying episclera and conjunctiva. But, once this barrier is breached, it is prone to infection by microorganisms.

Infectious scleritis is influenced by multiple predisposing factors, such as these (Pavesio, 2017):

1. Ocular trauma due to agricultural or farming activity (Figure 10.5a).
2. Spread to the sclera following microbial keratitis, recurrent viral keratitis, dry eye or infectious endophthalmitis.
3. After different types of ocular surgeries, such as pterygium surgery, scleral buckle, cataract removal, strabismus, glaucoma filtration surgery, and so forth.
4. Following treatment with antimetabolites such as mitomycin C and local radiotherapy.
5. Immunosuppression due to HIV infection, diabetes, or systemic immunosuppression.
6. Chronic use of topical corticosteroids.

In addition to the above predisposing factors two pathogenic mechanisms have been implicated in infectious scleritis (Ramenaden and Raiji, 2013):

1. Invasion of infectious organisms following the destruction of the tissue, and
2. Immune reactions induced by the infective microorganisms (Murthy et al., 2020).

10.5.2 Clinical Features of Infectious Scleritis

The clinical symptoms of infectious scleritis often overlap with non-infectious scleritis, which could be either diffuse, nodular, or necrotizing scleritis and thus differentiating infectious scleritis from non-infectious scleritis is a great challenge. A few of the differentiating features include the nature of the pain and time to presentation of

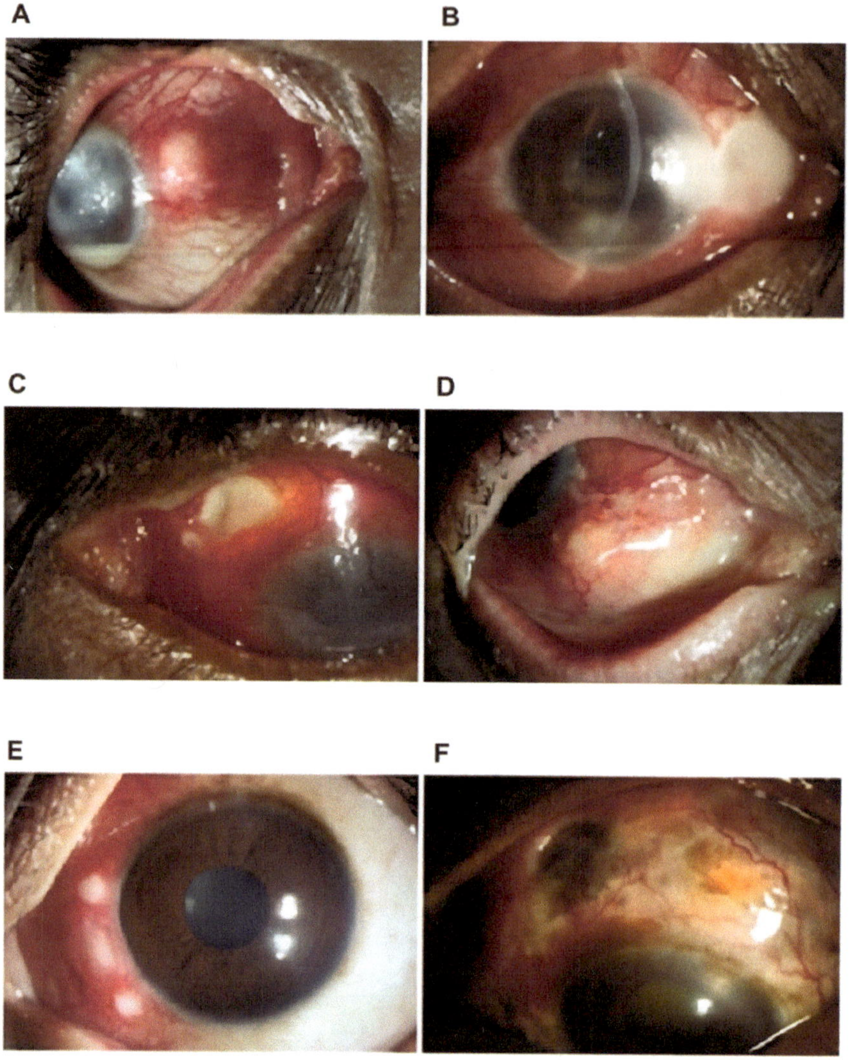

FIGURE 10.5 Etiological agents and confounding factors causing scleritis. (a) Fungal scleritis showing a scleral nodule following injury with vegetative material; (b) a scleral abscess following post-pterygium and cataract surgery; (c) ulcerated conjunctiva and scleral necrosis following *P. aeruginosa* scleritis; (d) a yellowish nodular lesion with underlying suppuration; (e) appearance of small multiple nodules in a case of bacterial scleritis; (f) multiple infective nodules after partial response to antifungals in a case of fungal scleritis. (Modified from Murthy et al. 2020. Published under the terms of the Creative Commons Attribution-Non-commercial-ShareAlike 4.0 License.

infection. Pain is mild to moderate in nodular scleritis, whereas it is severe in necrotizing scleritis (Remenaden and Raiji,2013; Kumar et al., 2012), and time to presentation of infection following an injury could take days to several years after an injury (Hodson et al., 2013; Cunningham et al., 2011). In a great majority of patients (94%) the most common predisposing factor for the infection was previous ocular surgery (83%), particularly excision of pterygium (57%) (Figure 10.5b) (Jan et al., 2022), prior accidental eye injury with organic matter, use of radiation, and the use of an antineoplastic like mitomycin (Paula et al., 2006; Hodson et al., 2013; Lin et al., 1997). Further, the period that lapses between trauma and onset of inflammation is shorter for post-traumatic cases compared to post-surgical cases (Ho et al., 2014; Sahu et al., 2013). Further time to resolution of scleritis was longer for infectious scleritis (Basti et al., 1994) and varied depending on the virulence of the etiological microorganism. *Mycobacteria*, *Nocardia* spp. or fungi can present as indolent infection developing slowly (Ho et al., 2014; Cunningham et al., 2011; Moorthy et al., 2012); but *Pseudomonas* spp., *Staphylococcal* spp. or viral infections have a shorter course of disease (Jain et al., 2009; Hodson et al., 2013; Huang et al., 1999; Feiz and Redline, 2007; Murthy et al., 2013). In developed countries *P. aeruginosa* is the most common etiological agent of scleritis (Ramenaden and Raiji, 2013; Hodson et al., 2013; Paula et al., 2006) (Figure 10.5c). The predominant clinical signs at presentation in 55 patients with infectious scleritis included episcleral and conjunctival hyperemia (98%) and scleral necrosis (93%) (Jain et al., 2009; Kumar et al., 2012). Though scleritis is restricted to the sclera it could also extend to the cornea, extraocular muscles and anterior chamber (Gonzalez-Gonzalez et al., 2012; Hsiao et al., 1998) and it could also be associated with endophthalmitis (Jain et al., 2009; Kumar et al., 2012). It was also observed that retinal surgeries predisposed patients with infectious scleritis to posterior scleral thickening and retinal or choroidal detachments (Margo and Pavan 2000; Lyne and Lloyd-Jones, 1979; Hsiao et al., 1998).

Among infectious scleritis, necrotizing scleritis is common and shows yellowish or reddish nodules due to underlying suppuration (Figure 10.5d). In some cases, numerous small subconjunctival abscesses are visible (Figure 10.5e) and at times these lesions may become large with mucopurulent discharge. Occasionally, the nodules may coalesce and result in scleral necrosis (Figure 10.5f). Hypopyon has also been strongly associated with infectious scleritis (Ho et al., 2014). Once the lesions heal, a fibrous tissue band indicative of infectious etiology is seen unlike immune lesions, which heal with only scleral thinning. Murthy et al. (2020) summarised the pathognomonic features of infectious scleritis as inclusive of purulent exudates, conjunctival ulceration and slough, visible pus points, abscesses and hypopyon or keratic precipitates. Jain et al. (2009) observed that corticosteroids and immunosuppressants reduce the inflammatory process and pain and congestion of an infectious nodule.

10.5.3 Types of Infectious Scleritis

10.5.3.1 Bacterial Scleritis

Bacterial scleritis is the most common type of scleritis, with a reported range of 53%–100% (Kumar et al., 2012; Hodson et al., 2013). Several bacteria have been identified as the etiological agents for infectious scleritis (Table 10.2). *P. aeruginosa* is the

TABLE 10.2

Bacteria, Fungi and Viruses that Cause Infectious Scleritis

Infectious organism
Bacteria
Pseudomonas aeruginosa, Streptococcus pneumoniae, methicillin resistant *Staphylococcus aureus* (MRSA), *Mycobacterium chelonae, M. marinum, M. tuberculosis,* and *Nocardia.*
Fungi
Aspergillus flavus, Scedosporium apiospermum, Pseudoallescheria boydii and *Rhinosporidium seeberi.*
Viruses
Varicella zoster virus, Herpes simplex virus type-1, Cytomegalovirus and Epstein Barr virus.
Protozoa
Acanthamoeba and *Toxoplasma*
Helminth parasite
Toxocara canis

Source: Compiled by author.

most common (51–81%) cause of bacterial scleritis in the Western world (Lin et al., 1997; Huang et al., 2000; Cunningham et al., 2011). Several features overlap between the bacteria causing scleritis but the bacteria causing scleritis may not be the same. For instance, scleritis following pterygium surgery is associated with *P. aeruginosa* (Huang et al., 2000) and *Streptococcus pneumoniae* (Altman et al., 1991) and scleritis following scleral buckle surgery is associated with *Pseudomonas* and MRSA (Ho et al., 2014; Basti et al., 1994). The most common Non-tuberculous Mycobacterial Scleritis (NTM) following scleral buckle surgery and pterygium excision is caused by *M. chelonae* (David et al., 1999) and *M. Marinum* (Kheir et al., 2015). Mycobacterial infection could be overcome by combination therapy involving amikacin eye drops (2.5% to 5%), with oral rifampicin, clarithromycin, or doxycycline from 4 weeks to 6 months after the resolution of the clinical signs. Surgical debridement is resorted to in cases with scleral abscesses (Murthy et al., 2020).

In contrast to NTM, *M. tuberculosis* (MTB) causes scleritis but more often it is an immune-mediated process (Bloomfield et al., 1976; Agarwal and Majumder, 2019). In fact, it is known that cell wall components of MTB proteins are immunogenic and have been reported to cause episcleritis and cause phlyctenular keratitis and phlyctenular keratoconjunctivitis often in conjunction with an active systemic disease (Murthy et al., 2020). Lane et al. (2019) observed that ocular tuberculosis accounted for 75% of infectious scleritis cases in India. Positive cases of tuberculosis should be treated using standard anti-tubercular therapy (Figure 10.6a-c) using multiple drugs: ethambutol (400 mg orally twice a day), isoniazid (300 mg orally once a day), rifampin (600 mg orally once a day), and pyridoxine (50 mg orally once a day) for 6–12 months (Murugan, 2019).

Nocardia an Actinomycetes bacterium is also known to be an etiological organism in 24% of patients with infectious scleritis (Jain et al., 2009; Richelle and Peacock, 2018) (Figure10.7a). These patients may have lung cavitation (Figure10.7b) and chronic diffuse scleritis with nodules and/or necrotic abscesses (Figure10.7b and 10.7c). Infection is controlled by surgical debridement of necrotic areas and therapy with Amikacin 2.5% eyedrops (Murthy et al., 2020).

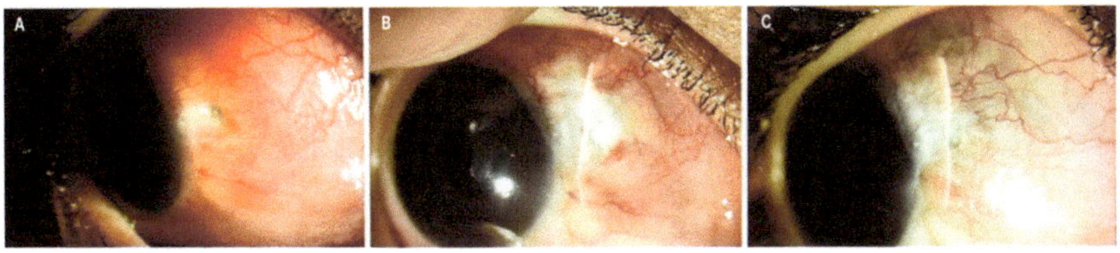

FIGURE 10.6 Treatment of necrotizing scleritis in a 19-year-old female patient (a) showing recovery signs after oral anti-tubercular therapy and topical corticosteroids (b) and complete resolution at two months follow up (c). (Modified from Murthy et al. (2020) published under the terms of the Creative Commons Attribution-Non-Commercial-ShareAlike 4.0 License.)

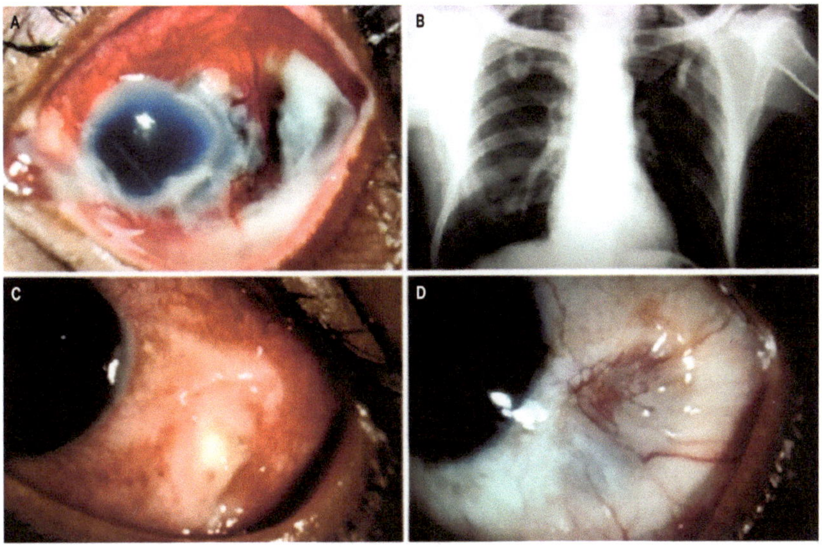

FIGURE 10.7 Nocardial scleritis in a patient. (a) Nocardial scleral abscess from endogenous spread; (b) lung with cavitatory lesion, which was the primary source of infection in the same patient. (c) A localized nodule with yellowish appearance, in case of injury with a thorn and *Nocardia* spp. infection. (d) Complete resolution two months later after debridement and therapy with amikacin 2.5% eyedrops. (Modified from Murthy et al. (2020) published under the terms of the Creative Commons Attribution-Non-Commercial-ShareAlike 4.0 License.)

10.5.3.2 Fungal Scleritis

Fungal scleritis is comparatively more severe compared to bacterial scleritis and could destroy the globe and cause loss of vision (Figure 10.8a and 10.8b). In tropical and developing countries, fungal scleritis ranged from 5 to 38% (Ho et al., 2014; Murthy et al., 2015) and constituted 25% of culture proven infectious scleritis (Pradhan and Jacob, 2013). Scleritis is known as the exogenous type when it originates from contamination following trauma or ocular surgeries and referred to as the endogenous

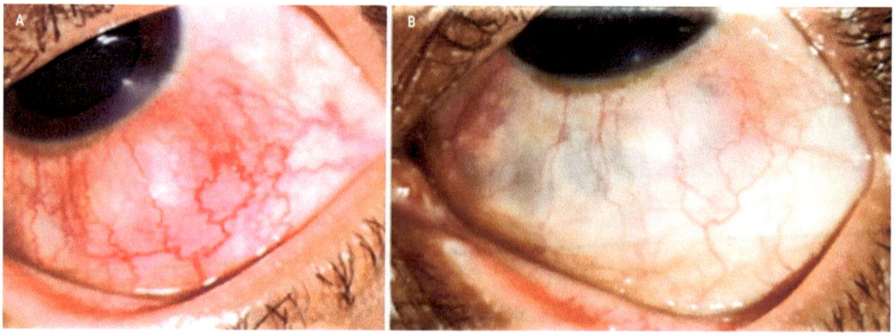

FIGURE 10.8 (a) Fungal scleritis showing a nodular lesion. (b) Resolution of scleritis after surgical debridement and treatment with topical and oral voriconazole. (Modified from Murthy et al. (2020) published under the terms of the Creative Commons Attribution-Non-commercial-ShareAlike 4.0 License.)

type when transmitted in immunocompromised patients or via intravenous drug use. The risk factors that are associated with fungal keratitis are many and include diseases of the eye or systemic diseases, use of contact lenses, addiction to intravenous use of narcotics and chronic use of topical corticosteroids. The common fungal etiological agents of scleritis include *Aspergillus flavus, Scedosporium apiospermum, Pseudoallescheria boydii* and *Rhinosporidium seeberi* (Cunningham et al., 2011; Todokoro et al., 2018). It is cautioned that since scleritis caused by *Mycobacterium* spp., *Acanthamoeba* spp. or anaerobes are clinically similar to fungal etiology, anti-fungal agents are recommended empirically as first line therapy in microbiologically negative cases. Visual prognosis in fungal scleritis is often poor due to various reasons such as delayed diagnosis, poor penetration of the antifungals and non-availability of antifungals. Fungal scleritis resolves after surgical debridement and treatment with topical and oral voriconazole and cyclosporine A (Sahu et al., 2013) and intravenous caspofungin (Fincher and Fulcher, 2007).

10.5.3.3 Viral Scleritis

Several different viruses, such as varicella zoster virus (VZV) (Livir-Rallatos et al., 1998), herpes simplex virus type-1 (HSV-1) (Gonzalez-Gonzalez et al., 2012), cytomegalovirus (Accorinti et al., 2013) and Epstein Barr virus have been implicated in scleritis (Furukawa et al., 1993). VZV is the most common cause of viral scleritis and presents as painful disease that is persistent and characterised by circumscribed nodules with translucent areas, with the risk of perforation or staphyloma formation (Sainz de la Maza et al., 2012a). Viral scleritis also may cause reduced corneal sensation in the affected area and additionally could induce sectorial iris atrophy. Scleritis is often associated with keratitis and anterior uveitis and diagnosis at times poses a challenge. For instance, in HSV anterior scleritis the condition is very similar to nodular immune mediated scleritis and may occur without corneal involvement. Further, scleritis may occur several months or several years after the onset of VZV infection. The additional complication is that detection of the virus by staining or immunofluorescence techniques, by culturing and by electron microscopy is not straight forward and may be difficult (Sainz de la Maza et al., 2012b). Corneal scars are indicative of viral disease and the scars are transient and resolve on anti-viral therapy.

10.5.3.4 Protozoal Scleritis

Compared to bacteria, fungi and viruses, the protozoans as etiological agents of scleritis are rare. *Acanthamoeba* and *Toxoplasma* infections of the sclera are uncommon (Lovieno et al., 2014; Kamath et al., 2013).

10.5.3.5 Infectious Posterior Scleritis

Infectious posterior scleritis is rare and may be secondary to systemic or local infectious process, surgical or accidental trauma involving an intraocular foreign body. Patients with infectious posterior scleritis present with deep-seated radiating pain around the orbit and when VZV is involved the pain becomes very severe and even extra-ocular movement becomes painful. Episcleral congestion may also be noted. The other features of posterior infectious scleritis include diminution of vision secondary to serous retinal detachment, macular edema or disc edema, sub-retinal fluid, sub-retinal mass, choroidal folds, and vitritis (Benson, 1988). A few cases may be associated with anterior uveitis. Diagnosis is based on B-scan ultrasonography to detect scleral mass and choroidal thickening with sub-tenons fluid, Fundus fluorescein angiography, computerized tomogram (CT) and magnetic resonance imaging (Diaz et al., 2016; Shields et al., 2014; Biswas et al., 1998). Lee et al. (2022) observed that most posterior scleritis patients were of non-infectious type but patients with a history of ocular surgery or trauma, especially with the presence of pus-containing nodules would be suspect for infectious posterior scleritis. It has been observed that impaired vision does not improve significantly after treatment of posterior scleritis thus prompt diagnosis and aggressive treatment are recommended.

10.5.4 Treatment of Infectious Scleritis

10.5.4.1 Medical Therapy

Treatment modalities for scleritis include both medical therapy and surgery and some of the strategies for medical therapy are as follows (Murthy et al., 2020):

1. Treatment with broad-spectrum antibiotics like ciprofloxacin, which is against *P. aeruginosa*, Gram-positive organisms, and *Nocardia* spp. is started even before the etiological agent is identified.
2. In case the microbiology report is negative but inflammation is observed auto-immune scleritis is suspected, and patients are immediately treated with oral corticosteroids, immunosuppressives, or biologics.
3. Empirical antifungal therapy with natamycin or voriconazole is started when the patient does not respond to antibiotics suspecting fungal infection.
4. When inflammation persists despite prolonged and adequate treatment with both antibiotics and antifungals topical corticosteroids are added to reduce tissue destruction and suppress host immune response.

5. In cases of acanthamoeba keratitis with scleritis, oral corticosteroids are recommended since the scleritis component is immune-mediated.

10.5.4.2 Surgical therapy

Apart from medical therapy surgical intervention is recommended for advanced cases of scleritis (Murthy et al., 2020).

1. In cases of severe scleral necrosis with abscesses refractive to medical treatment, the way forward is debridement of the scleral abscess and irrigation with either 2.5% betadine solution or with broad spectrum antibiotics or antifungals such as 2% voriconazole.
2. Surgical wound irrigation with antibiotics or antifungals is advisable.
3. Tectonic procedures such as amniotic membrane graft or tenon's patch graft or scleral graft and/or lamellar or penetrating keratoplasty with continued antibiotic coverage may be needed in cases of extensive scleral involvement (Murthy et al., 2020).

10.5.5 Pathogenesis of Scleritis

The pathogenesis of scleritis is still to be completely understood. In a review, Vergouwen et al. (2020) highlighted that non-infectious scleritis is immune system driven based on the local inflammation of the sclera and the increase in inflammatory factors in blood and tears. Further, the observation that immunosuppressive treatment is beneficial (Sainz de la Maza et al., 2016; Seo et al., 2006) and the fact that susceptibility genes overlap in systemic autoimmune diseases and idiopathic scleritis indicated the possibility of the involvement of the immune system in scleritis (Deng and Tsao, 2017; Karami et al., 2019; Li et al., 2019). Further, taking cue from the destructive power of matrix metalloproteinases (MMPs) in rheumatoid arthritis (Seo et al., 2006; Young et al., 2004; Burrage et al., 2006) the pathogenic role of MMPs (Burrage et al., 2006; Rose and Kooyman, 2016) in the pathogenesis of scleritis was implicated. MMPS could probably breakdown scleral collagen and proteoglycans and thus contribute to scleral necrosis (Wakefield et al., 2013). Increased levels of TNF-alpha and MMP-9 have been detected in tears of patients with necrotizing scleritis (Seo et al., 2006).

Vergouwen et al. (2020) in their review have flagged several pertinent questions such as: Why does sclera get inflamed? Why is scleritis associated with systemic autoimmune diseases? Why do some scleritis patients develop severe necrosis of scleral and adjacent corneal tissue? Explanations to these queries have been attempted (Watson and Romano, 2014; Orr et al., 2017) but are yet to be proved. Thus, to get a better understanding on the pathogenesis of scleritis future studies should focus on the identification and dynamics of autoantibodies and/or autoreactive T-cells in scleritis, understand the role of MMPs in scleritis, identify genes conferring susceptibility to scleritis, identify clinical biomarkers in scleritis and associated systemic autoimmune diseases, and improve imaging techniques to reveal additional information of the disease process. Finally, animal models would be highly relevant to plan studies on the molecular basis of scleritis. Overall, understanding the pathogenesis of scleritis would help develop strategies for treatment, improve therapy, and visual prognosis of this severe and disabling ocular disorder (Vergouwen et al., 2020).

10.5.6 Association of Scleritis with Ocular and Systemic Complications

Complications are common in scleritis patients and the complications appear to depend on the type of scleritis affecting the patient. A study in Japan indicated that all patients (100%) with either necrotizing or posterior scleritis had complications, whereas the number of patients with complications was only 44% for patients with nodular scleritis and 77% for patients with diffuse anterior scleritis (Keino et al., 2010). Elevated intraocular pressure as a complication was seen in up to 40% of patients with scleritis and was more frequently seen in patients with necrotizing scleritis and it caused decreased vision (Tanaka et al., 2018). Other complications include ocular hypertension in posterior and anterior necrotizing scleritis patients (Tanaka et al., 2018) and cataract (Jabs et al., 2000) with an incidence of 17% of cataract (Sainz de la Maza 1994) in necrotizing scleritis, diffuse and anterior scleritis and nodular scleritis patients (Jabs et al., 2000; Palkar et al., 2020). Complications that are sight-threatening such as perforation of the globe can also occur in patients with necrotizing scleritis (Tanaka et al., 2018). Bin Ismail et al. (2017) observed that 53.3% of patients with scleritis had ocular complications that included anterior uveitis (42.5%), raised intraocular pressure (12.5%), and corneal involvement (11.7%). Autoimmune causes were associated with 25.8% of patients, and 8.3% patients had an associated infective etiology, much higher than Caucasian studies (Bin Ismail et al., 2017).

Scleritis is also associated with several systemic infectious diseases like Tuberculosis (Yadav and Raval, 2015; Cheung and Chee, 2012), Leprosy, syphilis, Lyme's disease and viral diseases caused by HSV and VZV (Murthy et al., 2020). The type of scleritis in tuberculosis could be either anterior or posterior necrotising or non-necrotising tuberculous scleritis or nodular anterior scleritis (Paul et al., 2019; Patnaik et al., 2020). Anterior nodular scleritis is common in leprosy patients but at times necrotizing scleritis may occur. Scleritis has also been reported associated with syphilis patients and scleritis is diffuse but in tertiary syphilis patients scleritis is of the nodular type. Scleritis is also common in patients with Lyme's diseases but the time interval that lapses between Lyme's diseases and the detection of scleritis is prolonged (Doreen and Wenkel, 2002). Nodular posterior scleritis is rare in *Herpes Simplex Virus* disease but it is commonly associated with VZV (Table 10.1). Although rare, sometimes malignancies masquerade as scleritis (Kafkala et al., 2005). Choroidal melanoma can initially manifest as episcleritis (Fraser et al., 1979) and nodulo-ulcerative ocular surface squamous neoplasia can mimic necrotising scleritis (Kaliki et al., 2017).

10.6 Episcleritis

Episcleritis, also known as episcleritis periodica fugax, sub-conjunctivitis and phlegmatous conjunctivitis is an inflammatory disease associated with the conjunctival and superficial episcleral vascular plexi and the affected area appears bright red

(Okhravi et al., 2005). Episcleritis is referred to as being simple when the inflammation is diffuse, self-limiting and it resolves within days to weeks (Diaz et al., 2016). In contrast, nodular episcleritis lasts longer and appears as a well-defined area of elevation (Watson and Hayreh, 1976), which is more painful (Yanoff and Duker, 2014). Intraocular pressure may be elevated in episcleritis patients due to increased episcleral venous pressure (Pikkel et al., 2015). It is rare in children but more common in adults and roughly two-thirds of cases occur in females with a peak incidence in the fourth decade (Diaz et al., 2016; McGavin et al., 1976). Episcleritis patients complain of mild discomfort or irritation and may present with epiphora but acuity is normal (Diaz et al., 2016; Sims, 2012).

10.6.1 Diagnosis of Episcleritis

In addition to clinical examination, Optical coherence tomography (OCT), ultrasound bio-microscopy and B-scan ultrasound help to differentiate between episcleritis and scleritis. OCT could detect dilated deep episcleral vessels (Levison et al., 2016) and superficial layer involvement in episcleritis (Axman et al., 2016). Further ultrasound bio-microscopy can distinguish episcleral thickening from scleral involvement (Pavlin et al., 1993). Episcleritis is non-erythematous (Som and Curtin, 2011). Other dissimilarities with scleritis include common unilateral presentation, pain that does not correlate with the severity of inflammation (Diaz et al., 2016; Lavric et al., 2016), choroidal folds, serous retinal detachments, choroidal detachments, macular or optic disc edema, vitreous cells (Sims, 2012; Lavric et al., 2016) and fluid in the sub-Tenon's space (Lavric et al., 2016; Cunningham et al., 2016).

10.6.2 Treatment of Episcleritis

Episcleritis is self-limiting and about 20% of the patients recover without treatment (Diaz et al., 2016). Topical steroids have been used by many clinicians to treat episcleritis and it is known that NSAIDs provide no benefit over artificial tears (Williams et al., 2005). In case the inflammation persists, the dosing frequency is increased or a more potent topical steroid is used (Diaz et al., 2016; Jabs et al., 2000). It is known that more than 30% of episcleritis patients have an underlying systemic disease (Jabs et al., 2000; Diaz et al., 2016; Okhravi et al., 2005). Right diagnosis and prompt treatment of episcleritis patients leads to improved outcomes, as does adhering to a multidisciplinary approach (Williamson, 2017). In a recent study, a patient with Crohn's disease developed unilateral right episcleritis within hours following Moderna mRNA COVID-19 booster vaccination (Veenis et al., 2023).

10.7 Antibiotic Resistance in Scleritis

Much is known about the mechanism of resistance that is encountered in systemic infections (Tenover, 2001). But comparatively less information is available on antibiotic resistance in ophthalmic infections. Several factors contribute to the difficulty in assessing resistance in ophthalmology and include the following:

(i) Small case numbers have made it difficult to conclude about antibiotic resistance in ophthalmological practice (Asbell et al., 2008). In this context, the Ocular TRUST and ARMOR 2009 studies have provided valuable data on ocular isolates and their trends in resistance to antibiotics that could be statistically analyzed (Table 10.3).

(ii) Another limitation in assessing ocular bacterial resistance to antibiotics is using minimum inhibitory concentrations (MICs) of antibiotics determined for systemic drug administration (Haas et al., 2011; Asbell et al., 2008; 2015). This needs to be changed since MICs need to be determined when applied topically as in ophthalmology. Studies have indicated that topical application of antimicrobials, provides higher drug exposure over time than systemic use (Asbell et al., 2008). For instance, drug exposure time for 0.5% levofloxacin ophthalmic solution over 6 h was more than twice than that with oral or intravenous dose of 750 mg levofloxacin over 24 h (Asbell et al., 2008; Raizman et al., 2002). Drug exposure for azithromycin topical solution with 0.5%, 1% or 1.5% solution, resulted in two orders of magnitude higher than standard oral 3 or 5-day regimens or single-dose extended release respectively (Chiambaretta et al., 2007).

(iii) Need to consider antibiotic penetration as a factor since it is known to influence clinical success rates. For instance, topical moxifloxacin and gatifloxacin are known to penetrate the anterior chamber to a greater degree than ciprofloxacin, thus achieving much higher concentrations (Solomon et al., 2005).

(iv) Lack of a standardized framework for studying the susceptibility of ocular pathogens is yet another factor. Some used CLSI-defined breakpoints (Haas et al., 2011; Asbell et al., 2008; 2015), others used automatic microbial systems (Benz et al., 2004), and some reported only the method of testing (Ramakrishnan et al., 2008).

Thus, due to the above difficulties and the perceived threat from increasing antibiotic resistance worldwide in ophthalmology, the WHO and the United States Food and Drug Administration and others started two surveillance programs, namely the Ocular Tracking Resistance in the U.S. Today (TRUST) and Antibiotic Resistance Monitoring in Ocular Microorganisms (ARMOR) (Table 10.3). The former Ocular TRUST1 evaluated ophthalmic *S. aureus,* both methicillin-susceptible (MSSA) or methicillin-resistant (MRSA) isolates, *Streptococcus pneumoniae*, *Haemophilus influenzae* for *in vitro* susceptibility (Asbell et al., 2008; 2015). The ARMOR study extended the data to include analysis of *P. aeruginosa* and CoNS. The ARMOR 2013 study analyzed a total of 3237 isolates and is the largest study of its kind to date (Grzybowski et al., 2017).

10.7.1 Resistance in Ocular *Staphylococcus pneumoniae* from Scleritis Patients

Many ocular *S. pneumoniae* isolates (760) archived between 1999 and 2006 showed significant resistance to penicillin (34.1%) with cross-resistance to azithromycin and trimethoprim in about 75% of the isolates (Asbell et al., 2008). Subsequent prospective studies between 2005 and 2006 revealed that 18.3% of the *S. pneumoniae* isolates were resistant to penicillin, with concurrent resistance to trimethoprim, azithromycin, and

TABLE 10.3

Antibiotic Resistance Levels in the TRUST and ARMOR Studies

		Non-susceptible bacteria (%)							
Bacteria		**Penicillin**	**Azithromycin**	**Gatifloxacin**	**Moxifloxacin**	**Levofloxacin**	**Ofloxacin**	**Tobramycin**	**Ciprofloxacin**
Streptococcus Pneumoniae	TRUST-retro	34.1	33.4	0.3	0.1	0.1		95.1	9.7
	TRUST-pros	18.3	22.4	0	0	0		98	10.2
	ARMOR	31.8	34.8	0.4	0.3	0	0		
MS Staphyo-coccus aureus	TRUST-pros	90.2	45.7	18.9	18.9	18.9		7.3	20.1
	ARMOR		42.8	13.5	12	13.7	14.1	4.1	14.2
MR Staphyo-coccus aureus	TRUST-pros	100	93.9	84.8	84.8	84.9		63.6	84.8
	ARMOR		93.3	75.1	74	75.9	76.4	44.3	77.3
MS-CoNS	ARMOR		44.7	13.9	13.6	13.9	14.2	6.4	
MR-CoNS	ARMOR		78.3	55.7	51.2	56.8	56.9	23.1	58.5
H. influenzae	TRUST-retro	100	0.3	0.3	0.3	0.3		0	0.3
P. aeruginosa	ARMOR					6.9	10.1	3.1	7.7

Source: Modified from Grzybowski et al., 2017; published under the terms of the Creative Commons Attribution: Non-commercial–ShareAlike 4.0 License.

Note: MS-methicillin sensitive, MR-methicillin resistant, retro-retrospective, pros-prospective.

tobramycin (Asbell et al., 2008). A single isolate of the 760 isolates also showed resistance to third- and fourth-generation fluoroquinolones (levofloxacin, moxifloxacin, gatifloxacin), whereas resistance to ciprofloxacin was 9.7%. It was also observed that resistance of *S. pneumoniae* to fluoroquinolone during the 8 years of the study remained significantly unchanged (Asbell et al., 2008). The above studies carried out by Ocular TRUST 1 was confirmed by ARMOR 2013, which indicated that resistance to penicillin was 31.8%, to azithromycin 38.4%, and only a single isolate was resistant against any generation of fluoroquinolones (Asbell et al., 2015).

10.7.2 Resistance in Ocular *Pseudomonas aeruginosa* from Scleritis Patients

Resistance of *P. aeruginosa* isolates to antibiotics was tested in ARMOR 2013 and susceptibility was above 90% indicating only marginal resistance (Grzybowski et al., 2017). Meyer et al. (2008) described a case of necrotizing scleritis due to *P. aeruginosa* resistant to fourth-generation fluoroquinolones. The patient did not respond to empiric treatment and developed scleral edema with a nummular area of avascular sclera nasally. Histopathologic examination revealed necrotizing scleritis and cultures grew *P. aeruginosa* resistant to fourth-generation fluoroquinolones.

10.7.3 Resistance in Ocular *Haemophilus influenzae* from Scleritis Patients

Ocular TRUST1 (1999–2006) screened 356 archived isolates of *H. influenzae* and found that 37.4% were β-lactamase-positive. Further, majority of *H. influenzae* isolates (85.7%) were resistant to trimethoprim (Asbell et al., 2008). Subsequent prospective studies indicated that 44% were β-lactamase-producing (Asbell et al., 2008). The ARMOR study also showed that all isolates were resistant to the antibiotics tested except two isolates that were not (Haas et al., 2011; Asbell et al., 2015).

10.7.4 Resistance in Ocular *Staphylococcus aureus* from Scleritis Patients

The Ocular TRUST1, a prospective study, showed that a good majority of *S. aureus* isolates (83.2%) are methicillin susceptible (MSSA) and most of the MSSA isolates were susceptible to fluoroquinolones and resistance to the tested fluoroquinolones was less than 20% (Asbell et al., 2008). In contrast, 75–80% of methicillin-resistance *S. aureus* (MRSA) were resistant to fluoroquinolones (Asbell et al., 2008). A comparison of the methicillin resistance in *S. aureus* indicated that resistance increased from 16.8% in 2006 (Ocular TRUST1) to 39% in 2009 (the ARMOR study) (Haas et al., 2011; Asbell et al., 2008). Further, MRSA isolates exhibited multi-drug resistance to ciprofloxacin (79.5%), moxifloxacin (65.4%) and tobramycin (52.6%) (Haas et al., 2011). Comparison of the results of ARMOR 2009 with ARMOR 2013 indicated increase in resistance in *S. aureus* isolates to five drug classes from 11.5% to 86.8% (Haas et al., 2011). In addition, 77.3% of MRSA and methicillin-resistant CoNS, respectively also showed resistance to three or more drug classes.

10.7.5 Resistance in Ocular Coagulase-negative *Staphylococcus* from Scleritis Patients

S. epidermidis constitutes 75.9% of CoNS (Asbell et al., 2015) and the trend in resistance to antibiotics in CoNS was similar as in *S. aureus*, with comparable resistance rates. CoNS exhibited 49.7% resistance to methicillin (42.2% for *S. aureus*), 34.4% to ciprofloxacin (39.8% for *S. aureus*), and 61.3% to azithromycin (63.6% for *S. aureus*). The resistance profile during the ARMOR 2013 study (5-year surveillance) did not change (Asbell et al., 2015). But the subsequent ARMOR studies for 2014 and 2015 showed that resistance including MDR continue to be a reality and a challenge in ophthalmology (Bausch + Lomb 2016).

10.7.6 Resistance in Ocular *Mycobacterium tuberculosis* from Scleritis Patients

Mycobacterium tuberculosis (Watson and Hayreh, 1976) as an etiological agent of scleritis is rare, and occurs due to scleral invasion following pulmonary tuberculosis and at times it may occur post-trauma (Gupta et al., 2016) or due to infection from the conjunctiva, cornea, or uveal tissue. Scleral tuberculosis could also be due to circulating mycobacterial antigens in the absence of the bacterium in the eye (Bloomfield et al., 1976). Tuberculous scleritis commonly presents as anterior nodular scleritis and may involve the peripheral cornea causing sclero-keratitis (Arora et al., 2015; Shoughy et al .2016) or corneoscleral perforations (Wroblewski et al., 2011) despite systemic antitubercular therapy (ATT) and immunosuppressive therapy (Patel et al., 2013). Management of tuberculous scleritis is very challenging and normally includes ATT (rifampicin, isoniazid, ethambutol, pyrazinamide, and streptomycin) (Paul et al., 2019) with or without systemic corticosteroids. Sclero-keratitis was successfully treated with antitubercular drugs (Shoughy et al., 2016), whereas an HIV+ patient with anterior nodular scleritis was successfully treated using both the standard four drugs regimen of ATT along with antiretroviral treatment (Kariyawasam et al., 2020) and a patient with pan-uveitis preceded sclera inflammation was initially treated with only oral corticosteroid, and the patient showed poor response to ATT and immunosuppressive drugs. Globally, about 5% of tuberculosis patients have multidrug resistance (MDR) (resistant to isoniazid and rifampicin) or extensive drug resistance (XDR) (resistant to any fluoroquinolone, and one of the three second-line injectable drugs, capreomycin, kanamycin, and amikacin). Recent studies suggest that most MDR-TB is due to primary resistance, which results from infection with a drug resistant strain or secondary as an acquired resistance due to poor compliance and improper drug regimen (Dheda et al., 2017). Drug resistance in ocular TB has not been studied extensively. Kim and associates were the first to report antibiotic-resistant tuberculous choroiditis, which was sensitive to second-line ATT (Kim et al., 1993). Kesen et al. (2009) also reported a case of atypical multi-drug resistant tuberculous scleritis presenting as scleral nodules that subsequently progressed to necrotizing scleritis and sclera rupture after treatment with oral steroid and quadruple antituberculosis therapy. Subsequently, Sharma et al. (2015) reported rifampicin-resistant tubercular choroidal granuloma from the subretinal fluid and concluded that Multiplex PCR followed by rpoB sequencing are reliable for the detection of drug resistance (Sharma et al., 2014; 2019). Agarwal et al. (2022) reported the first case of culture-proven multidrug-resistant tuberculous sclero-keratitis in an immuno-competent patient. In this patient the visual acuity in the

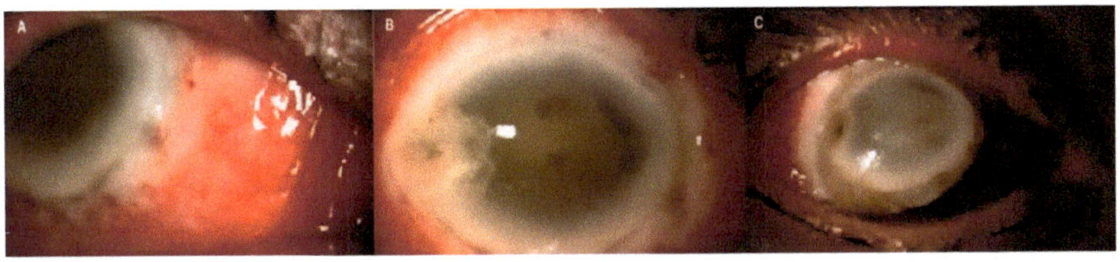

FIGURE 10.9 Slit lamp photograph of the right eye showing diffuse scleral congestion (A), 360 degrees sclero-corneal infiltrate and temporal corneal thinning (B), and scleral rupture with prolapse of the uveal tissue due to progression of scleral infiltrates at 3 weeks (C). (Courtesy Dr Mamta Agarwal, Uveitis & Cornea Services, Medical Research Foundation, Sankara Nethralaya, Chennai, India.)

affected right eye (RE) was only light perception and ocular movements were unhindered and painless. Slit-lamp examination of RE showed diffuse conjunctival and scleral congestion, chemosis, multiple yellowish scleral nodules, 360 degrees white limbal infiltrate with a 2 × 2 mm area of corneal thinning temporally, blood on the corneal endothelium, flat anterior chamber, and dense cataract (Figure 10.9 a-c). In this patient, prolonged immunosuppression with oral corticosteroids and azathioprine along with drug resistance to conventional ATT led to the dissemination of the tubercle bacilli from the intraocular tissues to the sclera and cornea. Analysis of the enucleated globe specimen after 3 weeks of treatment did not reveal any evidence of acid-fast bacilli (AFB), indicating that ATT was effective in controlling the infection but not the inflammation, which ultimately led to scleral perforation. The global treatment success rate in MDR-TB has been reported as 56%.

Studies have shown that in patients with intraocular tuberculosis, improvement with standard ATT is noticed mostly within 2 weeks to 3 months (Moimura et al., 2002). However, the emerging drug resistance, warrants an increase in immunosuppressive therapy. Agarwal et al. (2022) concluded that early identification of drug resistance can help improve the management of the disease.

10.8 Conclusions

Scleritis, an ocular inflammatory disease could be either of the non-infectious or infectious type. Non-infectious scleritis is immune mediated, whereas the etiological agents of infectious scleritis are bacteria, fungi, and viruses. The treatment modalities differ based on the type of scleritis but antibiotics are commonly used in both types. In infectious scleritis apart from medical therapy, surgical intervention is recommended. Scleritis could be severe and result in loss of vision and management is further compounded due to resistance to antibiotics. This menace of resistance warrants early identification of drug resistance to help improve the management of scleritis.

REFERENCES

Abdel-Aty A, Gupta A, Del Priore, L., Kombo N. Management of noninfectious scleritis. Ther Adv Ophthalmoln. 2022;14:1–12. DOI: 10.1177/25158414211070879

Abtahi MA, Abtahi SH, Fazel F, et al., Topiramate and the vision: A systematic review. *Clin Ophthalmol.* 2012;6:117–131.

Accorinti M, Abbouda A, Gilardi M, Zito R, Iannetti L. Cytomegalovirus-related scleritis. *Ocul Immunol Inflamm.* 2013;21:413–5.

Agarwal A, Majumder PD. Tubercular posterior scleritis: A case report and review of literature. *Indian J Ophthalmol.* 2019;67:1362-5.

Agrawal R, Lee CS, Gonzalez-Lopez JJ, Khan S, Rodrigues V, Pavesio C. Flurbiprofen: A nonselective cyclooxygenase (COX) inhibitor for treatment of non-infectious, non-necrotizing anterior Scleritis. *Ocul Immunol Inflamm.* 2016a;24(1):35–42.

Agrawal R, Lavric A, Restori M, Pavesio C, Sagoo MS. Nodular posterior scleritis: Clinico-sonographic characteristics and proposed diagnostic criteria. *Retina Phila Pa.* 2016b;36:392–401.

Agarwal M, Patnaik G, Agarwal S, Iyer G, Anand AR, Gayathri A, Biswas J, Zierhut M. Tuberculous Scleritis and multidrug resistance. *Ocul. Immunol. Inflamm.* 2022;30:4:915–924. DOI:10.1080/09273948.2020.1853176

Ahn SJ, Oh JY, Kim MK, Lee JH, Wee WR. Clinical features, predisposing factors, and treatment outcomes of scleritis in the Korean population. *Korean J Ophthalmol.* 2010;24:331-5.

Akhtar S, Babar S, Lightman S, et al., Adalimumab-induced remission of anterior scleritis: A very rare late manifestation of Takayasu arteritis. *Rheumatol.* 2015;54:2288–2290.

Aktas Z, Bektas C, Hasanreisoglu M. Pan-scleritis as an unusual complication of gonioscopy-assisted transluminal trabeculotomy. *J Glaucoma.* 2019;28:e21–e23.

Altman AJ, Cohen EJ, Berger ST, Mondino BJ. Scleritis and streptococcus pneumoniae. *Cornea.* 1991:10:341–5.

Amer R, Levinger N. Psoriasis-associated progressive necrotizing posterior scleritis: A 6-year follow-up. *Eur J Ophthalmol.* 2020;30(3):NP7–10. https://doi.org/10.1177/1120672118823121

Arora T, Sharma N, Shashni A, Titiyal JS. Peripheral ulcerative keratitis associated with chronicmal absorption syndrome and military tuberculosis in a child. *Oman J. Ophthalmol.* 2015;8(3):205–207. doi:10.4103/0974-620X.169888

Asbell PA, Colby KA, Deng S, McDonnell P, Meisler DM, Raizman MB, Sheppard JD, Sahm DF. Ocular TRUST: Nationwide antimicrobial susceptibility patterns in ocular isolates. *Am J Ophthalmol.* 2008;145(6):951–958. doi: 10.1016/j.ajo.2008.01.025.

Asbell PA, Sanfilippo CM, Pillar CM, DeCory HH, Sahm DF, Morris TW. Antibiotic resistance among ocular pathogens in the United States: five-year results from the antibiotic resistance monitoring in ocular microorganisms (ARMOR) surveillance study. *JAMA Ophthalmol.* 2015;133(12):1445–54

Axmann S, Ebneter A, Zinkernagel MS. Imaging of the sclera in patients with scleritis and episcleritis using anterior segment optical coherence tomography. *Ocul Immunol Inflamm.* 2016;24(1):29–34.

Basti S, Gopinathan U, Gupta S. Nocardial necrotizing scleritis after trauma. *Cornea.* 1994;13:274–5.

Bausch + Lomb (2016). Bausch + Lomb reports preliminary 2015 results of the antibiotic resistance monitoring in ocular microorganisms (ARMOR) study. www.bausch.com/our-company/recent-news/id/2262/542016-wednesday

Beardsley RM, Suhler EB, Rosenbaum JT, Lin P. Pharmacotherapy of Scleritis: Current paradigms and future directions. *Expert Opin Pharmacother.* 2013 March;14(4):411–424. doi:10.1517/14656566.2013.772982

Benson WE. Posterior scleritis. *Surv. Ophthalmol.* 1988;32:297–316.

Benz MS, Scott IU, Flynn HW, Unonius N, Miller D. Endophthalmitis isolates and antibiotic sensitivities: A 6-year review of culture-proven cases. *Am J Ophthalmol.* 2004;137(1):38–42. doi: 10.1016/S0002-9394(03)00896-1

Bhatt N, Dalal M, Tucker W, et al., Subconjunctival sirolimus in the treatment of autoimmune non-necrotizing anterior scleritis: Results of a phase I/II clinical trial. *Am J Ophthalmol.* 2015;159:601–606.

Bin Ismail MA, Lim, R., H., F., Fang, H. M., Wong, E., P. Y., Ho Su Ling, H. S. Wee Kiak Lim, W. K., Stephen C. Teoh, S., C., Agrawal, R. Ocular Autoimmune Systemic Inflammatory Infectious Study (OASIS)—report 4: Analysis and outcome of scleritis in an East Asian population. *J. Ophth Inflamm Infect.* 2017;7:6. DOI 10.1186/s12348-017-0124-5

Birolo C, Zannin ME, Arsenyeva S, et al., Comparable efficacy of abatacept used as first-line or secondline biological agent for severe juvenile idiopathic arthritis-related uveitis. *J Rheumatol.* 2016;43(11):2068–73. https://doi.org/10.3899/jrheum.151389

Biswas J, Mittal S, Ganesh SK, Shetty NS, Gopal. L. Posterior scleritis: Clinical profile and imaging characteristics. *Indian J Ophthalmol.* 1998;46:195–202.

Bloomfield SE, Mondino B, Gray GF. Scleraltuberculosis. *ArchOphthalmol.* 1976;94(6):954–956. doi:10.1001/archopht.1976.03910030482009

Botsios C, Sfriso P, Ostuni PA, Todesco S, Punzi L. Efficacy of the IL-1 Receptor antagonist, anakinra, for the treatment of diffuse anterior scleritis in rheumatoid arthritis. *Report of two cases. Rheumatol.* 2007;46:1042–1043.

Bottin C, Fel A, Butel N, et al., Anakinra in the treatment of patients with refractory scleritis: A pilot study. *Ocul Immunol Inflamm.* 2018; 26:915–920.

Burgos PI, Perkins EL, Pons-Estel GJ, et al., Risk factors and impact of recurrent lupus nephritis in patients with systemic lupus erythematosus undergoing renal transplantation: Data from a single US institution. *Arthritis Rheum* 2009; 60:2757–2766.

Burrage PS, Mix KS, Brinckerhoff CE. Matrix metalloproteinases: Role in arthritis. *Front. Biosci.* 2006;11:529–543.

Cao JH, Oray M, Cocho L, et al., Rituximab in the treatment of refractory noninfectious scleritis. *Am J Ophthalmol.* 2016;164:22–28.

Chandrakantan A, Ratanapanichkich P, Said M, et al., Recurrent IgA nephropathy after renal transplantation despite immunosuppressive regimens with mycophenolate mofetil. *Nephrol Dial Transplant.* 2005;20:1214–1221.

Cheung CMG, Chee SP. Posterior scleritis in children: Clinical features and treatment. *Ophthalmol.* 2012;119:59–65.

Chiambaretta F, Garraffo R, Elena P, Pouliquen P, Delval L, Rigal D, Dubray C, Goldschmidt P, Tabbara K, Cochereau I. Tear concentrations of azithromycin following topical administration of a single dose of azithromycin 0.5%, 1.0%, and 1.5% eyedrops (T1225) in healthy volunteers. *Eur J Ophthalmol.* 2007;18(1):13–20.

Choudhary A, Harding SP, Bucknall RC, et al., Mycophenolate mofetil as an immunosuppressive agent in refractory inflammatory eye disease. *J Ocul Pharmacol Ther.* 2006;22:168–175.

Cocho L, Gonzalez-Gonzalez LA, Molina-Prat N, Doctor P, Sainz-de-la-Maza M, Foster CS. Scleritis in patients with granulomatosis with polyangiitis (Wegener). *Br J Ophthalmol.* 2016;100:1062-5

Cunningham MA, Alexander JK, Matoba AY, Jones DB, Wilhemus KR. Management and outcome of microbial anterior scleritis. *Cornea.* 2011;30:1020-3.

Cunningham ET, McCluskey P, Pavesio C, et al., Scleritis. *Ocul Immunol Inflamm.* 2016;24(1):2–5.

Daniel E, Thorne JE, Newcomb CW, et al., Mycophenolate mofetil for ocular inflammation. Am *J Ophthalmol.* 2010;149:423–432.

Dastidar AG, Sugandha Goel, Debi Kundu, Jyotirmoy Biswas, Eesh Nigam, Preeti Sharma. A case of anterior scleritis in association with posterior scleritis — a diagnostic riddle GMS. *Ophthalmol. Cases.* 12. ISSN 2193-1496

David DB, Hirst LW, McMillen J, Whitby M. Mycobacterium marinum keratitis: Pigmentation a clue to diagnosis. *Eye.* 1999;13:377–9.

De Fidelix TS, Vieira LA, De Freitas D, et al., Biologic therapy for refractory scleritis: A new treatment perspective. *Int Ophthalmol.* 2015;35:903–912.

De Fidelix TS, Vieira LA, Trevisani VF. Management of necrotizing scleritis after pterygium surgery with rituximab. *Arq Bras Oftalmol.* 2016;79:339–341.

Deng Y, Tsao BP, 2017. Updates in lupus genetics. *Curr. Rheumatol. Rep.* 19:68.

Dheda K, Gumbo T, Maartens G, et al.,The epidemiology, pathogenesis, transmission, diagnosis, and management of multidrug-resistant, extensively drug-resistant, and incurable tuberculosis. *Lancet Respir Med.* 2017;5:291–360. doi:10.1016/S2213-2600(17)30079-6.

Diaz DJ, Sobol E, Gritz D. Treatment and management of scleraldisorders. *Surv Ophthalmol.* 2016;61(6):702–17.

Dick AD, Meyer P, James T, et al., Campath-1H therapy in refractory ocular inflammatory disease. *Br J Ophthalmol.* 2000;84:107–9.

Dominguez-Casas LC, Sánchez-Bilbao L, Calvo-Río V, et al., Biologic therapy in severe and refractory peripheral ulcerative keratitis (PUK). Multicenter study of 34 patients. *Semin Arthritis Rheum.* 2020;50(4):608–615. https://doi.org/10.1016/j

Doreen K, Wenkel H. Posterior scleritis associated with Borrelia burgdorferi (Lyme disease) infection. *Ophthalmol.* 2002;109:143-5.

Durrani K, Kempen JH, Ying GS, et al., Adalimumab for ocular inflammation. *Invest Ophthalmol Vis Sci.* 2012;53:5492.

Dutta-Majumder P, Jayshree G, David RL, Kaushik V. A case of idiopathic necrotizing scleritis with secondary glaucoma treated successfully with golimumab and Ahmed valve implantation. *Indian J Ophthalmol.* 2019a;67(8):1360–1362. https://doi.org/10.4103/ijo.IJO_2081_18

Dutta Majumder P, Ali S, George A, Ganesh S, Biswas J. Clinical Profile of Scleritis in Children. *Ocul Immunol Inflamm.* 2019b;27 :535–539.

Dutta Majumder P. Presumed tuberculous sclero-keratitis presenting with hypopyon. *Ocul Immunol Inflamm.* 2019c;27(7):1121–1123. doi:10.1080/09273948.2018.1511812

Dutta Majumder P, Agrawaly R, McCluskeyz P, Biswas J. Agrawal, Rupesh†; McCluskey, Peter‡; Biswas, Jyotirmay current approach for the diagnosis and management of noninfective Scleritis. *Asia-Pacific J Ophthalmol.* March-April 2021;10(2):212–223. | *DOI:* 10.1097/APO.0000000000000341

Fabiani C, Sota J, Sainz-de-la-Maza M, et al., New potential weapons for refractory scleritis in the era of targeted therapy. *Mediators Inflamm.* 2020;2020:8294560. https://doi.org/10.1155/2020/8294560.

Feiz V, Redline DE. Infectious scleritis after pars plana vitrectomy because of methicillin-resistant Staphylococcus aureus resistant to fourth-generation fluoroquinolones. *Cornea.* 2007;26:238–40.

Fincher T, Fulcher SFA. Diagnostic and therapeutic challenge of Aspergillus flavus scleritis. *Cornea.* 2007;26:618–20.

Foster CS, Kothari S, Anesi SD, et al., The Ocular Immunology and Uveitis Foundation preferred practice patterns of uveitis management. *Surv Ophthalmol.* 2016;61:1–17.

Fraser DJ Jr, Font RL. Ocular inflammation and haemorrhage as initial manifestations of uveal malignant melanoma: Incidence and prognosis. *Arch Ophthalmol.* 1979;97:1482–6.

Furukawa H, Hamada T, Nagaya K, Akiya S, Hamada T. A case of necrotizing scleritis associated with Epstein–Barr virus infection. *Nihon Ganka Gakkai Zasshi.* 1993;97:1337–42.

Galor A, Thorne JE. Scleritis and peripheral ulcerative keratitis. *Rheum Dis Clin North Am.* 2007;33:835–854.

Galor A, Jabs DA, Leder HA, et al., Comparison of antimetabolite drugs as corticosteroid-sparing therapy for noninfectious ocular inflammation. *Ophthalmol.* 2008;115:1826–32.

Gangaputra SS, Newcomb CW, Joffe MM, et al., Comparison between methotrexate and mycophenolate mofetil monotherapy for the control of non-infectious ocular inflammatory diseases. *Am J Ophthalmol.* 2019;208:68–75.

Gaujoux-Viala C, Giampietro C, Gaujoux T, et al., Scleritis: A paradoxical effect of etanercept? Etanercept-associated inflammatory eye disease. *J Rheumatol.* 2012;39:233–9.

Goldstein DA, Fontanilla FA, Kaul S, Sahin O, Tessler HH. Long-term follow-up of patients treated with short-term high-dose chlorambucil for sight-threatening ocular inflammation. *Ophthalmol.* 2002;109:370–7.

Goldstein DA, Patel SS, Tessler HH. Episcleritis and scleritis. In: Yanoff M, Duker J, ed. *Ophthalmology*. 4th ed. Philadelphia, PA: Elsevier, 2014:209–16.

Gonzalez-Gonzalez LA, Molina-Prat N, Doctor P, Tauber J, Sainz de la Maza MT, Foster CS. Clinical features and presentation of infectious scleritis from herpes viruses: A report of 35 cases. *Ophthalmol.* 2012;119(7):1460–1464.

González-López JJ, Lavric A, Dutta Majumder P, et al., Bilateral posterior scleritis: Analysis of 18 cases from a large cohort of posterior scleritis. *Ocul Immunol Inflamm*. 2016;24:16–23.

Gonzales JA, Shantha J, Acharya NR. Combination nivolumab- and cabiralizumab-associated acute bilateral anterior and posterior scleritis and anterior uveitis. *Am J Ophthalmol Case Rep*. 2018;10:117–118.

Grzybowski A, Brona P, Kim SJ. Microbial flora and resistance in ophthalmology: A review. *Graefes Arch Clin Exp Ophthalmol.* 2017 May;255(5):851–862. doi: 10.1007/s00417-017-3608-y. Epub 2017 Feb 22. PMID: 28229218; PMCID: PMC5394129

Gumus K, Mirza GE, Cavanagh HD, et al., Topical cyclosporine A as a steroid-sparing agent in steroid-dependent idiopathic ocular myositis with scleritis: A case report and review of the literature. *Eye Contact Lens.* 2009;35:275–278.

Gupta P, Singh R, Gupta S, Kumar A, Kakkar N. Tuberculosis presenting as posttraumatic pan-ophthalmitis. *Oman J Ophthalmol.* 2016;9(1):52–54.

Haas W, Pillar CM, Torres M, Morris TW, Sahm DF. Monitoring antibiotic resistance in ocular microorganisms: Results from the Antibiotic Resistance Monitoring in Ocular micRorganisms (ARMOR) 2009 surveillance study. *Am J Ophthalmol.* 2011;152(4):567–574. doi: 10.1016/j.ajo.2011.03.010

Hernandez-Illas M, Tozman E, Fulcher SF, Jundt JW, Davis J, Pflugfelder SC. Recombinant human tumor necrosis factor receptor Fc fusion protein (Etanercept): Experience as a therapy for sight-threatening scleritis and sterile corneal ulceration. *Eye Contact Lens*. 2004; 30:2–5.

Ho YF, Yeh LK, Tan HY, Chen HC, Yeong-Fong Chen YF, Lin HC, et al., Infectious scleritis in Taiwan—A 10-year review in a tertiary-care hospital. *Cornea.* 2014;33:838–43.

Hoang-XaunT, Foster CS, Rice BA. Scleritis in relapsing polychondritis. Response to therapy. *Ophthalmol.* 1990; 97:892–8.

Hodson KL, Galor A, Karp CL, et al., Epidemiology and visual outcomes in patients with infectious scleritis. *Cornea.* 2013;32(4):466–472.

Homayounfar G, Nardone N, Borkar DS, Vivien MT, Travis CP, Wayne TAE, et al., Incidence of scleritis and episcleritis: Results from the Pacific ocular inflammation study. *Am J Ophthalmol.* 2013;156:752–8.

Honik G, Wong IG, Gritz DC. Incidence and prevalence of episcleritis and scleritis in Northern California. *Cornea.* 2013;32:1562-6.

Hsiao CH, Chen JJ, Huang SC, Ma HK, Chen PY, Tsai RJ. Intrascleral dissemination of infectious scleritis following pterygium excision. *Br J Ophthalmol.* 1998;82(1):29–34.

Huang FC, Huang SP, Tseng SH. Management of infectious scleritis after pterygium excision. *Cornea.* 2000;19:34–9.

Huang SC, Lai HC, Lai IC. The treatment of Pseudomonas keratoscleritis after pterygium excision. *Cornea.* 1999;18:608–11.

Huynh N, Cervantes-Castaneda RA, Bhat P, et al., Biologic response modifier therapy forpsoriatic ocular inflammatory disease. *Ocul Immunol Inflamm.* 2008;16:89–93.

Hwang YS, Chen HC, Chen KJ, *et al.*, Enteric-coated mycophenolate sodium as a corticosteroid sparing agent for the treatment of autoimmune scleritis. *Cornea.* 2011;30:260–264.

Ingaralingam S., 1, Rauz1 S, Murray1 PI, Barry1 RJ,2*Effectiveness of pharmacological agents for the treatment of non-infectious scleritis: A systematic review protocol. *Sys. Rev.* 2020;9:54. https://doi.org/10.1186/s13643-020-01314-9

Jabs DA, Mudun A, Dunn JP, Marsh MJ. Episcleritis and scleritis: Clinical features and treatment results. *Am J Ophthalmol.* 2000;130:469–476.

Jabs D, Rosenbaum J. Guidelines for the use of immunosuppressive drugs in patients with ocular inflammatory disorders: Recommendations of an expert panel. *Am J Ophthalmol.* 2001;131(5):679.

Jain SS, Rao P, Kothari K, Bhatt D, Jain S. Posterior scleritis presenting as unilateral secondary angle-closure glaucoma. *Indian J Ophthalmol.* 2004; 52:241–244.

Jain V, Garg P, Sharma S. Microbial scleritis-experience from a developing country. *Eye (Lond).* 2009;23(2):255–261.

Jan R.-L., Ho C.-H, Wang J.-J, Tseng S.-H, Chang Y.-S. Associations between Sjögren Syndrome, Sociodemographic Factors, Comorbid Conditions, and Scleritis in a Taiwanese Population- Based Study. *J. Pers. Med.* 2022;12:105. https://doi.org/10.3390/jpm12010105

Johnson KS, Chu DS. Evaluation of sub-Tenon triamcinolone acetonide injections in the treatment of scleritis. *Am J Ophthalmol.* 2010;149:77–81.

Kafkala C, Daoud YJ, Paredes I, Foster CS. Masquerade scleritis. *Ocul Immunol Inflamm.* 2005;13:479-82.

Kaliki S, Freitag SK, Chodosh J. Nodulo-Ulcerative ocular surface squamous neoplasia in 6 patients: A rare presentation. *Cornea.* 2017;36:322-32.

Kamath YS, Rathinam SR, Kawali A. Ocular toxoplasmosis associated with scleritis. *Indian J Ophthalmol.* 2013;61:295–7.

Kamei RW. Posterior scleritis associated with central retinal vein occlusion and cystoid macular edema: Case report. *Arq Bras Oftalmol.* 2012;75:358–360.

Kaplan-Messas A, Barkana Y, Avni I, et al., Methotrexate as a first-line corticosteroid-sparing therapy in a cohort of uveitis and scleritis. *Ocul Immunol Inflamm.* 2003;11:131–139.

Karami J, Aslani S, Jamshidi A, Garshasbi M., Mahmoudi M. 2019. Genetic implications in the pathogenesis of rheumatoid arthritis; an updated review. *Gene.* 702:8–16.

Kariyawasam AGTA, Fonseka CL, DeSilva PUT, etal. Frequently relapsing anterior nodular scleritis as the initial presentation of a lethal systemic infection: Disseminated tuberculosis with HIVcoinfection.*CaseRep Infect Dis.* 2020;2020:9020864.

Keino H, Watanabe T, Taki W et al., Clinical features and visual outcomes of Japanese patients with scleritis. *Br J Ophthalmol.* 2010;94:1459–1463.

Kesen MR, Edwards DP, Rao NA. Atypical infectious nodular scleritis. *Arch Ophthalmol* 2009;127:1079–80. doi:10.1001/archophthalmol.2009.197

Kheir WJ, Sheheitli H, Fattah MA, Hamam RN. Nontuberculous mycobacterial ocular infections: A systematic review of the literature. *Biomed Res Int.* 2015;2015:164989. doi: 10.1155/2015/164989

Kim JY, Carroll CP, Opremcak EM. Antibiotic-resistant tubercu-lous choroiditis. *Am J Ophthalmol.* 1993;115:259–261. doi:10.1016/s0002-9394(14)73938–8

Kirtschig G, Murrell D, Wojnarowska F, Khumalo N. Interventions for mucous membrane pemphigoid and epidermolysis bullosa acquisita. *Cochrane Database Syst Rev.* 2003:CD004056.

Knickelbein JE, Tucker WR, Bhatt N, Armbrust K, Valent D, Obiyor D, Nussenblatt RB, Sen HN. Gevokizumab in the treatment of Autoimmune non-necrotizing anterior Scleritis: results of a phase I/II clinical trial. *Am. J. Ophthalmol.* 2016;172:104–110.

Kolomeyer AM, Ragam A, Shah K, *et al.,* Mycophenolate mofetil in the treatment of chronic non-infectious, non-necrotizing scleritis. *Ocul Immunol Inflamm.* 2012;20:113–118.

Kumar A, Ghose A, Biswas J, Dutta Majumder P. Clinical profile of patients with posterior scleritis: a report from Eastern India. *Indian J Ophthalmol.* 2018;66:1109–1112.

Kumar Sahu S, Das S, Sharma S, Sahu K. Clinico-microbiological profile and treatment outcome of infectious scleritis: Experience from a tertiary eye care center of India. *Int J Inflam.* 2012;2012:753–760.

Lane J, Nyugen E, Morrison J, et al., Clinical features of scleritis across the Asia-Pacific region. *Ocul Immunol Inflamm.* 2019;27:920–926.

Lavric A, Gonzalez-Lopez JJ, Majumder PD, et al., Posterior scleritis: Analysis of epidemiology, clinical factors, and risk of recurrence in a cohort of 114 patients. *Ocul Immunol Inflamm.* 2016;24(1):6–15.

Lawuyi LE, Gurbaxani A. Refractory necrotizing scleritis successfully treated with adalimumab. *J Ophthalmic Inflamm Infect.* 2016;6:37.

Lee DH, Hyuna Cho2, Eun Young Choi2^, Min Kim2^. Clinical features and long-term treatment outcome of posterior scleritis. Ann Transl Med. 2022;10(21):1162 | https://dx.doi.org/10.21037/atm-22-721

Lee YJ, Kim SW, Seo KY. Application for tacrolimus ointment in treating refractory inflammatory ocular surface diseases. *Am J Ophthalmol.* 2013;155:804–813.

Levison AL, Lowder CY, Baynes KM, et al., Anterior segment spectral domain optical coherence tomography imaging of patients with anterior scleritis. *Int Ophthalmol.* 2016;36(4):499–508.

Li F, Ma X, Du L, Shi L, Cao Q, Li N, Pang T, Liu Y, Kijlstra A, Yang P. Identification of susceptibility SNPs in CTLA-4 and PTPN22 for scleritis in Han Chinese. *Clin. Exp. Immunol.* 2019;197:230–236.

Lin CP, Shih MH, Tsai MC. Clinical experiences of infectious scleral ulceration: A complication of pterygium operation. *Br J Ophthalmol.* 1997;81(11):980–983.

Lin P, Bhullar SS, Tessler HH et al. Immunologic markers as potential predictors of systemic autoimmune disease in patients with idiopathic scleritis. *Am J Ophthalmol.* 2008;145:463–471.

Livir-Rallatos C, El-Shabrawi Y, Zatirakis P, Pellett PE, Stamey FR, Foster CS. Recurrent nodular scleritis associated with varicella zoster virus. *Am J Ophthalmol.* 1998;126:594–7.

Lovieno A, Gore DM, Carnt N, Dart JK. Acanthamoeba sclero- keratitis: Epidemiology, clinical features, and treatment outcomes. *Ophthalmol.* 2014;121:2340–7.

Lyne AJ, Lloyd-Jones D. Necrotizing scleritis after ocular surgery. *Trans Ophthalmol Soc U K.* 1979;99(1):146–149.

Madanagopalan VG, Shivananda N, Krishnan T. Surgically induced necrotizing scleritis after retinal detachment surgery masquerading as scleral abscess. *GMS Ophthalmol Cases.* 2019;9:Doc 18.

Margo CE, Pavan PR. Mycobacterium chelonae conjunctivitis and scleritis following vitrectomy. *Arch Ophthalmol.* 2000;118(8):1125–1128.

McCarthy JM, Dubord PJ, Chalmers A, et al., Cyclosporine A for the treatment of necrotizing scleritis and corneal melting in patients with rheumatoid arthritis. *J Rheumatol.* 1992;19:1358–1361.

McCluskey P, Wakefield D. Intravenous pulse methylprednisolone in scleritis. *Arch Ophth almol.*1987;105:793–7.

McCluskey PJ, Wakefield D. Prediction of response to treatment in patients with scleritis using a standardized scoring system. *Aust N Z J Ophthalmol.* 1991;19:211–5.

McCluskey PJ, Wakefield D (1996) Scleritis and Episcleritis. In: Pepose JS, Holland GN, Wilhelms KR (eds) *Ocular inflammation and immunity*. Mosby, St Louis; 642–647.

McCluskey PJ, Watson PG, Lightman S, Haybittle J, Restori M, Branley M. Posterior scleritis: clinical features, systemic associations, and outcome in a large series of patients. *Ophthalmol.* 1999;106:2380–2386.

McGavin DD, Williamson J, Forrester JV, et al., Episcleritis and scleritis. A study of their clinical manifestations and association with rheumatoid arthritis. *Br J Ophthalmol.* 1976;60(3):192–226.

Meyer J, Espandar L, Marx DP, Vitale A, Moshirfar M. Pseudomonas Scleritis resistant to fourth-generation Fluoroquinolones in a patient without prior trauma or surgery. *Ocul. Immunol. Inflamm.* 2008;16:3:127–129. DOI: 10.1080/09273940801947662

Miyazaki D, Tominaga T, Kakimaru-Hasegawa A, et al., Therapeutic effects of tacrolimus ointment for refractory ocular surface inflammatory diseases. *Ophthalmol.* 2008;115:988–992.

Moimura Y, Okada AA, Kawahara S, et al., Tuberculin skin testing in uveitis patients and treatment of presumed intra-ocular tuberculosis in Japan. *Ophthalmol.* 2002;109(5):851 857. doi:10.1016/S0161-6420(02)00973-9

Moorthy RS, Valluri S, Rao NA. Nontuberculous mycobacterial ocular infection. *Surv Ophthalmol.* 2012;57:202–35.

Moreno CB, João Carlos Gonçalves Cruz, Idiberto Jose Zotarelli-Filho, Maurélio de Lima Batista Ribeiro Junior, Marcello Novoa Colombo Barboza, Guilherme Novoa Colombo Barboza, Marta FabianeGouvêaBarioni. Necrotising scleritis and rheumatoid arthritis: A clinical case report supported by a brief review with risk of bias analysis. *Curr.* Rheumatol Rev. 2023 Februrary 23.

Murthy SI, Reddy JC, Sharma S, Sangwan VS. Infectious Scleritis. *CurrOphthalmol Rep.* 2015;3:147-57.

Murthy SI, Sabhapandit S, Balamurugan S, Subramaniam P, Sainz-de-la-Maza M, Agarwal M, et al., Scleritis: Differentiating infectious from non-infectious entities. *Indian J Ophthalmol.* 2020;68:1818–28.

Murthy SI, Sati A, Sangwan VS. Infectious scleritis mimicking severe ocular inflammation: Atypical initial presentation. *BMJ Case Rep.* 2013;2013:bcr2013008686. doi: 10.1136/bcr-2013-008686

Murugan SB. Commentary: Tubercular posterior scleritis-Nuances in practice! *Indian J Ophthalmol.* 2019;67:1366-7.

Naik PM, Lyon GM 3rd, Ramirez A, et al., Dapsone-induced hemolyticanemia in lung allograft recipients. *J Heart Lung Transplant.* 2008;27:1198–202.

Ng C CAileen Sy1,3 Emmett T. Cunningham Jr1,2,4,5 Rituximab for non-infectious Uveitis and Scleritis. *J Ophth Inflamm Infect.* 2021;11:23 https://doi.org/10.1186/s12348-021-00252-4

Okhravi N, Odufuwa B, McCluskey P et al., Scleritis. *SurvOphthalmol.* 2005;50:351–363.

Orr C, Vieira-Sousa E, Boyle DL, Buch MH, Buckley CD, Canete JD, Catrina AI, Choy EHS, Emery P, Fearon U, Filer A, Gerlag D, Humby F, Isaacs JD, Just SA, Lauwerys BR, Le Goff B, Manzo A, McGarry T, McInnes IB, Najm A, Pitzalis C, Pratt A, Smith M, Tak PP, Thurlings R, Fonseca JE, Veale DJ, Tas SW. Synovial tissue research: A state-of-the-art review. *Nat. Rev. Rheumatol.* 2017;13:463–475.

Paley MA, Karacal H, Rao PK, Margolis TP, Miner JJ. Tofacitinib for refractory uveitis and scleritis. *Am J Ophthalmol Case Rep.* 2018;13:53–5.

Palkar A, Sudharshan S, George AE, Ganesh SK, Biswas J, Dutta Majumder P. Cataract surgery in the setting of scleritis. *Ocul Immunol Inflamm.* 2020:1–4.

Pan Y, Yu-Xi Chen, Yao Lu, Yan-Yan Xie, Yi-Wen Xia, Dan Liang. Outcomes of escalating immunosuppressive treatments for recalcitrant *Noninfectious Posterior Scleritis Ophthalmol Ther.* 2022. https://doi.org/10.1007/s40123-022-00577-w

Patel SS, Saraiya NV, Tessler HH, Goldstein DA. Mycobacterial ocular inflammation: Delay in diagnosis and other factors impacting morbidity. *JAMA Ophthalmol.* 2013;131(6):752–758. doi:10.1001/jamaophthalmol.2013.71

Patnaik G, Sudharshan S, George AE, Ganesh SK,Biswas J, Dutta Majumder P. Clinical profile of patients with anterior nodularscleritis in India. *Indian J Ophthalmol.* 2020;68:1925–8.

Paul L, Agarwal M, Singh S, Katre P, Arora A S. Tubercular scleritis in a young Asian Indian girl-a case presentation and literature review. J *Ophthalmic Inflamm Infect.* 2019;9(1):22. doi:10.1186/s12348-019-0192-9

Pavlin C, Easterbrook M, Hurwitz J, et al., Ultrasound biomicroscopy in the assessment of anterior scleral disease. *Am J Ophthalmol.* 1993;116(5):628–35.

Pikkel J, Chassid O, Srour W, et al., Is episcleritis associated to glaucoma? *J Glaucoma.* 2015;24(9):669–71.

Paula JS, Simão ML, Rocha EM, Romão E, Velasco Cruz AA. Atypical pneumococcal scleritis after pterygium excision: Case report and literature review. *Cornea.* 2006 Jan;25(1):115–7.

Pavesio C, ed. *Scleritis.* Cham: Springer International Publishing; 2017.

Pavesio CE, Meier FM . Systemic disorders associated with episcleritis and scleritis. *CurrOpinOphthalmol.* 2001;12:471–478.

Piper JM, Ray WA, Daugherty JR, Griffin MR. Corticosteroid use and peptic ulcer disease: Role of nonsteroidal anti-inflammatory drugs. *Ann Intern Med.* 1991;114 :735–40.

Poelman HJ, Van Daele PLA , Rothova A. Successful tocilizumab treatment for scleritis. *Ocul Immunol Inflamm.* 2020; 28:285–287.

Pradhan Z, Jacob P. Infectious scleritis: Clinical spectrum and management outcomes in India. *Indian J Ophthalmol.* 2013;61:590-3

Promelle V, Vincent Goeb, Julie Gueudry. Rheumatoid Arthritis associated Episcleritis and Scleritis: An Update on Treatment Perspectives *J. Clin. Med.* 2021;10(10), 2118; https://doi.org/10.3390/jcm10102118

Pujari SS, Kempen JH, Newcomb CW, et al., Cyclophosphamide for ocular inflammatory diseases. *Ophthalmol.*. 2010;117:356–65.

Pyare R, Kaushik V, Dutta Majumder P, Biswas J. Tofacitinib in recalcitrant scleritis: First case report from India. *Indian J Ophthalmol.* 2020;68(9): 1988–90. https://doi.org/10.4103/ijo.IJO_534_20

Radis CD, Kahl LE, Baker GL, et al., Effects of cyclophosphamide on the development of malignancy and on long-term survival of patients with rheumatoid arthritis. A 20-year followup study. *Arthritis Rheum.* 1995; 38:1120–7.

Ragam A, Kolomeyer AM, Fang C, et al., Treatment of chronic, noninfectious, nonnecrotizing scleritis with tumor necrosis factor alpha inhibitors. *Ocul Immunol Inflamm.* 2014;22:469–477.

Raizman MB, Rubin JM, Graves AL, Rinehart M. Tear concentrations of levofloxacin following topical administration of a single dose of 0.5% levofloxacin ophthalmic solution in healthy volunteers. *Clin Ther.* 2002;24(9):1439–1450. doi: 10.1016/S0149-2918(02)80047-9

Ram R. Tectonic corneal lamellar grafting for surgically induced necrotizing scleritis after strabismus surgery: case report & literature review. *Am J Ophthalmol Case Rep.* 2018;11:28–31.

Ramakrishnan R, Bharathi M, Shivkumar C, Mittal S, Meenakshi R, Khadeer M, Avasthi A. Microbiological profile of culture-proven cases of exogenous and endogenous endophthalmitis: A 10-year retrospective study. *Eye.* 2009;23(4):945–956. doi: 10.1038/eye.2008.197

Ramenaden ER, Raiji VR. Clinical characteristics and visual outcomes in infectious scleritis: A review. *Clin Ophthalmol.* 2013;7:2113-22.

Restrepo JP, Molina MP. Successful treatment of severe nodular scleritis with adalimumab. *Clin Rheumatol.* 2010;29:559–561.

Reddy JC, Murthy SI, Reddy AK, Garg P. Risk factors and clinical outcomes of bacterial and fungal scleritis at a tertiary eye care hospital. *Middle East Afr J Ophthalmol.* 2015;22:203–11.

Richelle L, Guerrero-Wooley1, Peacock JE Jr. 2018. Infectious Scleritis: What the ID Clinician Should Know OFID 1-4. DOI: 10.1093/ofid/ofy140

Rose BJ, Kooyman DL. 2016. A tale of two joints: The role of matrix metalloproteases in cartilage biology. *Dis. Markers.* 2016, 4895050.

Roufas A, Jalaludin B, Gaskin C, McCluskey P. Subconjunctival triamcinolone treatment for nonnecrotizing anterior scleritis. *Br J Ophthalmol.* 2010;94:743–747.

Ryu SJ, Kang MH, Seong M, Cho H, Shin YU. Anterior scleritis following intravitreal injections in a patient with rheumatoid arthritis: a case report. *Medicine (Baltimore).* 2017;96:e8925.

Sahu SK, Das S, Sahani D, Sharma S. Fungal scleritis masquerading as surgically induced necrotizing scleritis: A case report. *J Med Case Reports.* 2013;7:288.

Sainz de la Maza M, Foster CS, Jabbur NS. Scleritis associated with systemic vasculitic diseases. *Ophthalmol.* 1995;102:687-92.

Sainz de la Maza M, Jabbur NS, Foster CS. Severity of scleritis and episcleritis. *Ophthalmol.* 1994;101:389–96.

Sainz de la Maza M, Molina N, Gonzalez-Gonzalez LA, Doctor PP, Tauber J, Foster CS. Clinical characteristics of a large cohort of patients with scleritis and episcleritis. *Ophthalmol.* 2012a;119:43–50.

Sainz de la Maza M, Tauber J, Foster CS. *The Sclera.* New York: Springer-Verlag; 2012b.

Sainz de la Maza MS, Molina N, Gonzalez- Gonzalez LA, et al., Scleritis therapy. *Ophthalmol.* 2012c;119:51–58.

Sainz de la Maza M, Molina N, Gonzalez-Gonzalez LA, Doctor PP, Tauber J, Foster CS. Scleritis associated with relapsing polychondritis. *Br J Ophthalmol.* 2016;100:1290–1294.

Sands DS, Chan SC, Gottlieb CC. Methotrexate for the treatment of noninfectious scleritis. *Can J Ophthalmol.* 2018;53:349–353.

Sen HN, Sangave AA, Goldstein DA, Suhler EB, Cunningham D, Vitale S, et al., A standardized grading system for scleritis. *Ophthalmol.* 2011;118:768–71.

Sen HN, Suhler EB, Al-Khatib SQ, *et al.,* Mycophenolate mofetil for the treatment of scleritis. *Ophthalmol.* 2003;110:1750–1755.

Sen HN, Ursea R, Nussenblatt RB, Buggage RR. Subconjunctival corticosteroid injection for the treatment of non-necrotising anterior scleritis. *Br J Ophthalmol.* 2005;89:917–8.

Seo KY, Lee HK, Kim EK, Lee JH. Expression of tumor necrosis factor alpha and matrix metalloproteinase-9 in surgically induced necrotizing scleritis. *Ophthalmic Res.* 2006;38, 66–70.

Sharma K, Bansal R, Sharma A, Gupta A, Fiorella PD. Successful treatmen to frifampicin-resistant intra ocular tuberculosis. *Ocul Immunol Inflamm.* 2015;23(1):93–96. doi:10.3109/09273948.2014.888084

Sharma K, Gupta A, Sharma M, et al., The emerging challenge ofdiagnosing drug-resistant tubercular uveitis:Experience of 110 eyes from north India. *Ocul Immunol Inflamm.* 2019;3:1–8.doi:10.1080/09273948.2019.1655581

Sharma K, Sharma A, Bansal R, Fiorella PD, Gupta A. Drug resistant tubercular uveitis. *Clin Microbiol.* 2014;52(11):4113–4114. doi:10.1128/JCM.01918-14

Sharma SM, Damato E, Hinchcliffe AE, et al., Long-term efficacy and tolerability of TNFα inhibitors in the treatment of non-infectious ocular inflammation: An 8-year prospective surveillance study. *Br J Ophthalmol.* 2021;105:1256–1262.

Shields CL, Pellegrini M, Ferenczy SR, Shields JA. Enhanced depth imaging optical coherence tomography of intraocular tumors: From placid to seasick to rock and rolling topography-- The 2013 Francesco Orzalesi Lecture. *Retina.* 2014;34:1495-512.

Shoughy SS, Jaroudi MO, Tabbara KF. Clinical manifestations and outcome of tuberculous sclera keratitis. *Br J Ophthalmol.* 2016;100(9):1301–1303. doi:10.1136/bjophthalmol-2015-307599

Sims J. Scleritis: Presentations, disease associations and management. *Postgrad Med J.* 2012;88(1046):713–8.

Smith JR, Levinson RD, Holland GN, et al., Differential efficacy of tumor necrosis factor inhibition in the management of inflammatory eye disease and associated rheumatic disease. *Arthritis & Rheumatism.* 2001; 45:252–257.

Smith JR, Mackensen F, Rosenbaum JT. Therapy insight: scleritis and its relationship to systemic autoimmune disease. *Nat Clin PractRheumatol.* 2007;3:219–226.

Sobrin L, Christen W, Foster CS. Mycophenolate mofetil after methotrexate failure or intolerance in the treatment of scleritis and uveitis. *Ophthalmol.* 2008;115:1416–1421.

Sohn EH, Wang R, Read R, et al., Long-term, multicenter evaluation of subconjunctival injection of triamcinolone for non-necrotizing, noninfectious anterior scleritis. *Ophthalmol.* 2011;118:1932–1937.

Solomon R, Donnenfeld ED, Perry HD, Snyder RW, Nedrud C, Stein J, Bloom A. Penetration of topically applied gatifloxacin 0.3%, moxifloxacin 0.5%, and ciprofloxacin 0.3% into the aqueous humor. *Ophthalmol.* 2005;112(3):466–469. doi: 10.1016/j.ophtha.2004.09.029.

Som PM, Curtin HD. *Head and Neck Imaging.* 5th ed. Boston:Mosby; 2011.

Sota J, Girolamo, M M., Tosi, B. F., G., M., Cantarini, L., Fabiani, C. Biologic Therapies and Small Molecules for the Management of Non-Infectious Scleritis: *A Narrative Review OphthalmolTher.* 2021;10:777–813 https://doi.org/10.1007/s40123-021-00393-8

Stem MS, Todorich B, Faia LJ. Ocular Pharmacology for Scleritis: Review of treatment and a practical perspective.. J Ocul Pharmacol Ther May 2017:240–246. http://doi.org/10.1089/jop.2016.0127

Taban M, Dupps WJ, Mandell B, Perez VL. Etanercept (enbrel)-associated inflammatory eye disease: Case report and review of the literature. *Ocul Immunol Inflamm.* 2006;14:145–50.

Tanaka R, Kaburaki T, Ohtomo K, et al., Clinical characteristics and ocular complications of patients with scleritis in Japanese. *Jpn J Ophthalmol.* 2018;62:517–524.

Tenover FC. Development and spread of bacterial resistance to antimicrobial agents: An overview. *Clin Infect Dis.* 2001;33(Supplement 3):S108–S115. doi: 10.1086/321834

Todokoro D, Hoshino J, Yo A, Makimura K, Hirato J, Akiyama H. Scedosporiumapiospermum infectious scleritis following posterior subtenon triamcinolone acetonide injection: A case report and literature review. *BMC Ophthalmol.* 2018;18:40.

Turgeon PW, Slamovits TL. Scleritis as the presenting manifestation of procainamide-induced lupus. *Ophthalmol.* 1989;96:68–71.

Ugurbas SH, Alpay A, Ugurbas SC. Posterior scleritis presenting with angle closure glaucoma. *Ocul Immunol Inflamm.* 2012; 20:218–220.

van der Maesen K, D'Haens EJ, van Rijn LJ, Pigeaud-Klessens MLE, Meenken C. Posterior scleritis in a 7-month-old infant. *J Pediatr Ophthalmol Strabismus.* 2007; 44:377–378.

Veenis A, Haghnegahdar M, Ajlan R. Unilateral Episcleritis Following COVID-19 Booster Vaccination of a Crohn's Disease Patient: A Case Report and Review of the Literature, *Int. Med. Case Rep. J.* 2023:91–96. DOI: 10.2147/ IMCRJ.S398502

Vergouwen DPC, Rothova A, Berge JCT, et al., Current insights in the pathogenesis of scleritis. *Exp Eye Res.* 2020;197:108078.

Wakefield D, Girolamo ND, Thurau S, Wildner G, McCluskey P. Scleritis: Immunopathogenesis and molecular basis for therapy. *Prog Retin Eye Res.* 2013;35:44–62.

Wakefield D, McCluskey P. Cyclosporin therapy for severe scleritis. *Br J Ophthalmol.* 1989; 73:743–746.

Watson PG. The diagnosis and management of scleritis. *Ophthalmol.* 1980;87:716–20.

Watson PG, Hayreh SS. Scleritis and Episcleritis. *Br J Ophthalmol.* 1976;60:163–191. doi:10.1136/bjo.60.3.163

Watson PG, Young RD. Scleral structure, organization and disease. A review. *Exp Eye Res* 2004;78:609–623.

Watson P, Romano A, 2014. The impact of new methods of investigation and treatment on the understanding of the pathology of scleral inflammation. *Eye*. 28:915–93.

Williams CPR, Browning AC, Sleep TJ, et al., A randomised, double-blind trial of topical ketorolac vs artificial tears for the treatment of episcleritis. *Eye (Lond).* 2005;19(7):739–42.

Williamson JODA. Red Eye: Scleritis or Episcleritis? Differentiating between the two is crucial to ensure you initiate the right treatment. *Review of Optometry*. November 15, 2017.

Wong MH, Su DH, Loh RS. Nodular scleritis and Sweet's syndrome. *Clin Experiment Ophthalmol*. 2007;35:858–860.

Wroblewski KJ, Hidayat AA, Neafie RC, Rao NA, Zapor M. Oculartuberculosis: A clinicopathologic and molecular study. *Ophthalmol*. 2011;118(4):772–777. doi:10.1016/j.ophtha.2010.08.011

Yadav S, Rawal G. Tubercular nodular episcleritis: A case report. *J Clin Diagn Res*. 2015;9:1–2.

Yalçindağ FN, Amer R, Forrester JV. Mycophenolate mofetil in the treatment of ocular inflammation in ANCA-associated vasculitis. *J Ocul Pharmacol Ther*. 2008;24:249–254.

Yang P, Ye Z, Tang J, et al., Clinical features and complications of scleritis in Chinese patients. *Ocul Immunol Inflamm*. 2018;26:387–396.

Yanoff M, Duker JS. *Ophthalmology*. Boston: Mosby; 2014.

Yazu H, Miyazaki D, Fujishima H. Experience with 0.1% tacrolimus eye drop for non-infectious, nonnecrotizing anterior scleritis. *Eye Contact Lens*. 2021;7(4):185–190. doi: 10.1097/ICL.0000000000000696. PMID: 32404650.

Yoo JH, Chodosh J, Dana R. Relapsing polychondritis: Systemic and ocular manifestations, differential diagnosis, management, and prognosis. *Semin Ophthalmol*. 2011;26:261–269.

Young TL, Scavello GS, Paluru PC, Choi JD, Rappaport EF, Rada JA, 2004. Microarray analysis of gene expression in human donor sclera. *Mol. Vis.* 10:163–176.

Young AL, Wong SM, Leung AT, Leung GY, Cheng LL, Lam DS. Successful treatment of surgically induced necrotizing scleritis with tacrolimus. *Clin Exp Ophthalmol*. 2005 Feb;33(1):98–9. doi: 10.1111/j.1442-9071.2005.00955.x. PMID: 15670089

Zamir E, Read RW, Smith RE, Wang RC, Rao NA. A prospective evaluation of subconjunctival injection of triamcinolone acetonide for resistant anterior scleritis. *Ophthalmol*. 2002;109:798–805.

11

Antimicrobial Resistance in the Diseased Human Eye: Uveitis

Shalem Raj Padakandla

11.1 Introduction

Uvea (derived from the Latin word for "grape") is the middle portion of the eye and comprises the iris, ciliary body, and choroid (Figure 11.1 A). Uveitis or intraocular inflammation involves the inflammation of the uveal tract or the adjacent intraocular structures including the retina, retinal blood vessels, optic nerve, vitreous, and sclera (Nussenblatt, 1990; Jabs et al., 2005; Maruyama et al., 2017; Joltikov and Lobo-Chan, 2021). It is the leading cause of inflammatory eye disease, and if left untreated could result in visual impairment due to retinal swelling (macular edema), retina scarring, glaucoma, cataracts, optic nerve damage, retinal detachment and permanent vision loss in over one-third of the patients (Wakefield and Chang, 2005; Rothova et al., 1996). Uveitis can occur in all age groups, often affecting young male adults with a mean age group of 35 to 45 years (Rothova et al., 1992; Tsirouki et al., 2018) and accounts for 10–15% of all cases of blindness (Durrani et al., 2004; Rosenbaum et al., 2019). With 14–17 per 100,000 cases of average annual incidence, the frequencies vary considerably by geographic location around the world, and it has been estimated roughly that the incidence is around 200 per 100,000 in the United States, 38 per 100,000 in France, and 730 per 100,000 in India (Rathinam and Namperumalsamy, 2007; Chang and Wakefield, 2002).

11.2 Classification of Uveitis

Various concurrent classification systems have described uveitis based on different characteristics. Among them, the most widely used was devised by the International Uveitis Study Group (IUSG) in 1987 and approved by the Standardization of Uveitis Nomenclature (SUN) Working Group in 2005 (Bloch-Michel and Nussenblatt, 1987; Deschenes et al., 2008). This classification is based on the anatomical location of the inflammation or the part of the uvea that is primarily affected and includes anterior uveitis (inflammation in the Iris, ciliary body, or both), intermediate uveitis (inflammation in vitreous humor or the pars plana), posterior uveitis (inflammation in choroid, retina, or both), and panuveitis (inflammation in anterior chamber, vitreous, retina, and choroid) (Table 11.1) (Figure 11.1 B, C, D). In addition, uveitis is also

DOI: 10.1201/9781003451105-11

TABLE 11.1
Classification of Uveitis Based on Primary Site of Inflammation as Categorised by SUN[a]

Type	Site of Inflammation[b]	Includes
Anterior uveitis	Anterior chamber (iris, ciliary body, or both)	Iritis, iridocyclitis, and anterior cyclitis.
Intermediate uveitis	Vitreous (vitreous humor or the pars plana)	Pars planitis, posterior cyclitis and hyalitis.
Posterior uveitis	Retina or choroid or both	Focal, multifocal, or diffuse choroiditis, chorioretinitis, retinochoroiditis, retinitis and neuroretinitis.
Panuveitis	Anterior chamber, vitreous, and retina or choroid	

Source: Data from Cunningham, 2000; London et al., 2010.

Notes:
[a] SUN = Standardization of uveitis nomenclature.
[b] Identified clinically and according to International Uveitis Study Group anatomic classification (Bloch-Michel and Nussenblatt, 1987).

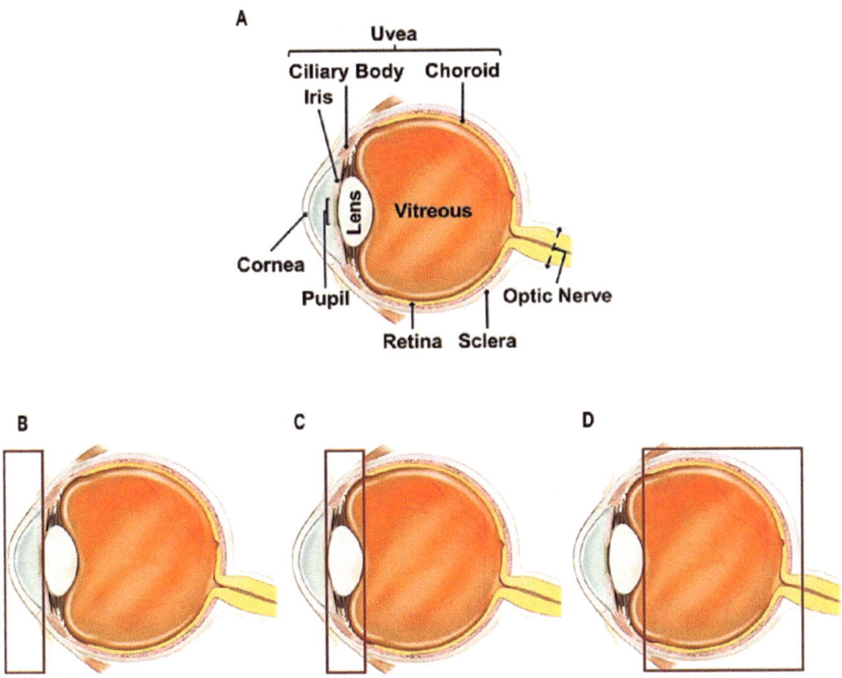

FIGURE 11.1 Anatomy of the healthy human eye (A) and the sites of inflammation (red box) in anterior uveitis (B), intermediate uveitis (C) and posterior uveitis (D). (Modified from www.nei.nih.gov/learn-about-eye-health/eye-conditions-and-diseases/uveitis.)

classified into acute or chronic, based on the duration of symptoms; granulomatous or non-granulomatous, according to histological features of inflammation; unilateral or bilateral, depending on the laterality of the condition and finally into infectious, autoimmune, drug-induced, or idiopathic with respect to the general underlying systemic causes of inflammation (Deschenes et al., 2008; Jabs et al., 2005; McCluskey et al., 2000; Gutteridge and Hall 2007; Guly and Forrester, 2010).

11.3 Etiology, Diagnosis, and Treatment of Uveitis

World-wide, uveitis is characterized by several variations, such as etiology, clinical presentations, prevalence and treatment outcomes that are usually influenced by geographic, racial and genetic conditions, socio-economic factors, prevalence of infections, and the availability of diagnosis and treatment facilities (de Smet et al., 2011) (Table 11.2). Infectious etiologies account for about 15–20% of all cases of uveitis in the United States. Significantly more proportion is observed world-wide, with at least 50% of total cases in developing countries (Acharya et al., 2013; Suhler et al., 2008, London et al., 2010; Forrester et al., 2018). In developing countries, infections that cause uveitis can range from tuberculosis (TB), toxoplasmosis, and herpesviruses which are quite common, to rare ones such as uveitis induced by trematode, dengue, chikungunya, onchocerciasis, cysticercosis, and trypanosomiasis and some endemic infections such as leprosy or leptospirosis which are almost unique to Asia-Pacific. Whereas, toxoplasmosis and viral infections are the most common ones in developed countries (Rathinam and Cunningham, 2000; London et al., 2010). Infectious agents can affect various anatomic locations ranging from anterior to posterior regions of the eye (Table 11.2). Infectious uveitis involving the posterior regions occurs usually after ocular trauma or surgery, or due to a systemic infection, or as a result of activation of a latent infection. Though uncommon, infectious posterior uveitis such as acute retinal necrosis caused by herpes family viruses, or toxoplasmosis retinochoroiditis, are usually sight-threatening (Mandelcorn, 2013) (Table 11.2). A variety of infectious agents can cause uveitis with identical clinical presentations, or single infectious agent resulting in multiple clinical presentations, and also are often similar to immune-mediated uveitis. Hence, an intimate knowledge of various types of infectious uveitis and identification of the causative agents through appropriate diagnostics is crucial in early identification and management of these conditions for visual recovery and sight preservation (Lin, 2015, Patel, 2021).

11.3.1 Tuberculous Uveitis

11.3.1.1 Etiology

Tuberculosis (TB) is caused by *Mycobacterium tuberculosis*, which following hematogenous dissemination may cause tuberculous seeding of other organs of the body, including the gastrointestinal system, cardiovascular system, genitourinary tract, central nervous system, musculoskeletal system, skin, and even the eye (Centers for Disease Control, 1997; Glassroth et al., 1980; Baydur, 1977; Gupta et al., 2007). Ocular TB or Tuberculous uveitis is an uncommon extrapulmonary

TABLE 11.2
Causes and Symptoms of Different Types of Uveitis

Type	Causes	Symptoms
Anterior uveitis	Idiopathic or postsurgical (most common causes), juvenile idiopathic arthritis, trauma, spondyloarthropathies, Herpes virus infection (HSV), varicella-zoster virus infection (VZV), and cytomegalovirus infection, tubulointerstitial nephritis and uveitis (TINU).	Acute cases are most symptomatic, usually manifesting with pain (ocular ache), redness, photophobia, and variable degree of decreased vision. Chronic cases have less dramatic symptoms and present with irritation or decreased vision.
Intermediate uveitis	Idiopathic (most common), multiple sclerosis, Tuberculosis (TB), Sarcoidosis, Syphilis and Lyme disease (in endemic regions).	Typically painless and manifests with floaters and decreased vision and macular edema. Primarily indicated by inflammatory cells in the vitreous humor, which often aggregate and condense as "snowball"-like appearance. "Snowbank"-like appearance caused by confluent and condensed vitreous cells and snowballs over the pars plana, usually associated with neovascularization of the retinal periphery.
Posterior uveitis	Idiopathic (most common), Toxoplasmosis, CMV (in immunocompromised patients), HSV/ VZV, Sarcoidosis, Syphilis and TB.	Diverse symptoms with floaters and decreased vision being the most common ones. Appearance of inflammatory cells in vitreous humor. White or yellow-white lesions in the retina (retinitis), underlying choroid (choroiditis), or both. Retinal vasculitis (with whitening or obliteration of blood vessels) and optic disk edema
Panuveitis	Idiopathic (most common), Sarcoidosis, TB, Vogt-Koynagi-Harada syndrome (VKH) and Syphilis.	Any combination of the previously mentioned symptoms and signs.

Source: Modified from www.msdmanuals.com/en-in/professional/eye-disorders/uveitis-and-related-disorders/overview-of-uveitis.

manifestation caused either through direct infection or indirect immune-mediated hypersensitivity response to mycobacterial antigens, usually with no defined active systemic lesion elsewhere or an inactive lesion (Rosen et al., 1990; Helm and Holland, 1993; Biswas et al., 1995; Sheu et al., 2001, Sakai et al., 2001). Typical

TABLE 11.3

Clinical Presentations in Tuberculous Uveitis

Anterior uveitis	Granulomatous, nongranulomatous, iris nodules and ciliary body tuberculoma.
Intermediate uveitis	Granulomatous, nongranulomatous with organizing exudates in the pars lana/peripheral uvea.
Posterior and panuveitis	Choroidal tubercle, choroidal tuberculoma, subretinal abscess and Serpiginous-like choroiditis

*Modified from Gupta et al. (2007).

symptoms include light sensitivity and blurry vision, along with redness of the eye, headache, floaters, or flashes. Patients may also be asymptomatic. Prevalence of tuberculous uveitis is highly variable in different regions of the world, ranging from 0.2–2.7% to 5.6–10.5% of the uveitis patients (Gupta et al., 2007; Ang and Chee, 2017; Lee et al., 2016; Abu El-Asrar et al., 2010). Tuberculous uveitis is considered a great mimicker of other uveitis entities and may affect all anatomical locations of the eye with posterior uveitis being the most common (35%–42%) clinical presentation, followed by cases of anterior uveitis (12%–36%), pan-uveitis (11–20%) and intermediate uveitis (11%) (Dalvin and Smith, 2017; Khochtali et al., 2015) (Table 11.3).

11.3.1.2 Diagnosis

Earlier, any granulomatous inflammation in the eye was considered to be of tuberculous origin (Guyton and Woods, 1941). Gupta et al. (2007), proposed that the diagnosis of tuberculous uveitis should be based on the exclusion of other known causes of uveitis, suggestive clinical signs, history and investigations like positive tuberculin skin test (TST), Quantiferon-Gold test, and chest X-ray to support the infection, response to standard antituberculosis treatment, and the presence of *M. tuberculosis* as evidenced by microscopic observation or a positive PCR reaction in the ocular fluids or tissue. Among these, TST is relatively nonspecific and can be influenced by atypical mycobacterial infection or by BCG vaccination. In contrast, the Quantiferon-Gold test, an interferon-gamma release assay (IGRA), can respond specifically to *M. tuberculosis* antigens and strongly indicate exposure to the bacterium. However, neither of these methods can distinguish an active disease from a latent tuberculous infection (Dinnes et al., 2007; Albini et al., 2008; Pai et al., 2008; Kurup et al., 2006). In addition, a low number of bacterial cells in the intraocular fluids and hosting of the bacterium primarily in the retinal pigment epithelium often leads to a negative microbiological or molecular assay for tuberculous uveitis (Alvarez and McCabe, 1984). With the involvement of mixed ocular tissue, variable clinical manifestations, and lack of international consensus on the investigations, and so forth, finally, a presumed diagnosis of tuberculous uveitis can only be made from the combination of various diagnostic assays (Lee et al., 2016; Ang and Chee, 2017; Gupta et al., 2007; Gupta et al., 2015; Rosenbaum, 2014; Kee et al., 2016).

11.3.1.3 Treatment

Ocular therapy alone cannot cure tuberculous uveitis and may even mask the disease symptoms. Hence the treatment primarily targets the underlying infection, be it a pulmonary or an extrapulmonary manifestation, and includes a four-drug therapy (Rifampicin, Isoniazid, Pyrazinamide, and Ethambutol) for 2 months, followed by an additional four-month course of isoniazid and rifampin. In conjugation, corticosteroids are often used to treat ocular manifestations such as the inflammatory consequences of the infection, and prevent ocular sequelae due to chronic uveitis (Lin 2015). In contrast, usage of anti-tuberculous drugs like Ethambutol (for 3–6 months) may also cause ocular complications such as dose-related optic neuritis, central scotomas, color vision deficiencies, and even retinal edema and foveal pigmentary changes (Koul 2015).

11.4 AMR in Uveitis

11.4.1 Drug Resistance of *M. tuberculosis* in Tuberculosis

Drug resistance in *M. tuberculosis* has been either intrinsic or acquired. The unusual structure of mycolic-acid containing cell wall that is less permeable to several antibiotics and chemotherapeutic agents and the efflux mechanisms are responsible for conferring intrinsic drug resistance (Jarlier and Nikaido, 1994; De Rossi et al., 2006). Unlike other bacteria where acquired drug resistance happens through mobile genetic elements mediating horizontal gene transfer, *M. tuberculosis* acquires resistance to one or more drugs due to spontaneous mutations like SNPs, insertions or deletions (indels) or sometimes large deletions in the genes that code for drug targets or drug-activating enzymes (Table 11.4) (Kochi et al., 1993; Gillespie 2002).

Streptomycin, the first antituberculosis agent was reported to cure a patient in 1944 (Daniel, 1997) and many cases of successful treatment soon followed (Hinshaw et al., 1946; Jellinek 1949). However, a substantial portion from a randomized clinical trial by British Medical Research Council in 1948 showed a relapse and the mycobacterial isolates cultured from these patients exhibited resistance to streptomycin (Crofton and Mitchison, 1948). Introduction of para-aminosalicylic acid and thiacetazone, two new antituberculosis agents in the same year resulted in the decline of acquired antibiotic resistance when they are used in combination with streptomycin (Medical Research Council., 1950). Later the testing of isonicotinic acid hydrazide (isoniazid) in 1951 dramatically improved the clinical outcomes of the treatment (Selikoff et al., 1952), and this was followed by the development of pyrazinamide (1952), cycloserine (1952), ethionamide (1956), rifampin (1957), and ethambutol (1962). Initially, the combination treatments included an 18-month regimen, but in 1957 the introduction and use of rifampin/rifampicin, a powerful anti-TB drug revolutionized the treatment of tuberculosis due to its high levels of efficiency (Daniel 1969; Vall-Spinosa et al., 1970; Cegielski, 2010). This led to the development of more effective "short-course chemotherapy" consisting of a combination of isoniazid- and rifampicin regimens (Seung et al., 2015). But the arrival of every new drug eventually led to resistance

TABLE 11.4

Genes in *M. tuberculosis* Responsible for Resistance to Classical, New and Repurposed Anti-TB Drugs

Drug	Mode of Action	Genetic mutation
Isoniazid	Cell wall mycolic acid synthesis inhibition	*kat*G, *inh*A, *kas*A
Rifampicin	RNA synthesis inhibition	*rpo*B
Ethambutol	Cell wall arabinogalactan biosynthesis inhibition	*emb*B, *ubi*A
Pyrazinamide	Membrane energy reduction, trans-translation inhibition and pantothenate and coenzyme A synthesis inhibition	*pnc*A, *rps*A, *pan*D
Streptomycin	Protein synthesis inhibition	*rps*L, *rrsV*, *gid*B
Fluoroquinolones	DNA synthesis inhibition	*gyr*A, *gyr*B
Capreomycin,	Inhibition of protein synthesis	*rrs*, *eis*, *tly*A
Ethionamide	Cell wall mycolic acid synthesis inhibition	*eth*A, *msh*A, *ndh*, *inh*A, *inh*A promoter
Para-aminosalicylic acid	Folic acid and thymine nucleotide metabolism inhibition	*thy*A, *fol*C, *rib*D
Bedaquiline	Mycobacterial ATP synthase inhibition	*rv*0678, *atp*E, *pep*Q
Clofazimine	Preferentially binds to mycobacterial DNA and inhibits growth. May also bind to potassium transporters and inhibit their function.	*rv*0678, *rv*1979c, *rv*2535c, *ndh*, *pep*Q
Delamanid/ pretonamid	Specific and selective inhibition of cell wall mycolic acid biosynthesis	*fgd*1, *fbi*A, *fbi*B, *fbi*C, *ddn*
Linezolid	Protein synthesis inhibition	*rpl*C, *rrl*

Source: Modified from Dookie et al., 2018.

in the bacterium through selection of mutations and rifampin was one among them (Manten and Van Wijngaarden, 1969). During monotherapy, rapid onset of isoniazid resistance was also observed whereas a combination of para-aminosalicylic acid or streptomycin suppressed the resistance. Based on these observations, British Medical Research Council designed a multidrug treatment that consists of a four-drug regimen for newly diagnosed tuberculosis, with the combination of isoniazid and rifampin as the backbone (Fox et al., 1999). However, a much longer treatment duration of 18–24 months and a combination of 8–10 drugs is required for multiple-drug resistant (MDR) (resistant to rifampin/rifampicin and isoniazid), and extensively drug-resistant (XDR) (resistant to rifampin/rifampicin, isoniazid, any fluoroquinolone, and any one of the injectable second-line drugs) strains of *M. tuberculosis* (Lin, 2015).

11.4.2 Drug Resistance of *M. tuberculosis* in Uveitis

Currently, the gold standard for *M. tuberculosis* drug susceptibility testing is by growing the bacterium on a solid medium and exposing it to various anti-tuberculosis drugs. However, given the slow-growing nature of the bacterium, requiring at least 4 to 8 weeks to produce visible colonies on Löwenstein-Jenson (egg-based) medium, the presence of bacterial cells in fewer numbers in intraocular fluids, and the difficulty in obtaining biopsy specimens from intraocular tissues, are the challenges in culturing the microorganism. In fact microbiological evidence in reported cases of tuberculous uveitis is extremely rare. Hence, the diagnosis of tuberculous uveitis is only presumptive, and treatment is often empirical, based on past medical or social history, or local prevalence of resistance, leading to a delay in the initiation of actual treatment (Dookie et al., 2018; Coll et al., 2015; Abu El-Asrar et al., 2009).

In recent years, use of molecular techniques like PCR, nucleic acid probes, restriction fragment length polymorphism and genetic-based susceptibility testing facilitate rapid and simultaneous detection of *M. tuberculosis* from intraocular samples and also identification of mutations conferring resistance to various antituberculous drugs, thereby preventing serious ocular sequelae (Su, 2002; Figueira et al., 2017). Several nucleic acid amplification tests (NAATs) have been reported to be useful in diagnosis of tuberculous uveitis in isolation, though with variable detection rates (Sharma et al., 2021). Sharma et al. (2013) demonstrated that multiplex PCR using three specific target genes, that is, *IS6110*, *MPB64*, and *protein b*, was able to diagnose the presence of *M. tuberculosis* from intraocular fluids, with a sensitivity and specificity of 77.77% and 100%, respectively and a positive predictive value of 100% and a negative predictive value of 88.88%. GeneXpert MTB/RIF (Xpert), a cartridge-based semi-automated format that uses real-time PCR was used to detect *M. tuberculosis* and rifampicin resistance from vitreous fluids and the overall sensitivity, specificity, positive and negative predictive values were 22.3%, 100%, 100% and 43.1%, respectively (Sharma et al., 2017a). MTBDRplus, a strip-based platform that simultaneously detects the presence of *M. tuberculosis* and mutations in rifampicin (rpoB gene) and isoniazid (katG and inhA gene) using reverse hybridization and biotinylated primers was tested on vitrous fluids, and the sensitivity, specificity, positive predictive value and negative predictive value were observed to be 36.36%, 100%, 100%, and 50.50%, respectively (Sharma et al., 2017b).

A recent comparative study on 110 eyes with drug-resistant tuberculous uveitis reported that MPCR had higher positivity rate of 71.8% followed by MTBDR plus (34.5%) and Xpert (17.2%) assays. Study also showed that the identification of both agent and mutated gene is possible through Xpert and MTBDR plus assay, but with a tradeoff of low sensitivity and false labeling of resistance. However, MPCR followed by rpoB gene sequencing was found to be a robust technique that offers a high yield of diagnosis of paucibacillary conditions like tuberculous uveitis, along with detection of drug resistance capabilities though foolproof identification of gene mutations (Sharma et al., 2021). Studies using whole genome sequencing were helpful in identifying the progressive development of drug resistance, that is, isoniazid resistance being the initial acquisition, followed by resistance to rifampicin or ethambutol, then to pyrazinamide and finally to second- and third-line drugs. In addition, using non-invasive samples for the diagnosis of *M. tuberculosis* and its drug resistance may be helpful in cases where intraocular samples are not available. Gaur et al. (2020) demonstrated the usage of PCR detection from stool samples of both pulmonary tuberculosis and extrapulmonary tuberculosis patients, with a diagnostic accuracy of 91% and 79% respectively. Xpert assay also revealed a positive predictive value of 95% and a negative predictive value of 100% (Gaur et al., 2020).

11.5 Conclusions

Definitive diagnosis of tuberculous uveitis, understanding the drug resistance patterns of *M. tuberculosis* followed by selection and compliance of appropriate treatment regimens still remain a challenge. This results in poor outcomes and further dissemination of MDR strains. Despite the fact that a number of genetic mutations are known to be associated with drug resistance in *M. tuberculosis*, several reports identified resistant strains without any mutation in the known genes. Development of newer diagnostic approaches, early and accurate detection of drug resistance, better knowledge of the mode of action of anti-TB drugs, and understanding the mechanisms of drug resistance development are key factors to contain the spread of resistance in tuberculous uveitis.

REFERENCES

Abu El-Asrar AM, Abouammoh M, Al-Mezaine HS. Tuberculous uveitis. *Middle East Afr J Ophthalmol.* 2009;16(4):188–201.

Abu El-Asrar AM, Abouammoh M, Al-Mezaine HS. Tuberculous uveitis. *Int Ophthalmol Clin.* 2010;50(2):19–39.

Acharya NR, Tham VM, Esterberg E, et al. Incidence and prevalence of uveitis: Results from the Pacific ocular inflammation study. *JAMA Ophthalmol.* Nov; 2013 131(11):1405–1412.

Albini TA, Karakousis PC, Rao NA. Interferon-gamma release assays in the diagnosis of tuberculous uveitis. *Am J Ophthalmol.* 2008;146(4):486–488.

Alvarez S, McCabe WR. Extrapulmonary tuberculosis revisited: A review of experience at Boston City and other hospitals. Medicine (Baltimore). 1984;63(1):25–55.

Ang M, Chee SP. Controversies in ocular tuberculosis. *Br J Ophthalmol.* 2017;101(1):6–9.

Baydur A. The spectrum of extrapulmonary tuberculosis. *West J Med.* 1977;126(4):253–262.

Biswas J, Madhavan HN, Gopal L, et al. Intraocular tuberculosis. Clinocopathologic study of five cases. *Retina.* 1995;15(6):461–468.

Bloch-Michel E, Nussenblatt RB. International Uveitis Study Group recommendations for the evaluation of intraocular inflammatory disease. *Am J Ophthalmol.* 1987;103(2):234–235.

Cegielski JP. Extensively drug-resistant tuberculosis: "there must be some kind of way out of here." *Clin Infect Dis.* 2010;50:Suppl 3:S195–S200.

Centers for Disease Control and Prevention. Case definitions for infectious conditions under public health surveillance. *MMWR Morb Mortal Wkly Rep.* 1997;46(RR-10):1–55.

Chang JH-M, Wakefield D. Uveitis: a global perspective. *Ocul Immunol Inflamm.* 2002;10(4):263–279.

Coll F, McNerney R, Preston MD et al. Rapid determination of anti-tuberculosis drug resistance from whole-genome sequences. *Genome Med.* 2015;7(1):51.

Crofton J, Mitchison DA. Streptomycin resistance in pulmonary tuberculosis. *Br Med J.* 1948;2(4588):1009–1015.

Dalvin LA, Smith WM. Intraocular manifestations of mycobacterium tuberculosis: A review of the literature. *J Clin Tuberc Other Microbact Dis.* 2017;7:13–21.

Daniel TM. Rifampin – a major new chemotherapeutic agent for the treatment of tuberculosis. *N Engl J Med.* 1969;280(11):615–616.

Daniel TM. *Captain of death: The story of tuberculosis.* Rochester, NY: University of Rochester Press, 1997:80–1.

De Rossi E, Aínsa JA, Riccardi G. Role of mycobacterial efflux transporters in drug resistance: An unresolved question. *FEMS Microbiol Rev*. 2006;30(1): 36–52.

Deschenes J, Murray PI, Rao NA, Nussenblatt RB, International Uveitis Study Group. "International Uveitis Study Group (IUSG): Clinical classification of uveitis". *Ocul Immunol Inflamm*. 2008;16(1):1–2.

de Smet MD, Taylor SR, Bodaghi B, et al. Understanding uveitis: The impact of research on visual outcomes. *Prog Retin Eye Res*. 2011;30(6):452–470.

Dinnes J, Deeks J, Kunst H, et al. A systematic review of rapid diagnostic tests for the detection of tuberculosis infection. *Health Technol Assess*. 2007;11(3):1–196.

Dookie N, Rambaran S, Padayatchi N, Mahomed S, Naidoo K. Evolution of drug resistance in Mycobacterium tuberculosis: A review on the molecular determinants of resistance and implications for personalized care. *J Antimicrob Chemother*. 2018;73(5):1138–1151.

Durrani OM, Tehrani NN, Marr JE, Moradi P, Stavrou P, Murray PI. Degree, duration, and causes of visual loss in uveitis. *Br J Ophthalmol*. 2004;88(9):1159–1162

Figueira L, Fonseca S, Ladeira I, Duarte R. Ocular tuberculosis: Position paper on diagnosis and treatment management. *Rev Port Pneumol (2006)*. 2017;23(1):31–38.

Forrester JV, Kuffova L, Dick AD. Autoimmunity, autoinflammation, and infection in uveitis. *Am J Ophthalmol*. 2018;189:77–85.

Fox W, Ellard GA, Mitchison DA. Studies on the treatment of tuberculosis undertaken by the British Medical Research Council tuberculosis units, 1946–1986, with relevant subsequent publications. *Int J Tuberc Lung Dis*. 1999;3(10 Suppl 2):S231-S279.

Gaur, M., Singh, A., Sharma, V. et al. Diagnostic performance of non-invasive, stool-based molecular assays in patients with paucibacillary tuberculosis. *Sci Rep*. 2020;10(1):7102.

Gillespie SH. Evolution of drug resistance in Mycobacterium tuberculosis: Clinical and molecular perspective. *Antimicrob Agents Chemother*. 2002;46(2):267–274.

Glassroth J, Robins AG, Snider DE Jr. Tuberculosis in the 1980s. *N Engl J Med*. 1980;302(26):1441–1450.

Guly CM, Forrester JV. Investigation and management of uveitis". *BMJ*. 2010;341:c4976.

Gupta V, Gupta A, Rao NA. Intraocular tuberculosis–an update. *Srv Opthalmol*. 2007;52(6):561–587.

Gupta A, Sharma A, Bansal R, Sharma K. Classification of intraocular tuberculosis. *Ocul Immunol Inflamm*. 2015;23(1):7–13.

Gutteridge IF, Hall AJ. "Acute anterior uveitis in primary care". *Clin Exp Optom*. 2007;90(2):70–82.

Guyton JS, Woods AC. Etiology of uveitis: A clinical study of 562 cases. *Arch Ophthalmol*. 1941;26(6):983–1018.

Helm CJ, Holland GN. Ocular tuberculosis. *Surv Ophthalmol*. 1993;38(3):229–256.

Hinshaw HC, Feldman WH, Pfuetze KH. Treatment of tuberculosis with streptomycin: A summary of observations on one hundred cases. *J Am Med Assoc*. 1946;132(13):778–782.

Jabs DA, Nussenblatt RB, Rosenbaum JT. Standardization of Uveitis Nomenclature (SUN) Working Group. Standardization of uveitis nomenclature for reporting clinical data. Results of the First International Workshop. *Am J Ophthalmol*. 2005; 140(3):509-516.

Jarlier V, Nikaido H. Mycobacterial cell wall: Structure and role in natural resistance to antibiotics. *FEMS Microbiol Lett*. 1994; 123(1-2): 11–18.

Jellinek K. Streptomycin in the treatment of extrapulmonary tuberculosis: early results. *N Engl J Med*. 1949;240(17):680–683.

Joltikov KA, Lobo-Chan A-M. Epidemiology and risk factors in non-infectious Uveitis: A systematic review. *Front. Med. (Lausanne)* 2021;8:695904.

Kee AR, Gonzalez-Lopez JJ, Al-Hity A, et al. Anti-tubercular therapy for intraocular tuberculosis: A systematic review and meta-analysis. *Surv Ophthalmol.* 2016; 61(5):628–653.

Khochtali S, Gargouri S, Abroug N, et al. The spectrum of presumed tubercular uveitis in Tunisia, North Africa. *Int Ophthalmol.* 2015;35(5):663–671.

Kochi A, Vareldzis B, Styblo K. Multidrug-resistant tuberculosis and its control. *Res Microbiol.* 1993;144(2):104–110.

Koul, PA. Ocular toxicity with ethambutol therapy: Timely recaution. *Lung India.* 2015;32(1):1–3.

Kurup SK, Buggage RR, Clarke GL, et al. Gamma interferon assay as an alternative to PPD skin testing in selected patients with granulomatous intraocular inflammatory disease. *Can J Ophthalmol.* 2006;41(6):737–740.

Lee C, Agrawal R, Pavesio C. Ocular tuberculosis–A clinical conundrum. *Ocul Immunol Inflamm.* 2016;24(2):237–242.

Lin P. Infectious Uveitis. *Curr Ophthalmol Rep.* 2015;3(3):170–183.

London NJ, Rathinam SR, Cunningham Jr. The epidemiology of uveitis in developing countries. *Int Ophthalmol Clin.* 2010;50(2):1–17.

Mandelcorn ED. Infectious causes of posterior uveitis. *Can J Ophthalmol.* 2013;48(1):31–9.

Manten A, Van Wijngaarden LJ. Development of drug resistance to rifampicin. *Chemotherapy.* 1969;14(2):93–100.

Maruyama K, Inaba T, Sugita S, Ichinohasama R, Nagata K, Kinoshita S, et al. Comprehensive analysis of vitreous specimens for uveitis classification: A prospective multicentre observational study. *BMJ Open.* 2017 Nov 16;7(11):e014549.

McCluskey PJ, Towler HM, Lightman S. "Management of chronic uveitis". *BMJ.* 2000;320(7234):555–8. PMC 1117601. PMID 10688564

Medical Research Council. Treatment of pulmonary tuberculosis with streptomycin and para-aminosalicylic acid: A Medical Research Council investigation. *Br Med J.* 1950;2(4688):1073–1085.

Nussenblatt RB. The natural history of uveitis. *Int Ophthalmol.* 1990 Oct;14(5–6):303–308.

Pai M, Zwerling A, Menzies D. Systematic review: T-cellbased assays for the diagnosis of latent tuberculosis infection: An update. *Ann Intern Med.* 2008;149(3):177–184.

Patel A, Kelgaonkar A, Kaza H, Tyagi M, Murthy S, Pathengay A, Basu S. Recent advances in diagnosis and treatment of infectious Uveitis prevalent in Asia-Pacific region. *Asia Pac J Ophthalmol (Phila).* 2021 Jan-Feb 01;10(1):99–108.

Rathinam SR, Cunningham Jr. Infectious causes of uveitis in the developing world. *Int Ophthalmol Clin.* 2000;40(2):137–152.

Rathinam, SR, Namperumalsamy, P. Global variation and pattern changes in epidemiology of uveitis. *Indian J Ophthalmol.* 2007;55(3):173–183.

Rosen PH, Spalton DJ, Graham EM. Intraocular tuberculosis. *Eye (Lond).* 1990;4(pt 3):486–492.

Rosenbaum JT. To be or not TB? *Br J Ophthalmol.* 2014;98(8):999–1000.

Rosenbaum JT, Bodaghi B, Couto C, Zierhut M, Acharya N, Pavesio C, et al. New observations and emerging ideas in diagnosis and management of non-infectious uveitis: A review. *Semin Arthritis Rheum.* 2019;49(3):438–445.

Rothova A, Buitenhuis HJ, Meenken C, Brinkman CJ, Linssen A, Alberts C, et al. Uveitis and systemic disease. *Br J Ophthalmol.* 1992;76(3):137–141.

Rothova A, Suttorp-van Schulten MS, Frits Treffers W, Kijlstrav A. Causes and frequency of blindness in patients with intraocular inflammatory disease. *Br J Ophthalmol.* 1996;80(4):332–336.

Sakai J, Matsuzawa S, Usui M, et al. New diagnostic approach for ocular tuberculosis by ELISA using the cord factor as antigen. *Br J Ophtalmol.* 2001;85(2):130–133.

Selikoff IJ, Robitzek EH, Ornstein GG. Treatment of pulmonary tuberculosis with hydrazide derivatives of isonicotinic acid. *J Am Med Assoc.* 1952;150(10):973–980.

Seung KJ, Keshavjee S, Rich ML. Multidrug-Resistant Tuberculosis and extensively drug-resistant Tuberculosis. *Cold Spring Harb Perspect Med.* 2015;5(9):a017863.

Sharma K, Gupta V, Bansal R, Sharma A, Sharma M, Gupta A. Novel multi-targeted polymerase chain reaction for diagnosis of presumed tubercular uveitis. *J Ophthalmic Inflamm Infect.* 2013;3(1):25.

Sharma K, Gupta V, Sharma A, et al. Gene Xpert MTB/RIF assay for the diagnosis of intra-ocular tuberculosis from vitreous fluid samples. *Tuberculosis (Edinb).* 2017a;102:1–2.

Sharma K, Gupta A, Sharma M, et al. MTBDRplus for the rapid diagnosis of ocular tuberculosis and screening of drug resistance. *Eye (Lond).* 2017b;32(2):451–456.

Sharma K, Gupta A, Sharma M, Sharma A, Bansal R, Sharma SP, Singh RD, Gupta V. The emerging challenge of diagnosing Drug-resistant Tubercular Uveitis: Experience of 110 Eyes from North India. *Ocul. Immunol Inflamm.* 2021;29(1):107–114.

Sheu SJ, Shyu JS, Chen LM, et al. Ocular manifestations of tuberculosis. *Ophthalmology* 2001;108(9):1580–1585.

Su WJ. Recent advances in the molecular diagnosis of tuberculosis. *J Microbiol Immunol Infect.* 2002;35(4):209–214.

Suhler EB, Lloyd MJ, Choi D, Rosenbaum JT, Austin DF. Incidence and prevalence of uveitis in Veterans Affairs Medical Centers of the Pacific Northwest. *Am J Ophthalmol.* 2008;146(6):890–896. e898.

Tsirouki T, Dastiridou A, Symeonidis C, Tounakaki O, Brazitikou I, Kalogeropoulos C, et al. A focus on the epidemiology of uveitis. *Ocul Immunol Inflamm.* 2018;26(1):2–16.

Vall-Spinosa A, Lester W, Moulding T, Davidson PT, McClatchy JK. Rifampin in the treatment of drug-resistant Mycobacterium tuberculosis infections. *N Engl J Med.* 1970;283(12):616–621.

Wakefield D, Chang, JH. Epidemiology of Uveitis. *Int. Ophthalmol. Clin.* 2005;45(2):1–13.

12

Antimicrobial Resistance in the Diseased Human Eye: Post-fever Retinitis

Arunasri Anakala Kotakonda

12.1 Introduction

Post-fever retinitis (PFR) is either an infectious or para-infectious uveitic entity usually manifesting between 2 to 4 weeks after fever in immunocompetent individuals—an entity that usually presents with a sudden diminution of vision. PFR may occur either due to direct invasion by a pathogen or indirectly mediated through an immune-mediated mechanism (Mahendradas et al., 2020). Based on clinical presentation alone, it is difficult to identify the exact etiology of PFR, which can be unilateral or bilateral and may present with focal or multifocal retinitis. Systemic symptoms such as joint pain and skin rash are common during the febrile stage. The other observation in PFR patients includes anterior uveitis, macular edema and retinal detachment at the macula (Figure 12.1). The localised involvement of the retinal vessel can manifest in the form of beading, tortuosity, and perivascular sheathing (Vishwanath et al., 2014). Among the clinical characteristics, the most common findings with retinitis were complete or incomplete macular star (95%), vessel wall exudation (84%) and retinal flame-shaped haemorrhages (62%) (Sundar et al., 2020). PFR is mostly prevalent in tropical and subtropical areas but can spread to other parts of the world mainly due to climate change and globalization. Bacterial causes of PFR would include rickettsia and typhoid, while PFR due to viral causes would include chikungunya, dengue, West Nile virus, and Zika virus (Mahendradas et al., 2020). The details on the disease, prevalence, diagnostics and the treatments are provided in Table 12.1.

Rickettsial disease affecting the eye is reported worldwide, while in India it is considered as a re-emerging disease (Rathi and Rathi 2010). Retinitis due to rickettsial fever shows primary involvement of retina unlike the full thickness involvement by other infectious agents in PFR (Mahendradas et al., 2020). Post-fever ocular involvement of *Salmonella typhi* has shown bilateral involvement, with extensive lesions defined as disc involvement, retinitis, and vasculitis (Acharya et al., 2018).

Dengue fever can lead to dengue maculopathy with the involvement of the posterior segment of the eye. The signs and symptoms observed in dengue maculopathy include retinal haemorrhages, vitreous haemorrhage, retinal vascular occlusion, foveolitis, macular oedema, and optic disc hyperaemia (Yip et al., 2012; Lim et al., 2004). Retinal changes are seen in 28%–34% of patients with dengue maculopathy (Su et al., 2007; Munk et al., 2016). In a cross-sectional observational study conducted in 2

 DOI: 10.1201/9781003451105-12

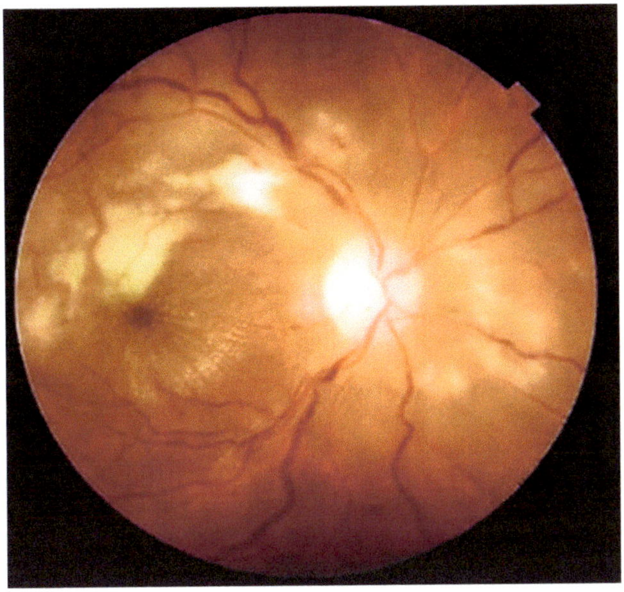

FIGURE 12.1 Retinitis lesions involving the outer retina in a post-fever retinitis patient. (Courtesy Dr. Mudit Tyagi, LV. Prasad Eye Institute, Hyderabad, India.)

general hospitals in Singapore on 197 hospitalised patients with dengue fever, dengue maculopathy was observed in 10% of patients (Su et al., 2007). Ocular involvement in patients with West Nile virus disease is very rare and a typical multifocal chorioretinitis was characterised as the most common ocular sign (Ng and Teoh 2015). Studies on the prevalence of Chikungunya virus (CHIKV) ocular manifestation are rare. In a recent retrospective study, it was reported that involvement of the ocular system in 26% of the CHIKV infected individuals (da Silva et al., 2022). From this study it was observed that ocular symptoms such as pain, inflammation and reduced visual acuity and optic neuritis were most common clinical symptoms in PFR patients (da Silva et al., 2022).

Zika is one of the diseases in the WHO-listed emergency pandemics. Ocular manifestation is observed in adults as well as in children. However, it is more devastating in children and referred to as congenital Zika syndrome (CZS) with a presentation rate of 21.4–55% (Martínez-Pulgarín et al., 2019). Out of the total CZS cases, ocular manifestations in patients present up to 70% of cases (de Oliveira Dias et al., 2018).

12.2 Diagnosis of PFR

The diagnosis of PFR is based on clinical signs and symptoms as well as based on the serological investigations, polymerase chain reaction and knowledge on the epidemics in the community. The clinical characteristics in PFR may be observed by using Fundus fluorescein angiography (FFA) and Optical coherence tomography (OCT). FFA in PFR reveal localized staining of vessels and infers disc leakage if any. OCT imaging in retinitis reveals the nerve fibre layer hyperreflectivity, cystic spaces in the

TABLE 12.1

The Causative Agents, Diagnosis and Treatment in Post-fever Retinitis

Sl. No.	Bacterium/ Virus	Epidemiological distribution	Diagnosis	Treatment
1	*Rickettsia*	World wide	Micro-immunofluorescence, latex agglutination, indirect hemagglutination, immunoperoxidase assay, enzyme-linked immunosorbent assay and immunofluorescence assay	Ciprofloxacin, azithromycin and intravitreal chloramphenicol.
2	*Salmonella typhi*	Indian subcontinent, Africa and South-east Asia	Blood culture, WIDAL test, culture-PCR assay	Steroids.
3	West Nile virus (WNV)	United States and Europe	Serologic testing, PCR assay	Ribavirin and interferon.
4	Dengue fever virus	Southeast Asia, South Asia and South America	NS I antigen assay, Real-Time (RT)-PCR assay	Treatment is unclear. Immunosuppressors are advised.
5	Chikungunya virus	Africa, Asia, and the America	PCR and serological tests	Timolol maleate 0.5% eye drops and oral acetazolamide.
6	Zika virus	Americas, Pacific, Asia, and Africa	PCR and serological tests	Symptomatic with topical steroids.

Sources: Data from Mahendradas et al., 2020; Vishwanath et al., 2014; da Silva et al., 2022; Kawali et al., 2020.

outer retina and at times detachment at the macula (Vishwanath et al., 2014). The specific diagnostic tests for different microorganisms are listed in Table 12.1.

Serological tests remained popular in the diagnosis of rickettsial infection, which would include micro-immunofluorescence, latex agglutination, indirect hemagglutination, immunoperoxidase assay, enzyme-linked immunosorbent assay (ELISA), immunofluorescence assay (IFA) and Weil Felix Test (Mahajan et al., 2006). Detection of *Salmonella typhi* is mostly by the blood culture method, which identifies 45–70% of confirmed cases (Zhou and Pollard 2010). Other tests would include the WIDAL test and culture-Polymerase Chain Reaction (PCR) assay (Zhou and Pollard 2010). The viral aetiologies are mostly confirmed by the serological and real-time PCR assay (Mahendrads et al., 2020).

12.3 Etiology of the Disease

Ascertaining the etiology of PFR is challenging, since the disease could occur following an infection as in isolated idiopathic conditions (Mahendradas et al., 2020; Panchal 2018). Despite similar clinical presentation, the underlying causes could be different (Karkhur and Soni 2022). Further, reports on PFR indicate an immune mediated ophthalmic manifestation of the disease with or without ocular infection, which may pose a threat to vision loss (Chan et al., 2006, Relhan et al., 2014; Velaitham and Vijayasingham 2016; Prabhushanker et al., 2017; Panchal 2018; Shinde et al., 2022). Infectious retinitis is caused due to viral and bacterial infections of the retina, whereas immune mediated retinitis involves a wide spectrum of retinal inflammation due to humoral or cellular immune reactions. About 21–39% of the cases remain idiopathic. The infectious causes of retinitis include viral infections (West Nile virus, Rift valley virus, Dengue virus, Chikungunya), bacterial infections (*Rickettsia conorii* (Mediterranean spotted fever), *Rickettsia rickettsii* (Rocky Mountain spotted fever) and *Salmonella typhi* (Typhoid fever)) (Mahendradas et al., 2020). In patients of PFR with an unknown etiology presence of virus cannot be completely ruled out (Shinde et al., 2022). In a recent study using metagenomic next generation sequencing, the microbiome comprising bacteria, fungi and viruses was studied in idiopathic PFR (Arunasri et al., 2020a; 2020b; 2021). This study identified no significant bacterial and fungal ocular pathogens in the vitreous fluid of patients, but indicated an association of Epstein Barr virus in idiopathic PFR.

12.4 Treatment of PFR

Literature is sparse in describing the clinical features, management options and treatment outcomes in PFR. Therefore, empirical treatment targeting the probable causative organism and the inflammatory process is followed. Treatment would comprise use of antibiotics, steroids, and intravitreal anti-vascular endothelial growth factor agents (Kawali et al., 2019; Vishwanath et al., 2014; Mahendradas et al., 2020; Sundar et al., 2020). Treatment with doxycycline remains the specific mode of management of PFR due to rickettsial infection and other alternative antibiotics would comprise ciprofloxacin and azithromycin (Kawali et al., 2020).

Treatment of other infectious etiologies may include use of steroidal, non-steroidal and immunotherapies (Mahendradas et al., 2020). In general, post-fever retinitis treatment is focused to relieve the symptoms by using analgesics, antipyretics, and anti-inflammatory agents. Nevertheless, for the confirmed viral etiology, antiviral agents, such as ribavirin and interferon may be helpful (Patel et al., 2019; van Tol et al., 2020).

12.5 AMR in PFR Patients

Antimicrobial resistance (AMR) in PFR patients has not been reported, though microorganisms in other human diseases have been reported to exhibit AMR. Since PFR occurs mostly due to the febrile illness caused by virus, the use of antiviral compounds is evident. Although there is no direct evidence of antiviral drug resistance in PFR patients it is predicted that prolonged drug use may lead to the development of resistant viral strains in immunocompromised patients (Strasfeld and Chou 2010). In patients, drug resistance due to the viral mutants may be determined by phenotypic assays. Flavipiravir-resistant Chikungunya virus variants with the acquired unique K291R mutation in the RNA-dependent RNA polymerase (RdRp) were reported in a cell culture-based study (Delang et al., 2014). In recent years repurposing of the antibiotics for controlling the viral infections is in use. Fluoroquinolone drugs are often repurposed as adjunct therapy for several viral infections including flavivirus infections notably for dengue virus and Zika virus infections (Scroggs et al., 2021). When fluoroquinolone drugs such as ciprofloxacin and difloxacin were used against Dengue virus (DENV-4), putative resistance conferring mutations were detected in the envelope genes of the resistant DENV-4 (Scroggs et al., 2021). As evident from the scantly available literature, antimicrobial resistance in chronic viral infections is of concern.

12.6 Trends in AMR in PFR

Reports on antimicrobial resistance in post-fever retinitis are none. Therefore, it is difficult to evaluate the trends in antimicrobial resistance in PFR. Updates on the trends in antimicrobial resistance are important for the ophthalmologists to deliver the appropriate treatment to the patients.

12.7 Multidrug Resistance in PFR

Salmonella typhi, one of the causative agents of PFR has been reported on numerous occasions to exhibit drug resistance in the Indian subcontinent, Southeast Asia, and Africa (Lee et al., 2016). An extensive drug-resistant (XDR) *Salmonella* typhi outbreak in Pakistan showed multidrug resistance to chloramphenicol, ampicillin, and trimethoprim-sulfamethoxazole (Chatham-Stephens et al., 2019). However direct report on the multidrug resistance in PFR is not evident.

12.8 Conclusions

Antimicrobial resistance in post-fever retinitis is poorly understood. Considering the recent literature on emerging antimicrobial resistance for fluoroquinolones, treatment modalities using such drugs need to be re-evaluated. It is also important to evaluate the antimicrobial resistance in rare ocular disease such as PFR time to time and understand the trends of drug resistance. Future research should focus on the evaluation of new antimicrobial agents for PFR and ways to reduce the burden of antimicrobial resistance.

REFERENCES

Acharya P, Ramamurthy LB, Venugopal KC, Manipur SR. Evaluation of posterior segment manifestations following typhoid fever-a clinical study *Indian J Clin Exp Ophthalmol.* 2018;4:421–425.

Arunasri K, Mahesh M, Sai Prashanthi G, Jayasudha R, Kalyana Chakravarthy S, Tyagi M, Pappuru RR, Shivaji S. Comparison of the vitreous fluid bacterial microbiomes between individuals with post fever retinitis and healthy controls. *Microorganisms* 2020a; 8, 751.

Arunasri K, Mahesh M, Sai Prashanthi G, Jayasudha R, Kalyana Chakravarthy S, Tyagi M, Pappuru RR, Shivaji S. Mycobiome changes in the vitreous of post fever retinitis patients. *PLoS ONE.* 2020b;15(11):e0242138.

Arunasri K, Sai Prashanthi G, Tyagi M, Pappuru RR, Shivaji, S. Intraocular viral communities associated with post fever retinitis. Frontiers in Medicine. *Front. Med.* 2021; 8:724195.

Chan DPL, Teoh SCB, Tan CSH, Nah GKM, Rajagopalan R, Prabhakaragupta MK, Chee CKL, Lim TH, Goh KY. Ophthalmic complications of dengue. *Emerg Infect Dis.* 2006;12:285–9.

Chatham-Stephens K, Medalla F, Hughes M, et al. Emergence of extensively drug-resistant *Salmonella* Typhi infections among travelers to or from Pakistan — United States, 2016–2018. *MMWR Morb Mortal Wkly Rep.* 2019;68:11–13. DOI: http://dx.doi.org/10.15585/mmwr.mm6801a3

Delang L, Segura Guerrero N, Tas A, Quérat G, Pastorino B, Froeyen M, Dallmeier K, Jochmans D, Herdewijn P, Bello F, Snijder EJ, de Lamballerie X, Martina B, Neyts J, van Hemert MJ, Leyssen P. Mutations in the chikungunya virus non-structural proteins cause resistance to favipiravir (T-705), a broad-spectrum antiviral. *J Antimicrob Chemother.* 2014 Oct;69(10):2770–84. doi: 10.1093/jac/dku209

de Oliveira Dias JR, Ventura CV, de Paula FB, Prazeres J, Ventura LO, Bravo-Filho V, et al. Ocular abnormalities in congenital Zika syndrome: Are the ophthalmoscopic findings "the top of the iceberg"? *Progress in Retinal and Eye Research.* 2018;66:85–106.

da Silva LCM, da Silva Platner F, da Silva Fonseca L, Rossato VF, de Andrade DCP, de Sousa Valente J, Brain SD, Fernandes ES. Ocular manifestations of Chikungunya Infection: A Systematic Review. *Pathogens.* 2022;11:412. https://doi.org/10.3390/pathogens11040412

Karkhur S, Soni D. Clinical spectrum, disease course and management outcome of post-fever retinitis cases: Experience from a tertiary eye institute in central India. *Int Ophthalmol.* 2022;42:2697–2709. https://doi.org/10.1007/s10792-022-02258-w

Kawali A, Mahendradas P, Mohan A, et al. Epidemic retinitis. *Ocular Immunol Inflam.* 2019;27:571–577.

Kawali A, Srinivasan S, Mohan A, Bavaharan B, Mahendradas P, Shetty B. Epidemic retinitis with macular edema –Treatment outcome with and without steroids. *Ocul Immunol Inflamm.* 2020;21:1–5.

Lee J-S, Mogasale VV, Mogasale V, Lee K. Geographical distribution of typhoid risk factors in low and middle income countries. *BMC Infect Dis.* 2016;16(1):732. doi: 10.1186/s12879-016-2074-1

Lim WK, Mathur R, Koh A, Yeoh R, Chee SP. Ocular manifestations of dengue fever *Ophthalmology.* 2004;111:2057–64.

Mahajan SK, Kashyap R, Kanga A, Sharma V, Prasher BS, Pal LS. Relevance of Weil-Felix test in diagnosis of scrub typhus in India. *J Assoc Physicians India.* 2006;54:619–21

Mahendradas P, Kawali A, Luthra S, et al. Post-fever retinitis–newer concepts. *Indian J Ophthalmol.* 2020;68:1775–1786.

Mahendradas P, Kawali A, Luthra S, Srinivasan S, Curi AL, Maheswari S, Ksiaa I, Khairallah M. Post-fever retinitis - Newer concepts. *Indian J Ophthalmol.* 2020 Sep;68(9):1775–1786. doi: 10.4103/ijo.IJO_1352_20. PMID: 32823394; PMCID: PMC7690479

Martínez-Pulgarín DF, Córdoba-Ortega CM, Padilla-Pantoja FD. The Eye and the Zika Virus. In: Rodriguez-Morales, A. J., editor. *Current Concepts in Zika Research [Internet]. London: IntechOpen;* 2019 [cited 2022 Oct 28]. Available from: www.intechopen.com/chapters/69790 doi: 10.5772/intechopen.89847

Munk MR, Jampol LM, Cunha Souza E, de Andrade GC, Esmili DD, Sarraf D, et al New associations of classic acute macular neuroretinopathy. *Br J Ophthalmol.* 2016;100:389–94.

Ng AW, Teoh SC. Dengue eye disease. *Surv Ophthalmol.* 2015;60:106–14.

Panchal B. Commentary: Post fever retinitis and vasculitis:A morphological conundrum. *Indian J Ophthalmol.* 2018;66:1844.

Patel S, Sinigaglia A, Barzon L, Fassan M, Sparber F, Landmann SL. Role of NS1 and TLR3 in pathogenesis and immunity of WNV. *Viruses.* 2019;11:603.

Prabhushanker M, Topiwalla TT, Ganesan G, Appandaraj S. Bilateral retinitis following typhoid fever. *Int. J. Retina Vitr.* 2017;3:11.

Rathi N, Rathi A. Rickettsial infections: Indian perspective. *Indian Pediatr.* 2010;47:157–164.

Relhan N, Pathengay A, Albini T, Priya K, Jalali S, Flynn HW. A case of vasculitis, retinitis and macular neurosensory detachment presenting post typhoid fever. *J. Ophth. Inflamm. Infect.* 2014;4:23.

Scroggs SLP, Gass JT, Chinnasamy R, Widen SG, Azar SR, Rossi SL, Arterburn JB, Vasilakis N, Hanley KA. Evolution of resistance to fluoroquinolones by dengue virus serotype 4 provides insight into mechanism of action and consequences for viral fitness. *Virol.* 2021 Jan 2;552:94–106. doi: 10.1016/j.virol.2020.09.004

Shinde P, Daigavane S, Sharma A. Post-Fever Retinitis. *Cureus.* 2022 Jun 29;14(6):e26429. doi:10.7759/cureus.26429

Strasfeld L, Chou S. Antiviral drug resistance: Mechanisms and clinical implications. *Infect Dis Clin North Am.* 2010 Jun;24(2):413–37. doi: 10.1016/j.idc.2010.01.001. Corrected and republished in: *Infect Dis Clin North Am.* 2010 Sep;24(3):809–33. PMID: 20466277; PMCID: PMC2871161

Su DH, Bacsal K, Chee SP, Flores JV, Lim WK, Cheng BC, Jap AH; Dengue Maculopathy Study Group. Prevalence of dengue maculopathy in patients hospitalized for dengue fever. *Ophthalmol.* 2007 Sep;114(9):1743–7. doi: 10.1016/j.ophtha.2007.03.054

Sundar MD, Chawla R, Balaji A, Garg I, Kalathil R, Hasan N, Vikas SJ, Kumar A. Clinical features, optical coherence tomography findings and treatment outcomes of post-fever retinitis. *Ther Adv Ophthalmol.* 2020; 12. doi: 10.1177/ 2515841420979111. PMID: 33415316; PMCID: PMC7750750.

van Tol S, Atkins C, Bharaj P, Johnson KN, Hage A, Freiberg AN. VAMP8 contributes to the TRIM6-mediated type I interferon antiviral response during West Nile virus infection. *J Virol.* 2020;94:e01454–19.

Velaitham P, Vijayasingham N. Central retinal vein occlusion concomitant with dengue fever. *Int. J. Retina Vitr.* 2016;2:1.

Vishwanath S, Badami K, Sriprakash KS, Sujatha BL, Shashidhar SD, Shilpa YD. Post-fever retinitis: A single center experience from south India. *Int Ophthalmol.* 2014, 34:851–7. doi: 10.1007/s10792-013-9891-7

Yip VC, Sanjay S, Koh YT. Ophthalmic complications of dengue fever: A systematic review. *Ophthalmol Ther.* 2012;1:2.

Zhou L, Pollard AJ. A fast and highly sensitive blood culture PCR method for clinical detection of Salmonella enterica serovar Typhi. *Ann Clin Microbiol Antimicrob.* 2010;9:14.

13

Antimicrobial Resistance of the Diseased Human Eye: Conjunctivitis

S. Shivaji

13.1 Introduction

Conjunctiva is the thin and translucent membrane that lines the sclera and inside of the eyelids. It is known as the bulbar conjunctiva when it covers the visible part of the sclera from the edge of the cornea, whereas it is referred to as the palpebral conjunctiva when it lines the inside of the eyelids. "Conjunctivitis" refers to Inflammation or infection of the conjunctiva and the characteristic features are hyperemia (excess of blood in the vessels) and conjunctival injection (redness in the white sclera) due to inflammation in the eye. Conjunctivitis is commonly referred to as "red eye" or "pink eye," and normally appears as a mild hyperemia with epiphora (tearing) to subconjunctival haemorrhage or chemosis (conjunctival edema) with purulent discharge and concomitant edema of the eyelid (Richards and Guzman-Cottrill, 2010) (Figure 13.1). The symptoms also include foreign body sensation, itching, burning and photophobia. At times the patients also present with matting of the lids and in severe cases it appears as a blister on the eyeball and may obstruct eyelid closure. At times conjunctivitis also results in the formation of papillae, follicles or granulomas, membranes, or pseudomembranes, and preauricular and submandibular lymphadenopathy (enlarged lymph nodes beneath the mandible) (Richards and Guzman- Cottrill, 2010).

13.2 Prevalence and Incidence of Conjunctivitis

Conjunctivitis occurs worldwide, and the annual economic burden in treating bacterial conjunctivitis is $857 million in USD (Smith and Waycaster, 2009). The incidence of viral conjunctivitis followed this order: adenovirus conjunctivitis (65,000 to 90,000 cases) > HSV conjunctivitis (1,300 to 4,800 cases) per 100,000 cases of acute conjunctivitis (Leibowitz, 2000). In contrast, the prevalence of the other types of conjunctivitis followed the order keratoconjunctivitis sicca (10,000 to 30,000 cases)

(Schaumberg et al., 2003) > allergic conjunctivitis (6,000 to 40,000 cases) (Leonardi et al., 2015)> bacterial conjunctivitis (1,350 cases) (Smith and Waycaster 2009) per 100,000 acute conjunctivitis cases annually. The occurrence of conjunctivitis is also influenced by age and the etiological agent involved. Viral conjunctivitis is

DOI: 10.1201/9781003451105-13

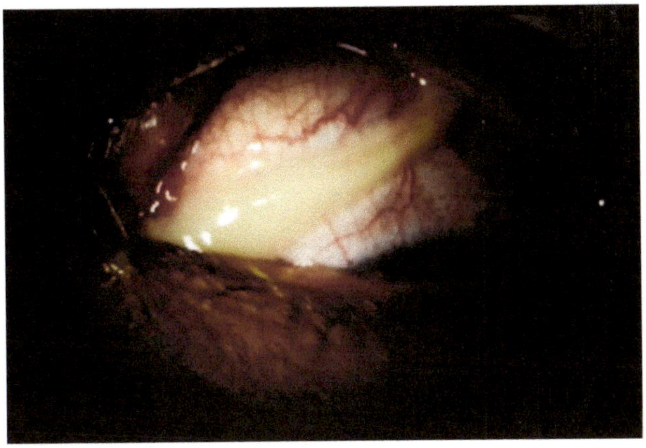

FIGURE 13.1 An acute bacterial conjunctivitis patient showing yellowish purulent discharge. (Adapted from Azari and Arabi (2020), published as an open access article and distributed under the terms of the Creative Commons Attribution-Non-Commercial-Share Alike 4.0 License.)

common in adults, whereas in children bacterial conjunctivitis is common (Johnson et al., 2022). Further, the incidence of infective conjunctivitis is higher in younger children (<1 year old) compared to the older children (>4 years of age) (Rose, 2007). VKC affects young males (Bonini et al., 2004), keratoconjunctivitis sicca affects patients older than 40 years (Schaumberg et al., 2009) and superior limbic keratoconjunctivitis (SLK) affects people around 60 years of age (Watson et al., 2002). Incidence of infective and neonatal conjunctivitis (Fitch et al., 1989) are not influenced by gender (Moore et al., 2015) unlike SLK, which commonly affects women (Nelson, 1989). Further, the occurrence of vernal keratoconjunctivitis and keratoconjunctivitis sicca shows geographical preferences with the former common in individuals with dark skin as in Africa and India (O'Brien et al., 2009), whereas keratoconjunctivitis sicca is common in Hispanic and Asian populations (Moss et al., 2000). Over the years neonatal conjunctivitis has decreased to 5 per 1,000 live births in developed countries (Azari and Barney 2013). Neonatal conjunctivitis due to gonococcal infection in Belgium and the Netherlands was estimated as 0.04 per 1,000 live births (Mallika et al., 2008), whereas in the United States it was 0.3 per 1,000 live births. In the United States it was also observed that the prevalence of neonatal conjunctivitis caused by chlamydial infection was higher 5 to 60 cases compared to 4 cases per 1,000 live births in UK. Further, in the United States incidence of bacterial conjunctivitis increased from 23,000 (in the 02 year age range), to 36,000 per 100,000 cases (in adults) (Epling, 2010). Trends related to the incidence of bacterial conjunctivitis indicated that it is decreasing in developing countries but in contrast in Africa, the incidence of neonatal conjunctivitis is still high (Schaller and Klauss, 2001). Neonatal conjunctivitis is still of global concern since worldwide, neonatal conjunctivitis or ophthalmia neonatorum is still the cause of blindness in approximately 10,000 babies annually (Isenberg et al., 1996).

13.3 Categorizing of Conjunctivitis

Conjunctivitis may be classified based on etiology as either infectious (caused by bacteria or viruses) or non-infectious conjunctivitis (caused by allergens or toxins). Further, conjunctivitis is designated acute when the onset is rapid and lasts for < four weeks and chronic when it lasts for >4 weeks. In severe conjunctivitis the affected individuals exhibit mucopurulent discharge and are referred to as blepharoconjunctivitis and viral keratoconjunctivitis when it involves the eyelid margins and cornea respectively (Azari and Arabi, 2020). Conjunctivitis is also known to be associated with several systemic syndromes and diseases like Reiter's syndrome, which is Reactive arthritis, Stevens-Johnson syndrome (SJS), rheumatoid arthritis, vitamin A deficiency, and Richner-Hanhart syndrome and porphyria (De laet et al., 2013; Sati et al., 2013).

13.3.1 Bacterial Conjunctivitis

Children are more prone to bacterial conjunctivitis than adults (Epling, 2010). Several factors contribute to the transmission of bacterial conjunctivitis such as contact with infected individuals, proliferation of the native conjunctival flora (Shields and Sloane, 1991), contaminated fingers, oculogenital spread, contaminated fomites (Karpecki et al., 2009), compromised tear production, corneal rupture due to trauma, abnormality of adnexal structures (eyebrow, eyelids, and lacrimal apparatus), and immunosuppressed status. In acute conjunctivitis the etiological bacterial agents are *Staphylococcus aureus, S. epidermidis, Staphylococcus* spp., *Haemophilus influenza, Streptococcus* spp., *Streptococcus pneumoniae, Moraxella catarrhalis,* and *Pseudomonas* (Høvding, 2008; Richards and Guzman- Cottrill, 2010). In contrast, in chronic conjunctivitis *Staphylococcus, M. catarrhalis,* and most commonly, *Chlamydia* (i.e., trachoma or inclusion conjunctivitis) are involved (Richards and Guzman- Cottrill, 2010). Symptoms of bacterial conjunctivitis include foreign body sensation, increased ocular secretion, conjunctival hyperemia (Figure 13.1), sticky eyelids and itching (Wall et al.1998), purulent secretion, ocular burning, acute otitis media and upper respiratory tract infection (Høvding, 2008). Conjunctival injection and irritation often occur in patients who have chronic conjunctivitis (Richards and Guzman-Cottrill, 2010). Bacterial conjunctivitis resolves spontaneously within one-two weeks in majority of the cases (>60%) (Rose et al., 2005) and very rarely serious complications may occur (Schiebel, 2003). Broad-spectrum topical antibiotics reduced the duration of the disease by 10% and topical antibiotics were more effective when patients had positive bacterial cultures (Azari and Barney, 2013; Tepedino et al., 2009; Hwang et al., 2003). The specific types of bacterial conjunctivitis are described below.

13.3.1.1 Methicillin-resistant S. aureus *Conjunctivitis*

Methicillin-resistant *S. aureus* (MRSA) conjunctivitis includes *S. aureus* resistant to methicillin and resistant to all β-lactam antimicrobials (Shanmuganathan et al., 2005a; b). The prevalence of MRSA in conjunctivitis is 3–64% of all *Staphylococcus* conjunctival infections (Azari and Barney, 2013; Zhang et al.,

2000; Shanmuganathan et al., 2005a; b). Suspected cases of MRSA conjunctivitis are effectively treated with fortified vancomycin eye drops or ointments (Gross et al., 2003; Thareja et al., 2019).

13.3.1.2 Ophthalmia Neonatorum

Ophthalmia neonatorum or neonatal conjunctivitis is a bacterial conjunctivitis in neonates occurring within 4 weeks of birth and the etiological agents include several different bacteria (*S. aureus, S epidermidis, S. pneumoniae, M. catarrhalis, C. trachomatis* and *N. gonorrhoeae*), and a virus (herpes simplex virus, HSV) (Richards and Guzman-Cottrill, 2010). *Chlamydia trachomatis* from a infected birth canal can cause neonatal conjunctivitis (Darville, 2005; Castro et al., 2020) between day 5 and 14 following vaginal delivery. The baby shows all the signs of conjunctivitis such as purulent discharge, erythema and edema of the eyelids and conjunctiva (Darville, 2005) and in untreated cases the cornea and/or conjunctiva develop a scar (Forster et al., 1970). Thus to avoid chlamydial conjunctivitis erythromycin or tetracycline ointment is applied within 1 hour of delivery (Rapoza and Chandler, 1988) or systemic erythromycin at 50 mg/kg per day for two weeks (Zikic et al., 2013). The later systemic treatment with erythromycin is superior but runs the risk of causing pyloric stenosis in neonates (Zikic et al., 2013). Additionally, it is suggested that azithromycin may improve compliance (Srivastava et al., 2013). It would be relevant to comment that *C. trachomatis* serotypes D-K in adults cause inclusion conjunctivitis (Burton, 2007; Granet et al., 2008).

13.3.1.3 Granulomatous Conjunctivitis

Unilateral granulomatous conjunctivitis is a characteristic feature of Parinaud oculoglandular syndrome (POS) when patients present with swollen ipsilateral preauricular or submandibular lymphadenopathy. Cat-scratch diseases caused by *Bartonella henselae* is the most common cause of POS. Several other microbes are known to cause POS and they include C. *trachomatis*, *Francisella tularensis*, *Mycobacterium tuberculosis, Sporothrix schenckii, Treponema pallidum, Actinomyces*, Epstein-Barr virus, and *Coccidioides immitis* (Richards and Guzman-Cottrill, 2010).

13.3.1.4 Gonococcal Conjunctivitis

Neisseria gonorrhoeae or *N. meningitidis* cause hyperacute gonococcal conjunctivitis (GC) in neonates and in sexually active adults (Shields and Sloane, 1991; Costumbrado et al., 2020; Richards and Guzman-Cottrill, 2010). GC individuals present with a high prevalence of corneal perforation (Høvding, 2008), a tender globe with periauricular lymphadenopathy, conjunctival injection, chemosis and copious mucopurulent discharge (Costumbrado et al., 2020). Both *N. gonorrhoeae* and *N. meningitidis* conjunctivitis should be treated with intramuscular or intravenous ceftriaxone (Richards and Guzman-Cottrill, 2010). The recommended therapy for neonates include single dose of ceftriaxone (25 to 50 mg/kg), or cefotaxime (100 mg/kg IV or IM) and hourly saline irrigation of the ocular surface (Costumbrado et al.,

2020; Pak et al., 2017). Non-neonates can be treated with combination of 1 gm of IM ceftriaxone given in a single dose and 1 gm of oral azithromycin (Costumbrado et al., 2020).

13.4 Diagnosis of Conjunctivitis

The appearance of 'red eye' is common to uveitis, keratitis, scleritis, conjunctivitis or subconjunctival haemorrhage, blepharitis, and so forth (Azari and Arabi, 2020). Thus, there is a need to differentiate conjunctivitis from other "red eye" diseases. Infectious conjunctivitis is characterised based on eye discharge, conjunctival injection, red eye(s), eyelashes matted, grittiness of the eye(s), edema, and also on the history of contact with conjunctivitis patients (Everitt and Little, 2002). On the other hand in allergic conjunctivitis patients complaint of itching, chemosis, redness and without significant corneal involvement and conjunctival swelling (Friedlaender, 2011). In difference to infectious and allergic conjunctivitis in vernal keratoconjunctivitis (VKC) the patients have severe itching and giant papillae in the superior tarsal conjunctiva (Bielory and

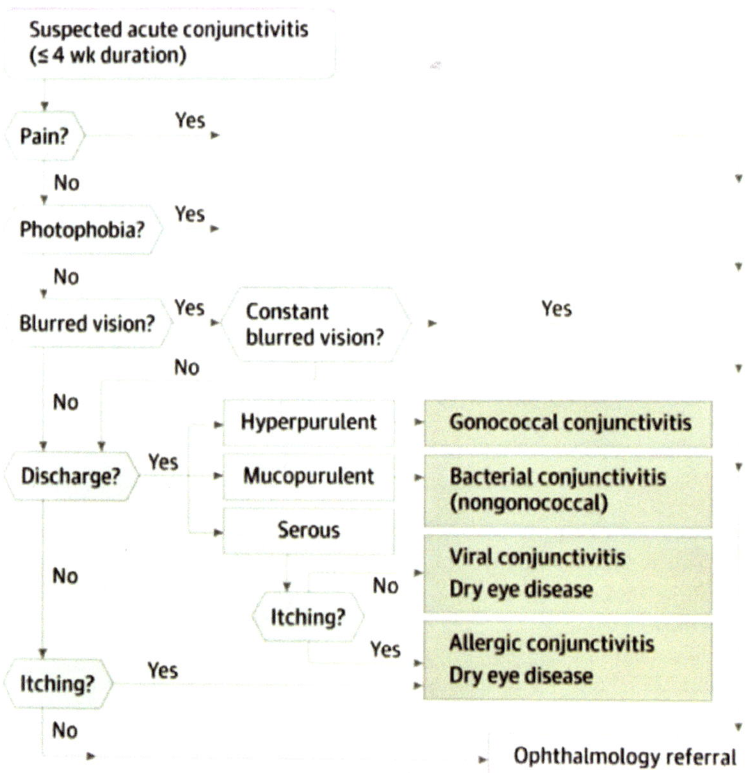

FIGURE 13.2 Differential diagnosis algorithm for suspected acute conjunctivitis. (Adopted and modified from Azari and Barney (2013).)

Frohman, 1992). Atopic VKC is also confirmed by the presence of conjunctival scar and anterior subcapsular cataract (Bielory and Bielory, 2010). Patients with chronic toxic conjunctivitis present with watery discharge, initial papillary conjunctival reaction followed by a follicular reaction, punctate epithelial erosion of the cornea, and eyelid dermatitis (Wilson-Holt and Dart, 1989; Soparkar et al., 1997). Overall, the diagnosis of viral, bacterial, and allergic conjunctivitis is difficult due to overlapping symptoms. Yeu & Hauswirth (2020) suggested an algorithm for differential diagnosis of acute conjunctivitis (Figure 13.2).

13.5 Treatment of Bacterial Conjunctivitis

Bacterial conjunctivitis is generally treated empirically with various classes of topical antibacterial products, which include the aminoglycosides, polymyxin B combinations, macrolides, and fluoroquinolones (Table 13.1) (Karpecki et al., 2009). But many of these can be irritating to the eye (Diamant and Hwang, 1999; Leeming, 1999) and are no longer widely prescribed due to toxicity concerns or due to bacterial resistance (Morrissey et al., 2004).

TABLE 13.1

Characteristics of Ophthalmic Antibiotic Classes Used to Treat Bacterial Conjunctivitis

Class	Bacteriostatic/ bactericidal	Effective against	Mechanisms of resistance
Aminoglycosides	Bactericidal	*Staphylococci*, Gram-negative pathogens	Enzymatic inhibition, membrane impermeability, and alteration of ribosomal target.
Polymyxin B plus bacitracin	Bacteriostatic/ bactericidal	Gram-positive pathogens	Membrane impermeability.
Polymyxin B plus neomycin	Bactericidal	Gram-negative pathogens	Enzyme inhibition.
Polymixin B plus trimethoprim	Bacteriostatic	Gram-negative and Gram-positive pathogens	Membrane impermeability, alteration and/or overproduction of enzymes.
Macrolides	Bacteriostatic	Gram-positive pathogens, *H. influenzae*	Enzyme inhibition, active efflux pump and alteration of ribosomal target.
Fluoroquinolones	Bactericidal	Gram-negative and Gram-positive pathogens	Alteration of target enzymes.

Source: Modified from Drug Facts and Comparisons (2008) and Karpecki et al. (2009).

13.5.1 Aminoglycosides for the Treatment of Bacterial Conjunctivitis

The aminoglycosides namely gentamicin and tobramycin (Table 13.1) were among the first topical antibiotics used to treat bacterial conjunctivitis (Gwon, 1992a; b). In patients with proven *H. influenzae* or *S. pneumoniae* conjunctivitis the cure rates with gentamicin 0.3% and polymyxin B/trimethoprim at days 3 to 6 of a 10-daycourse of therapy was very similar (46% and 45%, respectively); but polymyxin B/trimetho-prim combination (83%) proved to better than gentamicin alone (68%) after completion of therapy (Lohr et al., 1988). In a subsequent study the effect of gentamicin and tobramycin at the same concentration (0.3%) as an ophthalmic ointment in children and adolescents with bacterial conjunctivitis (Leibowitz et al., 1981; Timewell et al., 1983) revealed that the clinical cure rates were 39–55% for gentamicin and 46–61% for tobramycin and the bacteriological eradication rates were 77–84% for gentamycin and 88–91% for tobramycin 0.3% implying that tobramycin was better. These results with gentamicin and tobramycin compared favourably with the older fluoroquinolones including ciprofloxacin (Leibowitz, 1991), norfloxacin (Miller et al., 1992), ofloxacin (Gwon, 1992a; b) and lomefloxacin (Montero et al., 1998). Non-fortified aminoglycoside ophthalmic solutions were generally well tolerated (Miller et al., 1992; Cochereau et al., 2007) unlike the fortified aminoglycoside drops that have been associated with ocular irritation (O'Brien et al., 1995).

13.5.2 Macrolides for the Treatment of Bacterial Conjunctivitis

Erythromycin and azithromycin are macrolide antibiotics, which are bacteriostatic (Table 13.1). Erythromycin as an ophthalmic ointment has been used since 1982 for prophylaxis of *S. aureus* caused neonatal conjunctivitis and against Gram-negative organisms including *H. influenzae* (Drew and Gallis, 1992) in patients with bacterial conjunctivitis (Abelson et al., 2007; 2008). In fact bacteriological eradication rates with Azithromycin 1% was very effective and eradicated 83 to 94% of the most common pathogens (Abelson et al., 2007) and when azithromycin 1% and tobramycin 0.3% were compared for their potency, the rates of microbial eradication and bacterial infection recurrence were similar for both the antibiotics (Abelson et al., 2007; Protzko et al., 2007). Azithromycin does cause eye irritation, headache, worsening conjunctivitis, and conjunctival hyperaemia (Karpecki et al., 2009).

13.5.3 Fluoroquinolones for the Treatment of Bacterial Conjunctivitis

Fluoroquinolones (ciprofloxacin, ofloxacin, levofloxacin, moxifloxacin, gatifloxacin and besifloxacin) are broad-spectrum bactericidal agents effective against both Gram-positive (Staphylococci and Streptococci resistant to other antibiotics) and Gram-negative pathogens associated with bacterial conjunctivitis (Kaliamurthy et al., 2005; Hwang, 2004; Egger et al., 2001). All the listed fluoroquinolones have been used for treating bacterial conjunctivitis and are comparable in their effects with respect to clinical resolution rate (clinical improvement) (45.2%–86.5%), which increased with day of treatment and bacterial eradication rates (>90%) (Rose, 2007; Karpecki et al., 2009; Tepedino et al., 2009). However certain features appear to be specific to a particular fluoroquinolone. Overall, the fluoroquinolones were effective and the bacterial eradication rates were 84–94% for moxifloxacin treatment on days 5 to 6 (O'Brien, 2006) and eradication rates were 90.0–91.5% for besifloxacin ophthalmic suspension 0.6% on days 5 to 8 (Karpecki et al., 2009; Tepedino et al., 2009; O'Brien 2012). Comparing the effect of the drugs with respect to clinical resolution rate and bacterial eradication rates is confounded with many variables such as the concentration of the drug, the dosage, the regime and so forth and thus it is best done when the variables are reduced to a bare minimum.

In pediatric patients (aged 1 to 16 years) when ophthalmic solutions of levofloxacin 0.5% and ofloxacin 0.3% were compared for their potential use it was observed that both the antibiotics exhibited similar clinical cure rates (81 versus 86%, respectively) but levofloxacin was better with a higher microbial eradication rate (85 versus 68%, respectively), which improved in children aged 2 to 11 years (87 versus 62%) (Lichtenstein and Rinehart, 2003). Even, when all age groups (infants to adults) were compared levofloxacin exhibited higher microbial eradication rates (89 versus 80%) (Schwab et al., 2003). Further, when gatifloxacin 0.3% was compared with ofloxacin 0.3% ophthalmic solution the clinical cure rate with gatifloxacin was significantly higher than with ofloxacin (72.3 versus 63.5%) (Tepedino and Jensen, 2002). But the clinical resolution and bacterial eradication between 0.6% moxifloxacin and 0.6% besifloxacin ophthalmic suspensions were comparable (McDonald et al., 2009). Among the fluoroquinolones moxifloxacin is the most potent compared to ciprofloxacin, ofloxacin, levofloxacin, and gatifloxacin, which need to be administered more frequently (Vigamox, 2008; Gwon 1992a; b). But, fluoroquinolones are well tolerated (Hwang et al., 2003; Yee et al., 2005; Lichtenstein and Rinehart, 2003; Schwab et al., 2003; Comstock et al., 2010a; b) compared to older antibiotics (Gwon 1992a; b) and topical fluoroquinolones are mild and self-limiting (Gwon 1992a; b; Lichtenstein and Rinehart, 2003; Schwab et al., 2003; McDonald et al., 2009; Roberts, 2002; Comstock et al., 2010; Hwang et al., 2003; Yee et al., 2005).

13.6 Antimicrobial Resistance in Ophthalmic Infections

Assessing antibiotic resistance in ophthalmology is confounded by several factors such as small case numbers, single-centre studies, determining MICs based on systemic drug administration (Haas et al., 2011; Asbell et al., 2008; 2015), use of higher concentration of drug for topical applications (Raizman et al., 2002), absence of a standardized protocol for ascertaining resistance in ocular pathogens and analysing samples of cultured organisms (Asbell et al., 2008; 2015). Further, some studies use CLSI-defined breakpoints (Haas et al., 2011; Asbell et al., 2008; 2015), whereas others use automatic microbial systems (Benz et al., 2004), or a method used in a given laboratory (Ramakrishnan et al., 2009). Information on the resistance profile of ocular microorganisms would help in empirical treatment and facilitate effective pre-, peri and postoperative management (Grzybowski et al., 2017; Miño de Kaspar et al., 2001). It was also observed that prophylactic use of antibiotics causes increase in ocular resistant strains (Kim and Toma, 2011a; b; Dave et al., 2013; Milder et al., 2012) as observed in AMD patients treated with fluoroquinolone eye drops for 4 days (Milder et al., 2012). CoNS isolates treated with fluoroquinolones also showed increased resistance to ofloxacin, levofloxacin, gatifloxacin and moxifloxacin (Kim and Toma, 2011a; b) and so also resistance was evident in eyes treated with azithromycin (Kim and Toma, 2011a; b). Hsu et al. (2014) observed no significant change in the ocular flora resistance patterns following intravitreal injections with povidone-iodine antisepsis implying that resistance observed in the previous studies could be attributed to the use of antibiotics.

13.6.1 Emergence of Antimicrobial Resistance in Ocular Microorganisms in the United States

In the United States, two initiatives namely "Antibiotic Resistance Monitoring in Ocular Microorganisms' (ARMOR) and 'Ocular Tracking Resistance in the U.S. Today' (TRUST) tracked resistance in ocular microorganisms (Haas et al., 2011; Asbell et al., 2008; 2015). In the 3 ARMOR studies so far completed for the periods 2009-20013 (ARMOR 1) (Asbell et al., 2015), 2009–2016 (ARMOR 2) (Asbell and DeCory, 2018) and 2009–2018 (ARMOR 3) (Bispo et al., 2022) the following trends were obvious:

(a) Irrespective of the number of isolates that were evaluated methicillin resistance was observed by all the three studies (Asbell et al., 2015; Asbell and DeCory, 2018; Bispo et al., 2022). It was also reported that *S. aureus* and CoNS were resistant to several antibiotics including the fluoroquinolones, macrolides, and methicillin/oxacillin between 2009–2018 (Bispo et al., 2022).
(b) Methicillin resistance doubled the rate of multi-drug resistance (MDR) in MRSA and MDR-CoNS isolates (Asbell and DeCory, 2018., Asbell, 2015).
(c) Methicillin-resistant isolates exhibited concurrent resistance to fluoroquinolones, aminoglycosides, or macrolides (Asbell et al., 2015).
(d) MRSA prevalence peaked from 2005 to 2015 (Asbell et al., 2015) and subsequently showed a decreasing trend in recent years (Bispo et al. 2022).
(e) MDR was observed in *S. aureus* (86.8%) and CoNS isolates (77.3%) (Asbell et al., 2015) but in ARMOR 2 study multidrug resistance was less (30.2% and 39.0% respectively) compared to that observed in ARMOR 1(Asbell and DeCory, 2018); MDR was also reported by Bispo et al. (2022).
(f) Staphylococcal isolates were susceptible to vancomycin (Asbell et al., 2015), whereas *S. pneumoniae* exhibited highest resistance to azithromycin (Asbell et al., 2015).
(g) Among the isolates the resistance of *P. aeruginosa* and *H. influenzae* was low (Asbell et al., 2015).
(h) Small decrease in resistance to a number of antibiotics was observed in the three ARMOR studies and the trends were as follows: decrease in resistance to ciprofloxacin and tobramycin in ARMOR 1 (Asbell et al., 2015), decrease in resistance to azithromycin, ciprofloxacin, tobramycin, trimethoprim, and oxacillin in ARMOR 2 (Asbell and DeCory, 2018) and during 2009 and 2018 in ARMOR 3, a slight decreases in resistance were observed to several classes of antibiotics including fluoroquinolones, macrolides, methicillin/oxacillin, and aminoglycosides (Bispo et al., 2022). The microorganisms used for screening were different in these studies.
(i) In ARMOR 2 it was also observed that the resistance increased with increasing age among *S. aureus*, MRSA, and CoNS (Asbell and DeCory, 2018).

Overall, the ARMOR data was similar across the local and regional US centres and indicated that continued surveillance of ocular bacterial pathogens with respect to methicillin resistance and MDR prevalence would provide insight into emergence of new antibiotic resistance phenotypes.

13.6.2 Antimicrobial Resistance in Bacterial Isolates from Conjunctivitis Patients in the United States

Antimicrobial resistance was monitored in 1198 bacterial isolates from conjunctivitis patients across the United States during the period 2009–2016 (Asbell and DeCory, 2018). The isolates included 483 *S. aureus*, 305 CoNS, 208 *H. influenzae*, 118 *S. pneumoniae*, and 84 *P. aeruginosa* (Silvester et al., 2016). The data indicated that a large proportion of staphylococci were resistant to oxacillin and azithromycin, whereas the resistance of *S. pneumoniae*, *P. aeruginosa*, and *H. influenzae* was low against the antibiotics tested. Further, MDR (resistance to >3 antibiotic classes) was low in *S. aureus* and CoNS (30.2% and 39% respectively), whereas in MRSA, and MR-CoNS multi-drug resistance almost doubled (76.5% and 72.8% respectively) (Figure 13.3)

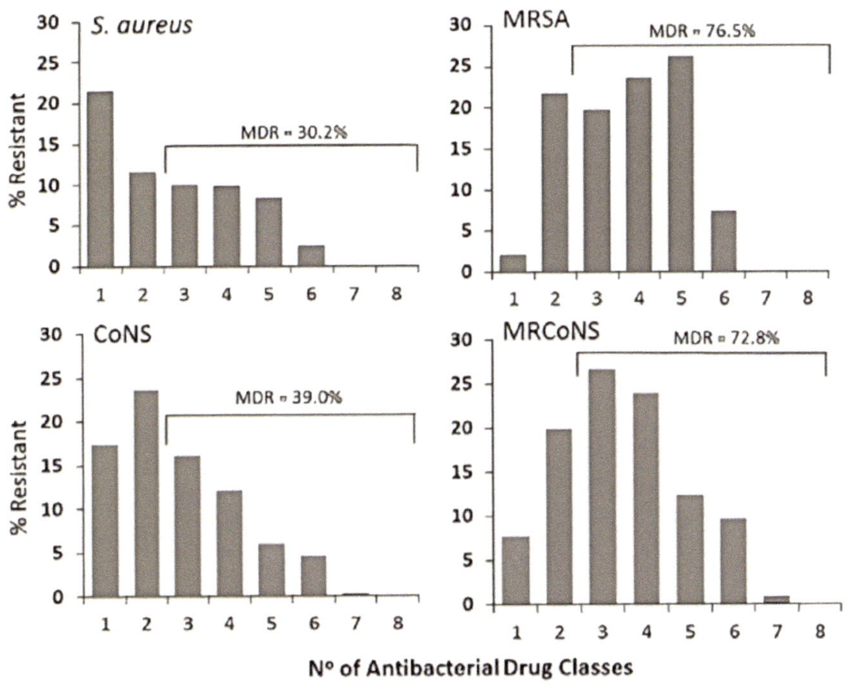

FIGURE 13.3 Multidrug resistance (MDR) among conjunctival isolates of *S. aureus*, coagulase-negative staphylococci (CoNS), methicillin-resistant *S. aureus* (MRSA) and methicillin-resistant coagulase-negative staphylococci (MRCoNS). Isolates were tested against ciprofloxacin, azithromycin, clindamycin, chloramphenicol, tobramycin, oxacillin, tetracycline, vancomycin, and trimethoprim. Percent resistance included intermediate resistance. (From Asbell and DeCory (2018). This is an open access article distributed under the terms of the Creative Commons Attribution License.)

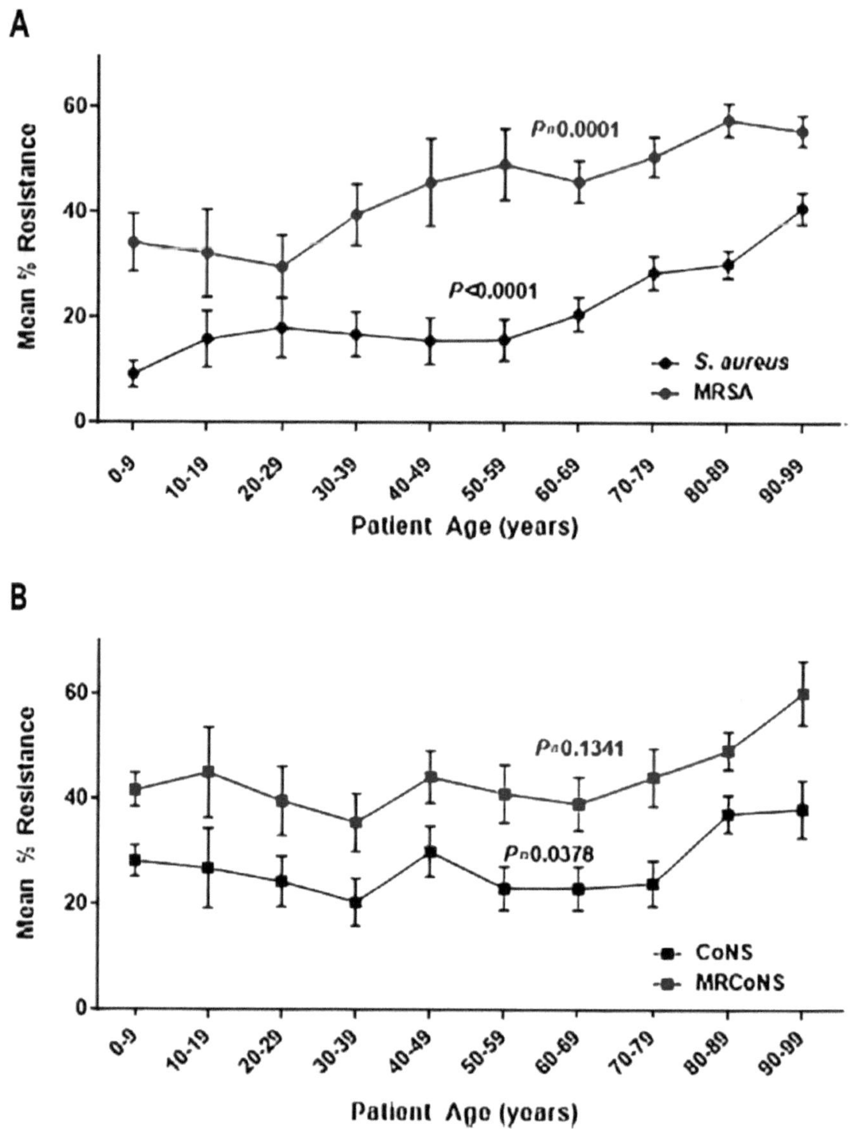

FIGURE 13.4 Increase in rate of resistance among *S. aureus,* coagulase-negative staphylococci (CoNS), methicillin-resistant *S. aureus* (MRSA) and methicillin-resistant coagulase-negative staphylococci (MRCoNS) from the conjunctiva with the age of the patient. (A) Mean ± SE percentage of resistance for *S. aureus* (black circles) and MRSA (blue circles); (B) mean ± SE percentage of resistance for CoNS (black squares) and MRCoNS (blue squares). In this study not all antibiotic classes were tested but the means of the percentage of drug classes to which each isolate of a species/species group was resistant was taken and analysed by ANOVA. $P<0.1$ criterion was considered statistical significant. (From Asbell and DeCory (2018). This is an open access article distributed under the terms of the Creative Commons Attribution License. Insert A and B.)

and resistance increased with increasing age among *S. aureus*, MRSA, and CoNS (Figure 13.4). During 2009–2018, *S. aureus*, CoNS and *P. aeruginosa* showed a significant decrease in resistance to different antibiotics. *S. aureus* showed decrease in resistance to azithromycin, ciprofloxacin, tobramycin, trimethoprim, and oxacillin, whereas CoNS and *P. aeruginosa* showed decrease in resistance to ciprofloxacin (Asbell and DeCory, 2018).

13.6.3 Emergence of Antimicrobial Resistance in Ocular Microorganisms in UK

A total of 2681 ophthalmic isolates were obtained from eye swabs (88.0%), conjunctival swabs (3.7%), contact lens swabs (2.0%), corneal scrapes (5.5%), and invasive samples (0.8%) from 2168 patients, between 2009 and 2015, from North-West London from primary, secondary, and tertiary care settings and their resistance to drugs was monitored (Lee et al., 2019). The pathogens isolated varied depending on the care facility and on the age of the individual with *Staphylococcus* spp., *S. aureus* and *Pseudomonas* spp. predominant in adults. In children, the most common pathogens included *Haemophilus* spp., *H. influenzae, Streptococcus spp., Staphylococcus spp., Moraxella spp.* and rarely *Pseudomonas* spp. The increased abundance of the pathogen *Staphylococcus spp.* in adults was indicative of increased use of contact lens in adults (Eltis, 2011). Further it was observed that the abundance of the isolates varied depending on the season and followed the trend spring (31.6%)> summer (24.8%) >, winter (23.9%), and autumn (19.7%) (Figure 13.5). Antimicrobial resistance increased across care settings. Resistance to chloramphenicol increased significantly from 11.8% to 33.8% from primary to tertiary care

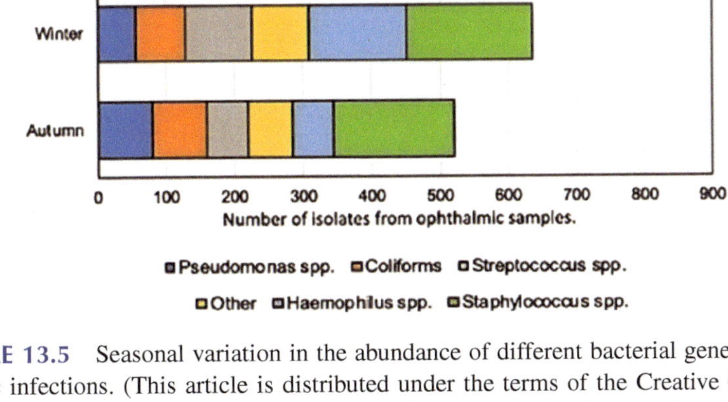

FIGURE 13.5 Seasonal variation in the abundance of different bacterial genera in ophthalmic infections. (This article is distributed under the terms of the Creative Commons Attribution 4.0 international license (http://creative commons.org/licenses/by/4.0/).)

and the intrinsically resistant *Pseudomonas* spp. was the most isolated chloramphenicol resistant pathogen. The observed increase in the intrinsically resistant *Pseudomonas* spp. may be attributed to increased contact lens-associated infections seen in tertiary versus primary care (Azari and Barney, 2013). This intrinsically resistant *Pseudomonas* spp. is also the most predominant moxifloxacin-resistant pathogen across all levels of care. The coliforms from all levels of care were also most frequently resistant to chloramphenicol. This contrasts with the findings of Silvester et al. (2016) on adult bacterial conjunctivitis who found lower resistance rate to chloramphenicol (8%). The resistant pathogen to fusidic varied between the centres with *Haemophilus* spp. dominant in both primary and secondary care, whereas *Pseudomonas* spp. was dominant in tertiary care. Lee et al. (2019) observed that resistance to fusidic acid among *Staphylococcus* spp., were similar between primary and secondary care (8.9% and 11.2% respectively) but increased significantly in tertiary care (27.0%).

13.6.4 Antimicrobial Resistance in Bacterial Isolates from Conjunctivitis Patients in UK

Conjunctival swabs from 8209 adults with suspected bacterial conjunctivitis from the Royal Liverpool University Hospital, UK, between 2001 and 2012 were analysed and 1300 (15.8%) were found to be culture positive for bacteria (Silvester et al., 2016). During this period resistance to chloramphenicol in the culture positive samples varied from a minimum of 3 to a maximum of 16.4%. Overall, the Gram-negative bacteria *Haemophilus influenzae, Pseudomonas sp., Moraxella catarrhalis, Morganella morganii, Neisseria gonorrhoeae and N. meningitidis,* were more resistant to chloramphenicol (8.1 to 37.3%). Resistance to ciprofloxacin and gentamicin between 2001–2012 was 16.4% and 14.0% among Gram-positive organisms *(Staphylococcus aureus and Streptococcus pneumoniae).* Further, a significant reduction in methicillin resistance among *S. aureus* isolates (mean 8.3%) over the 12-year period was observed (Figure 13.3). Based on these results it was suggested that chloramphenicol could be used particularly for conjunctivitis involving Gram-positive bacteria (Silvester et al., 2016).

13.6.5 Antimicrobial Resistance in Bacterial Isolates from Conjunctivitis patients in Ethiopia

In a prospective study carried out in Ethiopia, between September 1, 2018, to March 30, 2019, it was observed that 32.8% of individuals were diagnosed as having bacterial conjunctivitis (Mohammed et al., 2020). *S. aureus*, CoNS and *E. coli* were the most predominant bacteria (>10%) and were resistant to tetracycline and penicillin but majority of them were susceptible to ciprofloxacin. But the resistance data cannot be interpreted as that observed for conjunctivitis isolates since it was reported for all the ocular and periocular infections of the eye (Mohammed et al., 2020). It may be noted that though bacterial conjunctivitis knows no limits and is reported across the world the resistance of the isolates to antibiotics has not been studied in most cases.

13.7 Bacterial Resistance to Different Classes of Antibiotics

13.7.1 Bacterial Resistance to Aminoglycosides and Polymyxin B in Ocular Isolates from Conjunctivitis Patients

Several studies confirmed during different time periods 1990 to 1998 (Alexandrakis et al., 2000), 1994 to 2003 (Cavuoto et al., 2008), 1997–1998 (Block et al., 2000) and 2001 to 2002 (Morrissey et al., 2004) that several ocular isolates from conjunctivitis patients were resistant to aminoglycosides. For instance, conjunctival isolates like *S. aureus* (Cavuoto et al., 2008; Alexandrakis et al., 2000), *S. pneumoniae, S. aureus* (MSSA) and coagulase-negative *Staphylococcus* (Morrissey et al., 2004) both from adults and paediatric patients (Block et al., 2000) were resistant to topical gentamicin, tobramycin and polymyxin B (Morrissey et al., 2004; Alexandrakis et al., 2000; Cavuoto et al., 2008; Block et al., 2000). None of these studies showed any change in the pattern of resistance with time.

In Brazil, resistance to aminoglycosides was monitored among *S. pneumoniae* between 1989 to 1992 and it was observed that 100% of *S. pneumoniae* from the cornea or conjunctiva were susceptible to tobramycin and gentamicin (Chalita et al., 2004). But, subsequently from 1997 to 2000, in corneal and conjunctival cultures of *S. pneumoniae*, the susceptibility to tobramycin and gentamicin decreased substantially from 43.6–46%% for tobramycin and 42.3–56% respectively to gentamicin thus indicating an increased trend in resistance (Chalita et al., 2004). The most recent surveillance data (1999 to 2006) from the ocular TRUST study showed resistance rates of 73.1% to tobramycin and 100% to polymyxin B by *S. pneumoniae* but virtually no resistance to either antibiotic was observed in *H. influenzae* (Asbell et al., 2008).

13.7.2 Bacterial Resistance to Macrolides in Ocular Isolates from Conjunctivitis Patients

In October to April 2005 in the United States 34% of respiratory isolates of *S. pneumoniae* were resistant to the macrolide antibiotic azithromycin and in addition 80.6% of penicillin-resistant *S. pneumoniae* were also resistant to azithromycin (Critchley et al., 2007). The trends in resistance over a ten-year period (1994–2003) indicated that resistance to erythromycin in both *S. pneumoniae* and *S. aureus* increased two-fold from 22.1% (1994–1995) to 45.1% (2002–2003) and 23.8 to 48.9% respectively (Cavuoto et al., 2008). It was also demonstrated that four of seven strains of azithromycin resistant *Pneumococci* could be selected by using increasing concentrations of azithromycin (Asbell et al., 2008). Isolates from patients with bacterial conjunctivitis exhibited moderate to very high resistance to azithromycin and the MIC90 varied depending on the specific isolate. For instance, in *H. influenzae* the MIC90 to azithromycin was three-fold higher, for *S. pneumoniae* 16-fold higher, and for *S. aureus* and *S. epidermidis* were 128-fold higher (Ohnsman et al., 2007). Several other studies also confirmed that *S. pneumoniae* and *H. influenzae* were more than 70% resistant to azithromycin as observed in The TRUST surveillance study (1999 to 2006) and as reported by Block et al. (2000), Ohnsman et al. (2007) and Buznach et al. (2005). *H. influenzae* isolates were also resistant to erythromycin (78%) (Block et al.,

2000; Ohnsman et al., 2007; Buznach et al., 2005). Resistance to macrolides could be due to various reasons like its widespread usage, bacteriostatic effect, and persistence for a long period as in the case of azithromycin (Bergman et al., 2006; Dorfman et al., 2008).

13.7.3 Bacterial Resistance to Fluoroquinolones in Ocular Isolates from *Conjunctivitis Patients*

Resistance to fluoroquinolone antibiotics has been attributed to single or multiple mutations. The former confers "low-level" of resistance and is amenable to treatment (Egger et al., 2001). In contrast, when the bacterium has acquired multiple mutations (>2 mutations) it is critical and could result in treatment failure. Several ocular bacterial isolates including *S. aureus* and CoNS are resistant to ciprofloxacin and ofloxacin (8–36%) (Morrissey et al., 2004; Alexandrakis et al., 2000; Goldstein et al., 1999; Marangon et al., 2004) but in contrast resistance by *S. pneumoniae* and *H. influenzae* to fluoroquinolones in general including newer fluoroquinolones (gatifloxacin and moxifloxacin) is rare (Asbell et al., 2008; Ohnsman et al., 2007) although they may exhibit low resistance to ciprofloxacin (Asbell et al., 2008). Resistance of *S. aureus* from bacterial conjunctivitis and keratitis cases to ciprofloxacin was low but the resistance doubled between the periods 1990 to 1995 and 1996 to 2001 from 8–20.7% (Marangon et al., 2004), whereas in another study resistance to ciprofloxacin increased three-fold between 1994 to 1995 and 2002 to 2003 from 13.3–36.0% (Cavuoto et al., 2008). Over the years resistance to newer fluoroquinolones has been noticed and could be attributed to the extensive use of fluoroquinolones for systemic infections (Asbell et al., 2008; Jhanji et al., 2007; Kotlus et al., 2006; Miller et al., 2006; Brown, 2007). For instance, TRUST surveillance study data for the period October 2005 through June 2006 reported that 18.9 and 15.9% of methicillin-sensitive *S. aureus* isolates were resistant to gatifloxacin and moxifloxacin, respectively (Asbell et al., 2008). In the United States, a 10-year study, involving patients with proven bacterial conjunctivitis indicated that MRSA increased from 4.4% (1994–1995) to 42.9% (2002–2003) (Cavuoto et al., 2008) and this in contrast to 3% in United Kingdom (Shanmuganathan et al., 2005a; b) but comparable to 57% in Japan (Fukuda et al., 2002). The resistance of ocular isolates (including isolates from conjunctivitis patients) of MRSA to fluoroquinolones varied depending on the specific drug used and the country. For instance, resistance to gatifloxacin was 57.1% in six European countries from 2001 to 2002 (Morrissey et al., 2004), whereas in the United States the resistance rates were 68% for moxifloxacin, 71% for gatifloxacin, and 94% for ciprofloxacin and ofloxacin (Deramo et al., 2006). The Ocular TRUST study (October 2005 through June 2006) covering 19 states in the United States also reported high resistance rates by MRSA ranging from 63.6% to tobramycin to 100% to polymyxin B (100%). The resistance to other antibiotics was azithromycin 90.9%, ciprofloxacin 84.8%, levofloxacin 78.8%, gatifloxacin 81.8% and moxifloxacin 75.8% (Asbell et al., 2008). Further all the ciprofloxacin-resistant isolates including MRSA and MRSE were also resistant to the other fluoroquinolones (levofloxacin, ofloxacin, gatifloxacin, and moxifloxacin) (McDonald et al., 2008; Haas et al., 2009). Recent studies also indicated that 65% of MRSA were resistant to ophthalmic suspension of besifloxacin (0.6%) used for the treatment of bacterial conjunctivitis (Karpecki et al., 2009).

13.8 Antimicrobial Resistance in Specific Ocular Isolates from Conjunctivitis Patients

This aspect has already been mentioned above with emphasis on the resistance associated with different bacteria associated with conjunctivitis. But it does not include all the resistance profiles of a given ocular pathogen.

13.8.1 Resistance in *Pseudomonas aeruginosa*

The ARMOR 2013 study indicated that resistance of *P. aeruginosa* were low against all antibiotics tested except for polymyxin B, susceptibility was above 90% (Grzybowski et al., 2017).

13.8.2 Resistance in *Haemophilus influenzae*

In Ocular TRUST1 (1999–2006), it was observed that almost all *H. influenzae* isolates (99.7–100%) were susceptible to tobramycin, fluoroquinolones, azithromycin and polymyxin B with the exception of 14.3%, which were resistant to trimethoprim (Asbell et al., 2008). Subsequently, the Ocular TRUST1 study, which was carried out in October 2005–June 2006, indicated that β-lactamase producing isolates (Asbell et al., 2008) were susceptible to the above antibiotics tested except for trimethoprim (81.3% susceptible) (Asbell et al., 2008). These results were confirmed by the ARMOR study (2009–2013), which also showed that all the isolates were susceptible to all antibiotics tested in 2009 and in the following 4 years also a similar trend was observed except for one isolate (n = 284), which was resistant to chloramphenicol and one resistant to azithromycin (Haas et al., 2011; Asbell et al., 2015).

13.8.3 Resistance in CoNS and *Staphylococcus aureus*

Differential susceptibility to fluoroquinolones was reported for MSSA and MRSA in the Ocular TRUST1 study. Majority of the MSSA isolates (about 80%) were susceptible to ciprofloxacin, gatifloxacin, levofloxacin and moxifloxacin (Asbell et al., 2008), whereas in contrast majority of the MRSA isolates exhibited 75–85% resistance against the above fluoroquinolones. Further, trimethoprim was consistently active against both MSSA and MRSA (Asbell et al., 2008). The high resistance of MRSA isolates to ciprofloxacin (79.5%) and moxifloxacin (65.4%) was confirmed by Hass et al. (2011). This study also reported that MRSA isolates were multidrug-resistant since they were not only resistant to ciprofloxacin but were also resistant to tobramycin (Hass et al., 2011). The ARMOR 2009 and ARMOR 2013 studies indicated the emergence of multidrug resistance in *S. aureus* (Haas et al., 2011), in MRSA and in methicillin-resistant CoNS (*S. epidermidis*) (Asbell et al., 2015). Further, the resistance profile of CoNS and *S. aureus* followed similar trends with resistance rates of 49.7% and 42.2% to methicillin, 34.4% and 39.8% to ciprofloxacin and 61.3% and 63.6% to azithromycin respectively (Grzybowski et al., 2017) and resistance did not increase during the 5-year period of ARMOR 2013 study (Asbell et al., 2015). However, the subsequent ARMOR studies carried out in 2014 and 2015 showed high levels of resistance in ophthalmology, including multi-drug resistance (Bausch and Lomb, 2016).

13.8.4 Resistance in *Staphylococcus pneumoniae*

Archived samples, from ocular infections of 760 *S. pneumoniae* (1999 and 2006) exhibited 34.1% penicillin-resistance, with cross-resistance to azithromycin and trimethoprim in nearly three-quarters (Asbell et al., 2008). All the 760 isolates were susceptible to the fluoroquinolones gatifloxacin, levofloxacin and moxifloxacin (Asbell et al., 2008) and were resistant to polymyxin B (Asbell et al., 2008). Longitudinal analysis indicated that between 1999 to 2006 *S. pneumoniae* susceptibility rates to fluoroquinolones, azithromycin, penicillin, and trimethoprim were stable (Asbell et al., 2008).

13.9 Childhood Bacterial Conjunctivitis

Pichicherio (2011), identified several microorganisms as associated with acute conjunctivitis in children (Table 13.2) and non-typeable *H. influenzae* was the predominant organism followed by *S. pneumoniae* (Patel et al., 2007; Block et al., 2000). Treatment involved administration of oral broad spectrum bactericidal agent in the case of acute bacterial conjunctivitis with otitis media (Bodor, 1989), whereas topical ophthalmic agents were used in uncomplicated acute bacterial conjunctivitis. Wald et al. (2001) showed that oral cefixime and topical polymyxin B/bacitracin were

TABLE 13.2

Pathogens Associated with Acute Conjunctivitis in Children

Group	Age	Bacterium
Neonates	<1 week	*Neisseria gonorrhoeae*
Neonates	1–2 weeks	*Chlamydia trachomatis*
		Haemophilus influenzae
		Streptococcus pneumoniae
Infants and toddlers	1–5 years	*H. influenzae*
		S. pneumoniae
		Moraxella catarrhalis
		Staphylococcus aureus
Infants and Toddlers with otitis	1–5 years	*H. influenzae*
		S. pneumoniae
School-aged children/ adolescents		*S. aureus*
		H. influenzae
		S. pneumoniae

Source: Modified from Teoh and Reynolds (2003.

equally effective in achieving clinical cure in children below two and half years with acute bacterial conjunctivitis implying that an oral antibiotic is sufficient therapy and the addition of an ocular antibiotic is not needed. In a paediatric hospital in Greece, between 2013 to 2018, analysis of acute bacterial conjunctivitis in 191 cases of neonates under 28 days of life (Group A), children from 1 month to 2 years (Group B) and children from 2 years up to 14 years (Group C), revealed that *Staphylococcus* spp., *Haemophilus* spp. and *Streptococcus* spp. were the most prevalent pathogens in the three groups. Antibiotic resistance in the three groups was high against ampicillin, ceftriaxone, ceftazidime, and sulfamethoxazole, whereas resistance to ciprofloxacin were low and none of the isolates were resistant to vancomycin (Oikonomakou et al., 2019). Currently, aminoglycosides, polymyxin B combination therapies, macrolides, or fluoroquinolones are used as topical antimicrobials (Harbarth and Samore, 2005) for children 1 year and older.

13.9.1 Resistance to Aminoglycosides in Conjunctival Isolates from Children

Aminoglycosides (gentamicin and tobramycin) are active against ocular isolates of MSSA and *P. aeruginosa* (Gwon, 1992a; b) to a maximum extent of 77 and 70% respectively in patients of all ages (Rose, 2007) including children (<20 years) (Timewell et al., 1983). Resistance of MRSA and *S. pneumoniae* to tobramycin in 2005–2006 was comparable (63.6% and 65.3%) (Asbell et al., 2008). Further when archived isolates of *S pneumoniae* obtained between 1999 and 2006 were monitored they showed that their resistance to tobramycin was comparable to that seen by Asbell et al. (2008) in 2005–2006. Further it was observed that in the archived isolates the resistance to tobramycin was dependent on whether penicillin-sensitive *S. pneumoniae* (PSSP) *(59.9% resistance) or* penicillin-non-susceptible *S. pneumoniae* (PNSP) isolates were used (73.1% resistance) (Asbell et al., 2008).

13.9.2 Resistance to Polymyxin B Combination Therapy in Conjunctival Isolates from Children

Polymyxin B is bactericidal to Gram-negative organisms and therefore it is used in combination with other antibiotics for a broader spectrum of coverage. The common combinations include polymyxin B/trimethoprim, polymyxin B/bacitracin, and polymyxin B/neomycin/bacitracin (Pichichero, 2011), which are effective against both the Gram-negative and Gram-positive microorganisms even in childhood conjunctivitis. For instance, it is known that trimethoprim and bacitracin are active against most staphylococci and streptococci. But the polymyxin B combination regimens were not significantly effective compared to the single use antibiotic therapy when tested in paediatric patients (Block et al., 2000). Block et al. (2000) also showed that polymyxin B was not effective against *S. pneumoniae* and both penicillin-sensitive and penicillin-non-susceptible *S. pneumoniae* were not susceptible to polymyxin B. But when polymyxin B is used in combination with either neomycin or trimethoprim the combination was active against both *S. pneumoniae* and PSSP. Ocular TRUST 1 study also confirmed the resistance of *S. pneumoniae (100%)* and MSSA to polymyxin B. Trimethoprim was effective against PSSP, but 74% of PNSP were resistant to trimethoprim (Asbell et al., 2008).

13.9.3 Resistance to Macrolides in Conjunctival Isolates from Children

Macrolides like azithromycin and erythromycin are active against Gram-positive cocci and have been used as an ocular antibiotic for more than 25 years as a 0.5% ointment even for treatment of conjunctivitis from children. However, resistance among *Staphylococcus* species and poor activity against *H. influenzae* have relegated erythromycin to a marginal role in the treatment of bacterial conjunctivitis (Cavuoto et al., 2008). In the United States, *S. pneumoniae* and *S. aureus* isolates from conjunctivitis patients showed 20% and 30% resistance to azithromycin respectively (Ohnsman et al., 2007). Resistance to azithromycin among 76% of *H. influenzae* isolates was also reported (Pichichero, 2011).

13.9.4 Resistance to Fluoroquinolones in Conjunctival Isolates from Children

Fluoroquinolones (ofloxacin, ciprofloxacin, levofloxacin, moxifloxacin, and besifloxacin) are effective (Kaliamurthy et al., 2005) in both children and adults (66–96%) with bacterial conjunctivitis and the recently introduced fluoroquinolones (levofloxacin, moxifloxacin, and besifloxacin) were more effective (84% to 96%) (Keating et al., 2009). Besifloxacin 0.6% with appreciable clinical resolution rates (45% to 73%) and reasonably high bacterial eradication rates (88% to 91%) has been used both in children and adults to treat bacterial conjunctivitis (Karpecki et al., 2009; McDonald et al., 2009). But the emergence of resistance to both older and newer ophthalmic fluoroquinolones (ciprofloxacin, levofloxacin, and moxifloxacin) has discouraged their use. Resistance rates vary from organism to organism and ranged from 90% to 92% in MSSA isolates, but only 27% to 32% in MRSA isolates. This is further complicated by the observation that a consistent annual 2.5% increase in MRSA as a cause of ocular infections was identified (Wald et al., 2001). Another study reported an increase in resistance to ciprofloxacin by *S. aureus* isolates from 13.3% to 36.0% and the prevalence of methicillin resistance among these isolates increased concurrently from 4.4% to 42.9% (Cavuoto et al., 2008).

13.10 Other Effects of Topical Ophthalmic Antibiotics Used for the Treatment of Bacterial Conjunctivitis

Most topical ophthalmic antibiotics for treatment of bacterial conjunctivitis are safe and well tolerated except for a few adverse events, which are mild and transient. A few of the adverse effects are listed below:

(a) Aminoglycosides exert corneal and conjunctival toxicity when used frequently for instance tobramycin causes superficial punctate lesions and tobramycin, gentamicin, and neomycin cause ocular allergic reactions (Chou et al., 2004).
(b) Bacitracin a polypeptide antibiotic has been associated with contact dermatitis in the periocular area (Jacob and James, 2004).
(c) Polymyxin B/trimethoprim sulfate combination causes local irritation, whereas

Polymyxin B/bacitracin/neomycin combination cause allergic reactions.

(d) Macrolides are associated with minor ocular irritations, redness, and hypersensitivity reactions.

(e) Fluoroquinolone ophthalmic solutions are less toxic (eg, burning/stinging, chemosis, photophobia and so forth) compared to the other antibacterial classes (Roberts, 2002; Granet et al., 2007).

Pichichero (2011) suggested that an ophthalmic antibiotic for topical treatment of acute bacterial conjunctivitis should be non-toxic, well-tolerated, broad-spectrum, highly potent, exhibit rapid bactericidal activity and have convenient dosing regimens. Further AMR should be considered in selecting the appropriate antibiotic.

13.11 Overcoming the Dissemination of Resistance in Ophthalmology

Several strategies have been used to prevent the emergence and dissemination of resistance in ophthalmology such as avoiding the use of subtherapeutic dosage, using only the specific antibiotic, using only antibiotics with a tangible benefit, and using whenever possible alternatives to antibiotics.

13.11.1 Using Subtherapeutic Dose of the Drug

Using a subtherapeutic dose would fail to kill all the microorganisms thus allowing the surviving proportion of microbes to develop resistance, multiply and spread. It was demonstrated that subtherapeutic doses prescribed to about 50% of the children contributed to suboptimal clinical outcomes and an increased predominance of resistant pathogenic fungi (Lestner et al., 2015). Similarly, resistance can be promoted with antimicrobial treatment duration that is too short. It is in this context that it is important to prescribe the correct dosage for a given antibiotic.

13.11.2 Using the Specific Antibiotic

Antibiotic resistance could be overcome by prescribing the correct dose and the correct antibiotic. Time consuming methods (24–48 h) based on traditional and conventional methods of culturing and drug sensitivity tests based on disk diffusion or broth dilution resistance tests should be avoided (Goff et al., 2012). The more rapid modern DNA based methods like polymerase chain reaction (PCR), peptide nucleic acid fluorescence in situ hybridization (PNA-FISH) and bacterial cytological profiling, which are rapid should be employed (Quach et al., 2016).

13.11.3 Using Antibiotics only when Proven to Provide a Tangible Benefit

Antibiotics should be used for a specific and tangible benefit and not for antimicrobial prophylaxis (Parke et al., 2013). Antibiotics should also not be used topically prior to cataract surgery (Grzybowski et al., 2016) or for the treatment of non-target organisms like viruses or when allergic conjunctivitis is suspected (Parke et al., 2013; Azari and Barney, 2013). Moreover, topical antibiotics are contraindicated, as they do not protect against secondary infections, and may blur the clinical presentation by causing toxicity or allergic reactions (Azari and Barney, 2013).

13.11.4 Alternatives to Antibiotics

Several substitutes or adjuncts are emerging as substitutes to antibiotics (Czaplewski et al., 2016; Dossarps et al., 2015). Antiseptics like povidone iodine (PVI) and chlorhexidine act at the membrane level and cause cell death or have a bacteriostatic effect. PVI releases free iodine, which penetrates the microbial membrane, oxidise the proteins within and causes cell death (Lachapelle et al., 2013). Chlorhexidine, on the other hand destroys cell membranes and exerts a bacteriostatic effect. At higher concentrations, chlorhexidine is bactericidal causing leakage, coagulation, and precipitation of cellular contents (Lachapelle et al., 2013). Both antiseptics are effective against Gram-positive bacteria with PVI having a larger spectrum of activity against microorganisms. MRSA and other staphylococcal infections are known to be resistant to chlorhexidine (Horner et al., 2012). Isenberg et al. (2017) showed that PVI (1.25%) was an effective alternative to antibiotics in treating bacterial keratitis, suggesting that antiseptic use in ophthalmology has come to stay (Isenberg et al., 2017).

13.12 Viral Conjunctivitis

Viruses cause up to 80% of all cases of acute conjunctivitis and the viruses include adenoviruses (Stenson et al., 1982; Fitch et al., 1989), herpes simplex virus (HSV), and enterovirus (Newman and Gooding 2013).

13.12.1 Adenoviral Conjunctivitis

Human adenovirus (HAdV) genotypes have been classified into seven different species HAdV-A through HAdV-G (Singh et al., 2015; Kuo, 2019). These cause about 90% of viral conjunctivitis cases (O'Brien et al., 2009), which is very contagious (Udeh et al., 2008) and spreads through human contaminated body parts, medical devices, water or by sharing of personal items (Jackson et al., 2002). Spread of adenovirus is best overcome by washing of hands frequently, disinfecting instruments, and by isolating conjunctivitis patients (Bremond-Gignac et al., 2010). Fortunately, many of the clinical symptoms of adenoviral conjunctivitis could be prevented by frequently applying artificial tears, antihistamines eye drops, or cold compresses (Varu et al., 2019; Leibowitz, 1991; Richards and Guzman-Cottrill, 2010). Antiviral and antibiotic eye drops are not effective in the treatment of viral conjunctivitis but topical administration of PVI 2% four times a day for one week could lead to resolution of the disease in three-quarters of the eyes (Trinavarat and Atchaneeyasakul, 2012). Topical corticosteroids are not recommended (Kovalyuk et al., 2017); but combination therapies with corticosteroids and antibiotics are effective in treating both viral and bacterial conjunctivitis (Belfort et al., 2012; Kovalyuk et al., 2017; Clement et al., 2011; Pepose et al., 2019). A combination of a corticosteroid (eg, dexamethasone) and PVI has been used to treat both inflammatory and infectious conjunctivitis (Table 13.3). Other medications that have

TABLE 13.3

Combination Therapy Involving a Corticosteroid and an Anti-infective Agent for Treatment of Conjunctivitis

Corticosteroid Combination	Product Indication/Status	Reference
Tobramycin 0.3% plus dexamethasone 0.1% (TobraDex; Alcon Laboratories, Inc., Fort Worth, TX)	Corticosteroid prescribed for the inflammatory ocular condition and the antibiotic for superficial bacterial ocular infection or when a risk of bacterial ocular infection exists.	TobraDex (2009).
Tobramycin 0.3% plus loteprednol etabonate 0.5% (Zylet)		Zylet (2016).
Netilmicin 0.3% plus dexamethasone 0.1% (Netildex; Knight Therapeutics, Inc., Montreal, Quebec, Canada)		Faraldi et al. (2013).
Neomycin, polymyxin B sulfates plus dexamethasone (Maxitrol; Alcon Laboratories, Inc., Fort Worth,TX)		Maxitrol (2016).
Sulfacetamide plus prednisolone (Blephamide)		Blephamide (2017).
PVP-I 0.4% plus dexamethasone 0.1% (FST-100)	Tested for acute viral conjunctivitis	Pinto et al. (2015).
PVP-I 0.6% plus dexamethasone 0.1% (SHP640)	Ongoing phase 3 trials for treatment of bacterial and adenoviral conjunctivitis	ClinicalTrials.gov.https://clinicaltrials.gov/ct2/show/ NCT02998554 (2018). ClinicalTrials.gov. https://clinicaltrials.gov/ct2/show/ NCT02998541 (2018). ClinicalTrials.gov. https://clinicaltrials.gov/ct2/show/NCT03004924 (2018).
PVP-I 1.0% plus dexamethasone 0.1%	Tested for treatment of adenoviral keratoconjunctivitis	Kovalyuk et al. (2017).

Source: Modified from Holland et al. 2019.

been advocated include 1% and 2% cyclosporine-A (CsA) eye drops for the treatment of subepithelial infiltrates (SEIs), and for reducing the amounts of infiltrates (Levinger et al., 2010; Reinhard et al., 2000; Hillenkamp et al., 2001) and topical tacrolimus eye drops for the treatment of SEIs and to improve visual acuity of the patients (Berisa et al., 2017). Adenoviruses are known to cause Pharyngoconjunctival fever (PCF) and Epidemic keratoconjunctivitis (EKC) both involving the conjunctiva.

13.12.1.1 Pharyngoconjunctival Fever (PCF)

HAdV types 3, 4, and 7 cause pharyngoconjunctival fever (PCF) characterised by abrupt onset of high fever, pharyngitis and periauricular lymphadenopathy (Azari and Barney, 2013). In addition, they cause acute bilateral follicular conjunctivitis accompanied with edema, hyperemia, and petechial haemorrhages of the conjunctiva due to interaction between pro-inflammatory cytokines and conjunctival vasculature. PCF is not considered severe since it is self-limited and resolves spontaneously in two to three weeks without any treatment (Li et al., 2018a; Harley et al., 2001).

13.12.1.2 Epidemic Keratoconjunctivitis

Adenovirus serotypes 4, 8, 19, 37, D53 and D54 are responsible for the majority of EKC cases (Richards and Guzman-Cottrill, 2010). EKC disease affects both the conjunctiva and cornea and patients present with several ocular manifestations, which include conjunctival discharge, follicular conjunctivitis, corneal SEI, corneal scarring, development of conjunctival membranes and pseudo-membranes and symblepharon formation (Jhanji et al., 2015) (Figures 13.6 and 13.7), superficial punctate keratopathy and epithelial opacities (Chigbu and Labib, 2018). The opacities may last for years and

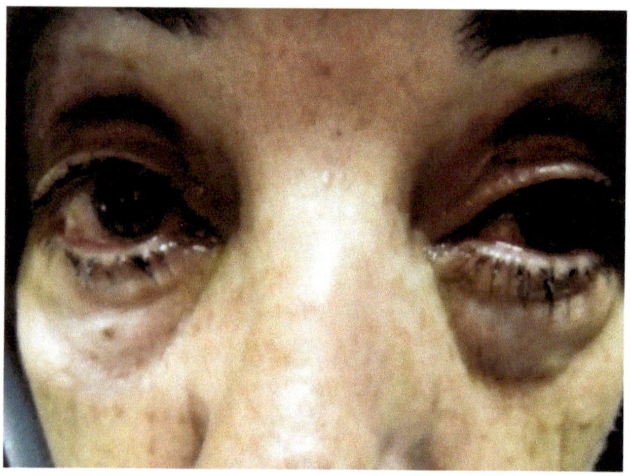

FIGURE 13.6 A patient with bilateral watery eyes due to adenoviral conjunctivitis. (Adapted from Azari and Arabi (2020) published as an open access article and distributed under the terms of the Creative Commons Attribution-Non-commercial-ShareAlike 4.0 License.)

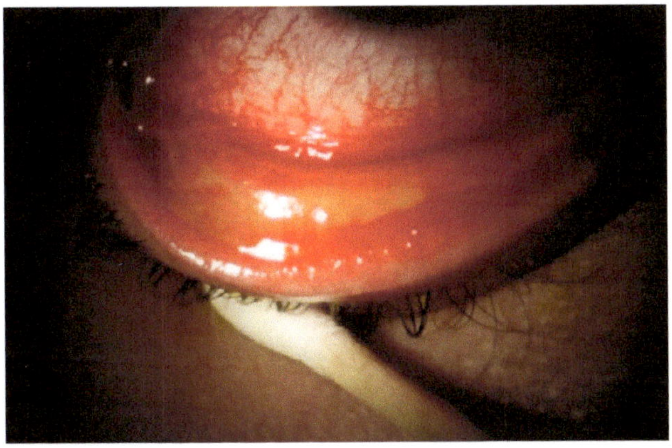

FIGURE 13.7 A patient with pseudo-membrane formation due to adenoviral conjunctivitis. (Adapted from Azari and Arabi (2020) published as an open access article and distributed under the terms of the Creative Commons Attribution-Non-commercial-ShareAlike 4.0 License.)

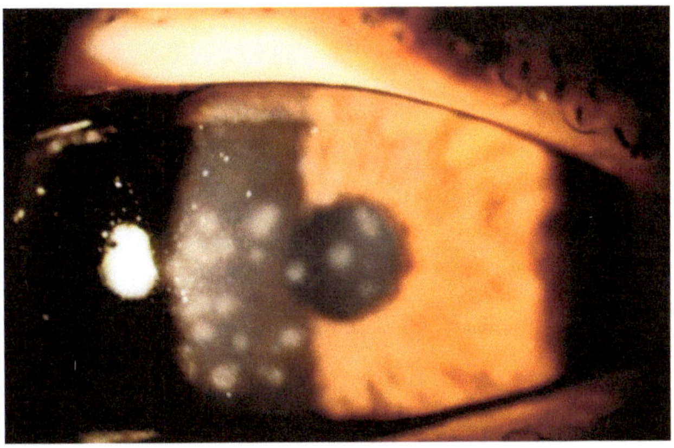

FIGURE 13.8 A patient with adenoviral conjunctivitis showing subepithelial infiltrations. (Adapted from Azari and Arabi (2020) published as an open access article and distributed under the terms of the Creative Commons Attribution-Non-commercial-ShareAlike 4.0 License.)

cause visual disturbance, photophobia, and astigmatism. Stromal SEI (in 49 to 80%) appear after a week of initiation of EKC (Richmond et al., 1984; Okumus et al., 2012) (Figure 13.8).

13.12.2 Herpetic Conjunctivitis

Compared to adenoviral conjunctivitis, which causes about 90% of viral conjunctivitis, HSV affects only a minor fraction (1.3–4.8%) of acute conjunctivitis individuals

(Sheikh et al., 2012; Papa et al., 2002). HSV conjunctivitis patients commonly presents with unilateral follicular conjunctivitis accompanied with a thin watery discharge and additionally vesicular lesions are seen on the eyelid skin. The antiviral agents (ganciclovir, idoxuridine, vidarabine, and trifluridine) are topically used as effective therapy. The other manifestations of HZV apart from conjunctivitis (41.1% of cases) include eyelid lesions (45.8%), uveitis (38.2%) and corneal lesions such as SEIs, pseudo-dendrites, and nummular keratitis (19.1%) (Puri et al., 2011).

13.12.3 Acute Haemorrhagic Conjunctivitis

Picornaviruses include the enterovirus 70 (EV70), coxsackievirus A24 variant (CA24v), and specific subtypes of adenoviruses and are the etiological agents of acute haemorrhagic conjunctivitis (AHC), a contagious form of conjunctivitis (Lim et al., 1973; Zhang et al., 2017; Richards and Guzman-Cottrill, 2010). AHC may also be transmitted primarily by hand-to eye-to-hand contact and infected fomites (Langford et al., 2015). AHC is a self-limiting disease and several symptoms that accompany the disease such as foreign body sensation, tearing, eyelid edema, dilatation of conjunctival vessels, chemosis, and subconjunctival haemorrhage diminish with time and completely resolve in two weeks (Langford et al., 2015). A few patients may also complaint of fever, fatigue, and leg pain. Affected individuals need to frequently wash hands and avoid contact with others to prevent spread into vulnerable groups including children, elderly, pregnant women, and immunocompromised individuals (Azari and Arabi, 2020).

13.12.4 Miscellaneous Viral Conjunctivitis

Several other viruses have been shown to cause conjunctivitis and these include P-2-virus that causes *Molluscum contagiosum* (MC), Ebola virus and coronaviruses. Transmission of the virus leading to conjunctivitis occurs due to skin-to-skin contact and sexual intercourse as in P-2-virus, through bodily fluids as in Ebola virus (Judson et al., 2015; Moshirfar et al., 2014) and from animal to humans as in coronavirus (Singla et al., 2020).

13.13 Allergic Conjunctivitis

Ocular allergic conditions based on IgE involvement have been classified into three categories: (i) the first category includes seasonal allergic conjunctivitis (SAC) and perennial allergic conjunctivitis (PAC), which are IgE-mediated reactions; (ii) the second category of allergic conjunctivitis includes VKC and AKC, which are mediated both by IgE and non-IgE-mediated reactions and (iii) the third category includes giant papillary conjunctivitis (GPC) and contact dermato-conjunctivitis (CDC), which are non-IgE-mediated reactions (Leonardi et al., 2007).

13.13.1 Seasonal Allergic Conjunctivitis

Seasonal Allergic Conjunctivitis (SAC) affects about 15–20% of the population and thus is considered as the most predominant allergic ocular conjunctivitis (Wong et al., 2014). SAC is a IgE-mediated hypersensitivity reaction, and IgE antibodies are common in both SAC and PAC (Bonini, 2004). The levels of histamine, prostaglandins, and leukotrienes are increased initially in the tear film and lasts for 20–30 min (La Rosa et al., 2013). The occurrence of SAC predominates during the spring and summer months and is associated with the release of pollen grains from plants and trees and it is also known as hay fever.

13.13.2 Perennial Allergic Conjunctivitis

Unlike SAC, which occurs during the pollen rich spring and summer months, perennial allergic conjunctivitis (PAC) occurs all through the year in response to allergens such as animal hair, mites, and feathers (Rathi and Murthy, 2017). Both SAC and PAC have similar clinical signs and symptoms and include itching of the eyes, burning of the eyes, tearing, and rhinorrhoea. The involvement of cornea in PAC is rare (Friedlaender, 2011).

13.14 Vernal Keratoconjunctivitis

Papillary hyperplasia, conjunctival injection, profuse tearing, severe itching, and photophobia are characteristic features of Vernal Keratoconjunctivitis (VKC) (Rathi and Murthy, 2017) thus differentiating it from SAC and PAC (Pucci et al., 2003). Limbal type of VKC could be differentiated based on limbal papillary reaction, thickening of the limbus and Horner-Trantas dots associated with the superior limbal margins (Rathi and Murthy, 2017). Palpebral VKC is distinct from limbal VKC and could be differentiated based on giant papillae and cobblestone appearance. In the mixed type of VKC both palpebral and limbal VKC features are seen (Figure 13.9).

According to Kumagai et al. (2006) mechanical trauma from the conjunctival papillae and the cytokines triggers the corneal changes. Ulceration of the cornea and plaques may develop in about 6% of patients leading to worsening of the vision (Cameron

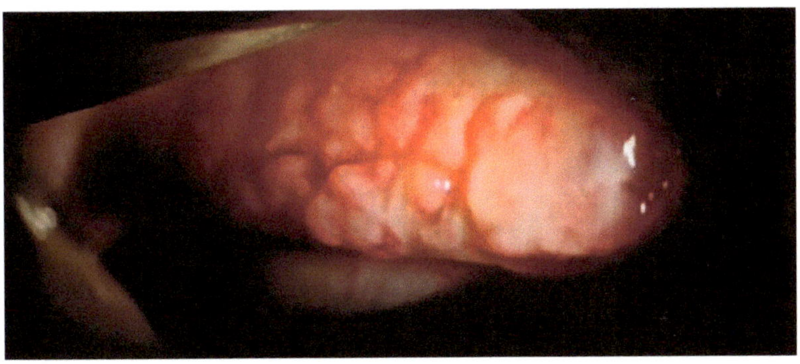

FIGURE 13.9 Cobblestone appearance of large conjunctival papillae in a patient with vernal keratoconjunctivitis. (Modified from Azari and Arabi (2020) published as an open access article and distributed under the terms of the Creative Commons Attribution-Non Commercial-ShareAlike 4.0 License.)

and Mullaney 1997). Keratoconus is highly associated with VKC (15%) (Totan et al., 2001). VKC is encountered both in children and adults in warmer climates (Sofi and Mufti, 2016; De Smedt et al., 2013) and it may be seasonal or perennial (Singhal et al., 2019). VKC is mediated by immune reactions due to IgE or non-IgE reactions due to wind, dust, and sunlight. Both T-helper 2 cells and IgE are implicated in the pathogenesis of VKC (Zicari et al., 2019). Zicari et al. (2013a; b; 19) have identified IL-17 serum levels as a marker for the severity of the disease. Further, VKC may be linked to auto-immune disorders like atopy (Occasi et al., 2015; Singhal et al., 2019).

13.14.1 Atopic Keratoconjunctivitis

Atopic Keratoconjunctivitis (AKC) is also an allergic disease affecting the eye (eyelid, cornea, and conjunctiva). It is chronic and approximately 95% of the AKC patients present with concomitant atopic dermatitis (La Rosa et al., 2013; Bielory and Bielory, 2010). It affects both the eyes (Li et al., 2018b) and occurs in early childhood and common in the second and third decades of life though it could be seen in the fifth decade of life (Chen et al., 2014). Clinicians distinguish AKC from VKC based on the age of the onset, duration of the disease, and clinical presentations of AKC (Chen et al., 2014). Clinical symptoms include epiphora, itching, redness, decreased vision, edematous eyelid, irregular skin with a sandpaper-like texture, conjunctival injection, chemosis and conjunctival scarring. Young AKC patients, are associated with "atopic cataracts" like shield-like cataracts, nuclear, cortical, and posterior subcapsular cataracts and approximately 50% of patients are negative for common allergens (La Rosa et al., 2013). In AKC cytokines released from the conjunctival epithelial cells and the inflammatory cells remodel the ocular surface connective tissue leading to mucus metaplasia, scar formation, and corneal neovascularization (Offiah and Calder, 2009).

13.15 Giant Papillary Conjunctivitis

Giant papillary conjunctivitis (GPC) like VKC also shows papillary hypertrophy of the superior tarsal conjunctiva (Elhers and Donshik, 2008). The symptoms include mucous secretion due to contact lens usage and this secretion occurs irrespective of the type of contact lens, which could be made with either hydroxyethyl methacrylate, silicone polymers, or the new gas permeable polymers (Elhers and Donshik, 2008) but occurs rarely when rigid contact lenses are used. In addition to its association with contact lens use GPC is also associated with corneal foreign bodies, filtering blebs, ocular prostheses, exposed sutures, limbal dermoids, and tissue adhesives (Forister et al., 2009; Donshik et al., 2008). In GPC patients, both Mast cells and eosinophils may be found in the conjunctiva; but GPC is not associated with increases in the levels of IgE or histamines in the tears of GPC patients (La Rosa et al., 2013).

13.16 Contact Dermato-conjunctivitis

Contact dermato-conjunctivitis (CDC) is a type-IV delayed hypersensitivity reaction (La Rosa et al., 2013; Niederkorn, 2008) that occurs in response to several low molecular weight allergens like poison ivy, poison oak, neomycin, nickel, latex, atropine and so forth (La Rosa et al., 2013). CDC pathogenesis is associated with IL-17-producing cells namely the Th cells and regulatory T cells (Oboki et al., 2008). CDC affects the conjunctiva, cornea, and eyelids and the symptoms include itching, lid swelling, follicular reaction, and cicatrisation. The corneal involvement of CDC may be manifested as punctate keratitis, pseudo-dendritic keratitis, and greyish stromal infiltrates (Dart et al., 2003).

13.17 Treatment of Allergic Conjunctivitis

Avoidance of allergens is an effective mode of therapy for allergic conjunctivitis. The other obvious strategies include the use of lubricating eye drops, antihistamines, and Mast cell stabilizers (Schmid and Schmid, 2000; Owen et al., 2004). Among the eye drops, preparations with dual action such as olopatadine, ketotifen, azelastine, and epinastine, which exhibit both antihistamine activity and Mast cell stabilizing effects are preferred for treatment of allergic conjunctivitis (Maziak et al., 2003). These agents apart from antagonising the histamine receptor also stabilize Mast-cell membranes and modify the action of eosinophils (Mishra et al., 2011). Though antihistamines are recommended they are known to induce ocular drying and thus worsen the symptoms of allergic conjunctivitis (Welch et al., 2002). Steroids should be sparingly used because corticosteroids could cause cataracts and elevate intraocular pressure (Rathi and Murthy, 2017) Thus non-steroidal anti-inflammatory drugs such as ketorolac and diclofenac and other steroid-sparing agents such as cyclosporine-A and tacrolimus are used in treating severe and chronic forms of ocular allergies. In addition, allergen-specific immunotherapy (Rathi and Murthy, 2017), which is like a targeted treatment inducing clinical tolerance to a specific allergen is suitable for allergic rhino-conjunctivitis which demonstrates specific IgE antibodies (Broide, 2009).

13.18 Parasitic Conjunctivitis

The human head-and-body louse may cause follicular conjunctivitis in one or both the eyes by infecting the eyelashes and causing itching of the eyelids, with conjunctival and lid margin injection. Manual removal of the lice and nits and using an ophthalmic ointment (e.g., erythromycin, or a bland ophthalmic ointment) for the lashes is an effective mode of treatment (Richards and Guzman-Cottrill, 2010).

13.19 Conjunctivitis Associated with Systemic Diseases

Conjunctivitis is known to be associated with many systemic diseases such as: reactive arthritis, Granulomatosis with polyangiitis, Systemic lupus erythematosus and so forth (Table 13.4).

13.19.1 Reactive Arthritis

Conjunctivitis is a common ocular manifestation of reactive arthritis, an inflammatory arthritis (Szamocki et al., 2016) and the characteristic features are conjunctival

TABLE 13.4

Conjunctivitis Associated with Systemic Diseases

Skin and mucous membrane diseases	Collagen-vascular diseases	Systemic autoimmunity
Acne rosacea	Reactive arthritis	Graft-versus-host disease (GVHD)
Xeroderma pigmentosum	Granulomatosis with polyangiitis	Stevens-Johnson syndrome (SJS) and toxic epidermal necrolysis (TEN)
Ichthyosis	Systemic lupus erythematosus (SLE)	Ocular cicatricial pemphigoid (OCP)
	Sjogren's disease	

Source: Compiled by the author.

hyperaemia with purulent discharge. Conjunctivitis is a part of the "Reiter's triad" (Keat, 1983) which consists of conjunctivitis, urethritis, and arthritis occurring after infection of the urogenital or gastrointestinal tract. Conjunctivitis usually presents early during reactive arthritis and may precede reactive arthritis. Conjunctivitis usually abates within one to four weeks; but at times conjunctivitis progresses and causes severe ocular surface problems (Hamdulay et al., 2006). Individuals with rheumatoid arthritis, are also more likely to develop conjunctivitis.

13.19.2 Granulomatosis with Polyangiitis

Granulomatosis with polyangiitis (GPA) causes granulomatous inflammation in the upper and lower respiratory tract and necrotizing glomerulonephritis. It occurs often in Caucasians (50–60 years of age) and could affect any organ of the human body including the eye. Studies have indicated that GPA could affect the eye in 16–78% of the individuals and in these individuals the anterior segment of the eye is affected (Říhová et al., 2018). Conjunctivitis with petechiae, granulomas, ulcerations and dry eye syndrome occur in approximately 16% of individuals (Fortney and Chodosh, 2002).

13.19.3 Systemic Lupus Erythematosus

Systemic Lupus Erythematosus (SLE), an autoimmune disease, affects women between ages 15 to 44 and causes inflammation of multiple organs in the body, either acutely or chronically. The common symptoms of SLE include skin rashes, arthritis, swelling in the feet, and around the eyes, fatigue, and fevers. Ocular manifestations of lupus are common, are sight-threatening and effect any part of the eye by inflammatory or thrombotic processes. In one third of the patients dry eye syndrome (keratoconjunctivitis sicca) is the most common ocular manifestation of SLE characterised by mild irritation and redness but pain is severe and vision may be compromised. Slit-lamp examination reveals a reduced tear film and corneal changes. SLE is accompanied with scleritis, severe retinopathy, episcleritis, anterior uveitis and dry eyes and these changes can usually be managed with eye drops (Sivaraj et al., 2007).

13.19.4 Sjögren's Syndrome

Dry mouth and dry eyes are the primary symptoms of Sjögren's syndrome (SS) which is a lifelong autoimmune disorder. In a recent study it was demonstrated that eight SS patients of mean age 79 years had cicatrizing conjunctivitis for more than a decade (between 2009 and 2020). Conjunctival biopsy revealed that all the patients had a negative immuno-reactant deposition. In three patients conjunctival scarring was seen and vision was compromised in 4 patients due to corneal ulceration, perforation and scarring (Sharon et al., 2021).

13.19.5 Rosacea

Rosacea is a chronic skin condition that commonly affects the face and eyes and at times it may affect other areas including the neck and chest and present with redness, pimples, and broken blood vessels. The cause of Rosacea is unknown. It normally affects patients between 30–60 years of age and more dominant in fair-skinned people and in women in menopause. Conjunctivitis is a common ocular manifestation of Rosacea and the patients present with follicular and papillary conjunctival reaction accompanied with interpalpebral conjunctival hyperaemia. In addition, cicatrization of the conjunctival tissue (Azari and Arabi, 2020) and conjunctival scarring secondary to entropion and trichiasis has been reported in 10% of the cases. Ocular Rosacea may also be associated with conjunctival granuloma, pinguecula, phlyctenule, and peripheral corneal infiltration (Faraj and Hoang-Xuan, 2001). Phlyctenule is a type IV hypersensitivity reaction to one or more antigens which could include commonly staphylococci but also occur in response to coccidiomycosis, *Candida*, HSV, Lymphogranuloma venereum, and tuberculosis. Phlyctenule normally presents as a painful round, bulbar peri-limbal conjunctival hyperaemia with sterile infiltrate on the limbal bulbar conjunctiva and the epithelium can break down. If Phlyctenule is untreated corneal vascularization may occur. Topical corticosteroids and antibiotics are recommended for this condition (Richards and Guzman-Cottrill, 2010).

13.19.6 Xeroderma Pigmentosum

Xeroderma pigmentosum (XP) is an autosomal recessive inherited disease of the skin where a person is highly sensitive to sunlight / ultraviolet light resulting in premature skin aging and a propensity to skin cancer development (Hossain et al., 2021). XP can also cause eye and neurological problems in some patients. A vast majority (80%) of XP patients suffer eye problems such as photophobia, conjunctivitis or frequent irritation, bloodshot and cloudy vision, or cancerous growths on the eyes. Neurological problems are less frequent (only 20%). XP affects all genders and races (Hossain et al., 2021).

13.19.7 Harlequin Ichthyosis

Harlequin ichthyosis (HI) is a rare autosomal recessive congenital disease caused by mutations of the ABCA12 gene (Elkhatib and Omar, 2022). HI is also known as 'ichthyosis fetalis', and patients show hyperkeratosis that is, thickening of the keratin in the skin and the stratum corneum resulting in dry "armor-like" scales covering the body surface. Infants with HI are more prone to infection and this is due to a compromised skin's protective barrier (Rathore et al., 2015; Jilumudi, 2012). Several other complications associated with HI include ectropion, recurrent conjunctivitis, and keratitis (Rajpopat et al., 2011; Shibata et al., 2015; Kün-Darbois et al., 2016).

13.19.8 Graft-versus-Host Disease

In Graft-versus-host disease (GVHD) patients the involvement of the conjunctiva could be categorised into 4 stages. Stage 1 is marked by simple conjunctival injection, stage 2 by exudative response, stage 3 by pseudo-membrane formation and stage 4 by scarring and cicatrization of the conjunctiva (Jabs et al., 1989; Nassar et al., 2013). Further, when GVHD is acute the following features could be observed: ulceration, conjunctival haemorrhage, exudative discharge, sterile purulent discharge, pseudo-membrane formation, and scarring (Jabs et al., 1989). Further, in the chronic patients of GVHD, dry eye disease may present in one-fourth to three-fourths of the patients (Tabbara et al., 2009). Frequently, keratoconjunctivitis sicca persists after remission of GVHD (Nassar et al., 2013).

13.19.9 Ocular Cicatricial Pemphigoid

Ocular cicatricial pemphigoid (OCP) is a chronic inflammation leading to desiccation of the ocular surface due to loss of goblet cells of the conjunctiva and abnormal mucosal epithelial turn-over (Wang et al., 2018). OCP affects patients in the age group of 50–60 years and more frequently in females (Bedell, 1964) and adhesion of the palpebral conjunctiva of the eyelid to the bulbar conjunctiva of the eyeball resulting in a Symblepharon formation is visible (Figure 13.10). Diagnosis is based on detection of immunoglobulins and/or complements at the epithelial basement membrane by direct immunofluorescence. OCP could be controlled by systemic immunosuppression and frequent lubrication.

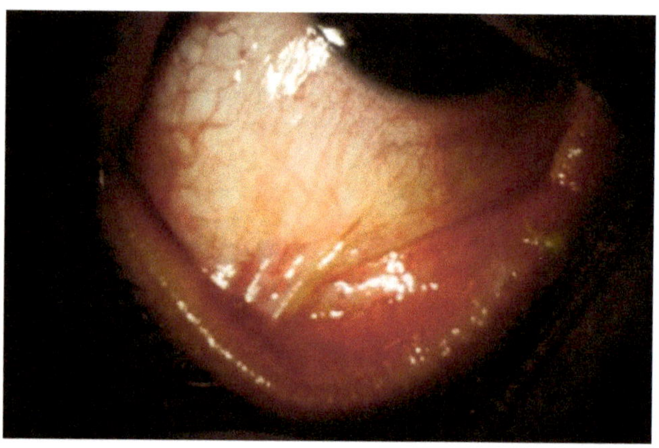

FIGURE 13.10 Symblepharon formation in a patient with ocular cicatricial pemphigoid. (Adapted from Azari and Arabi (2020).)

13.19.10 Stevens-Johnson Syndrome and Toxic Epidermal Necrolysis

Conjunctivitis in Stevens-Johnson syndrome (SJS) and toxic epidermal necrolysis (TEN) patients is characterised by conjunctival hyperaemia, a common sign of acute inflammation and in severe cases sloughing of palpebral conjunctiva and lid margins may be seen. Dry eyes, symblepharon formation, corneal limbal stem cell deficiency, and corneal scarring are observed in the acute stage. Acute ocular involvement occurs in up to 88% of the cases (Kohanim et al., 2016); but it is yet not clear whether the severity of ocular involvement is any different between SJS and TEN.

13.19.11 Kawasaki Disease and Conjunctivitis

Kawasaki disease (KD) is a systemic inflammatory disease of unknown cause. Patients of KD present with fever for at least 5 days with 4 of the 5 clinical findings: conjunctivitis, oropharyngeal changes (e.g., strawberry tongue, red fissured lips), unilateral cervical adenopathy, extremity changes of the limbs, and a polymorphous rash. Conjunctivitis is not purulent and is bilateral and conjunctival injection peri-limbally is absent (Richards and Guzman-Cottrill, 2010).

13.19.12 Ligneous Conjunctivitis

Ligneous conjunctivitis (LC) is a chronic pseudomembranous conjunctivitis (Richards and Guzman-Cottrill, 2010), which is rare and bilateral. Plasminogen deficiency causes LC and consequently it manifests as a pseudomembranous disease of mucous membranes in multiple organs including the eye. Ocular signs of LC include a "woody" induration of the tarsal conjunctiva associated with a yellowish, fibrinous, platelike pseudo-membrane that can be seen when the eyelids are everted. A mixture of fibrin, epithelial cells, and inflammatory cells make up these pseudo-membranes, which adhere firmly to the conjunctival surface. These pseudo-membranes recur even when removed. Successful treatment of LC has been achieved with intravenous lys-plasminogen (Kraft et al., 2000) or topical plasminogen drops (Watts et al., 2002).

13.20 Toxic Conjunctivitis

Studies have demonstrated that long-term and continuous use of topical eye medications could induce changes such as dry eyes, conjunctival inflammation, ocular surface fibrosis, scarring (Broadway et al., 1993), subclinical infiltration by inflammatory cells of the conjunctival epithelium and substantia propria (Baudouin et al., 1999). All these changes constitute the toxic effects of the drugs, which vary based on the drug used. Badouin et al. (2010) demonstrated that benzalkonium chloride (BAK), which is used as a preservative in eye drops has deleterious reactions such as conjunctival congestion, papillary conjunctivitis, and GPC (Badouin et al., 2010). These are the clinically noticeable side effect of the eye drops; but these side effects are less frequent and less harmful compared to their adverse toxic side effects. These toxic side effects manifest a few days after starting the treatment and are recognised as toxic side effects since they tend to resolve quickly when the medication is stopped. Zimmerman et al. (2009) analysed 300 patients in the United States between 2001 and 2004, and concluded that switching of eye drops is primarily to overcome the adverse side effects of the drug in use. Hyperaemia was the most common adverse effect in this study (Zimmerman et al., 2009).

13.21 Conclusions

Conjunctivitis a disease of the conjunctiva that occurs worldwide and the etiological agents of infectious conjunctivitis is either bacteria or virus. Non-infectious conjunctivitis is caused by allergy or toxins. Bacterial conjunctivitis is categorised into different types based on the bacterium involved like methicillin-resistant *S. aureus* conjunctivitis, Granulomatous conjunctivitis due to Chlamydia *trachomatis* and other bacteria and Gonococcal conjunctivitis caused by *Neisseria gonorrhoeae* or *N. meningitidis*. Bacterial conjunctivitis is treated empirically with aminoglycosides, polymyxin B combinations, macrolides, and fluoroquinolones. But the susceptibility of the bacteria (viz., *S. pneumoniae, P. aeruginosa, H. influenzae*, CoNS and *S. aureus*) from conjunctivitis patients to these antimicrobial agents changes with time and thus the emergence of resistance to antibacterial agents has been consistently reported across several countries including the United States, UK, Ethiopia and India. Viruses cause up to 80% of all cases of acute conjunctivitis and the viruses include adenoviruses, which cause Pharyngoconjunctival fever and Epidemic keratoconjunctivitis, herpes simplex virus that cause Herpetic conjunctivitis and enteric virus that cause Acute haemorrhagic conjunctivitis. Combination therapies with corticosteroids and antibiotics are effective in overcoming viral conjunctivitis. In addition, different types of conjunctivitis due to allergy have been reported, which includes Seasonal allergic conjunctivitis, Perennial allergic conjunctivitis, Vernal keratoconjunctivitis, Atopic Keratoconjunctivitis, Giant papillary conjunctivitis and Contact Dermatoconjunctivitis. Avoiding allergens is an effective strategy for the treatment for allergic conjunctivitis. Parasitic conjunctivitis and conjunctivitis associated with systemic diseases are also discussed in brief. The chapter concludes with methods to overcome the dissemination of antimicrobial resistance in ocular microorganisms.

REFERENCES

Abelson M, Protzko E, Shapiro A, Garces-Soldana A, Bowman L. A randomized trial assessing the clinical efficacy and microbial eradication of 1% azithromycin ophthalmic solution vs tobramycin in adult and pediatric subjects with bacterial conjunctivitis. *Clin Ophthalmol.* 2007;1:177–182.

Abelson MB, Heller W, Shapiro AM, Si E, Hsu P, Bowman LM. Clinical cure of bacterial conjunctivitis with azithromycin 1%: Vehicle- controlled, double-masked clinical trial. *Am J Ophthalmol.* 2008;145:959–965.

Alexandrakis G, Alfonso EC, Miller D. Shifting trends in bacterial keratitis in south Florida and emerging resistance to fluoroquinolones. *Ophthalmol.* 2000;107:1497–1502.

Asbell PA, Colby KA, Deng S, McDonnell P, Meisler DM, Raizman MB, Sheppard JD Jr, Sahm DF. Ocular TRUST: Nationwide antimicrobial susceptibility patterns in ocular isolates. *Am J Ophthalmol.* 2008;145:951–958.

Asbell PA, DeCory HH. Antibiotic resistance among bacterial conjunctival pathogens collected in the Antibiotic Resistance Monitoring in Ocular Microorganisms (ARMOR) surveillance study. *PLoS ONE.* 2018;13(10):e0205814. https://doi.org/10.1371/journal.pone.0205814

Asbell PA, Sanfilippo CM, Pillar CM, DeCory HH, Sahm DF, Morris TW. Antibiotic resistance among ocular pathogens in the United States: Five-year results from the antibiotic resistance monitoring in ocular microorganisms (ARMOR) surveillance study. *JAMA Ophthalmol.* 2015;133(12):1445–1454.

Azari AA, Arabi A. Conjunctivitis: A Systematic Review. *J Ophthalmic Vis Res.* 2020;15:372–395.

Azari AA, Barney NP. Conjunctivitis: A systematic review of diagnosis and treatment. *JAMA.* 2013;310(16):1721–1729. doi: 10.1001/jama.2013.280318

Baudouin C, Labbe A, Liang H, Pauly A, Brignole- Baudouin F. Preservatives in eyedrops: The good, the bad and the ugly. *Prog Retin Eye Res.* 2010;29:312–334.

Baudouin C, Pisella PJ, Fillacier K, et al., Ocular surface inflammatory changes induced by topical antiglaucoma drugs: Human and animal studies. *Ophthalmol.* 1999;106:556–563.

Baush + Lomb (2016) Bausch + Lomb reports preliminary 2015 results of the antibiotic resistance monitoring in ocular microorganisms (ARMOR) study. www.bausch.com/our-company/ recent-news/id/2262/542016-wednesday. Accessed 5 Aug 2016

Bedell AJ. Ocular pemphigus: A clinical presentation of kodachromes. *Trans Am Ophthalmol Soc.* 1964;62:109–122.

Belfort R, Jr., Gabriel L, Martins Bispo PJ, Muccioli C, Zacharias Serapicos PC, Clark L, et al., Safety and efficacy of moxifloxacin-dexamethasone eyedrops as treatment for bacterial ocular infection associated with bacterial blepharitis. *Adv Ther.* 2012;29:416–426.

Benz MS, Scott IU, Flynn HW, Unonius N, Miller D.. Endophthalmitis isolates and antibiotic sensitivities: A 6-year review of culture-proven cases. *Am J Ophthalmol.* 2004;137(1):38–42.

Bergman M, Huikko S, Huovinen P, Paakkari P, Seppälä H. Finnish Study Group for Antimicrobial Resistance (FiRe Network). Macrolide and azithromycin use are linked to increased macrolide resistance in *Streptococcus pneumoniae. Antimicrob Agents Chemother.* 2006;50:3646–3650.

Berisa Prado S, Riestra Ayora AC, Lisa Fernandez C, Chacon Rodriguez M, Merayo-Lloves J, Alfonso Sanchez JF. Topical tacrolimus for corneal subepithelial infiltrates secondary to adenoviral keratoconjunctivitis. *Cornea.* 2017;36:1102–1105.

Bielory L, Frohman LP. Allergic and immunologic disorders of the eye. *J Allergy Clin Immunol.* 1992;89:1–15.

Bielory B, Bielory L. Atopic dermatitis and keratoconjunctivitis. *Immunol Allergy Clin North Am.* 2010;30:323–336.

Bispo PJM, Sahm DF, Asbell PA. A systematic review of multi-decade antibiotic resistance data for ocular bacterial pathogens in the United States. *Ophthalmol Ther.* 2022;11:503–520. https://doi.org/10.1007/s40123-021-00449-9

Blephamide. Irvine, CA: Allergan;2017.

Block SL, Hedrick J, Tyler R, Smith A, Findlay R, Keegan E, Stroman DW. Increasing bacterial resistance in pediatric acute conjunctivitis (1997–1998). *Antimicrob Agents Chemother.* 2000;44:1650–1654.

Bodor FF. Systemic antibiotics for the treatment of the conjunctivitis otitis media syndrome. *Pediatr Infect Dis J.* 1989;8:287–290.

Bonini S. Atopic keratoconjunctivitis. *Allergy.* 2004;59:71–73.

Bonini S, Coassin M, Aronni S, Lambiase A. “Vernal keratoconjunctivitis”. *Eye (Lond).* 2004;18(4):345–351. doi:10.1038/sj.eye.6700675. PMID 15069427

Bremond-Gignac D, Mariani-Kurkdjian P, Beresniak A, ElFekih L, Bhagat Y, Pouliquen P, et al., Efficacy and safety of azithromycin 1.5% eye drops for purulent bacterial conjunctivitis in pediatric patients. *Pediatr Infect Dis J.* 2010;29:222–226.

Broadway D, Grierson I, Hitchings R. Adverse effects of topical antiglaucomatous medications on the conjunctiva. *Br J Ophthalmol.* 1993;77:590–596.

Broide DH. Immunomodulation of allergic disease. *Ann Rev Med.* 2009;60:279–291.

Brown L. Resistance to ocular antibiotics: An overview. *Clin Exp Optom.* 2007;90:258–262.

Burton MJ. Trachoma: An overview. *Br Med Bull* 2007;84:99–116.

Buznach N, Dagan R, Greenberg D. Clinical and bacterial characteristics of acute bacterial conjunctivitis in children in the antibiotic resistance era. *Pediatr Infect Dis J* 2005;24:823–828.

Cameron JA, Mullaney PB. Amblyopia resulting from shield ulcers and plaques of the cornea in vernal keratoconjunctivitis. *J Pediatr Ophthalmol Strabismus.* 1997;34:261–262.

Castro Ochoa KJ, Mendez MD. Ophthalmia Neonatorum. In: *Stat Pearls.* Treasure Island (FL): StatPearls Publishing LLC; 2020.

Cavuoto K, Zutshi D, Karp CL, Miller D, Feuer W. Update on bacterial conjunctivitis in South Florida. *Ophthalmol.* 2008;115:51–56.

Chalita MR, Hofling-Lima AL, Paranhos A Jr, Schor P, Belfort R Jr. Shifting trends in in vitro antibiotic susceptibilities for common ocular isolates during a period of 15 years. *Am J Ophthalmol.* 2004;137:43–51.

Chen JJ, Applebaum DS, Sun GS, Pflugfelder SC. Atopic keratoconjunctivitis: A review. *J Am Acad Dermatol.* 2014;70:569–575.

Chigbu DI, Labib BA. Pathogenesis and management of adenoviral keratoconjunctivitis. *Infect Drug Resist.* 2018;11:981–993.

Chou TM, Gigliotti F, Lichtenstein SJ. Bacterial conjunctivitis:setting a course for containment and cure. *Contemp Pediatr.* 2004;21:2–8.

Clement C, Capriotti JA, Kumar M, et al., Clinical and antiviral efficacy of an ophthalmic formulation of dexamethasone povidone-iodine in a rabbit model of adenoviral keratoconjunctivitis. *Invest ophthalmol & vis sci.* 2011;52:339–344.

Cochereau I, Meddeb-Ouertani A, Khairallah M, Amraoui A, Zaghloul K, Pop M, et al., 3-day treatment with azithromycin 1.5% eye drops versus 7-day treatment with tobramycin 0.3% for purulent bacterial conjunctivitis: Multicentre, randomised and controlled trial in adults and children. *Br J Ophthalmol.* 2007;91:465–469.

Comstock TL, Paterno MR, Lynch JA, Usner DW, Pichichero, ME. Efficacy and safety of besifloxacin ophthalmic suspension, 0.6% in pediatric patients with bacterial conjunctivitis. *Pediatr Drugs.* 2010a;12:105–112.

Comstock TL, Paterno MR, Decory HH, Usner DW. Safety and tolerability of besifloxacin ophthalmic suspension 0.6% in the treatment of bacterial conjunctivitis: Data from six clinical and phase I safety studies. *Clin Drug Investig.* 2010b;30:675–685.

Costumbrado J, Ng DK, Ghassemzadeh S. Gonococcal conjunctivitis. In: *StatPearls.* Treasure Island (FL): StatPearls Publishing; 2020.

Critchley IA, Brown SD, Traczewski MM, Tillotson GS, Janjic N. National and regional assessment of antimicrobial resistance among community-acquired respiratory tract pathogens identified in a 2005–2006 U.S. *Faropenem surveillance study. Antimicrob Agents Chemother.* 2007;51:4382–4389.

Czaplewski L, Bax R, Clokie M, Dawson M, Fairhead H, Fischetti VA, Foster S, Gilmore BF, Hancock RE, Harper D. Alternatives to antibiotics—a pipeline portfolio review. *Lancet Infect Dis.* 2016;16(2):P239–251, FEBRUARY 2016.

Dart J. Corneal toxicity: The epithelium and stroma in iatrogenic and factitious disease. *Eye.* 2003;17:886–892.

Darville T. Chlamydia trachomatis infections in neonates and young children. *Semin Pediatr Infect Dis.* 2005;16:235–244.

Dave SB, Toma HS, Kim SJ. Changes in ocular flora in eyes exposed to ophthalmic antibiotics. *Ophthalmol.* 2013;120(5):937–941.

de Laet C, Dionisi-Vici C, Leonard JV, McKiernan P, Mitchell G, Monti L, et al., Recommendations for the management of tyrosinaemia type 1. *Orphanet J Rare Dis.* 2013;8:8–8.

Deramo VA, Lai JC, Fastenberg DM, Udell IJ. Acute endophthalmitis in eyes treated prophylactically with gatifloxacin and moxifloxacin. *Am J Ophthalmol.* 2006;142:721 5.

De Smedt S, Wildner G, Kestelyn P. Vernal keratoconjunctivitis: An update. *Br J Ophthalmol.* 2013;97:9–14.

Diamant JI, Hwang DG. Therapy for bacterial conjunctivitis. *Ophthalmol Clin North Am.* 1999;12:15–20.

Donshik PC, Ehlers WH, Ballow M. Giant papillary conjunctivitis. *Immunol Allergy Clin North Am.* 2008;28:83–103, vi.

Dorfman MS, Wagner RS, Jamison T, Bell B, Stroman DW. The pharmacodynamic properties of azithromycin in a kinetics-of-kill model and implications for bacterial conjunctivitis treatment. *Adv Ther* 2008;25:208–217.

Dossarps D, Bron AM, Koehrer P, Aho-Glélé LS, Creuzot-Garcher C. Endophthalmitis after intravitreal injections: Incidence, presentation, management, and visual outcome. *Am J Ophthalmol.* 2015;160(1):17–25.

Drew RH, Gallis HA. Azithromycin, spectrum of activity, pharmacokinetics, and clinical applications. *Pharmacotherapy.* 1992;12:161–173.

Egger SF, Ruckhofer J, Alzner E, Hell M, Hitzl W, Huber-Spitzy V, Grabner G. In vitro susceptibilities to topical antibiotics of bacteria isolated from the surface of clinically symptomatic eyes. *Ophthalmic Res.* 2001;33:117–120.

Elhers WH, Donshik PC. Giant papillary conjunctivitis. *Curr Opin Allergy Clin Immunol.* 2008;8:445–449.

Elkhatib AM, Omar M. Ichthyosis Fetalis. [Updated 2022 Aug 8]. In: *StatPearls [Internet]. Treasure Island (FL).* StatPearls Publishing; 2022 Jan. Available from: www.ncbi.nlm.nih.gov/books/NBK560492/

Epling J. "Bacterial conjunctivitis". *BMJ Clin Evid.* 2010. PMC 2907624. PMID 21718563

Everitt H, Little P. How do GPs diagnose and manage acute infective conjunctivitis? A GP survey. *Fam Pract.* 2002;19:658–660.

Faraj HG, Hoang-Xuan T. Chronic cicatrizing conjunctivitis. *Curr Opin Ophthalmol.* 2001;12:250–257.

Faraldi F, Papa V, Rasà D, et al., Netilmicin/dexamethasone fixed combination in the treatment of conjunctival inflammation. *Clin Ophthalmol.* 2013;7:1239–1244.

Fitch CP, Rapoza PA, Owens S, Murillo-Lopez F, Johnson RA, Quinn TC, et al., Epidemiology and diagnosis of acute conjunctivitis at an inner-city hospital. *Ophthalmol.* 1989;96:1215–1220.

Forister JF, Forister EF, Yeung KK, Ye P, Chung MY, Tsui A, et al., Prevalence of contact lens-related complications: UCLA contact lens study. *Eye Contact Lens.* 2009;35:176– 180.

Forster RK, Dawson CR, Schachter J. Late follow-up of patients with neonatal inclusion conjunctivitis. *Am J Ophthalmol.* 1970;69:467–472.

Fortney AC, Chodosh J. Conjunctival ulceration in recurrent Wegener granulomatosis. *Cornea.* 21(6);2002:623–624.

Friedlaender MH. Ocular allergy. *Curr Opin Allergy Clin Immunol.* 2011;11:477–482.

Fukuda M, Ohashi H, Matsumoto C, Mishima S, Shimomura Y. Methicillin-resistant *Staphylococcus aureus* and methicillin-resistant coagulase-negative Staphylococcus ocular surface infection efficacy of chloramphenicol eye drops. *Cornea.* 2002;21:S86–89.

Goff DA, Jankowski C, Tenover FC. Using rapid diagnostic tests to optimize antimicrobial selection in antimicrobial stewardship programs. *Pharmacotherapy.* 2012;32(8):677–687.

Goldstein MH, Kowalski RP, Gordon YJ. Emerging fluoroquinolone resistance in bacterial keratitis: A 5-year review. *Ophthalmol.* 1999;106:1313–1318.

Granet D, Lichtenstein SJ, Onofrey B, Katz JA. An assessment of the tolerability of moxifloxacin 0.5% compared to nazithromycin 1.0% in DuraSite. *Clin Ophthalmol.* 2007;1:519–525.

Granet DB, Dorfman M, Stroman D, Cockrum P. A multicenter comparison of polymyxin B sulfate/trimethoprim ophthalmic solution and moxifloxacin in the speed of clinical efficacy for the treatment of bacterial conjunctivitis. *J Pediatr Ophthalmol Strabismus.* 2008;45:340–349.

Gross RD, Lichtenstein S, Schlech BA, Gower LA, Potts SL. Early clinical and microbiological responses in the treatment of bacterial conjunctivitis with moxifloxacin ophthalmic solution 0.5% (Vigamox*T M*) using B.I.D. dosing. *nToday Ther Trends.* 2003;21:227–237.

Grzybowski A, Brona, P., Kim S. J. Microbial flora and resistance in ophthalmology: A review Graefes Arch *Clin Exp Ophthalmol.* 2017;255:851–862. DOI 10.1007/s00417-017-3608-y

Grzybowski A, Kuklo P, Pieczynski J, Beiko G. A review of preoperative manoeuvres for prophylaxis of endophthalmitis in intraocular surgery: topical application of antibiotics, disinfectants, or both? *Curr Opin Ophthalmol.* 2016;27(1):9–23.

Gwon A. Ofloxacin vs tobramycin for the treatment of external ocular infection. Ofloxacin Study Group II. *Arch Ophthalmol.* 1992a;110:1234–1237.

Gwon A. Topical ofloxacin compared with gentamicin in the treatment of external ocular infection. Ofloxacin Study Group. *Br J Ophthalmol.* 1992b;76:714–718.

Haas W, Pillar CM, Torres M, Morris TW, Sahm DF. Monitoring antibiotic resistance in ocular microorganisms: results from the Antibiotic Resistance Monitoring in Ocular micRorganisms (ARMOR) 2009 surveillance study. *Am J Ophthalmol.* 2011;152(4):567–574.

Haas W, Pillar CM, Zurenko GE, Lee JC, Brunner LS, Morris TW. Besifloxacin, a novel fluoroquinolone, has broad-spectrum in vitro activity against aerobic and anaerobic bacteria. *Antimicrob Agents Chemother.* 2009;53:3552–3560.

Hamdulay SS, Glynne SJ, Keat A. When is arthritis reactive? *Postgrad Med J.* 2006;82:446–453.

Harbarth S, Samore MH. Antimicrobial resistance determinants and future control. *Emerg Infect Dis.* 2005 Jun;11(6):794–801. doi: 10.3201/eid1106.050167. PMID: 15963271; PMCID: PMC3367590

Harley D, Harrower B, Lyon M, Dick A. A primary school outbreak of pharyngoconjunctival fever caused by adenovirus type 3. *Comm Dis Intell.* 2001;25:9–12.

Hillenkamp J, Reinhard T, Ross RS, Böhringer D, Cartsburg O, Roggendorf M, et al., Topical treatment of acute adenoviral keratoconjunctivitis with 0.2% cidofovir and 1% cyclosporine: A controlled clinical pilot study. *Arch Ophthal.* 2001;119:1487–1491.

Holland EJ, Fingeret M, Francis S. Mah use of topical steroids in conjunctivitis: A review of the evidence. *Cornea.* May 2019;38(8):1. DOI: 10.1097/ICO.0000000000001982

Horner C, Mawer D, Wilcox M. Reduced susceptibility to chlorhexidine in staphylococci: Is it increasing and does it matter? *J Antimicrob Chemother.* 2012;67(11): 2547–2559, November 2012. https://doi.org/10.1093/jac/dks284

Hossain, M., Hasan A., Shawan, M. M. A. K., Banik, S., Jahan, I. Current therapeutic strategies of Xeroderma Pigmentosum. *Indian J Dermatol.* 2021 Nov-Dec;66(6):660–667. doi: 10.4103/ijd.ijd_329_21 PMCID: PMC8906321 PMID: 35283513

Høvding G. "Acute bacterial conjunctivitis". *Acta Ophthalmol.* 2008;86(1):5–17. doi:10.1111/j.1600-0420.2007.01006.x. PMID 17970823

Hsu J, Gerstenblith AT, Garg SJ, Vander JF. Conjunctival flora antibiotic resistance patterns after serial intravitreal injections without postinjection topical antibiotics. *Am J Ophthalmol.* 2014;157(3):514–518.

Hwang DG. Fluoroquinolone resistance in ophthalmology and the potential role for newer ophthalmic fluoroquinolones. *Surv Ophthalmol.* 2004;49(Suppl. 2):S79–83.

Hwang DG, Schanzlin DJ, Rotberg MH, Foulks G, Raizman MB. A phase III, placebo controlled clinical trial of 0.5% levofloxacin ophthalmic solution for the treatment of bacterial conjunctivitis. *Br J Ophthalmol.* 2003;87:1004– 1009.

Isenberg SJ, Apt L, Valenton M, Sharma S, Garg P, Thomas PA, Parmar P, Kaliamurthy J, Reyes JM, Ong D. Prospective,randomized clinical trial of povidone-iodine 1.25% solution versus topical antibiotics for treatment of bacterial keratitis. *Am J Ophthalmol.* 2017; 176, April 2017: 244–253. https://doi.org/10.1016/j.ajo.2016.10.004

Isenberg SJ, Apt L, Wood M. "The influence of perinatal infective factors on ophthalmia neonatorum". *J Pediatr Ophthalmol Strabismus.* 1996;33(3):185–188. PMID 8771523

Jabs DA, Wingard J, Green WR, Farmer ER, Vogelsang G, Saral R. The eye in bone marrow transplantation. III. Conjunctival graft-vs-host disease. *Arch Ophthalmol.* 1989;107:1343–1348.

Jackson WB, Low DE, Dattani D, Whitsitt PF, Leeder RG, MacDougall R. Treatment of acute bacterial conjunctivitis: 1% fusidic acid viscous drops vs. 0.3% tobramycin drops. *Can J Ophthalmol.* 2002;37:228–237. discussion 237.

Jacob SE, James WD. From road rash to top allergen in a flash: Bacitracin. *Dermatol Surg.* 2004;30(4 Pt 1): 521–524.

Jhanji V, Chan TC, Li EY, Agarwal K, Vajpayee RB. Adenoviral keratoconjunctivitis. *Surv Ophthalmol.* 2015;60:435–443.

Jhanji V, Sharma N, Satpathy G, Titiyal J. Fourth-generation fluoroquinolone- resistant bacterial keratitis. *J Cataract Refract Surg.* 2007;33:1488–1489.

Jilumudi UB. Harlequin ichthyosis: A medico legal case report & review of literature with peculiar findings in autopsy. *J Forensic Leg Med.* 2012 Aug;19(6):352-4.

Johnson D, Liu D, Simel D. Does this patient with acute infectious conjunctivitis have a bacterial infection? The Rational Clinical Examination Systematic Review. *JAMA.* 2022;327(22):2231–2237. doi:10.1001/jama.2022.7687

Judson S, Prescott J, Munster V. Understanding ebola virus transmission. *Viruses.* 2015 Feb 3;7(2):511–521. doi: 10.3390/v7020511. PMID: 25654239; PMCID: PMC4353901

Kaliamurthy J, Nelson Jesudasan CA, Geraldine P, Parmar P, Kalavathy CM, Thomas PA. Comparison of in vitro susceptibilities of ocular bacterial isolates to gatifloxacin and other topical antibiotics. *Ophthalmic Res.* 2005;37:117–122.

Karpecki P, Depaolis M, Hunter JA, White EM, Rigel L, nBrunner LS, et al., Besifloxacin ophthalmic suspension 0.6% in patients with bacterial conjunctivitis: A multicenter, prospective, randomized, double-masked, vehiclecontrolled, 5-day efficacy and safety study. *Clin Ther.* 2009;31:514–526.

Keat A. Reiter's syndrome and reactive arthritis in perspective. *New Engl J Med.* 1983;309:1606–1615.

Keating GM. Levofloxacin 0.5% ophthalmic solution: A review of its use in the treatment of external ocular infections and in intraocular surgery. *Drugs.* 2009;69:1267–1286.

Kim SJ, Toma HS. Ophthalmic antibiotics and antimicrobial resistance: A randomized, controlled study of patients undergoing intravitreal injections. *Ophthalmol.* 2011a;118(7):1358–1363.

Kim SJ, Toma HS. Antimicrobial resistance and ophthalmic antibiotics: 1-year results of a longitudinal controlled study of patients undergoing intravitreal injections. *Arch Ophthalmol.* 2011b;129(9):1180–1188.

Kohanim S, Palioura S, Saeed HN, et al., Acute and chronic ophthalmic involvement in Stevens-Johnson syndrome/toxic epidermal necrolysis - a comprehensive review and guide to therapy. *II. Ophthalmic disease. Ocul Surf.* 2016;14:168–188.

Kotlus BS, Wymbs RA, Vellozzi EM, Udell IJ. In vitro activity of fluoroquinolones, vancomycin, and gentamicin against methicillinresistant *Staphylococcus aureus* ocular isolates. *Am J Ophthalmol.* 2006;142:726–729.

Kovalyuk N, Kaiserman I, Mimouni M, Cohen O, Levartovsky S, Sherbany H, et al., Treatment of adenoviral keratoconjunctivitis with a combination of povidone-iodine 1.0% and dexamethasone 0.1% drops: A clinical prospective controlled randomized study. *Acta Ophthalmol.* 2017;95:e686–e692.

Kraft J, Lieb W, Zeitler P, Schuster V. Ligneous conjunctivitis in a girl with severe type I plasminogen deficiency. *Graefes Arch Clin Exp Ophthalmol.* 2000;238:797–800.

Kumagai N, Fukuda K, Fujitsu Y, Yamamoto K, Nishida T. Role of structural cells of the cornea and conjunctiva in the pathogenesis of vernal keratoconjunctivitis. *Prog Retin Eye Res.* 2006;25:165–187.

Kün-Darbois JD, Molin A, Jeanne-Pasquier C, Paré A, Bénateau H, Veyssière A. Facial features in Harlequin ichthyosis: Clinical findings about 4 cases. Rev Stomatol Chir *Maxillofac Chir Orale.* 2016 Feb;117(1):51–53.

Kuo IC. Adenoviral keratoconjunctivitis: Diagnosis, management, and prevention. *Curr Ophthalmol Rep.* 2019;7:118–127.

Lachapelle J-M, Castel O, Casado AF, Leroy B, Micali G, Tennstedt D, Lambert J. Antiseptics in the era of bacterial resistance: a focus on povidone iodine. *Clin Pract.* 2013;10(5):579–592.

Langford MP, Anders EA, Burch MA. Acute hemorrhagic conjunctivitis: Anti-coxsackievirus A24 variant secretory nimmunoglobulin A in acute and convalescent tear. *Clin Ophthalmol.* 2015;9:1665–1673.

La Rosa M, Lionetti E, Reibaldi M, et al., Allergic conjunctivitis: A comprehensive review of the literature. *Ital J Pediatr.* 2013;39:18.

Lee AE, Niruttan1 K, Rawson1 TM, Moore LSP 1,2,3,4* Antibacterial resistance in ophthalmic infections: A multi-centre analysis across UK care settings. *BMC Infect. Dis.* 2019;19:768.

Leeming JP. Treatment of ocular infections with topical antibacterials. *Clin Pharmacokinet.* 1999;37:351–360.

Leibowitz HM. Antibacterial effectiveness of ciprofloxacin 0.3% ophthalmic solution in the treatment of bacterial conjunctivitis. *Am J Ophthalmol.* 1991;112:29S–33S.

Leibowitz HM. "The red eye". *N Engl J Med.* 343(5): 345–351. doi:10.1056/NEJM200008033430507. PMID 10922425

Leibowitz HM, Hyndiuk RA, Smolin GR, Nozik RA, Hunter GJ, Cagle GD, Davis DS. Tobramycin in external eye disease: A doublemasked study vs. gentamicin. *Curr Eye Res.* 1981;1:259–266.

Leonardi A, Castegnaro A, Valerio AL, Lazzarini D. "Epidemiology of allergic conjunctivitis: clinical appearance and treatment patterns in a population-based study". *Curr Opin Allergy Clin Immunol.* 2015;15(5): 482–488. doi:10.1097/ACI.0000000000000204. PMID 26258920

Leonardi A, De Dominicis C, Motterle L. Immunopathogenesis of ocular allergy: A schematic approach to different clinical entities. *Curr Opin Allergy Clin Immunol.* 2007;7:429–435.

Lestner J, Versporten A, Doerholt K,Warris A, Roilides E, Sharland M, Bielicki J, Goossens H. Systemic antifungal prescribing in neonates and children: outcomes from the antibiotic resistance and prescribing in European children (ARPEC) study. *Antimicrob Agents Chemother.* 2015;59(2):782–789.

Levinger E, Slomovic A, Sansanayudh W, Bahar I, Slomovic AR. Topical treatment with 1% cyclosporine for subepithelial infiltrates secondary to adenoviral keratoconjunctivitis. *Cornea.* 2010;29:638–640.

Li J, Lu X, Sun Y, Lin C, Li F, Yang Y, et al., A swimming pool-associated outbreak of pharyngoconjunctival fever caused by human adenovirus type 4 in Beijing, China. *Int J Infect Dis.* 2018a;75:89–91.

Li A, Li S, Ruan F, Jie Y. Atopic keratoconjunctivitis: A diagnostic dilemma-a case report. *Medicine.* 2018b;97:e0372.

Lichtenstein SJ, Rinehart M. Levofloxacin Bacterial Conjunctivitis Study Group. Efficacy and safety of 0.5% levofloxacin ophthalmic solution for the treatment of bacterial conjunctivitis in pediatric patients. *J AAPOS.* 2003;7:317–324.

Lim KH, Yin-Murphy M. Epidemic conjunctivitis in Singapore in 1970 and 1971. *Singapore Med J.* 1973;14:86–89.

Lohr JA, Austin RD, Grossman M, Hayden GF, Knowlton GM, Dudley SM. Comparison of three topical antimicrobials for acute bacterial conjunctivitis. *Pediatr Infect Dis J.* 1988;7:626–629.

Mallika P, Asok T, Faisal H, Aziz S, Tan A, Intan G. "Neonatal conjunctivitis - a review". *Malays Fam Physician.* 2008;**3**(2):77–81. PMC 4170304. PMID 25606121

Marangon FB, Miller D, Muallem MS, Romano AC, Alfonso EC. Ciprofloxacin and levofloxacin resistance among methicillinsensitive *Staphylococcus aureus* isolates from keratitis and conjunctivitis. *Am J Ophthalmol.* 2004;137:453–458.

Mark E. Contact-lens-related microbial keratitis: Case report and review. *J Optom.* 2011;4(4):122–127. doi: 10.1016/S1888-4296(11)70053-X

Maxitrol. Fort Worth, TX: Alcon Laboratories, Inc.; 2016.

Maziak W, Behrens T, Brasky TM, Duhme H, Rzehak P, Weiland SK, et al., Are asthma and allergies in children and adolescents increasing? Results from ISAAC phase I and phase III surveys in Munster, Germany. *Allergy.* 2003;58:572–579.

McDonald MB, Blondeau JM, DeCory HH, Usner DW, Brunner LS, Haas W, Morris TW. *Multidrug-resistant strains in clinical trials of besifloxacin in the treatment of bacterial conjunctivitis.* Poster presented at the Annual Meeting of the American Academy of Ophthalmology, November 8–11, 2008. Atlanta, GA.

McDonald MB, Protzko EE, Brunner LS, Morris TW, Haas W, Paterno MR, Comstock TL, Usner DW. Efficacy and safety of besifloxacin ophthalmic suspension 0.6%

compared with moxifloxacin ophthalmic solution 0.5% for treating bacterial conjunctivitis. *Ophthalmol.* 2009;116:1615–1623.

Milder E, Vander J, Shah C, Garg S. Changes in antibiotic resistance patterns of conjunctival flora due to repeated use of topical antibiotics after intravitreal injection. *Ophthalmol.* 2012;119(7):1420–1424.

Miller D, Flynn PM, Scott IU, Alfonso EC, Flynn HW Jr. In vitro fluoroquinolone resistance in staphylococcal endophthalmitis isolates. *Arch Ophthalmol.* 2006;124:479–483.

Miller IM, Vogel R, Cook TJ, Wittreich J. Topically administered norfloxacin compared with topically administered gentamicin for the treatment of external ocular bacterial infections. *The Worldwide Norfloxacin Ophthalmic Study Group. Am J Ophthalmol.* 1992;113:638–644.

Miño de Kaspar H, Hoepfner AS, Engelbert M, Thiel M, Ta CN, Mette M, Schulze-Schwering M, Grasbon T, Sesma-Vea B, Casas JM, Iturralde-Goñi R, Klauss V, Kampik A. Antibiotic resistance pattern and visual outcome in experimentally-induced Staphylococcus epidermidis endophthalmitis in a rabbit model22. *Ophthalmol.* 2001;108(3):470–478. doi:10.1016/S0161-6420(00) 00545-5

Mishra GP, Tamboli V, Jwala J, Mitra AK. Recent patents and emerging therapeutics in the treatment of allergic conjunctivitis. *Recent Pat Inflamm Allergy Drug Discov.* 2011;5:26–36.

Mohammed AA, Ali MM, Zenebe MH. Bacterial etiology of ocular and periocular infections, antimicrobial susceptibility profile and associated factors among patients attending eye unit of Shashemene comprehensive specialized hospital, Shashemene, Ethiopia. *BMC Ophthalmol.* 2020 Mar 30;20(1):124. doi: 10.1186/s12886-020-01398-w. PMID: 32228528; PMCID: PMC7106738

Montero J, Casado A, Perea E, Vigo F, Pastor JC, Alio JL, Soler F, Escartin P, Clarke I, Juach A. A double-blind double-dummy comparison of topical lomefloxacin 0.3% twice daily with topical gentamicin 0.3% four times daily in the treatment of acute bacterial conjunctivitis. *J Clin Res.* 1998;1:29–39.

Moore DL, MacDonald NE, Canadian Paediatric Society, Infectious Diseases and Immunization Committee. "Preventing ophthalmia neonatorum". *Can J Infect Dis Med Microbiol.* 2015;**26**(3):122–125. PMC 4507834. PMID 26236350

Morrissey I, Burnett R, Viljoen L, Robbins M. Surveillance of the susceptibility of ocular bacterial pathogens to the fluoroquinolone gatifloxacin and other antimicrobials in Europe during 2001/2002. *J Infect.* 2004;49:109–114.

Moshirfar M., Fenz C, Li Z. What we know about ocular manifestations of *Ebola Clinical Ophthalmol.* 2014;8:2355–2357.

Moss SE, Klein R, Klein BE. "Prevalence of and risk factors for dry eye syndrome". *Arch Ophthalmol.* 2000;118(9):1264–1268. PMID 10980773

Nassar A, Tabbara KF, Aljurf M. Ocular manifestations of graft-versus-host disease. *Saudi J Ophthalmol.* 2013;27:215–222.

Nelson JD. "Superior limbic keratoconjunctivitis (SLK)". *Eye (Lond).* 1989;3(Pt 2):180–189. doi:10.1038/eye.1989.26. PMID 2695351

Newman H, Gooding C. Viral ocular manifestations: A broad overview. *Rev Med Virol.* 2013;23:281–294.

Niederkorn JY. Immune regulatory mechanisms in allergic conjunctivitis: Insights from mouse models. *Curr Opin Allergy Clin Immunol.* 2008;8:472–476.

O'Brien TP. Evidence-based review of moxifloxacin. *Int Ophthalmol Clin.* 2006;46:61–72.

O'Brien TP. Besifloxacin ophthalmic suspension, 0.6%: A novel topical fluoroquinolone for bacterial conjunctivitis. *Adv Ther.* 2012 Jun;29(6):473–490. doi: 10.1007/s12325-012-0027-7. Epub 2012 Jun 20. PMID: 22729919

O'Brien TP, Jeng BH, McDonald M, Raizman MB. "Acute conjunctivitis: truth and misconceptions". *Curr Med Res Opin*. 2009;25(8):1953–1961. doi:10.1185/03007990903038269. PMID 19552618

O'Brien TP, Maguire MG, Fink NE, Alfonso E, McDonnell P. Efficacy of ofloxacin vs cefazolin and tobramycin in the therapy for bacterial keratitis. *Report from the Bacterial Keratitis Study Research Group. Arch Ophthalmol.* 1995;113:1257–1265.

Oboki K, Ohno T, Saito H, Nakae S. Th17 and allergy. *Allergol Int*. 2008;57:121–134.

Occasi F, Zicari AM, Petrarca L, Nebbioso M, Salvatori G, Duse M. Vernal keratoconjunctivitis and immunemediated diseases: One unique way to symptom control? *Pediatr Allergy Immunol.* 2015;26:289–291.

Offiah I, Calder VL. Immune mechanisms in allergic eye diseases: What is new? *Curr Opin Allergy Clin Immunol.* 2009;9:477–481.

Ohnsman C, Ritterband D, O'Brien T, Girgis D, Kabat A. Comparison of azithromycin and moxifloxacin against bacterial isolates causing conjunctivitis. *Curr Med Res Opin* 2007;23:2241–2249.

Oikonomakou MZ, Makri OE, Panoutsou E, Kagkelaris K, Plotas P, Garatziotou D, Georgakopoulos CD, Eliopoulou MI. Bacteriology and antimicrobial susceptibility patterns of childhood acute bacterial conjunctivitis in Western Greece. *Med Hypothesis Discov Innov Ophthalmol.* 2019 Winter;8(4):266–271.

Okumus S, Coskun E, Tatar MG, Kaydu E, Yayuspayi R, Comez A, et al., Cyclosporine a 0.05% eye drops for the treatment of subepithelial infiltrates after epidemic keratoconjunctivitis. *BMC Ophthalmol.* 2012;12:42.

Owen CG, Shah A, Henshaw K, Smeeth L, Sheikh A. Topical treatments for seasonal allergic conjunctivitis: Systematic review and meta-analysis of efficacy and effectiveness. *Br J Gen Pract.* 2004;54:451–456.

Pak KY, Kim SI, Lee JS. Neonatal bacterial conjunctivitis in Korea in the 21st century. *Cornea* 2017;36:415–418.

Papa V, Aragona P, Scuderi AC, Blanco AR, Zola P, Alessandro Di B, et al., Treatment of acute bacterial conjunctivitis with topical netilmicin. *Cornea.* 2002;21:43–47.

Parke DW, Coleman AL, Rich WL, Lum F. Choosing wisely: Five ideas that physicians and patients can discuss. *Ophthalmol.* 120(3):443–444.

Patel PB, Diaz MC, Bennett JE, Attia MW. Clinical features of bacterial conjunctivitis in children. *Acad Emerg Med.* 2007;14:1–5.

Pepose JS, Ahuja A, Liu W, Narvekar A, Haque R. Randomized, controlled, phase 2 trial of povidoneiodine/ dexamethasone ophthalmic suspension for treatment of adenoviral conjunctivitis. *Am J Ophthalmol.* 2019;205:197.

Pichichero ME. Bacterial conjunctivitis in children: Antibacterial treatment options in an era of increasing drug resistance. *Clin Pediatr (Phila).* 2011 Jan;50(1):7–13. doi: 10.1177/0009922810379045. Epub 2010 Aug 19. PMID: 20724317

Pinto RD, Lira RP, Abe RY, et al., Dexamethasone/povidone eye drops versus artificial tears for treatment of presumed viral conjunctivitis: a randomized clinical trial. *Curr Eye Res.* 2015;40:870–877.

Protzko E, Bowman L, Abelson M, Shapiro A. Phase 3 safety comparisons for 1.0% azithromycin in polymeric mucoadhesive eye drops versus 0.3% tobramycin eye drops for bacterial conjunctivitis. *Invest Ophthalmol Vis Sci.* 2007;48:3425–3429.

Pucci N, Novembre E, Lombardi E, Cianferoni A, Bernardini R, Massai C, et al., Atopy and serum eosinophil cationic protein in 110 white children with vernal keratoconjunctivitis: Differences between tarsal and limbal forms. *Clin Exp Allergy.* 2003;33:325–330.

Puri LR, Shrestha GB, Shah DN, Chaudhary M, Thakur A. Ocular manifestations in herpes zoster ophthalmicus. *Nepal J Ophthalmol.* 2011;3:165–171.

Quach D, Sakoulas G, Nizet V, Pogliano J, Pogliano K. Bacterial Cytological Profiling (BCP) as a Rapid and accurate antimicrobial susceptibility testing method for Staphylococcus aureus. *EBioMedicine.* 2016;4:95–103.

Raizman MB, Rubin JM, Graves AL, Rinehart M. Tear concentrations of levofloxacin following topical administration of a single dose of 0.5% levofloxacin ophthalmic solution in healthy volunteers. *Clin Ther.* 2002;24(9):1439–1450. doi:10.1016/S0149- 2918(02)80047-9

Rajpopat S, Moss C, Mellerio J, Vahlquist A, Gånemo A, Hellstrom-Pigg M, Ilchyshyn A, Burrows N, Lestringant G, Taylor A, Kennedy C, Paige D, Harper J, Glover M, Fleckman P, Everman D, Fouani M, Kayserili H, Purvis D, Hobson E, Chu C, Mein C, Kelsell D, O'Toole E. Harlequin ichthyosis: A review of clinical and molecular findings in 45 cases. *Arch Dermatol.* 2011 Jun;147(6):681–686.

Ramakrishnan R, Bharathi M, Shivkumar C, Mittal S, Meenakshi R, KhadeerM, Avasthi A. Microbiological profile of cultureproven cases of exogenous and endogenous endophthalmitis: A 10- year retrospective study. *Eye.* 2009;23(4):945–956.

Rapoza PA, Chandler JW. Neonatal conjunctivitis: Diagnosis and treatment. In: *Focal Points 1988: Clinical Modules for Ophthalmologists.* San Francisco, Calif: American Academy of Ophthalmology; 1988:5–6.

Rathi VM, Murthy SI. Allergic conjunctivitis. *Commun Eye Health.* 2017;30:S7–S10.

Rathore S, David LS, Beck MM, Bindra MS, Arunachal G. Harlequin Ichthyosis: Prenatal Diagnosis of a Rare Yet Severe Genetic Dermatosis. *J Clin Diagn Res.* 2015 Nov;9(11):QD04-6.

Reinhard T, Godehardt E, Pfahl HG, Sundmacher R. Local cyclosporin A in nummuli after keratoconjunctivitis epidemica. a pilot study]. *Ophthalmologe.* 2000;97:764–768.

Richards A, Guzman-Cottrill JA. Conjunctivitis 196 pediatrics in review Vol.31 No.5 May 2010. (Downloaded from http://pedsinreview.aappublications.org/ at University of Michigan on February 20, 2015)

Richmond S, Burman R, Crosdale E, et al., A large outbreak of keratoconjunctivitis due to adenovirus type 8. *J. hyg.* 1984;93:285–291.

Říhová E, Svozílková P, Brichová M, Klímová A, Kuthan P, Diblík P. Ocular manifestations of granulomatosis with polyangiitis. *Čes. a slov. Oftal.*, 2018;74(5):167–174.

Roberts C. Comparison of the ocular comfort of ofloxacin 0.3% ophthalmic solution and lubricant eye drops. *Clin Pediatr.* 2002;41:151–154.

Rose P. "Management strategies for acute infective conjunctivitis in primary care: a systematic review". *Expert Opin Pharmacother.* 2007;**8** (12):1903–1921. doi:10.1517/14656566.8.12.1903. PMID 17696792

Rose PW, Harnden A, Brueggemann AB, Perera R, Sheikh A, Crook D, et al., Chloramphenicol treatment for acute infective conjunctivitis in children in primary care: a randomised double-blind placebo-controlled trial. *Lancet.* 2005;366:37–43.

Sati A, Sangwan VS, Basu S. Porphyria: Varied ocular manifestations and management. *BMJ Case Rep.* 2013;2013:bcr2013009496.

Schaller UC, Klauss V. "Is Credé's prophylaxis for ophthalmia neonatorum still valid?". *Bull World Health Organ.* 2001;79(3):262–263. PMC 2566367. PMID 11285676

Schaumberg DA, Dana R, Buring JE, Sullivan DA. "Prevalence of dry eye disease among US men: estimates from the Physicians' Health Studies". *Arch Ophthalmol.* 2009; 127(6):763–768. doi:10.1001/archophthalmol.2009.103

Schaumberg DA, Sullivan DA, Buring JE, Dana MR. "Prevalence of dry eye syndrome among US women". *Am J Ophthalmol.* 2003;136(2):318–326. PMID 12888056

Schiebel NE. Evidence-based emergency medicine/systematic review abstract. Use of antibiotics in patients with acute bacterial conjunctivitis. *Ann Emer Med.* 2003;41:407–409.

Schmid KL, Schmid LM. Ocular allergy: Causes and therapeutic options. *Clin Exp Optom.* 2000;83:257–270.

Schwab IR, Friedlaender M, McCulley J, Lichtenstein SJ, Moran CT. Levofloxacin bacterial conjunctivitis active control study group. A phase III clinical trial of 0.5% levofloxacin ophthalmic solution versus 0.3% ofloxacin ophthalmic solution for the treatment of bacterial conjunctivitis. *Ophthalmol.* 2003;110:457–465.

Shanmuganathan VA, Armstrong M, Buller A, Tullo AB. External ocular infections due to methicillin-resistant *Staphylococcus aureus* (MRSA). *Eye (Lond).* 2005a;19:284–291.

Shanmuganathan VA, Armstrong M, Buller A, Tullo AB. External ocular infections due to methicillin-resistant Staphylococcus aureus (MRSA). *Eye.* 2005b;19:284–291.

Sharon Y, Cui D, Akpek EK, Chu DS. Cicatrizing conjunctivitis as an uncommon manifestation of primary Sjögren's syndrome. *The Ocular Surface.* 2021;19:38–42. ISSN 1542-0124 https://doi.org/10.1016/j.jtos.2020.11.014

Sheikh A, Hurwitz B, van Schayck CP, McLean S, Nurmatov U. Antibiotics versus placebo for acute bacterial conjunctivitis. *Cochrane Database Syst Rev.* 2012; 19:Cd001211.

Shibata A, Akiyama M. Epidemiology, medical genetics, diagnosis and treatment of harlequin ichthyosis in Japan. *Pediatr Int.* 2015 Aug;57(4):516–522.

Shields T, Sloane PD. A comparison of eye problems in primary care and ophthalmology practices. *Fam Med.* 1991;23:544–546.

Silvester A, Neal T, Czanner G, Briggs M, Harding S, Kaye S. Adult bacterial conjunctivitis: resistance patterns over 12 years in patients attending a large primary eye care Centre in the UK. *BMJ Open Ophthalmol.* 2016;1:e000006.

Singh G, Zhou X, Lee JY, Yousuf MA, Ramke M, Ismail AM, et al., Recombination of the epsilon determinant and corneal tropism: Human adenovirus species D types 15, 29, 56, and 69. *Virol.* 2015;485:452–459.

Singhal D, Sahay P, Maharana PK, Raj N, Sharma N, Titiyal JS. Vernal keratoconjunctivitis. *Surv Ophthalmol.* 2019;64:289–311.

Singla R., Mishra A,[1] Joshi R,[1] Jha S,[2] Raj Sharma A,[3] Upadhyay S,[4] Sarma P,[1] Prakash A,[1] Medhi[1]B. Human animal interface of SARS-CoV-2 (COVID-19) transmission: A critical appraisal of scientific evidence. *Vet Res Commun.* 2020;44(3–4):119–130. doi: 10.1007/s11259-020-09781-0 PMCID: PMC7487339 PMID: 32926266

Sivaraj RR, Durrani OM, Denniston AK, Murray PI. Caroline Gordon Ocular manifestations of systemic lupus erythematosus. *Rheumatol.* December 2007;46(12):1757–1762. https://doi.org/10.1093/rheumatology/kem173

Smith AF, Waycaster C. "Estimate of the direct and indirect annual cost of bacterial conjunctivitis in the United States". *BMC Ophthalmol.* 2009;9:13. doi:10.1186/1471-2415-9-13. PMC 2791746. PMID 19939250

Sofi RA, Mufti A. Vernal keratoconjunctivitis in Kashmir: A temperate zone. *Int Ophthalmol.* 2016;36:875–879.

Soparkar CN, Wilhelmus KR, Koch DD, Wallace GW, Jones DB. Acute and chronic conjunctivitis due to over the- counter ophthalmic decongestants. *Arch Ophthalmol.* 1997;115:34–38.

Srivastava K, Arora A, Kataria A, Cappelleri JC, Sadosky A, Peterson AM. Impact of reducing dosing frequency on adherence to oral therapies: A literature review and metaanalysis. *Patient Prefer Adherence.* 2013;7:419–434.

Stenson S, Newman R, Fedukowicz H. Laboratory studies in acute conjunctivitis. *Arch Ophthalmol.* 1982;100:1275–1277.

Szamocki S, Martyn-Hemphill C, Green J. Reactive arthritis: Can't see, can't pee, can't climb a tree…. *Trends Urol Men Health.* 2016;7:17–20.

Tabbara KF, Al-Ghamdi A, Al-Mohareb F, Ayas M, Chaudhri N, Al-ShariF f, et al., Ocular findings after allogeneic hematopoietic stem cell transplantation. *Ophthalmol.* 2009;116:1624–1629.

Teoh DL, Reynolds S. Diagnosis and management of pediatric conjunctivitis. *Pediatr Emerg Care.* 2003;19:48–55.

Tepedino ME, Heller WH, Usner DW, Brunner LS, Morris TW, Haas W, et al., Phase III efficacy and safety study of besifloxacin ophthalmic suspension 0.6% in the treatment of bacterial conjunctivitis. *Curr Med Res Opin.* 2009;25:1159–1169.

Tepedino ME, Jensen HG. Efficacy and safety of gatifloxacin 0.3% ophthalmic solution compared with ofloxacin 0.3% for treatment of acute bacterial conjunctivitis. *Poster presented at the Annual Meeting of the American Academy of Ophthalmology (AAO),* October 20–23, 2002. Orlando, FL.

Thareja T, Kowalski R, Jhanji V, Kamyar R, Dhaliwal D. MRSA keratitis and conjunctivitis: What does it mean practically? *Curr Ophthalmol Rep.* 2019;7:110–117.

Timewell RM, Rosenthal AL, Smith JP, Cagle GD. Safety and efficacy of tobramycin and gentamicin sulfate in the treatment of external ocular infections of children. *J Pediatr Ophthalmol Strabismus.* 1983;20:22–26.

TobraDex. Fort Worth, TX: Alcon Laboratories, Inc.; 2009.

Totan Y, Hepsen IF, Cekic O, Gunduz A, Aydin E. Incidence of keratoconus in subjects with vernal keratoconjunctivitis: a videokeratographic study. *Ophthalmol.* 2001;108:824–827.

Trinavarat A, Atchaneeyasakul LO. Treatment of epidemic keratoconjunctivitis with 2% povidone-iodine: A pilot study. *J Ocul Pharmacol Ther.* 2012;28:53–58.

Udeh BL, Schneider JE, Ohsfeldt RL. Cost effectiveness of a point-of-care test for adenoviral conjunctivitis. *Am J Med Sci.* 2008;336:254–264.

Varu DM, Rhee MK, Akpek EK, Amescua G, Farid M, Garcia-Ferrer FJ, et al., Conjunctivitis preferred practice pattern®. *Ophthalmol.* 2019;126:P94–P169.

Vigamox. Fort Worth, TX: Alcon Laboratories, Inc; 2008.

Wald ER, Greenberg D, Hoberman A. Short-term oral cefixime therapy for treatment of bacterial conjunctivitis. *Pediatr Infect Dis J.* 2001;20:1039–1042.

Wall AR, Sinclair N, Adenis JP. Comparison of fucithalmic (fusidic acid viscous eyedrops 1%) and noroxin (norfloxacin ophthalmic solution 0.3%) in the treatment of acute bacterial conjunctivitis. *J Drug Assess.* 1998;1:549–558.

Wang K, Seitzman G, Gonzales JA. Ocular cicatricial pemphigoid. *Curr Opin Ophthalmol.* 2018;29:543–551.

Watson S, Tullo AB, Carley F. "Treatment of superior limbic keratoconjunctivitis with a unilateral bandage contact lens". *Br J Ophthalmol.* 2002;86(4): 485–486. PMC 1771108. PMID 11914237

Watts P, Suresh P, Mezer E, et al., Effective treatment of ligneousconjunctivitis with topical plasminogen. *Am J Ophthalmol.* 2002;133:451–455.

Welch D, Ousler GW, 3rd, Nally LA, Abelson MB, Wilcox KA. Ocular drying associated with oral antihistamines (loratadine) in the normal population-an evaluation of exaggerated dose effect. *Adv Exp Med Biol.* 2002;506:1051–1055.

Wilson-Holt N, Dart JK. Thiomersal keratoconjunctivitis, frequency, clinical spectrum and diagnosis. *Eye.* 1989;3:581–587.

Wolters Kluwer Health. Facts and Comparisons. *Drug Facts and Comparisons.* St. Louis, MO: Wolters Kluwer;2008.

Wong AH, Barg SS, Leung AK. Seasonal and perennial allergic conjunctivitis. *Recent Pat Inflamm Allergy Drug Discov.* 2014;8:139–153.

Yee RW, Tepedino M, Bernstein P, Jensen H, Schiffman R, Whitcup SM. A randomized, investigator-masked clinical trial comparing the efficacy and safety of gatifloxacin 0.3% administered BID versus QID for the treatment of acute bacterial conjunctivitis. *Curr Med Res Opin.* 2005;21:425–431.

Yeu E, Hauswirth S. A Review of the Differential Diagnosis of Acute Infectious Conjunctivitis: Implications for Treatment and Management. *Clin Ophthalmol.* 2020;805–813, DOI: 10.2147/OPTH.S236571

Zhang M, Hu Y, Chen F. [Clinical investigation of 0.3% levofloxacin eyedrops on the treatment of cases with acute bacterial conjunctivitis and bacterial keratitis]. *Yan Ke Xue Bao [Eye Sci].* 2000;16:146–148.

Zhang L, Zhao N, Huang X, Jin X, Geng X, Chan TC, et al., Molecular epidemiology of acute hemorrhagic conjunctivitis caused by coxsackie a type 24 variant in China, 2004–2014. *Sci Rep.* 2017;7:45202.

Zicari AM, Capata G, Nebbioso M, De Castro G, Midulla F, Leonardi L, et al., Vernal keratoconjunctivitis: An update focused on clinical grading system. *Ital J Pediatr.* 2019;45:64.

Zicari AM, Nebbioso M, Zicari A, Mari E, Celani C, Occasi F, et al., Serum levels of IL-17 in patients with vernal keratoconjunctivitis: A preliminary report. *Eur Rev Med Pharmacol Sci.* 2013a;17:1242–1244.

Zicari AM, Nebbioso M, Lollobrigida V, Bardanzellu F, Celani C, Occasi F, et al., Vernal keratoconjunctivitis: Atopy and autoimmunity. *Eur Rev Med Pharmacol Sci.* 2013b;17:1419–1423.

Zikic A, Schunemann H, Wi T, Lincetto O, Broutet N, Santesso N. Treatment of neonatal chlamydial conjunctivitis: A systematic review and meta-analysis. *J Pediatr Infect Dis Soc.* 2018;7:e107–e115.

Zimmerman TJ, Hahn SR, Gelb L, Tan H, Kim EE. The impact of ocular adverse effects in patients treated with topical prostaglandin analogs: Changes in prescription patterns and patient persistence. *J Ocul Pharmacol Ther.* 2009;25:145–152.

Zylet. Bridgewater, NJ: Bausch + Lomb;2016.

14

Antimicrobial Resistance of the Diseased Human Eye: Dacryocystitis

S. Shivaji

14.1 Introduction

Dacryocystitis is frequently caused by obstruction in the nasolacrimal duct (Assefa et al., 2015) due to a malformation or injury of the tear duct, due to eye infection, or trauma (Figure 14. 1). It is an inflammation and infection of the lacrimal sac. Such obstructions of the nasolacrimal duct could result in retention of tears and detritus in the lacrimal sac (Bale, 1987) and thus cause infection of the nasolacrimal sac. Dacryocystitis occurs in all age groups though it is more frequently seen in middle and elderly age individuals and predominantly occurs in women (Hartikainen et al., 1997; Bharathi et al., 2008; Assefa et al., 2015; Eslami et al., 2018; Aşgin and Çakmaklıoğulları, 2020). In China, the female-to-male ratio was 2.56 in adults consistent with the above studies (Luo et al., 2021) but in the paediatric group the female-to-male ratio was significantly lower (Eshraghi et al., 2014; Mills et al., 2007; Chung et al., 2019). In India the highest number of cases were seen in 51–60 years and females were more affected (74%) (Leelavathamma et al., 2018; Thomas et al., 2019) as in previous studies (Machin et al., 2003; Assefa et al., 2015; Ali et al., 2015; Chaudhry et al., 2010; Chung et al., 2019). The preponderance in female patients has been attributed to the smaller diameter of the nasolacrimal duct compared to men (Mccormick and Sloan, 2009). Distinct from dacryocystitis is dacryoadenitis which is an inflammation or infection of the lacrimal gland which secretes the tears (Figure 14.1).

14.2 Risk Factors of Dacryocystitis

Dacryocystitis is known to be influenced by several risk factors associated with the lifestyle of the individual. For instance, exposure to smoke (33.9%), use of eye cosmetics like kajal (18.3%), exposure to solvents (7.3%) and rubbing of eyes with contaminated hands (1.8%) are the common risk factors (Thomas et al., 2019). However, no correlation was observed between exposure to smoke and dacryocystitis (Thomas et al., 2019). Age is also a predisposing factor and particularly so in acute dacryocystitis patients which showed a greater prevalence with increasing age (Bale, 1987; Pinar-Sueiro et al., 2010). Dacryocystitis is rare in individuals below 30 years of age and individuals in this age group may be affected due to congenital dacryocystitis or following injury

DOI: 10.1201/9781003451105-14

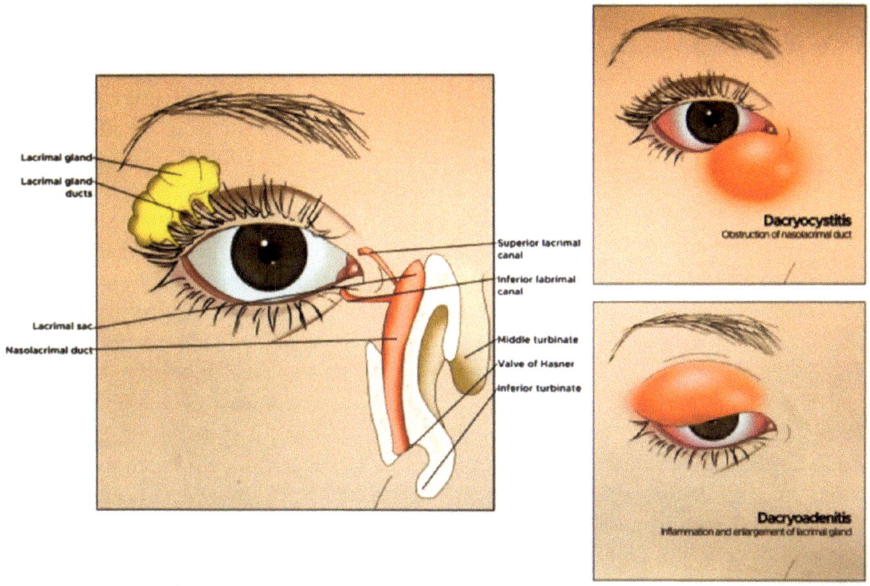

FIGURE 14.1 (a) Diagrammatic representation displaying the lacrimal gland and ducts, lacrimal sac, nasolacrimal duct, lacrimal canal (superior, inferior), middle turbinate, valve of Hasner and inferior turbinate of the eye; (b) Dacryocystitis and (c) Dacryoadenitis. (Adapted from Taylor and Ashurst (2022) contributed by Chelsea Rowe. This book is distributed under the terms of the Creative Commons Attribution-Non-Commercial-No-Derivatives 4.0 International) (CC BY-NC-ND 4.0) (http://creativecommons.org/licenses/by-nc-nd/4.0/).)

of the lacrimal drainage system (Bale, 1987; Pinar-Sueiro et al., 2010). In children dacryocystitis is rare and occurs associated with congenital nasolacrimal duct obstruction with up to 6% of healthy newborn (Ffooks, 1962; Petersen and Roob, 1978). Only 2.9% develop acute dacryocystitis (Pollard, 1991). Khatib et al. (2018) indicated that infantile dacryocystitis or dacryocystocele are almost invariably associated with an intranasal cyst and the percentage of dacryocystitis/dacryocystoceles in infants with nasal cyst ranged from 6% to 100% depending on the age (Levin et al., 2003; Lueder et al., 2004; 2012). Sex is also a risk factor with higher rates of both acute and chronic dacryocystitis in women (73% and 63.3%, respectively) (Summer and Jones, 1965; Sood et al., 1967). Additionally, 28.6% of adult dacryocystitis patients had nasal pathologies (Bale et al., 1987) and the most common being allergic rhinitis (6.4%), deviated nasal septum (1.8% to 69%), hypertrophied turbinate (2% to 41%) and atrophic rhinitis (9%-10%) (Patel et al., 2014; Mandal et al., 2008; Adlakha et al., 2015). Presence of dacryoliths in the lacrimal drainage system is also a known risk factor for the development of dacryocystitis (Bale, 1987; Jones, 1965; Yazici et al., 2001) and a higher proportion of men with dacryoliths was observed compared to women among those undergoing external dacryocystorhinostomy (Yazici et al., 2001).

14.3 Clinical Features of Dacryocystitis

Patients with dacryocystitis usually complain of pain, redness, swelling of the lower eyelid, epiphora (excessive tear production) and retention of lacrimal fluid (Huber et al., 1991; Leelavathamma et al., 2018). In India, Thomas et al. (2019) analysed 100 clinically diagnosed dacryocystitis patients and observed that apart from pain, redness and blurring of vision (8.3%) the other symptoms which were common were eye discharge (66.1%), epiphora (56.9%), swelling of the eye (15.6%) and fever (0.9%). The type of discharge was either serous discharge (42%), mucopurulent discharge (40%) or purulent discharge (18%); and there was no significant correlation between age and epiphora and between eye discharge and swelling of eye (Thomas et al., 2019). This study also agreed with earlier studies that had reported only epiphora in 50% to 100% of the patients, mucous or mucopurulent or purulent discharge in 10% to 71% and about 10% showed swelling and redness (Prakash et al., 2012; Sood et al., 1967; Siddiqui et al., 2013). Accumulation of microbes and severe inflammation were common in all dacryocystitis patients. Moreover, dacryocystitis was also observed to be associated with formation of draining fistulae and abscesses, recurrent conjunctivitis, orbital cellulitis and endophthalmitis in patients, who undergo intraocular surgery (Pinar-Sueiro et al., 2012). The presentation of dacryocystitis could also vary according to the geographical origin of the patients and also based on the etiological agents of dacryocystitis (Nigam et al., 2008).

14.4 Types of Dacryocystitis

14.4.1 Acute Dacryocystitis

Dacryocystitis is differentiated into acute and chronic types (Ataullah and Sloan, 2002; Leelavathamma et al., 2018). Acute dacryocystitis is common in adult women and it is associated with severe morbidity, lacrimal sac abscess (Janssen et al., 2000) and presents as a pre-septal infection which could be associated with orbital cellulitis, conjunctivitis, thrombosis of the superior ophthalmic vein, and cavernous sinus (Janssen et al., 2000; Maheshwari et al., 2009).The common symptoms include reddening, edema, presence of a painful area overlying the nasolacrimal sac, epiphora and discharge. When pressure is applied to the inflamed tear duct, purulent material may be expressed through the lacrimal punctum (Pinar-Sueiro et al., 2012).

Paediatric acute dacryocystitis (PAD) is a rare but potentially serious condition, characterized by acute inflammation of the lacrimal sac, which can rapidly evolve into lacrimal sac abscess or other serious life-threatening complications (Ali et al., 2015; Campolattaro et al., 1997; Cicco et al., 2020).

14.4.2 Chronic Dacryocystitis

The chronic form of dacryocystitis is more common than acute dacryocystitis and is associated with chronic tearing, thickening of the drainage system, conjunctival inflammation and majority of patients harbour multiple microorganisms (Grant, 2001). Like acute, the chronic dacryocystitis were also predominant in women (Mekonnen et al., 2009; Mills et al., 2007) and are a constant threat to the cornea and orbital tissue. It has been consistently observed that majority of the cases >65.9% had chronic dacryocystitis followed by acute dacryocystitis and congenital dacryocystitis (Bharathi et al., 2008; Eshraghi et al., 2014; Thomas et al., 2019). Complications of dacryocystitis include fistula, corneal ulcer, and orbital cellulitis and causes social embarrassment due to long-lasting epiphora (Badhu et al., 2006; Mandal et al., 2008; Huber-Spitzy et al., 1992; Chaudhry et al., 2005).

14.5 Microbiology of Dacryocystitis

14.5.1 Culture Positivity in Dacryocystitis

Isolation of bacteria from dacryocystitis patients is dependent on the nature of the sample collected (Hartikainen et al., 1997) which includes samples collected through the lacrimal punctum, posterior lacrimal flap or by swabbing the lower conjunctival cul-de-sac. The positive cultures ranged from 8.3% to 100% (Chaudhry et al., 2005; 2010; Assefa et al., 2015; Pornpanich et al., 2016; Kebede et al., 2010) and was observed to be more in chronic (95.34%) than in acute dacryocystitis (72.23%) patients (Kandati et al., 2015). Further the percentage of cases yielding single isolates was always consistently higher than isolation of multiple microorganisms (Thomas et al., 2019). Chung et al. (2019) observed polymicrobial infection in 33% of all cases and 38% in recurrent dacryocystitis, both slightly higher than the reported range of 7–30% (Mandal et al., 2008; Martins et al., 2008).The microbes associated with dacryocystitis differed between the two types of dacryocystitis; in severe acute dacryocystitis the infection is dominated by a single Gram-negative rod whereas in chronic dacryocystitis multiple species of bacteria were involved in the pathogenesis. According to Bharathi et al. (2008) in majority of patients polymicrobial infections are common and included bacteria, fungus, and virus.

14.5.2 Abundance of Gram-positive and Gram-negative Bacteria in Dacryocystitis

In both acute and chronic dacryocystitis Gram-positive organisms were predominant (Kuchar et al., 2000; Chaudhry et al., 2005; Huber-Spitzy et al., 1987; 1992; Mills et al., 2007; Prokosch et al., 2010; Ali et al., 2013; Assefa et al., 2015; Eslami et al., 2018), and constituted 61–90% of pathogens, whereas the remaining 2.5–39% included Gram-negative bacteria (Pornpanich et al., 2016; Chaudhry et al., 2005; 2010; Pradeep et al., 2013; Shah and Santani, 2011; Ali et al., 2015). The isolation rates of Gram-positive bacteria in dacryocystitis patients were 62% in Finland (Hartikainen et al., 1997), 54.4% in Australia (Sainju et al., 2005), 79.1% in Saudi Arabia (Begum et al., 2006), 71.2% in Iran (Razavi et al., 2010), 68.8% in the United States (Mills et al., 2007), 66.7% in Egypt (Negm et al., 2019) and 69.7%-75% in India (Bharathi et al., 2008; Delia et al., 2008). However, contrary to the above studies it was observed that the Gram-positive bacteria which were abundant did show a tendency to decrease (46%) and concomitantly the Gram-negative bacteria increased (40%) thus indicating a trend in increase in Gram-negative organisms in the lacrimal sac infections (Briscoe et al., 2005a; b; Kotlus et al., 2005; Kubo et al., 2002). More recently, Luo et al. (2021) also observed that Gram-positive isolates were not predominant in adults and both the Gram-positive and Gram-negative isolates were numerically equal; but in the paediatric group the Gram-positive bacteria were the major group (58.68%) and Gram-negative constituted a small proportion (26.45%).

14.5.2.1 Gram-positive Bacteria in Dacryocystitis

Overall, several studies indicated that bacteria of the genus *Staphylococcus* are the most predominant bacteria isolated from dacryocystitis patients but their abundance varied from study to study. The bacteria of the genus *Staphylococcus* included CoNS (29.0%), *S. aureus* (19.4%) and *S. pneumoniae* (3.9%) (Hartikainen et al., 1997; Madhusudhan et al., 2012; Assefa et al., 2015; Sivakrishna et al., 2018). Overall, the *Staphylococcus* species accounted for 30% to 88.68% (Huber-Spitzy et al., 1992; Sun et al., 2005; Chung et al., 2019) of the isolates in dacryocystitis patients with a predominance of *S. aureus* (54.61%) and *S. epidermidis* (45.39%) (Kandati et al., 2015) similar to the findings of Thicker and Buffam (1993) (73%), Huber-Spitzy et al. (1992) (51%), and Coden et al. (1993) (49%), Sainju et al. (2005), Delia et al. (2008) and Sivakrishna et al. (2018) (42.8%, *S. aureus*). In disagreement with these studies Aşgın and Çakmaklıoğulları (2020) reported that in patients with chronic dacryocystitis *S. epidermidis* (15.7%) and *S. aureus* (11.8%) were a minority and the predominant bacterium was a Gram-negative bacterium (*P. aeruginosa*, 45%). Among the *Staphylococcus* isolates, *S. pneumoniae* represented 2 to 23% of the bacterial isolates (Negm et al., 2019; Huber-Spitzy et al., 1992; Coden et al., 1993; Hartikainen et al., 1997) and 83.34% isolates were present in acute dacryocystitis and 16.66% in cases of chronic dacryocystitis which are higher than that reported by Huber-Spitzy et al. (1992) (2%), Coden et al. (1993) (2.3%), Hartikainen et al. (1997) (5%) and Bharathi et al. (2008) (10%). Further, 9.4% of *S. epidermidis* isolates were methicillin resistant and methicillin resistance among the *S. aureus* (MRSA) isolates varied from 3.1%-33% (Thomas et al., 2019; Aşgın and Çakmaklıoğulları, 2020). Studies also indicated that greater frequency of MRSA were seen in patients with acute dacryocystitis (Mills et al., 2007; Chung et al., 2019). Other species isolated from dacryocystitis patients included *Micrococcus* spp. only from acute dacryocystitis (3.77%) (Kandati et al., 2015) and *Propionibacterium acnes* from paediatric or adult patients with acute or chronic (10%) dacryocystitis (Chung et al., 2019).

14.5.2.2 Gram-negative Bacteria in Dacryocystitis

The Gram-negatives represented <50% of all isolates and several studies indicated that the most predominant are *Pseudomonas* species (10%-12%) (Bharathi et al., 2008; Chung et al., 2019) including *Pseudomonas aeruginosa* (9.7%-45%) (Assefa et al., 2015; Briscoe et al., 2005a; b; Aşgın and Çakmaklıoğulları, 2020). *E. coli* (34.64%) was the next most predominant according to Kandati et al. (2015) but differed with the findings of Bharathi et al. (2008) and Huber-Spitzy et al. (1992), who reported that *E. coli* constituted only 4.7% and 12% respectively. *Klebsiella pneumoniae* (6.5% to 14.38%) and *Haemophilus influenza* (4%-7.19%) (Kandati et al., 2015; Assefa et al., 2015) were the next predominant Gram-negatives in dacryocystitis patients in agreement with the observations of Huber-Spitzy et al. (1992), Coden et al. (1993), and Briscoe et al. (2005 a; b). A single study reported *Serratia marcescens* (Pinar-Sueiro et al., 2010) and another study reported *Acinetobacter* species (Thomas et al., 2019) in dacryocystitis patients. Thus, the diversity and abundance of the different types of bacteria between the studies seem to agree with respect to the dominant species but differ with respect to the actual abundance recorded in the different studies. The reasons for this discrepancy have not been addressed but could be attributed to multiple factors including the number of cases investigated, variation in the sociodemographic parameters of the patients, the geographical origins of the individuals and the type of dacryocystitis and so forth.

14.5.3 Diversity of Bacteria in Chronic and Acute Dacryocystitis

In adult dacryocystitis patients the common bacteria that are isolated include *S. epidermidis*, *S. aureus*, *S. pneumoniae*, and *P. aeruginosa* (Eslami et al., 2018; Eshraghi et al., 2014). But the prevalence of the above microorganisms varied between acute and chronic dacryocystitis patients. In 39 acute dacryocystitis patients *S. aureus* (33.3%), *S. epidermidis* (25.6%), and *S. pneumoniae* (17.9%) were predominant whereas in 58 chronic dacryocystitis patients both *S. aureus* (10.3%) and *S. epidermidis* (25.8%) as in acute dacryocystitis patients were dominant but in addition in these patients *Pseudomonas* (15.5%) and *C. albicans* (10.3%) were also dominant (Eslami et al., 2018). In yet another study *Pseudomonas* was reported in 2.5% of acute dacryocystitis patients whereas in chronic dacryocystitis patients *Pseudomonas* constituted 32.7% of the infections. Other bacteria isolated were *Burkholderia pseudomallei* (3.1%) and *Moraxella* species (1.5%) (Thomas et al., 2019; Kubo et al., 2002). Anaerobes have also been identified in 15.7% dacryocystitis cases and the frequently recovered anaerobic bacteria included *Peptostreptococcus* spp. (3.1%), *Propionibacterium* spp. (1.5%), anaerobic *Streptococcus* spp., *Fusobacterium* spp. and *Prevotella* spp. (Brook, 1998; Brook and Frazier, 1998; Hartikainen et al., 1997; Ali et al., 2013; Bekmez et al., 2019).

14.5.4 Diversity of Bacteria in Paediatric Patients with Dacryocystitis

In paediatric group, *Streptococcus pneumoniae* (24.79%), *S. aureus* (11.57%) (Assefa et al., 2015; Amin et al., 2013), *S. haemolyticus,* and *H. influenzae* were reported as the common pathogens in children (Eslami et al., 2018; Eshraghi et al., 2014). It was also reported that *Haemophilus influenza* was the predominant Gram-negative isolate in both adults (6.76%) and paediatric (5.78%) groups (Teweldemedhin et al., 2017; Kebede et al., 2010; Bekmez et al., 2019). However, in paediatric chronic dacryocystitis with nasolacrimal duct obstruction (NLDO), *S. pneumoniae* (28.43%) was predominant, whereas in paediatric acute dacryocystitis, *S. aureus* (42.11%) was predominant thus differentiating the bacterial spectrum between acute and chronic dacryocystitis patientts (Luo et al., 2021; Ali et al., 2013). Further fungus isolates were significantly more common in paediatric group (14.88%) compared to the adults (2.70%) (Pinar-Sueiro et al., 2012; Sun et al., 2005; Brook and Frazier 1998) and the common fungal pathogens that were identified were *Fusarium* spp., *Aspergillus* spp.,*Candida albicans*, *C. tropicalis* and *Mucor* spp. in dacroyocystitis patients (Sainju et al., 2005; Eshraghi et al., 2014).

14.6 Diagnosis of Dacryocystitis

14.6.1 Collection of Sample

This is a crucial step in the diagnosis of dacryocystitis. The mucopurulent sample is collected in a syringe from the infected area taken either from the conjunctival sac or by collecting the secretion by manually compressing the lacrimal sac. Antiseptic is applied prior to puncturing the lacrimal sac. The sample is immediately processed or stored in an anaerobic transport medium at 4 to 8°C until processing (within 24 h of collection) (Pinar-Sueiro et al., 2010). Several culture media that are routinely used in a clinical laboratory to culture both aerobic and anaerobic bacteria are recommended for culturing the samples (Etxebarria et al., 2009).

14.6.2 Etiological Agents of Dacryocystitis

Several studies have identified the flora of the lacrimal sac (Pinar-Sueiro et al., 2012) using samples such as the discharge expressed from the lacrimal sac (Chaudhry et al., 2005; Sun et al., 2005), the fluid collected following irrigating the lacrimal drainage system (Hartikainen et al., 1997; Sun et al., 2005), the fluid aspirated from a punctured lacrimal sac (Mills et al., 2007), or the sample obtained from lacrimal sac during dacryocystorhinostomy (Owji and Kahlili, 2009; Briscoe et al., 2005a; b). Not all the samples yielded positive cultures and the rate of positive cultures in the samples ranged from 8.3% to 100% (Owji and Kahlili, 2009; Delia et al., 2008). The samples also varied with respect to the percentage of Gram-positive to Gram-negative isolates, the specific bacteria and their predominance in acute and chronic dacryocystits patients and in in children and adults as presented separately above.

14.6.3 Blood Culture

Blood culture as part of microbiological diagnosis in dacryocystitis patients is extremely variable but used for paediatric patients. In 25 infants (<6 months of age) with dacryocystitis, 88% of the children yielded positive cultures and CoNS were identified in 22.7% of the sample (Baskin et al., 2008). Care should be taken to avoid contamination during blood culture by skin flora (Baskin et al., 2008). It could also be used to detect microorganisms in adult patients with immunosuppression and those with high fever (Pinar-Sueiro et al., 2012).

14.7 Treatment of Dacryocystitis

14.7.1 Treatment of Acute Dacryocystitis

Treatment of acute dacryocystitis in adults depends on whether the patients present with an abscess. Patients with an abscess go through a procedure of puncture followed by aspiration of the fluid from the lacrimal sac. The procedure could be painful and could result in incomplete drainage of the mucopurulent material from the sac. Such a condition could result in prolonged or recurrent inflammation and development of lacrimal cutaneous fistulae and formation of fibrous and granulation tissue in the sac and may thus cause adverse effects due to systemic antibiotics (Baskin et al., 2008). Surgical interventions like transcanalicular endoscopic laser-assisted dacryocystorhinostomy and nasal endoscopic surgery are suggested for the management of this type of infection (Wu et al., 2009) since these surgical procedures, especially nasal endoscopic dacryocystorhinostomy is more effective than conservative treatment and dacryocystitis resolves within a short span of time (Cahill and Burns, 1993; Wu et al., 2009). Nasal endoscopy facilitates higher rates of opening of the nasolacrimal ostium (Wu et al., 2009) and lowers the rate of stenosis. In patients devoid of an abscess dacryocystitis is manged with heat massage, antibiotics (oral or intravenous) and percutaneous drainage of the abscess (Cahill and Burns, 1993).

In paediatric patients with acute dacryocystitis, intubation of the nasolacrimal duct (Ffooks, 1962; Campolattaro et al., 1997) and intravenous antibiotics have been recommended. Sepsis is associated in up to 17.5% of pediatric patients subjected to the above procedure (Jones and Wobig, 1976; Walland and Rose, 1994) attributed to the immature immune system of children (Ffooks, 1962; Pollard 1991; Campolattaro et al., 1997; Strunk et al., 2007). In patients at risk of infectious endocarditis (Campolattaro et al., 1997; Walland and Rose, 1994) prior administration of antibiotics is statistically significantly associated with a higher success rate (Baskin et al., 2008).

14.7.2 Treatment of Chronic Dacryocystitis

Treatment of chronic dacryocystitis varies depending on the age of the patient. Adults with swelling of the lacrimal sac and suspected obstruction in the lacrimal drainage system probably due to tear stones are managed conservatively by massaging the lacrimal sac and irrigating the sac until symptoms improved (Traquair, 1941). Surgery was used as an option when patients were refractory to the conservative management approach. Surgical procedures like dacryocystorhinostomy carried out in the absence of antibiotic prophylaxis could expose the patient to the risk of soft-tissue infection (Huber-Spitzy et al., 1992). Thus, to curtail postoperative infection antibiotic therapy is recommended during surgery for chronic dacryocystitis. Pinar-Sueiro et al. (2012) recommended prophylactic antibiotics for dacryocystorhinostomy especially in patients, who have had prior episodes of mucocele, mucopyocele, or acute dacryocystitis. Pinar-Sueiro et al. (2012) used intravenous cefazolin (1 or 2 g) for this purpose.

Unlike in adults, in paediatric patients more than 90% with unilateral dacryoceles resolve spontaneously and the remaining could be managed using conservative treatment involving hydrostatic massage and bacitracin eyedrops. Surgery under general anaesthesia is indicated only if the patients had a dacryocele or acute dacryocystitis. The nasal cavity is decongested and visualized with a paediatric endoscope for the presence of nasolacrimal duct cysts. More recently a micro-debrider instrument has been used for the removal of the cyst (Khatib et al., 2018). In children (18 to 36 months age) dilation of the nasolacrimal duct or intubation is suggested but in older children (>3 years of age) dacryocystorhinostomy with or without antibiotic prophylaxis is the routinely used procedure. However, antibiotic prophylaxis is recommended under conditions of accumulation of mucopurulent fluid in the conjunctival sac. Thus, the use of antibiotics for the management of dacryocystitis (Assefa et al., 2015) would depend on several factors such as the age of the patient, disease status, etiological agent, and drug resistance profile (Coden et al., 1993; Assefa et al., 2015).

14.8 Antibacterial Resistance in Dacryocystitis

To put together all the information on antibiotic resistance in dacryocystitis is very complex considering that dacryocystitis could be either acute or chronic, effecting adult, paediatric groups or infants, individuals across several geographical regions, the diversity of bacteria involved, the diversity of fungi involved and further complicated by the innumerable number of antibiotics that have been used to treat dacryocystitis. Thus, the issue of antibiotic resistance may be best addressed with respect to resistance to a particular class of antibiotic irrespective of the other variables listed above.

14.8.1 Resistance to Fluoroquinolones

Over the years several studies have indicated that bacterial isolates from dacryocystitis patients were sensitive to fluoroquinolones. Some studies indicated that both Gram-positive and Gram-negative bacteria were sensitive to levofloxacin (Sun et al., 2005), ofloxacin (Kandati et al., 2015) and ciprofloxacin (Aşgın and Çakmaklıoğulları, 2020) with ofloxacin having better efficacy than ciprofloxacin (Kandati et al., 2015). Ciprofloxacin was preferentially active against Gram-negative bacteria (81.5%) and less active against Gram-positive bacteria (52.5%) (Aşgın and Çakmaklıoğulları, 2020). Some studies indicated that only a specific group of bacteria were sensitive to the fluoroquinolone like the Gram-negative isolates *Pseudomonas* spp., *P. aeruginosa, K. pneumonia, H. influenzae* and *E. coli* (50%), which were sensitive to ciprofloxacin (33.33% to 86%), oflaxacin and levofloxacin (Mandal et al., 2008). In contrast, Sun et al. (2005) reported that the fluoroquinolones levofloxacin, amikacin, ofloxacin and ciprofloxacin were most active against the Gram-positive bacteria and this was confirmed by Adlakha et al. (2015), who noticed that all Gram-positive isolates were 92.34% sensitive to ciprofloxacin and ofloxacin. In the same year *S. aureus* was also found to be sensitive to ciprofloxacin (Assefa et al., 2015). In 2019, Thomas et al. (2019) observed that *S. epidermidis*, was 50% sensitive to ciprofloxacin and levofloxacin. In China also Gram-positive bacteria showed high susceptibility to gatifloxacin (Chen et al., 2018). It was concluded that among the fluoroquinolones, gatifloxacin showed greater efficacy compared to ciprofloxacin and levofloxacin against all pathogens screened including Gram-positive, Gram-negative, and anaerobic isolates (Bharathi et al., 2008; Chen et al., 2018; Negm et al., 2019). Usha et al. (2006) also confirmed that Gram-positive organisms exhibited a high rate of sensitivity to Ofloxacin.

Despite all these studies on the use of fluoroquinolones to treat isolates of dacryocystitis patients only one study indicated the emergence of antibiotic resistance. Assefa et al. (2015) reported that CoNS were resistant to ciprofloxacin. Analysis of 51 dacryocystitis patients from January 2011-January 2012 in Ethiopia observed that 31 bacterial isolates were characterised as CoNS, *S. aureus*, *Pseudomonas* spp., *S. pneumoniae*, *Entrobacter* spp. *K. pneumoniae* and *H. influenzae* and these were resistant to amoxicillin (38.7%), ciprofloxacin (25.8%), chloramphenicol (25.8%), cotrimoxazole (25.8%), and ampicillin (19.4%). Further out of 31 bacterial isolates, 9, 5, 5, 5 and 4 isolates were resistant to 1, 2, 3, 4 or 5 antibiotics indicating that the maximum number of isolates (9) were resistant to at least one antibiotic and a small number of isolates (4 isolates) were resistant to all the 5 antibiotics (Assefa et al., 2015). *Citrobacter* spp. was the only isolate sensitive to all the antibiotics tested (Table 14.1).

14.8.2 Resistance to Nalidixic Acid

Nalidixic acid is not a typical quinolone because the 1,8-naphthyridine ring structure contains two nitrogen atoms in the place of a single nitrogen atom seen in quinolones (Emmerson and Jones, 2003). It is primarily active against Gram-negative bacteria, but also shows reduced activity against Gram-positive bacteria. Nalidixic acid has been used routinely to treating urinary tract infections caused by *E. coli*, *Proteus*, *Shigella*, *Enterobacter*, and *Klebsiella* (Pommier et al., 2010). Studies have indicated that different bacterial isolates from dacryocystitis patients were susceptible to nalidixic acid (87.1%), including *S. aureus* and CoNS (Assefa et al., 2015).

14.8.3 Resistance to Aminoglycosides

Aminoglycosides are a class of antibiotics used mainly to treat infections caused by aerobic Gram-negative bacilli but are also effective against *Staphylococcus* spp. and *Mycobacterium tuberculosis*. Aşgın and Çakmaklıoğulları (2020) confirmed that aminoglycosides were very effective against Gram-negative bacteria (>90%) and were preferred in the place of fluoroquinolones for empirical treatment of chronic dacryocystitis. A few studies also indicated that aminoglycosides were most effective against *S. epidermidis* (Patel et al., 2014; Thomas et al., 2019). The commonly used aminoglycosides for the treatment of dacryocystitis include amikacin, gentamicin and tobramycin, whereas reports on the use of neomycin, kanamycin and streptomycin are not common.

14.8.3.1 Amikacin

The aminoglycoside antibiotic amikacin was the most effective in treating dacryocystitis infection (Sun et al., 2005; Amin et al., 2013; Negm et al., 2019). The sensitivity to amikacin varied from 60% -100% (Shahraki et al., 2016; Thomas et al., 2019) depending on the species targeted for instance it was 75% in *Acinetobacter* spp. whereas it was 100% in the case of *S. epidermidis* (Thomas et al., 2019). Negm et al. (2019) confirmed that CoNS and *Klebsiella* spp are associated with dacryocystitis cases (60%) and were sensitive to amikacin (88%). Studies also indicated that bacterial pathogens isolated from acute dacryocystitis namely *S. aureus* (22.3%) and *P. aeruginosa* (21.1%) and those isolated from chronic dacryocystitis patients CoNS (44.2%), *S. aureus* (10.8%), and *Streptococcus pneumoniae* (10%) were all susceptible to amikacin (91.1%) (Bharathi et al., 2008) and it was concluded that amikacin could be used as a preferred drug against all common dacryocystitis pathogens except *Streptococcus pneumoniae*, which was sensitive to Cloxacillin (Kandati et al., 2015). Bharathi et al. (2008) also observed that the resistance of bacterial isolates to antibiotics was higher from chronic dacryocystitis patients compared to isolates in acute infection (Bharathi et al., 2008). As of date resistance to amikacin in dacryocystitis infection has not yet been reported.

14.8.3.2 Gentamicin

Gentamicin is yet another aminoglycoside antibiotic that is very effective (100%) against aerobic Gram-negative bacteria (Prakash et al., 2012). A few reports have also indicated that gentamicin is bactericidal to Gram-positive bacteria from dacryocystitis individuals (Aşgın and Çakmaklıoğulları, 2020). Eslami et al. (2018) confirmed that *S. aureus, S. epidermidis,* and *S. pneumoniae* the predominant Gram-positive bacteria both from acute and chronic dacryocystitis were susceptible to gentamicin. Bharathi et al. (2008) based on antibiotic susceptibility tests demonstrated that the resistance of bacterial isolates recovered from chronic dacryocystitis – CoNS (44.2%), *S. aureus* (10.8%) and *S. pneumoniae* (8.7%) – to gentamicin was higher (45.7%), than the resistance of bacteria isolates from acute dacryocystitis patients – *S. aureus* (22.3%), *Pseudomonas* spp. (21.1%) and *S. pneumoniae* (16%) – to gentamicin (24.6%).

14.8.3.3 Tobramycin

Tobramycin is a natural aminoglycoside isolated from *Streptomyces tenebrarius* effective against Gram-negative infections especially caused by *Pseudomonas.* Tobramycin is used frequently to eliminate *Pseudomonas aeruginosa* and *Klebsiella pneumoniae.* Gram-negative isolates from dacryocystitis patients were most sensitive to tobramycin (100%) (Prakash et al. (2012). Further in 2008, it was demonstrated that the resistance of bacterial isolates to tobramycin (50.8%), were found to be higher from chronic dacryocystitis patients compared to tobramycin (35%), in acute infection (Bharathi et al., 2008).

14.8.4 Resistance to Cephalosporins (β-lactam antibiotics)

Cephalosporins are also known as β-lactam antibiotics and are grouped into five generations. The first generation cephalosporins (cefazolin, cephalexin, cephalosporin C, cephalothin and cefapirin) are used to treat soft tissue infections, second-generation cephalosporins (cefuroxime and cefoxitin) have several uses such as to treat respiratory tract infections, UTIs, skin infections, soft tissue infections, and Lyme disease, third-generation cephalosporins (ceftriaxone and ceftazidime) are used to overcome serious infections, including meningitis and sepsis, fourth-generation cephalosporin (cefepime) to treat staphylococcal infections and fifth-generation cephalosporin (ceftaroline) is effective against MRSA (methicillin-resistant *Staphylococcus aureus*) and community-acquired pneumonia. Some of these have also been used to treat he etiological agents of dacryocystitis.

14.8.4.1 Cefazolin

Pinar-Sueiro et al. (2012) recommended intravenous cefazolin (1 or 2 g) as a prophylactic antibiotic prior to dacryocystorhinostomy in patients with prior episodes of mucocele, mucopyocele, or acute dacryocystitis. Adlakha et al. (2015) noticed that all Gram-positive isolates were 94.59% sensitive to cefazolin but subsequently Thomas et al. (2019) observed that Gram negative isolates were least sensitive to cefazolin.

14.8.4.2 Cephalexin

Briscoe et al. (2005a) monitored the antibiotic sensitivities of purulent dacryocystitis patients and reported that microorganisms showed high sensitivity to the third-generation cephalosporins (ceftazidime and cefuroxime, 95% and 50% respectively) but the susceptibility to cephalexin, a first-generation cephalosporin, was low (14%). Negm et al. (2019) also demonstrated that cephalexin showed only 58% efficacy against Gram-positive isolates, which was less compared with ceftazidime (66%), and cefotaxime (66%) the third generation of cephalosporins. Further, it was also observed that Gram-negative isolates were also more sensitive to cefotaxime (80%) and ceftazidime (60%) compared to cephalexin (40%). It was also demonstrated that 59% of the isolates were resistant to cephalexin (Bharathi et al., 2008; Kandati et al., 2015; Delia et al., 2008).

14.8.4.3 Cefuroxime

Cefuroxime is a second-generation cephalosporin and studies on the sensitivity of isolates from purulent dacryocystitis had indicated that Gram-negative isolates were more sensitive to ceftazidime (95%) and ciprofloxacin (86%) compared to cefuroxime (50%). But the susceptibility to cefuroxime (50%) was higher than that observed for cefalexin and ampicillin (< 30%) (Briscoe et al. (2005a; b). Thomas et al. (2019) also indicated that *E coli* were only 16.7% sensitive to cefuroxime and the most sensitive antibiotic for Gram-negative isolates were colistin and least sensitive was cefazolin.

14.8.4.4 Ceftriaxone

Ceftriaxone, a third-generation cephalosporin, has been used to treat bacterial infections such as meningitis, gonorrhoea and community-acquired pneumonia. It has also been used to treat dacryocystitis. Pinar-Sueiro et al. (2010) while evaluating the prophylactic use of antibiotics in external dacryocystorhinostomy for the prevention of postoperative complications reported that bacteria were sensitive to ceftriaxone. This observation was confirmed by Eslami et al. (2018), who observed that bacteria isolated from acute and chronic dacryocystitis patients were sensitive to ceftriaxone (73.6%) and other antibiotics. In an earlier study Assefa et al. (2015) also indicated that bacterial isolates from dacryocystitis patients were susceptible to ceftriaxone (95.3%), but CoNS were not susceptible to ceftriaxone.

14.8.4.5 Ceftazidime

Briscoe et al. (2005a; b) studied the sensitivity of isolates from purulent dacryocystitis patients and observed that Gram-negative isolates were most sensitive to ceftazidime (95%), a third-generation cephalosporin, in addition to being sensitive to ciprofloxacin (86%), cefuroxime (50%) and less than 30% to Cephalexin and Ampicillin. Pornpanich et al. (2016) and Thomas et al. (2019) also found that all *P. aeruginosa* isolates were susceptible to ceftazidime and other antibiotics like ciprofloxacin, gentamicin, and meropenem. Several of these observations were confirmed by Negm et al. (2019), who showed that the third generation of cephalosporins, ceftazidime (66%), and cefotaxime (66%) were more effective compared to cephalexin a first generation of cephalosporins, which showed only 58% efficacy against Gram-positive isolates. Further the susceptibility of Gram-negative isolates was maximum to cefotaxime (80%) followed by ceftazidime (60%) and cephalexin (40%). Resistance to ceftazidime by dacryocystitis isolates has not yet been reported.

14.8.4.6 Cefepime and Ceftaroline

Cefepime and ceftaroline are fourth-generation and fifth-generation cephalosporin antibiotics respectively used very scarcely in treating dacryocystitis. Prakash et al. (2012) in a bacteriology study of dacryocystitis showed that Gram-negative organisms were sensitive to cefepime (98.79%). However as yet there is no report on the effects of ceftaroline though ceftaroline is effective against MRSA (methicillin-resistant *S. aureus)* (Reese et al., 2017).

14.8.5 Penicillins

Penicillins are β-lactam antibiotics produced from *P. chrysogenum* and *P. rubens*, which are moulds. Penicillin G and Penicillin V were used as the first line of treatment against bacterial infections caused by staphylococci and streptococci. Even to this day these are still used for different bacterial infections. But over the years there have been several reports on the emergence of resistance to penicillins. In one report based on antibiotic susceptibility in isolates from dacryocystitis patients it was observed that CoNS and *S. aureus* showed 33% resistance to penicillin (Assefa et al., 2015).

Clindamycin is used to treat bacterial infections when penicillin is not an option. Sivakrishna et al. (2018) confirmed that Gram-positive isolates from dacryocystitis patients were more sensitive to vancomycin but were equally sensitive to linezolid (100%) followed by Clindamycin (86%).

14.8.5.1 Ampicillin

Ampicillin is a "broad spectrum" penicillin antibiotic used to treat bacterial infections caused by Gram-positive (*S. pneumoniae*, *S. pyogenes*, and *S. aureus)* and Gram-negative bacteria like *Enterococcus faecalis* and *E. faecium* (Magdesian, 2017), *Neisseria meningitidis*, some *Haemophilus influenzae*, and some Enterobacteriaceae (though most Enterobacteriaceae and *Pseudomonas* are resistant) (Magdesian, 2017; Papich, 2015). Some isolates from dacryocystitis patients showed least sensitivity to ampicillin (Adlakha et al., 2015; Chen et al., 2018; Sivakrishna et al., 2018; Thomas et al., 2019) like *S. pneumoniae*, which was completely resistant to ampicillin in contrast to earlier finding by Kebede et al. (2010) and *S. epidermidis*, which showed only 15.4% sensitivity to ampicillin (Thomas et al., 2019). But CoNS from dacryocystitis patients were sensitive to ampicillin (Assefa et al., 2015). Gram-positive isolates were extremely susceptible to vancomycin, teicoplanin and linezolid and exhibited least susceptibility to ampicillin (Thomas et al., 2019; Pornpanich et al., 2016). Briscoe et al. (2005a; b) demonstrated that Gram-negative bacteria from dacryocystitis patients exhibited less than 30% sensitivity to ampicillin. The variation in resistance pattern observed by different investigators has been attributed to empirical and irrational use of antibiotics (Assefa et al., 2015).

14.8.5.2 Methicillin

Methicillin is a β-lactam antibiotic used to treat infections caused by Gram-positive bacteria like *S. aureus*, *S. epidermidis*, *Streptococcus pyogenes*, and *S. pneumoniae*. But this antibiotic is no longer effective due to resistance conferred by activation of the *mecA gene* that codes for a bacterial penicillin binding protein. In 2015 it was reported that all isolates of CoNS were sensitive to methicillin and other antibiotics (ampicillin, methicillin, nalidixic acid, and erythromycin) (Assefa et al., 2015). But in the subsequent years it was observed that 9.4% of *S. epidermidis* and 3.1%–33% of *S. aureus* were methicillin resistance (Aşgın and Çakmaklıoğulları, 2020). Thomas et al. (2019) also observed that the methicillin resistant *S. epidermidis* was 50% sensitive to ciprofloxacin and levofloxacin and 16.7% sensitive to azithromycin.

14.8.5.3 Piperacillin

Pipercillin is a broad-spectrum β-lactam antibiotic (Tan and File, 1995) and is active against the hospital pathogen *P. aeruginosa*. Pornpanich et al. (2016) and Thomas et al. (2019) observed that all *P. aeruginosa* isolates from dacryocystitis patients showed less than 50% sensitivity to aztreonam, piperacillin and tazobactam.

14.8.5.4 Amoxicillin-clavulanic

Amoxicillin is a β-lactam antibiotic, which is bacteriolytic against both Gram-positive and Gram-negative bacteria. *Streptococcus, Bacillus subtilis, Enterococcus, Haemophilus, Helicobacter,* and *Moraxella* are susceptible to amoxicillin, whereas *Citrobacter, Klebsiella* and *P. aeruginosa* are resistant to it (Huttner et al., 2020). Some strains of *E. coli* and most strains of *S. aureus* have developed resistance to amoxicillin (Tadesse et al., 2017). In acute dacryocystitis, Amoxicillin-clavulanic acid combination was very effective against *S. aureus* (80%) and *H. influenzae* (100%) (Pinar-Sueiro et al., 2010). But in a study, covering the period January 2011-January 2012, 38.7% bacterial isolates from dacryocystitis patients in Ethiopia—which were identified as CoNS, *S. aureus*, *Pseudomonas* spp., *S. pneumoniae*, *Entrobacter* spp. *K. pneumoniae* and *H. influenzae*—were resistant to amoxicillin (Assefa et al., 2015). *Citrobacter* spp. was the only isolate sensitive to all the antibiotics tested (Table 14.1). The variation in resistance pattern may be due to inappropriate use of antibiotics (Assefa et al., 2015) and due to variations depending on the region and availability (Chung et al., 2019). For instance, the most effective treatment for dacryocystitis in Iran was amoxicillin-clavulanic acid and third-generation cephalosporin or ciprofloxacin (Razavi et al., 2010) whereas in other countries it was different such as ciprofloxacin in Thailand (Pornpanich et al., 2016), gatifloxacin, ofloxacin, and amikacin in Egypt (Amin et al., 2013), and norfloxacin in India (Shah and Santani, 2011). These findings highlight regional variation in antibiotic resistance and sensitivities, as well as variation in the antibiotics that may be commercially available.

14.8.5.5 Carbapenems

The carbapenems are members of the β-lactam antibiotics used for known or suspected MDR bacterial infections. These exhibit a broader spectrum of activity and unlike

TABLE 14.1

Antibiotic Resistance Patterns[a] of Bacterial Pathogens Isolated from Dacryocystitis Patients from Ethiopia from 2011 to 2012

Bacterial isolates	No. (%)	S/R	Common antibiotics tested, No. (%)											
			AMP	AMO	CIP	CN*	MET*	NA	CAF	ERY*	PEN*	TTC	SXT	CEF
S. aureus	6 (19.4)	S	5 (83.3)	5 (83.3)	6 (100.0)	-	5 (83.3)	6 (100)	6 (100)	4 (66.7)	4 (66.7)	3 (50.0)	4 (66.7)	6 (100)
		R	1 (16.7)	1 (16.7)	0 (0.0)	-	1 (16.7)	0 (0.0)	0 (0.0)	2 (33.3)	2 (33.3)	3 (50.0)	2 (33.3)	0 (0.0)
CNS	9 (29.0)	S	9 (100)	7 (77.8)	5 (55.6)	-	9 (100)	9 (100)	7 (77.8)	9 (100)	6 (66.7)	7 (77.8)	6 (66.7)	8 (88.9)
		R	0 (0.0)	2 (22.2)	4 (44.4)	-	0 (0.0)	0 (0.0)	2 (22.2)	0 (0.0)	3 (33.3)	2 (22.2)	3 (33.3)	1 (11.1)
S. pneumoniae	2 (6.5)	S	0 (0.0)	2 (100)	1 (50.0)	-	2 (100)	2 (100)	1 (50.0)	2 (100)	2 (100)	0 (0.0)	2 (100)	2 (100)
		R	2 (100.0)	0 (0)	1 (50.0)	-	0 (0.0)	0 (0.0)	1 (50.0)	0 (0.0)	0 (0.0)	2 (100)	0 (0.0)	0 (0.0)
Entero-bacter spp.	2 (6.5)	S	2 (100.0)	1 (50.0)	1 (50.0)	-	-	0 (0.0)	1 (50.0)	1 (50.0)	1 (50.0)	2 (100)	0 (0.0)	2 (100)
		R	0 (0.0)	1 (50.0)	1 (50.0)	-	-	2 (100)	1 (50.0)	1 (50.0)	1 (50.0)	0 (0.0)	2 (100)	0 (0.0)
E. coli	1 (3.2)	S	1 (100.0)	0 (0.0)	1 (100)	1 (100)	-	0 (0.0)	1 (100)	-	-	0 (0.0)	1 (100)	1 (100)
		R	0 (0.0)	1 (100)	0 (0.0)	0 (0.0)	-	1 (100)	0 (0.0)	-	-	1 (100)	0 (0.0)	0 (0.0)

K. pneumoniae	2 (6.5)	S	1 (50.0)	1 (50.0)	2 (100)	2 (100)	-	1 (50)	2 (100)	-	-	1 (50.0)	2 (100.)	1 (50.0)
		R	1 (50.0)	1 (50.0)	0 (0.0)	0 (0.0)	-	1 (50)	0 (0.0)	-	-	1 (50.0)	0 (0.0)	1 (50.0)
Pseudo-monas spp.	3 (9.7)	S	2 (66.7)	1 (33.3)	2 (66.7)	2 (66.7)	-	2 (66.7)	2 (66.7)	-	-	2 (66.7)	2 (66.7)	3 (100.)
		R	1 (33.3)	2 (66.7)	1 (33.3)	1 (33.3)	-	1 (33.3)	1 (33.3)	-	-	1 (33.3)	1 (33.3)	0 (0.0)
Citrobacter spp.	1 (3.2)	S	1 (100)	1 (100)	1 (100)	1 (100)	-	1 (100)	1 (100)	-	-	1 (100)	1 (100)	1 (100)
		R	0 (0.0)	0 (0.0)	0 (0.0)	0 (0.0)	-	0 (0.0)	0 (0.0)	-	-	0 (0.0)	0 (0.0)	0 (0.0)
H. influenzae	2 (6.5)	S	2 (100.0)	0 (0.0)	2 (100)	2 (100)	-	2 (100)	1 (50)	-	-	1 (50.0)	1 (50.0)	2 (100)
		R	0 (0.0)	2 (100)	0 (0.0)	0 (0.0)	-	0 (0.0)	1 (50)	-	-	1 (50.0)	1 (50.0)	0 (0.0)
P. vulgaris	1 (3.2)	S	1 (100.0)	0 (0.0)	1 (100)	1 (100)	-	1 (100)	1 (100)	-	-	1 (100)	1 (100)	1 (100)
		R	0 (0.0)	1 (100)	0 (0.0)	0 (0.0)	-	0 (0.0)	0 (0.0)	-	-	0 (0.0)	0 (0.0)	0 (0.0)
P. mirabilis	1 (3.2)	S	0 (0.0)	0 (0.0)	1 (100)	1 (100.)	-	1 (100)	0 (0.0)	-	-	1 (100)	1 (100)	1 (100)
		R	1 (100)	1 (100)	0 (0.0)	0 (0.0)	-	0 (0.0)	1 (100)	-	-	0 (0.0)	0 (0.0)	0 (0.0)

(*continued*)

TABLE 14.1 (Continued)

Antibiotic Resistance Patterns[a] of Bacterial Pathogens Isolated from Dacryocystitis Patients from Ethiopia from 2011 to 2012

Bacterial			Common antibiotics tested, No. (%)											
isolates	**No. (%)**	**S/R**	**AMP**	**AMO**	**CIP**	**CN***	**MET***	**NA**	**CAF**	**ERY***	**PEN***	**TTC**	**SXT**	**CEF**
Providencia spp.	1 (3.2)	S	1 (100)	0 (0.0)	1 (100)	0 (0.0)	-	1 (100)	1 (100)	-	-	0 (0.0)	0 (0.0)	1 (100)
		R	0 (0.0)	1 (100)	0 (0.0)	1 (100)	-	0 (0.0)	0 (0.0)	-	-	1 (100)	1 (100)	0 (0.0)
Total	31 (100)	S	25 (80.6)	19 (61.3)	23 (74.2)	10 (83.3)	17 (89.5)	27 (87.1)	23 (74.2)	16 (84.2)	13 (68.4)	22 (71.0)	23 (74.2)	29 (93.5)
		R	6 (19.4)	12 (38.7)	8 (25.8)	2 (16.7)	2 (10.5)	4 (12.9)	8 (25.8)	3 (15.8)	6 (31.6)	6 (29.0)	8 (25.8)	2 (6.5)

Source: Data from Assefa et al., 2015.

Notes:

[a] CAF – chloramphenicol; CL – cephalexin; CAZ – ceftazidime; CTX – cefotaxime; CN – gentamycin; AK – amikacin; TOB – tobramycin; GAT – gatifloxacin; CIP – ciprofloxacin; LEV – levofloxacin.

Keys: AMP – Ampicillin; AMO – Amoxicillin; CIP – Ciprofloxacin; CN – Gentamycin; MET – Methicillin; NA – Nalidic Acid; CAF – Chloramphinicol; ERY – Erythromycin; VAN – Vancomycin; PEN – penicillin; TTC – Tetracycline; SXT – Co-trimoxazole; CEF – Ceftriaxone.

other β-lactams, resistance to this group is not common. The carbapenems namely imipenem and meropenem are preferred for patients showing late-onset hospital-acquired or ventilator-associated pneumonia, caused by *Pseudomonas*, *Acinetobacter*, or extended spectrum beta-lactamase producing *Enterobacteriaceae* . Combination therapy including a carbapenem and an aminoglycoside, is recommended for *Pseudomonas* infections to avoid development of antibiotic resistance (Infectious Diseases Society of America, 2005).

14.8.5.6 Imipenem

Thomas et al. (2019) observed that the common Gram-negative bacteria from dacryocystitis patients namely *P. aeruginosa* and *Acinetobacter* spp., were 100% and 50% sensitive to imipenem respectively apart from their sensitivity to other antibiotics like colistin and cefazolin. This observation in dacryocystitis patients was surprising since the Gram-negative isolates *K. pneumoniae*, *Pseudomonas* spp., and *H.* influenzae were resistant to most antibiotics tested (Kebede et al., 2010; Assefa et al., 2015) including imipenem prior to 2019, (Sivakrishna et al., 2018).

14.8.5.7 Meropenem

Pornpanich et al. (2016) and Thomas et al. (2019) also found that all *P. aeruginosa* isolates from dacryocystitis patients were 100% susceptible to ceftazidime, gentamicin, and meropenem but showed less than 50% sensitivity to aztreonam, piperacillin and tazobactam. Acinetobacter spp. also showed only 50% sensitive to imipenem and 25% sensitive to amoxyclav and cefotaxime.

14.8.6 Resistance to Macrolides

Macrolides belong to the polyketide class of natural antibiotics and are bacteriostatic rather than bactericidal. Among this group the most popularly used antibiotics used for treatment of dacryocystitis include erythromycin and azithromycin.

14.8.6.1 Erythromycin and Azithromycin

Assefa et al. (2015) reported that CoNS from dacryocystitis patients were sensitive to erythromycin whereas *S. sureus* were less susceptible to erythromycin and 33.3% were resistant to erythromycin and other antibiotics like penicillin, tetracycline and co-trimoxazole, The results on the sensitivity to different antibiotics was also confirmed by Eslami et al. (2018), who observed that various strains like *S. aureus, S. epidermidis,* and *S. pneumoniae* the common strains *in* acute and chronic dacryocystitis patients were sensitive to erythromycin (52.9%) and several other antibiotics including ciprofloxacin (75.9%), ceftriaxone (73.6%), vancomycin (67.8%), chloramphenicol (60.9%) and gentamicin (56.3%).

Compared to erythromycin, the use of azithromycin in dacryocystitis is very scanty though azithromycin has been used to treat several systemic bacterial infections caused by *S. aureus* (Methicillin-sensitive only), *H. influenzae* and others not involved in dacryocystitis. There is only one report that observed that the most common Gram-positive bacterium from dacryocystitis patients *S. epidermidis* (methicillin resistant) was only 16.7% sensitive to azithromycin (Thomas et al., 2019).

14.8.7 Resistance to Glycopeptides

14.8.7.1 Vancomycin

Vancomycin is glycopeptide antibiotic effective principally against Gram-positive cocci (Sarkar et al., 2017) like methicillin-resistant *Staphylococcus aureus* (MRSA). Several studies confirmed that Gram-positive bacteria from dacryocystitis individuals were highly sensitive to vancomycin. Usha et al. (2006) confirmed that Gram-positive organisms from congenital nasolacrimal duct obstruction were susceptible to vancomycin. Further, Adlakha et al. (2015), Pornpanich et al. (2016), Chen et al. (2018), Sivakrishna et al. (2018) and Thomas et al. (2019) observed that Gram-positive dacryocystitis isolates were most sensitive to vancomycin in addition to being very sensitive to teicoplanin and linezolid (Pornpanich et al. 2016). Eslami et al. (2018) also observed that *S. aureus, S. epidermidis,* and *S. pneumoniae* from dacryocystitis patients, both acute and chronic, were most susceptible to vancomycin and chloramphenicol. Interestingly it was also observed that both the under and above 10year age groups were sensitive to ciprofloxacin and vancomycin and thus could be used for empirical therapy of dacryocystitis (Eslami et al., 2018). Several other glycopeptide antibiotics including teicoplanin, telavancin, ramoplanin, decaplanin, corbomycin, complestatin and bleomycin have been in use (Sarkar et al., 2017) but not with respect to dacryocystitis. The resistance profile of dacryocystitis isolates to vancomycin is also not reported.

14.8.8 Resistance to Other Antibiotics

14.8.8.1 Linezolid

Linezolid is in a class of antibacterials called oxazolidinones used for the treatment of infections caused by Gram-positive bacteria including streptococci, vancomycin-resistant enterococci (VRE), and MRSA. Pornpanich et al. (2016) demonstrated that Gram-positive isolates were very sensitive to linezolid and this was confirmed subsequently by Thomas et al. (2019).

14.8.8.2 Chloramphenicol

Chloramphenicol has been used to treat several ocular diseases (conjunctivitis, endophthalmitis and blepharitis) caused by bacteria like *S. aureus, Streptococcus pneumoniae*, and *E. coli*. It is not effective against *P. aeruginosa* (Edwards et al., 2009; Gower et al., 2017). Usha et al. (2006) observed that Gram-positive microorganisms were susceptible to chloramphenicol and it was subsequently demonstrated that Gram-positive isolates were 95.94% sensitive to chloramphenicol (Adlakha et al., 2015). Prakash et al. (2012) in a bacteriology study of dacryocystitis patients showed that Gram-negative organisms were 97.14% sensitive to chloramphenicol and ciprofloxacin (61.64%) was the least effective antibiotic. Resistance to chloramphenicol was reported by Assefa et al. (2015) in dacryocystitis patients from Ethiopia.

14.8.8.3 Tetracycline

Tetracycline, is an oral, broad spectrum antibiotic effective against systemic infections like cholera, plague, malaria and so forth. (Todar, 2012). Tetracycline effects both Gram-positive and Gram-negative bacteria but *P. aeruginosa* and *Proteus* spp., are resistant, attributed to their intrinsic resistance. Resistance has now been reported in previously sensitive microorganisms like *Staphylococcus* spp., *Streptococcus* spp., *Neisseria gonorrhoeae*, anaerobic members of the Enterobacteriaceae implying the emergence of resistance to antibiotics. The molecular mechanism of resistance acquisition to tetracycline in bacteria is due to horizontal transfer of a gene (Chopra and Roberts, 2001; Connell et al., 2003). Resistance to tetracycline in streptococci has been reported and Assefa et al. (2015) observed that 33.3% of the isolates of *S. aureus* and 100% *S. pneumoniae* respectively were resistant to tetracycline in contrast to earlier findings by Kebede et al. (2010).

14.8.8.4 Rifampicin

Rifampicin, is an antibiotic used to treat tuberculosis (TB), leprosy, and Legionnaires' disease apart from several other diseases of bacteria (Sensi, 1983). Pinar-Sueiro et al. (2010) studied the effectiveness of antibiotic prophylaxis in external dacryocystorhinostomy and observed that both Gram-positive and Gram-negative bacteria were most sensitive to rifampicin.

14.8.8.5 Polymixin B

Polymyxin B is an antibiotic specific against Gram-negative infections and commonly used to treat patients with meningitis, pneumonia, sepsis, and urinary tract infections. Thomas et al. (2019) showed that the common Gram-negative bacterium from dacryocystitis patients *P. aeruginosa* was 100% sensitive to polymixin B.

14.8.8.6 Cotrimoxazole

Cotrimoxazole, a combination of Trimethoprim and sulfamethoxazole, is a fixed-dose combination antibiotic used to treat urinary tract infections, MRSA, skin infections, travelers' diarrhoea, respiratory tract infections, cholera, pneumocystis pneumonia and toxoplasmosis and so forth. (Wormser et al., 1982). A few studies have indicated that the dacryocystitis isolates namely *S. epidermidis*, was 100% sensitive to cotrimoxazole whereas *Acinetobacter* spp., was 75% sensitive to cotrimoxazole (Thomas et al., 2019).

14.8.8.7 Tazobactam

Tazobactam is a drug that inhibits the action of bacterial β-lactamases and thus when added to certain antibiotics it makes them less vulnerable to the β-lactamases. Tazobactam along with piperacillin is used to treat infections caused by *P. aeruginosa* (Lodise et al., 2007). Pornpanich et al. (2016) and Thomas et al. (2019) observed that all *P. aeruginosa* isolates from dacryocystitis patients showed less than 50% sensitivity to tazobactam.

TABLE 14.2

Antibiotic Susceptibility of Bacteria (Disc Diffusion Method) Isolated from Lacrimal Sacs of Chronic Adult Dacryocystitis Patients

Antibiotic tested	Susceptibility to antibiotics (%)*									
Name of organism	**CAF**	**CL**	**CAZ**	**CTX**	**CN**	**AK**	**TOB**	**GAT**	**CIP**	**LEV**
Staphylococcus aureus (3)	100	66.6	66.6	66.6	66.6	66.6	66.6	100	66.6	66.6
Staphylococcus epidermidis (4)	75	50	75	75	75	100	100	100	75	75
Streptococcus pneumonia (4)	50	75	50	50	50	75	50	75	50	50
Streptococcus viridans (1)	100	0	100	100	100	100	100	100	100	100
Total Gram-positive (12)	75	58	66.6	66.6	66.6	83.3	75	91.6	66.6	66.6
Klebsiella (3)	33.3	33.3	33.3	66.6	66.6	100	66.6	66.6	66.6	66.6
Pseudomonas (2)	50	50	100	100	50	100	50	100	50	50
Total Gram-negative (5)	40	40	60	80	60	100	60	80	60	60
Total 17	65	53	65	70.5	65	88.8	70.5	88.8	65	65

Source: Modified from Negm et al., 2019.

14.8.8.8 Aztreonam

Azetronam is a monbactam antibiotic used to treat infections by *Pseudomonas aeruginosa*. It is active against *Acinetobacter anitratus*, *E. coli*, *P. aeruginosa*, and *Proteus mirabilis* while *S. aureus*, *S. haemolyticus* and *Xanthomonas maltophilia* are resistant to aztreonam (Brogden et al., 1986). Pornpanich et al. (2016) and Thomas et al. (2019) observed that all *P. aeruginosa* isolates from dacryocystitis patients showed less than 50% sensitivity to aztreonam, piperacillin, tazobactam.

14.8.8.9 Colistin

Colistin or polymyxin E, is a last-resort treatment for MDR Gram-negative infections including pneumonia (Pogue et al., 2017) caused by the etiological agents *P. aeruginosa*, *K. pneumoniae*, or *Acinetobacter* (Falagas et al., 2008). Colistin is most effective against *P. aeruginosa* and *Acinetobacter* spp., (Thomas et al., 2019).

14.9 Dacryocystitis in Different Countries

The microbiology profile of chronic dacryocystitis patients is known to vary according to geographical origin of the patient, probably due to variation in environmental factors

(Badhu et al.2006). In general, Gram-positive bacteria in dacryocystitis infection accounted for higher isolation rate. For instance, it was 62.6% in Ethiopia (Kebede *et al.*, 2010), 62% in Finland (Hartikainen et al., 1997), 54.4% in Australia (Sainju et al., 2005), 79.1% in Saudi Arabia (Begum et al., 2006), 71.2% in Iran (Razavi et al., 2010) and 68.8% in the United States (Mills et al., 2007). In contrast, in Israel 61% of the bacteria were Gram-negative, particularly *P. aeruginosa* (22%) (Briscoe et al., 2005a; b). The predominant species varied from region to region. *S. epidermidis* and S. aureus were dominant in Saudi Arabia (Chaudhry et al., 2005), China (Sun et al., 2005), Austria (Huber Spitzy et al., 1987), Australia (Sainju et al., 2005), and Spain (Pinar-Sueiro et al., 2010; 2012). The diversity of bacteria associated with dacryocystitis is described for a few countries below.

14.9.1 Ethiopia

Assefa et al. (2015) identified the bacterial etiological agents of dacryocystitis in patients attending University of Gondar Eye Specialized Hospital, in Northwest Ethiopia between January 2011 and January 2012 and evaluated their antibiotic susceptibility. Of the 51 dacryocystitis cases bacteria were the etiological agents in 60.8% cases dominated by CoNS (29.0%), *S. aureus* (19.4%), and *Pseudomonas* spp. (9.7%) and the other isolates were identified as *S. pneumoniae*, *Entrobacter* spp., *K. pnemoniae* and *H. influenzae* and each of these accounted for 6.5% of the cases. Resistance of the bacteria to antibiotics followed the trend amoxicillin (38.7%) > ciprofloxacin, chloramphenicol and co-trimoxazole when tested separately (25.8%) > ampicillin (19.4%). *Citrobacter* was the only species susceptible to the antibiotics tested. The percentage of bacteria that were resistant to one antibiotic was 29.0% and this percentage represented the maximum percentage of resistance. Further, the resistance to more than one antibiotic (two, three or four antibiotics) was less and accounted for only 16.1% (Assefa et al., 2015).

14.9.2 Egypt

Negm et al. (2019) studied the clinic-microbiological profile of 25 adults with chronic dacryocystitis and observed that majority of the cases (60%) were infected with bacteria with 48% showing the involvement of a single bacterial isolate while 12% had mixed bacterial isolates. Gram positives constituted 66.7% of the cases and 27.7% had Gram-negative isolates with one person showing a fungal etiological agent. CoNS were the most predominant Gram-positive bacteria (22.2%) whereas *Klebsiella* spp. was the predominant Gram-negative bacteria (16.6%). Majority of the isolates were sensitive to gatifloxacin (88%) and amikacin (88%) (Table 14.2) and 59% were resistant to cephalexin.

14.9.3 India

Dacryocystitis was seen in the age group of 30–60 years, and females were more affected (51.43–74%) (Sivakrishna et al., 2018; Delia et al., 2008). Chronic dacryocystitis was more common (57–75%) compared to acute dacryosystitis (25–42.29%) and the clinical symptoms included only epiphora in 35–100% of cases, epiphora with muco-purulent discharge in 40% of the cases and epiphora with swelling in the medial canthus region in 25% of the cases (Kandati et al., 2015; Patel et al., 2014; Leelavathamma et al., 2018). Dacryocystitis affecting one eye was observed in 61.75% cases and in the remaining both the eyes (38.25%) were affected (Kandati et al, 2015). The spectrum of bacteria associated with dacryocystitis differed in their diversity and abundance and varied between acute and chronic dacryocystitis patients. Culture-positivity was high in chronic dacryocystitis (90–95.34%) compared to acute dacryocystitis (57.4–72.3%) and in majority of the eyes (74.5% to >90%) single bacterial species was isolated, and only in 6.2–16.39% of eyes, two bacterial species were isolated (Bharathi et al., 2008; Kandati et al., 2015; Delia et al., 2008). In acute dacryocystitis, the predominant bacterium that was isolated was *S. aureus* (22.3%) followed by *P. aeruginosa* (21.1%) whereas in chronic dacryocystitis CoNS (44.2%), *S. aureus* (10.8%), and *S. pneumoniae* (10%) were the frequently isolated bacteria. More than 90% of the isolates were susceptible to gatifloxacin (96.5%), ofloxacin (94.8%), and amikacin (91.1%). The resistance profiles of the bacterial isolates to antibiotics varied between the isolates from chronic and acute dacryocystitis patients. Overall, the resistance to gentamicin (45.7%), tobramycin (50.8%), norfloxacin (50.7%), and ciprofloxacin (30.4%) were found to be higher from chronic dacryocystitis patients compared to isolates from acute dacryocystitis patients, which showed decreased resistance to gentamicin (24.6%), tobramycin (35%), norfloxacin (36.5%), and ciprofloxacin (19.9%) (Bharathi et al., 2008). *S. aureus* was the most abundant (Patel et al., 2014; Leelavathamma et al., 2018) with 55.84% in acute and 44.16% in chronic dacryocystitis (Kandati et al., 2015). This was followed by S. epidermidis 59.37% in acute and 40.63% in chronic dacryocystitis and then *Streptococcus pneumoniae* 83.34% in acute and 16.67% in chronic dacryocystitis (Kandati et al., 2015). *Streptococcus* and *Klebsiella* were the least abundant (3.4%) in the study by Leelavathamma et al. (2018) whereas in the study by Sivakrishna et al. (2018) *P. aeruginosa* were the least represented (2.85%). This contrasts with the findings of Kandati et al. (2015), who reported *P. aeruginosa* as the common pathogen (20.9% in acute, and 79.1% in chronic dacryocystitis). The other bacteria isolated include *E. coli*, CoNS, *Pneumococcus,* and *H. influenzae*. Most of the isolates *of S. aureus* were sensitive to ciprofloxacin (82.9%) (Patel et al., 2014). Antibiotic susceptibility studies indicated that amikacin was the most effective antibiotic and was the drug of choice as against all common pathogens except *S. pneumoniae*, which was sensitive to cloxacillin (Kandati et al., 2015). In Nepal, the most common microorganism was *S. pneumoniae* (Badhu et al., 2006).

14.9.4 Iran

Nasolacrimal duct discharges from 129 patients in Iran with acute (34.1%) and chronic (65.9%) dacryocystitis were analysed during 2016–2017. The data indicated that culture growth was positive in 75.2% of patients and the predominant bacteria were *S. aureus, S. epidermidis*, *Streptococcus pneumoniae* and *Pseudomonas* spp. One yeast i.e., *Candida albicans* was also identified. The diversity reported was like that reported from other countries (Negm et al., 2019; Leelavathamma et al., 2018). Antibiotic susceptibility tests indicated that these organisms were sensitive to ciprofloxacin (75.9%), ceftriaxone (73.6%), vancomycin (67.8%), and chloramphenicol (60.9%). Interestingly it was observed that patients below and above 10 years were sensitive to ciprofloxacin and vancomycin and could be used for empirical therapy of both acute and chronic dacryocystitis (Eslami et al., 2018).

14.9.5 China

In a retrospective study, sixty-four adults and one hundred and five paediatric dacryocystitis patients in central China were investigated to identify the microbiology profile for the clinical treatment of dacryocystitis (Luo et al., 2021). All the adults had chronic dacryocystitis with nasolacrimal duct obstruction (NLDO) whereas the paediatric patients included 89 with chronic dacryocystitis with NLDO and 16 had only acute dacryocystitis. When the percentage of bacteria were analysed both the Gram-positive and Gram-negative isolates were numerically equal (48.65%) in the adults whereas in the paediatric group Gram-positive isolates constituted the major group (58.68%). Between the adults and paediatric group *Streptococcus pneumonia* was the most predominant isolate in the dacryocystitis patients. Further between the two paediatric subgroups Gram-positive isolates were the major organisms with *Streptococcus pneumonia* (28.43%) being most common in chronic dacryocystitis with NLDO and *S. aureus* (42.11%) in acute dacryocystitis. Moreover, virulent organisms were more common in acute dacryocystitis than in chronic dacryocystitis with NLDO (Luo et al., 2021).

14.10 Wrap up on Antimicrobial Resistance in Isolates from Dacryocystitis Patients

Antibiotics belonging to all the different classes have been used to treat bacterial infections in dacryocystitis. The list of antibiotics is very lengthy and some have been more popular and commonly used and in all cases the sensitivity of the isolates to the antibiotics and the patient's recovery from the diseases has been recorded. But the development of antibiotic resistance has been reported only for some of the antibiotics and there are many that have the potential to be used for dacryocyctitis therapy. This chapter reviews the sensitivity /resistance of dacryocystitis isolates to antibiotics and the effects are characterised with respect to the efficacy of the candidate antibiotic against the dacryocystitis etiological agent.

REFERENCES

Adlakha N, Shakya D, Gandhi S, Kujur R, Chauhan HS, Ranjan K. A clinico-bacteriological study of lacrimal regurgitate in cases of chronic dacryocystitis in a referral hospital in Madhya Pradesh. *Int J Med Public Health.* 2015;5(4):270–270.

Ali MJ, Joshi SD, Naik MN, Honavar SG. Clinical profile and management outcome of acute dacryocystitis: Two decades of experience in a tertiary eye care center. *Semin Ophthalmol.* 2015;30(2):118–123. doi:10.3109/08820538.2013.833269

Ali MJ, Motukupally SR, Joshi SD, Naik MN. The microbiological profile of lacrimal abscess: Two decades of experience from a tertiary eye care center. *J Ophthalmic Inflamm Infect.* 2013;3(1):57.

Amin RM, Hussein FA, Idriss HF, Hanafy NF, Abdallah DM. Pathological, immunohistochemical and microbiologicalal analysis of lacrimal sac biopsies in patients with chronic dacrocystitis. *Int J Ophthalmol.* 2013;6(6):817–826.

Aşgın N, Çakmaklıoğulları M. Bacterial profile and antibiotic susceptibility pattern in patients with chronic dacryocystitis. *J Surg Med.* 2020;4(3):217–221.

Assefa Y, Moges F, Endris M. et al., Bacteriological profile and drug susceptibility patterns in dacryocystitis patients attending Gondar University teaching hospital, Northwest Ethiopia. *BMC Ophthalmol.* 2015;15:34. https://doi.org/10.1186/s12886-015-0016-0

Ataullah S, Sloan B. Acute dacryocystitis presenting as an orbital abscess. *Clin Exp Ophthalmol.* 2002;30:44–6.

Badhu BP, Karki BS, Khanal B, Dulal S, Das H. Microbiological patterns of chronic dacryocystitis. *Ophthalmol.* 2006;113(12):2377.e1-2

Bale RN. Dacryocystitis: Bacteriological study and its relation with nasal pathology. *Indian J Ophthalmol.* 1987;35(4):178–82.

Baskin DE, Ashvini KR, Chu YI, et al. The timing of antibiotic administration in the management of infant dacryocystitis. *J AAPOS* 2008; 12:456–9.

Begum NN, Al-Khattaf AS, Al-Mansouri SM, Yeboah EA, Kambal AM. A study of bacterial isolates from corneal specimens and their antibiotic resistance profile. *Saudi Med J.* 2006;27(1):41–5.

Bekmez S, Eriş E, Altan EV, Dursun V. The role of bacterial Etiology in the tear duct infections secondary to congenital Nasolacrimal duct obstructions. *J Craniofac Surg.* 2019;30(7):2214–6.

Bharathi MJ, Ramakrishnan R, Maneksha V, Shivakumar C, Nithya V, Mittal S. Comparative bacteriology of acute and chronic dacryocystitis. *Eye.* 2008;22:953–60.

Briscoe D, Rubowitz A, Assia EI, Changing MD. Bacterial isolates and antibiotic sensitivities of purulent dacryocystitis. *Orbit.* 2005a;24:29–32.

Briscoe D, Rubowitz A, Assia EI. Changing bacterial isolates and antibiotic sensitivities of purulent dacryocystitis. *Orbit.* 2005b;24:95–98. doi:10.1080/01676830590926585.

Brogden RN, Heel RC. Aztreonam. A review of its antibacterial activity, pharmacokinetic properties and therapeutic use. *Drugs.* 1986 February;31(2):96–130. doi: 10.2165/00003495-198631020-00002. PMID: 3512234

Brook I. Dacryocystitis caused by anaerobic bacteria in the newborn. *Pediatr Infect Dis J.* 1998;17(2):172–3.

Brook I, Frazier EH. Aerobic and anaerobic microbiology of dacryocystitis. *Am J Ophthalmol.* 1998;125(4):552–520.

Cahill KV, Burns JA. Management of acute dacryocystitis in adults. *Ophthal Plast Reconstr Surg.* 1993;9:38–41.

Campolattaro BN, Lueder GT, Tychsen L. Spectrum of pediatric dacryocystitis: Medical and surgical management of 54 cases. *J Pediatr Ophthalmol Strabismus.* 1997;34:143–53.

Chaudhary M, Bhattarai A, Adhikari SK, Bhatta DR. Bacteriology and antimicrobial susceptibility of adult chronic dacryocystitis. *Nepal J Ophthalmol.* 2010;2(2):105–113.

Chaudhry IA, Shamsi FA, Al-Rashed W. Bacteriology of chronic dacryocystitis in a tertiary eye care center. *Ophthal Plast Reconstr Surg.* 2005;21(3):207–210. doi:10.1097/01.IOP.0000161718.54275.7D

Chen L, Fu T, Gu H, Jie Y, Sun Z, Jiang D, Yu J, Zhu X, Xu J, Hong J. Trends in dacryocystitis in China: A STROBE-compliant article. *Medicine (Baltimore).* 2018 June;97(26):e11318. doi: 10.1097/MD.0000000000011318. PMID: 29953020; PMCID: PMC6039673

Chopra I, Roberts M. "Tetracycline antibiotics: Mode of action, applications, molecular biology, and epidemiology of bacterial resistance". *Microbiol. Mol. Biol. Rev.* 2001 June;65(2):232–60.

Chung SY, Rafailov L, Turbin RE, Langer PD. The microbiologic profile of dacryocystitis. *Orbit*. 2019;38(1):72–8.

Cicco MD, Bellino EM, Marabotti A, Luti L, Peroni DG, Baroncelli GI. Acute dacryocystitis with giant lacrimal abscess: A case report. *Ital. J. Pediatri*. 2020;46:15. https://doi.org/10.1186/s13052-020-0779-7

Coden DJ, Hornblass A, Haas BD. Clinical bacteriology of dacryocystitis in adults. *Ophthal Plast Reconstr Surg*. 1993;9(2):125–31.

Connell SR, Tracz DM, Nierhaus KH, Taylor DE. "Ribosomal protection proteins and their mechanism of tetracycline resistance". *Antimicrob. Agents Chemother*. 2003 December;47(12):3675–3681. doi:10.1128/AAC.47.12.3675-3681.2003. PMC 296194. PMID 14638464

Delia, A, Ganesh Ch, Uuri C, Battacharjee K, Das D, Gogoi U. Bacteriology of chronic Dacryocystitis in adult population of Northeast India. Orbit, 27:243–247, 2008. ISSN: 0167-6830 print / 1744-5108 online. doi: 10.1080/01676830802224668

Edwards KH, Sherman J, Portello JK, Rosenfield M. (2009). Examination of the anterior segment of the eye. In Mark Rosenfield and Nicola Logan (Eds.). *Optometry: Science, techniques and clinical management*. Elsevier Health Sciences. p. 102. ISBN 978-0750687782

Emmerson AM, Jones AM. "The quinolones: Decades of development and use" (PDF). *J. Antimicrob. Chemother*. 2003 May;51(Suppl 1):13–20. doi:10.1093/jac/dkg208. PMID 12702699

Eshraghi B, Abdi P, Akbari M, Fard MA. Microbiologic spectrum of acute and chronic dacryocystitis. *Int J Ophthalmol*. 2014;7(5):864–867.

Eslami F, Basir HR, Moradi A, Farah S. Microbiological study of dacryocystitis in northwest of Iran. *Clin Ophthalmol*. 2018;12:1859–1864.

Etxebarría J, López-Cerero L, Mensa J. Diagnóstico microbiológico de las infecciones oculares. *Enferm Infecc Microbiol Clin*. 2009;27(9):531–5.

Falagas ME, Grammatikos AP, Michalopoulos A. "Potential of old-generation antibiotics to address current need for new antibiotics". *Expert Rev Anti-Infect. Ther*. 2008 October;6(5):593–600. doi:10.1586/14787210.6.5.593

Ffooks OO. Dacryocystitis in infancy. *Br J Ophthalmol*. 1962;46:422–34.

Gower EW, Lindsley K, Tulenko SE, Nanji AA, Leyngold I, McDonnell PJ. "Perioperative antibiotics for prevention of acute endophthalmitis after cataract surgery". *Cochrane Database Syst Rev*. 2017 Feb 13;2(2):CD006364. doi: 10.1002/14651858.CD006364.pub3. PMID: 28192644; PMCID: PMC5375161.

Grant GG. Dacryocystitis. *J Ophthalmol*. 2001;2:1–10.

Hartikainen J, Lehtonen O, Saari KM. Bacteriology of lacrimal duct obstruction in adults. *Br J Ophthalmol*. 1997;81:37–40.

Huber E, Steinkogler FJ, Huber-Spitzy V. A new antibiotic in the treatment of dacryocystitis. *Orbit*. 1991;10(1):33–5.

Huber-Spitzy V, Steinkogler FJ, Haselberger C. The pathogen spectrum in neonatal dacryocystitis. *Klin Monastsbl Augeenheilkd*. 1987;190:445–6.

Huber-Spitzy V, Steinkogler FJ, Huber E, Arocker-Mettinger E, Schiffbänker M. Acquired dacryocystitis: Microbiology and conservative therapy. *Acta Ophthalmol (Copenh)* 1992;70(6):745–749. https://doi.org/10.1111/j.1755-3768.1992.tb04880.x

Huttner A, Bielicki J, Clements MN, Frimodt-Møller N, Muller AE, Paccaud J-P, Mouton JW, Oral amoxicillin and amoxicillin–clavulanic acid: Properties, indications and usage. *Clin. Microbiol Infect*. 2020;26(7):871–879. ISSN 1198-743X

Infectious Diseases Society of America. "Guidelines for the management of adults with hospital-acquired, ventilator-associated, and healthcare-associated pneumonia". *Am. J. Respir. Crit. Care Med.* 2005;171(4):388–416. doi:10.1164/rccm.200405-644ST. PMID 15699079

Janssen AG, Mansour K, Bos JJ, Manoliu RA, Castelijns JA. Abscess of the lacrimal sac due to chronic or subacute dacryocystitis: Treatment with temporary stent placement in the nasolacrimal duct. *Radiology*. 2000;215(1):300–4. 107515

Jones LT. Tear-sac foreign bodies. *Am J Ophthalmol*. 1965;60:111–3.

Jones LT, Wobig JL, eds. *Surgery of the eyelids and lacrimal system*. Birmingham, AL: Aesculapius Publishing Co, 1976: 185–93.

Kandati J, Kumar GK, Avanish G, Buchineni M, Pathapati RM, Srinivas P. "Microbial Surveillance of acute and Chronic Dacryocystitis in a tertiary care hospital". *J. Evol. Med. Dent. Sci.* 2015 January 08;4(03): 408–415. DOI: 10.14260/jemds/2015/62

Kebede A, Adamu Y, Bejiga A. Bacteriological study of dacryocystitis among patients attending in Menelik II Hospital, Addis Ababa, Ethiopia. *Ethiop Med J*. 2010;48(1):29–33.

Khatib L, Lebanon B, Katowitz WR. Managing Dacryoceles and Dacryocystitis. *Rev Ophthalmol*. 10 April 2018.

Kotlus BS, Rodgers IR, Udell IJ. Dacryocystitis caused by community-onset methicillin-resistant Staphylococcus aureus. *Ophthal Plast Reconstr Surg*. 2005;21:371–375. doi:10.1097/01.iop.0000175035.22953.71

Kubo M, Sakuraba T, Arai Y, Nakazawa M. Dacyocystorhinostomy for dacryocystitis caused by methicillin-resistant Staphylococcus aureus: Report of four cases. *Jpn J Opthalmol*. 2002;46:177–182. doi:10.1016/S0021-5155(01)00496-8

Kuchar A, Lukas J, Steinkogler FJ. Bacteriology and antibiotic therapy in congenital nasolacrimal duct obstruction. *Acta Ophthalmol Scand*. 2000;78:694–8.

Leelavathamma T, Penchalaiah T, Sandhya CS, Subhahani MA. A clinico microbiological study of Dacryocystitis in Adults. *IOSR Journal of Dental and Medical Sciences* (IOSR-JDMS) 2018 Ver. IV January;17(01):11–15. e-ISSN: 2279-0853, p-ISSN: 2279-0861. www.iosrjournals.org

Levin AV, Wygnanski-Jaffe T, Forte V, et al., Nasal endoscopy in the treatment of congenital lacrimal sac mucoceles. *Int J Pediatr Otorhinolaryngol* 2003;67:3:255–61.

Lodise TP Jr, Lomaestro B, Drusano GL. Piperacillin-tazobactam for Pseudomonas aeruginosa infection: Clinical implications of an extended-infusion dosing strategy. *Clin Infect Dis.* 2007 February 1;44(3):357–63. doi: 10.1086/510590. Epub 2007 Jan 2. PMID: 17205441

Lueder GT. Endosopic treatment of intranasal abnormalities associated with nasolacrimal duct obstruction. *J AAPOS*. 2004;8:2:128–32.

Lueder GT. The association of neonatal dacryocystoceles and infantile dacryocystitis with nasolacrimal duct cysts (an American ophthalmological society thesis). *Trans Am Ophthalmol Soc*. 2012;110:74–93.

Luo B, Li M, Xiang N, Hu W, Liu R, Yan X. The microbiologic spectrum of dacryocystitis. *BMC Ophthalmol.* 2021;21:29.

Machin SJ, et al., Lacrimal duct obstruction treated with lacrimonasal stent. *Arch Soc Esp opthalomol*. 2003; 78(6):315–318.

Madhusudhan MY, Ismail N, Hussein A. Microbiological aetiology of acute dacryocystitis in hospital University Sains Malaysia, Kelantan Malaysia. *J Acute Dis*. 2012;1:31–4.

Magdesian KG (2017). Equine pharmacology, an issue of veterinary clinics of North America: Equine practice, e-book. *Elsevier Health Sciences*. p. 59. ISBN 978-0-323-52438-4

Maheshwari R, Maheshwari S, Shah T. Acute dacryocystitis causing orbital cellulitis and abscess. *Orbit*. 2009;28:196–9.

Mandal R, Banerjee A, Biswas M. Clinicobacteriological study of chronic dacryocystitis in adults. *J Indian Med Assoc*. 2008;106(5):296–304.

Martins MC, Ricardo JR, Akaishi PM, et al., Orbital abscess secondary to acute dacryocystitis: Case report. *Arq Bras Oftalmol*. 2008;71(4):576–578. doi:10.1590/S0004-27492008000400020

Mccormick A, Sloan B. The diameter of the nasolacrimal canal measured by computed tomography: Gender and racial differences. *Clin Exp Ophthalmol*. 2009;37(4):357–361.

Mekonnen W, Adamu Y. Outcome of external dacryocystorhinostomy in Ethiopian patients. *Ethiop Med J*. 2009;47(3):221–6.

Mills DM, Bodman MG, Meyer DR, Morton AD 3rd, ASOPRS Dacryocystitis Study Group. The microbiologic spectrum of dacryocystitis: A national study of acute versus chronic infection. *Ophthalmic Plast Reconstr Surg*. 2007;23(4):302–6. doi:10.1097/IOP.0b013e318070d237

Negm S, Aboelnour A, Saleh T, Yasser M, Hassanin O. Clinicobacteriological study of chronic dacryocystitis in *Egypt Bulletin of the National Research Centre* 2019;43, Article number: 35.

Nigam E, Chandrakar AK, Garg ML, Pandey N, Jaiswal V, Radke N, et al., Age wise microbiological profile in chronic dacryocystitis. In: *AIOC 2008 Proceedings*. 2008.

Owji N, Kahlili MR. Normalization of conjunctival flora after dacryocystorhinostomy. *Ophthal Plast Reconstr Surg*. 2009;25:136–8.

Papich MG (2015). Saunders handbook of veterinary drugs: Small and large animal. *Elsevier Health Sciences*. pp. 43–47. ISBN 978-0-323-24485-5. www.papichdrugformulary.com

Patel K, Pradhan A, Sethia S, Lune A, Magdum R, Misra R. A clinicobateriological study of chronic dacryocystitis. *Sudanese J Ophthalmol*. 2014;6(1):1–5.

Petersen RA, Roob RM. The natural course of congenital obstruction of the nasolacrimal duct. *J Pediatr Ophthalmol Strabismus*. 1978;15:246–50.

Pinar-Sueiro S, Fernández-Hermida RV, Gibelalde A, et al. Study on the effectiveness of antibiotic prophylaxis in external dacryocystorhinostomy: A review of 697 cases. *Ophthal Plast Reconstr Surg*. 2010; 26(6): 467–72. doi:10.1097/IOP.0b013e3181d644cf

Pinar-Sueiro S, Sota M, Lerchundi T, Gibelalde A, Berasategui B, Vilar B. Dacryocystitis: Systematic approach to diagnosis and therapy. *Curr Infect Dis Rep*. 2012;14(2):137–146. doi:10.1007/ s11908-012-0238-8

Pogue JM, Ortwine JK, Kaye KS. "Clinical considerations for optimal use of the polymyxins: A focus on agent selection and dosing". *Clinical Microbiology and Infection*. 2017 April;23(4):229–233. doi:10.1016/j.cmi.2017.02.023. PMID 28238870

Pollard ZF. Treatment of acute dacryocystitis in neonates. *J Pediatr Ophthalmol Strabismus*. 1991;28:341–3.

Pommier Y, Leo E, Zhang H, Marchand C. "DNA topoisomerases and their poisoning by anticancer and antibacterial drugs". *Chem. Biol*. 2010;17(5):421–433. doi:10.1016/j.chembiol.2010.04.012. PMC 7316379. PMID 20534341

Pornpanich K, Luemsamran P, Leelaporn A, et al., Microbiology of primary acquired nasolacrimal duct obstruction: Simple epiphora, acute dacryocystitis, and chronic dacryocystitis. *Clin Ophthalmol*. 2016;10:337–342. doi:10.2147/OPTH.S100280

Pradeep AV, Patil SS, Koti SV, et al., Clinico-bacteriological study of chronic dacryocystitis cases in Northern Karnataka, India. *J Clin Diagn Res*. 2013 November;7(11):2502–2504.

Prakash R, Girish Babu RJ, Nagaraj ER, Prashanth HV, Jayashree S, Shah A. Bacteriological study of Dacryocystitis. *J Clin Diagn Res*. 2012;6(4):652–655.

Prokosch V, Prokosh JE, Promesberger J, et al., Bacteriology of occluded nasolacrimal ducts in infants. *Klin Monbl Augenheilkd*. 2010;227(7):585–8.

Razavi EM, Ansari-Astaneh MR, Farzadnia M, Rahmaniyan H, Moghiman T. Bacteriological evaluation of adult dacryocystitis in Iran. *Orbit*. 2010;29(5):286–90. doi:10.3109/01676831003664293

Reese A, Cosimi, Beik N, Kubiak DW, Johnson JA. Ceftaroline for severe Methicillin-Resistant Staphylococcus aureus Infections: A systematic review. *Open Forum Infect. Dis*. Spring 2017;4(2), ofx084. https://doi.org/10.1093/ofid/ofx084

Sainju R, Franzco AA, Shrestha MK, Ruit S. Microbiology of dacryocystitis among adults population in southern Australia. *Nepal Med Coll J*. 2005;7(1):18–20.

Shah CP, Santani D. A comparative bacteriological profile and antibiogram of dacryocystitis. *Nepal J Ophthalmol*. 2011;3(2):134–139.

Sarkar P, Yarlagadda V, Ghosh C, Haldar, J. "A review on cell wall synthesis inhibitors with an emphasis on glycopeptide antibiotics". *MedChemComm*. 1 2017 March;8(3):516–533. doi:10.1039/c6md00585c. PMC 6072328. PMID 30108769

Sensi P. "History of the development of rifampin". *Rev. Infect. Dis*. 1983;5(Suppl 3):S402–S406. doi:10.1093/clinids/5.supplement_3. s402. JSTOR 4453138. PMID 6635432

Shahraki K, Makateb A, Shirzadi K, Khosravifard K. Frequency of bacterial samples from patients with chronic acquired nasolacrimal duct obstruction. *Med Arch*. 2016;70(6):433–436.

Siddiqui AP, Bandil SB, Sukhadeve S. Chronic Dacryocystitis - Its evaluation and management by various investigative and diagnostic test. *IOSR J Pharm (IOSRPHR)*. 2013;03(10):28–33.

Sivakrishna A, Jithendhra K, Sreenivasulu Reddy P, Avinash G, vasundhara P, Deepa T, Ravikumar T, Prasad Naidu M. "A bacteriological study of Dacryocystitis in a tertiary care hospital" *Am. J. Microbiol Res*. 2018;6(2):38–41. doi: 10.12691/ajmr-6-2-1

Sood N, Ratnaraj A, Balaraman G, Madhavan H. Chronic dacryocystitis - a clinic- bacteriological study. *Indian J Ophthalmol*. 1967;15(3):107–110.

Strunk T, Richmond P, Simmer K, et al., Neonatal immune responses to coagulase-negative staphylococci. *Curr Opin Infect Dis*. 2007;20:370–5.

Summer Skill, WA, Jones LT. Tear-sac foreign bodies. *Am J Ophthalmol*. 1965;60:111–3.

Sun X, Liang Q, Luo S, Wang Z, Li R, Jin X. Microbiological analysis of chronic dacryocystitis. *Ophthalmic Physiol Opt*. 2005;25(3):261–263.

Tadesse BT, Ashley EA, Ongarello S, Havumaki J, Wijegoonewardena M, González IJ, Dittrich S. "Antimicrobial resistance in Africa: A systematic review". *BMC Infect. Dis*. 2017 September;17(1):616. doi:10.1186/s12879-017-2713-1. PMC 5594539. PMID 28893183

Tan JS, File TM. "Antipseudomonal penicillins". *Med. Clin. North Am.*. 1995 July;79(4):679–93. doi:10.1016/s0025-7125(16)30032-3. PMID 7791416

Taylor RS, Ashurst JV. (2022). StatPearls Publishing LLC as contributed by Chelsea Rowe.

Teweldemedhin M, Gebreyesus H, Atsbaha AH, Asgedom SW, Saravanan M. Bacterial profile of ocular infections: A systematic review. *BMC Ophthalmol*. 2017;17(1):212.

Thicker JA, Buffam FV. Lacrimal sac, conjunctival, and nasal culture results in dacryocystorhinostomy patients. *Ophthal Plast Reconstr Surg*. 1993; 9:43–46.

Thomas T, Meena Dias1, Bichu Joseph Maliakal2. Microbiological profile and antimicrobial susceptibility pattern of the isolates indacryocystitis: A prospective study in a tertiary care hospital. *Indian J. Microbiol. Res*. 2019;6(4):328–335.

Todar K. (2012). "Antimicrobial agents in the treatment of infectious disease.". *Todars Online Text Book of Bacteriology*. Archived from the original on 2013-10-08

Traquair H. Chronic dacryocystitis. Its causation and treatment. *Arch Ophthalmol.* 1941;26:165–80.

Usha K. et al spectrum nd the suspectibilities of microbial isolates in case of congenital nasolacrimal duct obstruction *JAAPOS.* 2006;10(5):469–478.

Walland MJ, Rose GE. Soft tissue infections after open lacrimal surgery. *Ophthalmol.* 1994;101:608–11.

Wormser GP, Keusch GT, Heel RC. "Co-trimoxazole (trimethoprim-sulfamethoxazole): An updated review of its antibacterial activity and clinical efficacy". *Drugs.* 1982 December;24(6):459–518. doi:10.2165/00003495-198224060-00002. PMID 6759092. S2CID 209121818

Wu W, Yan W, MacCallum JK, et al. Primary treatment of acute dacryocystitis by endoscopic dacryocystorhinostomy with silicone intubation guided by a soft probe. *Ophthalmol.* 2009;116:116–122.

Yazici B, Hammad AM, Meyer DR. Lacrimal sac dacryoliths. Predictive factors and clinical characteristics. *Ophthalmol.* 2001;108:1308–12.

15

Antimicrobial Resistance of the Human Eye: A Way Forward

Ponnari Gottipati

15.1 Introduction

Antimicrobial Resistance (AMR) is a major public health issue that affects people worldwide, regardless of their nationality or geographic location. The spread of AMR is not limited by borders, and the emergence of drug-resistant infections in one part of the world can easily spread to other parts of the world, threatening public health globally (Molton et al., 2013). An illustration of this is the emergence of resistance to colistin, a last resort antibiotic identified only in 2015, which is now present in over 50 countries spanning 5 continents (Wang et al., 2018). Bacteria harbouring the New Delhi metallo-beta-lactamase 1 (NDM1) resistance gene identified first in clinical samples in India in 2018 has already spread to the rest of the world (Khan et al., 2017), including Arctic soils (McCann et al., 2019). In fact, detection of antibiotic-resistant bacteria in the Arctic and Antarctic regions (Hernåndez et al., 2016), some of the most remote and isolated areas on the planet, is a stark example of how easily AMR can spread. Studies have shown that AMR bacteria have been found in various environments in the Arctic, including soil, water, and wildlife (McCann et al., 2019). Nearly 100% of the bacterial isolates from Arctic soil samples were found to be resistant to at least one antibiotic, and a significant percentage were multidrug-resistant (Mogrovejo et al., 2020). Similarly, studies reported the detection of antibiotic-resistant bacteria in wastewater from an Antarctic research station, bird faeces as well as sea waters (Rabbia et al., 2016; Tam et al., 2015). These examples reiterate the need for a global concerted effort in tackling the issue, as AMR in one part of the world is a problem elsewhere.

15.2 Why Study Antimicrobial Resistance of the Human Eye?

AMR in one part of the body can have significant consequences for the overall health of an individual. Limiting their effectiveness has broader implications for antimicrobial treatments in other parts of the body, leading to longer hospital stays, increased healthcare costs, and poorer health outcomes for patients. As such, it is essential to study AMR, not only in systemic infections, but also in site-specific infections such as

 DOI: 10.1201/9781003451105-15

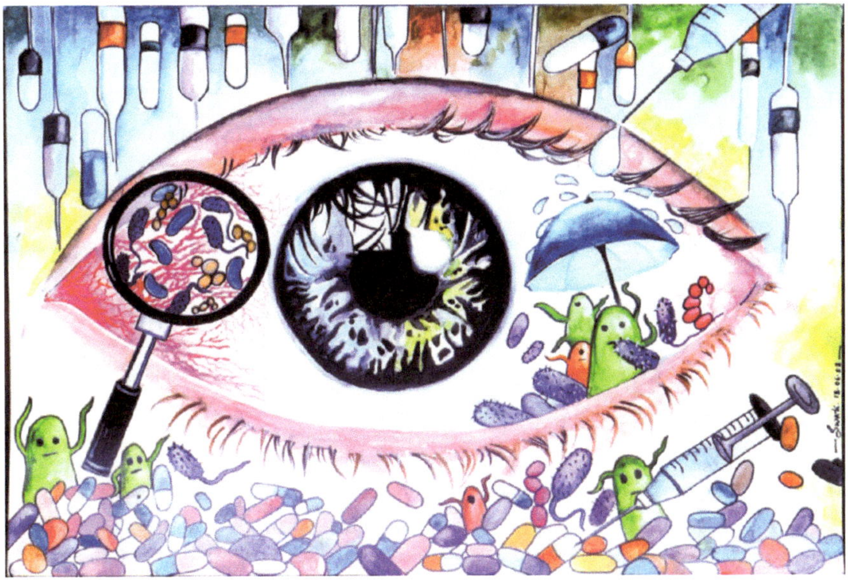

FIGURE 15.1 An artist's view of antimicrobial resistance in the human eye. (Courtesy Swatilekha Hazra, PhD Scholar, L.V. Prasad Eye Institute, Hyderabad, India.)

the eye (Figure 15.1), to understand the magnitude of the problem and devise strategies to manage it.

15.3 Systemic versus Ocular Infections

There is a great deal of evidence to show how systemic infections lead to complications in the eye (Lynn et al., 2004). Systemic infections caused by bacterial, fungal, viral, and parasitic pathogens can spread to the eye and lead to various ocular complications. Reduced immunity due to ageing or underlying diseases such as diabetes mellitus and rheumatoid arthritis may lead to systemic infections that can spread to ocular tissues, especially endophthalmitis (Wong et al., 2000; Falavarjani et al., 2012; Wiskur et al., 2008) and keratitis (Das et al., 2006). The same could be the case with clinical procedures that suppress the immune system or prophylactic measures (Henry et al., 2012). Certain systemic infections can spread to the eyes through the bloodstream. For example, bacteria that cause endocarditis can travel to the eyes and cause endophthalmitis (Chen et al., 2019). In fact, in addition to the direct involvement of the eye, some infections cause intraocular damage through indirect mechanisms, leading to opportunistic infections such as cytomegalovirus infection of the eye or leprosy-related periocular nerve involvement (Lynn et al., 2004). Severe malaria may result in the development of Malarial Retinopathy (Beare, 2006). Eye symptoms can also indicate the outcome of an underlying infection, such as poor vision after congenital or acquired toxoplasmosis (Park et al., 2013).

In the opposite scenario, some ocular infections in rare cases may lead to more serious life-threatening conditions. Infectious conjunctivitis can potentially spread to other areas of the body, leading to more severe secondary infections like meningitis (Barquet et al., 1990). Conditions like Orbital Cellulitis, if not handled correctly can very quickly lead to severe conditions like meningitis and osteomyelitis and even death (Chaudhry et al., 2012).

15.4 Microbiome and Ocular Infections

Changes in the normal ocular microbiota can contribute significantly to the pathogenesis of ophthalmic diseases. As discussed in Chapter 1, until the advent of genomics, Gram-positive and a few Gram-negative bacteria, as well as some fungal strains, were thought to dominate the ocular surface microbiota. However, recent genomics research has revealed a much wider range of microorganisms on the ocular surface, which are described in detail in earlier chapters of the book. The most found bacteria in the ocular microbiota include *Pseudomonas*, *Propionibacterium*, *Acinetobacter*, *Staphylococcus*, *Corynebacterium* and so forth (Petrillo et al., 2020). However, this balanced microbiota can be easily disrupted by environmental factors, use of antibiotics, infections such as conjunctivitis, pathological conditions like dry eye syndrome or personal habits like improper use of contact lenses which may provide a route for opportunistic pathogens, leading to different kinds of ocular infections (Xue et al., 2021).

Recent studies indicated that changes in the microbiota of other body sites can sometimes lead to ophthalmic pathologies. While the microbiome of the eye is different from the rest of the body, and the diversity of the commensal bacterial flora in the eye might be less compared to other organs, emerging research indicates that the microbiota present in regions other than the eyes, such as the gut and oral microbiota, may contribute to the development of certain ophthalmic diseases (Xue et al., 2021). Specifically, there is increasing evidence of a link between oral pathogens and several eye diseases, including Age-related Macular Degeneration (AMD), Glaucoma, Uveitis, and Diabetic Retinopathy (Arjunan, 2022).

There is therefore a great deal of interconnectedness between eye pathogenesis and systemic infections, and the state of the microbiome has a major influence on ocular diseases. In the context of AMR, systemic infections may be more challenging to treat than eye infections. This is because systemic infections are often more severe, and the use of antibiotics is usually required to control the infection. Additionally, systemic infections may be caused by more diverse microorganisms, making treatment more complicated. Furthermore, the use of antibiotics to treat systemic infections may have a greater impact on the overall microbiome of the patient, leading to the emergence of AMR bacteria.

On the other hand, eye infections are usually caused by a smaller range of microorganisms, making treatment more straightforward. Additionally, topical antibiotics, which have a lower risk of promoting AMR than systemic antibiotics, are often used to treat eye infections. However, the ocular microbiome is highly susceptible to AMR because of its ability to form biofilms (Konduri et al., 2021). Biofilms are resistant to antibiotics, and ocular biofilms are particularly challenging to treat. Knowledge of resistant strains and their mechanisms in site-specific infections will invariably help in advancements in tackling the issue at large.

15.5 AMR in Ocular Infections

As demonstrated in this book in much detail, the issue of Antimicrobial Resistance has gained more attention in the context of eye infections in recent years, with certain microorganisms exhibiting resistance to multiple classes of antibiotics. There is a sharp rise in bacterial keratitis (Egrilmez et al., 2020) and endophthalmitis (Wu et al., 2022) infections due to AMR, and more and more fungal infections are starting to show resistance, adding to the load of AMR infections in countries like India, the United States, Australia, Scandinavian countries and so forth. (Biswas et al., 2022). Resistant bacteria strains like MRSA are increasingly found in ocular infections (Ritterband, 2013). Multi-drug-resistant strains of *Pseudomonas aeruginosa* and *Acinetobacter baumannii* are also reported in keratitis (Chatterjee et al., 2016) and in endophthalmitis cases (Chen et al., 2008).

Ocular infections are a common cause of morbidity, especially in developing countries. If left untreated or if treatment is delayed due to AMR, the infection can spread and lead to severe complications such as corneal scarring, blindness, and even death. Although the incidence of AMR in ocular infections is on the rise, there is a lack of information on the prevalence and susceptibility patterns of ocular pathogens and their resistance strains. Therefore, studies aimed at understanding AMR in ocular pathogens can help not only for proper management of ocular infections and preserving visual function but also to reduce the prevalence of drug resistance in ocular infections. Conducting large-scale microbiological surveillance and antibiotic susceptibility studies for ocular infections can provide valuable information on the prevalence and distribution of AMR in ocular pathogens. This information can be used to develop targeted treatment strategies and establish antimicrobial prescription guidelines and policies for ophthalmic infections based on regional resistance patterns.

Antimicrobial Resistance is a growing public-health concern worldwide, and the eye is not exempt from this issue. A Lancet study estimated the death toll of AMR to be 4.95 million in 2019, and AMR was the direct cause of death in 1.27 million of those cases (Antimicrobial Resistance Collaborators, 2022). It is forecast that the global death toll due to AMR could increase up to 10 million per year by 2050, in the absence of immediate action to curtail the issue (O'Neill, 2016). The rampant and injudicious use of antibiotics is a daunting challenge in managing ocular infections (Luyt et al., 2014; Ventola, 2015). This issue is particularly augmented due to topical administration of antibiotics in ocular infections. Furthermore, research has revealed that 30% to 60% of antibiotics prescribed in intensive care units (ICUs) are either unnecessary, unsuitable, or suboptimal (Luyt et al., 2014). Post-marketing surveillance studies on antibiotics, both in outpatient settings and ICUs, hold the key to analyzing real-world usage patterns to issue appropriate prescription guidelines.

15.6 The Solutions

The drivers of AMR are multifaceted and include human, animal, plant, food, and environmental factors. In humans, overuse, and misuse of antibiotics, as well as poor infection control, contribute to the problem. In animal agriculture, the routine use of antibiotics for growth promotion and disease prevention is a significant driver of AMR. Antibiotics in plant agriculture can also lead to resistant plant pathogens, and contaminated food can transmit resistant bacteria to humans. Finally, the environment can play a role in the spread of AMR. Addressing these drivers through coordinated efforts and adopting a "One Health" approach is advocated as crucial to combat AMR and preserve the effectiveness of antibiotics for future generations (www.who.int/news-room/fact-sheets/detail/antimicrobial-resistance).

The Global Action Plan for AMR containment, which has been adapted by many nations through national action plans, aims to strengthen awareness, reduce the incidence of infection, optimize the use of antimicrobial agents, and promote investments in AMR activities while stressing the need for the countries to strengthen leadership on AMR (WHO, 2015).

A coordinated, multi-sectoral global response is urgently needed to address the problem of AMR, including efforts to improve surveillance and monitoring of drug-resistant infections, promote responsible use of antibiotics, develop new antimicrobial drugs, vaccines, alternate treatments (example: phage therapy) and other antimicrobial agents (example: antimicrobial peptides), as well as investments in faster diagnostics, as advocated in the 2019 report by the Inter-Agency Coordination Group (IACG) on AMR. The report stresses the importance of global political commitment and leadership, as well as greater awareness and education at all levels of society. The IACG recommends the establishment of national AMR action plans and the strengthening of international collaboration to tackle the growing threat of AMR. IACG gives a stark warning that failure to act swiftly could result in disastrous consequences within a single generation (WHO, 2019). To counter this, all stakeholders must share a sense of urgency and unite in their commitment to combat AMR.

As the world increasingly adopts a One Health approach to tackle AMR, it is crucial to address not only the human drivers of resistance, but also to promote global and regional cooperation in combating both systemic and site-specific infections such as that of the eye. Similarities or differences in resistance patterns in systemic versus site-specific infections can reveal important information to help in this cause. In addition, understanding the epidemiology of AMR in various infections will add to the knowledge in the field and help understand variations. Urgent action is needed to reduce transmission rates, and prevent the emergence of new resistant strains, which can significantly reduce the human and healthcare costs associated with AMR. This includes reducing antibiotic use in human and animal medicine, improving infection control practices, and monitoring resistance patterns in both the clinical samples and the environment. Understanding Antimicrobial Resistance in various disease conditions is a crucial first step, and this book provides a detailed account of this in the context of the eye.

REFERENCES

Antimicrobial Resistance Collaborators. Global burden of bacterial antimicrobial resistance in 2019: A systematic analysis. *Lancet*. 399 (10325) (2022), p. 629–655.

Arjunan P, Swaminathan R. Do oral pathogens inhabit the eye and play a role in ocular diseases? *J Clin Med*. 11(10) 2022 May 23, p. 2938. doi: 10.3390/jcm11102938. PMID: 35629064; PMCID: PMC9146391

Barquet N, Gasser I, Domingo P, Moraga FA, Macaya A, Elcuaz R. Primary meningococcal conjunctivitis: Report of 21. *Patients and Review, Rev. Infect. Dis.* 12(5) (1990), p. 838–847.

Beare NA, Taylor TE, Harding SP, Lewallen S, Molyneux ME. Malarial retinopathy: A newly established diagnostic sign in severe malaria. *Am J Trop Med Hyg.* 75(5) (2006), p. 790–7.

Biswas P, Batra S, Gurha N, Maksane N. Emerging antimicrobial resistance and need for antimicrobial stewardship for ocular infections in India: A narrative review. *Indian J Ophthalmol.* 70(5) (2022) p. 1513–1521.

Chatterjee S, Agrawal D. Multi-drug resistant Pseudomonas aeruginosa keratitis and its effective treatment with topical colistimethate. *Indian J Ophthalmol.* 64(2) (2016) p. 153–7.

Chaudhry IA, Al-Rashed W, Arat YO. The hot orbit: Orbital cellulitis. *Middle East Afr J Ophthalmol.* 19(1) (2012) p. 34–42.

Chen K-J, Hou CH, Sun MH, Lai CC, Sun CC, Hsiao CH. Endophthalmitis caused by Acinetobacter baumannii: Report of two cases. *J Clin Microbiol.* 46(3) (2008) p. 1148–50.

Chen K-J, Ming-Hui Sun, Yen-Po Chen, Nan-Kai Wang, Wei-Chi Wu, Chi-Chun Lai. *Ophthalmol. Retina.* 3(4), (2019) p. 382–384.

Das S, Constantinou M, Daniell M, Taylor HR. Moraxella keratitis: Predisposing factors and clinical review of 95 cases. *Br J Opthalmol.* 90 (2006) p. 1236–8.

Egrilmez S, Yildirim-Theveny Ş. Treatment-resistant bacterial keratitis: Challenges and solutions. *Clin Ophthalmol.* 29(14) (2020) p. 287–297.

Falavarjani KG, Nekoozadeh S, Modarres M, Parvaresh M, Hashmi M, Soodi R, Alemzadeh SA. Isolates and antibiotic resistance of culture-proven Endophthalmitis cases presented to a referral Center in Tehran. *Middle East Afr J Ophthalmol.* 19(4) (2012) p. 361–3.

Henry CR, Flynn W, Miller D, Forster RK, Alfonso EC. Infectious keratitis progressing to Endophthalmitis: A 15-year-study of microbiology, associated factors, and clinical outcomes. *Ophthalmol.* 119(12) 2012 p. 2443–9.

Hernández J, González-Acuña D. Anthropogenic antibiotic resistance genes mobilization to the polar regions. *Infect Ecol Epidemiol.* 12(6) (2016), p. 32112.

Khan AU, Maryam L, Zarrilli R. Structure, genetics and worldwide spread of New Delhi Metallo-β-lactamase (NDM): A threat to public health. *BMC Microbiol.* 17 (2017) p. 101.

Konduri R, Sharma S, Shivaji S. Microbes of the human eye: Microbiome, antimicrobial resistance and biofilm formation. *Exp Eye Res.* (2021) p. 205:108476. doi: 10.1016/j.exer.2021.108476

Luyt CE, Brechot N, Trouillet JL, Chastre J. Antibiotic stewardship in the intensive care unit. *Crit Care.*18(5) (2014) p. 480.

Lynn WA, Lightman S. The eye in systemic infection. *Lancet.* 364(9443) (2004) p. 1439–50.

McCann CM, Christgen B, Roberts JA, Su J-Q, Arnold KE, Gray ND, Zhu Y-G, Graham DW. Understanding drivers of antibiotic resistance genes in high Arctic soil ecosystems. *Environ. Int.* 125 (2019) p. 497–504.

Mogrovejo DC, Perini L, Gostinčar C, Sepčić K, Turk M, Ambrožič-Avguštin J, Brill FHH, Gunde-Cimerman N. Prevalence of antimicrobial resistance and hemolytic phenotypes in culturable arctic bacteria. *Front. Microbiol.* (2020) 11 Apr 3;11:570. doi: 10.3389/fmicb.2020.00570. PMID: 32318045; PMCID: PMC7147505.

Molton JS, Tambyah PA, Ang BSP, Ling ML, Fisher DA. The global spread of healthcare-associated multidrug-resistant bacteria: A perspective from Asia. *Clin. Infect. Dis.* 56 (9) (2013), p. 1310–1318.

O'Neill J. *Tackling Drug-Resistant Infections Globally: Final Report and Recommendations.* Wellcome Trust, HM Government; 2016.

Park YH, Nam HW. Clinical features and treatment of ocular toxoplasmosis. *Korean J Parasitol.* 51(4) (2013) p. 393–9.

Petrillo F, Pignataro D, Lavano MA, Santella B, Folliero V, Zannella C, Astarita C, Gagliano C, Franci G, Avitabile T, Galdiero M. Current evidence on the ocular surface microbiota and related diseases. *Microorganisms.* 8(7) (2020) p. 1033.

Rabbia, V., Bello-Toledo H, Jiménez S, Quezada M, Domínguez M, Vergara L, Gómez-Fuentes C, Calisto-Ulloa N, González-Acuña D, López J, González-Rocha G. Antibiotic resistance in Escherichia coli strains isolated from Antarctic bird faeces, water from inside a wastewater treatment plant, and seawater samples collected in the Antarctic Treaty area. *Polar Sci.* 10(2) (2016) p. 123–131.

Ritterband D. Methicillin-Resistant Staphylococcus aureus and the Eye: Current concepts and management strategies. *Curr Ophthalmol Rep.* 1 (2013) p. 151–160.

Tam HK, Wong CMVL, Yong ST, Blamey J, González M. Multiple-antibiotic-resistant bacteria from the maritime Antarctic. *Polar Biol.* 38 (2015) p. 1129–1141.

Ventola CL. The antibiotic resistance crisis: Part 1: Causes and threats. *P T.* 40(4) (2015) p. 277–283.

Wang, R., van Dorp L, Shaw LP, Bradley P, Wang Q, Wang X, Jin L, Zhang Q, Liu Y, Rieux A, Dorai-Schneiders T, Weinert L.A, Iqbal Z, Didelot X, Wang H, Balloux F. The global distribution and spread of the mobilized colistin resistance gene mcr-1. *Nat. Commun.* 9(1) (2018) p. 1179.

WHO. Global action plan on antimicrobial resistance. (2015). www.who.int/publications/i/item/9789241509763

WHO. No time to Wait: Securing the future from drug-resistant infections; Report to the Secretary-General of the United Nations (2019). www.who.int/publications/i/item/no-time-to-wait-securing-the-future-from-drug-resistant-infections

Wiskur BJ, Hunt JJ, Callegan MC. Hypermucoviscosity as a virulence factor in experimental Klebsiella Pneumoniae Endophthalmitis. *Invest Ophthalmol Vis Sci.* 49(11) (2008) p. 4931–8.

Wong JS, Chan TK, Lee HM, Chee SP. Endogenous bacterial endophthalmitis: An east Asian experience and a reappraisal of a severe ocular affliction. *Ophthalmol.* 107(8) (2000) p. 1483–91.

Wu X-N, Chen Y-H, Sharief L, Al-Janabi A, Al Qassimi N, Lightman S, Tomkins-Netzer O. Emerging antibiotic resistance patterns affect visual outcome treating acute Endophthalmitis. *Antibiotics.* 11(7) (2022) p. 843.

Xue W, Li JJ, Zou Y, Zou B, Wei L. Microbiota and ocular diseases. *Front Cell Infect Microbiol.* 21(11) (2021) p. 759333.

Index

Note: Figures are indicated by *italics*. Tables are indicated by **bold**.